D1217036

HARCOURT
Math

Harcourt School Publishers
Orlando • Boston • Dallas • Chicago • San Diego
www.harcourtschool.com

Printed in the United States of America

ISBN 0-15-320747-7

7 8 9 10 032 10 09 08 07 06 05 04

Senior Author

Evan M. Maletsky
Professor of Mathematics
Montclair State University
Upper Montclair, New Jersey

Mathematics Advisor

Richard Askey
Professor of Mathematics
University of Wisconsin
Madison, Wisconsin

Authors

Angela Giglio Andrews
Math Teacher, Scott School
Naperville District #203
Naperville, Illinois

Jennie M. Bennett
Instructional Mathematics Supervisor
Houston Independent School District
Houston, Texas

Grace M. Burton
Chair, Department of Curricular Studies
Professor, School of Education
University of North Carolina
at Wilmington
Wilmington, North Carolina

Howard C. Johnson
Dean of the Graduate School
Associate Vice Chancellor for Academic Affairs
Professor, Mathematics and Mathematics Education
Syracuse University
Syracuse, New York

Lynda A. Luckie
Administrator/Math Specialist
Gwinnett County Public Schools
Lawrenceville, Georgia

Joyce C. McLeod
Visiting Professor
Rollins College
Winter Park, Florida

Vicki Newman
Classroom Teacher
McGaugh Elementary School
Los Alamitos Unified School District
Seal Beach, California

Janet K. Scheer
Executive Director
Create A Vision
Foster City, California

Karen A. Schultz
College of Education
Georgia State University
Atlanta, Georgia

Program Consultants and Specialists

Janet S. Abbott
Mathematics Consultant
California

Elsie Babcock
Director, Mathematics and Science Center
Mathematics Consultant
Wayne Regional Educational Service Agency
Wayne, Michigan

William J. Driscoll
Professor of Mathematics
Department of Mathematical Sciences
Central Connecticut State University
New Britain, Connecticut

Lois Harrison-Jones
Education and Management Consultant
Dallas, Texas

Arax Miller
Curriculum Coordinator and English Department Chairperson
Chamlian School
Glendale, California

Rebecca Valbuena
Language Development Specialist
Stanton Elementary School
Glendora, California

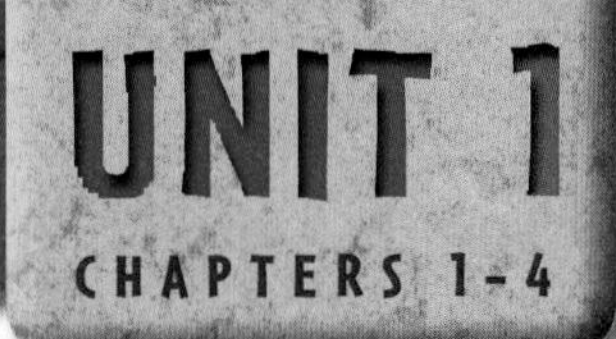

Understand Numbers and Operations

1 PLACE VALUE AND NUMBER SENSE

2 COMPARE, ORDER, AND ROUND NUMBERS . . 16

Technology Link

Harcourt Math Newsroom Video: *Chapter 2, p. 21*

E-Lab:
Chapter 1, p. 3
Chapter 3, p. 40
Chapter 4, p. 57

Mighty Math Calculating Crew:
Chapter 3, p. 47
Chapter 4, p. 63

Multimedia Math Glossary: *The Learning Site* at
www.harcourtschool.com/mathglossary

UNIT 2
CHAPTERS 5-6

Money and Time

5

Technology Link

Harcourt Math Newsroom Video: *Chapter 6, p. 100*

E-Lab:
Chapter 5, p. 81
Chapter 6, p. 95
Chapter 6, p. 98

Mighty Math Calculating Crew:
Chapter 5, p. 87

Multimedia Math Glossary: *The Learning Site* at
www.harcourtschool.com/mathglossary

6 UNDERSTAND TIME 92

UNIT WRAPUP

UNIT 3
CHAPTERS 7–10

Multiplication Concepts and Facts

Harcourt Math Newsroom Video: *Chapter 10, p. 165*

E-Lab:
Chapter 7, p. 116
Chapter 9, p. 150

Mighty Math Carnival Countdown:
Chapter 7, p. 124
Chapter 8, p. 139

Multimedia Math Glossary: *The Learning Site* at
www.harcourtschool.com/mathglossary

9 MULTIPLICATION FACTS AND STRATEGIES . . 146

10 MULTIPLICATION FACTS AND PATTERNS 162

UNIT WRAPUP

UNIT 4
CHAPTERS 11-13

Division Concepts and Facts

11

12

Technology Link

Harcourt Math Newsroom Video: *Chapter 13, p. 216*
E-Lab:
Chapter 11, p. 188
Multimedia Math Glossary: *The Learning Site* at
www.harcourtschool.com/mathglossary

13

UNIT WRAPUP

Data, Graphing, and Probability

14

15

Technology Link

Harcourt Math Newsroom Video: *Chapter 15, p. 264*

E-Lab: *Chapter 14, p. 242*

Mighty Math Number Heroes: *Chapter 16, p. 270*

Multimedia Math Glossary: *The Learning Site* at
www.harcourtschool.com/mathglossary

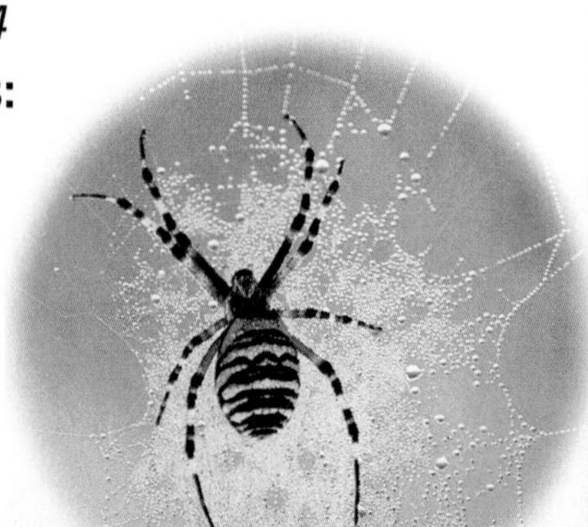

16

UNIT WRAPUP

UNIT 6

CHAPTERS 17–19

Geometry

17

18

Technology Link

Harcourt Math Newsroom Video: *Chapter 17, p. 301*

E-Lab:
Chapter 19, p. 333

Mighty Math Number Heroes:
Chapter 18, p. 317
Chapter 19, p. 338

Multimedia Math Glossary: *The Learning Site* at
www.harcourtschool.com/mathglossary

19 CONGRUENCE AND SYMMETRY

UNIT WRAPUP

UNIT 7

CHAPTERS 20-22

Measurement

20

21

Harcourt Math Newsroom Video: *Chapter 20, p. 353*

E-Lab:
Chapter 21, p. 373
Chapter 22, p. 395

Mighty Math Carnival Countdown:
Chapter 22, p. 391

Multimedia Math Glossary: *The Learning Site* at
www.harcourtschool.com/mathglossary

22 PERIMETER, AREA, AND VOLUME 386

UNIT WRAPUP

UNIT 8

CHAPTERS 23-26

Fractions and Decimals

Harcourt Math Newsroom Video: *Chapter 26, p. 474*

E-Lab:
Chapter 23, p. 418
Chapter 25, p. 457

Mighty Math Number Heroes:
Chapter 24, p. 438
Chapter 25, p. 452

Multimedia Math Glossary: *The Learning Site* at
www.harcourtschool.com/mathglossary

UNIT 9
CHAPTERS 27–30

Multiply and Divide by 1-Digit Numbers

Technology Link

Harcourt Math Newsroom Video: *Chapter 27, p. 491*

E-Lab:
Chapter 27, p. 489
Chapter 28, p. 503

Mighty Math Calculating Crew:
Chapter 29, p. 529
Chapter 30, p. 543

Multimedia Math Glossary: *The Learning Site* at
www.harcourtschool.com/mathglossary

29

30

UNIT WRAPUP

Student Handbook

Welcome!

The authors of *Harcourt Math* want you to enjoy learning math and to feel confident that you can do it. We invite you to share your math book with family members. Take them on a guided tour through your book!

The Guided Tour

Choose a chapter you are interested in. Show your family some of these things in the chapter that will help you learn.

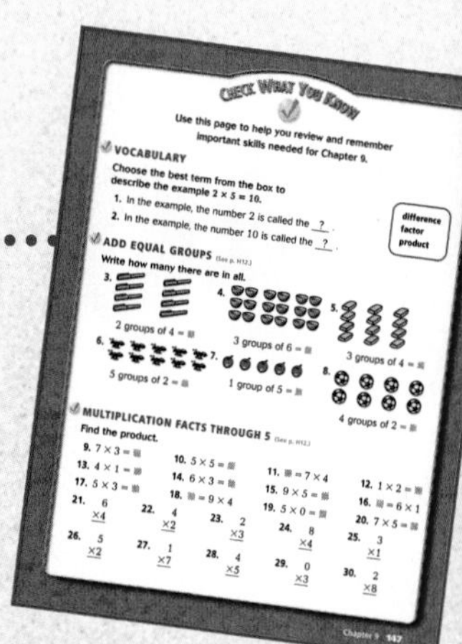

Check What You Know

Do you need to review any skills before you begin the next chapter? If you do, you will find help in the Handbook in the back of your book.

The Math Lessons

- **Quick Review** to check the skills you need for the lesson.
- **Learn section** to help you study problems, models, examples, and questions that give you different ways to learn.
- **Check** to make sure you understood the lesson.
- **Practice and Problem Solving** to practice what you have just learned.
- **Mixed Review and Test Prep** to keep your skills sharp and to prepare you for important tests. Look back at the pages shown next to each problem to get help if you need it.

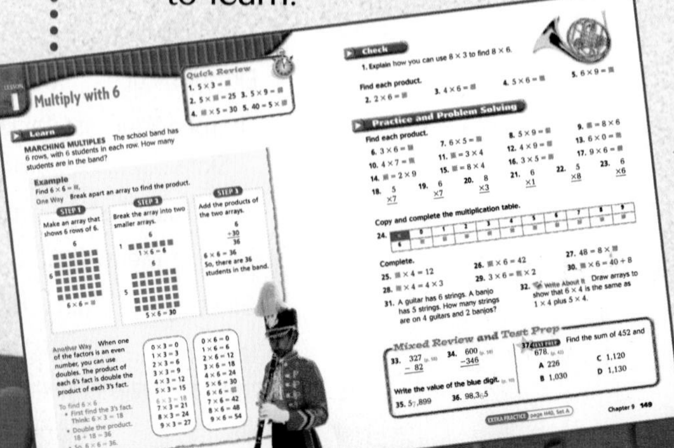

Student Handbook

Now show your family the **Student Handbook** in the back of your book. The sections will help you in many different ways.

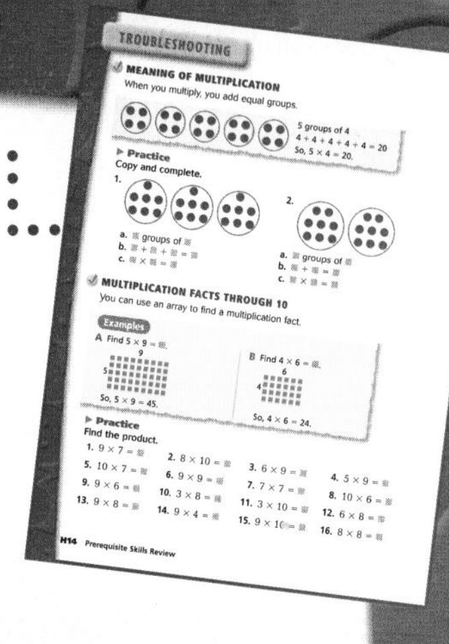

TROUBLESHOOTING

MEANING OF MULTIPLICATION

When you multiply, you add equal groups.

5 groups of 4
$4 + 4 + 4 + 4 + 4 = 20$
So, $5 \times 4 = 20$.

Practice

Copy and complete.

1.
a. ■ groups of ■
b. ■ + ■ + ■ = ■
c. ■ × ■ = ■

2.
a. ■ groups of ■
b. ■ + ■ = ■
c. ■ × ■ = ■

MULTIPLICATION FACTS THROUGH 10

You can use an array to find a multiplication fact.

Examples

A Find $5 \times 9 =$ ■.
So, $5 \times 9 = 45$.

B Find $4 \times 6 =$ ■.
So, $4 \times 6 = 24$.

Practice

Find the product.

1. $9 \times 7 =$ ■
2. $8 \times 10 =$ ■
3. $6 \times 9 =$ ■
4. $5 \times 9 =$ ■
5. $10 \times 7 =$ ■
6. $9 \times 9 =$ ■
7. $7 \times 7 =$ ■
8. $10 \times 6 =$ ■
9. $9 \times 6 =$ ■
10. $3 \times 8 =$ ■
11. $3 \times 10 =$ ■
12. $6 \times 8 =$ ■
13. $9 \times 8 =$ ■
14. $9 \times 4 =$ ■
15. 9 × 1[illegible] = ■
16. $8 \times 8 =$ ■

H14 Prerequisite Skills Review

- **Troubleshooting** will help you review and remember skills from last year.
- **Extra Practice** can be used to make sure that you are ready to move on to the next lesson.
- **Sharpen Your Test-Taking Skills** will help you feel confident that you can do well on a test.
- **Basic Facts Tests** will check whether you have memorized all of the basic facts and will show you which facts you still need to practice.

Invite your family members to

- talk with you about what you are learning.
- help you correct errors you have made on completed work.
- help you set a time and find a quiet place to do math homework.
- help you memorize the addition, subtraction, multiplication, and division facts.
- solve problems as you play together, shop together, and do household chores.
- visit **The Learning Site** at **www.harcourtschool.com**
- have **Fun with Math!**

Have a great year!

The Authors

Be a Good Problem Solver!

You need to organize your thinking. You can use problem-solving steps to stay on track.

Use these problem-solving steps. They can help you think through a problem.

UNDERSTAND the problem.

What are you asked to find?	Restate the question in your own words.
What information will you use?	List all the information in the problem.
Is there information you will not use? If so, what?	You may not need all the information given.

PLAN a strategy to solve.

What strategy can you use to solve the problem?	Think about some problem solving strategies you can use. Then choose one.

SOLVE the problem.

How can you use the strategy to solve the problem?	Follow your plan. Show your solution.

CHECK your answer.

Look back at the problem. Does the answer make sense? Explain.	Be sure you answered the question that is asked.
What other strategy could you use?	Solving the problem by another method is a good way to check your work.

PROBLEM SOLVING STRATEGIES

- Draw a Diagram or Picture
- Make a Model or Act it Out
- Make an Organized List
- Find a Pattern
- **Make a Table or Graph**
- Predict and Test
- Work Backward
- Solve a Simpler Problem
- Write a Number Sentence or an Equation
- Use Logical Reasoning

Try It

Here's how you can use the problem-solving steps to solve a problem.

Make a Table

PROBLEM The children in Ms. Ling's class wanted to care for an animal. They could choose a fish, a rabbit, or a hamster. The children wrote their votes on slips of paper. The votes are shown here. Which animal did they choose? How can the class keep a record of their votes?

hamster rabbit rabbit fish hamster
rabbit fish rabbit hamster rabbit
hamster rabbit fish fish rabbit
hamster

UNDERSTAND the problem.

I need to find which animal was chosen. I also need to keep a record of the votes. The votes are shown on the slips of paper.

PLAN a strategy to solve.

I can *make a table* to show tallies of the votes. Then I can count the tallies to see which animal was chosen.

SOLVE the problem.

OUR VOTES	
Animal	**Tallies**
Fish	||||
Rabbit	𝍸 ||
Hamster	𝍸

The rabbit has 7 votes. So, it was chosen by the class.

CHECK your answer.

I can check that the vote tallies in the table match the slips of paper shown in the problem. Since *Rabbit* has the most votes, this solves the problem.

Getting Ready!

Remember the Properties of Addition

These rules, called properties, can help you recall basic facts and compute mentally.

ORDER PROPERTY OF ADDITION

Changing the order of the addends does not change the sum.

5 + 3 = 8	3 + 5 = 8
addends sum	addends sum

This means that $5 + 3 = 3 + 5$.

ZERO PROPERTY OF ADDITION

When you add zero to a number, the sum is that number.

$6 + 0 = 6 \qquad 0 + 5 = 5$

GROUPING PROPERTY OF ADDITION

When you group addends in different ways, the sum is the same.

$$5 + (3 + 2) = (5 + 3) + 2$$
$$5 + 5 = 8 + 2$$
$$10 = 10$$

▶ Practice

Copy and complete. Write the property shown.

1. $4 + 9 = ■ + 4$

2. $0 + 10 = ■$

3. $3 + (7 + 2) = ■$
$(3 + 7) + 2 = ■$

4. $4 + (5 + 6) = ■$
$(4 + 5) + 6 = ■$

5. $7 + 0 = ■$

6. $9 + ■ = 7 + 9$

Practice Addition and Subtraction Facts

Use addition properties and fact families to review addition and subtraction facts.

A **FACT FAMILY** is a set of related addition and subtraction number sentences that use the same numbers.

8, 6, 14		**5, 5, 10**
$8 + 6 = 14$	$6 + 8 = 14$	$5 + 5 = 10$
$14 - 6 = 8$	$14 - 8 = 6$	$10 - 5 = 5$

▶ Practice

Copy. Find the sum or difference.

1. $3 + 4 =$ ■ **2.** $7 + 6 =$ ■ **3.** $0 + 8 =$ ■ **4.** $5 + 5 =$ ■

5. $3 + 9 =$ ■ **6.** $5 + 7 =$ ■ **7.** $9 + 9 =$ ■ **8.** $7 + 8 =$ ■

9. $4 + 6 =$ ■ **10.** $15 - 9 =$ ■ **11.** $12 - 7 =$ ■ **12.** $15 - 8 =$ ■

13. $11 - 6 =$ ■ **14.** $17 - 8 =$ ■ **15.** $16 - 9 =$ ■ **16.** $14 - 7 =$ ■

17. $9 - 0 =$ ■ **18.** $13 - 6 =$ ■ **19.** $15 - 6 =$ ■ **20.** $12 - 4 =$ ■

21. $\begin{array}{r} 8 \\ +5 \\ \hline \end{array}$ **22.** $\begin{array}{r} 16 \\ -\ 8 \\ \hline \end{array}$ **23.** $\begin{array}{r} 4 \\ +7 \\ \hline \end{array}$ **24.** $\begin{array}{r} 10 \\ -\ 9 \\ \hline \end{array}$ **25.** $\begin{array}{r} 11 \\ -\ 2 \\ \hline \end{array}$

26. $\begin{array}{r} 6 \\ +3 \\ \hline \end{array}$ **27.** $\begin{array}{r} 15 \\ -\ 6 \\ \hline \end{array}$ **28.** $\begin{array}{r} 5 \\ +6 \\ \hline \end{array}$ **29.** $\begin{array}{r} 12 \\ -\ 7 \\ \hline \end{array}$ **30.** $\begin{array}{r} 8 \\ -8 \\ \hline \end{array}$

31. $\begin{array}{r} 3 \\ +8 \\ \hline \end{array}$ **32.** $\begin{array}{r} 14 \\ -\ 8 \\ \hline \end{array}$ **33.** $\begin{array}{r} 7 \\ +7 \\ \hline \end{array}$ **34.** $\begin{array}{r} 18 \\ -\ 9 \\ \hline \end{array}$ **35.** $\begin{array}{r} 5 \\ +9 \\ \hline \end{array}$

36. $\begin{array}{r} 6 \\ +6 \\ \hline \end{array}$ **37.** $\begin{array}{r} 13 \\ -\ 5 \\ \hline \end{array}$ **38.** $\begin{array}{r} 4 \\ +9 \\ \hline \end{array}$ **39.** $\begin{array}{r} 17 \\ -\ 9 \\ \hline \end{array}$ **40.** $\begin{array}{r} 8 \\ +8 \\ \hline \end{array}$

CHAPTER 1

Place Value and Number Sense

There are more than 150 kinds of horses. Sometimes horses are used on farms to help with the work. Horses need hay, oats, and fresh water to stay healthy.

PROBLEM SOLVING Use the pictograph. Skip-count to find how much water a horse needs each week and each month.

DATA LINK

HOW MUCH WATER A HORSE NEEDS

Time	Water
1 day	🪣
1 week	🪣🪣🪣🪣🪣🪣🪣
1 month	🪣🪣🪣🪣🪣🪣🪣🪣🪣🪣🪣🪣🪣🪣🪣 🪣🪣🪣🪣🪣🪣🪣🪣🪣🪣🪣🪣🪣🪣🪣

Water

Key: Each 🪣 = 10 gallons.

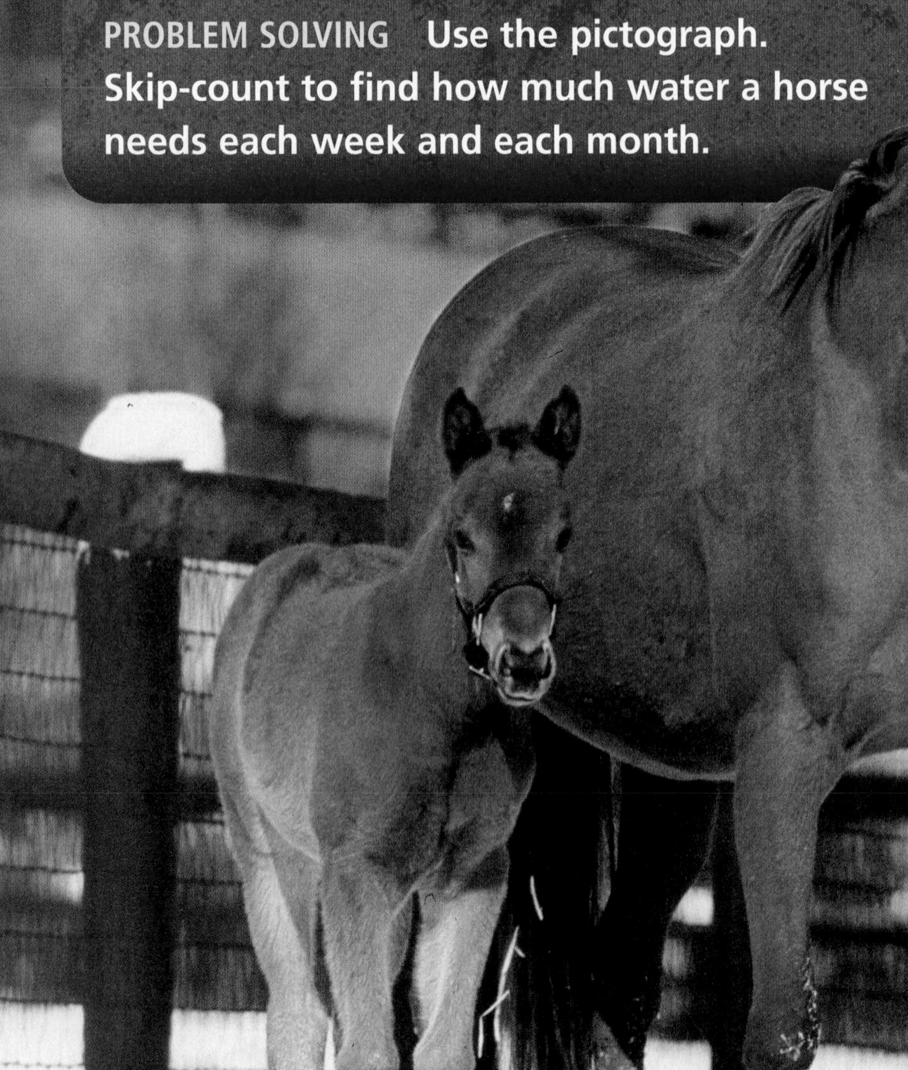

Use this page to help you review and remember important skills needed for Chapter 1.

ORDINAL NUMBERS (For Intervention, see p. H2.)

For 1–6, use the list of names.

1. Kelly is first on the list. Who is third?

2. Who is ninth on the list?

3. Who is fifth on the list?

4. In which position is Juan on the list?

5. Who is eighth on the list?

6. In which position is Julie on the list?

Kelly
Tom
Sally
Susan
Timothy
Julie
Juan
Matt
Maria

READ AND WRITE ONES, TENS, HUNDREDS (For Intervention, see p. H2.)

Write the number.

7. **8.** **9.** **10.**

11. fifty-eight

12. three hundred twenty-one

13. six hundred forty-five

14. nine hundred four

PLACE VALUE WITH ONES, TENS, HUNDREDS (For Intervention, see p. H3.)

Write the number.

15. 6 tens 5 ones

16. 4 tens 9 ones

17. 9 tens 0 ones

18. 3 hundreds 6 tens 1 one

19. 1 hundred 7 tens 5 ones

20. 5 hundreds 2 tens 9 ones

21. 4 hundreds 2 tens 7 ones

22. 8 hundreds 0 tens 3 ones

23. 6 hundreds 8 tens 0 ones

24. 7 hundreds 3 tens 8 ones

LESSON 1

Patterns on a Hundred Chart

Quick Review

1. 8 + 2
2. 6 + 3
3. 9 + 3
4. 10 + 5
5. 25 + 5

VOCABULARY
even odd

MATERIALS
hundred chart, crayons

Explore

Use the hundred chart. Start at 2. Skip-count by twos.

STEP 1
Use a hundred chart.

STEP 2
Start at 2. Shade that box.

STEP 3
Skip-count by twos, and shade each box you land on.

1	2	3	4	5	6	7	8	9	10
11	12	13	14	15	16	17	18	19	20
21	22	23	24	25	26	27	28	29	30
31	32	33	34	35	36	37	38	39	40
41	42	43	44	45	46	47	48	49	50
51	52	53	54	55	56	57	58	59	60
61	62	63	64	65	66	67	68	69	70
71	72	73	74	75	76	77	78	79	80
81	82	83	84	85	86	87	88	89	90
91	92	93	94	95	96	97	98	99	100

MATH IDEA **Even** numbers have a 0, 2, 4, 6, or 8 in the ones place. **Odd** numbers have a 1, 3, 5, 7, or 9 in the ones place.

Try It

a. Start at 2. Skip-count by twos. What numbers do you land on? Are they even or odd? How do you know?

b. Start at 3. Skip-count by threes. What numbers do you land on? Are they even or odd? How do you know?

REASONING **What if** you want to count by twos and name *odd* numbers? On what number should you start?

2, 4, 6, 8, . . . what number do we land on next?

Connect

Look at your shaded chart. What pattern do you see?

All the even numbers are shaded. Even numbers have a 0, 2, 4, 6, or 8 in the ones place.

The odd numbers are not shaded. Odd numbers have a 1, 3, 5, 7, or 9 in the ones place.

1	2	3	4	5	6	7	8	9	10
11	12	13	14	15	16	17	18	19	20
21	22	23	24	25	26	27	28	29	30
31	32	33	34	35	36	37	38	39	40
41	42	43	44	45	46	47	48	49	50
51	52	53	54	55	56	57	58	59	60
61	62	63	64	65	66	67	68	69	70
71	72	73	74	75	76	77	78	79	80
81	82	83	84	85	86	87	88	89	90
91	92	93	94	95	96	97	98	99	100

Practice and Problem Solving

Technology Link

More Practice: Use E-lab, *Number Patterns.*

www.harcourtschool.com/elab2002

Use the hundred chart. Tell whether the number is *odd* or *even*.

1. 7 **2.** 8 **3.** 12 **4.** 30

5. 27 **6.** 98 **7.** 19 **8.** 45

9. 44 **10.** 81 **11.** 76 **12.** 100

Use the hundred chart.

13. Start at 5. Skip-count by fives. Move 4 skips. What number do you land on? Is it odd or even?

14. Start at 10. Skip-count by tens. Move 6 skips. What number do you land on? Is it odd or even?

15. The houses on Quinn's street are numbered 4, 8, 12, 16, and 20. What are the next three house numbers? Explain.

16. **REASONING** Marcos skip-counted. He started at 2. He landed on 15. Could he be skip-counting by twos? Why?

Mixed Review and Test Prep

17. $\begin{array}{r} 64 \\ +21 \\ \hline \end{array}$

18. $\begin{array}{r} 78 \\ -54 \\ \hline \end{array}$

19. $25 + 11 = \blacksquare$

20. $37 - 10 = \blacksquare$

21. **TEST PREP** Jon had 12 red marbles and 17 blue marbles. How many marbles did he have in all?

A 30 **B** 29 **C** 19 **D** 5

LESSON 2

Understand Place Value

Quick Review

Write the number.

1. 5 tens 1 one
2. 4 tens 3 ones
3. 7 tens 0 ones
4. 1 ten 9 ones
5. 2 hundreds 6 tens 8 ones

VOCABULARY

digits
standard form
expanded form
word form

Learn

FARM FACTS The symbols 0, 1, 2, 3, 4, 5, 6, 7, 8, and 9 are **digits**. Numbers are made up of digits.

On Mr. Sam's farm there are 248 chickens. What does the number 248 mean?

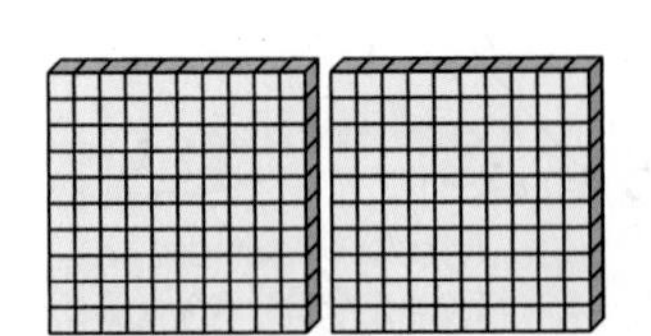

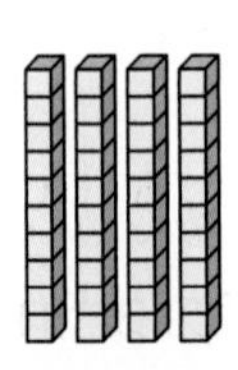

HUNDREDS	TENS	ONES
2	4	8

So, 248 means 2 hundreds + 4 tens + 8 ones or 200 + 40 + 8.

MATH IDEA You can write a number in different ways: standard form, expanded form, and word form.

Standard form: 248

Expanded form: 200 + 40 + 8

Word form: two hundred forty-eight

- What does the number 527 mean?

Check

USE DATA For 1–2, use the table.

1. **Tell** the value of the digit 3 in the number of cows on the farm.
2. What is the expanded form for the number of pigs on the farm?

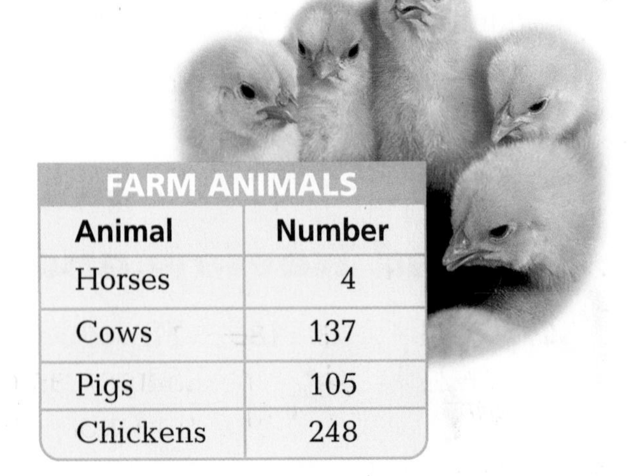

FARM ANIMALS

Animal	Number
Horses	4
Cows	137
Pigs	105
Chickens	248

Write each number in standard form.

3. 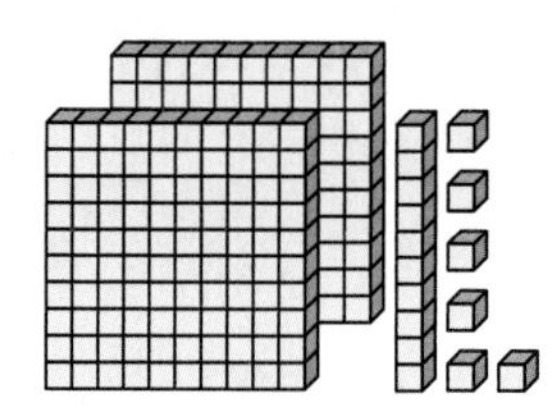

4. 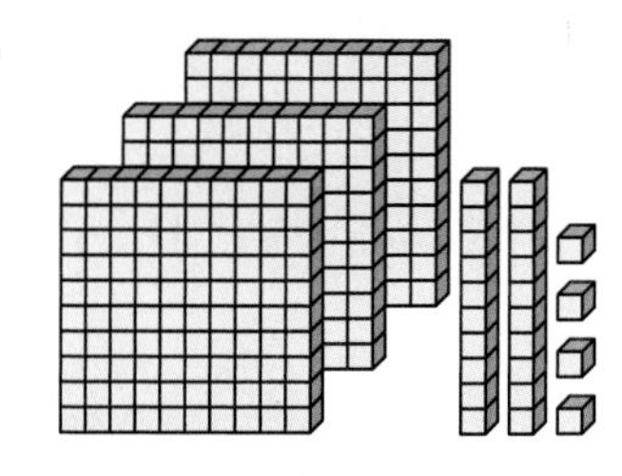

5. 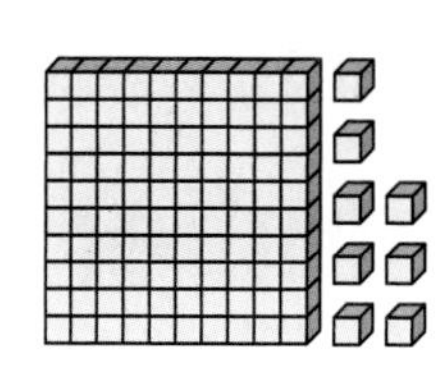

Practice and Problem Solving

Write each number in standard form.

6. 100 + 50 + 3
7. 400 + 70 + 6
8. 600 + 30 + 9
9. 900 + 2
10. 4 hundreds 2 tens 1 one
11. 6 hundreds 8 tens 3 ones
12. 7 hundreds 2 tens 3 ones
13. 4 hundreds 5 ones
14. one hundred three
15. three hundred forty-five
16. six hundred eleven
17. nine hundred seventy-one

Write the value of the blue digit.

18. 846
19. 267
20. 493
21. 923

22. Mr. Sam put 297 bales of hay in one barn. There are still 86 bales of hay in the field. How many more bales of hay are in the barn than in the field?

23. There are 100 cows in one field and 30 cows in another field. There are 7 cows in the barn. How many cows are there in all?

24. **REASONING** What is the greatest 3-digit number you can write using the digits 8, 9, and 4?

25. **REASONING** I am a digit in each of the numbers 312, 213, and 132. My value is different in all three numbers. What digit am I? What is my value in each number?

Mixed Review and Test Prep

26. 24 + 24 = ■
27. 24 − 24 = ■

28. Pablo came in second, Jacob came in third, and Lauren came in ahead of Pablo. Who won the race?

29. What number continues the pattern? (p. 2)

3, 6, 9, 12, 15, ■

30. **TEST PREP** Which number is greater than 54?

A 34 B 45 C 53 D 55

EXTRA PRACTICE page H32, Set A

LESSON 3

Understand Greater Numbers

Quick Review

Write the value of the blue digit.

1. 246 2. 394 3. 714

4. 520 5. 802

Learn

BUILDING BLOCKS You can show 100 with a base-ten hundreds block or with a 10-by-10 paper grid. How can you show 1,000?

Activity

Materials: base-ten blocks, 10-by-10 grid paper, paste, stapler

Model 1,000. Stack hundreds blocks until you have built a cube of 1,000. Then make a book of 1,000 squares.

STEP 1

Paste 10-by-10 paper grids onto pieces of paper. Use one grid for each piece of paper. Number each page at the bottom. Staple the pages together.

STEP 2

Number the squares of the grids from 1 to 1,000.

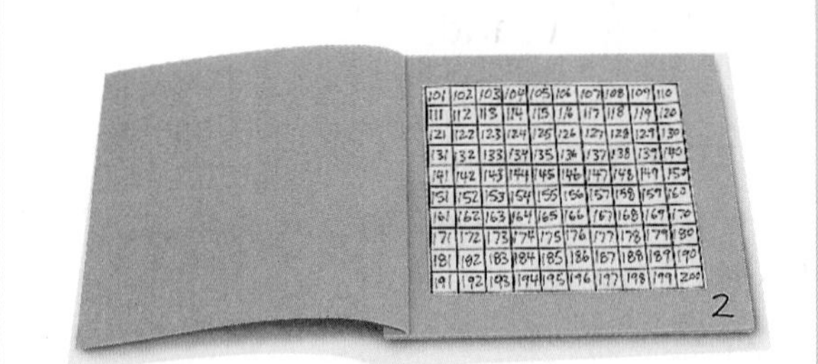

A thousands cube and a book of 1,000 squares are models for 1,000.

- How many 10-by-10 grids did it take to make the 1,000 book? Explain.

Building with Thousands

What if you put 3 books of one thousand squares together? How many squares will you have in all?

You can add the squares.

$1,000 + 1,000 + 1,000 = 3,000$

You can skip-count by thousands.

1,000 2,000 3,000

So, the 3 books equal 3,000 squares.

- How many books would you need to have 5,000 squares?

Examples

Here are some ways to show 2,346.

A With base-ten blocks

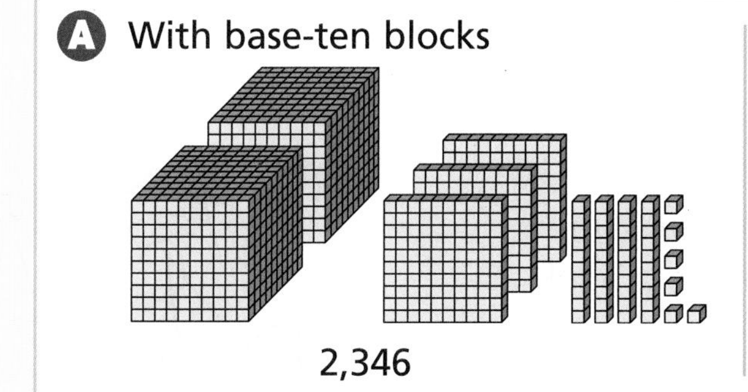

2,346

B On a place-value chart

Thousands	Hundreds	Tens	Ones
2,	3	4	6

C In standard form:
2,346
↑
A comma separates the thousands and hundreds.

D In expanded form:
2,000 + 300 + 40 + 6

E In word form:
two thousand, three hundred forty-six

Check

1. **Tell** how to write the expanded form for 5,403.

Write in standard form.

2.

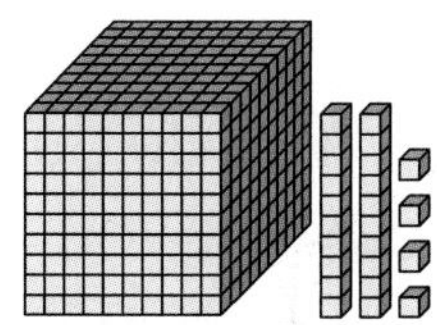

3.

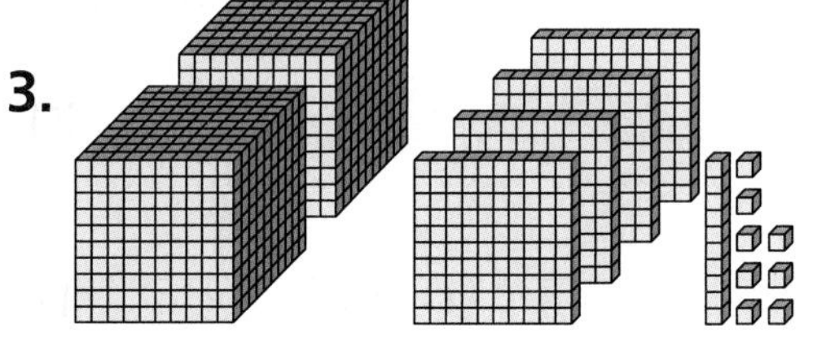

Write in expanded form.

4. 5,632 5. 7,401 6. 8,011 7. 3,462

LESSON CONTINUES

Practice and Problem Solving

Write in standard form.

8.

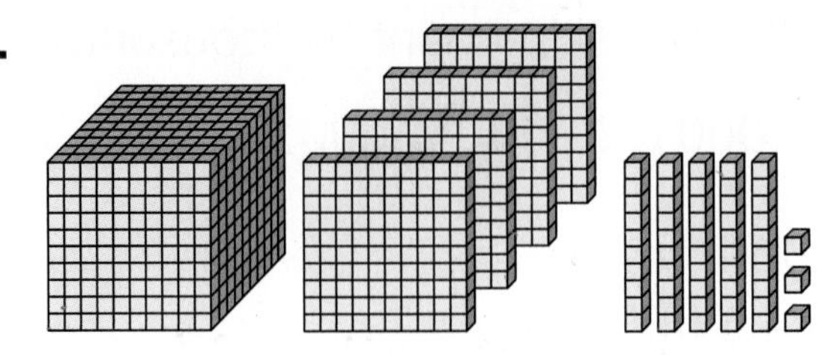

9.

10. $5,000 + 400 + 50$

11. $2,000 + 300 + 90 + 7$

12. $3,000 + 700 + 20 + 3$

13. $1,000 + 10 + 8$

14. two thousand, four hundred eighty-three

15. six thousand, one hundred ninety-four

Write in expanded form.

16. 1,234 **17.** 4,321 **18.** 3,016 **19.** 8,367

Write in word form.

20. 87 **21.** 148 **22.** 317 **23.** 599

Complete.

24. $1,000 + 500 + \blacksquare + 8 = 1,548$

25. $3,000 + \blacksquare + 90 + 7 = 3,897$

26. **USE DATA** The pictograph shows what Molly saw at the farm. How many animals did Molly see in all?

FARM ANIMALS

Horses	
Lambs	
Cows	

Key: Each = **2 Animals.**

27. **REASONING** I am a 3-digit number. My hundreds digit equals the sum of my tens digit and my ones digit. My tens digit is 4 more than my ones digit. My ones digit is 1. What number am I?

28. What is the least possible number you can write with the digits 2, 9, 4, and 7? Use each digit only once.

29. **Write About It** Why do you have to use a zero when you write one thousand, six hundred four in standard form?

30. The number 124 is an even number. Write 5 more even numbers including one with 3 digits.

31. Show that each of the even numbers 6, 8, 10, and 12 can be written as the sum of a group of 2's.

Mixed Review and Test Prep

Tell whether the number is *even* or *odd*. (p. 2)

32. 61 **33.** 18 **34.** 58

35. 97 **36.** 102 **37.** 183

Find the sum.

38. $6 + 3$ **39.** $4 + 7$ **40.** $12 + 4$

41. $9 + 8$ **42.** $10 + 5$ **43.** $11 + 10$

Find the difference.

44. $14 - 5$ **45.** $16 - 8$ **46.** $13 - 7$

47. $17 - 7$ **48.** $18 - 6$ **49.** $11 - 6$

Find the sum.

50. $6 + 5 + 3$ **51.** $9 + 7 + 1$ **52.** $8 + 3 + 7$

53. TEST PREP Which number is eight hundred four in standard form? (p. 4)

A 940 C 804
B 840 D 84

54. TEST PREP Which number is the expanded form for 432? (p. 4)

F $40 + 2$ H $400 + 30$
G $400 + 2$ J $400 + 30 + 2$

PROBLEM SOLVING Thinker's Corner

REASONING

Materials: number cards 0–9, paper and pencil

A. Play the game with a partner. Choose 3 number cards. Make a 3-digit number. Record it.

B. Use these same number cards to make and record as many *different* 3-digit numbers as you can.

C. Players get one point for each different 3-digit number.

D. Players take turns choosing 3 cards and making all the possible numbers.

The player who scores 25 points first wins the game.

CHALLENGE: Play the game again, choosing 4 number cards and making 4-digit numbers.

EXTRA PRACTICE page H32, Set B

LESSON 4

Understand Ten Thousands

Quick Review

Write in expanded form.

1. 384
2. 51
3. 677
4. 9,240
5. 3,818

Learn

BUILDING BOOKS How many 1,000 books do you need to show 10,000 squares?

Activity

MATERIALS: books of 1,000 squares

STEP 1

Put 2 books of 1,000 together.

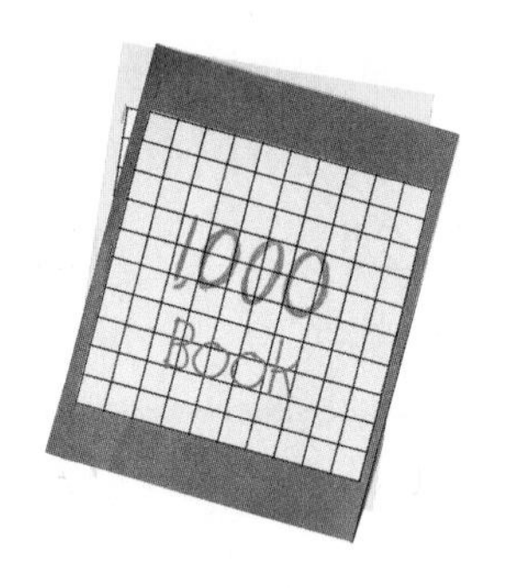

STEP 2

Continue putting books of 1,000 together until they show 10,000.

- How many books of 1,000 squares would you need to have 15,000 squares? to have 23,000 squares?

Remember

Put a comma between the thousands place and the hundreds place.

23,680
↑
comma

What does each digit in 23,680 mean?

Use a place-value chart.

TEN THOUSANDS	THOUSANDS	HUNDREDS	TENS	ONES
2	3,	6	8	0

Standard form: 23,680

Expanded form: 20,000 + 3,000 + 600 + 80

Word form: twenty-three thousand, six hundred eighty

- What is the value of the 3 in 34,152?

Check

1. **Tell** the expanded form for 21,694.

Write in standard form.

2. 20,000 + 6,000 + 700 + 30 + 4
3. thirty-five thousand, nine hundred forty-seven

Write in expanded form.

4. 16,723
5. 52,019

Practice and Problem Solving

Write in standard form.

6. 20,000 + 6,000 + 700 + 30 + 4
7. 10,000 + 400 + 8
8. forty-two thousand, three hundred fifteen
9. eighteen thousand, nine hundred

Write in expanded form.

10. 16,723
11. 55,119
12. 11,012
13. 49,207

Write the value of the blue digit.

14. 81,465
15. 26,817
16. 43,912
17. 19,273

Complete.

18. 10,000 + 2,000 + ■ + 50 + 1 = 12,651
19. 60,000 + ■ + 300 + 10 + 9 = 62,319

20. **REASONING** I am an even number between 51,680 and 51,700. The sum of my digits is 23. What number am I?

21. **What's the Error?** Karla wrote eleven thousand, forty-five as 1,145. Explain her error. Write the number correctly in standard form.

Mixed Review and Test Prep

22. $\begin{array}{r} 77 \\ -52 \\ \hline \end{array}$

23. $\begin{array}{r} 72 \\ +19 \\ \hline \end{array}$

24. 25 + ■ = 29

25. 43 + ■ = 49

26. **TEST PREP** What is the value of the blue digit in 5,789? (p. 6)

A 7
B 70
C 700
D 7,000

EXTRA PRACTICE page H32, Set C

LESSON 5

Problem Solving Strategy
Use Logical Reasoning

Quick Review

1. 400 + ■ + 9 = 499
2. ■ + 700 + 40 + 5 = 8,745
3. ■ + 60 + 3 = 763
4. 900 + 30 + ■ = 936
5. ■ + 400 + 7 = 10,407

PROBLEM I am a 2-digit number. The sum of my digits is 17. The tens digit is odd. The ones digit is even. What number am I?

UNDERSTAND

- What are you asked to find?
- Is there information you will not use? If so, what?

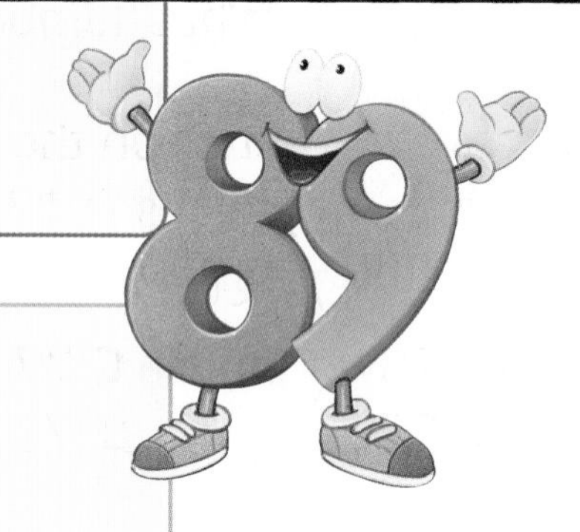

PLAN

- What strategy can you use to solve the problem?

You can *use logical reasoning.*

SOLVE

- How can you use the strategy to solve the problem?

Use a hundred chart. Find the numbers with digits whose sum equals 17. The numbers are 89 and 98.
Then decide which number has a tens digit that is odd and a ones digit that is even.
That number is 98.

So, the number is 98.

1	2	3	4	5	6	7	8	9	10
11	12	13	14	15	16	17	18	19	20
21	22	23	24	25	26	27	28	29	30
31	32	33	34	35	36	37	38	39	40
41	42	43	44	45	46	47	48	49	50
51	52	53	54	55	56	57	58	59	60
61	62	63	64	65	66	67	68	69	70
71	72	73	74	75	76	77	78	79	80
81	82	83	84	85	86	87	88	89	90
91	92	93	94	95	96	97	98	99	100

CHECK

- Look at the problem. Does your answer make sense? Explain.
- Explain how you know that 89 and 98 are the only 2-digit numbers whose digits have a sum that equals 17.

Problem Solving Practice

- Draw a Diagram or Picture
- Make a Model or Act It Out
- Make an Organized List
- Find a Pattern
- Make a Table or Graph
- Predict and Test
- Work Backward
- Solve a Simpler Problem
- Write a Number Sentence
- **Use Logical Reasoning**

Use logical reasoning and solve.

1. **What if** the riddle said that the tens digit is even, and the ones digit is odd and the sum of the digits is 5? What would be the answer to this riddle?

2. I am a number on the hundred chart. Both of my digits are odd. Both of my digits are the same. What numbers can I be?

3. I am a number on the hundred chart. The sum of my digits is 12. My tens digit is 7. What number am I?

   A 72 **C** 77
   B 75 **D** 79

4. I am a number in the third row on the hundred chart. My ones digit is 5 more than my tens digit. What number am I?

   F 22 **H** 27
   G 24 **J** 30

Mixed Strategy Practice

5. Write the greatest possible four-digit number using the digits 3, 4, 5, and 6. Write the least possible four-digit number.

6. **REASONING** A number has the same number of ones, tens, and hundreds. If the sum of the digits is 9, what is the number?

7. **USE DATA** The pictograph shows Tim's pets. How many more lambs than ponies does Tim have? How many animals does Tim have in all?

8. **Write a problem** about a number riddle. Use the hundred chart to help you. Tell how to solve the problem.

TIM'S PETS

Ponies	
Calves	
Lambs	

Key: Each [symbol] = 1 Animal.

Problem Solving Strategy

CHAPTER 1

Review/Test

CHECK VOCABULARY AND CONCEPTS

Choose the best term from the box.

odd
even
digits
expanded form

1. A number that ends in 0, 2, 4, 6, or 8 is an __?__ number. (p. 2)
2. The symbols 0, 1, 2, 3, 4, 5, 6, 7, 8, and 9 are called __?__. (p. 4)

Write each number in standard form. (pp. 4–9)

3. 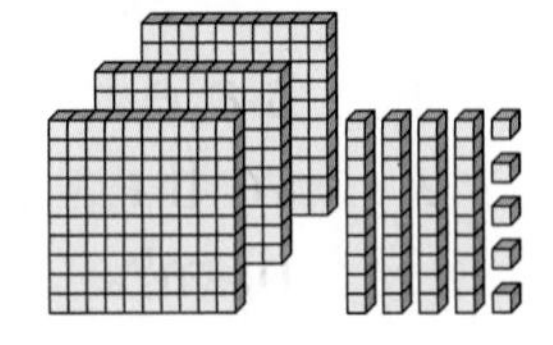

4. 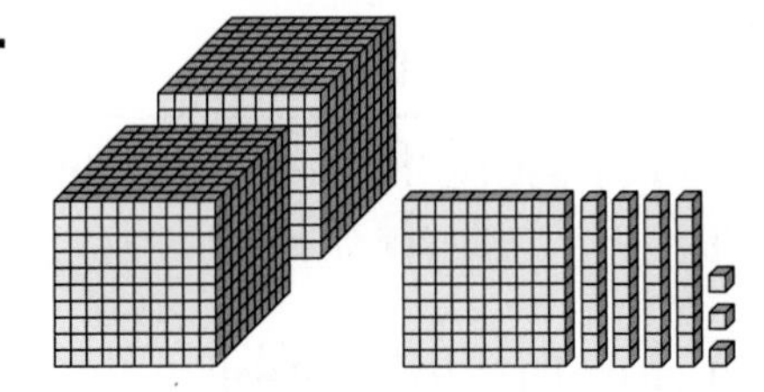

CHECK SKILLS

Write whether the number is *odd* or *even*. (pp. 2–3)

5. 317
6. 200
7. 1,348
8. 12,999

Write in standard form. (pp. 4–11)

9. 800 + 60 + 9
10. 3,000 + 700 + 10 + 1
11. 8,000 + 500 + 20 + 2
12. 30,000 + 4,000 + 700 + 5
13. two thousand, thirty-nine
14. fifteen thousand, sixty-five

Write the value of the blue digit. (pp. 4–5, 10–11)

15. 863
16. 9,845
17. 12,053
18. 32,859

CHECK PROBLEM SOLVING

Solve. Use a hundred chart. (pp. 12–13)

19. I am a 2-digit number. My tens digit is two more than my ones digit. My ones digit is between 4 and 6. What number am I?
20. I am a 2-digit number. I am greater than 40 but less than 60. My tens and ones digits are the same. I am an odd number. What number am I?

Standardized Test Prep

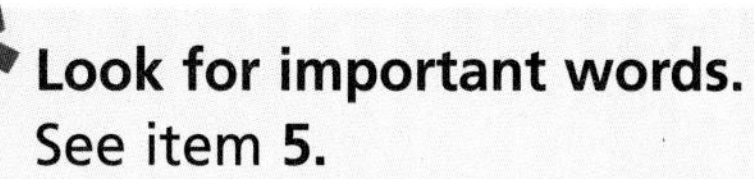

TIP! **Look for important words.**
See item **5.**

Not is an important word. **Not** odd means a number that is even.

Also see problem **2,** p. H62.

For 1–9, choose the best answer.

1. What does the 6 mean in 461?

A 6 **B** 60 **C** 600 **D** 6,000

---

2. How is 6 hundreds 2 tens 3 ones written in standard form?

F 263 **G** 326 **H** 623 **J** 6,023

---

3. How is 2,047 written in expanded form?

A $2{,}000 + 40 + 7$
B $2{,}000 + 400 + 7$
C $20 + 40 + 7$
D $2{,}000 + 47$

---

4. Which number makes the number sentence true?

$2{,}000 + 400 + ■ + 5 = 2{,}475$

F 7 **G** 70 **H** 700 **J** 7,000

---

5. Which number is **not** odd?

A 35 **B** 41 **C** 50 **D** 77

6.

$$\begin{array}{r} 43 \\ 21 \\ +19 \\ \hline \end{array}$$

F 73 **G** 83 **H** 713 **J** NOT HERE

---

7. The sum of the digits of a number is 10. Both of the digits are even. The ones digit is 2 more than the tens digit. What is the number?

A 19 **B** 28 **C** 37 **D** 46

---

8. What is the value of the 2 in 23,790?

F 20 **G** 200 **H** 2,000 **J** NOT HERE

---

9. A number has the same number of ones, tens, and hundreds. The sum of the digits is 24. What is the number?

A 699 **B** 777 **C** 888 **D** NOT HERE

Write What You Know

10. What is the value of the 3 in 25,381 and in 39,047? In which number does the 3 have a greater value? Explain your answer.

11. These numbers follow a pattern.
4, 8, 12, 16, 20
Describe the pattern and tell how you can find the next three numbers.

Compare, Order, and Round Numbers

Crater Lake in the Cascade Mountains of Oregon is the deepest lake in the United States.

PROBLEM SOLVING Use the graph. Which lakes have depths that round to 800 feet when rounded to the nearest 100?

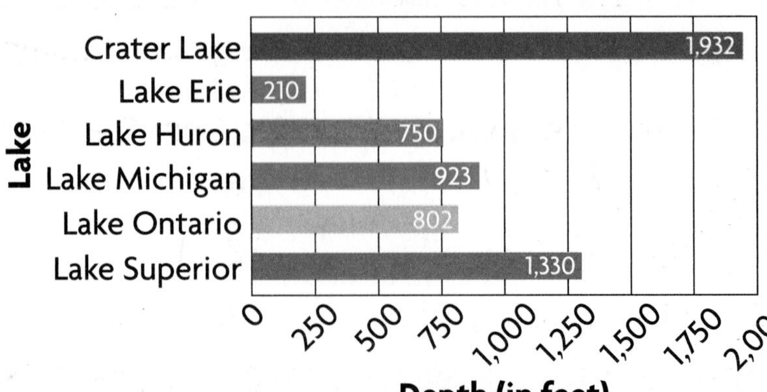

CHECK WHAT YOU KNOW

Use this page to help you review and remember important skills needed for Chapter 2.

COMPARE 2-DIGIT NUMBERS (For Intervention, see p. H3.)

Write the words *greater than* or *less than*.

1. 78 is ? 41.
2. 35 is ? 55.
3. 41 is ? 45.
4. 63 is ? 68.
5. 37 is ? 31.
6. 56 is ? 58.
7. Which number is greater, 98 or 89?

ORDER NUMBERS (For Intervention, see p. H4.)

Write the number that is just after, just before, or between.

8. 7, ■
9. ■, 45
10. 21, ■, 23
11. 307, ■
12. ■, 768
13. 871, ■, 873
14. 454, ■
15. ■, 133
16. 189, ■, 191

Write the number each model shows.

17. 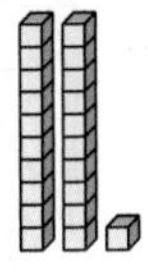
18. 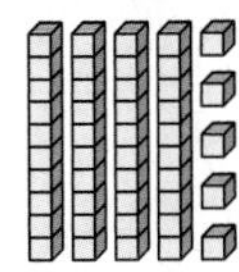
19. 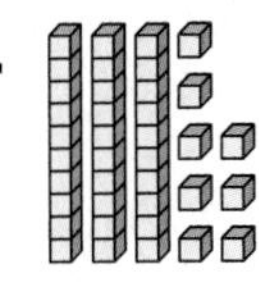
20. Which model above shows the greatest number?

Write the number each model shows.

21. 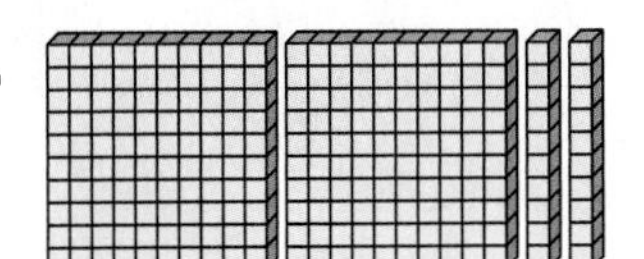
22. 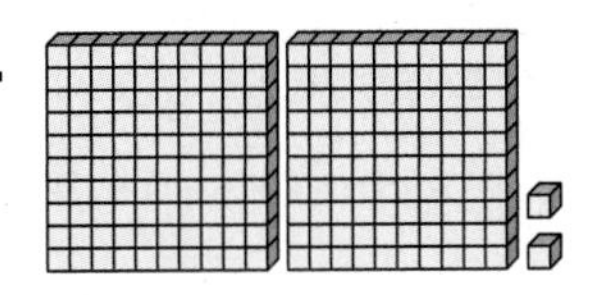
23. 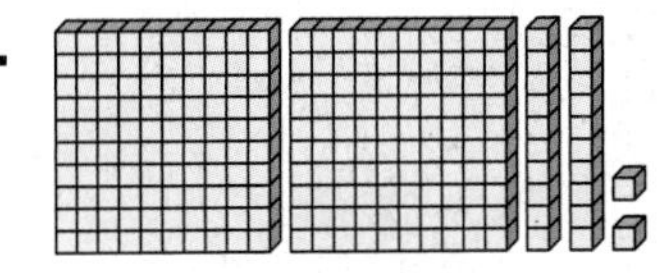
24. Which model shows the greatest number?

Write the numbers in order from least to greatest.

25. 74, 79, 73
26. 19, 16, 18
27. 49, 57, 51
28. 117, 127, 111
29. 181, 178, 176
30. 342, 349, 344
31. 248, 255, 250
32. 75, 69, 84
33. 428, 423, 425

LESSON 1

Size of Numbers

Quick Review

What number is just before each?

1. 9 **2.** 16

3. 20 **4.** 103

5. 330

VOCABULARY

benchmark numbers

Learn

HOW MANY? Numbers that help you estimate the number of objects without counting them are called **benchmark numbers**. Any useful number can be a benchmark.

About how many beans are in Jar B?
You can use 25 as a benchmark to estimate.

There are 25 beans in Jar A.

There are ■ beans in Jar B.

There are about two times as many beans in Jar B.

So, there are about 50 beans in Jar B.

Think about the number of students in your class, your grade, and your school. Which has about 20 students? Which has about 100 students? Which has about 500 students?

BENCHMARK	NUMBER TO BE ESTIMATED
20	students in your class
100	students in your grade
500	students in your school

- There are 5 third-grade classes in Nora's school. About how many students are in the third grade? What benchmark can you use to help you?

Check

1. **Explain** the benchmark you would use to estimate the number of girls in your class.

Estimate the number of beans in each jar. Use Jars A and B as benchmarks.

Jar A has 10 beans.

Jar B has about 50 beans.

2. 

10 or 50?

3. 

25 or 50?

4. 

100 or 200?

Practice and Problem Solving

5. Estimate the number of beans in the jar at the right. Use Jars A and B as benchmarks.

Choose a benchmark of 10, 100, or 500 to estimate each.

6. the number of players on a soccer team
7. the number of pretzels in a large bag
8. the number of sheets in a package of notebook paper
9. the number of leaves on a tree in summer
10. Juan has 30 blocks. He gives 18 to Rick but gets 14 from Ron. How many blocks does Juan have now?
11. **Write a problem** in which a benchmark is used to estimate. Solve.

Mixed Review and Test Prep

Write each number in expanded form. (p. 4)

12. 268 **13.** 354 **14.** 420 **15.** 679

16. **TEST PREP** Choose the value of the blue digit in 15,688. (p. 10)

A 6 **B** 60 **C** 600 **D** 700

EXTRA PRACTICE page H33, Set A

LESSON

2 Compare Numbers

Quick Review

Write how many tens each number has.

1. 15 **2.** 20 **3.** 54
4. 37 **5.** 182

VOCABULARY

compare
is less than <
is greater than >
is equal to =

Learn

HOW NEAR? HOW FAR? Beth lives 262 miles from Homer and 245 miles from Lakewood. Which city does she live closer to?

Compare numbers to decide which of two numbers is greater. Use these symbols.

is less than	**is greater than**	**is equal to**
<		=

Use base-ten blocks to compare 262 and 245.

STEP 1

Show 262 and 245 with base-ten blocks.

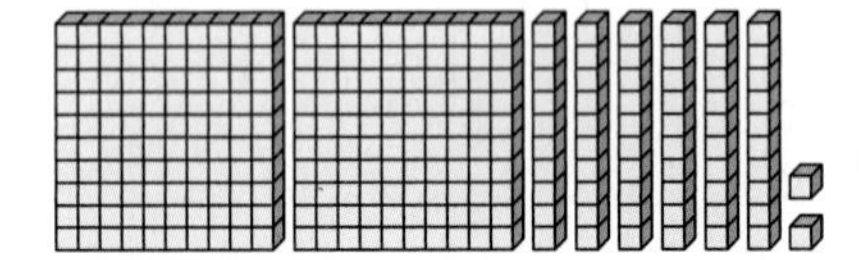

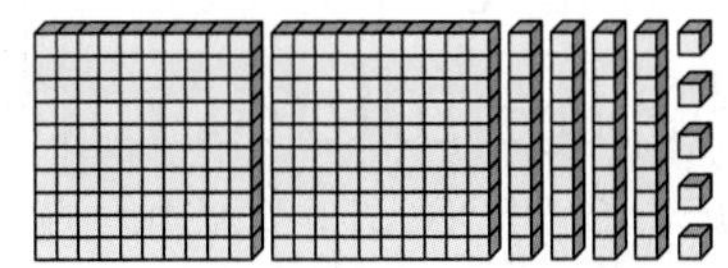

STEP 2

Compare from left to right. First compare the hundreds. Since they are the same, compare the tens.

6 tens is greater than 4 tens. So, 262 is greater than 245.

Beth lives closer to Lakewood since $262 > 245$.

You can use a number line to compare numbers.

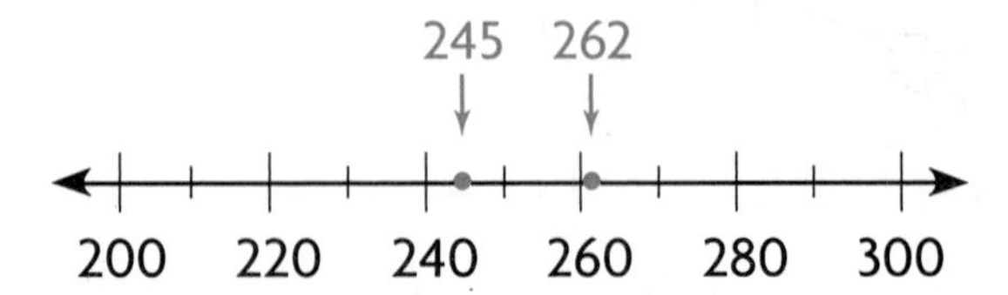

From left to right, the numbers on a number line are in order from *least* to *greatest*.

So, $245 < 262$ or $262 > 245$.

Comparing Thousands

You can use a place-value chart to compare numbers.

Compare 3,165 and 3,271 starting from the left.

THOUSANDS	HUNDREDS	TENS	ONES
3,	1	6	5
3,	2	7	1

↑ Thousands are the same.

↑ 200 > 100

Technology Link
To learn more about *Comparing Numbers*, watch the **Harcourt Math Newsroom Video,** *Chicago Skyscraper.*

So, 3,271 > 3,165 or 3,165 < 3,271.

MATH IDEA Compare numbers by using base-ten blocks, a number line, or a place-value chart.

Check

1. **Explain** how to use base-ten blocks to compare 341 and 300 + 40 + 1. What do you notice?
2. Use the number line on page 20 to compare 268 and 279. Which number is greater? Explain.

Compare the numbers. Write <, >, or = for each ●.

3. 68 ● 98

4. 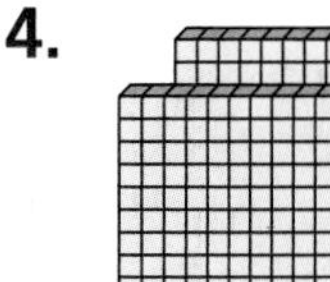 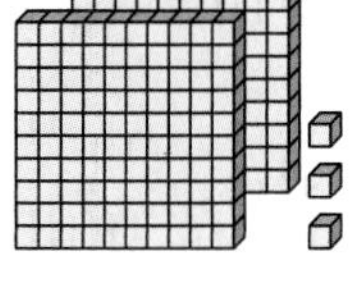 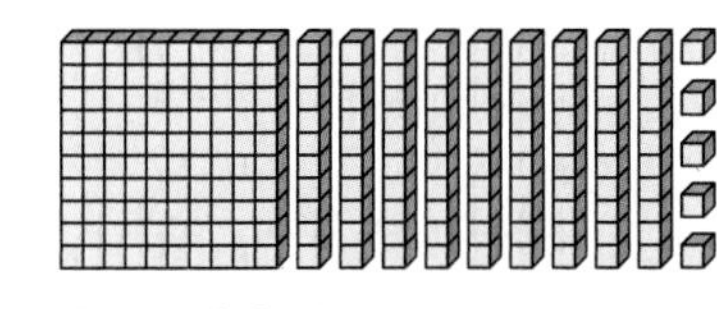

203 ● 195

5. 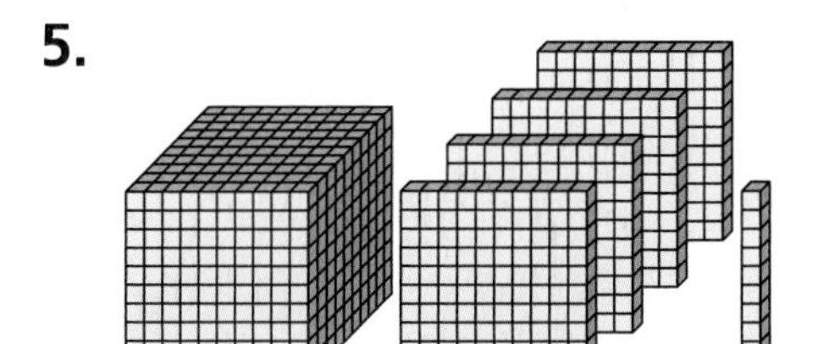 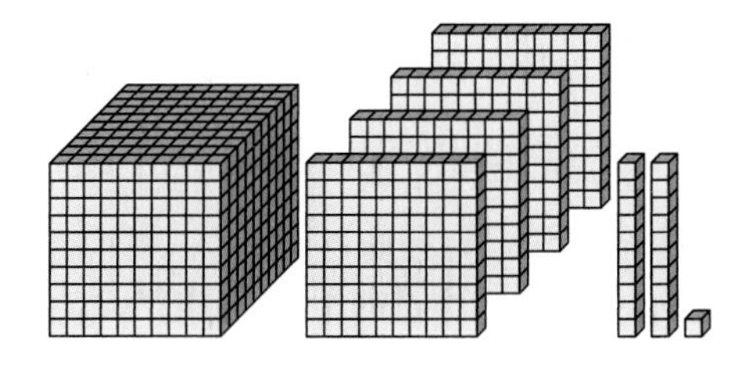

1,411 ● 1,421

LESSON CONTINUES

Practice and Problem Solving

Compare the numbers. Write <, >, or = for each ●.

6.

T	O
9	2
8	3

92 ● 83

7.

H	T	O
1	0	1
1	1	0

101 ● 110

8.

H	T	O
4	2	8
4	3	8

428 ● 438

9. 629 ● 631

10. 758 ● 750

11. 439 ● 438

12. 3,425 ● 3,799

13. 5,712 ● 5,412

14. 2,412 ● 2,412

15. 894 ● 2,139

16. 348 ● 348

17. 7,393 ● 7,396

18. 151 + 200 ● 350

19. 274 + 128 ● 402

20. 475 + 52 ● 537

21. 344 − 103 ● 244

22. 786 − 231 ● 555

23. 696 − 418 ● 296

24. What is the greatest place-value position in which the digits of 831 and 819 are different? Compare the numbers.

25. Compare the numbers 5,361 and 3,974. How can you tell which number is greater?

For 26–28, use the numbers on the box.

26. List all the numbers that are less than 575.

27. List all the numbers that are greater than 830.

28. List all the numbers that are greater than 326 and less than 748.

29. **What's the Question?** Louis read 125 pages. Tom read 137 pages. The answer is 12.

30. **Write About It** You have 3 four-digit numbers. The digits in the thousands, hundreds, and ones places are the same. Which digit would you use to compare the numbers? Explain.

31. The numbers 456 and 564 have the same digits in a different order. Do they both have the same value? Explain.

32. The sum of three addends is 24. One addend is 5. Another addend is 3 more than 7. What is the missing addend?

Mixed Review and Test Prep

Tell whether the number is odd or even. (p. 2)

33. 13 **34.** 46 **35.** 187

36. 35 **37.** 2,721 **38.** 544

39. 736 **40.** 4,922 **41.** 6,571

Write the number in standard form. (p. 10)

42. 50,000 + 8,000 + 300

43. 30,000 + 700 + 5

Choose the letter for the number in standard form. (p. 10)

44. TEST PREP fifty-three thousand, six hundred seventy-two

A 53,072 **C** 53,627
B 53,602 **D** 53,672

45. TEST PREP thirty-four thousand, five hundred twenty

F 3,452 **H** 34,502
G 34,052 **J** 34,520

PROBLEM SOLVING

Thinker's Corner

STRATEGY • COMPARE Early settlements were often built near rivers. Rivers provided people with food, water, and easy trade routes. Even today, many goods are shipped on riverboats rather than by truck or train.

When you *compare* things, you decide how they are alike. When you *contrast* things, you decide how they are different. To solve some problems, you need to compare and contrast the information.

LENGTHS OF RIVERS

Name	Miles
Mississippi	2,348
Missouri	2,315
Yukon	1,979
Rio Grande	1,885
Snake	1,083
Red	1,018

USE DATA For 1–4, use the table.

1. The Mississippi River is the longest river in the United States. In which place-value positions are the lengths of the Mississippi and Missouri Rivers different?
2. Compare the lengths of the Rio Grande and the Yukon River. Which river is longer?
3. How much longer is the Mississippi River than the Missouri River?
4. Explain how to compare the lengths of the Snake River and the Red River.

EXTRA PRACTICE page H33, Set B

LESSON 3

Order Numbers

Quick Review

Tell which number is greater.

1. 37 or 29
2. 21 or 32
3. 58 or 65
4. 120 or 99
5. 235 or 253

Learn

HOW TALL? HOW SHORT? The table lists the heights of three mountains in the United States.

Which mountain is the tallest?

Use a number line to order the numbers.

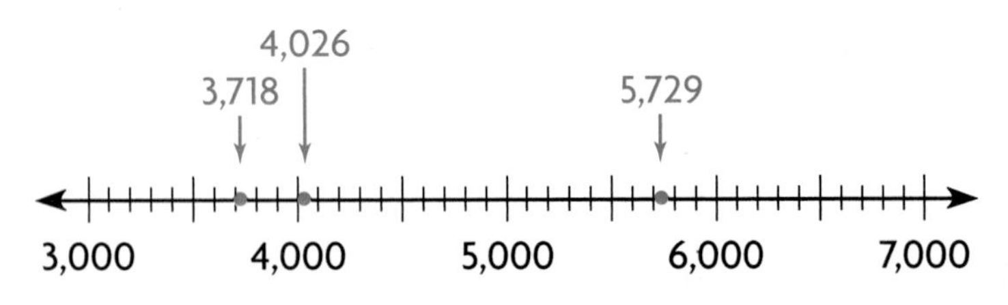

$3{,}718 < 4{,}026 < 5{,}729$

So, Mount Rogers is the tallest.

You can order numbers by comparing the digits in the same place-value positions.

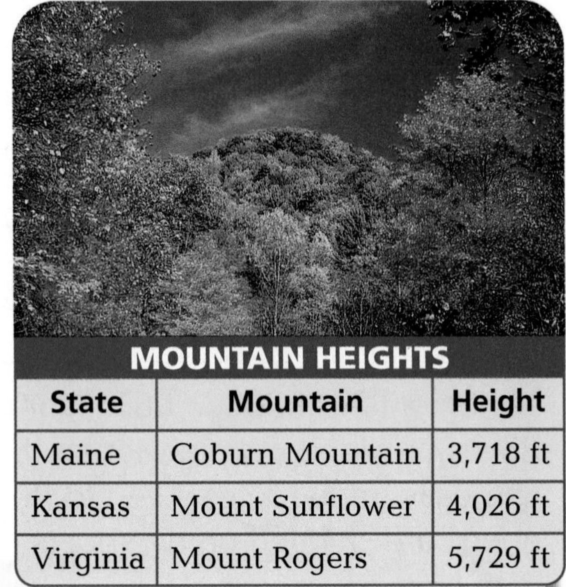

MOUNTAIN HEIGHTS

State	Mountain	Height
Maine	Coburn Mountain	3,718 ft
Kansas	Mount Sunflower	4,026 ft
Virginia	Mount Rogers	5,729 ft

Example

Order 7,613; 7,435; and 7,551.

STEP 1

Compare the thousands.

7,613
7,435
7,551

The digits are the same.

STEP 2

Compare the hundreds.

7,613
7,435
7,551

They are not the same.
$6 > 5 > 4$

STEP 3

Write the numbers in order from greatest to least.

$7{,}613 > 7{,}551 > 7{,}435$

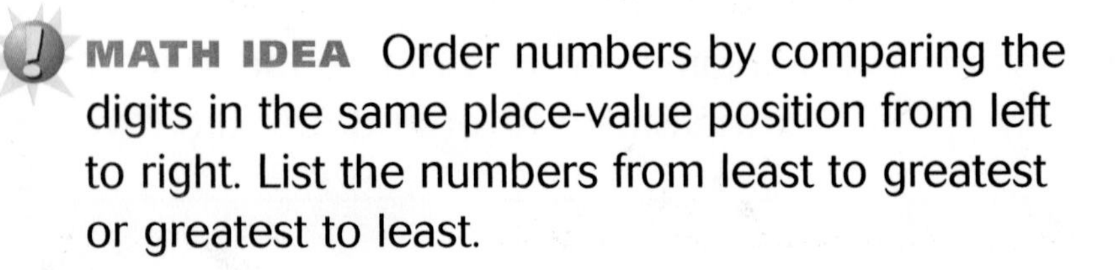

MATH IDEA Order numbers by comparing the digits in the same place-value position from left to right. List the numbers from least to greatest or greatest to least.

Check

1. **Explain** how you can order the numbers 1,432; 1,428; and 1,463 from greatest to least.

Write the numbers in order from least to greatest.

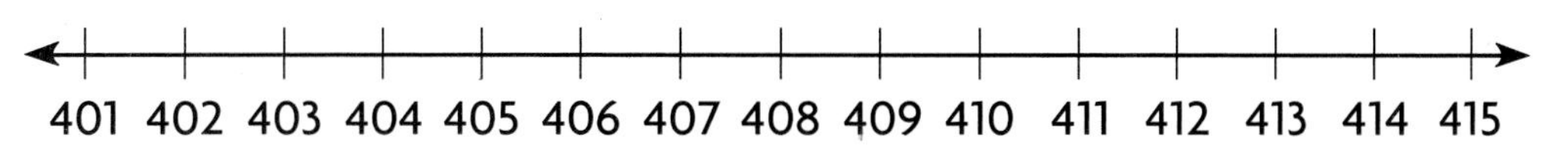

2. 408, 413, 411
3. 403, 410, 407
4. 415, 405, 409

Practice and Problem Solving

Write the numbers in order from least to greatest.

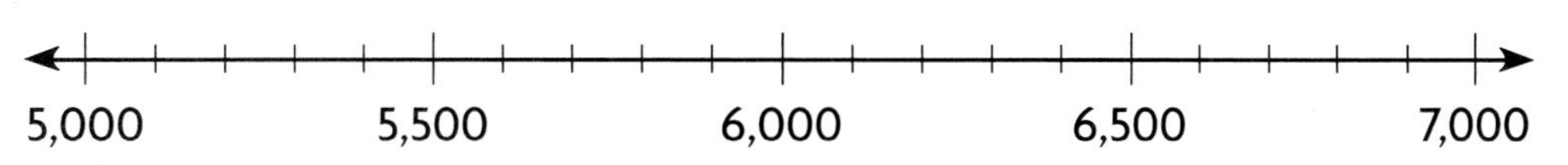

5. 5,200; 6,500; 5,900
6. 6,750; 6,125; 6,500

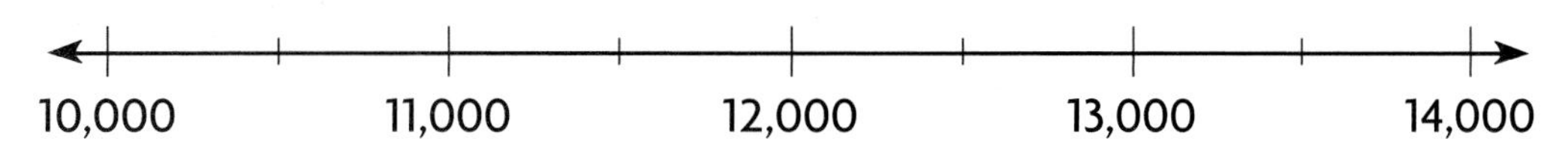

7. 10,500; 13,000; 12,500
8. 12,240; 11,845; 13,156

Write the numbers in order from greatest to least.

9. 384, 526, 431
10. 793, 728, 756
11. 37, 18, 24
12. 583, 791, 476
13. 7,837; 5,126; 3,541
14. 17,655; 22,600; 9,860
15. **REASONING** Write in order from least to greatest the six numbers whose digits are 2, 8, and 9.
16. **What's the Error?** Jason put the following numbers in order from least to greatest: 3,545; 3,556; 3,554. What was his error?

Mixed Review and Test Prep

17. What number continues the pattern? 27, 23, 19, 15, ■ (p. 2)
18. Write the standard form for 60,000 + 500 + 9. (p. 10)
19. 45 + 10
20. 56 + 20
21. **TEST PREP** Which number is greater than 567? (p. 20)
    - **A** 560
    - **B** 549
    - **C** 562
    - **D** 657

EXTRA PRACTICE page H33, Set C

LESSON 4

Problem Solving Skill
Identify Relationships

UNDERSTAND · PLAN · SOLVE · CHECK

FOLLOW THE TRAIL Nancy and Emilio have been studying Mount Hood, Mount Jefferson, North Sister, and Broken Top in the Oregon Cascade Mountains. They want to hike to the mountain that is higher than North Sister but not as high as Mount Hood. Which one should they choose?

Quick Review

Compare. Write <, >, or = for each ●.

1. 124 ● 118
2. 229 ● 232
3. 244 ● 184
4. 156 ● 156
5. 371 ● 372

Knowing how the numbers are related can help you solve the problem.

MOUNTAINS IN THE OREGON CASCADES

Name	Height
Mount Hood	11,235
Mount Jefferson	10,495
North Sister	10,085
Broken Top	9,152

Example

STEP 1

Identify how the numbers are related.

$11,235 > 10,495 > 10,085 > 9,152$

The mountain heights in the table are listed in order from greatest to least.

STEP 2

Find all the mountains that are higher than North Sister.

Mount Jefferson and Mount Hood

STEP 3

Find all the mountains that are not as high as Mount Hood.

Mount Jefferson, North Sister, and Broken Top

STEP 4

Find the mountain that is listed in both Step 2 and Step 3.

Mount Jefferson is the only mountain listed in both steps.

So, Nancy and Emilio chose Mount Jefferson.

Talk About It

- How does the height of Broken Top compare to that of North Sister?

Problem Solving Practice

1. **What if** Emilio and Nancy hiked to the mountain that is higher than Broken Top but not as high as Mount Jefferson? To which mountain did they hike?
2. What is the highest mountain under 11,000 feet that Emilio and Nancy studied?
3. **What if** the mountains were listed in order of height from least to greatest? What would be the order of the mountains?

USE DATA For 4–5, use the table at the right.

4. Which state is larger than Connecticut but smaller than New Jersey?
   - **A** Hawaii
   - **B** Rhode Island
   - **C** New Jersey
   - **D** Delaware
5. Name the three smallest states.
   - **F** Delaware, Rhode Island, Hawaii
   - **G** New Jersey, Hawaii, Connecticut
   - **H** Hawaii, Rhode Island, New Jersey
   - **J** Rhode Island, Delaware, Connecticut

SMALLEST STATES IN THE U.S.

Name	Size in Square Miles
Connecticut	5,006
New Jersey	7,790
Rhode Island	1,213
Hawaii	6,459
Delaware	2,026

Problem Solving Skill

Mixed Applications

6. Louis had base-ten blocks that showed 6 hundreds, 7 tens, 3 ones. Tom gave him 2 hundreds, 1 ten, 5 ones. Using standard form, write the number that shows the value of Louis's blocks now.
7. Celia lives in a town with a population of 12,346. Last year there were 1,000 fewer people living in the town. How many people lived in the town last year?
8. Tony had 24 postcards. Arlo gave him some more postcards. Now Tony has 46 postcards. How many postcards did Arlo give to Tony?
9. There were 53 students that went to the Science Museum. Of them, 27 were girls. How many were boys?
10. **Write About It** Explain how you would compare 4,291; 4,921; and 4,129 to put them in order from greatest to least.

LESSON 5

Round to Nearest 10 and 100

Learn

HOW CLOSE? There are 43 third graders and 47 fourth graders going on a field trip to the San Diego Zoo. About how many students in each grade are going to the zoo?

Rounding is one way to estimate when you want to know *about how many*.

A number line can help you.

Quick Review

Write the numbers in order from least to greatest.

1. 17, 87, 57
2. 23, 19, 16
3. 37, 31, 23
4. 29, 33, 32
5. 59, 57, 58

VOCABULARY

rounding

Example 1

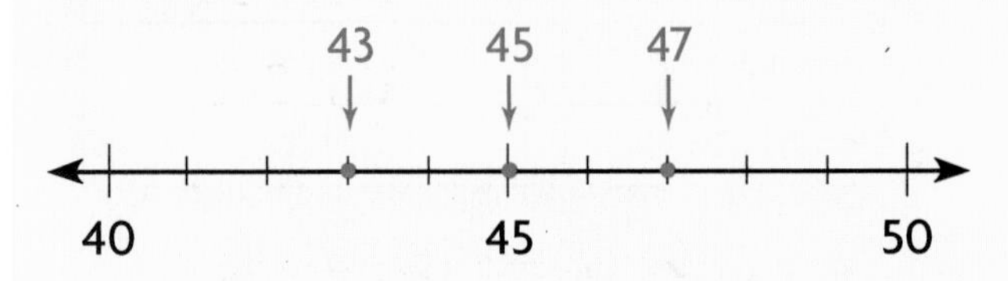

43 is closer to 40 than to 50.
43 rounds to 40.

47 is closer to 50 than to 40.
47 rounds to 50.

45 is halfway between 40 and 50.
If a number is halfway between two tens, round to the greater ten. 45 rounds to 50.

So, about 40 third graders and about 50 fourth graders are going to the zoo.

Example 2

Round 3-digit numbers to the nearest ten or hundred.

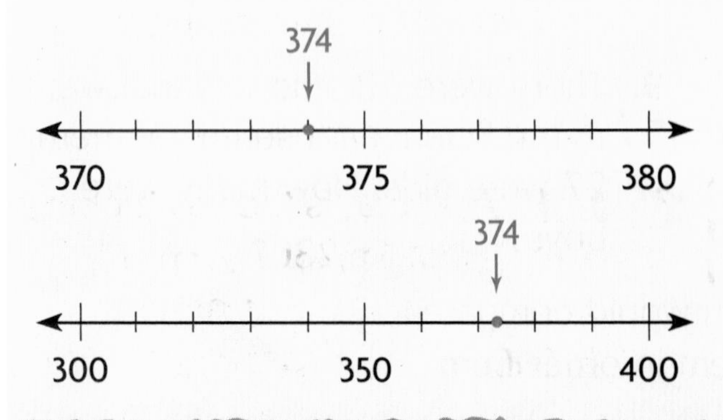

Round to the nearest 10.
374 is closer to 370 than to 380.
374 rounds to 370.

Round to the nearest 100.
374 is closer to 400 than to 300.
374 rounds to 400.

Check

1. **Explain** how you can round 350 to the nearest hundred using the number line.

Round to the nearest hundred and the nearest ten.

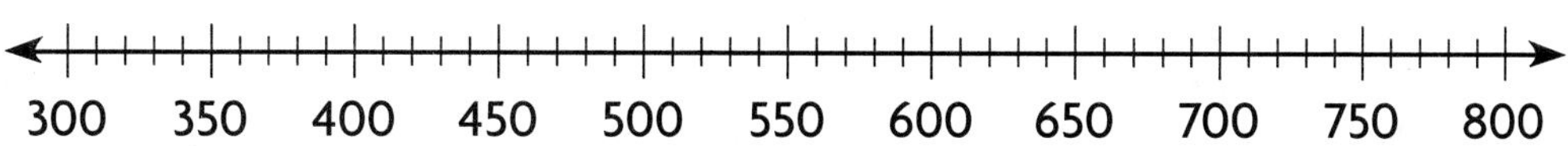

2. 643 3. 377 4. 445 5. 518 6. 750

Practice and Problem Solving

Round to the nearest ten.

7. 16 8. 74 9. 53 10. 5 11. 78

12. 37 13. 44 14. 78 15. 94 16. 98

Round to the nearest hundred and the nearest ten.

17. 363 18. 405 19. 115 20. 165 21. 952

22. 237 23. 917 24. 385 25. 456 26. 883

USE DATA For 27–28, use the table.

ZOO ANIMALS	
Type	**Number**
Mammals	214
Birds	428
Reptiles	174

27. To the nearest hundred, about how many kinds of birds does the zoo have?

28. **REASONING** The number of __?__ + the number of __?__ < the number of __?__ .

29. Kim rounded 348 to the nearest ten and said it was 350. She rounded 348 to the nearest hundred and said it was 400. Was this correct? Explain.

30. **Write a problem** about animals. Use rounding to the nearest ten or to the nearest hundred in your problem.

Mixed Review and Test Prep

Find 100 more. (p. 6)

31. 877 32. 352

33. 461 34. 208

35. **TEST PREP** What is the value of the blue digit in 16,230? (p. 10)

A 10 **C** 1,000
B 100 **D** 10,000

EXTRA PRACTICE page H33, Set D

LESSON 6

Round to Nearest 1,000

Quick Review

Round to the nearest ten.

1. 52 (50)
2. 15 (20)
3. 22 (20)
4. 68 (70)
5. 54 (50)

Learn

ABOUT HOW MANY? When the Bronx Zoo in New York City first opened in 1899, it had 843 animals. In the fall of 2000, the Bronx Zoo had 8,861 animals.

To the nearest thousand, how many animals are in the zoo?

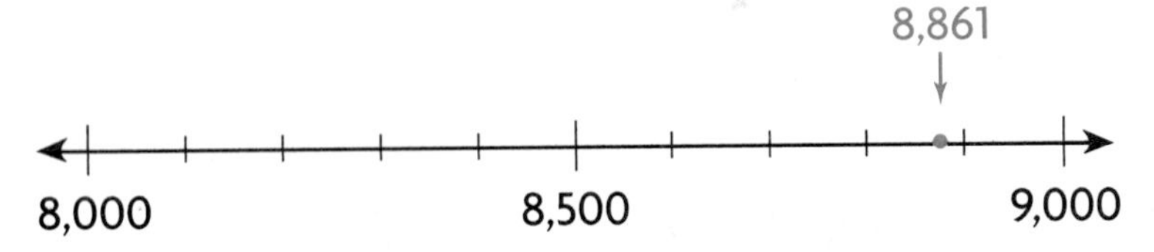

8,861 is closer to 9,000 than to 8,000.
8,861 rounds to 9,000.

So, there are about 9,000 animals in the zoo.

You can use rounding rules to round numbers.

Examples

A Round 2,641 to the nearest *thousand*.

2,641
↑

Look at the hundreds digit. Since 6 > 5, the 2 thousands digit rounds to 3 thousands. So, 2,641 rounds to 3,000.

B Round 2,641 to the nearest *hundred*.

2,641
↑

Look at the tens digit. Since 4 < 5, the 6 hundreds digit stays the same. So, 2,641 rounds to 2,600.

Rounding Rules

- Find the place to which you want to round.
- Look at the digit to its right.
- If the digit is less than 5, the digit in the rounding place stays the same.
- If the digit is 5 or more, the digit in the rounding place increases by 1.

Check

1. **Tell** how you would use the rounding rules to round 2,641 to the nearest ten.

Round to the nearest thousand.

2. 6,427 3. 2,500 4. 4,526 5. 1,670

Practice and Problem Solving

Round to the nearest thousand.

6. 8,312 7. 4,500 8. 674 9. 9,478
10. 1,611 11. 5,920 12. 2,543 13. 4,444

Round to the nearest thousand, the nearest hundred, and the nearest ten.

14. 3,581 15. 6,318 16. 2,350 17. 8,914
18. 4,624 19. 5,337 20. 1,273 21. 2,845

USE DATA For 22–24, use the table.

22. To the nearest thousand pounds, about how much does the African elephant weigh?
23. Round the weights of the giraffe and rhinoceros to the nearest thousand pounds. About how many giraffes would it take to equal the weight of the rhinoceros?
24. **Write About It** Tell how to round the weight of the hippopotamus to the nearest thousand, hundred, and ten.

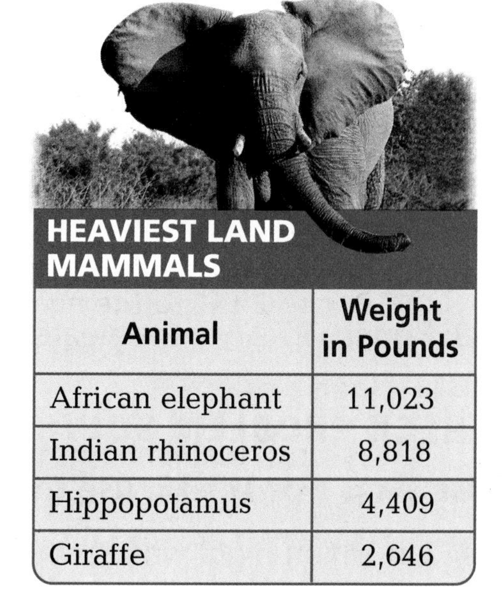

HEAVIEST LAND MAMMALS

Animal	Weight in Pounds
African elephant	11,023
Indian rhinoceros	8,818
Hippopotamus	4,409
Giraffe	2,646

Mixed Review and Test Prep

Write the value of the blue digit. (p. 10)

25. 8,251 26. 87,668

Write in expanded form. (p. 10)

27. 337 28. 12,982

29. **TEST PREP** $68 + 42 + 36 = \blacksquare$

A 136 C 142
B 138 D 146

EXTRA PRACTICE page H33, Set E

Review/Test

✓ CHECK VOCABULARY AND CONCEPTS

Choose the best term from the box.

greatest to least
compare
round

1. You can use $<, >,$ or $=$ to __?__ numbers. (p. 20)
2. One way to estimate is to __?__ numbers. (p. 28)

Suppose you want to round 371 to the nearest hundred. (pp. 28–29)

3. Which hundreds is 371 between?
4. Which hundred is 371 closer to?

✓ CHECK SKILLS

Compare the numbers. Write $<, >,$ or $=$ for each ●. (pp. 20–23)

5. 532 ● 523
6. 3,246 ● 325
7. 7,583 ● 7,583

Write the numbers in order from least to greatest. (pp. 24–25)

8. 143, 438, 92
9. 7,304; 7,890; 7,141
10. 23,256; 23,161; 23,470
11. Round 85 to the nearest ten. (pp. 28–29)
12. Round 824 to the nearest hundred. (pp. 28–29)
13. Round 3,721 to the nearest thousand and hundred. (pp. 30–31)

✓ CHECK PROBLEM SOLVING

USE DATA For 14–15, use the table. (pp. 26–27)

14. On which night was the number of tickets sold greater than the number sold on Monday but less than the number sold on Wednesday?
15. On which night was the number of tickets sold less than the number sold on Friday but greater than the number sold on Wednesday?

TICKET SALES

Day	Number
Monday	1,079
Tuesday	1,580
Wednesday	1,493
Thursday	1,208
Friday	2,112

Standardized Test Prep

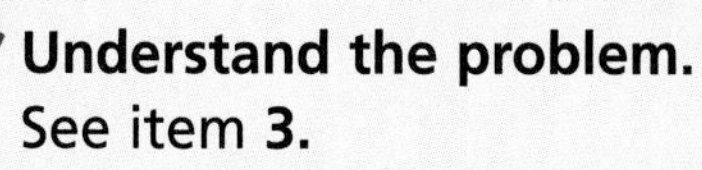

Understand the problem.
See item **3**.

There is more than one number greater than 240 and less than 250. Find digits that add up to 7 from these numbers.

Also see problem **1**, p. H62.

For 1–8, choose the best answer.

1. $\begin{array}{r} 38 \\ -13 \\ \hline \end{array}$

A 24 C 51
B 25 D NOT HERE

2. Which number is greater than 672 and less than 683?

F 671 H 685
G 682 J 687

3. A number is greater than 240 and less than 250. The sum of its digits is 7. What is the number?

A 142 C 241
B 214 D 242

4. Which shows the numbers in order from greatest to least?

F 13,456; 43,165; 14,653
G 13,456; 14,653; 43,165
H 43,165; 14,653; 13,456
J 43,165; 13,456; 14,653

For 5–6, use the table.

RIVER LENGTHS		
Continent	**River Name**	**Length (miles)**
North America	Mississippi	2,340
Asia	Ob	2,268
Europe	Volga	2,290

5. What is the Volga River's length rounded to the nearest hundred miles?

A 2,200 C 2,300
B 2,290 D 2,390

6. What is the Ob River's length rounded to the nearest ten miles?

F 2,258 H NOT HERE
G 2,260 J 2,270

7. On Friday, 8,656 people attended the football game. What is that number rounded to the nearest thousand?

A 9,000 C 8,600
B 8,700 D 8,000

8. Which number is **not** even?

F 753 H 3,558
G 1,350 J 6,904

Write What You Know

9. Janet rounded 418 to 420. Lorna rounded 418 to 400. Explain why 418 can be rounded to both 420 and 400.

10. Use the table from Exercises 5–6. List the order of the rivers from greatest length to least length. Explain your answer.

CHAPTER 3
Addition

Dogs were first used as watchdogs, herding dogs, and hunting dogs. Now, more dogs are pets than workers. Some dogs are still trained to help disabled people.

PROBLEM SOLVING Look at the chart. How many dogs in all graduated in 1997 and 1998?

Use this page to help you review and remember important skills needed for Chapter 3.

VOCABULARY

Choose the best term from the box.

1. The answer to an addition problem is called the ? .
2. In $8 + 4 = 12$, the 8 and 4 are ? .

addends
difference
sum

ADDITION FACTS (For Intervention, see p. H4.)

Add.

3. $2 + 7$
4. $9 + 4$
5. $6 + 5$
6. $3 + 8$
7. $8 + 7$
8. $5 + 7$
9. $8 + 2$
10. $7 + 9$
11. $5 + 3$
12. $4 + 3$
13. $7 + 7$
14. $3 + 9$
15. $6 + 6$
16. $7 + 3$
17. $8 + 8$

2-DIGIT ADDITION (For Intervention, see p. H5.)

Add.

18.

tens	ones
1	2
+ 1	5

19.

tens	ones
1	6
+ 1	8

20.

tens	ones
4	3
+ 3	6

21. $11 + 65$
22. $53 + 18$
23. $27 + 19$
24. $26 + 35$
25. $15 + 45$

MENTAL MATH: ADD 2-DIGIT NUMBERS (For Intervention, see p. H5.)

Use mental math to find the sum.

26. $23 + 10 =$ ■
27. $24 + 11 =$ ■
28. $32 + 14 =$ ■
29. $15 + 25 =$ ■
30. $22 + 35 =$ ■
31. $45 + 21 =$ ■

LESSON 1

Column Addition

Quick Review

1. $3 + 2 + 3$
2. $4 + 5 + 6$
3. $8 + 2 + 3$
4. $6 + 1 + 9$
5. $8 + 7 + 3$

VOCABULARY

Grouping Property of Addition

Learn

PET PORTIONS Maria bought 5 pounds of cat food, 2 pounds of birdseed, and 8 pounds of dog food. How many pounds of pet food did she buy?

$5 + 2 + 8 = ■$

MATH IDEA The **Grouping Property of Addition** states that you can group addends in different ways. The sum is always the same.

$5 + (2 + 8) = ■$ → $5 + 10 = 15$

$(5 + 2) + 8 = ■$ → $7 + 8 = 15$

Add the numbers in ( ) first.

So, Maria bought 15 pounds of pet food.

Example

Find $13 + 18 + 27$.

STEP 1

Add ones.
$10 + 8 = 18$ ones

$$\begin{array}{r} {}^{1} \\ 13 \\ 18 \\ +27 \\ \hline 8 \end{array}$$

Group 3 and 7 to make a ten. (3 and 7 → 10)

STEP 2

Add tens.
$1 + 4 = 5$ tens

$$\begin{array}{r} {}^{1} \\ 13 \\ 18 \\ +27 \\ \hline 58 \end{array}$$

Check

1. **Explain** what happens to the sum when you group addends in different ways.

Find the sum.

2. $(5 + 5) + 6 = ■$
3. $8 + (2 + 8) = ■$
4. $(7 + 13) + 2 = ■$

Practice and Problem Solving

Find the sum.

5. $(7 + 3) + 6 = ■$
6. $7 + (5 + 5) = ■$
7. $(8 + 2) + 4 = ■$
8. $8 + (5 + 3) = ■$
9. $(1 + 4) + 11 = ■$
10. $(9 + 8) + 20 = ■$

Use the Grouping Property to find the sum.

11. $\begin{array}{r} 8 \\ 9 \\ +2 \\ \hline \end{array}$
12. $\begin{array}{r} 5 \\ 5 \\ +8 \\ \hline \end{array}$
13. $\begin{array}{r} 4 \\ 9 \\ +8 \\ \hline \end{array}$
14. $\begin{array}{r} 3 \\ 9 \\ +7 \\ \hline \end{array}$
15. $\begin{array}{r} 21 \\ 45 \\ +32 \\ \hline \end{array}$
16. $\begin{array}{r} 34 \\ 27 \\ +11 \\ \hline \end{array}$

17. $8 + 2 + 6 = ■$
18. $5 + 18 + 5 = ■$
19. $12 + 8 + 9 = ■$

USE DATA For 20-21, use the table.

20. How many pounds did the puppy gain in three months?
21. How many more pounds did the puppy gain in September and October than in November?

POUNDS GAINED		
September	October	November
6	4	5

22. **Write About It** Does $18 + 4 + 3 = 5 + 2 + 10 + 8$? Explain.
23. **Write a problem** with three addends. Use the Grouping Property to help you solve.

Mixed Review and Test Prep

24. Add 20 and 15.
25. Write 16,072 in expanded form. .(p. 10)
26. What is the least number you can write with the digits 4, 8, 7, and 2? (p. 6)
27. What is the place-value position of the digit 6 in 16,083? (p. 10)
28. **TEST PREP** Which is greater than 7,855? (p. 20)
    - **A** 6,999
    - **B** 7,755
    - **C** 7,847
    - **D** 7,901

EXTRA PRACTICE page H34, Set A

LESSON 2

Estimate Sums

Quick Review

1. 5 + 25
2. 30 + 20
3. 70 + 20
4. \$4 + \$6
5. 20 + 55

VOCABULARY

estimate

Learn

MANATEE MEALS Two manatees live at a sea park. In one day, Manny ate 91 pounds of food and Marsha ate 88 pounds. About how much did they eat in all?

To find *about* how much, you can **estimate**.

Example

STEP 1

Round each number to the nearest ten.

$$\begin{array}{r} 88 \\ +91 \\ \hline \end{array} \rightarrow \begin{array}{r} 90 \\ +90 \\ \hline \end{array}$$

STEP 2

Find the estimated sum.

$$\begin{array}{r} 90 \\ +90 \\ \hline 180 \end{array}$$

So, they ate about 180 pounds of food.

More Examples

A Round to the nearest hundred.

$$\begin{array}{r} 174 \\ +123 \\ \hline \end{array} \rightarrow \begin{array}{r} 200 \\ +100 \\ \hline 300 \end{array}$$

B Round to the nearest dollar.

$$\begin{array}{r} \$7.80 \\ +\$4.35 \\ \hline \end{array} \rightarrow \begin{array}{r} \$8 \\ +\$4 \\ \hline \$12 \end{array}$$

C Round to the nearest thousand.

$$\begin{array}{r} 3,260 \\ +2,755 \\ \hline \end{array} \rightarrow \begin{array}{r} 3,000 \\ +3,000 \\ \hline 6,000 \end{array}$$

▲ A manatee will eat from 32 to 108 pounds of plants every day, depending on its size.

MATH IDEA When you do not need an exact answer, you can estimate.

Check

1. **Explain** why it makes sense for the sea park to estimate how much food the manatees need.

Estimate the sum.

2. $\begin{array}{r} 67 \\ +19 \\ \hline \end{array}$

3. $\begin{array}{r} 410 \\ +890 \\ \hline \end{array}$

4. $\begin{array}{r} \$5.30 \\ +\$3.80 \\ \hline \end{array}$

Practice and Problem Solving

Estimate the sum.

5. $\begin{array}{r} 12 \\ +23 \\ \hline \end{array}$

6. $\begin{array}{r} 14 \\ +28 \\ \hline \end{array}$

7. $\begin{array}{r} 24 \\ +78 \\ \hline \end{array}$

8. $\begin{array}{r} 518 \\ +305 \\ \hline \end{array}$

9. $\begin{array}{r} 206 \\ +668 \\ \hline \end{array}$

10. $\begin{array}{r} \$6.38 \\ +\$1.04 \\ \hline \end{array}$

11. $\begin{array}{r} 480 \\ +240 \\ \hline \end{array}$

12. $\begin{array}{r} 379 \\ +325 \\ \hline \end{array}$

13. $\begin{array}{r} 109 \\ +515 \\ \hline \end{array}$

14. $\begin{array}{r} 922 \\ +111 \\ \hline \end{array}$

15. $\begin{array}{r} \$5.90 \\ +\$2.99 \\ \hline \end{array}$

16. $\begin{array}{r} 2,610 \\ +3,397 \\ \hline \end{array}$

17. $\begin{array}{r} 7,800 \\ +1,620 \\ \hline \end{array}$

18. $\begin{array}{r} 1,340 \\ +5,600 \\ \hline \end{array}$

19. $\begin{array}{r} \$4.70 \\ +\$2.90 \\ \hline \end{array}$

20. **REASONING** Josh put 21 heads of lettuce in each basket to feed to the manatees. About how many heads of lettuce would there be in 3 baskets?

21. Erica earned $2.90 on Monday. If she earns about the same amount Tuesday and Wednesday, can she buy a $13.00 CD? Explain.

For 22–25, use the numbers at the right.

Choose two numbers whose sum is about:

22. 70. 23. 500. 24. 8,000.

25. **Write a problem** in which you estimate the sum of two or more numbers from the list.

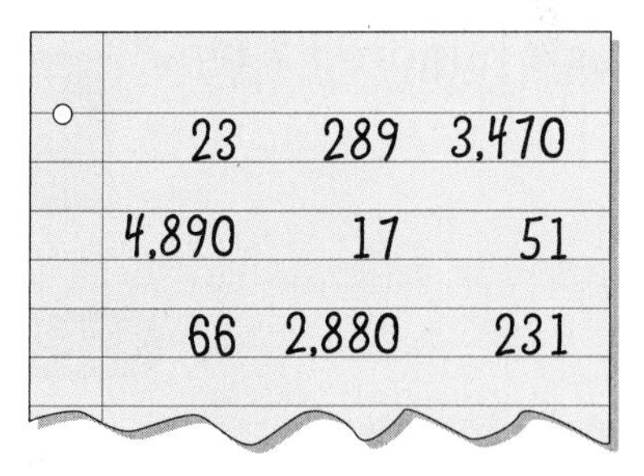

23	289	3,470
4,890	17	51
66	2,880	231

Mixed Review and Test Prep

Write <, >, or = for each ●. (p. 20)

26. 27 ● 38

27. 723 ● 726

28. Write in order from least to greatest: 3,291; 3,245; 3,311. (p. 24)

29. Write 40,000 + 3,000 + 700 + 9 in standard form. (p. 10)

30. **TEST PREP** Which is sixty thousand, two hundred forty written in standard form? (p. 10)

A 60,024
B 60,204
C 60,240
D 62,040

EXTRA PRACTICE page H34, Set B

LESSON 3

Add 3-Digit Numbers

Quick Review

1. 12 + 10
2. 14 + 15
3. 41 + 39
4. 25 + 36
5. 21 + 57

MATERIALS
base-ten blocks

Explore

Make a model to add 134 and 279.

$$\begin{array}{r} 134 \\ +279 \\ \hline \end{array}$$

Activity

STEP 1

Add ones.
4 + 9 = 13 ones
Regroup 13 ones
as 1 ten 3 ones.

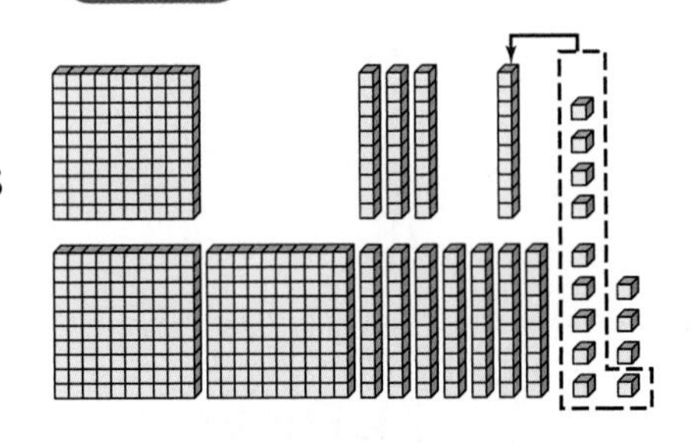

STEP 2

Add tens.
1 + 3 + 7 = 11 tens
Regroup 11 tens
as 1 hundred 1 ten.

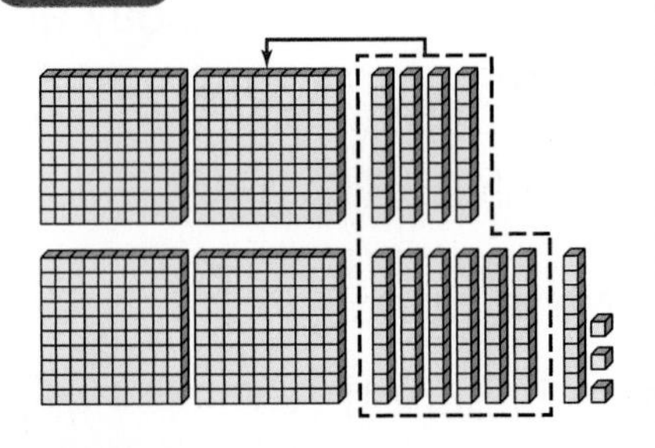

STEP 3

Add hundreds.
1 + 1 + 2 = 4 hundreds

So, 134 + 279 = 413.

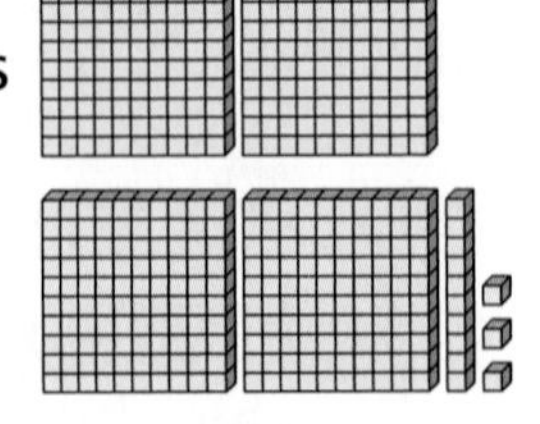

Technology Link
More Practice:
Use E-Lab, Addition of Three-Digit Numbers.
www.harcourtschool.com/elab2002

We are adding 116 and 144. We now have 10 ones. What should we do next?

Try It

Use base-ten blocks to find each sum.

a. 116 + 144 = ■ **b.** 269 + 358 = ■

Connect

Here is a way to record addition. To add 137 and 264, first line up hundreds, tens, and ones.

Example

STEP 1

Add the ones.
Regroup.
11 ones = 1 ten 1 one

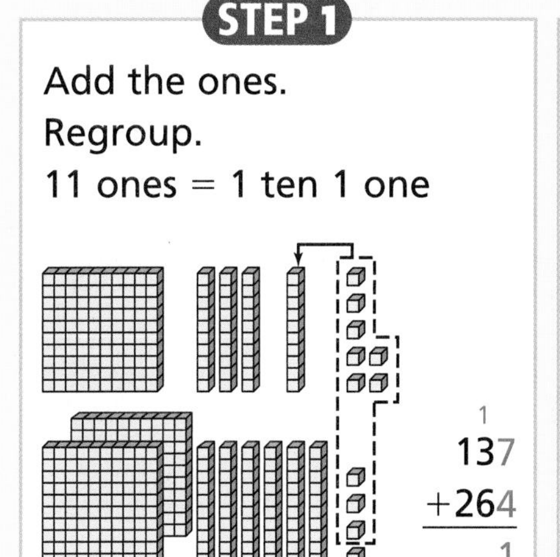

$$\begin{array}{r} {}^{1}\\ 137 \\ +264 \\ \hline 1 \end{array}$$

STEP 2

Add the tens.
Regroup.
10 tens = 1 hundred 0 tens

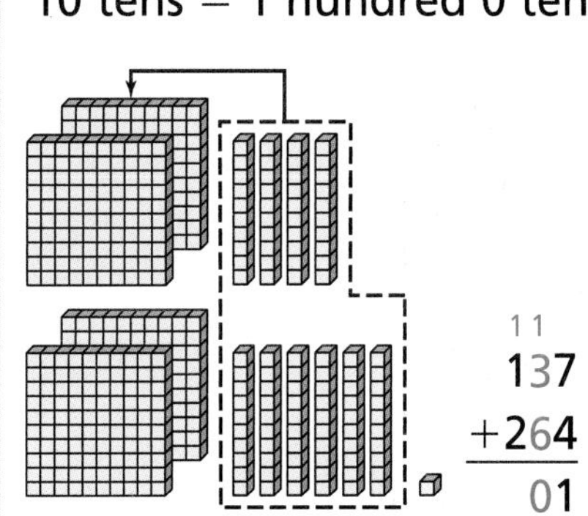

$$\begin{array}{r} {}^{1\,1}\\ 137 \\ +264 \\ \hline 01 \end{array}$$

STEP 3

Add the hundreds.

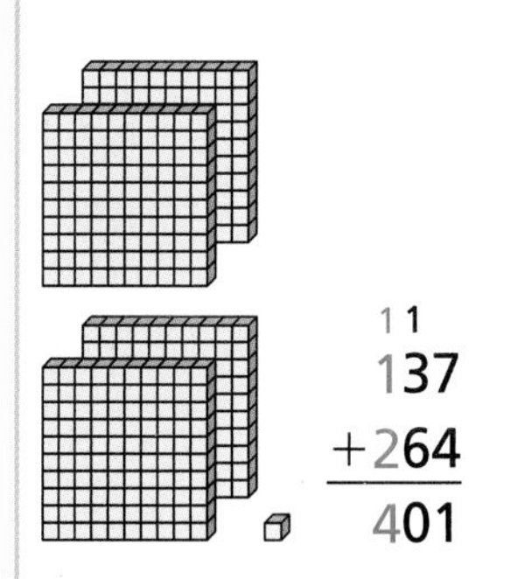

$$\begin{array}{r} {}^{1\,1}\\ 137 \\ +264 \\ \hline 401 \end{array}$$

Practice and Problem Solving

Use base-ten blocks to find each sum.

1. 134 + 217 = ■ **2.** 265 + 423 = ■ **3.** 368 + 416 = ■

4. 333 + 128 = ■ **5.** 295 + 382 = ■ **6.** 192 + 439 = ■

7. 493 + 256 = ■ **8.** 563 + 139 = ■ **9.** 612 + 308 = ■

USE DATA For 10, use the table.

10. REASONING Theo drove from Pittsburgh to Allentown. David drove from State College to Harrisburg. Who drove farther? How much farther?

DISTANCES BETWEEN PENNSYLVANIA CITIES	
State College to Pittsburgh	132 miles
Pittsburgh to Philadelphia	302 miles
Philadelphia to Harrisburg	104 miles
Pittsburgh to Harrisburg	203 miles
Harrisburg to Allentown	80 miles

Mixed Review and Test Prep

Compare. Use <, >, or = for each ●. (p. 20)

11. 89 ● 78 **12.** 324 ● 342

13. 142 ● 412 **14.** 8 + 3 ● 20 − 7

15. TEST PREP Which group of numbers is in order from least to greatest? (p. 24)

A 394, 379, 380 **C** 427, 450, 431
B 521, 539, 540 **D** 201, 263, 229

LESSON 4

Add 3-Digit Numbers

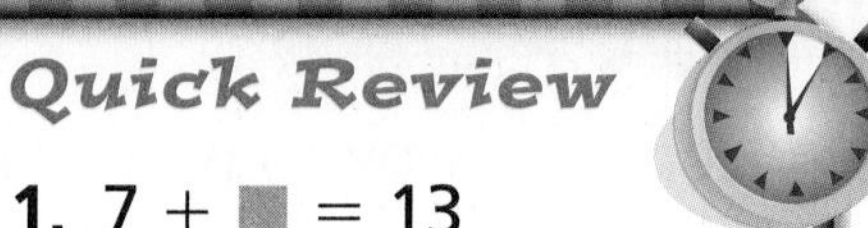

Quick Review

1. 7 + ■ = 13
2. ■ + 9 = 18
3. 5 + 6
4. 8 + 7
5. 6 + 8

Learn

BUNCHES OF BOOKS How many books did Mr. Chi's and Mrs. Garcia's classes read in all?

198 + 165 = ■

READ-A-THON RESULTS	
Mr. Chi's class	198 books
Mrs. Garcia's class	165 books
Mrs. Miller's class	203 books

Estimate.

$$\begin{array}{r} 198 \\ +165 \\ \hline \end{array} \rightarrow \begin{array}{r} 200 \\ +200 \\ \hline 400 \end{array}$$

Example

STEP 1

Add the ones. Regroup.
13 ones = 1 ten 3 ones.

$$\begin{array}{r} {}^{1} \\ 198 \\ +165 \\ \hline 3 \end{array}$$

STEP 2

Add the tens. Regroup.
16 tens = 1 hundred 6 tens.

$$\begin{array}{r} {}^{1\,1} \\ 198 \\ +165 \\ \hline 63 \end{array}$$

STEP 3

Add the hundreds.

$$\begin{array}{r} {}^{1\,1} \\ 198 \\ +165 \\ \hline 363 \end{array}$$

So, the two classes read 363 books in all. Since 363 is close to 400, the answer is reasonable.

More Examples

A

$$\begin{array}{r} {}^{1} \\ 325 \\ +\ 67 \\ \hline 392 \end{array}$$

B

$$\begin{array}{r} {}^{2} \\ 591 \\ 173 \\ +290 \\ \hline 1,054 \end{array}$$

C

$$\begin{array}{r} {}^{1} \\ \$4.83 \\ +\$2.74 \\ \hline \$7.57 \end{array}$$

↑
decimal point

- Add money like whole numbers.
- Then use a decimal point to separate dollars and cents.

MATH IDEA Estimate to see if your answer is reasonable.

Check

1. **Explain** whether you would regroup to find how many books Mr. Chi's and Mrs. Miller's classes read.

Find the sum. Estimate to check.

2. $\begin{array}{r} 224 \\ +511 \\ \hline \end{array}$
3. $\begin{array}{r} 298 \\ +172 \\ \hline \end{array}$
4. $\begin{array}{r} \$9.07 \\ +\$1.25 \\ \hline \end{array}$
5. $\begin{array}{r} 468 \\ +\ 89 \\ \hline \end{array}$

Practice and Problem Solving

Find the sum. Estimate to check.

6. $\begin{array}{r} 321 \\ +268 \\ \hline \end{array}$
7. $\begin{array}{r} 505 \\ +228 \\ \hline \end{array}$
8. $\begin{array}{r} 173 \\ +368 \\ \hline \end{array}$
9. $\begin{array}{r} 561 \\ +246 \\ \hline \end{array}$
10. $\begin{array}{r} 299 \\ +\ 66 \\ \hline \end{array}$
11. $\begin{array}{r} 284 \\ +325 \\ \hline \end{array}$
12. $\begin{array}{r} \$7.44 \\ +\$5.02 \\ \hline \end{array}$
13. $\begin{array}{r} 629 \\ +\ 67 \\ \hline \end{array}$
14. $\begin{array}{r} \$1.42 \\ +\$5.61 \\ \hline \end{array}$
15. $\begin{array}{r} 152 \\ +339 \\ \hline \end{array}$
16. $\begin{array}{r} 297 \\ +580 \\ \hline \end{array}$
17. $\begin{array}{r} 579 \\ +486 \\ \hline \end{array}$
18. $\begin{array}{r} \$7.39 \\ +\$2.51 \\ \hline \end{array}$
19. $\begin{array}{r} 896 \\ +424 \\ \hline \end{array}$
20. $\begin{array}{r} 177 \\ +695 \\ \hline \end{array}$

21. $219 + 316 + 222 = \blacksquare$

22. $267 + 741 + 109 = \blacksquare$

23. **ALGEBRA** Write the missing addend. $230 + \blacksquare + 50 = 282$

24. Manuel spent \$3.65 for lunch on Monday and \$2.78 on Tuesday. How much did he spend in all?

25. **What's the Question?** Eva read to page 112 in her book. There are 67 more pages in the book. The answer is 179 pages.

26. **What's the Error?** Sharon added 458 and 83 like this.

$\begin{array}{r} 458 \\ +83 \\ \hline 1{,}288 \end{array}$

Describe her error and solve.

Mixed Review and Test Prep

27. $19 + 24 = \blacksquare$

28. $30 + 62 = \blacksquare$

29. Explain the pattern. (p. 2)
86, 81, 76, 71, 66

30. What is the value of the blue digit in 34,241? (p. 10)

31. **TEST PREP** $50 - 27 = \blacksquare$
   - **A** 21
   - **B** 23
   - **C** 32
   - **D** 33

EXTRA PRACTICE page H34, Set C

LESSON 5

Problem Solving Strategy
Predict and Test

Quick Review

1. 21 + 6
2. 45 + 15
3. 9 + 36
4. 72 + 8
5. 12 + 13

PROBLEM The third-grade classes bought 75 containers of food for the animal shelter. They had 15 more cans than bags of food. How many bags and cans did the classes buy?

UNDERSTAND

- What are you asked to find?
- What information will you use?
- Is there any information you will not use?

PLAN

- What strategy can you use to solve the problem?

You can *predict and test* to find the number of bags and cans the classes bought.

SOLVE

- How can you use the strategy to solve the problem?

Predict the number of bags the classes bought. Add 15 to that number for the number of cans. Then test to see if the sum is 75.

BAGS	CANS	TOTAL	NOTES
20	20+15=35	20+35=55	too low
50	50+15=65	50+65=115	too high
30	30+15=45	30+45=75	just right

So, the classes bought 30 bags and 45 cans of food.

CHECK

- How can you use the first two predictions to make a better prediction?

Problem Solving Practice

Use *predict and test* to solve.

1. **What if** the classes bought 120 containers and had 30 more cans than bags? How many bags and how many cans did they buy?
2. Pilar has 170 stamps in her collection. Her first book of stamps has 30 more stamps in it than her second book. How many stamps are in each book?

PROBLEM SOLVING STRATEGIES

- Draw a Diagram or Picture
- Make a Model or Act It Out
- Make an Organized List
- Find a Pattern
- Make a Table or Graph
- **Predict and Test**
- Work Backward
- Solve a Simpler Problem
- Write a Number Sentence
- Use Logical Reasoning

Two numbers have a sum of 27. Their difference is 3. What are the two numbers?

3. Which is a reasonable prediction for one of the numbers?

   A 3 **C** 27
   B 10 **D** 30

4. What solution answers the question?

   F 3 and 27 **H** 10 and 17
   G 10 and 13 **J** 12 and 15

Mixed Strategy Practice

USE DATA For 5–6, use the table.

5. The number of pounds used in Week 2 was greater than in Week 1, but less than in Week 3. The number of pounds used in Week 2 is an odd number that does not end in 5. How many pounds were used in Week 2?

DOG FOOD USED AT SHELTER	
February	**Pounds**
Week 1	73
Week 2	■
Week 3	79
Week 4	81

6. **Write a problem** about the dog food used at the shelter in which the difference is greater than 5.
7. The sum of two numbers is 55. Their difference is 7. What are the numbers?
8. There are 4 students in line. Max is before Keiko but after Liz. Adam is fourth. Who is first?

Problem Solving Strategy

LESSON

6 Choose a Method

Quick Review

1. 350 + 40
2. 150 + 212
3. 560 + 161
4. 205 + 52
5. 90 + 215

Learn

You can find a sum by using mental math, paper and pencil, or a calculator.

PADDLE POWER Tom and Eli paddled from White Rock to Bear Corner to Raccoon Falls. How many yards did they paddle in all?

$$4{,}365 + 3{,}852 = \blacksquare$$

Estimate. $4{,}000 + 4{,}000 = 8{,}000$

Use Paper and Pencil The numbers are large. The problem involves regrouping. So, paper and pencil is a good choice.

STEP 1

Add the ones.

$$\begin{array}{r} 4{,}365 \\ +3{,}852 \\ \hline 7 \end{array}$$

STEP 2

Add the tens. Regroup. 11 tens = 1 hundred 1 ten

$$\begin{array}{r} ^{1} \\ 4{,}365 \\ +3{,}852 \\ \hline 17 \end{array}$$

STEP 3

Add the hundreds. Regroup. 12 hundreds = 1 thousand 2 hundreds

$$\begin{array}{r} ^{1\ 1} \\ 4{,}365 \\ +3{,}852 \\ \hline 217 \end{array}$$

STEP 4

Add the thousands.

$$\begin{array}{r} ^{1\ 1} \\ 4{,}365 \\ +3{,}852 \\ \hline 8{,}217 \end{array}$$

So, Tom and Eli paddled 8,217 yards. Since 8,217 is close to 8,000, the answer is reasonable.

Use a Calculator

$$4{,}365 + 3{,}852 + 3{,}978 = \blacksquare$$

The numbers are large. The problem involves regrouping. So, a calculator is a good choice.

     12195.

Use Mental Math

$9.30 + $5.60 = ■

There is no regrouping. You can add the dollar and cents amounts in your head. So, mental math is a good choice.

Think: Add the dollar amounts. $9.00 + $5.00 = $14.00
Then add the cents. $0.30 + $0.60 = $0.90
Find the sum. $14.00 + $0.90 = $14.90

So, $9.30 + $5.60 = $14.90.

Examples

A
```
   11
  373
 +497
 ----
  870
```

B
```
   21
2,094
  167
+5,041
------
 7,302
```

C
```
 $5.10
+$2.20
------
 $7.30
```

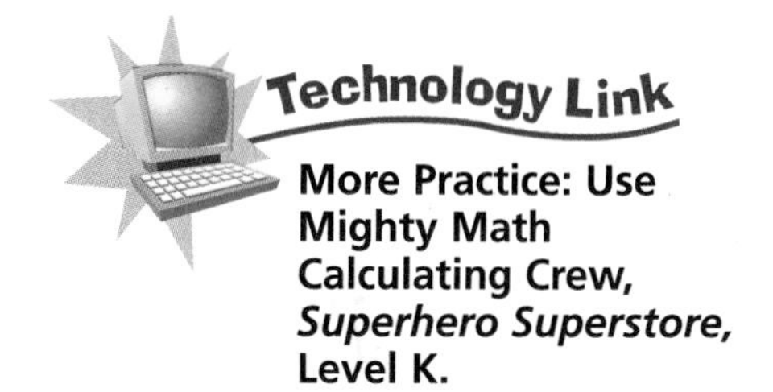

Technology Link
More Practice: Use Mighty Math Calculating Crew, *Superhero Superstore*, Level K.

- Which example can you solve by using mental math? Explain.
- Which method would you choose to solve Example B? Explain.

MATH IDEA You can find a sum by using mental math, paper and pencil, or a calculator. Choose the method that works best with the numbers in the problem.

Check

1. **Explain** how you can use mental math to add 747 and 242.

Find the sum. Tell what method you used.

2. $347 + 91$
3. $1,348 + 1,231$
4. $919 + 489$
5. $1,625 + 350$
6. $5.80 + $5.25

7. 1,032 + 5,198 = ■
8. $69.81 + $23.11 = ■
9. 3,035 + 989 + 4,918 = ■
10. 2,354 + 4,526 + 831 = ■

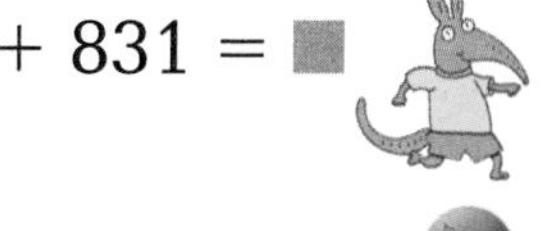

LESSON CONTINUES

Practice and Problem Solving

Find the sum. Tell what method you used.

11. 709 + 226

12. $2.78 + $5.01

13. 821 + 744

14. $3.58 + $2.65

15. 259 + 74

16. 458 + 221

17. $7.35 + $2.44

18. 624 + 347

19. 769 + 347

20. $4.11 + $3.48

21. 641 + 989

22. 329 + 110

23. 5,492 + 1,205

24. 1,895 + 1,700

25. 9,294 + 2,104

26. 2,164 + 6,235

27. 4,921 + 8,000

28. 7,493 + 2,020

29. 4,733 + 3,256

30. 8,284 + 1,213

Use mental math. Find the sum.

31. 320 + 433 = ■

32. $4.50 + $3.25 = ■

33. 910 + 120 = ■

34. 429 + 640 = ■

35. 565 + 424 = ■

36. $14.40 + $10.20 = ■

Use a calculator. Find the sum.

37. 798 + 987 = ■

38. $12.79 + $3.49 + $6.98 = ■

39. 1,647 + 897 + 3,467 = ■

40. 2,491 + 1,109 + 743 = ■

41. **NUMBER SENSE** Write a number less than 3,425 + 8,630 but greater than 7,614 + 4,429.

42. **ALGEBRA** Write the missing addend. 4,020 + ■ = 4,222

43. **ESTIMATION** Allie estimates that 5,109 + 4,995 is about 1,000. Do you agree or disagree? Explain.

44. **What's the Error?** Sergio used paper and pencil to find this sum. Describe his error. Find the sum.

```
  1 11
  8,235
+   986
  9,211
```

45. **USE DATA** Use the price list. If Craig mows and rakes 2 lawns, how much has he earned?

CRAIG'S PRICE LIST	
Weed Garden	$5.00
Mow Lawn	$7.50
Rake Lawn	$4.50

46. Can you add two 3-digit numbers and get a sum greater than 2,000? Explain.

Mixed Review and Test Prep

47. $6 + 8 + 4 = ■$ (p. 36)

48. $24 + 36 + 19 = ■$ (p. 36)

49. $\$15.76 + \$22.19 + \$3.00 = ■$ (p. 36)

Choose <, >, or = for each ●. (p. 20)

50. $305 + 281$ ● 550

51. $416 + 966$ ● $1,600$

52. $223 + 100$ ● $50 + 50 + 223$

53. The sum of two numbers is 20. The difference of the numbers is 10. What are the numbers? (p. 44)

A 10, 10 **B** 20, 10 **C** 12, 8 **D** 15, 5

For 54, use the graph.

FAVORITE SEASONS

Summer	☀ ☀ ☀ ☀ ☀
Winter	☀ ☀ ☀
Spring	☀ ☀ ☀ ☀
Fall	☀ ☀

Key: Each ☀ = 2 votes.

54. TEST PREP How many students did NOT vote for summer?

F 12 **G** 18 **H** 20 **J** 24

PROBLEM SOLVING  Thinker's Corner

Try to make a greater sum than your partner's.

MATERIALS: index cards numbered 0–9

A. Player 1 chooses 4 cards and uses the digits to write two 4-digit addends. Each digit should be used twice. Player 1 replaces the cards.

B. Player 2 repeats Step A.

C. Both players find the sum. The player with the greatest sum wins. Play this game several times. See if you can find a winning strategy.

D. Repeat this game. Try to make the least sum.

1. When making the greatest sum, where is the best place to put a 9?
2. When making the least sum, where is the best place to put your highest digit?

EXTRA PRACTICE page H34, Set D

Review/Test

CHECK VOCABULARY AND CONCEPTS

Choose the best term from the box.

regroup
Grouping Property of Addition
estimate

1. To find *about* how much, you can ? . (p. 38)
2. The ? states that the sum is the same no matter how you group the addends. (p. 36)

For 3, think of how to model 120 + 138. (pp. 40-41)

3. Do you need to regroup to find the sum 120 + 138? Explain.

CHECK SKILLS

Use the Grouping Property to find the sum. (pp. 36–37)

4. 47 + 39 + 21
5. 26 + 34 + 18
6. 65 + 32 + 55
7. 87 + 91 + 63

Estimate the sum. (pp. 38–39)

8. 76 + 17
9. 267 + 193
10. $5.92 + $3.25
11. 3,200 + 1,900

Find the sum. (pp. 42–43, 46–49)

12. 419 + 451 = ■
13. 321 + 683 = ■
14. 127 + 315 + 299 = ■
15. $6.33 + $2.98
16. 5,436 + 7,695
17. 4,782 + 3,917
18. 3,764 + 8,109

CHECK PROBLEM SOLVING

Solve. (pp. 44–45)

19. Two numbers have a sum of 47. Their difference is 5. What are the two numbers?
20. Mr. Samuel has 150 pennies in two jars. There are 40 more pennies in one jar than in the other. How many pennies are in each jar?

Standardized Test Prep

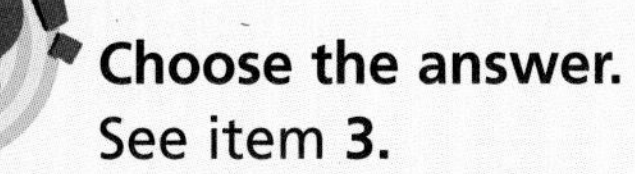

Choose the answer.
See item **3.**

If your answer doesn't match one of the choices, check your computation. If your computation is correct, mark NOT HERE.

Also see problem **6,** p. H64.

For 1–8, choose the best answer.

1. There were 596 people at the fair on Saturday and 246 people at the fair on Sunday. How many people in all were at the fair on these days?

A 742 **C** 842
B 832 **D** NOT HERE

2. Which number is less than 4,562?

F 4,565 **H** 4,572
G 4,560 **J** 4,625

3. Which number has the least value?

A 1,198 **C** 1,527
B 2,571 **D** NOT HERE

4. What is the sum of 637 and 295?

F 822 **G** 832 **H** 931 **J** 932

5. Which two numbers have a sum of about 300?

A 144 + 66
B 175 + 123
C 127 + 88
D 134 + 116

6. The sum of two numbers is 54. Their difference is 8. What are the two numbers?

F 20 and 34
G 23 and 31
H 24 and 32
J 50 and 4

7. Barry put 23 red marbles, 46 green marbles, and 34 blue marbles in a bag. How many marbles did he put in the bag in all?

A 103 **B** 104 **C** 107 **D** 113

8. A number is between 10 and 20 and it is even. The difference between the digits is 5. What is the number?

F 14 **G** 15 **H** 16 **J** 27

Write What You Know

9. Explain how to find the sum of 2,563 and 4,798. Show the addition.

10. Use the table. How many students voted in all? Explain how you grouped the numbers.

VOTES FOR CLASS TRIP	
Trip	**Number of Votes**
TV Station	16
Nature Park	8
Museum	4

CHAPTER 4 Subtraction

Mammoth Cave National Park in Kentucky is the longest known cave system in the world. More than 330 miles of caves have been mapped.

PROBLEM SOLVING Use the table to find the longest tour and the shortest tour. What is the difference between the lengths of these tours?

MAMMOTH CAVE TOURS

Tour	Length (in feet)	Minimum Number of Stairs
Great Onyx Lantern	5,280	40
Gothic Lantern	7,920	200
Mammoth Passage	3,960	120
Mobility Impaired	5,280	140
Travertine	1,320	36

Use this page to help you review and remember important skills needed for Chapter 4.

VOCABULARY

Choose the best term from the box.

difference
estimate
regroup

1. When you want to know *about* how many, you can find an ? .
2. In 15 − 8 = 7, the 7 is the ? .

SUBTRACTION FACTS (For Intervention, see p. H6.)

3. $12 - 9$
4. $15 - 7$
5. $11 - 5$
6. $14 - 6$
7. $13 - 5$
8. $10 - 3$
9. $10 - 6$
10. $12 - 7$
11. $13 - 4$
12. $16 - 7$
13. $11 - 4$
14. $18 - 9$
15. $17 - 8$
16. $16 - 9$
17. $11 - 3$

2-DIGIT SUBTRACTION (For Intervention, see p. H6.)

Subtract.

18.

tens	ones
4	6
−1	4

19.

tens	ones
8	2
−2	8

20.

tens	ones
7	0
−5	6

21. $33 - 18$
22. $90 - 48$
23. $98 - 17$
24. $57 - 27$
25. $25 - 19$

MENTAL MATH: SUBTRACT 2-DIGIT NUMBERS (For Intervention, see p. H7.)

Use mental math to find the difference.

26. $38 - 10 = ■$
27. $43 - 11 = ■$
28. $52 - 12 = ■$
29. $28 - 15 = ■$
30. $36 - 15 = ■$
31. $65 - 14 = ■$
32. $48 - 24 = ■$
33. $75 - 25 = ■$

LESSON 1

Estimate Differences

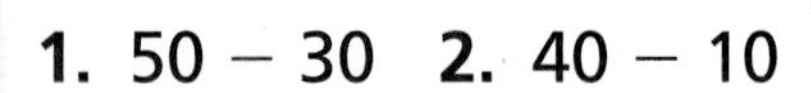

Quick Review

1. 50 − 30
2. 40 − 10
3. 35 − 10
4. 60 − 20
5. 50 − 40

Learn

TRACKING TURTLES Every summer, scientists keep track of loggerhead sea turtle nests. The table shows how many nests were seen on some South Carolina beaches in July 2000. About how many more nests were on Cape Island than on Pritchards Island?

LOGGERHEAD SEA TURTLE NESTS

Beach	Nests Found
Cape Island	615
Hilton Head Island	129
Kiawah Island	205
Pritchards Island	106

To find *about* how many more, you can estimate.

Example

STEP 1

Round each number to the nearest hundred.

$$\begin{array}{r} 615 \\ -106 \\ \hline \end{array} \rightarrow \begin{array}{r} 600 \\ -100 \\ \hline \end{array}$$

STEP 2

Find the estimated difference.

$$\begin{array}{r} 600 \\ -100 \\ \hline 500 \end{array}$$

So, about 500 more nests were on Cape Island.

More Examples

A Round to the nearest ten.

$$\begin{array}{r} 61 \\ -48 \\ \hline \end{array} \rightarrow \begin{array}{r} 60 \\ -50 \\ \hline 10 \end{array}$$

B Round to the nearest dollar.

$$\begin{array}{r} \$8.95 \\ -\$3.35 \\ \hline \end{array} \rightarrow \begin{array}{r} \$9 \\ -\$3 \\ \hline \$6 \end{array}$$

C Round to the nearest thousand.

$$\begin{array}{r} 6,860 \\ -4,655 \\ \hline \end{array} \rightarrow \begin{array}{r} 7,000 \\ -5,000 \\ \hline 2,000 \end{array}$$

Sometimes it makes sense to round to a different place. Try estimating 341 − 265.

$$\begin{array}{r} 341 \\ -265 \\ \hline \end{array} \rightarrow \begin{array}{r} 300 \\ -300 \\ \hline 0 \end{array}$$

Round to the nearest ten to get a closer estimate.

$$\begin{array}{r} 341 \\ -265 \\ \hline \end{array} \rightarrow \begin{array}{r} 340 \\ -270 \\ \hline 70 \end{array}$$

MATH IDEA To estimate a difference, round to the place value that makes sense.

Check

1. **Tell** how you would estimate 203 − 181.

Estimate the difference.

2. $\begin{array}{r} 87 \\ -32 \\ \hline \end{array}$ 3. $\begin{array}{r} 478 \\ -115 \\ \hline \end{array}$ 4. $\begin{array}{r} 813 \\ -491 \\ \hline \end{array}$ 5. $\begin{array}{r} \$9.01 \\ -\$2.60 \\ \hline \end{array}$ 6. $\begin{array}{r} 5{,}020 \\ -1{,}750 \\ \hline \end{array}$

Practice and Problem Solving

Estimate the difference.

7. $\begin{array}{r} 42 \\ -19 \\ \hline \end{array}$ 8. $\begin{array}{r} 51 \\ -29 \\ \hline \end{array}$ 9. $\begin{array}{r} 84 \\ -28 \\ \hline \end{array}$ 10. $\begin{array}{r} 69 \\ -43 \\ \hline \end{array}$ 11. $\begin{array}{r} 91 \\ -23 \\ \hline \end{array}$

12. $\begin{array}{r} 470 \\ -110 \\ \hline \end{array}$ 13. $\begin{array}{r} 613 \\ -371 \\ \hline \end{array}$ 14. $\begin{array}{r} \$9.08 \\ -\$3.80 \\ \hline \end{array}$ 15. $\begin{array}{r} 880 \\ -114 \\ \hline \end{array}$ 16. $\begin{array}{r} 625 \\ -489 \\ \hline \end{array}$

17. $\begin{array}{r} \$5.17 \\ -\$1.01 \\ \hline \end{array}$ 18. $\begin{array}{r} 322 \\ -199 \\ \hline \end{array}$ 19. $\begin{array}{r} 3{,}288 \\ -1{,}255 \\ \hline \end{array}$ 20. $\begin{array}{r} 1{,}970 \\ -1{,}050 \\ \hline \end{array}$ 21. $\begin{array}{r} 4{,}819 \\ -1{,}766 \\ \hline \end{array}$

Estimate the difference. Round to the place value that makes sense.

22. $\begin{array}{r} 422 \\ -394 \\ \hline \end{array}$ 23. $\begin{array}{r} 201 \\ -178 \\ \hline \end{array}$ 24. $\begin{array}{r} 4{,}411 \\ -3{,}509 \\ \hline \end{array}$

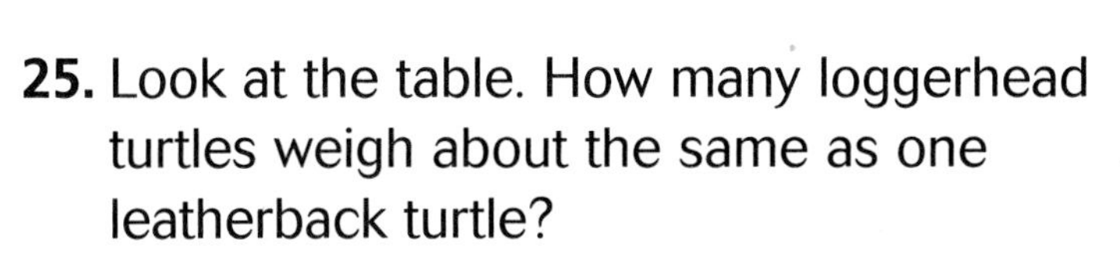

25. Look at the table. How many loggerhead turtles weigh about the same as one leatherback turtle?

26. **Write a problem** about estimating. Use the table at the right.

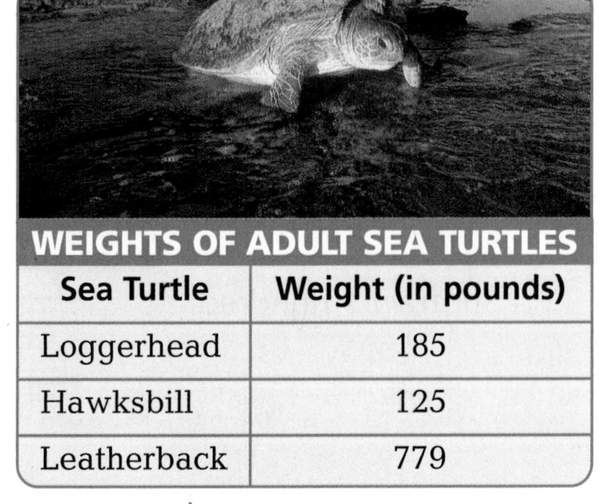

WEIGHTS OF ADULT SEA TURTLES

Sea Turtle	Weight (in pounds)
Loggerhead	185
Hawksbill	125
Leatherback	779

Mixed Review and Test Prep

Write the value of the blue digit.

(pp. 6 and 10)

27. 2,051   28. 1,321

29. 37,820   30. 52,902

31. TEST PREP Amy had 34 shells. On Friday she found more and had a total of 45 shells. How many shells did she find on Friday?

A 9   **B** 11   **C** 55   **D** 79

EXTRA PRACTICE page H35, Set A

LESSON 2

HANDS ON

Subtract 3-Digit Numbers

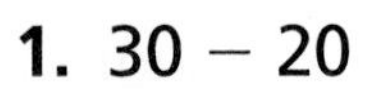

Quick Review

1. 30 − 20
2. 120 − 10
3. 62 − 32
4. 92 − 31
5. 35 − 15

MATERIALS
base-ten blocks

Explore

Use models to subtract 195 from 324.

$$\begin{array}{r} 324 \\ -195 \\ \hline \end{array}$$

Example 1

STEP 1

Show 324.

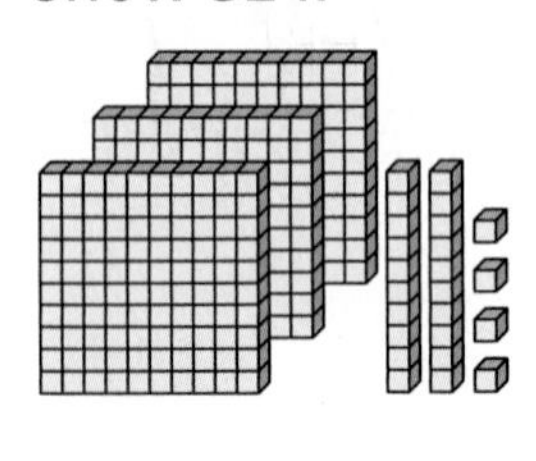

STEP 2

Try to subtract 5 ones. Since there are not enough ones, regroup 1 ten as 10 ones. Subtract 5 ones.

STEP 3

Try to subtract 9 tens. Since there are not enough tens, regroup 1 hundred as 10 tens. Subtract 9 tens.

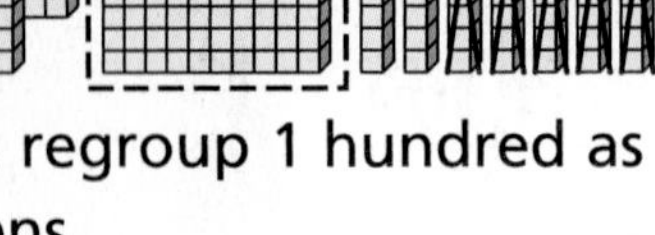

STEP 4

Subtract 1 hundred.

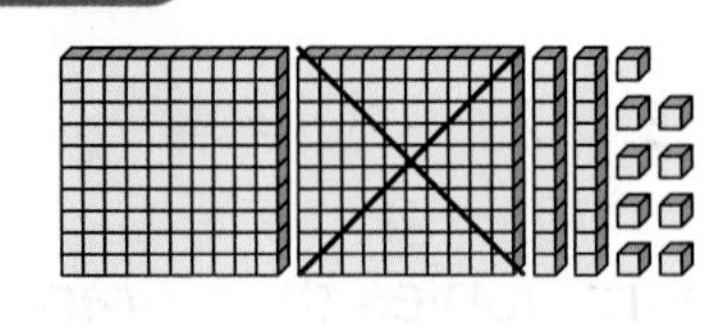

So, 324 − 195 = 129.

- **Explain** why you regroup 1 ten as 10 ones in Step 2.

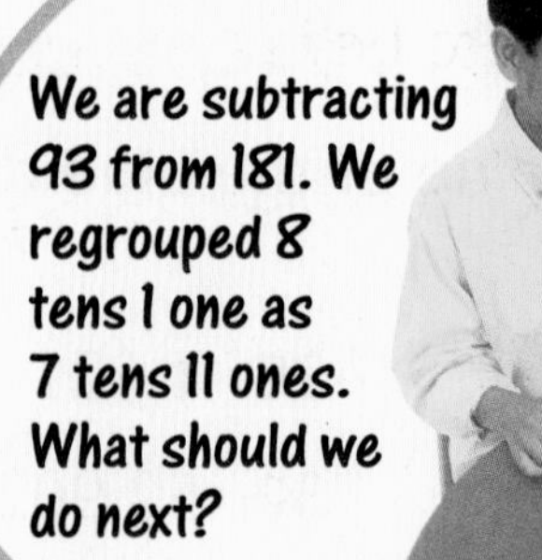

We are subtracting 93 from 181. We regrouped 8 tens 1 one as 7 tens 11 ones. What should we do next?

Try It

Use base-ten blocks to find each difference.

a. 181 − 93 = ■ **b.** 369 − 140 = ■

Connect

Technology Link

More Practice: Use E-Lab, *Subtraction of Three-Digit Numbers.*

www.harcourtschool.com/elab2002

Here is a way to record subtraction. To subtract 126 from 215, first line up the hundreds, tens, and ones.

Example 2

STEP 1

Subtract the ones. 6 > 5
Regroup.
1 ten 5 ones = 0 tens 15 ones

$$\begin{array}{r} \overset{0\ 15}{2\ \cancel{1}\ \cancel{5}} \\ -1\ 2\ 6 \\ \hline 9 \end{array}$$

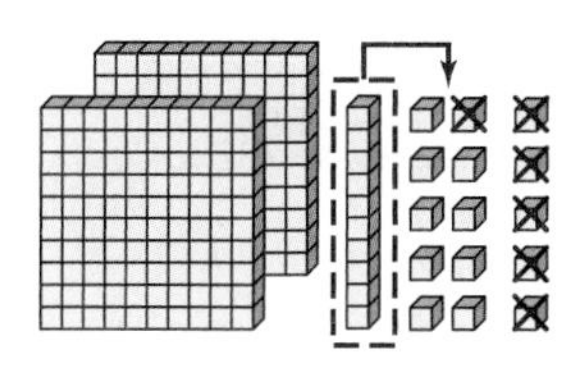

STEP 2

Subtract the tens. 2 > 0
Regroup.
2 hundreds 0 tens =
1 hundred 10 tens

$$\begin{array}{r} \overset{1\ \overset{10}{\cancel{0}}\ 15}{\cancel{2}\ \cancel{1}\ \cancel{5}} \\ -1\ 2\ 6 \\ \hline 8\ 9 \end{array}$$

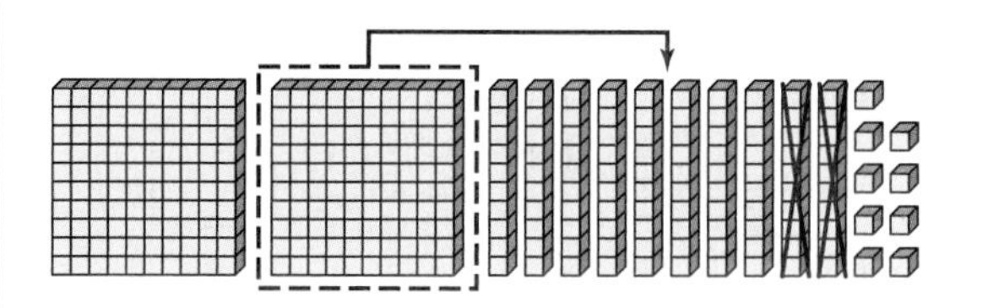

STEP 3

Subtract the hundreds.

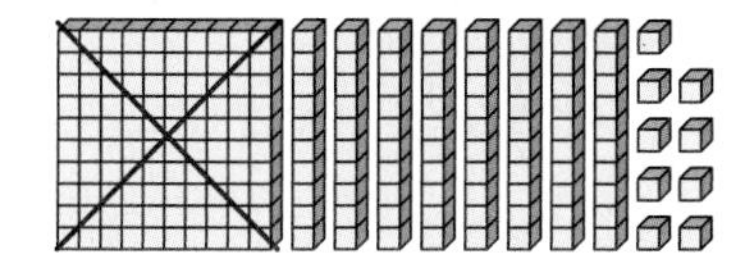

$$\begin{array}{r} \overset{1\ \overset{10}{\cancel{0}}\ 15}{\cancel{2}\ \cancel{1}\ \cancel{5}} \\ -1\ 2\ 6 \\ \hline 8\ 9 \end{array}$$

Practice and Problem Solving

Use base-ten blocks to find each difference.

1. $94 - 28 = \blacksquare$
2. $223 - 122 = \blacksquare$
3. $437 - 243 = \blacksquare$
4. $183 - 159 = \blacksquare$
5. $329 - 87 = \blacksquare$
6. $223 - 135 = \blacksquare$
7. **REASONING** There are 32 more apples than oranges at a fruit stand. How many apples and oranges could there be?

Mixed Review and Test Prep

8. $9 - 5$
9. $17 - 8$
10. $6 + 7$
11. $8 + 4$
12. **TEST PREP** $3 + (8 + 12)$ (p. 36)

A 11
B 20
C 23
D 26

LESSON 3

Subtract 3-Digit Numbers

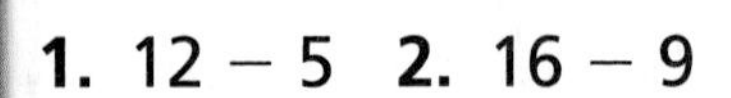

Quick Review

1. 12 − 5 2. 16 − 9
3. 11 − 7 4. 15 − 6
5. 15 − 8

Learn

TURTLE TALK A sea park has two green sea turtles, Tara and Tom. Tom weighs 332 pounds and Tara weighs 198 pounds. How much more does Tom weigh than Tara?

▲ A green sea turtle weighs between 150 and 410 pounds.

$332 - 198 = \blacksquare$

Estimate. $332 \rightarrow 300$, $-198 \rightarrow -200$; $300 - 200 = 100$

Example 1

STEP 1

Subtract the ones. 8 > 2
Regroup.
3 tens 2 ones = 2 tens 12 ones

$$\begin{array}{r} \scriptstyle 2\ 12 \\ 3\,\cancel{3}\,\cancel{2} \\ -1\,9\,8 \\ \hline 4 \end{array}$$

STEP 2

Subtract the tens. 9 > 2
Regroup.
3 hundreds 2 tens = 2 hundreds 12 tens

$$\begin{array}{r} \scriptstyle 12 \\ \scriptstyle 2\ \cancel{2}\ 12 \\ \cancel{3}\,\cancel{3}\,\cancel{2} \\ -1\,9\,8 \\ \hline 3\,4 \end{array}$$

STEP 3

Subtract the hundreds.

$$\begin{array}{r} \scriptstyle 12 \\ \scriptstyle 2\ \cancel{2}\ 12 \\ \cancel{3}\,\cancel{3}\,\cancel{2} \\ -1\,9\,8 \\ \hline 1\,3\,4 \end{array}$$

So, Tom weighs 134 pounds more than Tara. Since 134 is close to 100, the answer is reasonable.

More Examples

A

$$\begin{array}{r} \scriptstyle 2\ 15 \\ \cancel{3}\,\cancel{5}\,9 \\ -\ \ 8\,4 \\ \hline 2\,7\,5 \end{array}$$

B

$$\begin{array}{r} \scriptstyle 13 \\ \scriptstyle 7\ \cancel{3}\ 16 \\ \cancel{8}\,\cancel{4}\,\cancel{6} \\ -6\,9\,8 \\ \hline 1\,4\,8 \end{array}$$

C

$$\begin{array}{r} \scriptstyle 6\ 12 \\ \$4.\,\cancel{7}\,\cancel{2} \\ -\$1.\,1\,5 \\ \hline \$3.\,5\,7 \end{array}$$

Use a decimal point to separate dollars and cents.

Subtract Across Zeros

What if Tom weighed 300 pounds and Tara weighed 178 pounds? How much more would Tom weigh?

$300 - 178 = \blacksquare$

Estimate. $300 - 180 = 120$

Example 2

STEP 1

8 > 0. Since there are 0 tens, regroup hundreds.
3 hundreds 0 tens =
2 hundreds 10 tens

$$\begin{array}{r} {\scriptstyle 2\ 10} \\ \not{3}\not{0}0 \\ -178 \\ \hline \end{array}$$

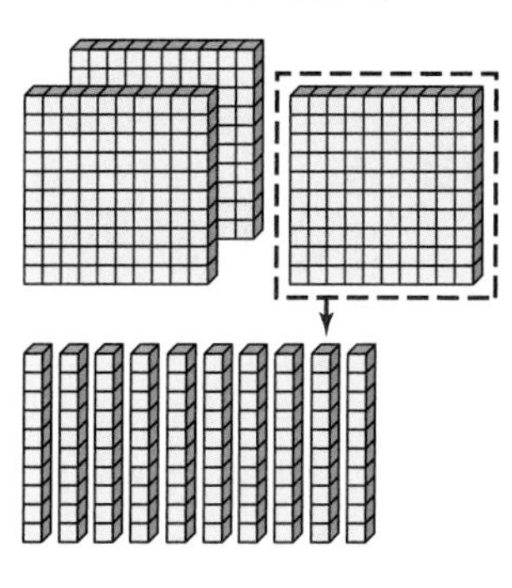

STEP 2

Regroup tens.
10 tens 0 ones =
9 tens 10 ones

$$\begin{array}{r} {\scriptstyle 9} \\ {\scriptstyle 2\ \not{10}\ 10} \\ \not{3}\not{0}\not{0} \\ -178 \\ \hline \end{array}$$

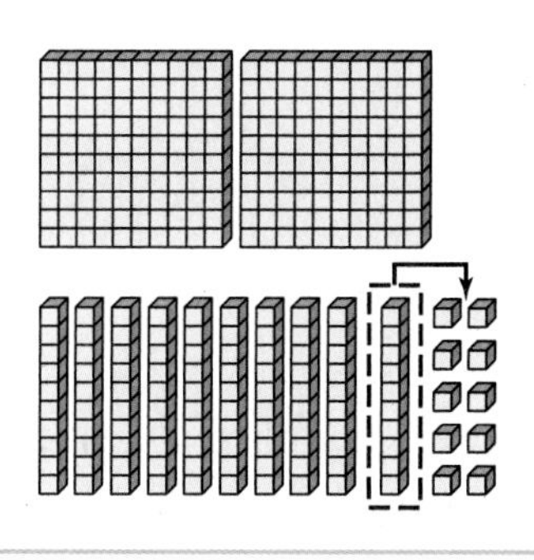

STEP 3

Subtract the ones.
Subtract the tens.
Subtract the hundreds.

$$\begin{array}{r} {\scriptstyle 9} \\ {\scriptstyle 2\ \not{10}\ 10} \\ \not{3}\not{0}\not{0} \\ -178 \\ \hline 122 \end{array}$$

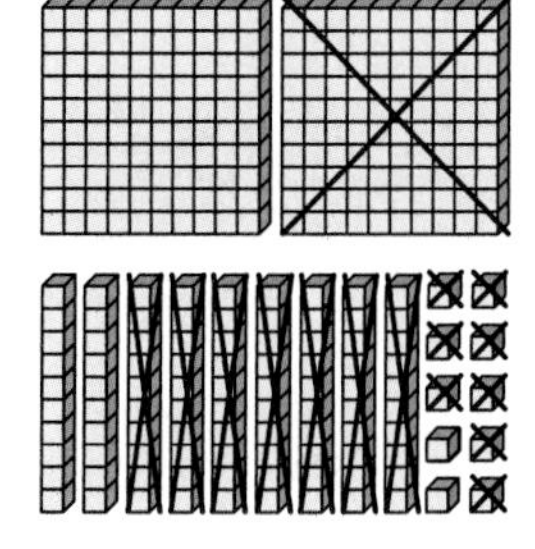

So, Tom would weigh 122 pounds more than Tara. Since 122 is close to 120, the answer is reasonable.

More Examples

A

$$\begin{array}{r} {\scriptstyle 9} \\ {\scriptstyle 1\ \not{10}\ 14} \\ \not{2}\not{0}\not{4} \\ -\ 87 \\ \hline 117 \end{array}$$

B

$$\begin{array}{r} {\scriptstyle 9} \\ {\scriptstyle 4\ \ \not{10}\ 13} \\ \$\not{5}.\not{0}\not{3} \\ -\ \$1.24 \\ \hline \$3.79 \end{array}$$

C

$$\begin{array}{r} {\scriptstyle 9} \\ {\scriptstyle 6\ \not{10}\ 10} \\ \not{7}\not{0}\not{0} \\ -481 \\ \hline 219 \end{array}$$

CHECK ✓

$$\begin{array}{r} 219 \\ +481 \\ \hline 700 \end{array}$$

You can add to check your answer.

MATH IDEA When you subtract across zeros, you may need to regroup the hundreds first.

Check

1. **Explain** why 9 is written above the crossed-out 10 in Example A.

LESSON CONTINUES

Find the difference. Estimate to check.

2. $\begin{array}{r} 595 \\ -242 \\ \hline \end{array}$ 3. $\begin{array}{r} 336 \\ -191 \\ \hline \end{array}$ 4. $\begin{array}{r} 300 \\ -\ 84 \\ \hline \end{array}$ 5. $\begin{array}{r} \$4.00 \\ -\$2.83 \\ \hline \end{array}$ 6. $\begin{array}{r} 607 \\ -349 \\ \hline \end{array}$

Practice and Problem Solving

Find the difference. Estimate to check.

7. $\begin{array}{r} 574 \\ -412 \\ \hline \end{array}$ 8. $\begin{array}{r} 584 \\ -250 \\ \hline \end{array}$ 9. $\begin{array}{r} 438 \\ -119 \\ \hline \end{array}$ 10. $\begin{array}{r} 672 \\ -468 \\ \hline \end{array}$ 11. $\begin{array}{r} \$9.63 \\ -\$4.05 \\ \hline \end{array}$

12. $\begin{array}{r} 294 \\ -137 \\ \hline \end{array}$ 13. $\begin{array}{r} 891 \\ -\ 86 \\ \hline \end{array}$ 14. $\begin{array}{r} \$6.57 \\ -\$4.98 \\ \hline \end{array}$ 15. $\begin{array}{r} 372 \\ -196 \\ \hline \end{array}$ 16. $\begin{array}{r} 730 \\ -217 \\ \hline \end{array}$

17. $\begin{array}{r} 800 \\ -585 \\ \hline \end{array}$ 18. $\begin{array}{r} 506 \\ -439 \\ \hline \end{array}$ 19. $\begin{array}{r} \$7.00 \\ -\$2.11 \\ \hline \end{array}$ 20. $\begin{array}{r} 805 \\ -\ 99 \\ \hline \end{array}$ 21. $\begin{array}{r} \$9.06 \\ -\$4.08 \\ \hline \end{array}$

22. $\begin{array}{r} 354 \\ -148 \\ \hline \end{array}$ 23. $\begin{array}{r} 942 \\ -817 \\ \hline \end{array}$ 24. $\begin{array}{r} 647 \\ -435 \\ \hline \end{array}$ 25. $\begin{array}{r} 461 \\ -178 \\ \hline \end{array}$ 26. $\begin{array}{r} 724 \\ -536 \\ \hline \end{array}$

Subtract. Use addition to check.

27. $308 - 149 = \blacksquare$ 28. $900 - 312 = \blacksquare$ 29. $604 - 485 = \blacksquare$

30. $401 - 173 = \blacksquare$ 31. $304 - 255 = \blacksquare$ 32. $300 - 92 = \blacksquare$

33. **ALGEBRA** Write the missing addend. $255 + \blacksquare = 305$

34. **What's the Error?** Michael wrote a subtraction problem like this. Describe his error. Find the difference.

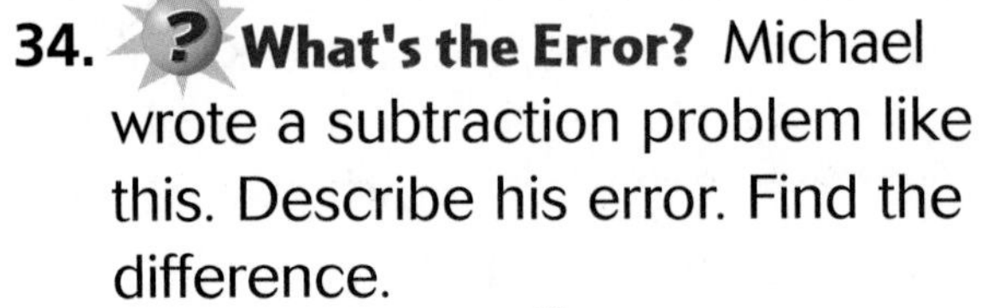

35. Sherrie is thinking of a number. It is 128 less than 509. What number is she thinking of?

36. **REASONING** Would you have to regroup to find $255 - 125$? Explain.

USE DATA For 37–38, use the graph.

37. How much do Tim, Ted, and Talia weigh altogether?

38. If Tim Turtle gained 100 pounds, how much more would he weigh than Talia Turtle?

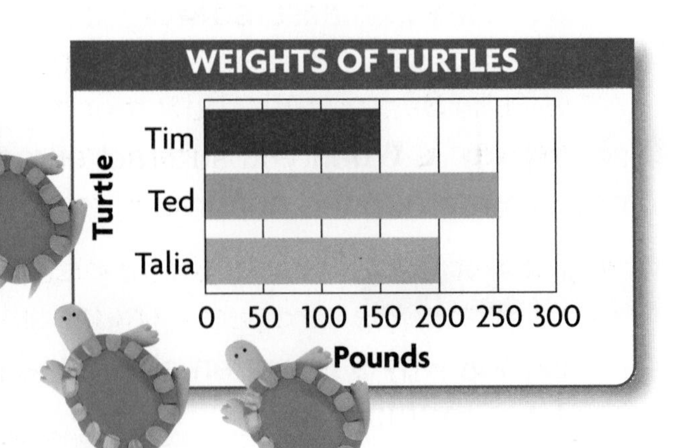

Mixed Review and Test Prep

Write <, >, or = for each ●. (p. 20)

39. 72 ● 27

40. 321 ● 213

41. $66 - 32$

42. (p. 42) $148 + 134$

43. (p. 42) $399 + 722$

44. TEST PREP Find 95 + 37.

A 58
B 132
C 217
D 300

45. Carlos found 112 green bottles, 115 soda cans, and 129 plastic cups. How many cups and cans did he find? (p. 42)

46. (p. 42) $182 + 28$

47. (p. 42) $325 + 149$

48. (p. 42) $456 + 344$

49. TEST PREP Find the sum of 342 and 160. (p. 42)

F 182
G 402
H 500
J 502

PROBLEM SOLVING LINKUP . . . to Science

Turtles are the only reptiles with shells. Most turtles can pull their head, legs, and tail into their shells. The female turtle digs a hole on land, lays her eggs, and covers them. The heat from the sun hatches the eggs.

1. There are about 250 kinds of turtles. About 50 kinds live in the United States and in Canada. About how many kinds don't live in the United States and Canada?
2. One of the largest green sea turtles ever measured weighed 871 pounds. If a green sea turtle weighs 395 pounds, how much heavier is the largest green sea turtle?
3. If a flatback sea turtle swam 752 miles in the spring and 374 miles in the summer, how much farther did the turtle swim in the spring?
4. If a leatherback sea turtle traveled 989 miles one year and 873 miles the next year, how far did it travel in the two years?

EXTRA PRACTICE page H35, Set B

LESSON 4

Choose a Method

Quick Review

1. 135 − 20
2. 500 − 50
3. 395 − 30
4. 671 − 421
5. 105 − 95

Learn

You can find a difference by using mental math, paper and pencil, or a calculator.

UP, UP AND AWAY A hot-air balloon was at 1,025 feet above the ground. Then it rose to 1,920 feet above the ground. How much higher was it then?

1,920 − 1,025 = ■

Estimate. 2,000 − 1,000 = 1,000

Use Paper and Pencil

The numbers are large. The problem involves regrouping. So, using paper and pencil is a good choice.

STEP 1

Subtract the ones. 5 > 0
Regroup.
2 tens 0 ones =
1 ten 10 ones

$$\begin{array}{r} {\scriptstyle 1\ 10} \\ 1,9\not{2}\not{0} \\ -1,025 \\ \hline 5 \end{array}$$

STEP 2

Subtract the tens. 2 > 1
Regroup.
9 hundreds 1 ten =
8 hundreds 11 tens

$$\begin{array}{r} {\scriptstyle 11} \\ {\scriptstyle 8\ \not{1}\ 10} \\ 1,\not{9}\not{2}\not{0} \\ -1,025 \\ \hline 95 \end{array}$$

STEP 3

Subtract the hundreds.
Subtract the thousands.

$$\begin{array}{r} {\scriptstyle 11} \\ {\scriptstyle 8\ \not{1}\ 10} \\ 1,\not{9}\not{2}\not{0} \\ -1,025 \\ \hline 895 \end{array}$$

So, the hot-air balloon was 895 feet higher. Since 895 is close to 1,000, the answer is reasonable.

REASONING How can you add to check your answer?

Use a Calculator

$38.94 − $25.96 = ■

The money amounts are large. The problem involves regrouping. So, using a calculator is a good choice.

12.98

Use Mental Math

$\$6.80 - \$3.10 = \blacksquare$

There is no regrouping. You can subtract the dollars and cents amounts in your head. So, using mental math is a good choice.

Think: Subtract the dollar amounts. $6.00 − $3.00 = $3.00
Subtract the cents amounts. $0.80 − $0.10 = $0.70
The difference is $3.00 + $0.70, or $3.70.

Examples

A	B	C
$39.85 −$25.65 $14.20	8 17 (regrouping) 897 −689 208	11 12 / 1 2 16 (regrouping) 1,236 − 847 389

- Which problem can you solve using mental math? Explain.
- Which method would you use to solve Example B? Explain.

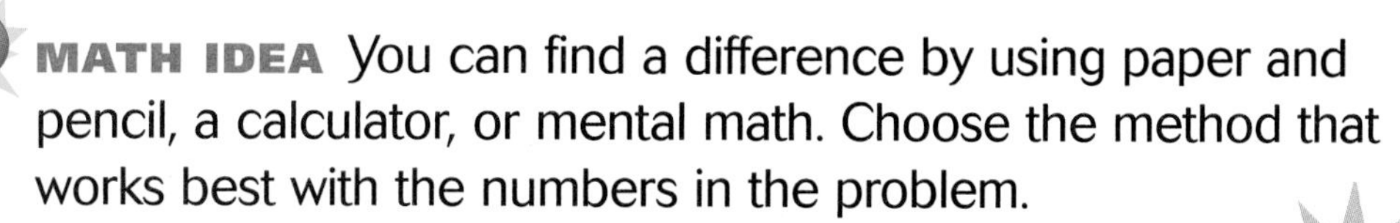

MATH IDEA You can find a difference by using paper and pencil, a calculator, or mental math. Choose the method that works best with the numbers in the problem.

Technology Link
More Practice: Use Mighty Math Calculating Crew, *Superhero Superstore*, Levels G and H.

▶ Check

1. **Explain** how you can use mental math to subtract 656 from 987.

Find the difference. Tell what method you used.

2. 287 − 178
3. $7.98 − $3.56
4. 127 − 94
5. 2,165 − 1,084
6. $35.98 − $15.46

LESSON CONTINUES ▶

Practice and Problem Solving

Find the difference. Tell what method you used.

7. $365 - 104$
8. $884 - 282$
9. $951 - 148$
10. $821 - 631$
11. $930 - 821$
12. $211 - 120$
13. $545 - 438$
14. $921 - 642$
15. $760 - 479$
16. $397 - 254$
17. $5{,}673 - 1{,}294$
18. $7{,}116 - 2{,}005$
19. $4{,}690 - 3{,}282$
20. $3{,}050 - 1{,}422$
21. $2{,}860 - 750$
22. $8{,}907 - 5{,}605$
23. $9{,}437 - 6{,}420$
24. $7{,}884 - 3{,}802$
25. $8{,}932 - 4{,}613$
26. $4{,}507 - 1{,}602$

Subtract. Use mental math.

27. 8,000 − 5,000 = ■
28. 4,500 − 1,400 = ■
29. 5,850 − 4,420 = ■
30. 2,000 − 900 = ■

Use a calculator to solve.

31. 4,665 − ■ = 3,962
32. ■ − 978 = 396
33. $42.98 − ■ = $24.50
34. ■ − 1,324 = 687

USE DATA For 35–38, use the table.

35. What is the difference between the lengths of the Ohio River and the Rio Grande?
36. **ALGEBRA** The Ohio River is 998 miles shorter than which one of these rivers?
37. **Write a problem** about the difference between river lengths. Exchange with a partner. Solve.
38. **What's the Question?** The answer is 1,367 miles.

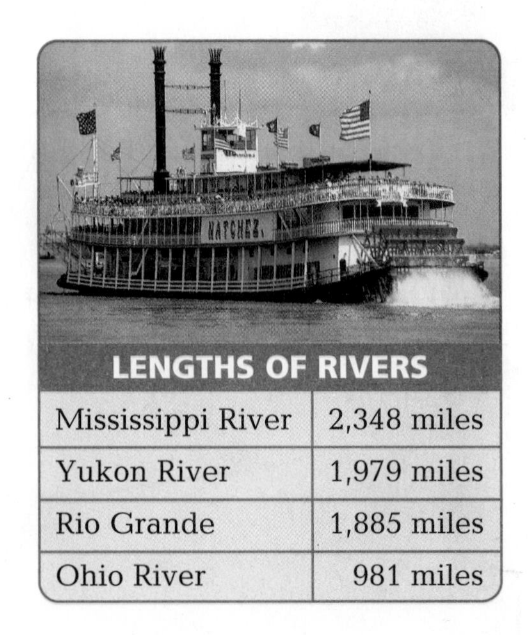

LENGTHS OF RIVERS	
Mississippi River	2,348 miles
Yukon River	1,979 miles
Rio Grande	1,885 miles
Ohio River	981 miles

39. Sheli's mother was born 25 years before 1987. How old was Sheli's mother in 2000?

40. Write a 3-digit number you could subtract from 274 without regrouping.

Mixed Review and Test Prep

Find each sum or difference.

41. $28 + 7$

42. $45 - 8$

43. $67 - 29$

44. $85 - 59$

Write the missing number.

45. $80 - ■ = 71$

46. $45 + ■ = 52$

47. $■ - 20 = 30$

48. $■ - 25 = 40$

Use the Grouping Property to find the sum. (p. 36)

49. $7 + 3 + 16$

50. $21 + 8 + 9$

51. **TEST PREP** Mrs. Jimenez baked 75 cookies and sold 58 of them at the bake sale. How many cookies were left?

A 17 **B** 23 **C** 27 **D** 133

52. **TEST PREP** Which number is between 3,478 and 4,309? (p. 24)

F 3,380 **H** 4,041
G 3,409 **J** 4,330

PROBLEM SOLVING

Thinker's Corner

SOLVE IT!

Find the sum or difference.

51 − 23 **O**	36 + 49 **N**	70 − 53 **H**		
64 − 15 **Y**	91 + 59 **P**	647 + 178 **A**	313 + 448 **U**	500 − 195 **E**
200 − 77 **C**	464 + 446 **T**	384 + 165 **J**	675 − 179 **S**	853 − 194 **M**

To answer the riddle, match the letters from the sums and differences above to the numbers below.

Who can jump higher than a house? __ __ __ __ __ __.
825 85 49 28 85 305

__ __ __ __ __ __ __ __ __ ' __ __ __ __ __!
825 17 28 761 496 305 123 825 85 910 549 761 659 150

LESSON 5

Problem Solving Skill
Estimate or Exact Answer

UNDERSTAND | PLAN | SOLVE | CHECK

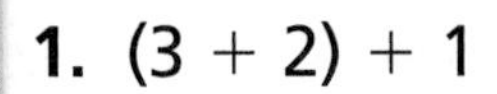

Quick Review

1. (3 + 2) + 1
2. 4 + (5 + 5)
3. 7 + (3 + 4)
4. 2 + 3 + 4
5. 32 + 6 + 14

PLANNING AHEAD Alicia invited 9 boys, 19 girls, and 8 adults to a meeting with the park ranger. She must decide whether to hold the meeting in the classroom, the library, or the cafeteria.

Since she doesn't need to know exactly how many people are coming, she estimates.

$$\begin{array}{rcr} 9 & \rightarrow & 10 \\ 19 & \rightarrow & 20 \\ +\ 8 & \rightarrow & +10 \\ \hline & & 40 \end{array}$$

MEETING ROOMS	
Place	**Number of Seats**
Classroom	25
Library	50
Cafeteria	100

So, she needs a room with about 40 seats. She can hold the meeting in the library.

What if Alicia must make a name tag for each person at the meeting? Alicia must decide how many name tags she needs to make.

Since she doesn't want to make too many or too few name tags, she finds an exact answer.

$$9 + 19 + 8 = 36$$

So, she must make 36 name tags.

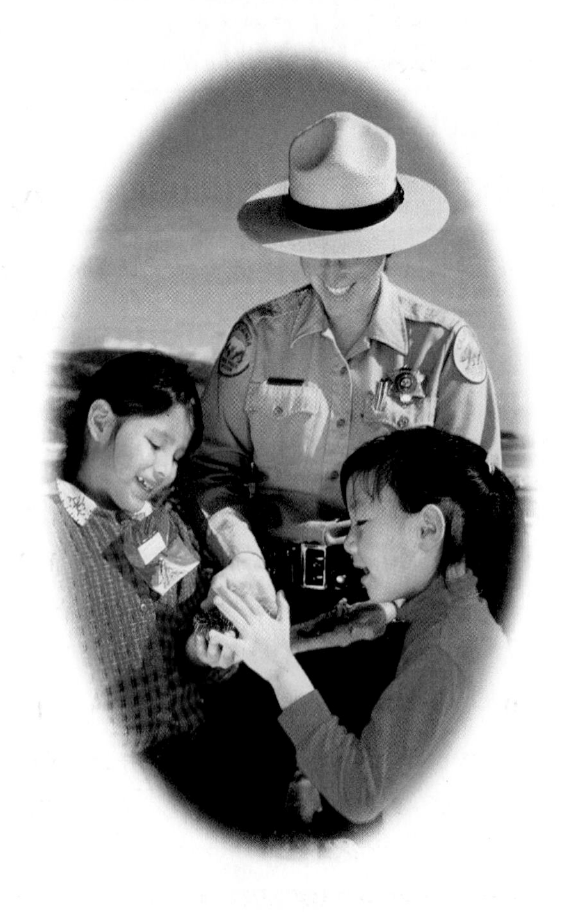

Talk About It

- Why was an estimate good enough to decide which room to use?
- Why did Alicia need an exact answer to decide how many name tags to make?

Problem Solving Practice

USE DATA Use the table for 1–2. Write whether you need an exact answer or an estimate. Then solve.

SUPPLIES	
Item	**Price**
Glitter	$3.50
Glue	$1.79
Hole punch	$3.99

1. Roberta has $15. Can she buy glitter and a hole punch? Explain.
2. Lorenzo pays for glue with a $5 bill. How much change will he get?

Clarissa is planning a picnic for two scout troops. There are 29 scouts in one troop and 17 scouts in the other troop.

3. Clarissa must make a name tag for each scout. Which sentence shows how many name tags she must make?

   A $30 + 20 = 50$
   B $29 + 17 = 46$
   C $29 + 17 = 36$
   D $29 - 17 = 12$

4. A package of cupcakes holds 10 cupcakes. If Clarissa wants each scout to have at least 1 cupcake, how many packages should she buy?

   F 1
   G 2
   H 4
   J 5

Problem Solving Skill

Mixed Applications

5. Wesley has 4 more hockey cards than baseball cards. If he has 28 cards in all, how many hockey cards does he have?
6. Each person at a picnic needs about 2 cups of punch. There will be 19 girls and 29 boys at the picnic. About how many cups should Jeff make?
7. I am a number greater than 110 and less than 120. The sum of my digits is 9. What number am I?
8. Mitch has 43 bottle caps and 12 buttons. If he finds 12 more bottle caps, how many caps will he have in all?
9. **? What's the Question?** Last week Luann ran 50 miles and Patrice ran 15 miles. The answer is 35 miles.
10. **REASONING** Joel had 32 inches of ribbon. He cut 10 inches off each end of the ribbon. What is the length of the ribbon he has left?

LESSON 6

Algebra: Expressions and Number Sentences

Quick Review

1. 14 + 14
2. 12 + 21
3. 42 + 25
4. 90 + 18
5. 25 + 49

VOCABULARY
expression

Learn

LUNCH LINE In the morning, visitors bought 34 packets of food for the animals in the petting zoo. In the afternoon, visitors bought 58 packets. How many food packets were bought in all?

You can write an expression for this problem.

34 packets plus 58 packets
↓ ↓ ↓
34 + 58

An **expression** is part of a number sentence. It combines numbers and operation signs. It does not have an equal sign.

34 + 58 = 92 is a number sentence.

92 is the number of food packets bought in all.

MATH IDEA A number sentence can be true or false.

4 + 3 = 7 is true.
4 − 3 = 7 is false.

Mike spent $12 for a basketball and $18 for a soccer ball. How much more did the soccer ball cost?

$18 ● $12 = $6

Which symbol will make the sentence *true*?

Try + $18 + $12 = $6 *Not* true.
Try − $18 − $12 = $6 **True**.

So, the correct operation symbol is −.

Check

Write an expression for each.

1. Takeo had 273 cards. He gave away 35. How many cards does he have left?
2. Mia had 13 apples. She bought 7 more. How many does she have in all?

Write + or − to make the number sentence true.

3. 12 ● 2 = 10 4. 37 ● 11 = 48 5. 126 ● 79 = 47 6. 367 ● 43 = 410

Practice and Problem Solving

Write an expression for each.

7. Gwen bought 12 red pencils, 2 blue pencils, and 22 yellow pencils. How many blue and red pencils did she buy?
8. Ned has 17 crayons. He has 15 pens. How many more crayons than pens does he have?

Write + or − to make the number sentence true.

9. 4 ● 3 = 1
10. 28 ● 9 = 37
11. 329 ● 87 = 242
12. 559 ● 50 = 609
13. 74 ● 47 = 17 + 10
14. 444 ● 6 = 460 − 10

Write the missing number that makes the number sentence true.

15. ■ + 3 = 14
16. 140 + 5 = ■
17. 45 − ■ = 25
18. 309 − ■ = 209
19. 215 − ■ = 120
20. ■ − 125 = 318
21. **REASONING** Blair says, "12 + 3 + 1 = 5 + 11 is a true number sentence." Do you agree or disagree? Explain.
22. Selma wrote 18 > 27. Is this true? If not, rewrite her sentence to make it true.

Mixed Review and Test Prep

Find each sum. (p. 36)

23. 64 + 91 + 26
24. 15 + 73 + 22
25. 41 + 66 + 19
26. Find 1,035 − 887. (p. 62)
27. **TEST PREP** Benjy had 62 marbles. He gave 15 to Lisa, 14 to Lenny, and 9 to Jake. He kept the rest. Who has the most marbles?
   - **A** Lisa
   - **B** Lenny
   - **C** Benjy
   - **D** Jake

EXTRA PRACTICE page H35, Set D

Review/Test

CHECK VOCABULARY AND CONCEPTS

Choose the correct term from the box.

estimate
expression

1. A part of a number sentence that combines numbers and operation signs is an ? . (p. 68)

Tell whether you need to regroup to find each difference. Explain. (pp. 56–57)

2. $342 - 214$
3. $312 - 181$
4. $162 - 51$

CHECK SKILLS

Estimate the difference. (pp. 54–55)

5. $67 - 29$
6. $967 - 283$
7. $748 - 599$
8. $4,175 - 1,832$
9. $8,596 - 3,714$

Find the difference. (pp. 58–65)

10. $341 - 133 = \blacksquare$
11. $837 - 247 = \blacksquare$
12. $\$3.73 - \$2.08 = \blacksquare$
13. $645 - 347 = \blacksquare$
14. $\$4.02 - \$2.56$
15. $800 - 364$
16. $602 - 531$
17. $200 - 95$
18. $703 - 155$
19. $4,250 - 2,872$
20. $9,308 - 5,970$
21. $8,029 - 6,047$
22. $7,300 - 1,074$
23. $5,789 - 898$

CHECK PROBLEM SOLVING

24. Abe wants each person at his party to have about 1 cup of punch. If he invites 18 children and 9 adults, about how many cups of punch should he make? (pp. 66–67)

25. Barry collected 29 cans on Monday, 12 cans on Tuesday, and 17 cans on Friday. Write an expression to show how to find how many cans Barry collected. (pp. 68–69)

Standardized Test Prep

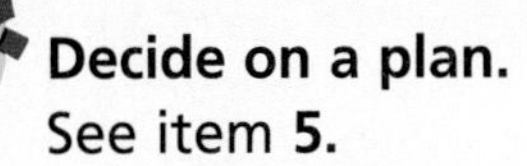

Decide on a plan.
See item **5**.

The question asks how many in all. Look for an expression that shows the correct operation for putting groups together.

Also see problem **4**, p. H63.

For 1–7, choose the best answer.

1. Which two numbers have a difference of about 500?

A 647 and 374
B 521 and 110
C 782 and 367
D 659 and 188

2. 528 − 265 = ■

F 263 **H** 363
G 343 **J** 793

3. What is 4,612 rounded to the nearest thousand?

A 3,000 **C** 5,000
B 4,000 **D** 6,000

4. What is the sum of 539 and 649?

F 1,177 **H** 1,187
G 1,178 **J** 1,188

5. Jay has 147 stamps from the United States and 64 stamps from other countries. Which expression tells how many stamps he has in all?

A 147 − 64
B 147 + 147
C 147 + 64
D 64 − 64

6. A number is odd and the sum of its digits is 15. The number is greater than 80 and less than 96. What is the number?

F 86 **H** 89
G 87 **J** NOT HERE

7. $\begin{array}{r} 6{,}003 \\ -2{,}336 \\ \hline \end{array}$

A 3,667 **C** 4,773
B 3,763 **D** NOT HERE

Write What You Know

8. Cedric invited 9 friends from his soccer team to a party. He invited 18 friends from school. He wants to buy enough party favors so that each friend gets one party favor. Does he need an estimate or an exact answer? How do you know?

9. Use mental math to find the missing number.

92 − 38 = ■

Explain what you did.

PROBLEM SOLVING
MATH DETECTIVE

Putting It Together

For each case, copy the squares on grid paper. Then fill in the missing numbers.

Remember

A hundred chart has numbers from 1 to 100 arranged in ten rows and ten columns.

Case 1

1. If the hundred chart were cut apart into the pieces shown, what numbers would go in the blank squares?

A 57

B 66

C 34

Case 2

2. **What if** you made a two hundred chart by showing twenty rows and ten columns? What numbers would go in the blank squares?

A 127

B 131

C 146

STRETCH YOUR THINKING If this square were cut out of a hundred chart, how would you use the value of the number in box E to figure out the other numbers?

A	B	C
D	E	F
G	H	I

Challenge

Understand 100,000

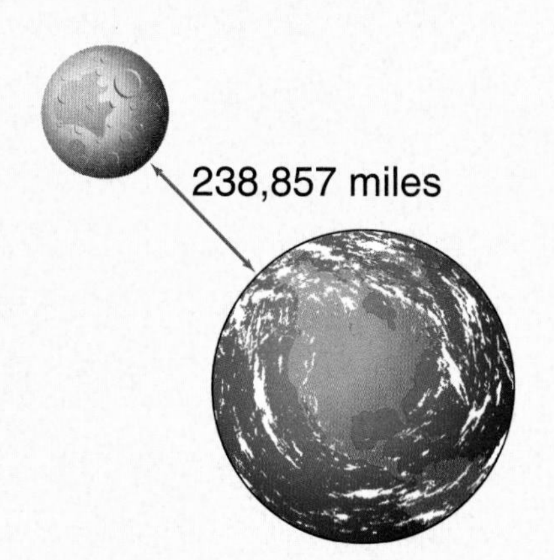

The distance from Earth to the moon is about 238,857 miles. Use a place-value chart to show the value of each digit in this number.

Hundred Thousands	Ten Thousands	Thousands	Hundreds	Tens	Ones
2	3	8,	8	5	7

Standard form: 238,857

Expanded form:
200,000 + 30,000 + 8,000 + 800 + 50 + 7

Word form:
two hundred thirty-eight thousand, eight hundred fifty-seven

Talk About It

- What is the value of the 2 in the number 238,857?

Try It

Write in standard form.

1. 300,000 + 20,000 + 6,000 + 700 + 40 + 4
2. 800,000 + 5,000 + 400 + 90 + 2
3. seven hundred forty-six thousand, one hundred eighteen
4. five hundred twelve thousand, twenty-two

Write in expanded form.

5. 312,456
6. 605,127

Write the value of the blue digit.

7. 312,876 8. 897,546 9. 185,254 10. 302,170

Study Guide and Review

STUDY AND SOLVE

Chapter 1

Identify numbers.

Even numbers end in 0, 2, 4, 6, or 8.
Odd numbers end in 1, 3, 5, 7, or 9.

Tell whether the number is *odd* or *even*. (pp. 2-3)

1. 9 **2.** 23 **3.** 40

Understand place value.

Ten Thousands	Thousands	Hundreds	Tens	Ones
4	2,	1	0	5

Standard form: 42,105
Expanded form:
40,000 + 2,000 + 100 + 5
Word form:
forty-two thousand, one hundred five

Write in standard form. (pp. 4-11)

4. seven hundred twenty-eight
5. two hundred eighty
6. 3,000 + 200 + 20 + 5
7. 10,000 + 2,000 + 500 + 20 + 2

Write in expanded form. (pp. 4-11)

8. 4,542 **9.** 71,061

Chapter 2

Compare and order numbers.

Write in order from greatest to least.
2,761; 1,793; 5,219

Compare thousands.
5 > 2 > 1

5,219; 2,761; 1,793

Write <, >, or = for each ●. (pp. 20-23)

10. 739 ● 728 **11.** 461 ● 461
12. 3,125 ● 452 **13.** 1,203 ● 1,209

Write in order from greatest to least. (pp. 24-25)

14. 423, 432, 417
15. 9,005; 5,009; 5,010

Round numbers.

Round 4,483 to the nearest thousand.
4,483 is between 4,000 and 5,000.
It is closer to 4,000.
So, 4,483 rounded to the nearest thousand is 4,000.

Round to the nearest hundred. (pp. 28-29)

16. 144 **17.** 653 **18.** 247

Round to the nearest thousand. (pp. 30-31)

19. 3,609 **20.** 1,289

Chapter 3

Add two or more numbers.

6 + (3 + 1) = 10 and (6 + 3) + 1 = 10

Add 3- and 4-digit numbers.

```
   1          1 1
  437        3,987
 +155       +2,532
  592        6,519
```

Find the sum. (pp. 36–37)

21. (5 + 3) + 8 = ■

22. (4 + 6) + 3 = ■

Find the sum. Estimate to check. (pp. 42–43, 46–49)

23. 192 + 432 **24.** 643 + 289 **25.** 534 + 846

26. 4,276 + 1,071 **27.** 2,008 + 6,439 **28.** 5,976 + 8,668

Chapter 4

Estimate differences.

```
 689 →  700
-408 → -400
        300
```

- Round to the nearest hundred.

Subtract 3- and 4-digit numbers.

```
 8 13          9
 8 9 3      2 1010
             3,0 0 0
-5 0 8      -1,6 5 0
 3 8 5       1,3 5 0
```

Estimate the difference. (pp. 54–55)

29. 58 − 19 **30.** 311 − 196 **31.** 3,032 − 1,114

Subtract. (pp. 58–65)

32. 562 − 313 **33.** 430 − 287 **34.** 406 − 89

35. 6,314 − 2,509 **36.** 2,006 − 1,734 **37.** 3,508 − 2,779

PROBLEM SOLVING PRACTICE

Solve. (pp. 12–13, 44–45)

38. I am a number less than 100. My ones digit is 6. The sum of my digits is 11. What number am I?

39. Susan has 10 more goldfish than Gary. Together, they own 50 goldfish. How many goldfish does each have?

PERFORMANCE ASSESSMENT

TASK A • MAKE A GOOD GUESS

Abe, Beth, Carla, and Devon guessed the number of jellybeans in a jar. Beth guessed that there were 2,459. Abe, Carla, and Devon gave clues for their guesses.

a. Copy and complete the table. Use the clues to write possible numbers for the other three guesses.

JELLYBEAN GUESSES	
Abe	
Beth	2,459
Carla	
Devon	

Abe's clue: My guess is less than 2,900, but greater than Beth's guess.

Carla's clue: My guess is 100 more than Abe's guess.

Devon's clue: My guess rounded to the nearest hundred is 2,800.

b. Write the guesses in order from the least number to the greatest number.

TASK B • TRIP TO THE BEACH

For their summer vacation, Jeremy and his family are driving from their home to the beach. The map shows the roads they can take and the distances in miles.

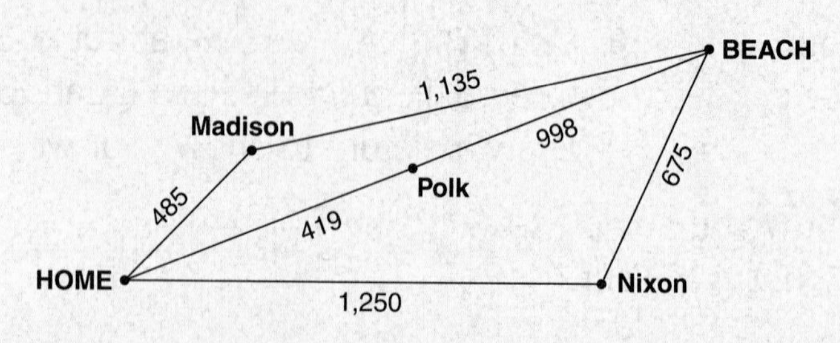

a. Write the name of one city they could drive through. Then estimate the distance they will drive if they choose this city.

b. Use the distances shown on the map to find the total number of miles they will drive if they choose this city.

E-Lab • Addition of Three-Digit Numbers

There were 295 people watching the dog show on Saturday and 147 people watching on Sunday. How many people in all watched the dog show?

You can use E-Lab to add three-digit numbers.

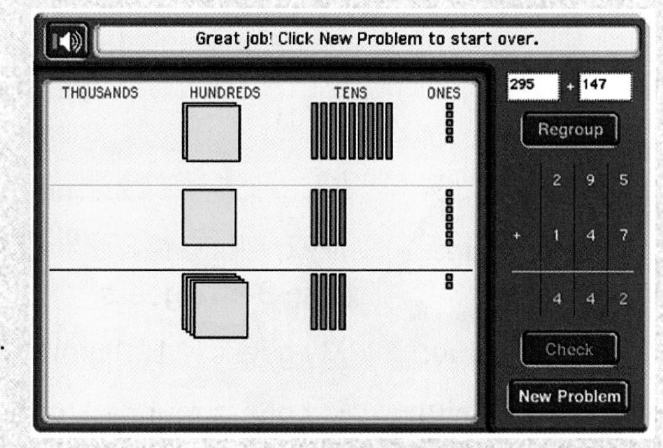

- Click *Addition of Three-Digit Numbers.*
- Click *New Problem* to begin.
- Type 295. Press *Enter.*
- Type 147. Press *Enter.*
- Click *Regroup* to regroup 10 ones as 1 ten.
- Click *Regroup* to regroup 10 tens as 1 hundred.
- Click *Check.* Record the sum.

So, 442 people watched the dog show.

How many times would you need to *regroup* to find 609 + 386? What is the sum?

Practice and Problem Solving

Use E-Lab to add.

1. 319 + 245 **2.** 621 + 190 **3.** 411 + 499 **4.** 523 + 188

5. Jamie biked 124 miles in June and 137 miles in July. How many miles in all did she bike?

6. **WRITE A PROBLEM** about adding three-digit numbers. Use E-Lab to solve. Draw a picture to show your work.

7. **STRETCH YOUR THINKING** Explain how you could use E-Lab to find 119 + 233 + 308. What is the sum?

Multimedia Math Glossary **www.harcourtschool.com/mathglossary**

8. **Vocabulary** Visit the Multimedia Math Glossary to find the symbols for *less than* and *greater than.* Use < to compare two numbers in *expanded form.* Use > to compare two numbers in *word form.*

at the
Statue of Liberty

LIBERTY ISLAND, NEW YORK

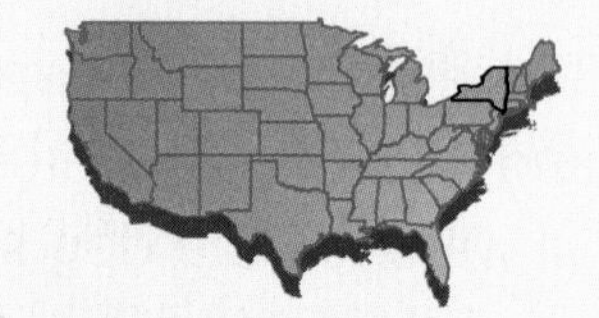

People from all over the world visit the Statue of Liberty in New York City.

USE DATA For 1–5, use the table below.

1. The table lists the average number of visitors per day. What is this number rounded to the nearest thousand?
2. The Statue of Liberty is about 300 feet high. Is this number rounded to the nearest ten or hundred? Explain.
3. What is the total weight of the steel frame and copper skin?
4. One day 9,503 people visited the statue. How many more visitors was this than the average?
5. **REASONING** The Statue of Liberty Monument's total height is the sum of the heights of the foundation, the pedestal, and the statue from base to torch. Which height is missing from the chart? Find that height.

Statue of Liberty Facts	
Average visitors per day	6,614 people
Weight of copper "skin"	200,000 pounds
Weight of steel frame	250,000 pounds
Height from base to torch	151 feet
Height of granite pedestal	89 feet
Total height	305 feet

The Statue of Liberty was a gift from the people of France to the United States for its 100th birthday on October 28, 1886.

PARIS, FRANCE

A small version of the Statue of Liberty stands on a small island in the Seine River in Paris.

1. In 1885, U.S. citizens living in Paris gave a model of the Statue of Liberty to France. How long ago was that?

2. The Statue of Liberty in Paris is about 38 feet high. The one in New York is 151 feet high. Compare the heights of the two statues.

3. The Statue of Liberty in Paris weighs about 28,000 pounds. If this weight has been rounded to the nearest thousand, could the actual weight be 28,500 pounds? Explain.

4. It took 214 crates to ship the Statue of Liberty to New York. It took 6 crates to ship France's Statue of Liberty to Japan for a festival. How many more crates did it take to ship the Statue of Liberty to New York than it did to ship France's statue to Japan?

5. **REASONING** Two art students are touring Paris. They each buy a one-day museum pass for $14. Each student also buys a ticket to the Eiffel Tower for $11 and a boat ticket for $3. How much do the two students spend altogether? Explain.

The Statue of Liberty in Paris faces west toward New York, while the Statue of Liberty in New York faces east toward Paris.

CHAPTER 5 Use Money

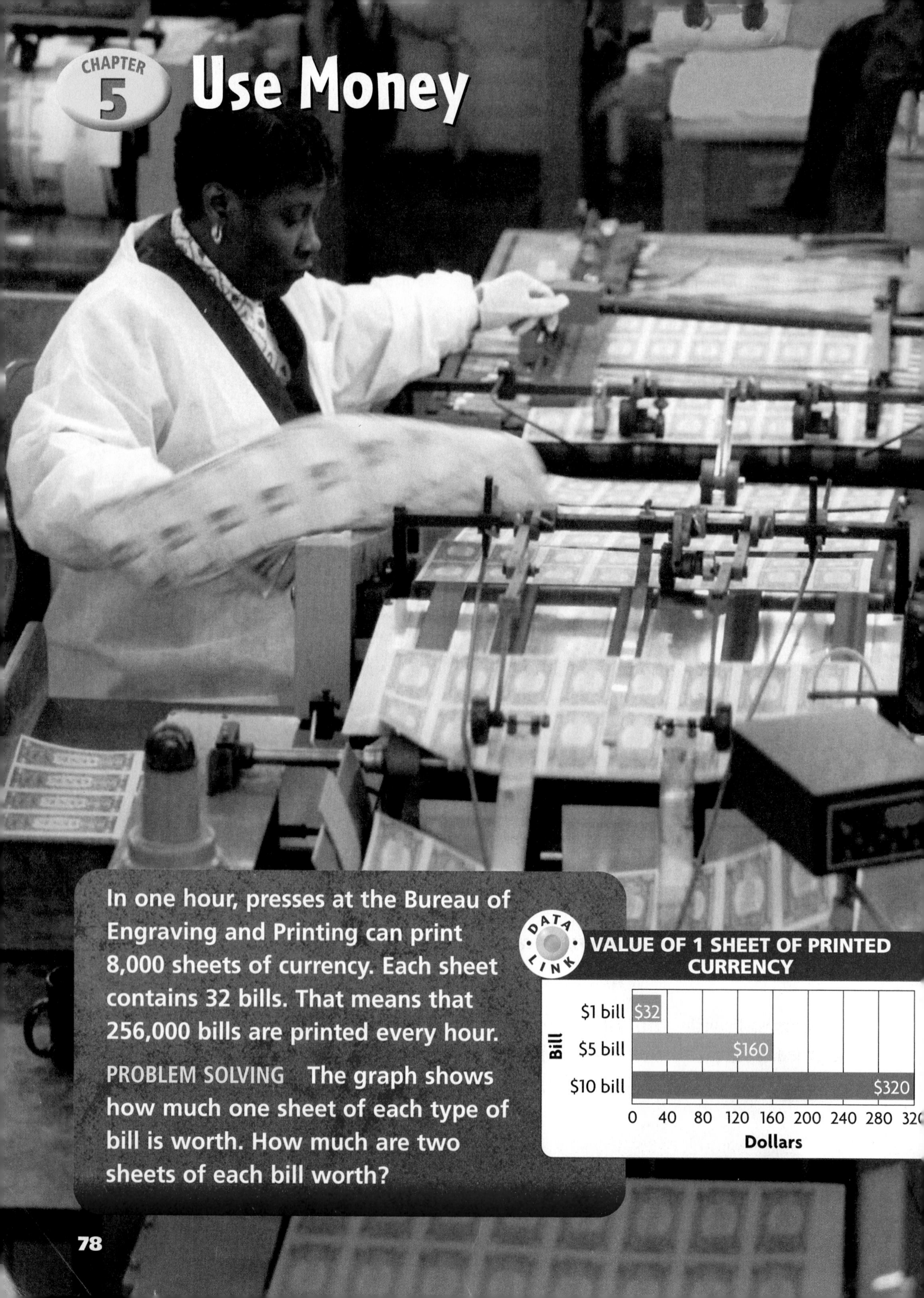

In one hour, presses at the Bureau of Engraving and Printing can print 8,000 sheets of currency. Each sheet contains 32 bills. That means that 256,000 bills are printed every hour.

PROBLEM SOLVING The graph shows how much one sheet of each type of bill is worth. How much are two sheets of each bill worth?

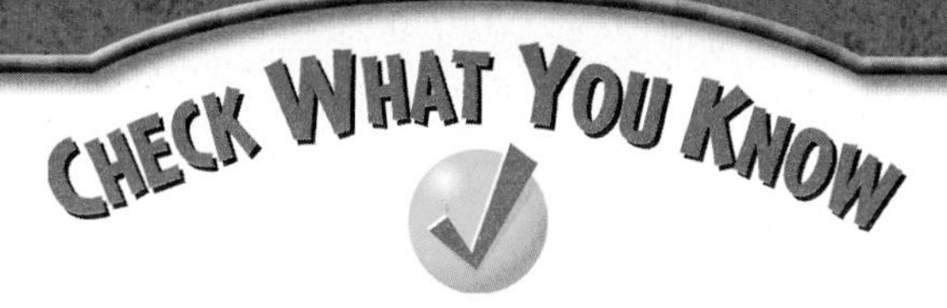

Use this page to help you review and remember important skills needed for Chapter 5.

VOCABULARY

Choose the best term from the box.

quarter
dime
nickel

1. The value of a __?__ is $0.10.
2. The value of a __?__ is $0.25.

MONEY: COUNT BILLS AND COINS (For Intervention, see p. H7.)

Count and write the amount.

3. 

4. 

5. 

6. 

7. 2 quarters, 1 dime, 3 nickels, and 2 pennies
8. 1 quarter, 3 dimes, 2 nickels, and 4 pennies

List the coins you would use to make each amount.

9. 39¢
10. 72¢
11. 68¢
12. 46¢

Count and write the amount. List the coins you would use to make the same amount with the fewest coins.

13. 

14. 

LESSON 1

Make Equivalent Sets

Quick Review

List the coins to make each amount.

1. 11¢ **2.** 35¢ **3.** 55¢
4. 75¢ **5.** 80¢

VOCABULARY
equivalent

MATERIALS
play bills and coins

Explore

Sets of money that have the same value are **equivalent**.

Examples

Examples A, B, C, and D are equivalent sets. Each set of coins has a value of 10¢.

Activity

Use play money to make equivalent sets. This set of coins has a value of $1.25.

$0.25 $0.50 $0.75 $1.00 $1.25

- Make at least three equivalent sets with a value of $1.25. One set should use the fewest bills and coins. Draw pictures to record each set.

What are two other ways we could show $2.40?

Try It

Find two equivalent sets for:

a. $2.40 **b.** $1.85

Connect

Make two equivalent sets with a value of $6.13. The first set uses the fewest bills and coins.

THINK:		NOW I HAVE:
one $5 bill	→	$5.00
plus one $1 bill	→	$6.00
plus 1 dime	→	$6.10
plus 3 pennies	→	$6.13

THINK:		NOW I HAVE:
six $1 bills	→	$6.00
plus 2 nickels	→	$6.10
plus 3 pennies	→	$6.13

MATH IDEA You can make equivalent sets of money by using different combinations of bills and coins.

Technology Link

More Practice: Use E-Lab, *Equivalent Sets of Coins.*

www.harcourtschool.com/elab2000

Practice and Problem Solving

Make two equivalent sets for each amount. List the bills and coins you used.

1. $1.35 **2.** $5.50 **3.** $2.46 **4.** $6.92

5. $3.75 **6.** $5.03 **7.** $2.25 **8.** $8.04

9. If Rosa has 5 nickels, how many pennies could she trade them for? Copy and complete the table.

nickels	1	2	3	4	5
pennies	5	■	■	■	■

10. REASONING Fiona has 2 quarters and 1 nickel. Jake has an equal amount in dimes and nickels. Jake has 8 coins. How many dimes and nickels does he have?

11. List the fewest bills and coins you can use to make $2.37.

Mixed Review and Test Prep

Round to the nearest hundred. (p. 28)

12. 87 **13.** 267 **14.** 142

15. Which digit is in the thousands place of 5,723? (p. 6)

16. TEST PREP Which digit is in the hundreds place of 6,295? (p. 6)

A 2 **C** 6

B 5 **D** 9

LESSON 2

Problem Solving Strategy
Make a Table

Quick Review

1. 2 quarters = ■ pennies
2. 10 dimes = ■ quarters
3. 4 nickels = ■ dimes
4. 1 quarter = ■ nickels
5. 10 pennies = ■ nickels

PROBLEM Patty has four $1 bills, 3 quarters, 5 dimes, 1 nickel, and 5 pennies. How many different equivalent sets of bills and coins can she use to pay for a magazine that costs $4.75?

UNDERSTAND

- What are you asked to find?
- What information will you use?

PLAN

- What strategy can you use?

You can *make a table* to find sets of bills and coins with a value of $4.75.

SOLVE

- How can you use the strategy to solve the problem?

Make a table to show equivalent sets of money.

$1 BILLS	QUARTERS	DIMES	NICKELS	PENNIES	VALUE
4	3				$4.75
4	2	2	1		$4.75
4	2	2		5	$4.75
4	1	5			$4.75
4	1	4	1	5	$4.75

So, there are 5 equivalent sets.

CHECK

- How can you decide if your answer is correct?

Problem Solving Practice

PROBLEM SOLVING STRATEGIES

- Draw a Diagram or Picture
- Make a Model or Act It Out
- Make an Organized List
- Find a Pattern
- **Make a Table or Graph**
- Predict and Test
- Work Backward
- Solve a Simpler Problem
- Write a Number Sentence
- Use Logical Reasoning

Make a table to solve.

1. **What if** Patty's magazine costs $5.25? How many different equivalent sets of bills and coins can she use?

2. Tyler has one $1 bill, 5 quarters, 1 dime, and 2 nickels. How many ways can he pay for a goldfish that costs $1.35?

Kevin has 7 quarters, 4 dimes, and 1 nickel. He wants to buy a bookmark that costs $1.80.

3. Kevin wants to keep 1 quarter. Which set of coins should he use?
   - **A** 6 quarters, 2 dimes, 1 nickel
   - **B** 6 quarters, 3 dimes
   - **C** 7 quarters, 1 nickel
   - **D** 6 quarters, 3 dimes, 1 nickel

4. If Kevin uses the fewest coins, which type of coin will he NOT use?
   - **F** quarters
   - **G** dimes
   - **H** nickels
   - **J** none of the above

Mixed Strategy Practice

USE DATA For 5–7, use the table.

Camping Equipment

Item	Price
Flashlight	$5.99
Canteen	$4.65
Can Opener	$1.05
Bug Spray	$2.49

5. Laura has one $5 bill, four $1 bills, 7 quarters, 2 dimes, 2 nickels, and 4 pennies. How many different ways can she pay for the flashlight?

6. Paco has only quarters and nickels in his pocket. If he uses 9 coins to buy the can opener, what coins does he use?

7. Fran, Geri, Harold, and Ivan each buy a different item. Use the clues to decide what each person buys.

   Fran pays with one $1 bill and 1 nickel. Ivan pays without using pennies. Geri pays with three $1 bills.

8. **Write About It** Betty has three $1 bills, 5 quarters, 7 dimes, and 2 nickels. Explain how Betty can trade some of her bills and coins for a $5 bill.

LESSON 3

Compare Amounts of Money

Quick Review

Write <, >, or = for each ●.

1. 463 ● 643 **2.** 510 ● 105

3. 874 ● 748 **4.** 296 ● 926

5. 713 ● 713

Learn

MONEY MATTERS Ming and Ben have these sets of bills and coins. Who has more money?

Count each amount and compare.

Ming has \$5.75.
Ben has \$5.50.

$\$5.75 > \$5.50$.
So, Ming has more money.

Ming's money

Ben's money

Examples Compare. Which amount is greater?

A

Since $\$2.73 = \$2.73$, the amounts are equal.

B

Since $\$3.54 < \$4.12$, then \$4.12 is the greater amount.

- **REASONING** Is a set of bills and coins always worth more than a set that has fewer bills and coins? Explain.

MATH IDEA To compare amounts of money, count each set and decide if one is greater than, less than, or equal to the other.

Check

1. **Explain** how you can use what you know about comparing whole numbers to compare amounts of money.

Use > or < to compare the amounts of money.

2. a. 

   b.   

Practice and Problem Solving

Use > or < to compare the amounts of money.

3. a. 

   b.  

4. a. 

   b. 

5. a. 

   b. 

6. Setsuo sells lemonade for 25¢ a glass. He has 9 quarters, 6 dimes, and 3 nickels. How many glasses of lemonade did he sell? Draw a picture to explain.

7. **What's the Error?** Janice says that \$4.87 is greater than \$6.21 because 87 cents is greater than 21 cents. Describe her error. Explain which is greater.

Mixed Review and Test Prep

8. (p. 36)

$$\begin{array}{r} 21 \\ 98 \\ +45 \\ \hline \end{array}$$

9. (p. 36)

$$\begin{array}{r} 18 \\ 24 \\ +42 \\ \hline \end{array}$$

10. (p. 42)

$$\begin{array}{r} 256 \\ +148 \\ \hline \end{array}$$

11. Describe and continue the pattern.
13, 23, 33, 43, ■, ■, ■ (p. 2)

12. TEST PREP What is the value of the blue digit? 15,271 (p. 10)

A 5 **B** 50 **C** 500 **D** 5,000

EXTRA PRACTICE page H36, Set A

LESSON 4

Make Change

Quick Review

Add 10¢ to each amount.

1. 55¢ **2.** 70¢ **3.** 83¢
4. 29¢ **5.** 41¢

MATERIALS
play coins

PET SUPPLIES
Dog Leash.................$5.99
Dog Shampoo..........$3.68
Kitty Toy....................$0.76
Fish Food..................$0.89
Chew Bone...............$0.59
Bird Seed Bell..........$0.63

Explore

Jessica buys a kitty toy at Pal's Pet Store. She pays with a $1 bill. How much change will she get?

To find Jessica's change, **count on** from the cost of the kitty toy to the amount paid.

$0.77

$0.78

$0.79

$0.80

$0.90

$1.00

4 pennies and 2 dimes equal $0.24.
So, Jessica will get $0.24 in change.

Try It

Each person pays with a $1 bill. Use play money to make change. Draw a picture to show the change each person will get.

a. Tony buys a bird seed bell.

b. Marian buys fish food. Show her change, using the fewest coins.

c. Emma buys a chew bone. Show at least two different ways to make change.

- Why does it help to count on with pennies first when making change for $0.63 from a $1 bill?

$0.64, $0.65, $0.75 . . . What should I count next to make change?

Connect

Dog shampoo costs $3.68. Anton pays with a $5 bill. How much change will he get?

You can count on from the cost of the dog shampoo to the amount paid.

$3.69 $3.70 $3.75 $4.00 $5.00

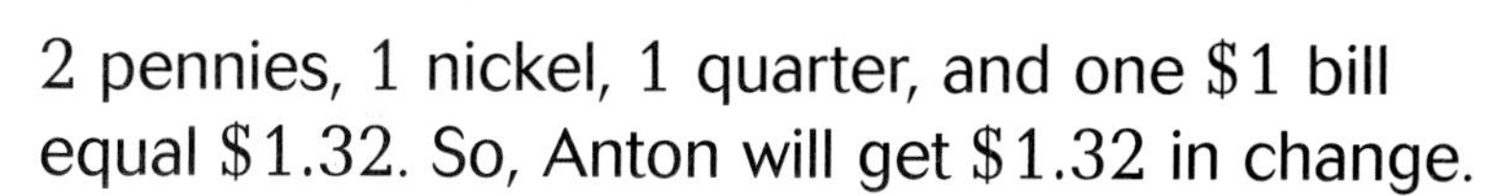

2 pennies, 1 nickel, 1 quarter, and one $1 bill equal $1.32. So, Anton will get $1.32 in change.

Technology Link

Mighty Math
More Practice: Use Mighty Math Calculating Crew, *Superhero Superstore* Levels B and C.

Practice and Problem Solving

Copy and complete the table. Use play money.

	COST OF ITEM	AMOUNT PAID	CHANGE IN COINS AND BILLS	TOTAL AMOUNT OF CHANGE
1.	$0.54	$1.00	■	■
2.	$3.23	$5.00	■	■
3.	$2.69	$3.00	■	■

4. **REASONING** Dana buys a rock for her fish tank for $0.65. She pays with a $1 bill. The clerk doesn't have any quarters for change. What is the least number of coins Dana can get? List the coins.

5. **? What's the Question?** Evan bought a dog bowl for $2.85. He paid with a $5 bill. The answer is $2.15.

Mixed Review and Test Prep

6. $\begin{array}{r} 375 \\ +499 \\ \hline \end{array}$ (p. 42)

7. $\begin{array}{r} 4,359 \\ -2,276 \\ \hline \end{array}$ (p. 62)

8. Order 2,743; 987; and 8,143 from least to greatest. (p. 24)

9. What is one hundred more than 3,420? (p. 6)

10. **TEST PREP** What is the standard form of seven thousand, nine hundred twenty-three? (p. 6)

A 7,092 **C** 7,923
B 7,903 **D** 70,923

LESSON 5

Add and Subtract Money

Quick Review

1. $863 + 219$
2. $673 - 482$
3. $457 + 361$
4. $1{,}073 - 845$
5. $920 + 549$

Learn

CHECK YOUR CHANGE Ryan bought a dog collar for \$3.95 and a leash for \$4.64. How much money did Ryan spend?

Example 1

Add. \$3.95 + \$4.64 = ■

STEP 1

Estimate the sum. Round to the nearest dollar.

$$\begin{array}{r} \$3.95 \\ +\$4.64 \\ \hline \end{array} \rightarrow \begin{array}{r} \$4.00 \\ +\$5.00 \\ \hline \$9.00 \end{array}$$

STEP 2

Add money amounts as you add whole numbers.

$$\begin{array}{r} \$3.95 \\ +\$4.64 \\ \hline \end{array} \rightarrow \begin{array}{r} \overset{1}{3}95 \\ +464 \\ \hline 859 \end{array}$$

STEP 3

Write the sum in dollars and cents.

$$\begin{array}{r} \$3.95 \\ +\$4.64 \\ \hline \$8.59 \end{array}$$

So, Ryan spent \$8.59. Since \$8.59 is close to \$9.00, the answer is reasonable.

Ryan paid for the collar and leash with a \$10 bill. How much change should he get?

Example 2

Subtract. \$10.00 − \$8.59 = ■

STEP 1

Estimate the difference. Round to the nearest dollar.

$$\begin{array}{r} \$10.00 \\ -\$\ 8.59 \\ \hline \end{array} \rightarrow \begin{array}{r} \$10.00 \\ -\$\ 9.00 \\ \hline \$1.00 \end{array}$$

STEP 2

Subtract money amounts as you subtract whole numbers.

$$\begin{array}{r} \$10.00 \\ -\$\ 8.59 \\ \hline \end{array} \rightarrow \begin{array}{r} \overset{0}{\not1},\overset{\overset{9}{\not10}}{\not0}\overset{\overset{9}{\not10}}{\not0}\overset{10}{\not0} \\ -\ \ 859 \\ \hline 141 \end{array}$$

STEP 3

Write the difference in dollars and cents.

$$\begin{array}{r} \$10.00 \\ -\$\ 8.59 \\ \hline \$\ 1.41 \end{array}$$

So, Ryan should get \$1.41 in change. \$1.41 is close to \$1.00, so the answer is reasonable.

Check

1. **Explain** how you can check the subtraction to be sure Ryan got the correct change.

Find the sum or difference. Estimate to check.

2. $\begin{array}{r} \$1.45 \\ +\$2.32 \\ \hline \end{array}$ **3.** $\begin{array}{r} \$3.67 \\ +\$1.19 \\ \hline \end{array}$ **4.** $\begin{array}{r} \$4.83 \\ +\$22.51 \\ \hline \end{array}$ **5.** $\begin{array}{r} \$1.76 \\ -\$0.45 \\ \hline \end{array}$

Practice and Problem Solving

Find the sum or difference. Estimate to check.

6. $\begin{array}{r} \$2.63 \\ +\$1.24 \\ \hline \end{array}$ **7.** $\begin{array}{r} \$4.55 \\ +\$10.38 \\ \hline \end{array}$ **8.** $\begin{array}{r} \$4.64 \\ -\$1.80 \\ \hline \end{array}$ **9.** $\begin{array}{r} \$1.85 \\ -\$0.73 \\ \hline \end{array}$

10. $\begin{array}{r} \$3.52 \\ -\$2.34 \\ \hline \end{array}$ **11.** $\begin{array}{r} \$8.26 \\ +\$4.81 \\ \hline \end{array}$ **12.** $\begin{array}{r} \$7.69 \\ +\$5.95 \\ \hline \end{array}$ **13.** $\begin{array}{r} \$10.00 \\ -\$\ \ 5.25 \\ \hline \end{array}$

14. $4.28 + $2.59 = ■ **15.** $6.72 − $3.94 = ■ **16.** $3.26 − $1.09 = ■

Write <, >, or = for each ●.

17. $5.00 ● $3.94 + $1.06 **18.** $4.57 − $1.14 ● $5.71

USE DATA For 19–21, use the table.

CAT SUPPLIES	
Cat Bed	$7.58
Bowl	$3.42
Kitty Perch	$9.45
Ball	$0.83

19. How much more does the kitty perch cost than the cat bed?

20. Do a cat bed and a bowl cost more or less than a kitty perch and a ball? Explain.

21. Estimation Tara has $5.00. Does she have enough money for a bowl and a ball? Explain.

22. **What's the Error?** Roberto wrote $7.40 + $5.60 = $1.80. Describe his error. What should the sum be?

Mixed Review and Test Prep

Write <, >, or = for each ●. (p. 20)

23. 127 ● 171 **24.** 682 ● 659

25. Round 463 to the nearest hundred. (p. 28)

26. 723 − 581 = ■ (p. 58)

27. **TEST PREP** Which shows the numbers in order from least to greatest? (p. 24)

A 481, 419, 527 **C** 521, 518, 509
B 519, 531, 656 **D** 704, 698, 549

EXTRA PRACTICE page H36, Set B

Review/Test

CHECK VOCABULARY AND CONCEPTS

Choose the best term from the box.

equivalent
change

1. Sets of money that have the same value are ___?___ . (p. 80)

Find two equivalent sets for each amount. List the bills and coins you can use. (pp. 80–81)

2. $0.78 **3.** $3.65 **4.** $5.17 **5.** $8.42

Copy and complete the table. (pp. 86–87)

	COST OF ITEM	AMOUNT PAID	CHANGE IN COINS AND BILLS	AMOUNT OF CHANGE
6.	$0.62	$1.00	■	■
7.	$3.49	$5.00	■	■

CHECK SKILLS

Use > or < to compare the amounts of money. (pp. 84–85)

8. a.

b.

Find the sum or difference. (pp. 88–89)

9. $\begin{array}{r} \$2.46 \\ +\$3.37 \\ \hline \end{array}$ **10.** $\begin{array}{r} \$6.39 \\ -\$1.81 \\ \hline \end{array}$ **11.** $\begin{array}{r} \$6.74 \\ +\$4.46 \\ \hline \end{array}$ **12.** $\begin{array}{r} \$8.00 \\ -\$4.72 \\ \hline \end{array}$ **13.** $\begin{array}{r} \$9.05 \\ -\$2.88 \\ \hline \end{array}$

CHECK PROBLEM SOLVING

Solve. (pp. 82–83)

14. Michelle has two $1 bills, 5 quarters, 3 dimes, and 4 nickels. How many ways can she make $3.50?

15. Brian has one $5 bill, one $1 bill, 5 quarters, 3 dimes, and 3 nickels. How many ways can he make $7.35?

Standardized Test Prep

TIP! **Get the information you need.** See item **2.**

Before you can find which is greatest, you need to find the amounts to compare. After you find the amount in cents for each answer choice, you can compare and order them.

Also see problem **3,** p. H63.

For 1–7, choose the best answer.

1. Jonathan wants to buy a book that costs $1.95. Which set of bills and coins makes exactly $1.95?

- **A** $1 bill, 3 quarters, 2 dimes
- **B** $1 bill, 9 quarters, 5 nickels
- **C** $1 bill, 5 dimes, 6 nickels
- **D** $1 bill, 5 dimes, 2 nickels

2. Which is the greatest amount of money?

- **F** 3 quarters, 5 dimes, 1 nickel
- **G** 3 quarters, 3 dimes, 3 nickels
- **H** 4 quarters, 3 nickels
- **J** 5 quarters

3. What is the sum of 762 and 498?

- **A** 1,150
- **B** 1,160
- **C** 1,250
- **D** 1,260

4. Find the sum.

$$\begin{array}{r} \$8.45 \\ +\$3.87 \\ \hline \end{array}$$

- **F** $11.22
- **G** $11.32
- **H** $12.32
- **J** NOT HERE

5. Emma has $9.32. Jenelle has $6.65. How much more money does Emma have?

- **A** $2.63
- **B** $2.67
- **C** $3.33
- **D** NOT HERE

6. Mindy's school has collected 872 cans for the food drive. What is that number rounded to the nearest hundred?

- **F** 700
- **G** 800
- **H** 900
- **J** 1,000

7. Tara practiced dancing for 45 minutes. She talked on the phone for 15 minutes and read for 22 minutes. How much time did she spend on these activities?

- **A** 80 minutes
- **B** 82 minutes
- **C** 85 minutes
- **D** 87 minutes

Write What You Know

8. List a set of coins that is worth 46 cents. Can you make the same amount using fewer coins? Explain.

9. How many sets of coins can be made with a value of 16 cents? Make a list to show all of the possible combinations.

Understand Time

Living things grow at different rates. The time it takes flowers to grow depends on many things, such as temperature, soil, and light.

PROBLEM SOLVING Look at the chart. Which flower sprouts the fastest? Which flower takes the longest to bloom?

DATA LINK

GROWTH TIMES OF FLOWERS

Flower	Days to Sprout	Days to Bloom
Black-Eyed Susan	10–15	60–90
Snapdragon	8–14	55–100
Sunflower	10–18	120
Petunia	4–10	45–80
Marigold	7–18	70
Shasta Daisy	14–21	180–300

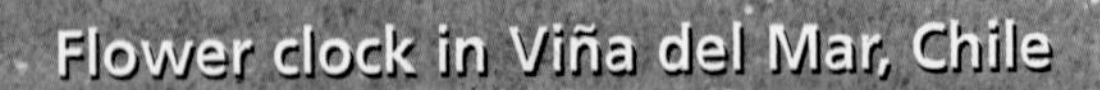
Flower clock in Viña del Mar, Chile

Use this page to help you review and remember important skills needed for Chapter 6.

VOCABULARY

Choose the best term from the box.

1. The short hand on the clock is the ? .
2. The long hand on the clock is the ? .

time
minute hand
hour hand

TELL TIME (For Intervention, see p. H8.)

Read and write the time.

3. 
4. 
5. 
6. 
7. 
8. 
9. 
10. 
11. 

CALENDAR (For Intervention, see p. H8.)

For 12–14, use the calendar.

12. There are ? Fridays in November.
13. The second Monday in November is ? .
14. The third Wednesday in November is ? .

November						
Sun	Mon	Tue	Wed	Thu	Fri	Sat
					1	2
3	4	5	6	7	8	9
10	11	12	13	14	15	16
17	18	19	20	21	22	23
24	25	26	27	28	29	30

LESSON 1

Time to the Minute

Quick Review

Add 5 to each number.

1. 25 **2.** 10

3. 15 **4.** 5

5. 30

VOCABULARY

minute

MATERIALS

clocks with movable hands

Explore

The hands, numbers, and marks on a clock help you tell what time it is.

In five minutes, the minute hand moves from one number to the next.

In one **minute**, the minute hand moves from one mark to the next.

To find the number of minutes after the hour, count by fives and ones to where the minute hand is pointing.

5 minutes
10 minutes
15 minutes
20 minutes
25 minutes
26 minutes

Read: nine twenty-six or 26 minutes after nine

Write: 9:26

Remember

In one *hour*, the hour hand moves from one number to the next. There are 60 minutes in 1 hour.

- Show 7:42 and 9:07 on your clock.

Try It

Read and write each time.

a.

b.

Show each time on your clock. Then read each time in two ways.

c.

d.

Connect

When a clock shows 31 or more minutes *after* the hour, you can read the time as a number of minutes *before* the next hour.

Count back by fives and ones to where the minute hand is pointing.

Read: 18 minutes before two

Write: 1:42

Technology Link

More Practice:
Use E-Lab, *Time to the Minute.*

www.harcourtschool.com/elab2002

Practice and Problem Solving

Read and write each time.

1. 

2. 

3. 

4. 

Show each time on your clock. Then write two ways you can read each time.

5. 

6. 

7. 

8. 

9. **REASONING** Does it take about 1 minute or about 5 minutes to tie your shoe? to make your lunch?

10. **REASONING** It takes Jan 18 more minutes to walk to school than Rob. It takes Rob 15 minutes. How many minutes does it take Jan?

Mixed Review and Test Prep

11. $\begin{array}{r} 600 \\ -343 \\ \hline \end{array}$ (p. 58)

12. $\begin{array}{r} 3,191 \\ +\ 980 \\ \hline \end{array}$ (p. 46)

Round to the nearest thousand. (p. 30)

13. 5,439

14. 3,572

15. **TEST PREP** What is the value of the 2 in 56,297? (p. 10)

A 2
B 20
C 200
D 2,000

LESSON

2 A.M. and P.M.

Quick Review

Write the time 10 minutes later.

1. 3:00
2. 12:30
3. 9:45
4. 5:15
5. 11:55

VOCABULARY

A.M.
P.M.
noon
midnight

Learn

IT'S ABOUT TIME Using A.M. and P.M. helps you know what time of the day or night it is. Times from midnight to noon are in the **A.M.** Times from noon to midnight are in the **P.M.**

12:00 in the day is **noon**.
12:00 at night is **midnight**.

Here are some ways to read and write times.

quarter to midnight
eleven forty-five P.M.
11:45 P.M.

quarter past seven
seven fifteen A.M.
7:15 A.M.

half past three
three thirty P.M.
3:30 P.M.

MATH IDEA The hours between midnight and noon are A.M. hours. The hours between noon and midnight are P.M. hours.

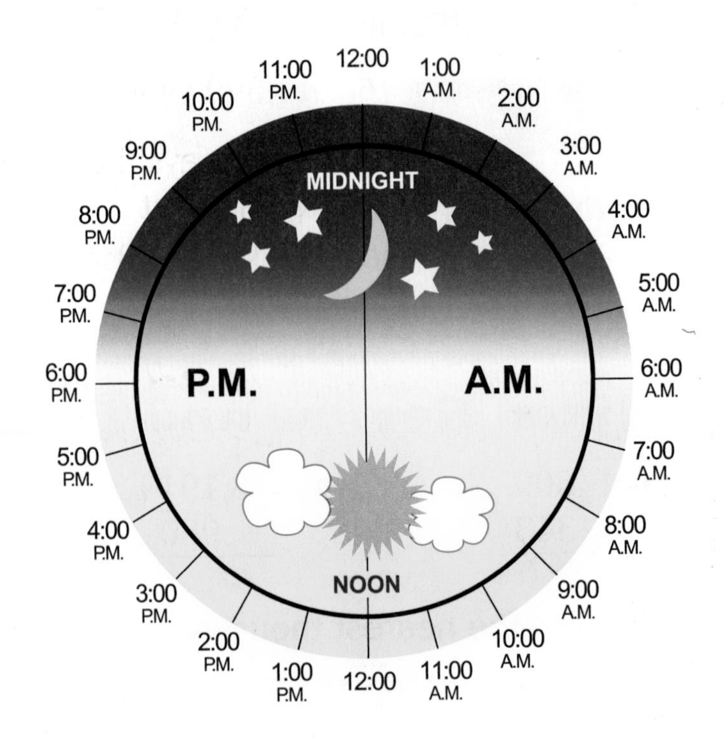

Check

1. **List** three things that you do in the A.M. hours.
2. **Name** something you do at 9:00 A.M. and something you do at 9:00 P.M. Explain how you know these times are not the same.

Write the time, using A.M. or P.M.

3. 
school starts

4. 
eat lunch

5. 
do homework

6. 

sun sets

Practice and Problem Solving

Write the time, using A.M. or P.M.

7. 
get ready for school

8. 
go to the store

9. 
recess

10. 

go to bed

Write two ways you can read each time.
Then write the time, using A.M. or P.M.

11. 
play softball

12. 
moon shines

13. 
sun rises

14. 

eat dinner

15. **What's the Error?** Ty says that 11:45 A.M. is close to midnight. Explain his error. Then give a time that is close to midnight.

16. **REASONING** Are you awake during more A.M. or P.M. hours? Explain.

Mixed Review and Test Prep

Write + or − to make the sentence true. (p. 68)

17. 5 ● 3 = 2 18. 15 ● 9 = 24

19. 129 ● 5 = 134

20. 637 ● 42 = 595

21. **TEST PREP** Which set of coins is equivalent to $0.31? (p. 80)

A 1 dime, 3 pennies
B 1 quarter, 1 dime, 1 penny
C 1 quarter, 1 nickel, 1 penny
D 3 dimes, 1 nickel

EXTRA PRACTICE page H37, Set A

Elapsed Time

Quick Review

Write the time 2 hours later.

1. 1:00
2. 6:30
3. 3:15
4. 8:55
5. 11:25

VOCABULARY

elapsed time

MATERIALS

clocks with movable hands

Explore

Cameron read a book from 8:15 P.M. to 8:45 P.M. How long did Cameron read?

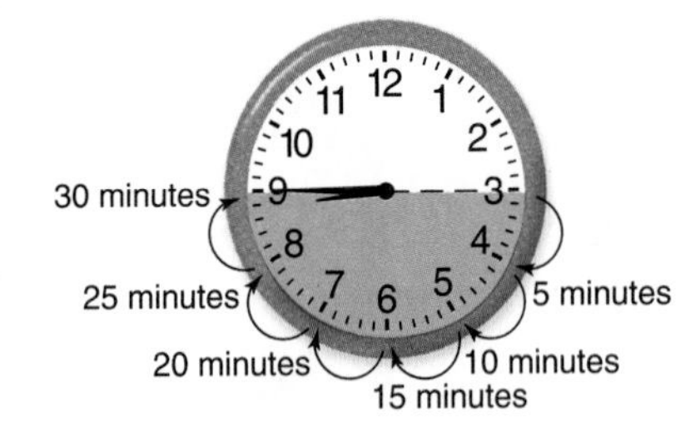

Start:
8:15

Count the minutes:
30 minutes

So, Cameron read for 30 minutes.

Elapsed time is the amount of time that passes from the start of an activity to the end of that activity.

- Find the elapsed time from 1:15 P.M. to 1:45 P.M.

Technology Link

More Practice: Use E-Lab, *Elapsed Time: Minutes and Hours.*

www.harcourtschool.com/elab2002

Try It

Use a clock to find the elapsed time.

a. start: 9:15 A.M.
end: 10:00 A.M.

b. start: 3:45 P.M.
end: 4:30 P.M.

c. start: 10:00 A.M.
end: 11:15 A.M.

d. start: 11:45 A.M.
end: 1:15 P.M.

e. start: 5:10 P.M.
end: 5:20 P.M.

f. start: 11:30 P.M.
end: 12:30 A.M.

The start time was 9:15. The end time was 10:00. How much time has elapsed?

Connect

Soccer practice starts at 11:00 A.M. It lasts 2 hours 15 minutes. What time does practice end?

Start: 11:00

Count the hours.

Count the minutes.

So, practice ends at 1:15 P.M.

MATH IDEA If you know when an activity starts and how long it takes, you can find the time it ends.

Practice and Problem Solving

Use a clock to find the elapsed time.

1. start: 7:00 A.M.
   end: 7:45 A.M.
2. start: 5:15 P.M.
   end: 6:20 P.M.
3. start: 11:30 A.M.
   end: 12:15 P.M.

Use a clock to find the end time.

4. start: 4:15 P.M.
   elapsed time: 1 hour 15 minutes
5. start: 11:45 A.M.
   elapsed time: 20 minutes
6. **What's the Question?** The basketball game started at 11:30 A.M. It ended at 1:15 P.M. The answer is 1 hour 45 minutes.
7. **Write About It** Explain how you can use a clock to find the elapsed time from 11:30 A.M. to 1:15 P.M.

Mixed Review and Test Prep

Describe the pattern. Then continue the pattern. (p. 2)

8. 3, 6, 9, 12, ■
9. 7, 14, 21, 28, ■
10. 4, 8, 12, 16, ■

11. Order 297, 179, and 253 from greatest to least. (p. 24)
12. **TEST PREP** Which is 100 less than 461? (p. 4)
    A 361 **B** 460 **C** 451 **D** 561

Use a Schedule

Learn

RIGHT ON TIME A **schedule** is a table that lists activities or events and the times they happen.

You can use what you know about elapsed time to finish Stacy's schedule.

STACY'S SCHEDULE		
Activity	**Time**	**Elapsed Time**
Eat snack	3:45 P.M.–4:05 P.M.	20 minutes
Do homework	4:05 P.M.–5:10 P.M.	■
Walk dog	5:10 P.M.–■	25 minutes

How long will Stacy do homework?

Think: Find the elapsed time.
4:05 P.M. to 5:05 P.M. 1 hour
5:05 P.M. to 5:10 P.M. 5 minutes

So, Stacy will do homework for 1 hour 5 minutes.

When will Stacy walk her dog?

Think: Find the end time.
Start: 5:10 P.M.
Count on 25 minutes to 5:35 P.M.

So, Stacy will walk her dog from 5:10 P.M. to 5:35 P.M.

MATH IDEA You can use a schedule to find elapsed times of events. If you know the elapsed times, you can find start or end times on a schedule.

Check

1. **Explain** why Stacy won't start walking her dog at 5:00 P.M.

USE DATA **For 2–3, use Stacy's schedule above.**

2. What time does Stacy finish her homework?
3. How long does it take Stacy to do her homework and walk the dog?

Quick Review

1. 15 + 5 = ■
2. 15 + ■ = 25
3. ■ + 15 = 30
4. 15 + 30 = ■
5. 30 minutes after 1:15 is ■.

VOCABULARY
schedule

Technology Link
To learn more about schedules, watch the **Harcourt Math Newsroom Video** ***Kangaroo Foster Mom.***

Practice and Problem Solving

USE DATA For 4–6, use the class schedule.

4. Which activities last 45 minutes each?

5. Which activity is the longest?

6. **ESTIMATION** About how long are the reading and math activities altogether?

MORNING CLASS SCHEDULE	
Activity	**Time**
Reading	8:30 A.M. – 9:15 A.M.
Math	9:15 A.M. – 10:15 A.M.
Recess	10:15 A.M. – 10:35 A.M.
Music	10:35 A.M. – 11:20 A.M.
Art	11:20 A.M. – 12:05 P.M.

Copy and complete the schedule.

	THE SCIENCE CHANNEL SCHEDULE		
	Program	**Time**	**Elapsed Time**
7.	Animals Around Us	6:00 P.M. – 7:00 P.M.	■
8.	Wonderful Space	7:00 P.M. – ■	25 minutes
9.	Weather in Your Town	7:25 P.M. – 7:30 P.M.	■
10.	Earthly Treasures	7:30 P.M. – ■	30 minutes

For 11–12, use the schedule you completed.

11. *Weather in Your Town* begins ■ minutes after 7:00 P.M.

12. *Earthly Treasures* begins ■ hour and ■ minutes after *Animals Around Us* begins.

13. **REASONING** Sean needs at least 30 minutes to get ready for school. If he leaves for school at 8:05 A.M., what is the latest he can start getting ready?

14. **Write About It** Think about activities you do on a school day and how much time each takes. Make a schedule. Be sure to include start and end times.

Mixed Review and Test Prep

15. Kate has 5 quarters, 1 dime, and 2 nickels. Jim has one $1 bill and 1 half dollar. Who has more money? (p. 84)

16. (p. 42)
$$\begin{array}{r} 345 \\ 102 \\ +593 \\ \hline \end{array}$$

17. (p. 42)
$$\begin{array}{r} 199 \\ 255 \\ +175 \\ \hline \end{array}$$

18. Write 3,000 + 400 + 8 in standard form. (p. 6)

19. **TEST PREP** The difference between $4.35 and $1.67 is ■. (p. 88)

A $6.02 **C** $3.32

B $3.78 **D** $2.68

LESSON

5 Use a Calendar

Quick Review

1. 7 pennies = ■¢
2. 3 nickels = ■¢
3. 4 dimes = ■¢
4. 1 quarter = ■¢
5. ? is the month after September.

VOCABULARY

calendar

Learn

DAYS AND DATES A **calendar** shows the days, weeks, and months of a year.

MATH IDEA You can use a calendar to find elapsed time in days, weeks, and months.

The Youngs are taking a trip. They left on July 21 and will return on August 6. How long will they be gone?

July						
Sun	Mon	Tue	Wed	Thu	Fri	Sat
	1	2	3	4	5	6
7	8	9	10	11	12	13
14	15	16	17	18	19	20
21	22	23	24	25	26	27
28	29	30	31			

August						
Sun	Mon	Tue	Wed	Thu	Fri	Sat
				1	2	3
4	5	6	7	8	9	10
11	12	13	14	15	16	17
18	19	20	21	22	23	24
25	26	27	28	29	30	31

Think: Start: Sunday, July 21.
Move down 2 weeks to August 4.
Then count on 2 days to August 6.

Remember

Units of Time
7 days = 1 week
12 months = 1 year

So, the Youngs will be gone for 2 weeks and 2 days, or 16 days.

- **What if** the Youngs are gone for 2 months? In what month will they return?

Check

1. **Explain** how you know 1 week and 2 days is the same amount of time as 9 days.

For 2, use the calendars above.

2. How many weeks are there from July 30 to August 20? How many days?

Practice and Problem Solving

For 3–6, use the calendars.

February						
Sun	Mon	Tue	Wed	Thu	Fri	Sat
					1	2
3	4	5	6	7	8	9
10	11	12	13	14	15	16
17	18	19	20	21	22	23
24	25	26	27	28		

March						
Sun	Mon	Tue	Wed	Thu	Fri	Sat
					1	2
3	4	5	6	7	8	9
10	11	12	13	14	15	16
17	18	19	20	21	22	23
24/31	25	26	27	28	29	30

April						
Sun	Mon	Tue	Wed	Thu	Fri	Sat
	1	2	3	4	5	6
7	8	9	10	11	12	13
14	15	16	17	18	19	20
21	22	23	24	25	26	27
28	29	30				

3. Doug went on vacation from February 16 to March 9. How many weeks was his vacation?

4. Chantel started her art project on March 1. She worked for 3 weeks and 4 days. When did she finish?

5. **ESTIMATION** Ms. Horner's class rehearsed for a play from February 11 to March 9. About how long did the class rehearse?

6. Mr. Todd left for a 4-week trip on February 4. He came home for a week and then left again for 10 days. Did he return by March 19? Explain.

USE DATA For 7–9, use the calendars above and the pictograph.

7. "Raisins in the Sun" first hit number 1 on March 13. When did it lose its first-place spot?

8. Which song was number 1 for the longest time?

9. **REASONING** How many more days was "Dance and Sing" number 1 than "Rabbit in the Moon?"

TOP-OF-THE-CHART SONGS

Dance and Sing	♪ ♪ ♪ ♪
Raisins in the Sun	♪ ♪
Rabbit in the Moon	♪ ♪ ♪

Key: Each ♪ = 1 week.

Mixed Review and Test Prep

10. Round 4,812 to the nearest thousand. (p. 30)

11. (p. 58) $457 - 186$

12. (p. 88) $\$7.83 + \$5.09$

13. (p. 88) $\$1.62 - \$0.85$

14. **TEST PREP** Jay buys a book for $4.65. He pays with a $5 bill. Which coins can he get for change? (p. 86)

A 1 quarter, 1 nickel
B 3 dimes, 1 nickel
C 3 nickels, 5 pennies
D 6 nickels

EXTRA PRACTICE page H37, Set C

LESSON 6

Problem Solving Skill
Sequence Events

UNDERSTAND | PLAN | SOLVE | CHECK

Quick Review

1. $24 + 5 = ■$
2. $17 - 9 = ■$
3. $2 + 14 = ■$
4. $31 - 11 = ■$
5. 2 weeks = ■ days

GOING CAMPING Mr. Cob is planning a camping trip. Use the list of things to do. Help him decide what to do first.

To Do List

Today's Date: October 22
Trip Date: November 16

Things to do:
- 3 days before the trip, check the weather.
- Tell mail carrier about the trip 1 week from today.
- Rent camping gear 2 weeks before the trip.

October

Sun	Mon	Tue	Wed	Thu	Fri	Sat
		1	2	3	4	5
6	7	8	9	10	11	12
13	14	15	16	17	18	19
20	21	22	23	24	25	26
27	28	29	30	31		

November

Sun	Mon	Tue	Wed	Thu	Fri	Sat
					1	2
3	4	5	6	7	8	9
10	11	12	13	14	15	16
17	18	19	20	21	22	23
24	25	26	27	28	29	30

Use the calendars to find the date for each.

Find the date 3 days before the trip.

Start: November 16.
Count back 3 days.

On November 13, check the weather.

Find the date 1 week from today.

Start: October 22.
Count on 1 week.

Tell mail carrier on October 29.

Find the date 2 weeks before the trip.

Start: November 16.
Count back 2 weeks.

Rent gear on November 2.

Write the things to do in order.

Things to Do	Date
Tell mail carrier.	October 29
Rent gear.	November 2
Check weather.	November 13

So, Mr. Cob should start by telling the mail carrier about the trip.

Talk About It

- How can you find the date in 3 weeks?

Problem Solving Practice

For 1–4, use the calendars and the list.

April						
Sun	Mon	Tue	Wed	Thu	Fri	Sat
	1	2	3	4	5	6
7	8	9	10	11	12	13
14	15	16	17	18	19	20
21	22	23	24	25	26	27
28	29	30				

May						
Sun	Mon	Tue	Wed	Thu	Fri	Sat
			1	2	3	4
5	6	7	8	9	10	11
12	13	14	15	16	17	18
19	20	21	22	23	24	25
26	27	28	29	30	31	

To Do List

Today's Date: April 25
Date of Party: May 18

Things to do:
- Rent tables and chairs 2 weeks before the party.
- Send invitations in 5 days.
- Order cake 10 days after sending the invitations.

1. Use the list of things to do to help plan the party. Write what needs to be done, in order, and tell the date for each.

2. **What if** the date of the party changes to May 11? List what needs to be done, in order, and tell the date for each.

Mr. Hulin's class will go to the science museum on April 26, to the history museum on April 5, and to the art museum on May 9.

3. Which shows the museum field trips in order?
   - **A** science, history, art
   - **B** history, science, art
   - **C** science, art, history
   - **D** history, art, science

4. How many days are there between the history trip and the science trip?
   - **F** 14 days
   - **G** 21 days
   - **H** 31 days
   - **J** 35 days

Mixed Applications

5. Terry will need chairs for 8 adults and 14 children at her party. Can Terry estimate to decide how many chairs to rent? Explain.

6. Ms. Tate leaves on May 15. She must reserve a room 10 days before she leaves and a flight 2 weeks before she leaves. Which should she do first?

7. The sum of two numbers is 72. Their difference is 24. What are the numbers?

8. **Write a problem** you can solve by using a calendar. Trade problems with a partner and solve.

CHAPTER 6

Review/Test

CHECK VOCABULARY

Choose the best term from the box.

midnight
noon
elapsed time

1. 12:00 at night is __?__. (p. 96)
2. The amount of time that passes from the start of an activity to the end of that activity is __?__. (p. 98)

CHECK SKILLS

Write two ways you can read each time. Then write the time, using A.M. or P.M. (pp. 94–97)

3. 

dance class

4. 

bedtime

5. 

eat lunch

6. 

eat breakfast

USE DATA **For 7–9, use the schedule.** (pp. 98–101)

RAMON'S SATURDAY SCHEDULE	
Activity	**Time**
Breakfast	7:30 A.M. – 8:00 A.M.
Clean room	8:00 A.M. – 8:45 A.M.
Read book	8:45 A.M. – 9:15 A.M.
Play softball	9:15 A.M.–11:15 A.M.
Lunch	11:15 A.M.–11:35 A.M.

7. At what time does Ramon start cleaning his room?
8. How long does Ramon read a book?
9. Which activity is the shortest?

CHECK PROBLEM SOLVING

For 10, use the calendar. (pp. 102–105)

10. A soap-box derby will be on January 30. It will take Ken 2 weeks to build his car and 1 week to paint it. What is the latest date Ken should begin work on his car?

January						
Sun	Mon	Tue	Wed	Thu	Fri	Sat
		1	2	3	4	5
6	7	8	9	10	11	12
13	14	15	16	17	18	19
20	21	22	23	24	25	26
27	28	29	30	31		

Standardized Test Prep

TIP! Get the information you need.
See item **6.**

Use the schedule to find the number of minutes each group is in class. Then compare and order them.

Also see problem **3,** p. H63.

For 1–7, choose the best answer.

1. A number has 2 digits. The tens digit is 4 more than the ones digit. The sum of the digits is 8. What is the number?

   A 53 **B** 62 **C** 73 **D** 84

2. What time does the clock show?

   F 6:40 **G** 7:36 **H** 7:41 **J** 8:41

3. Which of these activities would **not** likely happen at 7:30 P.M.?

   A get ready for bed
   B eat dinner
   C study for a test
   D eat breakfast

4. $8.69 − $4.78 = ■

   F $3.91 **H** $4.91
   G $4.11 **J** NOT HERE

For 5–6, use the schedule.

SWIMMING LESSONS	
Group	**Schedule**
Ages 4–6	9:40 A.M.–10:05 A.M.
Ages 7–9	10:05 A.M.–10:35 A.M.
Ages 10–12	10:35 A.M.–11:10 A.M.
Ages 13–15	11:10 A.M.–11:50 A.M.

5. How long is the lesson for Ages 10–12?

   A 40 minutes **C** 30 minutes
   B 35 minutes **D** 25 minutes

6. Which age group has the longest lesson?

   F Ages 4–6 **H** Ages 10–12
   G Ages 7–9 **J** Ages 13–15

7. Julie saved $5.86 in April and $8.77 in May. How much did she save in all?

   A $14.63 **C** $14.73
   B $14.65 **D** $15.63

Write What You Know

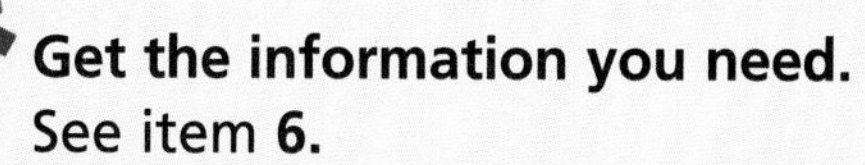

8. Josh worked on a project for 1 hour 45 minutes. He started work at 4:15 P.M. Find the time that he stopped working. Explain how you found your answer.

9. July and August each have 31 days. Suppose the last day in July this year is on a Wednesday. Make a calendar for August. How many Mondays are in August? What are their dates?

Making Cents of It

Use the clues to solve the puzzles.

Puzzle 1

You have four coins.
The value of the coins is 30¢.
What coins do you have?

Puzzle 2

You have three coins.
The value of the coins is 40¢.
What coins do you have?

Puzzle 3

You have 60¢.
You have six coins.
You have no nickels.
What coins do you have?

Puzzle 4

You have 40¢.
You have four coins.
You have no dimes.
What coins do you have?

Puzzle 5

You have five coins.
The value of the coins is greater than 75¢ and less than 90¢.
You have no half dollars.
What coins could you have?

Puzzle 6

You have 90¢.
You have four coins.
One of the coins is a half dollar.
What other coins do you have?

Talk About It

- John has $1.19 in coins. John cannot trade his coins evenly for a $1 bill. What coins could John have?

Challenge

Time: Year, Decade, Century

You know how to use minutes, hours, days, weeks, and months to measure time. You can use *years, decades*, and *centuries* to measure longer periods of time.

UNITS OF TIME

1 year = 365 days, or about 52 weeks

1 decade = 10 years

1 century = 100 years, or 10 decades

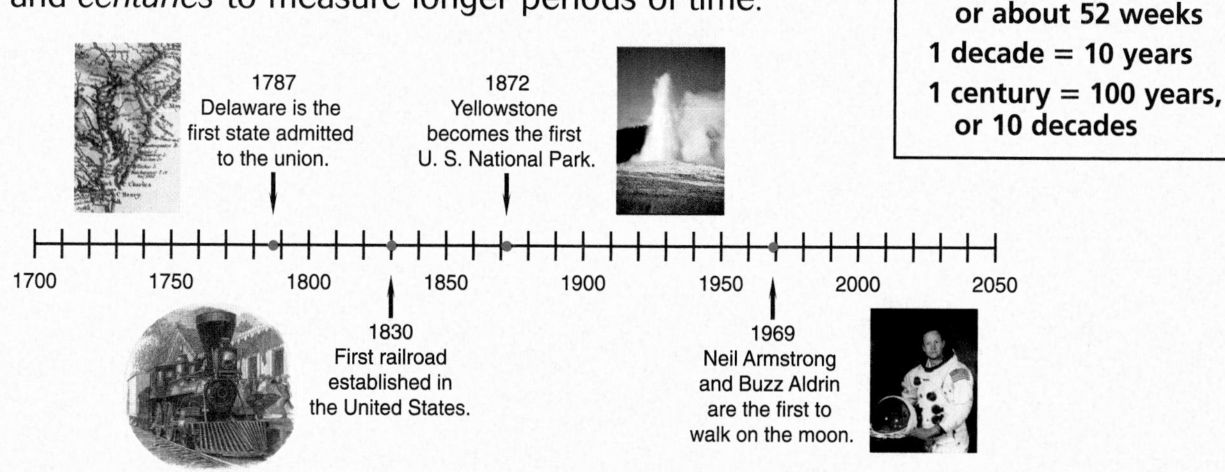

1787
Delaware is the first state admitted to the union.
1872
Yellowstone becomes the first U. S. National Park.
1700
1750
1800
1850
1900
1950
2000
2050
1830
First railroad established in the United States.
1969
Neil Armstrong and Buzz Aldrin are the first to walk on the moon.

A time line shows when events happened. Time lines can help you understand how years, decades, and centuries are related.

Talk About It

- How many centuries are there from the year 1800 to the year 2000? How many decades? How many years?
- How many years passed from the date of the first railroad in the United States to the first time a person walked on the moon?

Try It

Make a time line to show the year, decade, and century you were born. Use your time line for 1–2.

1. What year were you born? How do you know where to mark your birth date on a time line?
2. Mark today's year on your time line. About how much time has passed from your birth date to today?
3. **STRETCH YOUR THINKING** How many centuries are there in 1,000 years? Explain how you know.

Study Guide and Review

VOCABULARY

Choose the best term from the box.

1. A table that lists activities or events and the times they happen is called a __?__. (p. 100)
2. The hours between midnight and noon are __?__ hours. (p. 96)

calendar
schedule
A.M.
P.M.

STUDY AND SOLVE

Chapter 5

Make equivalent sets of money.

Examples A and B are equivalent sets with a value of $1.40.

Find two equivalent sets for each amount. List the bills and coins you can use. (pp. 80–81, 84–85)

3. $1.65
4. $3.70
5. $5.37
6. $5.82
7. $2.13
8. $0.49

Count on to make change.

A comb costs $0.69. Nikki pays with a $1 bill. How much change should she get?

$0.69 $0.70 $0.75 $1.00

1 penny, 1 nickel, and 1 quarter equal $0.31. So, Nikki will get $0.31 in change.

Copy and complete the table. (pp. 86–87)

	COST OF ITEM	AMOUNT PAID	AMOUNT OF CHANGE
9.	$1.25	$2.00	■
10.	$3.72	$5.00	■
11.	$2.34	$3.00	■
12.	$0.17	$1.00	■

Add or subtract money amounts.

Add or subtract as with whole numbers. Then write the sum or difference in dollars and cents.

$\begin{array}{r} \scriptstyle 1 \\ \$2.59 \\ +\$3.17 \\ \hline \$5.76 \end{array}$	$\begin{array}{r} \scriptstyle 3\,12 \\ \$7.\not{4}\not{2} \\ -\$2.23 \\ \hline \$5.19 \end{array}$

Find the sum or difference. Estimate to check. (pp. 88–89)

13. $\begin{array}{r} \$3.58 \\ +\$2.21 \\ \hline \end{array}$ **14.** $\begin{array}{r} \$8.64 \\ -\$7.75 \\ \hline \end{array}$ **15.** $\begin{array}{r} \$2.46 \\ +\$6.54 \\ \hline \end{array}$

16. $\begin{array}{r} \$5.75 \\ -\$2.30 \\ \hline \end{array}$ **17.** $\begin{array}{r} \$5.16 \\ +\$3.95 \\ \hline \end{array}$ **18.** $\begin{array}{r} \$8.07 \\ -\$6.49 \\ \hline \end{array}$

Chapter 6

Tell time to the minute.

Read:
three fifteen
15 minutes after three
quarter past three

Write: 3:15

Write two ways you can read each time. (pp. 94–97)

19.

20.

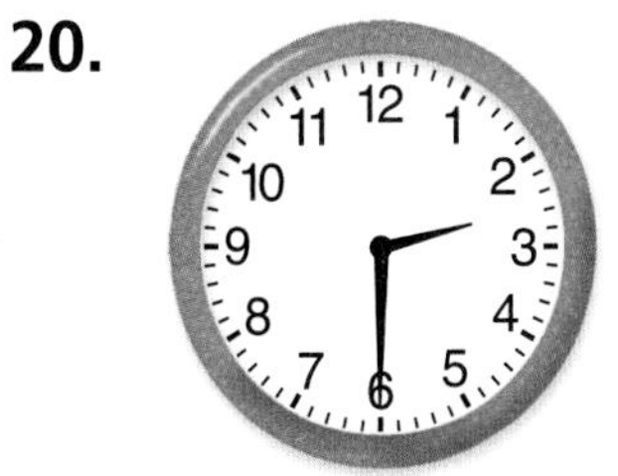

Find elapsed time.

Jerry played basketball from 11:30 A.M. to 1:45 P.M. How long did Jerry play?

Think:
From 11:30 A.M. to 1:30 P.M. is 2 hours.
From 1:30 P.M. to 1:45 P.M. is 15 minutes.

So, Jerry played for 2 hours 15 minutes.

Copy and complete the schedule. (pp. 98–101)

	Activity	Time	Elapsed Time
21.	Reading	11:45 A.M.–12:30 P.M.	■
22.	Lunch	12:30 P.M.–1:00 P.M.	■
23.	Soccer	1:00 P.M.–■	1 hour 30 minutes

PROBLEM SOLVING PRACTICE

Solve. (pp. 82–83, 104–105)

24. Ann has one $1 bill, 8 dimes, and 7 nickels. She wants to buy a notebook for $1.35. List the fewest bills and coins she can use.

25. Jim's recital is on November 12. His soccer game is on October 19. His class project is due October 6. List Jim's activities in order.

PERFORMANCE ASSESSMENT

TASK A • SAVING MONEY

Materials: bills and coins

Becka and Martin are saving money to buy a soccer ball.

a. Becka has saved $5.63. Use bills and coins to show $5.63 in different ways. Copy and complete the table.

b. Martin has saved more than $3.00, but less than Becka. Write an amount Martin could have saved.

c. How much have Becka and Martin saved in all? Is this enough to buy a soccer ball that costs $10.00? Explain.

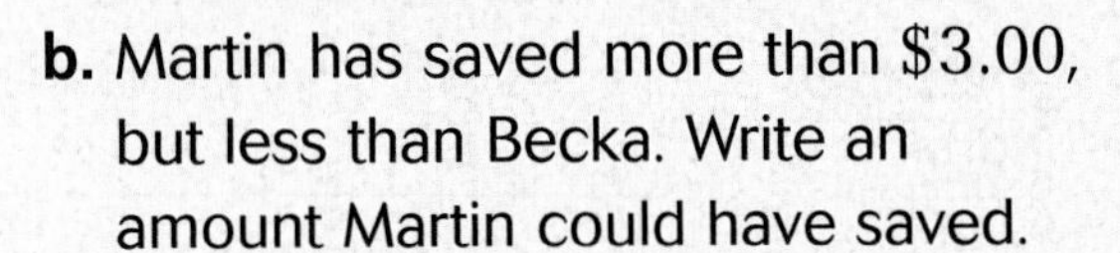

$1 bills	Quarters	Dimes	Nickels	Pennies
4				
4				
3				
3				

TASK B • SOCCER TIME

Materials: clockface

The Kickers soccer team meets every Saturday morning at 8:00 A.M. The table shows the amount of time spent for each activity.

ACTIVITY	ELAPSED TIME
Kicking goals	30 minutes
Dribbling balls	45 minutes
Running laps	40 minutes
Team meeting	30 minutes

a. Copy and complete the practice schedule below. The breaks last 10 minutes.

b. How long is the practice from the start time to the end time?

c. It takes Doug about 45 minutes to get home and eat lunch. Then he has 45 minutes of chores. At what time could he schedule a piano lesson? Tell why you chose that time.

KICKERS SOCCER PRACTICE SCHEDULE

Activity	Start Time	Elapsed Time	End Time
Kicking	8:00 A.M.	30 minutes	
Break			
Dribbling			
Break			
Running			
Break			
Team meeting			

Mighty Math Calculating Crew Counting Coins

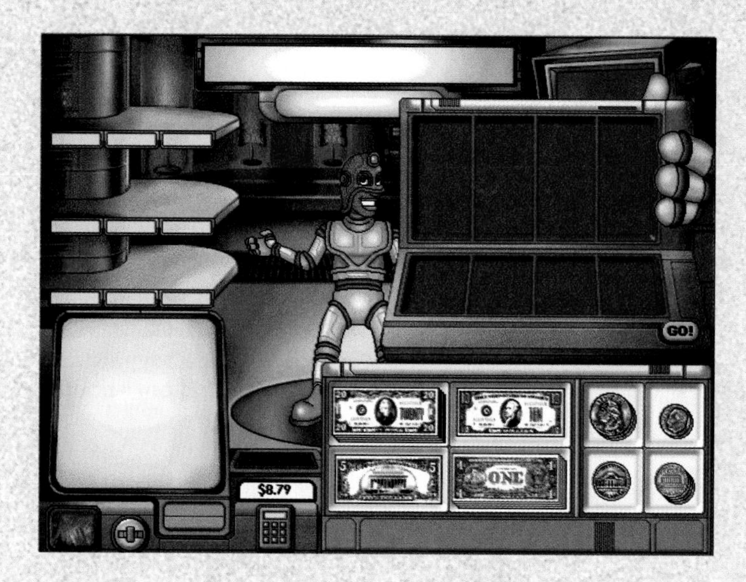

You can use Mighty Math Calculating Crew to help you count money amounts.

Choose *Superhero Superstore*.

- Click [button]. Choose Level C. Click *ok*. Complete the activity by counting coins and bills to $9.99.

Practice and Problem Solving

Write the amount of money.

1. Click [button]. Choose Level E. Complete the activity by counting coins and bills to $39.99.
2. Click [button]. Choose Level F. Complete the activity by counting coins and bills to $99.99.

Solve.

3. Antonio has two $1 bills, 5 quarters, 3 dimes, and 5 pennies. How much money does he have?
4. Brianna buys a notebook for $1.79 and paper for $0.99. She pays with a $5 bill. How much change does she get?
5. **REASONING** Janell has 7 coins that equal $1.30. What coins does she have?

Multimedia Math Glossary www.harcourtschool.com/mathglossary

6. **Vocabulary** Look up *A.M.*, *P.M.*, and *schedule* in the Multimedia Math Glossary. Write a paragraph to explain these terms to a second grader.

The Castle is now the administration building. You can find out about museum programs here.

PROBLEM SOLVING ON LOCATION

in Washington, D.C.

IMAX at the Planetarium	
Moviegoer	**Price per Ticket**
Adult	$5.50
Youth/Senior	$4.25
School Groups	$3.75

Times for *Cosmic Voyage*	
Show 1	9:45 A.M.
Show 2	10:30 A.M.
Show 3	12:45 P.M.
Show 4	3:00 P.M.

SMITHSONIAN INSTITUTION

The Smithsonian museums are called our nation's attic. You can see rockets in the Air and Space Museum. In the Natural History Museum you can see dinosaurs.

USE DATA For 1–4, use the table and schedule.

1. You and your family get to the Castle at 9:00 A.M. You check the show times for *Cosmic Voyage*, and 20 minutes later you are in your seats. How much longer is it before the movie begins?

2. You buy 1 adult ticket and 1 youth ticket and pay with the exact amount. What is the fewest number of bills and coins you could use? List the bills and coins.

3. *Cosmic Voyage* lasts 45 minutes. How much time is there between the end of show 2 and the start of show 3?

4. **REASONING** Planetarium shows run every 40 minutes starting at 11:00 A.M. What is the last show you can go to and still meet your friends at 3:30 P.M?

THE NATIONAL GALLERY OF ART

The National Gallery of Art is just across the Mall from the Smithsonian. The mall is a large grassy area surrounded by many museums and other national buildings.

A family of two adults and two children is planning a day trip to Washington, D.C. This chart shows what they want to do.

Event	Place	Time	Cost
Tour of the East Building Art Collection (45 minutes)	National Gallery of Art	11:30 A.M., 1:30 P.M., 3:30 P.M.	Free
IMAX film: *Galapagos* (40 minutes)	Natural History Museum	10:20 A.M., 1:00 P.M., 2:00 P.M., 3:45 P.M., 5:45 P.M.	Adult: \$6.50 Youth/Senior \$5.50
Ebony Angels Double Dutch Team Performances	National Museum of American History	12:00 P.M.–4:00 P.M.	Free
Highlights Tour (45 minutes)	Air and Space Museum	10:15 A.M., 1:00 P.M.	Free

I. M. Pei designed the East Building of the National Gallery of Art.

USE DATA For 1–2, use the information from the table.

1. Make a schedule for the family's visit.
   a. Show the events in the order the family will attend them. List the name and place of each event.
   b. Show the beginning and ending times for each event. Allow at least 1 hour to see the performance at the National Museum of American History.
   c. Allow at least 1 hour for the family to have lunch and at least 15 minutes between events.
2. Find the total cost for the events the family will attend.

Understand Multiplication

Millions of pounds of fruit are grown each year. Knowing about how much different fruits weigh can help you decide how much fruit to buy.

PROBLEM SOLVING Use the pictograph. About how much more would 3 pears weigh than 3 peaches?

DATA LINK

WEIGHT OF FRUITS

Apple	🛍 🛍 🛍 🛍 🛍 🛍
Orange	🛍 🛍 🛍 🛍 🛍
Peach	🛍 🛍 🛍
Banana	🛍 🛍 🛍 🛍
Kiwi	🛍 🛍 🛍
Plum	🛍 🛍
Pear	🛍 🛍 🛍 🛍 🛍 🛍

Key: Each 🛍 = 1 ounce.

Apple orchard, Hope, Maine

CHECK WHAT YOU KNOW

Use this page to help you review and remember important skills needed for Chapter 7.

SKIP-COUNT (For Intervention, see p. H9.)

1. Skip-count by twos.

2, 4, 6, ■, ■, ■, ■

2. Skip-count by fives.

5, 10, 15, ■, ■, ■

Skip-count to find the missing numbers.

3. 2, 4, 6, ■, ■, ■, 14, 16, ■, ■

4. 3, 6, ■, 12, ■, ■, 21

5. 5, 10, ■, ■, 25, ■, ■, ■, ■, 50

6. 10, 20, ■, ■, 50, ■, ■, ■, 90

EQUAL GROUPS (For Intervention, see p. H9.)

Write how many there are in all.

7.

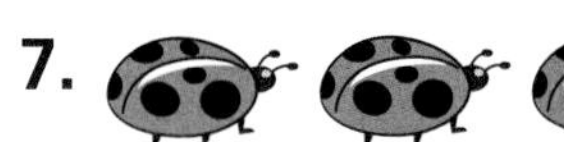

3 groups of 3 = ■

8.

5 groups of 2 = ■

9.

3 groups of 4 = ■

Find how many in all. You may wish to draw a picture.

10. 2 groups of 5

11. 3 groups of 5

12. 4 groups of 2

13. 2 groups of 2

14. 1 group of 3

15. 2 groups of 3

COLUMN ADDITION (For Intervention, see p. H10.)

Find the sum.

16. $\begin{array}{r} 1 \\ 4 \\ +6 \\ \hline \end{array}$

17. $\begin{array}{r} 2 \\ 1 \\ +9 \\ \hline \end{array}$

18. $\begin{array}{r} 6 \\ 6 \\ +6 \\ \hline \end{array}$

19. $\begin{array}{r} 5 \\ 4 \\ +7 \\ \hline \end{array}$

20. $\begin{array}{r} 8 \\ 5 \\ +2 \\ \hline \end{array}$

LESSON 1

Algebra: Connect Addition and Multiplication

Quick Review

1. 3 + 3 + 3
2. 5 + 5 + 5
3. 2 + 2 + 2
4. 2 + 2 + 2 + 2
5. 4 + 4 + 4

VOCABULARY

multiply

Learn

SLURP! There are 3 juice boxes in a package. If Cara buys 5 packages, how many juice boxes will she have?

You can add to find how many in all.

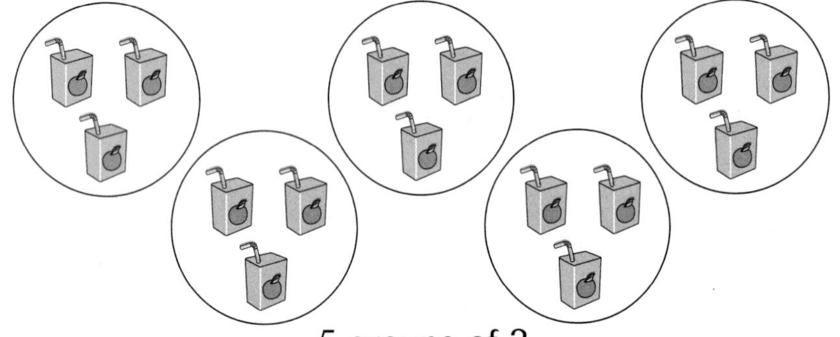

5 groups of 3

Write: $3 + 3 + 3 + 3 + 3 = 15$

Say: 5 threes equal 15.

You can **multiply** to find how many in all.

Write: $5 \times 3 = 15$

Say: 5 times 3 equals 15.

So, Cara will have 15 juice boxes.

MATH IDEA When the groups have the same number, you can multiply to find how many in all.

REASONING Can you use multiplication to find 2 + 3 + 2? Why or why not?

Remember

You can use a number sentence to show addition.
$2 + 2 + 2 = 6$

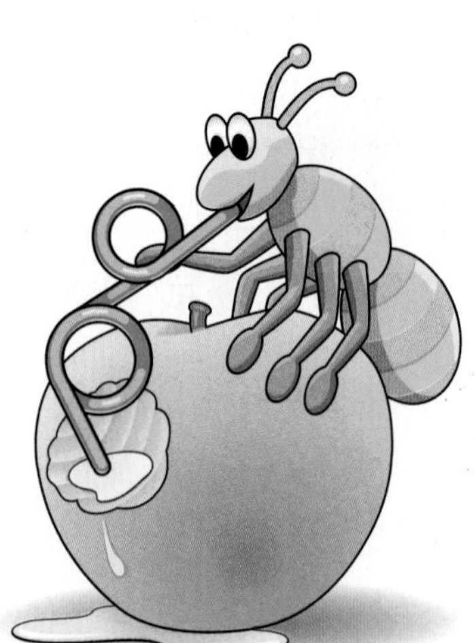

Technology Link

More Practice:
Use E-Lab, *Exploring Multiplication.*
www.harcourtschool.com/elab2002

Check

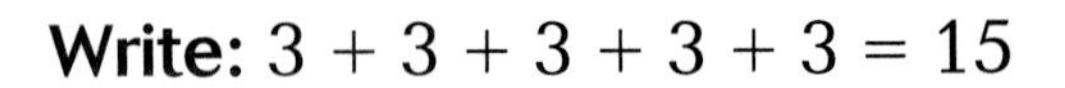

1. **Tell** two ways to find the total if the juice boxes come in packages of 4, and Cara buys 3 packages.

Copy and complete.

2.   

a. ■ groups of ■ = ■
b. ■ + ■ + ■ = ■
c. ■ × ■ = ■

3. 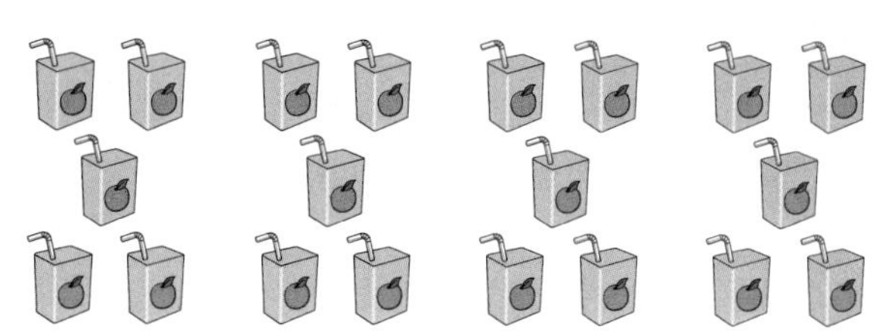

a. ■ groups of ■ = ■
b. ■ + ■ + ■ + ■ = ■
c. ■ × ■ = ■

Practice and Problem Solving

Copy and complete.

4.     

a. ■ groups of ■ = ■
b. ■ + ■ + ■ = ■
c. ■ × ■ = ■

5.   

a. ■ groups of ■ = ■
b. ■ + ■ = ■
c. ■ × ■ = ■

For 6–9, choose the letter of the number sentence that matches.

a. $6 \times 2 = 12$	**b.** $3 \times 8 = 24$	**c.** $3 \times 4 = 12$	**d.** $6 \times 4 = 24$

6. $4 + 4 + 4 = 12$
7. $2 + 2 + 2 + 2 + 2 + 2 = 12$
8. $8 + 8 + 8 = 24$
9. $4 + 4 + 4 + 4 + 4 + 4 = 24$

10. **REASONING** Can you multiply to find 10 + 10 +10 + 10? Explain.

11. **REASONING** Yogurts come in packages of 6. How many packages are needed to give 23 students each a yogurt?

Mixed Review and Test Prep

12. $\begin{array}{r} 437 \\ -229 \\ \hline \end{array}$ (p. 58)

13. $\begin{array}{r} 684 \\ +321 \\ \hline \end{array}$ (p. 42)

14. $\begin{array}{r} 9{,}239 \\ +1{,}605 \\ \hline \end{array}$ (p. 46)

15. $\begin{array}{r} 3{,}030 \\ -1{,}923 \\ \hline \end{array}$ (p. 62)

16. **TEST PREP** What is the elapsed time from 3:15 P.M. until 4:30 P.M.? (p. 98)

A 15 minutes
B 45 minutes
C 1 hour 15 minutes
D 2 hours

EXTRA PRACTICE page H38, Set A

LESSON 2

Multiply with 2 and 5

Quick Review

How many are in all?

1. 1 row of 8
2. 3 rows of 2
3. 2 rows of 5
4. 4 rows of 2
5. 3 rows of 3

VOCABULARY

factors
product

Learn

SMART ROCKS The chips that run computers are made from a mineral found in rocks. Mrs. Frank asked 5 students to bring in 2 rocks each for a science project. How many rocks does she need?

Use counters.

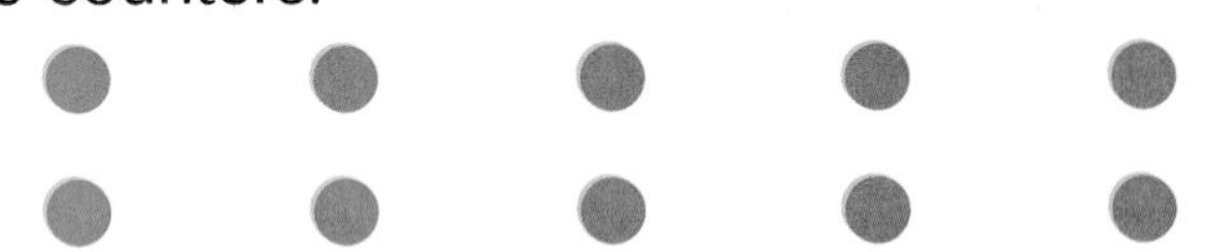

There are 5 groups with 2 in each group.

Since each group has the same number, you can multiply to find how many in all.

$$\underset{\text{factor}}{5} \times \underset{\text{factor}}{2} = \underset{\text{product}}{10}$$

$$\begin{array}{r} 2 \\ \times 5 \\ \hline 10 \end{array} \quad \begin{array}{l} \leftarrow \text{factor} \\ \leftarrow \text{factor} \\ \leftarrow \text{product} \end{array}$$

So, Mrs. Frank needs 10 rocks in all.

Computer chip

Crystal

MATH IDEA The numbers that you multiply are **factors**. The answer is the **product**.

- Name the factors and product in $3 \times 2 = 6$.

Check

1. **Find** the products 1×2 through 9×2. What do you notice about the products? Are they always even or odd numbers? Why?

Find the product.

2. 
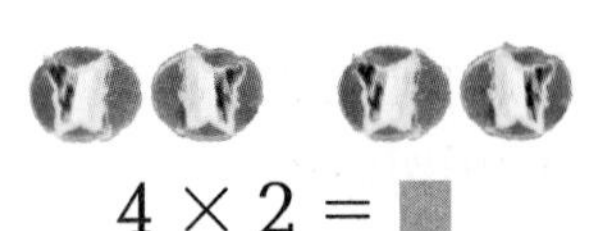
$4 \times 2 = \blacksquare$

3. 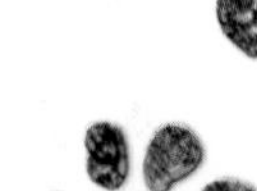
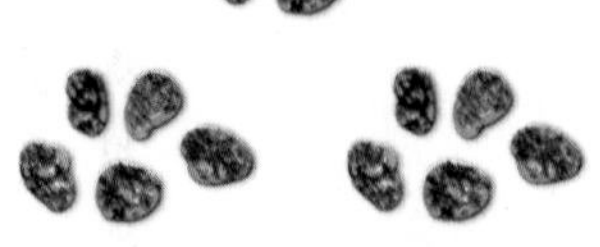
$3 \times 5 = \blacksquare$

4. 
$6 \times 2 = \blacksquare$

Practice and Problem Solving

Find the product.

5.  

$2 \times 2 = ■$

6. 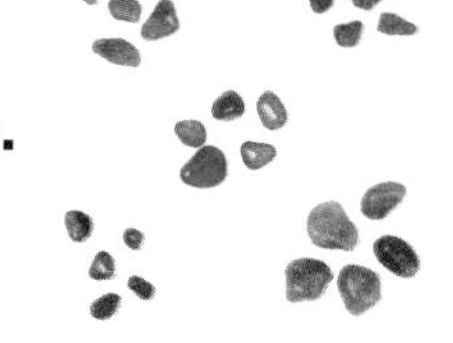

$5 \times 5 = ■$

7. 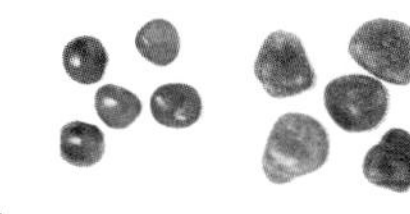

$4 \times 5 = ■$

Copy and complete the multiplication table.

	×	1	2	3	4	5	6	7	8	9
8.	2	■	■	■	■	■	■	■	■	■
9.	5	■	■	■	■	■	■	■	■	■

Complete.

10. $8 \times 2 = ■$

11. $■ = 4 \times 5$

12. $6 \times 2 = ■$

13. $■ = 7 \times 5$

14. $■ = 9 \times 5$

15. $8 \times 5 = ■$

16. $■ = 7 \times 2$

17. $6 \times 5 = ■$

18. $\begin{array}{r} 2 \\ \times 8 \\ \hline \end{array}$

19. $\begin{array}{r} 5 \\ \times 8 \\ \hline \end{array}$

20. $\begin{array}{r} 2 \\ \times 9 \\ \hline \end{array}$

21. $\begin{array}{r} 5 \\ \times 9 \\ \hline \end{array}$

22. $3 + 3 = 2 \times ■$

23. $4 \times ■ = 4 + 4 + 4$

24. $2 \times 5 = ■ + 5$

25. Sal bought 8 packages of rocks at the science center. There are 2 rocks in each package. How many rocks did Sal buy?

26. **REASONING** Drew has 5 pairs of white socks and 2 pairs of black socks. How many more white socks than black socks does Drew have?

27. **? What's the Error?** Midville School has 5 classrooms with 6 computers in each. Jan said that there are 11 computers in all. Describe Jan's error.

Mixed Review and Test Prep

28. 5 dimes = ■ quarters (p. 80)

29. 9 nickels = ■ pennies (p. 80)

30. Round 4,787 to the nearest hundred. (p. 28)

31. $\begin{array}{r} \$4.56 \\ +\$2.98 \\ \hline \end{array}$ (p. 88)

32. **TEST PREP** Jim went to play basketball at 2:15 P.M. and returned at 4:45 P.M. How long was he gone? (p. 98)

A 2 hours
B 2 hours 15 minutes
C 2 hours 30 minutes
D 2 hours 45 minutes

EXTRA PRACTICE page H38, Set B

Arrays

Quick Review

Find how many in all.

1. 2 groups of 2
2. 1 group of 9
3. 4 groups of 2
4. 3 groups of 4
5. 2 groups of 3

VOCABULARY
array
Order Property of Multiplication

MATERIALS
square tiles

Explore

An **array** shows objects in rows and columns.

Activity

Make an array to find how many are in 3 rows of 5.

STEP 1

Make an array with 3 rows and 5 columns.

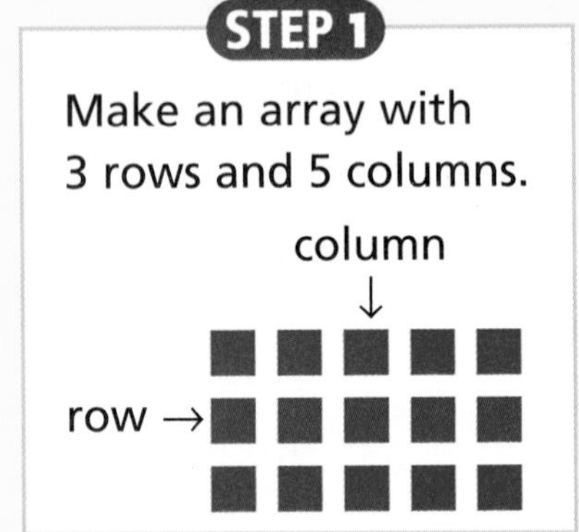

STEP 2

Count the tiles.

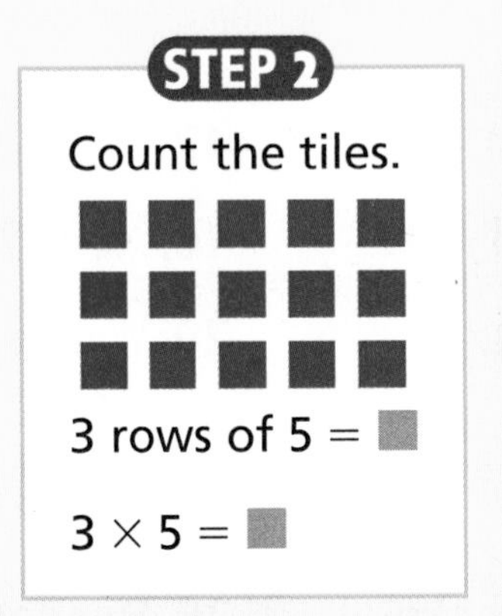

3 rows of 5 = ___

$3 \times 5 =$ ___

- How many tiles are in the 3 rows of 5?
- What multiplication sentence can you write to find the number of tiles?
- Make an array with 3 rows of 3. What shape is formed by this array? What multiplication sentence can you write to find the number of tiles?

Try It

Copy and complete.

a.

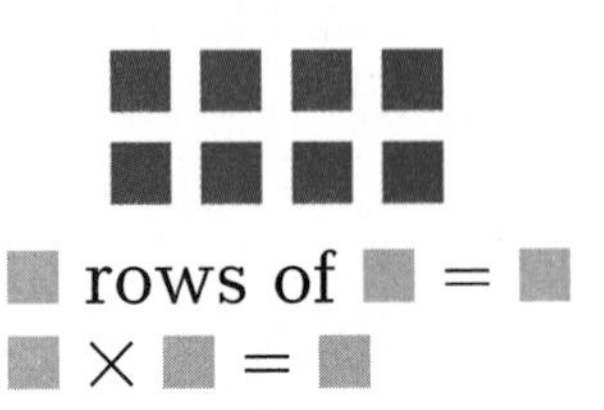

___ rows of ___ = ___

___ $\times$ ___ = ___

b.

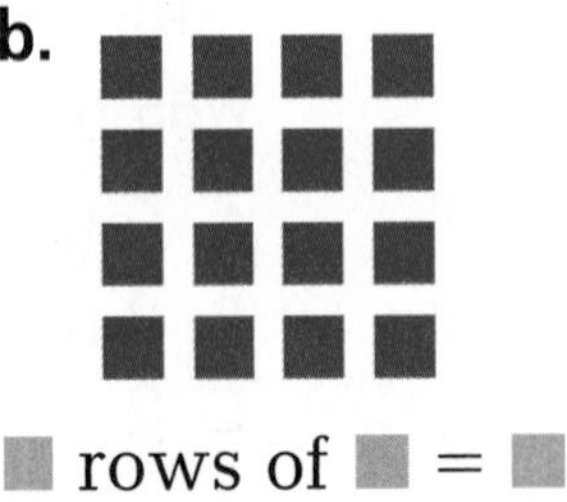

___ rows of ___ = ___

___ $\times$ ___ = ___

I have 2 rows of 4. How many are there in all?

Connect

The **Order Property of Multiplication** means that two numbers can be multiplied in any order. The product is the same.

Use arrays to show the Order Property of Multiplication.

Examples

A

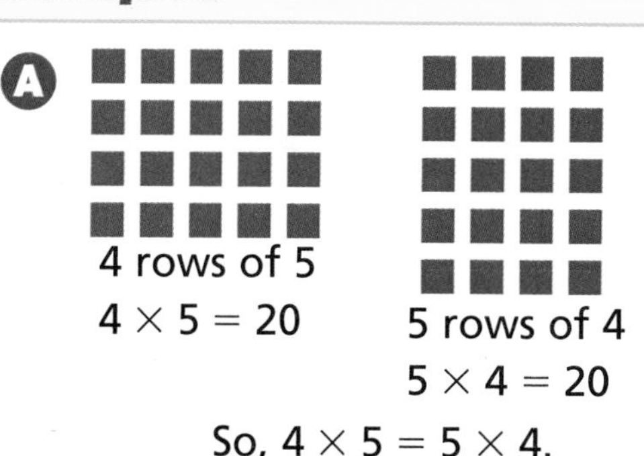

4 rows of 5
$4 \times 5 = 20$

5 rows of 4
$5 \times 4 = 20$

So, $4 \times 5 = 5 \times 4$.

B

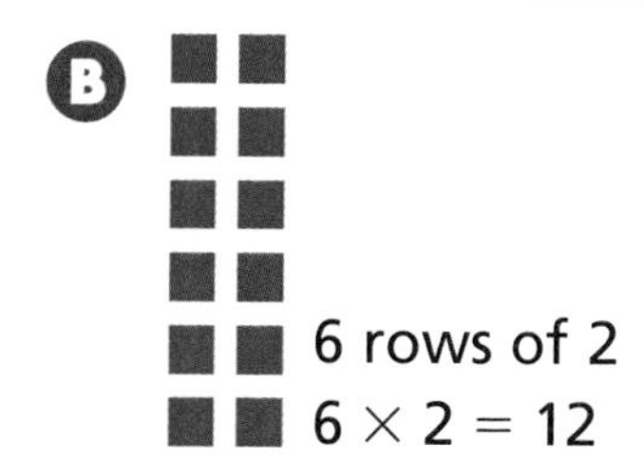

6 rows of 2
$6 \times 2 = 12$

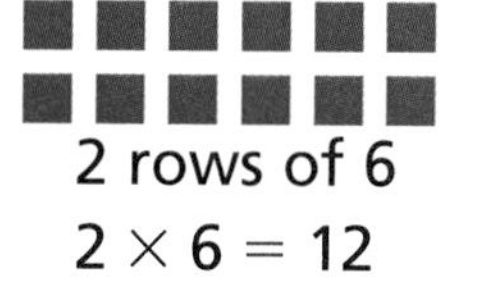

2 rows of 6
$2 \times 6 = 12$

So, $6 \times 2 = 2 \times 6$.

Practice and Problem Solving

Copy and complete.

1.

■ rows of ■ = ■
■ × ■ = ■

2.

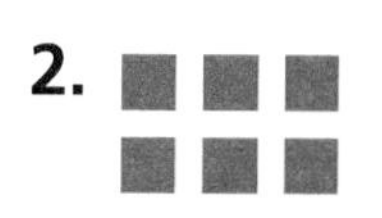

■ rows of ■ = ■
■ × ■ = ■

3.

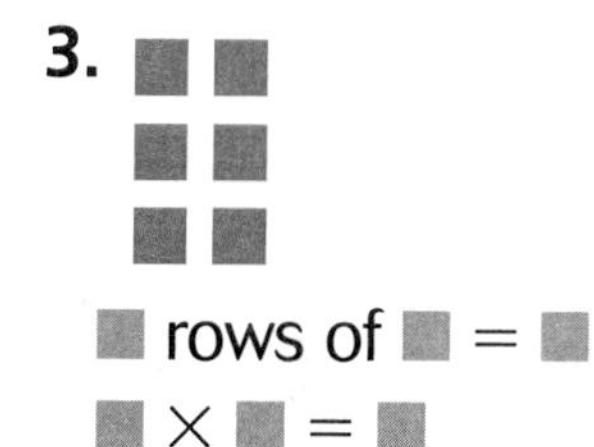

■ rows of ■ = ■
■ × ■ = ■

Find the product. You may wish to draw an array.

4. $2 \times 3 =$ ■

5. $6 \times 4 =$ ■

6. $8 \times 3 =$ ■

7. $3 \times 5 =$ ■

8. **Write About It** Miguel needs a book cover that costs \$1.99 and a package of markers that costs \$2.79. He has \$5.00. Does he have enough money to buy both items?

9. The sum of Jarrod's age and Kayla's age is 21. Kayla is 5 years older than Jarrod. How old are Kayla and Jarrod?

Mixed Review and Test Prep

10. 24 + 46 + 93 (p. 36)

11. 446 − 267 (p. 58)

12. 34 − 15

13. Write the value of the 7 in 67,409. (p. 10)

14. **TEST PREP** What is the total value of 3 dimes and 4 nickels?

A 34¢
B 50¢
C 55¢
D 70¢

LESSON 4

Multiply with 3

Quick Review

1. $2 + 2 = 2 \times$ ■
2. $3 \times$ ■ $= 5 + 5 + 5$
3. $2 \times 6 = 6 +$ ■
4. ■ $\times 5 = 10$
5. $2 \times 4 =$ ■

Learn

PRACTICE, PRACTICE, PRACTICE

Pat practiced soccer 2 hours each day for 3 days. How many hours did he practice in all?

For 2 hours, move 2 spaces. For 3 days, make 3 jumps of 2 spaces.

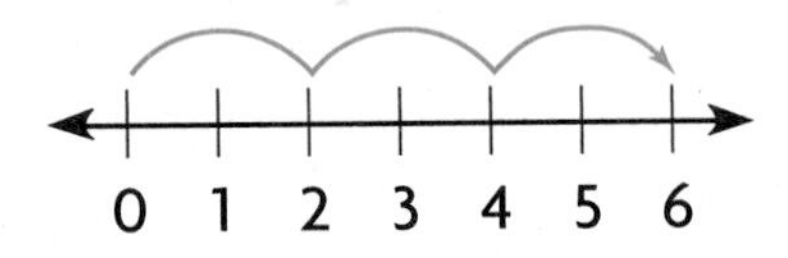

Multiply: $3 \times 2 = 6$

Val practiced soccer 3 hours each day for 2 days. How many hours did she practice in all?

For 3 hours, move 3 spaces. For 2 days, make 2 jumps of 3 spaces.

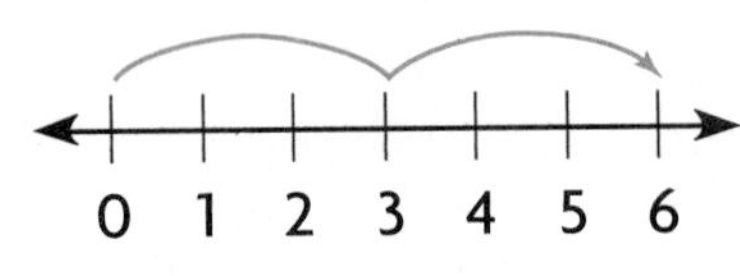

Multiply: $2 \times 3 = 6$

So, both Pat and Val practiced for 6 hours.

A number line can help you understand the Order Property of Multiplication.

Example 1

Multiply. $3 \times 5 = 15$

0 1 2 3 4 5 6 7 8 9 10 11 12 13 14 15

Multiply. $5 \times 3 = 15$

0 1 2 3 4 5 6 7 8 9 10 11 12 13 14 15

So, $3 \times 5 = 5 \times 3$.

- **REASONING** Use the factors 3 and 6 to explain the Order Property of Multiplication.

Multiplication Practice

What if Pat scored 4 goals in each of 3 games and Val scored 3 goals in each of 4 games? Which number line shows Pat's goals? Val's goals? How many goals did each player score?

Example 2

A

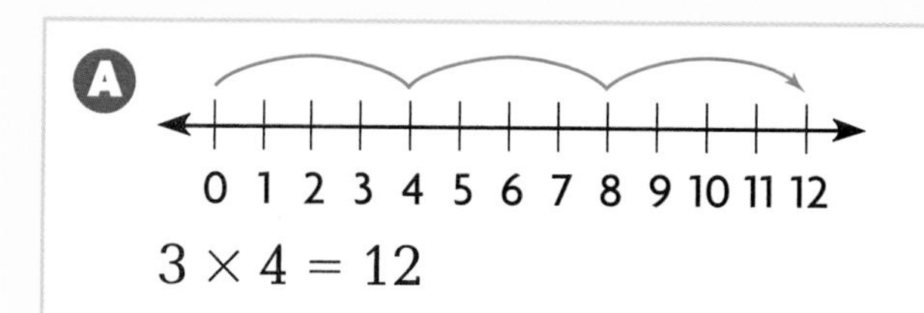

$3 \times 4 = 12$

B

0 1 2 3 4 5 6 7 8 9 10 11 12

$$\begin{array}{r} 4 \\ \times 3 \\ \hline 12 \end{array}$$

Number line A shows Pat's goals.

Number line B shows Val's goals.

Each player scored 12 goals.

- **What if** Val scored 8 goals in each of 3 games and Pat scored 3 goals in each of 8 games? How many goals did each player score?

Check

1. **Explain** how knowing the product 7×3 can help you find the product 3×7.

2. Is the product even or odd when you multiply 3 by an even number? by an odd number?

Use the number line to find the product.

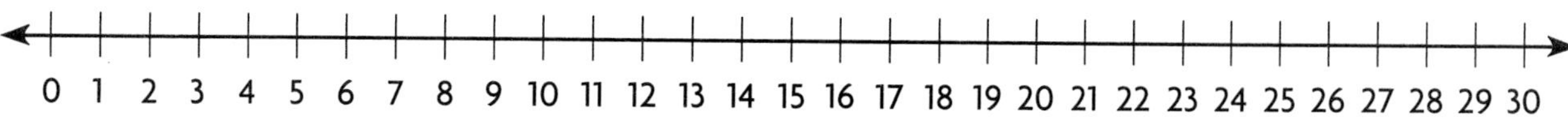

3. $4 \times 3 = ■$ **4.** $■ = 3 \times 4$ **5.** $7 \times 3 = ■$ **6.** $■ = 3 \times 7$

7. $■ = 6 \times 3$ **8.** $■ = 3 \times 6$ **9.** $3 \times 8 = ■$ **10.** $■ = 8 \times 3$

11. $\begin{array}{r} 5 \\ \times 3 \\ \hline \end{array}$ **12.** $\begin{array}{r} 3 \\ \times 5 \\ \hline \end{array}$ **13.** $\begin{array}{r} 3 \\ \times 9 \\ \hline \end{array}$ **14.** $\begin{array}{r} 9 \\ \times 3 \\ \hline \end{array}$

LESSON CONTINUES

Practice and Problem Solving

Use the number line to find the product.

15. $3 \times 6 = \blacksquare$

16. $\blacksquare = 5 \times 5$

17. $3 \times 9 = \blacksquare$

18. $\blacksquare = 9 \times 3$

19. $\blacksquare = 9 \times 2$

20. $7 \times 3 = \blacksquare$

21. $\blacksquare = 4 \times 5$

22. $2 \times 8 = \blacksquare$

23. $7 \times 2 = \blacksquare$

24. $\begin{array}{r} 3 \\ \times 8 \\ \hline \end{array}$

25. $\begin{array}{r} 1 \\ \times 5 \\ \hline \end{array}$

26. $\begin{array}{r} 3 \\ \times 9 \\ \hline \end{array}$

Copy and complete the multiplication table.

27.

×	1	2	3	4	5	6	7	8	9
3	■	■	■	■	■	■	■	■	■

Write the missing factor.

28. $2 \times 3 = \blacksquare \times 2$

29. $3 \times \blacksquare = 7 \times 3$

30. $7 \times 3 = \blacksquare \times 7$

31. $6 \times 3 = \blacksquare \times 2$

32. $4 \times 3 = 6 \times \blacksquare$

33. $8 \times 3 = \blacksquare \times 8$

More Practice: Use Mighty Math Carnival Countdown, *Snap Clowns*, Level P.

34. **What's the Error?** Sam played soccer for 2 hours a day, 3 days a week, for 4 weeks. Sam said he played soccer for a total of 6 hours during the month. What is Sam's error?

35. **REASONING** If you add 3 to an odd number, is the sum even or odd? If you multiply an odd number by 3, is the product even or odd?

USE DATA For 36–39, use the bar graph.

36. How many more goals did Matt's soccer team score in Game 4 than in Game 1?

37. How many goals did Matt's team score in the four games?

38. If Matt's team scores twice as many goals in Game 5 as in Game 3, how many goals will it score?

39. **Write a problem** using the bar graph.

40. Last week, the team practiced 2 days for 2 hours each day. This week, they practiced 3 days for 2 hours each day. How many hours did they practice in all?

41. Tim needs 25 table tennis balls. There are 5 tennis balls in 1 package. If he buys 4 packages, will he have enough tennis balls? Explain.

Mixed Review and Test Prep

Write *A.M.* **or** *P.M.* **for each.** (p. 96)

42. see the moon: 10:00 __?__

43. eat lunch: 11:45 __?__

44. play at park: 4:30 __?__

Find the product. (p. 118)

45. $6 \times 2 = ■$ **46.** $8 \times 2 = ■$

47. $4 \times 5 = ■$ **48.** $6 \times 5 = ■$

49. TEST PREP $35 + 14 + 26$ (p. 36)

A 65 **B** 75 **C** 76 **D** 85

50. TEST PREP Tina went to the mall at 3:00 P.M. She returned home at 4:15 P.M. How long was she gone? (p. 98)

F 15 minutes
G 1 hour
H 1 hour 15 minutes
J 1 hour 30 minutes

PROBLEM SOLVING  Thinker's Corner

SOLVE THE RIDDLE! **Find the product. To answer the riddle, match the letters to the products below.**

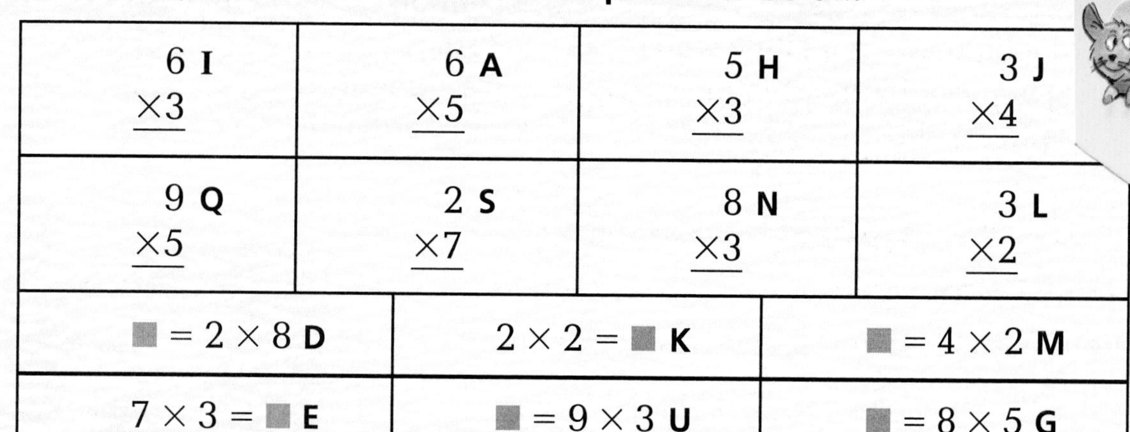

6 **I** ×3	6 **A** ×5	5 **H** ×3	3 **J** ×4
9 **Q** ×5	2 **S** ×7	8 **N** ×3	3 **L** ×2

$■ = 2 \times 8$ **D**	$2 \times 2 = ■$ **K**	$■ = 4 \times 2$ **M**
$7 \times 3 = ■$ **E**	$■ = 9 \times 3$ **U**	$■ = 8 \times 5$ **G**

What is a mouse's favorite game?

?/15 ?/18 ?/16 ?/21   ?/30 ?/24 ?/16   ?/14 ?/45 ?/27 ?/21 ?/30 ?/4

EXTRA PRACTICE page H38, Set C

LESSON 5

Problem Solving Skill
Too Much/Too Little Information

Quick Review

1. 3×5
2. 4×3
3. 3×3
4. 9×3
5. 3×6

UNDERSTAND | PLAN | SOLVE | CHECK

FIND THE FACTS Three students walked 6 blocks to the craft store. Each one bought 5 pieces of poster board to make posters for the school book fair. They stayed at the store for 45 minutes. How many pieces of poster board in all did the students buy?

Example

STEP 1

Find what the problem asks.
- How many pieces of poster board did the students buy in all?

STEP 2

Find what facts are needed to solve the problem.
- the number of students
- the number of pieces of poster board each one buys

STEP 3

Look for extra information.
- how far they walked
- how long they were at the store

Do you need this information to solve the problem?

STEP 4

Solve the problem.
- multiply

3 students × 5 pieces = 15 pieces

So, the students bought 15 pieces of poster board in all.

Talk About It

- Is there too much or too little information in the problem above?
- How do you know which information you need to solve a problem?
- Three students went to a restaurant. They each bought a sandwich. How much did they spend in all? Does this problem have too much, too little, or the right amount of information?

Practice and Problem Solving

USE DATA For 1–4, use the table. Write *a*, *b*, or *c* to tell whether the problem has

a. too much information.
b. too little information.
c. the right amount of information.

Solve those with too much or the right amount of information. Tell what is missing for those with too little information.

SCHOOL SUPPLIES	
Pack of Paper	$1
Backpack	$9
Pack of Pencils	$3
Lunch Box	$4

1. Felix wants to buy a backpack and a box of crayons. How much will he spend?
2. Marisa bought 2 packs of pencils. She was second in line to pay for her supplies. How much did Marisa spend?
3. Sam bought 2 backpacks and a lunch box. He received $3 change. How much money had he given the clerk?
4. Sally had $15. She bought 5 packs of paper and a lunch box. How much did she spend?

Mixed Applications

USE DATA For 5–6, use the table above.

You have $15 to spend on school supplies.

5. Which items can you buy?
   - **A** a backpack, 3 packs of pencils
   - **B** a backpack, a pack of pencils, a lunch box
   - **C** a lunch box, 2 packs of paper, a backpack
   - **D** 4 packs of pencils, a backpack
6. How much more money do you need if you choose to buy 2 packs of pencils, a lunch box, and a backpack?
   - **F** $1
   - **G** $2
   - **H** $3
   - **J** $4
7. Joe bought two tapes at the music store. They cost $7.28 and $7.71. How much change did he receive from $20.00?
8. **What's the Question?** There are 4 people in the Tamura family. Movie tickets cost $6 each. The answer is $24.

Review/Test

CHECK VOCABULARY AND CONCEPTS

Choose the best term from the box.

array
factors
multiply
product

1. When groups have the same number, you can __?__ to find how many in all. (p. 116)
2. The numbers you multiply are __?__. (p. 118)
3. The answer to a multiplication problem is the __?__. (p. 118)

Find the product. You may wish to draw an array. (pp. 120–121)

4. $4 \times 5 = \blacksquare$
5. $3 \times 2 = \blacksquare$
6. $3 \times 6 = \blacksquare$
7. $7 \times 3 = \blacksquare$

CHECK SKILLS

For 8–9, choose the letter of the number sentence that matches. (pp. 116–117)

8. $3 + 3 + 3 + 3 = 12$
9. $4 + 4 + 4 + 4 + 4 = 20$

a. $5 \times 4 = 20$
b. $6 \times 2 = 12$
c. $4 \times 3 = 12$

Find the product. (pp. 118–125)

10. $3 \times 8 = \blacksquare$
11. $7 \times 3 = \blacksquare$
12. $\blacksquare = 5 \times 6$
13. $4 \times 2 = \blacksquare$
14. 3×9
15. 6×3
16. 8×5
17. 7×5
18. 5×9

CHECK PROBLEM SOLVING

Write *a*, *b*, or *c* to tell whether the problem has

a. too much information.
b. too little information.
c. the right amount of information.

Solve those with too much or the right amount of information. Tell what is missing for those with too little information. (pp. 126–127)

19. Pete practices 3 hours a day, Monday through Friday. How many hours does he practice each week?

20. Ramiro worked on his science project 3 hours longer than Sue. How much time did each of them spend on the project?

Standardized Test Prep

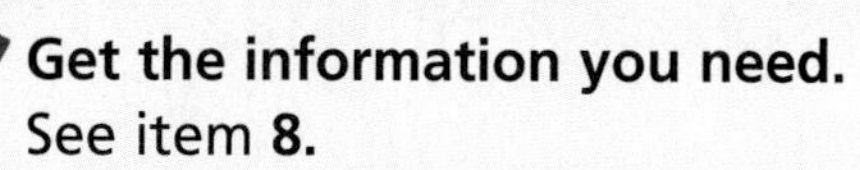

TIP! **Get the information you need.**
See item **8.**

Don't be confused by information that is not needed. You need the cost of the packs and how many Ben bought to find how much he spent.

Also see problem **3**, p. H63.

For 1–8, choose the best answer.

1. What time does the clock show?

A 6:08   C 7:08
B 6:40   D 7:40

2. What is the value of the 1 in 17,405?

F 10   H 1,000
G 100   J 10,000

3. $3 \times 9 = \blacksquare$

A 6   C 27
B 12   D NOT HERE

4. $3 \times 8 = \blacksquare$

F 11   H 20
G 16   J NOT HERE

5. Libby bought 8 packages of fishing bait for \$5 each. How much money did she spend on the fishing bait?

A \$30   C \$38
B \$35   D \$40

6. Shelly made 6 glasses of grape drink. She used 2 scoops of powder in each glass. How many scoops did she use in all?

F 12   H 4
G 8   J NOT HERE

7. Ramon has 64¢. A pack of gum costs 25¢ and a snack bar costs 35¢. How much money will Ramon have left if he buys a snack bar?

A 60¢   C 29¢
B 39¢   D 4¢

8. Ben had \$25. He bought 7 packs of cards that cost \$2 each. How much money did he spend?

F \$9   H \$14
G \$11   J \$25

Write What You Know

9. Write a multiplication sentence that shows another way to write $5 + 5 + 5 + 5 = 20$.
Explain your answer.

10. Abdul planted 3 rows of trees. There were 6 trees in each row. Draw a picture of this. How many trees did Abdul plant?

CHAPTER 8

Multiplication Facts Through 5

Mantis is a stand-up roller coaster in Sandusky, Ohio. Each train has eight rows of four riders.

There are about 175 roller coasters in the United States today. Some can reach speeds of 70 miles per hour!

PROBLEM SOLVING Compare the number of riders on the roller coasters named on the pictograph with the coaster in the photo. What is a quick way to find the number of riders on Mantis?

ROLLER COASTER RIDERS

Hercules (Pennsylvania)	■■■ ■■■
The Psyclone (California)	■■■ ■■■■
The Viper (Georgia)	■■■ ■■■■
Mamba (Missouri)	■■■■ ■■■■■

Key: Each ■ = 4 riders.

Use this page to help you review and remember important skills needed for Chapter 8.

VOCABULARY

Choose the best term from the box.

addends
factors
product

1. The numbers that you multiply are called _?_ .
2. The answer to a multiplication problem is the _?_ .

ADDITION (For Intervention, see p. H10.)

Complete.

3. $3 + 3 + 3 + 3 = \blacksquare$
4. $8 + \blacksquare = 8$
5. $1 + 1 + 1 + 1 + 1 = \blacksquare$
6. $4 + 4 + 4 + 4 = \blacksquare$
7. $\blacksquare = 1 + 9$
8. $6 + 0 = \blacksquare$
9. $0 + 0 + 0 = \blacksquare$
10. $4 + 4 + 4 = \blacksquare$
11. $\blacksquare + 7 = 15$

ORDER PROPERTY OF ADDITION (For Intervention, see p. H11.)

Complete.

12. $4 + 7 = 7 + \blacksquare$
13. $8 + 5 = 5 + \blacksquare$
14. $9 + 6 = \blacksquare + 9$
15. $\blacksquare + 9 = 9 + 8$
16. $3 + \blacksquare = 7 + 3$
17. $\blacksquare + 4 = 4 + 9$

MULTIPLICATION FACTS (For Intervention, see p. H11.)

Find the product.

18. $1 \times 2 = \blacksquare$
19. $4 \times 5 = \blacksquare$
20. $2 \times 5 = \blacksquare$
21. $\blacksquare = 3 \times 3$
22. $3 \times 6 = \blacksquare$
23. $\blacksquare = 2 \times 9$
24. $7 \times 3 = \blacksquare$
25. $4 \times 2 = \blacksquare$
26. $\blacksquare = 9 \times 3$
27. $8 \times 5 = \blacksquare$
28. $\blacksquare = 6 \times 2$
29. $8 \times 2 = \blacksquare$
30. $9 \times 5 = \blacksquare$
31. $\blacksquare = 5 \times 6$
32. $2 \times 7 = \blacksquare$
33. $\blacksquare = 3 \times 8$

LESSON 1

Multiply with 0 and 1

Quick Review

1. $2 \times 5 = \blacksquare$
2. $3 \times 4 = 4 + 4 + \blacksquare$
3. $6 \times 2 = \blacksquare$
4. $5 + 5 + 5 = \blacksquare \times 5$
5. $3 \times 2 = \blacksquare$

Learn

ALL OR NOTHING Tina saw 5 cars. One clown sat in each car. How many clowns were there in all?

Example

STEP 1

Count the cars.

STEP 2

Count the clowns in the cars.

STEP 3

Write the multiplication sentence.

5	×	1	=	5
↑		↑		↑
number of groups		number in each group		number in all

So, there were 5 clowns in all.

Suppose Tina saw 3 cars with 0 clowns in each car. How many clowns were there in all?

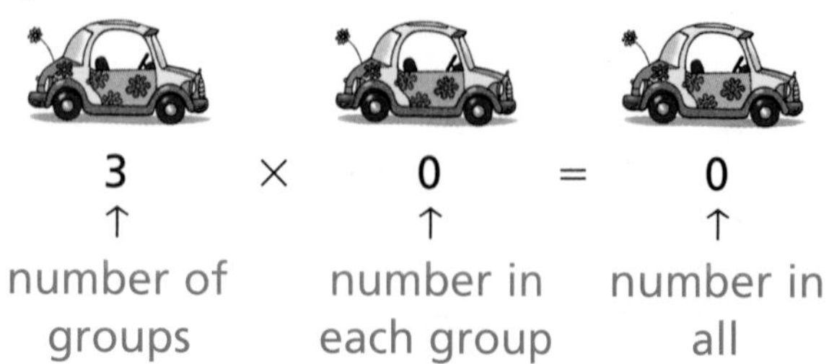

3	×	0	=	0
↑		↑		↑
number of groups		number in each group		number in all

So, there were 0 clowns in all.

MATH IDEA The product of 1 and any number equals that number. The product of 0 and any number equals 0.

REASONING What is 498×1? 498×0? How do you know?

Check

1. **Explain** what happens when you multiply by 1. What happens when you multiply by 0?

Find the product.

2. $4 \times 1 = ■$ 3. $5 \times 0 = ■$ 4. $1 \times 3 = ■$

Practice and Problem Solving

Find the product.

5. $2 \times 1 = ■$
6. $4 \times 0 = ■$
7. $0 \times 5 = ■$
8. $■ = 9 \times 1$
9. $1 \times 6 = ■$
10. $■ = 0 \times 9$
11. $7 \times 1 = ■$
12. $2 \times 4 = ■$
13. $■ = 0 \times 7$
14. $3 \times 3 = ■$
15. $■ = 1 \times 8$
16. $0 \times 0 = ■$
17. 9×0
18. 1×3
19. 5×4
20. 8×5
21. Multiply 4 by 1.
22. Find the product of 0 and 8.
23. Find the product of 1 and 7.
24. What is 3 times 9?

Complete.

25. $■ = 9 \times 5$
26. $3 + 9 = 3 \times ■$
27. $3 \times 6 = ■ \times 9$
28. $8 + 7 = ■ \times 5$
29. $0 \times 8 = ■ \times 9$
30. $9 \times ■ = 3 \times 3$
31. **REASONING** Ann is younger than Rick. Rick is older than Tracy. Tracy is older than Ann. Who is the oldest?
32. **Write About It** Which is less, the product of your age and 1 or the product of your age and 0? Explain.
33. **Write a problem that has zero as the product.**

Mixed Review and Test Prep

Find the value of the blue digit. (p. 10)

34. 92,348
35. 38,965
36. Find the sum of 652 and 738. (p. 42)
37. Put the numbers in order from least to greatest. 43, 24, 34, 42 (p. 24)
38. **TEST PREP** Carol bought a pencil for \$0.32. She paid with a \$1 bill. How much change did she get? (p. 86)

   A \$0.32 **C** \$0.78
   B \$0.68 **D** \$1.32

EXTRA PRACTICE page H39, Set A

LESSON

2 Multiply with 4

Quick Review

1. 0×3
2. 4×2
3. 1×9
4. 8×1
5. 2×5

Learn

TWISTS AND TURNS There are 6 cars in the Twister ride at the amusement park. Each car holds 4 people. How many people does the Twister ride hold?

$$6 \times 4 = ■$$

- How could you use an array to find 6×4?

You can use a multiplication table to find the product.

Multiplication Table

column ↓ (4); row → (6)

×	0	1	2	3	4	5	6	7	8	9
0	0	0	0	0	0	0	0	0	0	0
1	0	1	2	3	4	5	6	7	8	9
2	0	2	4	6	8	10	12	14	16	18
3	0	3	6	9	12	15	18	21	24	27
4	0	4	8	12	16	20	24	28	32	36
5	0	5	10	15	20	25	30	35	40	45
6	0	6	12	18	24	30	36	42	48	54
7	0	7	14	21	28	35	42	49	56	63
8	0	8	16	24	32	40	48	56	64	72
9	0	9	18	27	36	45	54	63	72	81

The product is found where row 6 and column 4 meet.

6	×	4	=	24
↑		↑		↑
factor		factor		product

$$\begin{array}{r} 4 \\ \times 6 \\ \hline 24 \end{array}$$

4 ← factor
×6 ← factor
24 ← product

So, the Twister ride holds 24 people.

MATH IDEA All multiplication facts that have 4 as a factor have an even product.

REASONING How do you know that 2×5 plus 2×5 is the same as finding 4×5?

Check

1. **Explain** how you can use the multiplication table to find 4×8.

Find the product.

2. $2 \times 4 = ■$
3. $9 \times 4 = ■$
4. $■ = 4 \times 5$
5. $4 \times 3 = ■$

Practice and Problem Solving

Find the product.

6. $4 \times 2 = ■$
7. $9 \times 0 = ■$
8. $4 \times 5 = ■$
9. $■ = 5 \times 8$
10. $4 \times 4 = ■$
11. $1 \times 9 = ■$
12. $■ = 4 \times 7$
13. $3 \times 0 = ■$
14. $■ = 8 \times 4$
15. 2×3
16. 5×1
17. 4×6
18. 0×4
19. 7×4
20. 5×5
21. 3×6
22. 4×9
23. 8×2
24. 7×5
25. 4×8
26. 9×4

Copy and complete.

27.

×	0	1	2	3	4	5	6	7	8	9
4	■	■	■	■	■	■	■	■	■	■

28. $9 \times 0 = ■ \times 4$
29. $2 \times 9 = ■ \times 3$
30. $4 \times 4 = ■ \times 2$
31. Multiply 5 by 4.
32. Find the product of 4 and 9.
33. Find the product of 7 and 0.
34. What is 1 times 12?
35. Each ride at the amusement park costs 4 tickets. If Tonya went on 7 different rides, how many tickets did she use?
36. Ahmed has 3 packs of 8 baseball cards and 11 extra cards. How many cards does he have in all?
37. Is the product 6×4 greater than, less than, or equal to 3×8? Explain.
38. Since $9 \times 4 = 36$ and $10 \times 4 = 40$, what is 11×4? How do you know?

Mixed Review and Test Prep

39. Ida has one \$1 bill, 4 quarters, 2 dimes, and 1 nickel. How much money does she have? (p. 84)
40. \$4.89 + \$7.77 (p. 88)
41. \$9.42 − \$5.54 (p. 88)
42. Round 547 to the nearest ten. (p. 28)
43. **TEST PREP** Which shows the numbers in order from least to greatest? (p. 24)
   - **A** 468, 486, 648
   - **B** 468, 648, 486
   - **C** 486, 468, 648
   - **D** 648, 486, 468

EXTRA PRACTICE page H39, Set B

LESSON 3

Problem Solving Strategy
Find a Pattern

THE PROBLEM Emily is playing a number pattern game. She says the numbers 3, 5, 8, 10, 13, 15, 18, 20, and 23. What is the rule for her pattern? What are the next four numbers she will say?

Quick Review

1. $3 \times \blacksquare = 6$
2. $3 \times \blacksquare = 9$
3. $\blacksquare \times 4 = 12$
4. $3 \times \blacksquare = 15$
5. $3 \times \blacksquare = 18$

UNDERSTAND

- What are you asked to find?
- What information will you use?
- Is there information you will not use? If so, what?

PLAN

- What strategy can you use to solve the problem?

  You can *find a pattern.*

SOLVE

- How can you use the strategy to solve the problem?

  Use a number line to find the pattern. Then write the rule and the next four numbers.

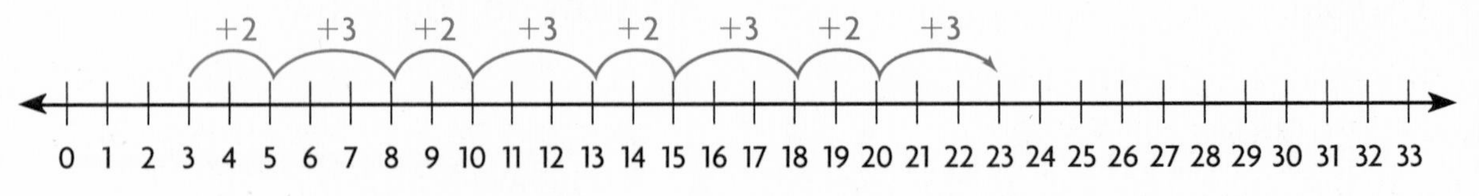

So, the rule is to add 2 and then add 3. The next four numbers in Emily's pattern will be 25, 28, 30, and 33.

CHECK

- How do you know if your answer is correct?

Problem Solving Practice

Use *find a pattern* to solve.

1. **What if** Emily's pattern is 5, 4, 9, 8, 13, 12, 17, 16, and 21? What are the rule and the next four numbers in her pattern?
2. Albert's pattern is 3, 6, 9, and 12. What are the rule and the next four numbers?

PROBLEM SOLVING STRATEGIES

- Draw a Diagram or Picture
- Make a Model or Act It Out
- Make an Organized List
- **Find a Pattern**
- Make a Table or Graph
- Predict and Test
- Work Backward
- Solve a Simpler Problem
- Write a Number Sentence
- Use Logical Reasoning

Karen is thinking of a number pattern. The first four numbers are 4, 8, 12, and 16.

3. What are the next three numbers in Karen's pattern?

   A 16, 20, 24 **C** 18, 20, 22
   B 17, 19, 21 **D** 20, 24, 28

4. Which number cannot be in Karen's pattern?

   F 20 **H** 32
   G 28 **J** 35

Mixed Strategy Practice

5. Bo bicycled 4 miles a day last week. He did not bicycle on Saturday or Sunday. How far did he bicycle last week?
6. **What's the Error?** Look at this pattern. Describe the error and tell how to correct it.
   11, 21, 31, 41, 51, 60, 71
7. **REASONING** Write the greatest possible 4-digit number using 4 different digits. Write the smallest possible 4-digit number using 4 different digits.
8. If this is the time now, what time will it be in 2 hours 35 minutes?

CARL'S JUMPING JACKS

Monday	
Tuesday	
Wednesday	

Key: Each = 3 jumping jacks.

9. **USE DATA** If Carl continues the pattern above, how many jumping jacks will he do on Saturday?
10. **Write a problem** about a number pattern. Tell how you would explain to a second grader how to find the next number in your pattern.

LESSON 4

Practice Multiplication

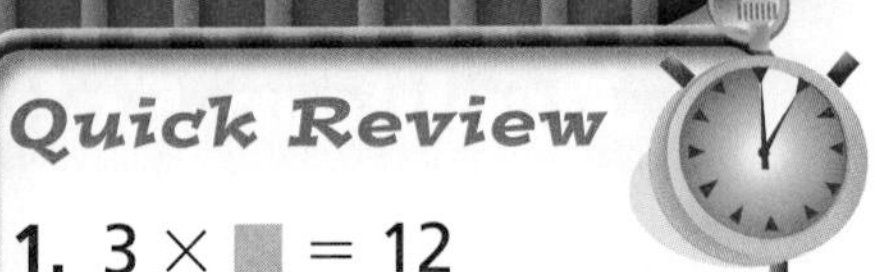

Quick Review

1. $3 \times ■ = 12$
2. $2 \times 6 = ■$
3. $0 \times 8 = ■$
4. $4 \times 5 = ■$
5. $9 \times ■ = 9$

Learn

FACTS IN FLIGHT At the airport, Nicole saw 6 jets waiting to take off. Each jet had 3 engines. How many engines were there in all?

$$6 \times 3 = ■$$

There are many ways to find a product.

A. You can make equal groups or arrays.

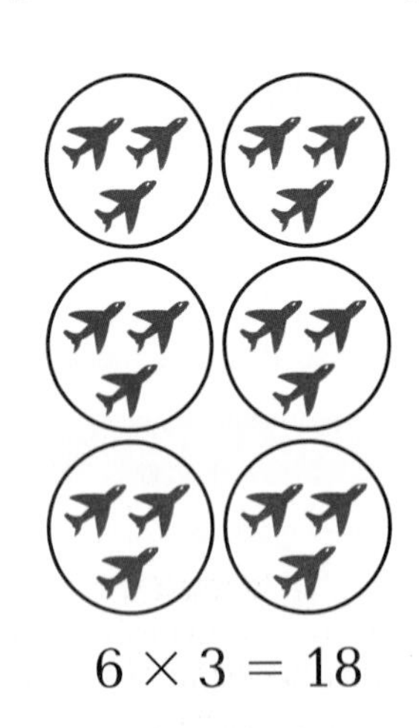

$6 \times 3 = 18$

$6 \times 3 = 18$

B. You can skip-count on a number line.

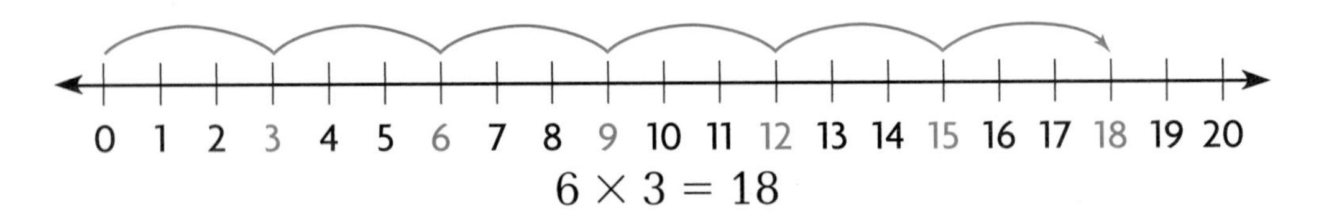

$6 \times 3 = 18$

C. You can double a fact that you already know.

Think: $3 \times 3 = 9$ and $9 + 9 = 18$, so,
$6 \times 3 = 18$.

D. You can use the Order Property of Multiplication.

Think: $3 \times 6 = 6 \times 3 = 18$.

Ways to Find a Product

E. You can use a multiplication table.

column ↓ (column 3); row → (row 6)

×	0	1	2	3	4	5	6	7	8	9
0	0	0	0	0	0	0	0	0	0	0
1	0	1	2	3	4	5	6	7	8	9
2	0	2	4	6	8	10	12	14	16	18
3	0	3	6	9	12	15	18	21	24	27
4	0	4	8	12	16	20	24	28	32	36
5	0	5	10	15	20	25	30	35	40	45
row→ 6	0	6	12	18	24	30	36	42	48	54
7	0	7	14	21	28	35	42	49	56	63
8	0	8	16	24	32	40	48	56	64	72
9	0	9	18	27	36	45	54	63	72	81

Think: The product is found where row 6 and column 3 meet.

$$6 \times 3 = 18$$

So, there are 18 engines in all.

MATH IDEA You can use arrays, skip-counting, doubles, the Order Property of Multiplication, or a multiplication table to help you find products.

Technology Link

More Practice: Use Mighty Math Carnival Countdown, *Snap Clowns,* Level W.

Check

1. **Explain** two ways to find 4×8.

Write a multiplication sentence for each.

2.

3.

4.

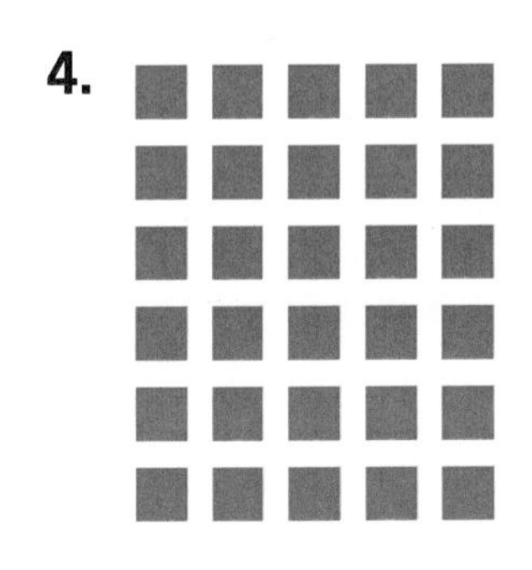

Find the product.

5. $2 \times 6 = ■$
6. $5 \times 3 = ■$
7. $■ = 1 \times 7$
8. $■ = 7 \times 3$

LESSON CONTINUES

Practice and Problem Solving

Find the product.

9. $7 \times 4 =$ ■
10. $2 \times 8 =$ ■
11. $7 \times 5 =$ ■
12. $4 \times 2 =$ ■
13. ■ $= 5 \times 4$
14. $9 \times 4 =$ ■
15. ■ $= 2 \times 5$
16. $9 \times 1 =$ ■
17. ■ $= 1 \times 9$
18. $6 \times 2 =$ ■
19. $5 \times 6 =$ ■
20. ■ $= 8 \times 4$
21. $5 \times 1 =$ ■
22. $3 \times 0 =$ ■
23. $2 \times 9 =$ ■
24. $6 \times 5 =$ ■
25. $2 \times 4 =$ ■
26. ■ $= 3 \times 7$
27. $8 \times 5 =$ ■
28. ■ $= 9 \times 3$

29. $\begin{array}{r} 9 \\ \times 3 \\ \hline \end{array}$
30. $\begin{array}{r} 0 \\ \times 6 \\ \hline \end{array}$
31. $\begin{array}{r} 5 \\ \times 7 \\ \hline \end{array}$
32. $\begin{array}{r} 8 \\ \times 3 \\ \hline \end{array}$
33. $\begin{array}{r} 4 \\ \times 9 \\ \hline \end{array}$
34. $\begin{array}{r} 4 \\ \times 4 \\ \hline \end{array}$

Copy and complete.

35.

×	2	4	7	8	9
2	■	■	■	■	■

36.

×	3	5	7	8	9
3	■	■	■	■	■

37.

×	2	7	5	3	8
4	■	■	■	■	■

38.

×	1	6	9	7	8
5	■	■	■	■	■

Compare. Write $<$, $>$, or $=$ for each ●.

39. 3×2 ● 4×1
40. 7×4 ● 4×8
41. 5×8 ● $35 + 6$
42. 4×6 ● 8×3
43. 3×6 ● 5×4
44. 7×5 ● 8×3

45. **REASONING** Jenny baked some cookies. She put 4 chocolate chips and 2 pecans on each cookie. If she used 24 chocolate chips in all, how many pecans did she use?

46. Pedro and Jon have 20 toy cars altogether. If Jon buys another toy car, he will have twice as many toy cars as Pedro. How many toy cars does Pedro have?

47. **What's the Error?** To find the product 5×6, Ellen made this array. What did Ellen do wrong?

48. **REASONING** Look at this number pattern. What is the rule? What are the missing numbers?

8, 11, 14, ■, ■, ■, 26

49. Three vans are going to the airport. There are 9 people in each van. How many people are going to the airport?

50. Marie has 4 loose stamps and 5 sheets of 8 stamps each. How many stamps does she have in all?

Mixed Review and Test Prep

51. How many minutes are between 12:25 P.M. and 12:30 P.M.? (p. 96)

52. $\begin{array}{r} 3{,}509 \\ +4{,}737 \\ \hline \end{array}$ (p. 46)

53. $\begin{array}{r} 7{,}324 \\ -2{,}195 \\ \hline \end{array}$ (p. 62)

54. What time will it be in 1 hour 45 minutes? (p. 98)

55. $17 + 88 + 71 = \blacksquare$ (p. 36)

56. $\blacksquare + 14 = 23$

57. $51 - \blacksquare = 39$

58. $872 + 208 = \blacksquare$ (p. 42)

59. TEST PREP Which shows the numbers in order from greatest to least? (p. 24)

A 998, 989, 999
B 989, 998, 999
C 999, 989, 998
D 999, 998, 989

60. TEST PREP Leo bought a comic book for \$2.59. He paid with a \$5 bill. How much change did he get back? (p. 86)

F \$2.41 **H** \$3.41
G \$2.51 **J** \$3.51

PROBLEM SOLVING

Thinker's Corner

USE DATA For 1–4, use the pictograph.

DAILY FLIGHTS FROM PORTLAND TO NEW YORK CITY

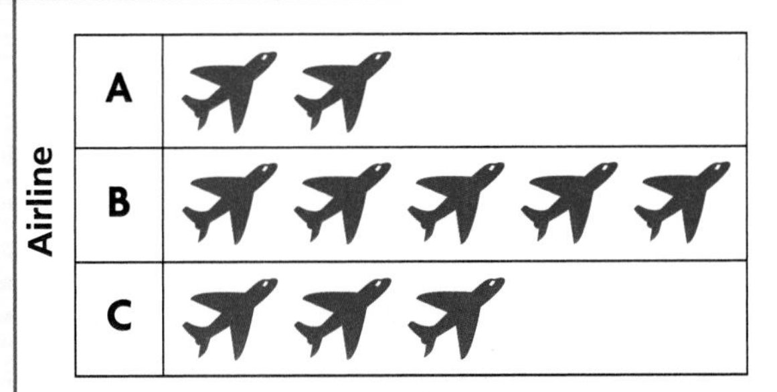

1. How many flights does Airline A have to New York City each day? Airline B? Airline C?

2. Sara's mother is flying from Portland to New York City for business. How many flights does she have to choose from?

3. During a week, how many more flights does Airline C have than Airline A?

4. How many more flights does Airline B have to New York City each day than Airline A? than Airline C?

LESSON 5

Algebra: Find Missing Factors

Quick Review

1. $4 \times 6 = ■$
2. $2 \times 7 = ■$
3. $■ = 3 \times 0$
4. $6 \times 1 = ■$
5. $3 \times 6 = ■$

Learn

BLUE RIBBON BAKING Mike's muffins won first prize at the county fair. Each plate held 5 muffins. He made 35 muffins. How many plates did he use?

A multiplication table can help you find the missing factor.

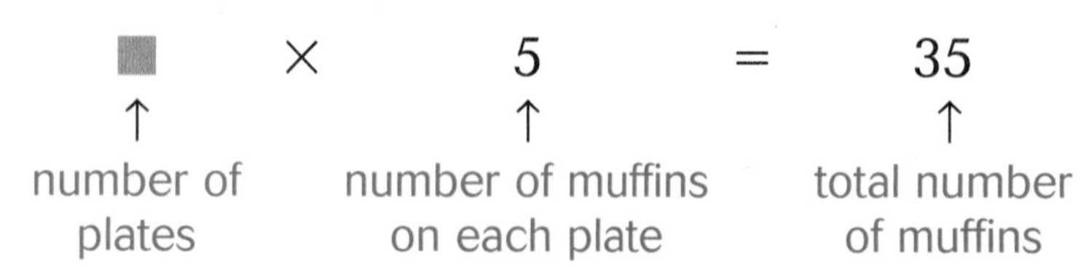

$$■ \times 5 = 35$$

- ■ ↑ number of plates
- 5 ↑ number of muffins on each plate
- 35 ↑ total number of muffins

Start at the column for 5.
Look down to the product, 35.
Look left across the row from 35.
The missing factor is 7.

$$7 \times 5 = 35$$

- 7 ↑ factor *row*
- 5 ↑ factor *column*
- 35 ↑ product *row 7 column 5*

So, Mike used 7 plates.

column ↓ (5); row → (7)

×	0	1	2	3	4	5	6	7	8	9
0	0	0	0	0	0	0	0	0	0	0
1	0	1	2	3	4	5	6	7	8	9
2	0	2	4	6	8	10	12	14	16	18
3	0	3	6	9	12	15	18	21	24	27
4	0	4	8	12	16	20	24	28	32	36
5	0	5	10	15	20	25	30	35	40	45
6	0	6	12	18	24	30	36	42	48	54
7	0	7	14	21	28	35	42	49	56	63
8	0	8	16	24	32	40	48	56	64	72
9	0	9	18	27	36	45	54	63	72	81

Example

Find the missing factor.
$4 \times ■ = 12$
Start at the row for 4.
Look right, to the product, 12.
Look up the column from 12.
The missing factor is 3.

So, $4 \times 3 = 12$.

column ↓ (3); row → (4)

×	0	1	2	3	4	5	6	7	8	9
0	0	0	0	0	0	0	0	0	0	0
1	0	1	2	3	4	5	6	7	8	9
2	0	2	4	6	8	10	12	14	16	18
3	0	3	6	9	12	15	18	21	24	27
4	0	4	8	12	16	20	24	28	32	36
5	0	5	10	15	20	25	30	35	40	45
6	0	6	12	18	24	30	36	42	48	54
7	0	7	14	21	28	35	42	49	56	63
8	0	8	16	24	32	40	48	56	64	72
9	0	9	18	27	36	45	54	63	72	81

MATH IDEA When you know the product and one factor, a multiplication table can help you find the missing factor.

Check

1. **Explain** how to use the table to find the missing factor in $\blacksquare \times 6 = 18$.

Find the missing factor.

2. $\blacksquare \times 2 = 8$
3. $3 \times \blacksquare = 9$
4. $5 \times \blacksquare = 20$

Practice and Problem Solving

Find the missing factor.

5. $\blacksquare \times 4 = 12$
6. $\blacksquare \times 3 = 21$
7. $5 \times \blacksquare = 0$
8. $2 \times \blacksquare = 12$
9. $1 \times \blacksquare = 9$
10. $8 \times \blacksquare = 24$
11. $\blacksquare \times 6 = 30$
12. $\blacksquare \times 4 = 32$
13. $\blacksquare \times 2 = 18$
14. $4 \times \blacksquare = 16$
15. $5 \times \blacksquare = 15$
16. $\blacksquare \times 6 = 24$
17. $\blacksquare \times 4 = 36$
18. $\blacksquare \times 3 = 27$
19. $6 \times \blacksquare = 18$
20. $8 \times \blacksquare = 40$
21. $4 \times 6 = \blacksquare \times 3$
22. $9 \times \blacksquare = 50 - 5$
23. $7 \times \blacksquare = 32 - 4$

24. The product of 4 and another factor is 28. What is the other factor?

25. If you multiply 9 by a number, the product is 27. What is the number?

26. There are 2 chairs at each table. If there are 14 chairs, how many tables are there? Write a multiplication sentence to solve.

27. There are 4 oatmeal cookies and 3 sugar cookies on each plate. How many cookies are on 5 plates?

28. Isabel has $1.29 and Luke has $5.95. How much more does Luke have than Isabel?

29. **? What's the Question?** Pies are on sale for $3 each. Carly spent $12 on pies. The answer is 4 pies.

Mixed Review and Test Prep

30. Tony has two $1 bills and 1 dime. Benita has one $1 bill and 4 quarters. Who has the greater amount of money? (p. 84)

31. $400 - 137$ (p. 58)

32. $453 + 487$ (p. 42)

33. Write <, >, or =. (p. 20)
5,450 ● 5,405

34. **TEST PREP** Which number means $10,000 + 1,000 + 10 + 1$? (p. 10)
 - A 10,111
 - B 11,011
 - C 11,101
 - D 11,110

EXTRA PRACTICE page H39, Set D

Review/Test

CHECK VOCABULARY

Choose the best term from the box.

array
zero
product
one
factor

1. The product of ? and any number equals that number. (p. 132)
2. The product of ? and any number equals zero. (p. 132)
3. On a multiplication table, the ? is found where the factor row and the factor column meet. (p. 134)

CHECK SKILLS

Find the product. (pp. 132-133)

4. $5 \times 1 = \blacksquare$
5. $0 \times 6 = \blacksquare$
6. $\blacksquare = 1 \times 8$
7. $\blacksquare = 9 \times 0$

Find the product. (pp. 134-135)

8. $\blacksquare = 4 \times 3$
9. $2 \times 4 = \blacksquare$
10. $5 \times 4 = \blacksquare$
11. $\blacksquare = 8 \times 4$
12. $4 \times 9 = \blacksquare$
13. $4 \times 4 = \blacksquare$
14. $\blacksquare = 6 \times 4$
15. $4 \times 0 = \blacksquare$

Find the missing factor. (pp. 142-143)

16. $2 \times \blacksquare = 8$
17. $\blacksquare \times 5 = 30$
18. $1 \times \blacksquare = 9$
19. $\blacksquare \times 6 = 18$
20. $\blacksquare \times 1 = 9$
21. $4 \times \blacksquare = 32$
22. $\blacksquare \times 3 = 18$
23. $7 \times \blacksquare = 0$

CHECK PROBLEM SOLVING

Solve. (pp. 136-137)

24. The first four numbers in the pattern are 4, 8, 12, 16. What is the rule? What are the next three numbers?

25. Lin saw this number pattern: 10, 13, 16, 19, 22, 25, and 28. What is the rule? What are the next three numbers?

Standardized Test Prep

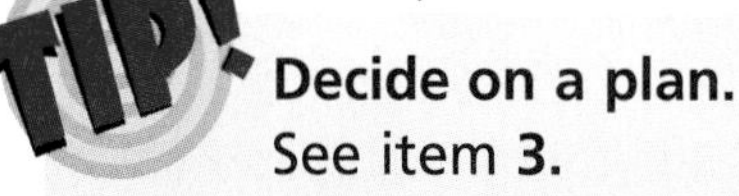

Decide on a plan.
See item **3.**
The problem tells you to multiply a number by 4. You should start by writing a number sentence from the words. Then solve the number sentence.

Also see problem **4,** p. H63.

For 1–8, choose the best answer.

1. Maria had 8 plates. She put 1 cookie on each plate. How many cookies did she use?

A 0 **B** 4 **C** 6 **D** 8

2. A number multiplied by 4 is 36. What is the number?

F 40 **G** 32 **H** 9 **J** 8

3. Matt has 7 folders for his school work. Each folder has 4 assignments in it. How many assignments are there altogether?

A 28 **B** 24 **C** 21 **D** 11

4. Justin put 5 jelly beans in each of 8 cups. How many jelly beans are in the cups altogether?

F 13 **G** 32 **H** 33 **J** 40

5. These numbers follow a pattern.

5, 9, 13, 17

Which numbers continue the pattern?

A 18, 19, 20
B 21, 25, 29
C 19, 21, 23
D 20, 23, 26

6. Bradley has 4 each of 3 kinds of marbles. How many marbles does he have in all?

F 7
G 12
H 15
J NOT HERE

7. Which number makes this equation true?

$$■ \times 1 = 5$$

A 6
B 5
C 4
D 2

8. Pete is setting the table for 9 people. Each person will get 1 fork, 1 knife, and 1 spoon. How many forks, knives and spoons will Pete put on the table in all?

F 45
G 42
H 27
J 14

Write What You Know

9. Find the products and explain how you found them. Which product is greater?

$8 \times 1 = ■$
$9 \times 0 = ■$

10. Tonya wrote a pattern. She started with the number 1 and used a rule. Her rule uses multiplication. What is her rule? Explain how the rule works.

1, 2, 4, 8, 16

Multiplication Facts and Strategies

A marching band plays music and moves in formation. You may see bands march in parades and at sporting events, such as football games.

PROBLEM SOLVING Look at the pictograph. Draw an array that each band could use as a formation for its brass players.

DATA LINK

MARCHING BANDS

High School	Brass Players
Loch Raven	🎺 🎺 🎺
Patapsco	🎺 🎺 🎺 🎺 🎺 🎺 🎺 🎺
Severna Park	🎺 🎺 🎺 🎺 🎺 🎺 🎺

Key: Each 🎺 = 4 Brass Players.

Marching band,
Severna Park High School
Baltimore, Maryland

CHECK WHAT YOU KNOW

Use this page to help you review and remember important skills needed for Chapter 9.

VOCABULARY

Choose the best term from the box to describe the example $2 \times 5 = 10$.

difference
factor
product

1. In the example, the number 2 is called a __?__ .
2. In the example, the number 10 is called the __?__ .

EQUAL GROUPS (For Intervention, see p. H12.)

Write how many there are in all.

3. 

   2 groups of 4 = ■

4. 

   3 groups of 6 = ■

5. 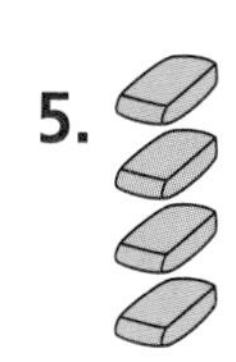 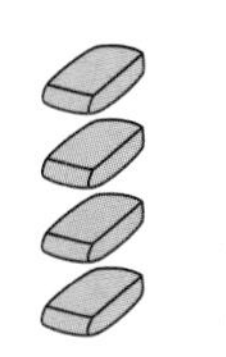

   3 groups of 4 = ■

6.  

   5 groups of 2 = ■

7. 

   1 group of 5 = ■

8.     

   4 groups of 2 = ■

MULTIPLICATION FACTS THROUGH 5 (For Intervention, see p. H12.)

Find the product.

9. $7 \times 3 =$ ■
10. $5 \times 5 =$ ■
11. ■ $= 7 \times 4$
12. $1 \times 2 =$ ■
13. $4 \times 1 =$ ■
14. $6 \times 3 =$ ■
15. $9 \times 5 =$ ■
16. ■ $= 6 \times 1$
17. $5 \times 3 =$ ■
18. ■ $= 9 \times 4$
19. $5 \times 0 =$ ■
20. $7 \times 5 =$ ■
21. $\begin{array}{r} 6 \\ \times 4 \\ \hline \end{array}$
22. $\begin{array}{r} 4 \\ \times 2 \\ \hline \end{array}$
23. $\begin{array}{r} 2 \\ \times 3 \\ \hline \end{array}$
24. $\begin{array}{r} 8 \\ \times 4 \\ \hline \end{array}$
25. $\begin{array}{r} 3 \\ \times 1 \\ \hline \end{array}$
26. $\begin{array}{r} 5 \\ \times 2 \\ \hline \end{array}$
27. $\begin{array}{r} 1 \\ \times 7 \\ \hline \end{array}$
28. $\begin{array}{r} 4 \\ \times 5 \\ \hline \end{array}$
29. $\begin{array}{r} 0 \\ \times 3 \\ \hline \end{array}$
30. $\begin{array}{r} 2 \\ \times 8 \\ \hline \end{array}$

LESSON 1

Multiply with 6

Quick Review

1. $5 \times 3 = ■$
2. $5 \times ■ = 25$
3. $5 \times 9 = ■$
4. $■ \times 5 = 30$
5. $40 = 5 \times ■$

Learn

MARCHING MULTIPLES The school band has 6 rows, with 6 students in each row. How many students are in the band?

Example

Find $6 \times 6 = ■$.

One Way Break apart an array to find the product.

STEP 1

Make an array that shows 6 rows of 6.

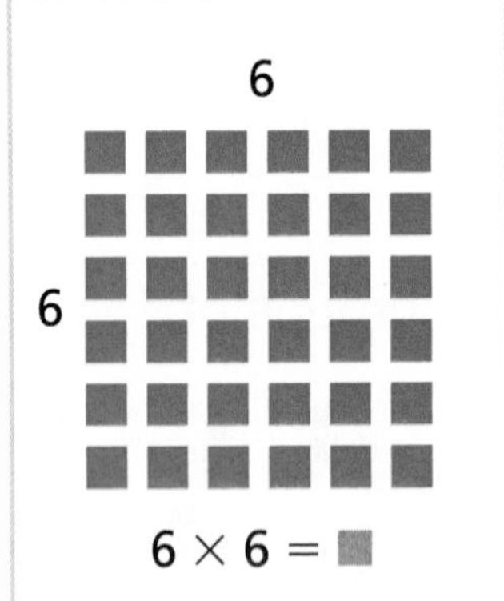

$6 \times 6 = ■$

STEP 2

Break the array into two smaller arrays.

$1 \times 6 = \mathbf{6}$

$5 \times 6 = \mathbf{30}$

STEP 3

Add the products of the two arrays.

$$\begin{array}{r} 6 \\ +30 \\ \hline 36 \end{array}$$

$6 \times 6 = 36$

So, there are 36 students in the band.

Another Way When one of the factors is an even number, you can use doubles. The product of each 6's fact is double the product of each 3's fact.

To find 6×6

- First find the 3's fact. **Think:** $6 \times 3 = 18$
- Double the product. $18 + 18 = 36$
- So, $6 \times 6 = 36$.

3's facts	6's facts
$0 \times 3 = 0$	$0 \times 6 = 0$
$1 \times 3 = 3$	$1 \times 6 = 6$
$2 \times 3 = 6$	$2 \times 6 = 12$
$3 \times 3 = 9$	$3 \times 6 = 18$
$4 \times 3 = 12$	$4 \times 6 = 24$
$5 \times 3 = 15$	$5 \times 6 = 30$
$6 \times 3 = 18$	$6 \times 6 = ■$
$7 \times 3 = 21$	$7 \times 6 = 42$
$8 \times 3 = 24$	$8 \times 6 = 48$
$9 \times 3 = 27$	$9 \times 6 = 54$

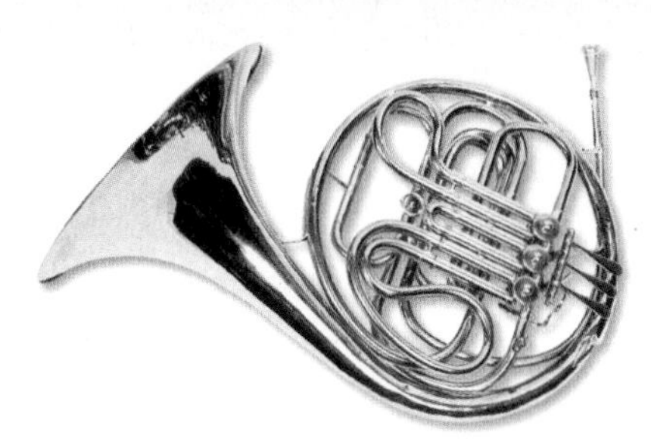

Check

1. **Explain** how you can use 8×3 to find 8×6.

Find each product.

2. $2 \times 6 = \blacksquare$ 3. $4 \times 6 = \blacksquare$ 4. $5 \times 6 = \blacksquare$ 5. $6 \times 9 = \blacksquare$

Practice and Problem Solving

Find each product.

6. $3 \times 6 = \blacksquare$ 7. $6 \times 5 = \blacksquare$ 8. $5 \times 9 = \blacksquare$ 9. $\blacksquare = 8 \times 6$

10. $4 \times 7 = \blacksquare$ 11. $\blacksquare = 3 \times 4$ 12. $4 \times 9 = \blacksquare$ 13. $6 \times 0 = \blacksquare$

14. $\blacksquare = 2 \times 9$ 15. $\blacksquare = 8 \times 4$ 16. $3 \times 5 = \blacksquare$ 17. $9 \times 6 = \blacksquare$

18. $\begin{array}{r} 5 \\ \times 7 \\ \hline \end{array}$ 19. $\begin{array}{r} 6 \\ \times 7 \\ \hline \end{array}$ 20. $\begin{array}{r} 8 \\ \times 3 \\ \hline \end{array}$ 21. $\begin{array}{r} 6 \\ \times 1 \\ \hline \end{array}$ 22. $\begin{array}{r} 5 \\ \times 8 \\ \hline \end{array}$ 23. $\begin{array}{r} 6 \\ \times 6 \\ \hline \end{array}$

Copy and complete the multiplication table.

24.

×	0	1	2	3	4	5	6	7	8	9
6	■	■	■	■	■	■	■	■	■	■

Complete.

25. $\blacksquare \times 4 = 12$ 26. $\blacksquare \times 6 = 42$ 27. $48 = 8 \times \blacksquare$

28. $\blacksquare \times 4 = 4 \times 3$ 29. $3 \times 6 = \blacksquare \times 2$ 30. $\blacksquare \times 6 = 40 + 8$

31. A guitar has 6 strings. A banjo has 5 strings. How many strings are on 4 guitars and 2 banjos?

32. **Write About It** Draw arrays to show that 6×4 is the same as 1×4 plus 5×4.

Mixed Review and Test Prep

33. $\begin{array}{r} 327 \\ -\ 82 \\ \hline \end{array}$ (p. 58) 34. $\begin{array}{r} 600 \\ -346 \\ \hline \end{array}$ (p. 58)

Write the value of the blue digit. (p. 10)

35. 57,899 36. 98,365

37. **TEST PREP** Find the sum of 452 and 678. (p. 42)

A 226 C 1,120

B 1,030 D 1,130

EXTRA PRACTICE page H40, Set A

LESSON 2

Multiply with 7

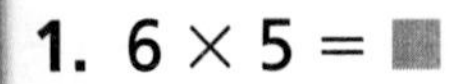

Quick Review

1. $6 \times 5 = ■$
2. $6 \times 4 = ■$
3. $3 \times 6 = ■$
4. $2 \times 6 = ■$
5. $5 \times 6 = ■$

Learn

PARADE! PARADE! Students from the local high school built a float for a parade. They worked on the float for 8 weeks. How many days did they work on the float?

Example

Find $8 \times 7 = ■$.
Break apart an array to find the product.

STEP 1

Make an array that shows 8 rows of 7.

7

8

$8 \times 7 = ■$

STEP 2

Break the array into two smaller arrays.

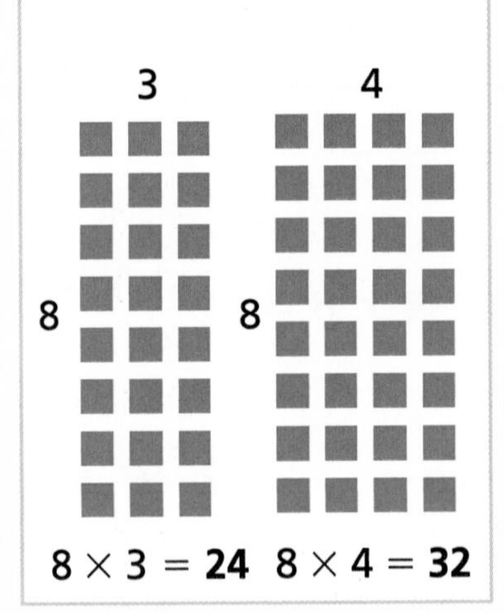

3 4

8 8

$8 \times 3 = \mathbf{24}$ $8 \times 4 = \mathbf{32}$

STEP 3

Add the products of the two arrays.

$$\begin{array}{r} 24 \\ +32 \\ \hline 56 \end{array}$$

$8 \times 7 = 56$

So, the students worked 56 days on the float.

Check

1. **Tell** how you could break apart 7×5 into two facts to help you find the product.

Technology Link

More Practice: Use E-Lab, *Multiplication Arrays.*

www.harcourtschool.com/elab2002

Find each product.

2. $7 \times 4 = ■$
3. $0 \times 7 = ■$
4. $7 \times 5 = ■$
5. $7 \times 6 = ■$

Practice and Problem Solving

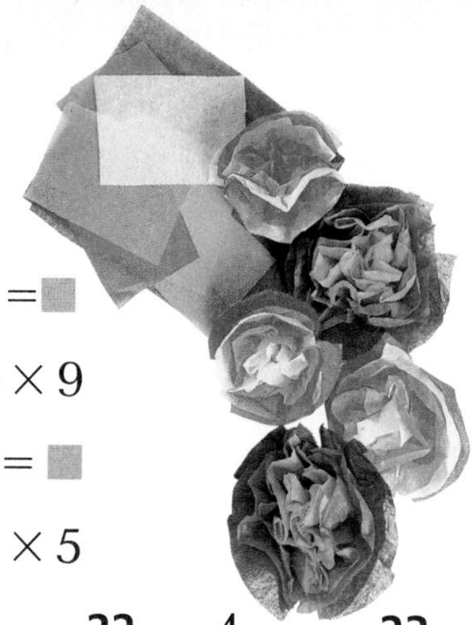

Find each product.

6. $2 \times 7 =$ ■
7. $2 \times 9 =$ ■
8. $3 \times 7 =$ ■
9. $6 \times 6 =$ ■
10. $7 \times 6 =$ ■
11. ■ $= 6 \times 9$
12. ■ $= 5 \times 9$
13. $7 \times 7 =$ ■
14. $4 \times 7 =$ ■
15. $3 \times 6 =$ ■
16. ■ $= 4 \times 9$
17. ■ $= 5 \times 5$
18. $\begin{array}{r} 1 \\ \times 7 \\ \hline \end{array}$
19. $\begin{array}{r} 7 \\ \times 9 \\ \hline \end{array}$
20. $\begin{array}{r} 6 \\ \times 8 \\ \hline \end{array}$
21. $\begin{array}{r} 5 \\ \times 8 \\ \hline \end{array}$
22. $\begin{array}{r} 4 \\ \times 5 \\ \hline \end{array}$
23. $\begin{array}{r} 5 \\ \times 7 \\ \hline \end{array}$

Copy and complete the multiplication table.

24.

×	0	1	2	3	4	5	6	7	8	9
7	■	■	■	■	■	■	■	■	■	■

Complete.

25. $7 \times 6 =$ ■ $+ 21$
26. ■ $\times 4 = 30 - 2$
27. $8 + 6 = 7 \times$ ■

28. **REASONING** How can you tell without multiplying that 7×9 is less than 9×8?

29. **REASONING** Explain how you can use $6 \times 2 = 12$ to find 7×2.

30. Shayla was on vacation for 7 weeks. She spent 3 weeks at band camp and the rest of the time at home. How many days did she spend at home?

31. Break apart the array. Then write the multiplication fact.

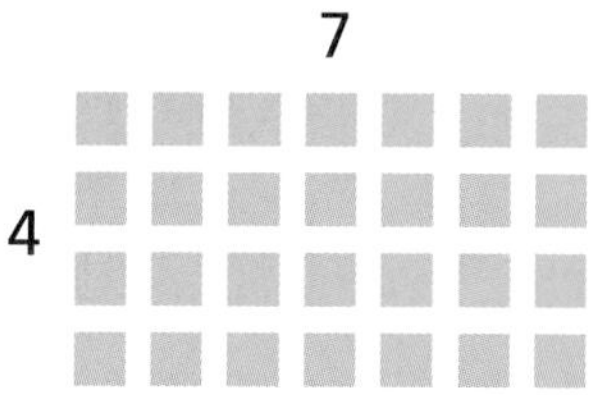

32. ALGEBRA Write a one-digit number to make this number sentence true.
■ $\times 7 + 10 > 67 - 9$

Mixed Review and Test Prep

33. $\begin{array}{r} 3{,}458 \\ +1{,}679 \\ \hline \end{array}$ (p. 46)
34. $\begin{array}{r} 2{,}008 \\ +1{,}256 \\ \hline \end{array}$ (p. 46)
35. $\begin{array}{r} 2{,}814 \\ -1{,}680 \\ \hline \end{array}$ (p. 62)
36. $\begin{array}{r} 8{,}093 \\ -5{,}934 \\ \hline \end{array}$ (p. 62)

37. **TEST PREP** Subtract 49 from 201. (p. 58)

A 152 **B** 162 **C** 250 **D** 252

EXTRA PRACTICE page H40, Set B

LESSON

3 Multiply with 8

Quick Review

1. $35 = ■ \times 7$
2. $2 \times 7 = ■$
3. $7 \times 3 = ■$
4. $7 \times ■ = 49$
5. $6 \times 7 = ■$

Learn

BAKE-OFF Mr. Lee baked 6 peach pies for the state fair. He used 8 peaches in each pie. How many peaches did he use in all?

Example

Find $6 \times 8 = ■$.

One Way Break apart an array to find the product.

STEP 1 

Make an array that shows 6 rows of 8.

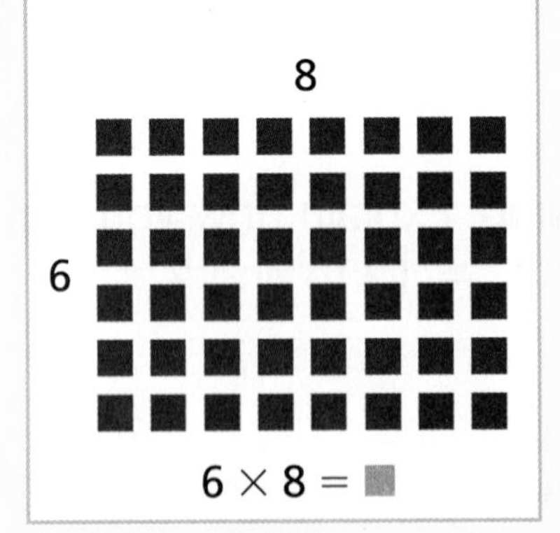

$6 \times 8 = ■$

STEP 2

Break the array into two smaller arrays.

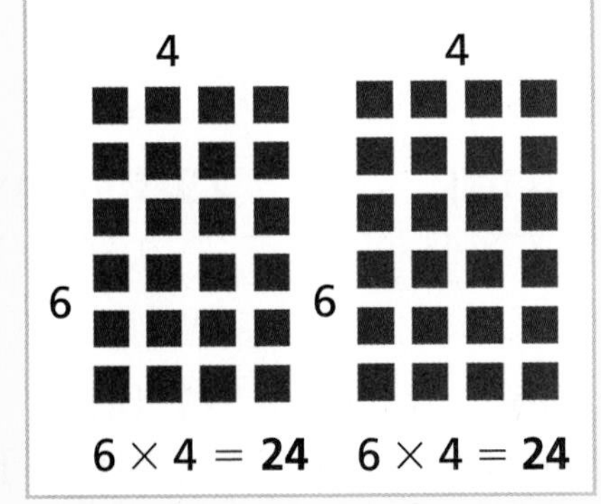

$6 \times 4 = \mathbf{24}$ $\quad 6 \times 4 = \mathbf{24}$

STEP 3

Add the products of the two arrays.

$$\begin{array}{r} 24 \\ +24 \\ \hline 48 \end{array}$$

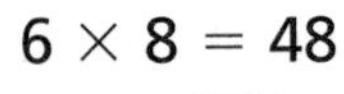

$6 \times 8 = 48$

So, Mr. Lee used 48 peaches in all.

Another Way When one of the factors is an even number, you can use doubles. The product of each 8's fact is double the product of each 4's fact.

To find 6×8

- First find the 4's fact. **Think:** $6 \times 4 = 24$
- Double the product. $24 + 24 = 48$
- So, $6 \times 8 = 48$.

4's facts	8's facts
$0 \times 4 = 0$	$0 \times 8 = 0$
$1 \times 4 = 4$	$1 \times 8 = 8$
$2 \times 4 = 8$	$2 \times 8 = 16$
$3 \times 4 = 12$	$3 \times 8 = 24$
$4 \times 4 = 16$	$4 \times 8 = 32$
$5 \times 4 = 20$	$5 \times 8 = 40$
$6 \times 4 = 24$	$6 \times 8 = ■$
$7 \times 4 = 28$	$7 \times 8 = 56$
$8 \times 4 = 32$	$8 \times 8 = 64$
$9 \times 4 = 36$	$9 \times 8 = 72$

Check

1. **Explain** how you can use $4 \times 5 = 20$ to find 8×5.

Find each product.

2. $8 \times 4 = \blacksquare$
3. $8 \times 2 = \blacksquare$
4. $5 \times 8 = \blacksquare$
5. $1 \times 8 = \blacksquare$

Practice and Problem Solving

Find each product.

6. $5 \times 4 = \blacksquare$
7. $8 \times 3 = \blacksquare$
8. $9 \times 8 = \blacksquare$
9. $4 \times 7 = \blacksquare$
10. $8 \times 6 = \blacksquare$
11. $3 \times 4 = \blacksquare$
12. $5 \times 7 = \blacksquare$
13. $8 \times 8 = \blacksquare$
14. $2 \times 9 = \blacksquare$
15. $7 \times 8 = \blacksquare$
16. $5 \times 9 = \blacksquare$
17. $6 \times 6 = \blacksquare$
18. 4×8
19. 7×7
20. 9×6
21. 7×9
22. 3×7
23. 6×8

Complete the multiplication table.

24.

×	0	1	2	3	4	5	6	7	8	9
8	■	■	■	■	■	■	■	■	■	■

Compare. Write <, >, or = for each ●.

25. 2×3 ● 2×4
26. 5×8 ● 8×5
27. 5×5 ● 4×6

28. **ALGEBRA** Hal has 7 bags of 8 green apples and 1 bag of red apples. He has 60 apples in all. How many red apples does he have?

29. **What's the Error?** Robin says, "I can find 8×7 by thinking of $3 \times 7 = 21$ and doubling it."

30. **REASONING** If you know $9 \times 4 = 36$, how can you find 8×4?

31. **What's the Question?** Joanna has 9 boxes of pears. She has 72 pears in all. The answer is 8 pears.

Mixed Review and Test Prep

32. $172 + 781$ (p. 42)
33. $399 + 421$ (p. 42)
34. $2,523 + 1,607$ (p. 46)
35. $3,627 + 4,482$ (p. 46)

36. **TEST PREP** What is another way to show $2 + 2 + 2 + 2$? (p. 116)
   A 2×2 B 4×2 C 2×8 D 8×4

EXTRA PRACTICE page H40, Set C

LESSON 4

Problem Solving Strategy
Draw a Picture

Quick Review

1. $2 \times 8 = \blacksquare$
2. $8 \times \blacksquare = 24$
3. $64 = 8 \times \blacksquare$
4. $\blacksquare \times 8 = 32$
5. $6 \times 8 = \blacksquare$

PROBLEM Jacob's grandmother has made 16 squares for a quilt. How can she arrange the squares so the quilt is the same number of squares wide as it is long?

UNDERSTAND

- What are you asked to find?
- What information will you use?
- Is there information you will not use?

PLAN

- What strategy can you use to solve the problem?
  Draw a picture to show how to arrange the squares.

SOLVE

- How can you use the strategy to solve the problem?

Try 3 squares wide and 3 squares long.

$3 \times 3 = 9$

This quilt is 9 squares in all. Try again.

Try 4 squares wide and 4 squares long.

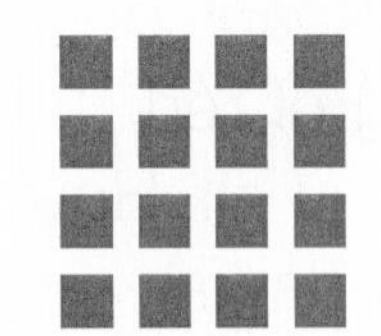

$4 \times 4 = 16$

This quilt is 16 squares in all.

So, her quilt should be 4 squares wide and 4 squares long.

CHECK

- Look at the Problem again. Explain why this answer makes sense.

Problem Solving Practice

Use *draw a picture* to solve.

1. **What if** Jacob's grandmother had a total of 25 squares? How could she arrange the squares so the quilt is the same number of squares long as it is wide?

2. Amrita has completed 32 squares. How many more does she have to make to be able to arrange them so that the quilt is the same number of squares long as it is wide?

PROBLEM SOLVING STRATEGIES

- **Draw a Diagram or Picture**
- Make a Model or Act It Out
- Make an Organized List
- Find a Pattern
- Make a Table or Graph
- Predict and Test
- Work Backward
- Solve a Simpler Problem
- Write a Number Sentence
- Use Logical Reasoning

Mrs. Adams is hanging her students' pictures. She has 24 pictures to arrange.

3. Which arrangement can Mrs. Adams **not** use?

   A 3 rows of 8 **C** 4 rows of 6
   B 6 rows of 4 **D** 4 rows of 8

4. Mrs. Adams decides to put 4 pictures in a row. How many rows does she need?

   F 3 **G** 4 **H** 6 **J** 8

Mixed Strategy Practice

5. **REASONING** I am a one-digit number. If you multiply me by 3 the product has a 1 in the ones place. What number am I?

6. List all the ways you can use coins to make 21¢.

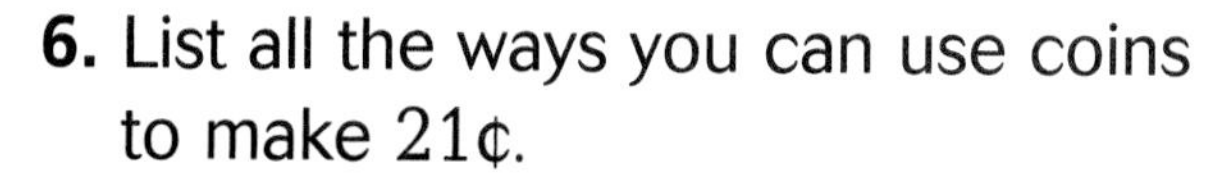

USE DATA For 7–9, use the graph.

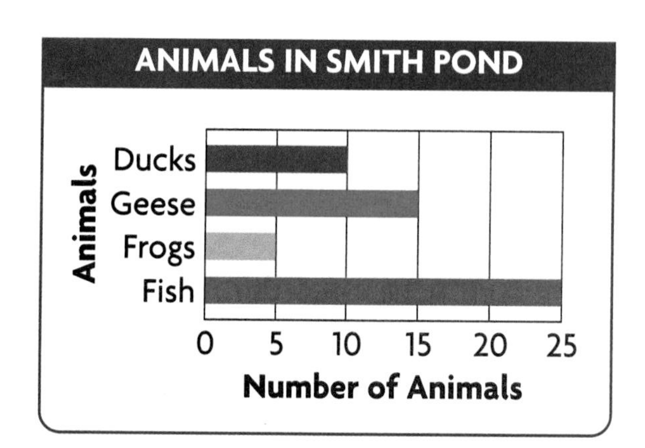

7. If 3 frogs leave the pond, how many animals in all are in the pond?

8. If 5 ducks come to the pond and 2 geese leave the pond, how many ducks and geese are in the pond?

9. If 5 more frogs come to the pond, how should the bar graph change?

LESSON

5 Algebra: Practice the Facts

Quick Review

1. $6 \times 3 = \blacksquare$
2. $5 \times 4 = \blacksquare$
3. $2 \times 8 = \blacksquare$
4. $4 \times 9 = \blacksquare$
5. $7 \times 8 = \blacksquare$

Learn

Splash! Each instructor teaches a group of 6 children. If there are 7 instructors, how many children are taking swimming lessons?

$$7 \times 6 = \blacksquare$$

You have learned many ways to find 7×6.

A. Break an array into known facts.

6

2 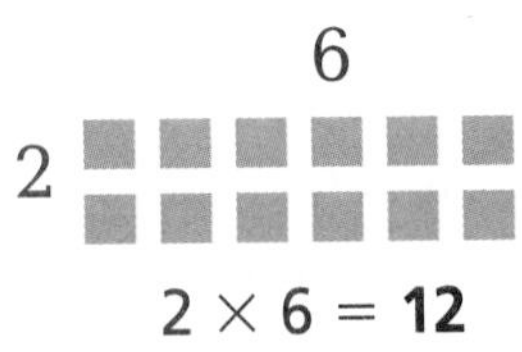

$2 \times 6 = \mathbf{12}$

6

5 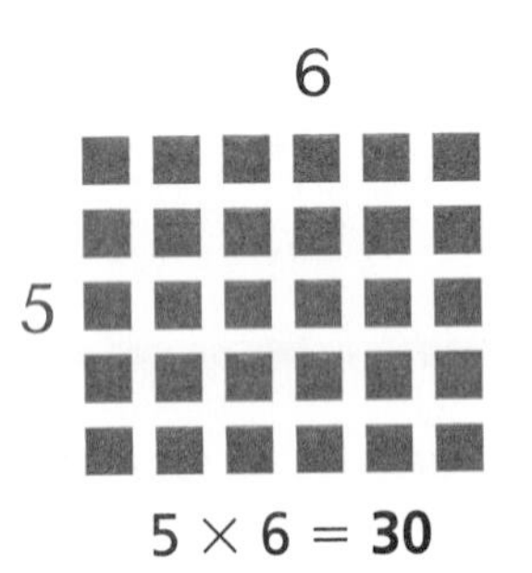

$5 \times 6 = \mathbf{30}$

$12 + 30 = 42$, so $7 \times 6 = 42$.

B. Use a multiplication table.

×	0	1	2	3	4	5	6	7	8	9
0	0	0	0	0	0	0	0	0	0	0
1	0	1	2	3	4	5	6	7	8	9
2	0	2	4	6	8	10	12	14	16	18
3	0	3	6	9	12	15	18	21	24	27
4	0	4	8	12	16	20	24	28	32	36
5	0	5	10	15	20	25	30	35	40	45
6	0	6	12	18	24	30	36	42	48	54
7	0	7	14	21	28	35	42	49	56	63
8	0	8	16	24	32	40	48	56	64	72
9	0	9	18	27	36	45	54	63	72	81

$7 \times 6 = 42$

C. Use the Order Property of Multiplication.

Try changing the order of the factors:
Think: If $6 \times 7 = 42$, then $7 \times 6 = 42$.

D. When one of the factors is an even number, you can use doubles.

To find a 6's fact, you can double a 3's fact.

- First find the 3's fact.
  Think: $7 \times 3 = 21$
- Double the product. $21 + 21 = 42$
  $7 \times 6 = 42$.

So, 42 children are taking lessons.

Ways to Find a Product

What if there are 8 instructors with 5 swimmers each? How many children are taking lessons?

$$8 \times 5 = \blacksquare$$

David and Niam use different ways to find 8×5.

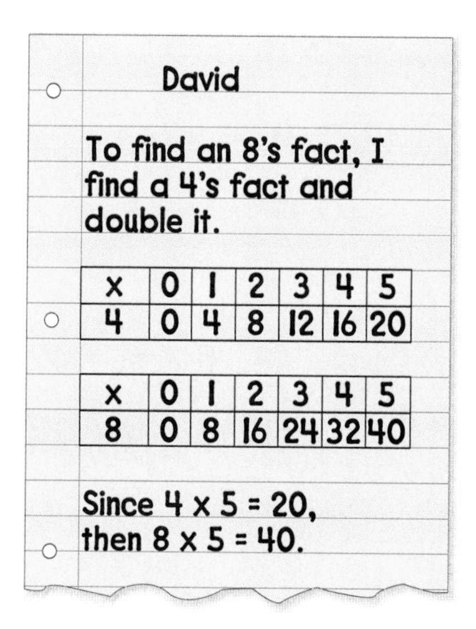
David

To find an 8's fact, I find a 4's fact and double it.

x	0	1	2	3	4	5
4	0	4	8	12	16	20

x	0	1	2	3	4	5
8	0	8	16	24	32	40

Since 4 x 5 = 20, then 8 x 5 = 40.

Niam

I can use the Order Property of Multiplication.

I know that 8 x 5 is the same as 5 x 8.

5 x 8 = 8 x 5 = 40

- What is another way that David or Niam could find 8×5?
- If you can't remember the fact 7×8, what strategy can you use?

Check

1. **Explain** how you could use $9 \times 5 = 45$ to find 8×5.
2. **Describe** how you could use doubles to find 6×9.

Find each product.

3. $4 \times 5 = \blacksquare$
4. $3 \times 7 = \blacksquare$
5. $6 \times 4 = \blacksquare$
6. $7 \times 6 = \blacksquare$
7. $2 \times 5 = \blacksquare$
8. $6 \times 6 = \blacksquare$
9. $\blacksquare = 4 \times 3$
10. $2 \times 2 = \blacksquare$
11. $1 \times 8 = \blacksquare$
12. $\blacksquare = 5 \times 3$
13. $9 \times 2 = \blacksquare$
14. $5 \times 5 = \blacksquare$
15. 5×9
16. 6×3
17. 7×7
18. 4×8
19. 0×4
20. 3×3

Practice and Problem Solving

Find each product.

21. $8 \times 5 = ■$ **22.** $0 \times 6 = ■$ **23.** $9 \times 3 = ■$ **24.** $5 \times 6 = ■$

25. $3 \times 2 = ■$ **26.** $8 \times 3 = ■$ **27.** $■ = 1 \times 8$ **28.** $8 \times 8 = ■$

29. $6 \times 8 = ■$ **30.** $■ = 4 \times 9$ **31.** $■ = 2 \times 8$ **32.** $8 \times 7 = ■$

33. $\begin{array}{r} 2 \\ \times 6 \\ \hline \end{array}$ **34.** $\begin{array}{r} 4 \\ \times 1 \\ \hline \end{array}$ **35.** $\begin{array}{r} 8 \\ \times 9 \\ \hline \end{array}$ **36.** $\begin{array}{r} 5 \\ \times 7 \\ \hline \end{array}$ **37.** $\begin{array}{r} 5 \\ \times 1 \\ \hline \end{array}$ **38.** $\begin{array}{r} 4 \\ \times 4 \\ \hline \end{array}$

39. $\begin{array}{r} 2 \\ \times 9 \\ \hline \end{array}$ **40.** $\begin{array}{r} 0 \\ \times 3 \\ \hline \end{array}$ **41.** $\begin{array}{r} 2 \\ \times 4 \\ \hline \end{array}$ **42.** $\begin{array}{r} 4 \\ \times 7 \\ \hline \end{array}$ **43.** $\begin{array}{r} 9 \\ \times 6 \\ \hline \end{array}$ **44.** $\begin{array}{r} 7 \\ \times 9 \\ \hline \end{array}$

Find each missing factor.

45. $■ \times 4 = 20$ **46.** $8 \times ■ = 56$ **47.** $■ \times 6 = 0$ **48.** $6 \times ■ = 42$

49. $8 \times ■ = 24$ **50.** $■ \times 5 = 40$ **51.** $4 \times ■ = 16$ **52.** $3 \times ■ = 12$

Write $<$, $>$, or $=$ for each ●.

53. 3×2 ● 6 **54.** 4×2 ● $5 + 2$ **55.** 6×3 ● 7×2

56. 4×9 ● 6×6 **57.** 3×4 ● 18 **58.** 5×9 ● 6×8

59. 7×4 ● $30 - 4$ **60.** 3×8 ● 6×4 **61.** 5×7 ● 6×6

62. $8 + 9$ ● 8×9 **63.** 7×7 ● 50 **64.** 9×3 ● $9 + 9$

65. **REASONING** Jenna says, "The product of any number with a factor of 7 is an odd number." Do you agree or disagree? Explain.

66. ALGEBRA Write *true* or *false* for each.

a. $1 \times 8 = 9$ **b.** $0 \times 7 = 0$

c. $0 \times 0 = 0$ **d.** $1 + 9 = 9$

USE DATA For 67–69, use the table.

BAKE SALE
Cake - $3
Brownie - $1
Cupcake - $0.50
Loaf of Bread - $2

67. Olga buys 4 cakes and 1 loaf of bread at the bake sale. How much does she pay?

68. Greg buys 2 cakes and one cupcake. How much change does he get from a $10 bill?

69. **Write a problem** about the bake sale. Use multiplication. Exchange with a partner and solve.

70. Mr. Wu taught 3 lessons each day for 6 days. Then he taught 2 lessons each day for 3 days. How many lessons did he teach?

71. Ed arranged 5 rows of 6 pennies each. He had one more coin in his pocket. If he had a total of $0.35, what coin was in his pocket?

Mixed Review and Test Prep

Choose <, >, or = for each ●. (p. 20)

72. 67 ● 76 **73.** 254 ● 257

74. 1,007 ● 985 **75.** 4,902 ● 4,092

Write + or − to make each sentence true. (p. 68)

76. 39 ● 14 = 25 **77.** 47 ● 7 = 54

Write the missing addend. (p. 68)

78. 350 + ■ = 402

79. ■ + 120 = 135

80. TEST PREP Tameo bought some tape for $1.39 and a sheet of stickers for $1.53. How much did he spend? (p. 88)

A $0.14 C $2.92
B $1.81 D $3.51

PROBLEM SOLVING

Thinker's Corner

ALL SQUARED OFF When both factors are the same, the product is called a *square number*.

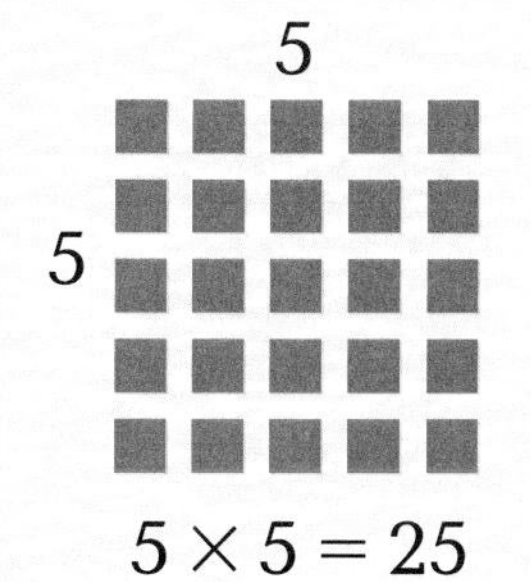

$5 \times 5 = 25$

Notice that the arrays have equal sides. They are square.

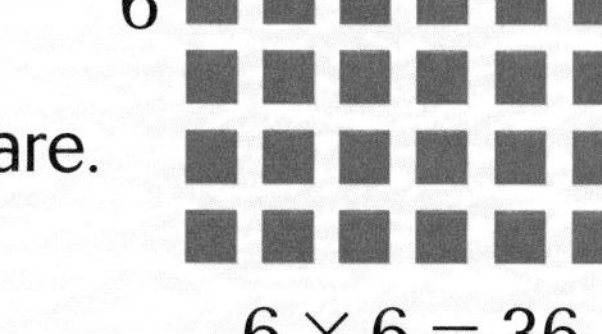

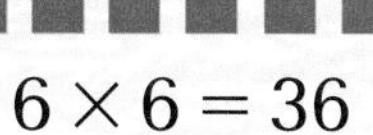

$6 \times 6 = 36$

Both factors are the same, so the products are square numbers.

Write the multiplication fact for each array.
Write *yes* or *no* to tell if the product is a square number.

1. 

2. 

3. 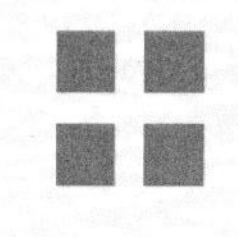

4. Write a multiplication fact where the product is a square number. Then draw an array to show your fact.

EXTRA PRACTICE page H40, Set D

Review/Test

CHECK CONCEPTS

Name a way to break apart each array. Then write the product. (pp. 148–153)

1. 

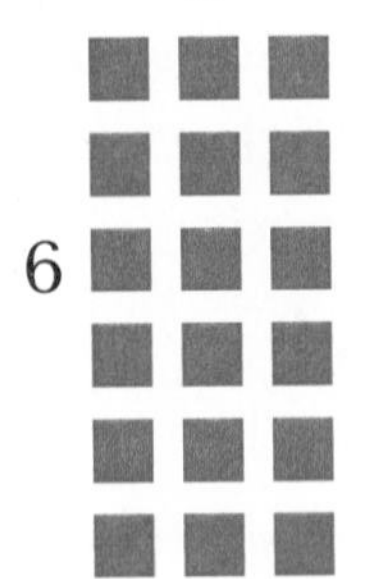

2. 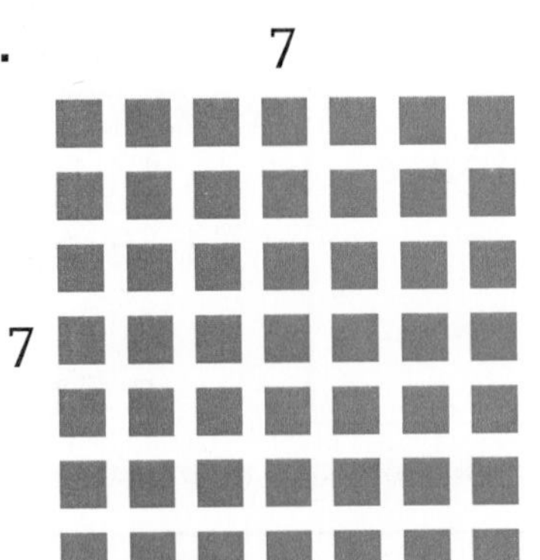

3. 

CHECK SKILLS

Find each product. (pp. 148–153, 156–159)

4. $6 \times 7 = ■$ **5.** $■ = 8 \times 7$ **6.** $4 \times 6 = ■$ **7.** $9 \times 7 = ■$

8. $■ = 5 \times 8$ **9.** $6 \times 0 = ■$ **10.** $3 \times 8 = ■$ **11.** $5 \times 7 = ■$

12. $6 \times 6 = ■$ **13.** $8 \times 9 = ■$ **14.** $8 \times 6 = ■$ **15.** $6 \times 9 = ■$

16. 8×8 **17.** 8×4 **18.** 2×7 **19.** 7×3 **20.** 6×2 **21.** 7×4

CHECK PROBLEM SOLVING

Solve. (pp. 154–155)

22. Hisako has 2 rows of 6 cookies. What is another way she could arrange the cookies in equal rows?

23. Calvin had 3 rows of 7 cookies. He gave away 8 cookies. How many were left?

24. Rita has 49 squares for a quilt. How can she arrange the squares so the quilt is the same number of squares long as it is wide?

25. Brad had 6 equal rows of stamps. He then bought 3 more stamps. If he now has 33 stamps, how many were in each row?

Standardized Test Prep

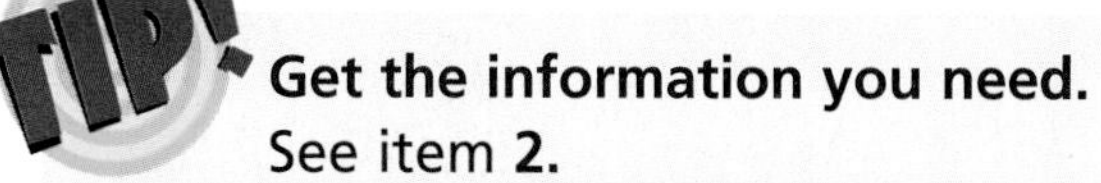

TIP! Get the information you need.
See item **2.**

You need to remember how many days are in a week to solve this problem.

Also see problem **3,** p. H63.

For 1–8, choose the best answer.

1. Betty baked some cookies. She put 6 rows of cookies, with 5 cookies in each row, on each pan. How many cookies did she put on each pan?

A 30 **B** 25 **C** 24 **D** 11

2. Evan attended camp for 4 weeks. How many days are in 4 weeks?

F 30 **G** 28 **H** 24 **J** 20

3. Which number makes this number sentence true? $7 \times ■ = 42$

A 5 **B** 6 **C** 7 **D** 8

4. What is another way to show $7 + 7 + 7$?

F 7×7
G 21×3
H 3×7
J $3 + 7$

5. $\begin{array}{r} 8 \\ \times 8 \\ \hline \end{array}$

A 56
B 64
C 72
D NOT HERE

6. Marcie bought 3 colors of yarn to knit a sweater. She bought 6 packages of each color. Which number sentence tells how many packages of yarn she bought?

F $6 \times 1 = 6$
G $3 \times 3 = 9$
H $3 \times 6 = 18$
J $6 \times 6 = 36$

7. Mrs. Kahn chose 7 students to make bookmarks for the book fair. Each student made 9 bookmarks. How many bookmarks did they make in all?

A 16 **B** 56 **C** 63 **D** 72

8. Andy wants to buy colored pencils for \$0.59, paper for \$0.85, and an eraser for \$0.29. How much money does Andy need?

F \$1.73
G \$2.27
H \$3.27
J \$4.73

Write What You Know

9. Kim wants to make a quilt that is the same number of squares long as it is wide. She wants to use more than 20 squares. She has made 21 squares so far. How many more squares does she need? Draw a picture to solve.

10. Mr. Perez made a display with 4 rows of 9 apples. Ted's mother bought 15 apples. How many apples were left in the display? Draw a picture to solve.

Multiplication Facts and Patterns

Dogsled teams were used to deliver mail in places such as Michigan, Minnesota, Wisconsin, and Alaska. Large teams of dogs could pull 400 to 500 pounds of mail.

PROBLEM SOLVING A full-grown Samoyed sled dog weighs about 50 pounds. A team of 10 Samoyeds would weigh 500 pounds. What would 10 newborn Samoyed puppies weigh?

DATA LINK

WEIGHT OF A NEWBORN PUPPY

Siberian Husky	🐶🐶🐶🐶🐶🐶
Alaskan Malamute	🐶🐶🐶🐶🐶🐶🐶🐶
American Eskimo	🐶🐶🐶🐶🐶
Samoyed	🐶🐶🐶🐶🐶

Key: Each 🐶 = 2 ounces.

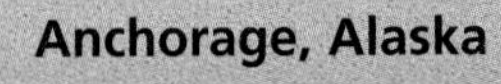

CHECK WHAT YOU KNOW

Use this page to help you review and remember important skills needed for Chapter 10.

SKIP-COUNT BY TENS (For Intervention, see p. H13.)

Continue the pattern.

1. 10, 20, 30, 40, ■, ■

2. 30, 40, 50, 60, 70, ■, ■

Skip-count by tens to find the missing numbers.

3. 3, 13, 23, ■, ■, 53, ■, ■, 83, ■

4. 7, 17, 27, ■, ■, 57, ■, 77, ■, ■

5. 5, 15, 25, ■, ■, 55, ■, ■, ■

6. 64, 54, 44, ■, ■, ■, ■

MULTIPLICATION FACTS THROUGH 8 (For Intervention, see p. H13.)

Find the product.

7. $5 \times 1 = ■$ **8.** $■ = 6 \times 3$ **9.** $■ = 5 \times 2$ **10.** $3 \times 7 = ■$

11. $■ = 8 \times 4$ **12.** $5 \times 6 = ■$ **13.** $7 \times 4 = ■$ **14.** $■ = 8 \times 2$

15. $4 \times 5 = ■$ **16.** $8 \times 6 = ■$ **17.** $5 \times 8 = ■$ **18.** $0 \times 6 = ■$

19. $3 \times 8 = ■$ **20.** $■ = 6 \times 7$ **21.** $■ = 1 \times 4$ **22.** $■ = 3 \times 2$

23. $■ = 6 \times 6$ **24.** $5 \times 3 = ■$ **25.** $2 \times 7 = ■$ **26.** $3 \times 1 = ■$

27. $\begin{array}{r} 9 \\ \times 6 \\ \hline \end{array}$ **28.** $\begin{array}{r} 7 \\ \times 8 \\ \hline \end{array}$ **29.** $\begin{array}{r} 8 \\ \times 8 \\ \hline \end{array}$ **30.** $\begin{array}{r} 9 \\ \times 0 \\ \hline \end{array}$ **31.** $\begin{array}{r} 8 \\ \times 9 \\ \hline \end{array}$

32. $\begin{array}{r} 4 \\ \times 2 \\ \hline \end{array}$ **33.** $\begin{array}{r} 5 \\ \times 7 \\ \hline \end{array}$ **34.** $\begin{array}{r} 5 \\ \times 5 \\ \hline \end{array}$ **35.** $\begin{array}{r} 1 \\ \times 7 \\ \hline \end{array}$ **36.** $\begin{array}{r} 3 \\ \times 3 \\ \hline \end{array}$

LESSON 1

Multiply with 9 and 10

Quick Review

1. $4 \times 8 = \blacksquare$
2. $7 \times 5 = \blacksquare$
3. $8 \times 0 = \blacksquare$
4. $3 \times 6 = \blacksquare$
5. $6 \times 7 = \blacksquare$

Learn

TIMBER! Beavers cut down trees with their large front teeth. If 5 beavers each cut down 10 trees, how many trees did they cut down in all?

$$5 \times 10 = \blacksquare$$

You can skip-count by tens 5 times.

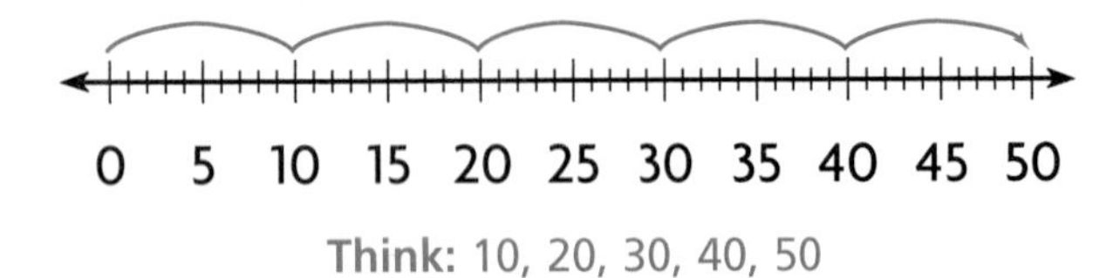

Think: 10, 20, 30, 40, 50

So, $5 \times 10 = 50$.

You can also use what you know about place value.

5×10 is the same as 5 tens.

5 tens $= 50$

So, $5 \times 10 = 50$.

So, the beavers cut down 50 trees in all.

▲ **When beavers cut down trees, they eat the bark and use the branches to build homes in the water.**

- Describe a pattern in the list of 10's facts.

$0 \times 10 = 0$
$1 \times 10 = 10$
$2 \times 10 = 20$
$3 \times 10 = 30$
$4 \times 10 = 40$
$5 \times 10 = 50$
$6 \times 10 = 60$
$7 \times 10 = 70$
$8 \times 10 = 80$
$9 \times 10 = 90$
$10 \times 10 = 100$

×	0	1	2	3	4	5	6	7	8	9	10
0	0	0	0	0	0	0	0	0	0	0	0
1	0	1	2	3	4	5	6	7	8	9	10
2	0	2	4	6	8	10	12	14	16	18	20
3	0	3	6	9	12	15	18	21	24	27	30
4	0	4	8	12	16	20	24	28	32	36	40
5	0	5	10	15	20	25	30	35	40	45	50
6	0	6	12	18	24	30	36	42	48	54	60
7	0	7	14	21	28	35	42	49	56	63	70
8	0	8	16	24	32	40	48	56	64	72	80
9	0	9	18	27	36	45	54	63	72	81	90
10	0	10	20	30	40	50	60	70	80	90	100

Multiply with 9

Lynn's class made 7 animal posters. The students drew 9 animals on each poster. How many animals did they draw in all?

$$7 \times 9 = \blacksquare$$

Lynn and Jeff use different ways to find 7×9.

Lynn

I'll think of the 10's fact first.

$7 \times 10 = 70$

Next, I'll subtract the first factor, 7.

$70 - 7 = 63$

Since $70 - 7 = 63$, $7 \times 9 = 63$.

Jeff

I'll use a pattern in the products.

- **The tens digit is 1 less than the factor 7.**
- **The sum of the digits is 9.**

$7 \times 9 = 63$

$0 \times 9 = 0$
$1 \times 9 = 9$
$2 \times 9 = 18$
$3 \times 9 = 27$
$4 \times 9 = 36$
$5 \times 9 = 45$
$6 \times 9 = 54$
$7 \times 9 = \blacksquare$
$8 \times 9 = 72$
$9 \times 9 = 81$

So, Lynn's class drew 63 animals.

MATH IDEA You can use facts you already know or a pattern to find 9's facts.

Technology Link

To learn more about multiplying, watch the Harcourt Math Newsroom Video, *How to Snag Baseballs.*

Check

1. **Explain** how to use a 10's fact to find 9×3.
2. **Explain** how to use a pattern to find 6×9.

Find the product.

3. $3 \times 10 = \blacksquare$
4. $10 \times 6 = \blacksquare$
5. $\blacksquare = 7 \times 9$
6. $1 \times 9 = \blacksquare$
7. $8 \times 9 = \blacksquare$
8. $\blacksquare = 9 \times 2$
9. $4 \times 10 = \blacksquare$
10. $2 \times 10 = \blacksquare$

Find the missing factor.

11. $\blacksquare \times 7 = 63$
12. $5 \times \blacksquare = 30$
13. $10 \times \blacksquare = 60$
14. $\blacksquare \times 6 = 48$
15. $\blacksquare \times 9 = 90$
16. $6 \times \blacksquare = 24$
17. $\blacksquare \times 8 = 64$
18. $9 \times \blacksquare = 45$

LESSON CONTINUES

Practice and Problem Solving

Find the product.

19. $\blacksquare = 10 \times 4$ **20.** $4 \times 8 = \blacksquare$ **21.** $9 \times 8 = \blacksquare$ **22.** $\blacksquare = 8 \times 6$

23. $9 \times 9 = \blacksquare$ **24.** $\blacksquare = 7 \times 10$ **25.** $10 \times 10 = \blacksquare$ **26.** $2 \times 8 = \blacksquare$

27. $\blacksquare = 5 \times 5$ **28.** $5 \times 9 = \blacksquare$ **29.** $\blacksquare = 6 \times 6$ **30.** $\blacksquare = 10 \times 0$

31. $\begin{array}{r} 10 \\ \times\ 8 \\ \hline \end{array}$ **32.** $\begin{array}{r} 9 \\ \times 4 \\ \hline \end{array}$ **33.** $\begin{array}{r} 5 \\ \times 10 \\ \hline \end{array}$ **34.** $\begin{array}{r} 8 \\ \times 8 \\ \hline \end{array}$ **35.** $\begin{array}{r} 9 \\ \times 3 \\ \hline \end{array}$ **36.** $\begin{array}{r} 7 \\ \times 8 \\ \hline \end{array}$

37. $\begin{array}{r} 4 \\ \times 3 \\ \hline \end{array}$ **38.** $\begin{array}{r} 10 \\ \times\ 1 \\ \hline \end{array}$ **39.** $\begin{array}{r} 8 \\ \times 3 \\ \hline \end{array}$ **40.** $\begin{array}{r} 6 \\ \times 9 \\ \hline \end{array}$ **41.** $\begin{array}{r} 9 \\ \times 7 \\ \hline \end{array}$ **42.** $\begin{array}{r} 10 \\ \times\ 6 \\ \hline \end{array}$

Find the missing factor.

43. $\blacksquare \times 6 = 0$ **44.** $10 \times \blacksquare = 20$ **45.** $\blacksquare \times 7 = 28$

46. $\blacksquare \times 5 = 50$ **47.** $6 \times \blacksquare = 54$ **48.** $7 \times \blacksquare = 42$

49. $\blacksquare \times 3 = 9$ **50.** $8 \times \blacksquare = 64$ **51.** $3 \times 6 = \blacksquare \times 9$

52. $5 \times \blacksquare = 4 \times 10$ **53.** $\blacksquare \times 2 = 12 + 8$ **54.** $6 \times 6 = \blacksquare \times 4$

Compare. Write <, >, or = for each ●.

55. $10 \times 6 \bullet 75 - 15$ **56.** $9 \times 9 \bullet 10 \times 8$ **57.** $7 \times 9 \bullet 10 \times 7$

58. $8 \times 9 \bullet 9 \times 8$ **59.** $16 + 40 \bullet 9 \times 6$ **60.** $10 \times 10 \bullet 50 + 50$

61. **Write About It** Sydney says, "The problem 3×10 is the same as $10 + 10 + 10$." Do you agree or disagree? Explain.

62. Malcolm cut 3 pies into 10 pieces each and 2 pies into 8 pieces each. How many pieces of pie did he have in all?

63. **What's the Error?** Describe Mike's error. Then solve the problem correctly.

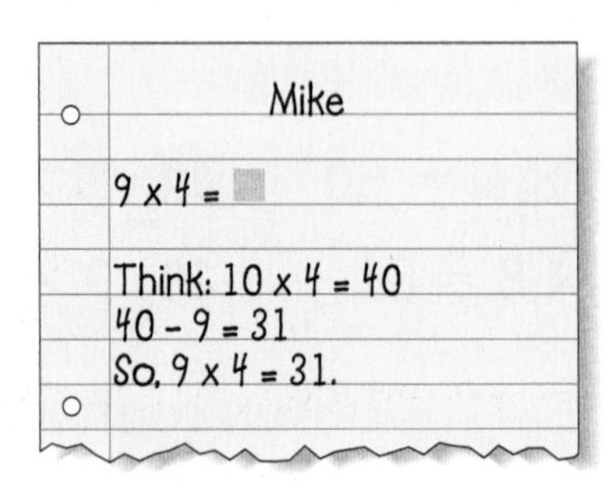
Mike

9 x 4 = ■

Think: 10 x 4 = 40
40 - 9 = 31
So, 9 x 4 = 31.

64. Emiko had 4 sheets with 10 animal stickers on each. After she gave some stickers away, she had 37 left. How many stickers did she give away?

Mixed Review and Test Prep

Order each group of numbers from least to greatest. (p. 24)

65. 243, 536, 144

66. 1,390; 1,039; 1,930

67. 6,321; 6,967; 5,644

Find the sum. (p. 46)

68. 1,568 + 3,215 = ■

69. 2,513 + 874 = ■

70. TEST PREP Meg went to the store at 11:15 A.M. She got home 2 hours later. At what time did she get home? (p. 98)

A 9:15 A.M. **C** 1:15 P.M.
B 12:15 P.M. **D** 2:15 P.M.

71. TEST PREP Jesse has 3 rows of 9 stamps. Lou has 2 rows of 6 stamps. How many stamps do they have in all? (p. 122)

F 27 **G** 39 **H** 56 **J** 60

PROBLEM SOLVING LINKUP ... to Reading

Strategy • Analyze Information The information in a problem can offer clues about how to solve it. *Analyze,* or look carefully at, each part of the problem. Read the following problems carefully.

ANIMAL FAMILIES AT THE POND

Ducks	🐾🐾🐾🐾🐾🐾🐾🐾🐾
Beavers	🐾🐾🐾🐾🐾
Turtles	🐾🐾🐾🐾🐾🐾🐾🐾
Deer Mice	🐾🐾🐾🐾

Key: Each [symbol] = 1 family.

USE DATA For 1–4, use the graph.

1. A beaver community is made up of several beaver families. If two of the families have 4 members and the rest of the families have 5 members, how many beavers are there in all?

2. Deer mice eat berries, leaves, nuts, seeds, and insects. If there are 10 deer mice in each family, how many are there in all?

3. Ducks build homes in clumps of grass or reeds. If each duck family has 8 ducklings, how many ducklings are there in all?

4. Otters make their homes in burrows near water or under rocky ledges. If a family of otters moves to the pond, how many animal families live at the pond in all?

EXTRA PRACTICE page H41, Set A

LESSON 2

Algebra: Find a Rule

Quick Review

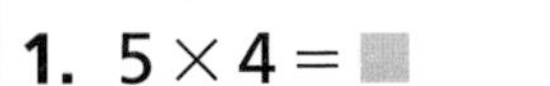
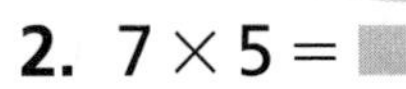

1. $5 \times 4 =$ ■
2. $7 \times 5 =$ ■
3. $6 \times 8 =$ ■
4. $7 \times 2 =$ ■
5. $8 \times 6 =$ ■

Learn

CLIP CLOP Horses wear a horseshoe on each of their 4 hooves. How many horseshoes are needed for 6 horses?

Think: 1 horse needs 4 horseshoes.
2 horses need 8 horseshoes.
3 horses need 12 horseshoes, and so on.

Look for a pattern. Write a rule.

Horses	1	2	3	4	5	6
Horseshoes	4	8	12	16	20	■

Pattern: The number of horseshoes equals the number of horses times 4.

Rule: Multiply the number of horses by 4.

Since $6 \times 4 = 24$, then 24 horseshoes are needed for 6 horses.

MATH IDEA You can write a rule to describe a number pattern in a table.

Example

Write a rule to find the cost of the bread.

Loaves of bread	1	2	4	5	7	9
Cost	\$3	\$6	\$12	\$15	\$21	\$27

Rule: Multiply the number of loaves of bread by \$3.

- **Explain** how to use the rule to find the cost of 3 loaves of bread.

Check

1. **Describe** how you could use a rule to find the cost of 10 loaves of bread.

2. Write a rule for the table. Then copy and complete the table.

Nickels	1	2	3	4	5	6	7	8	9	10
Pennies	5	10	15	■	■	■	■	■	■	■

Practice and Problem Solving

Write a rule for each table. Then copy and complete the table.

3.

Spiders	1	2	3	4	5	6
Legs	8	16	24	■	■	■

4.

Toy cars	1	2	3	4	5	6
Cost	$2	$4	$6	■	■	■

5.

Tables	3	4	5	7	8	9
Legs	12	16	20	■	■	■

6.

Guitars	2	3	5	6	7	8
Strings	12	18	30	■	■	■

For 7-9, use the table below.

Dimes	1	2	3	4	5	6	7	8	9	10
Nickels	2	4	6	■	■	■	■	■	■	■

7. Write a rule to find the number of nickels. Copy and complete the table.

8. How many nickels can you trade for 8 dimes?

9. **REASONING** How many dimes can you trade for 18 nickels?

10. Each pudding pack costs $4. How much would 5 packs cost? Make a table and write a rule to find your answer.

Mixed Review and Test Prep

For 11–14, find the elapsed time. (p. 98)

11. 8:00 A.M. to 3:00 P.M.

12. 11:30 A.M. to 1:05 P.M.

13. 3:00 A.M. to 5:45 A.M.

14. 11:00 A.M. to midnight

15. **TEST PREP** Nieta had 314 stickers. She gave 54 away and collected 32 more. How many does she have now? (p. 58)

A 400   C 260

B 292   D 228

EXTRA PRACTICE page H41, Set B

LESSON 3

Algebra: Multiply with 3 Factors

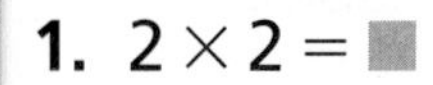

Quick Review

1. $2 \times 2 = \blacksquare$
2. $4 \times 7 = \blacksquare$
3. $3 \times 2 = \blacksquare$
4. $6 \times 5 = \blacksquare$
5. $9 \times 4 = \blacksquare$

VOCABULARY

Grouping Property of Multiplication

Learn

PRACTICE, PRACTICE . . . Julia has been taking horseback riding lessons for 3 months. At each lesson she rides for 2 hours. If she has 4 lessons each month, for how many hours has she ridden?

$$3 \times 2 \times 4 = \blacksquare$$

MATH IDEA The **Grouping Property of Multiplication** states that when the grouping of factors is changed, the product remains the same.

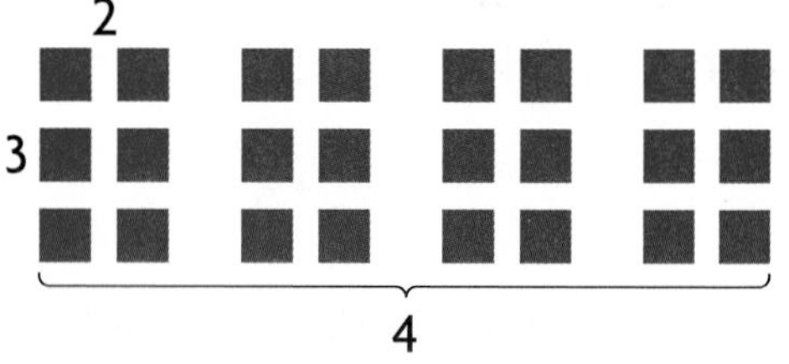

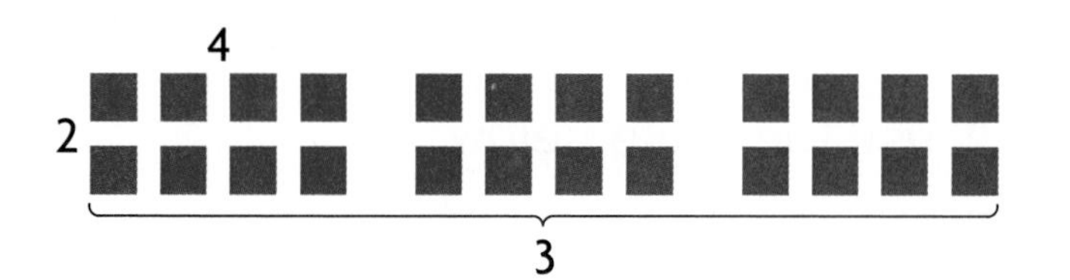

$(3 \times 2) \times 4 = \blacksquare$

$\downarrow$

$6 \times 4 = 24$

$3 \times (2 \times 4) = \blacksquare$

$\downarrow$

$3 \times 8 = 24$

Multiply the numbers in ( ) first.

So, Julia has ridden for 24 hours.

Check

1. **Tell** which numbers you would multiply first to find $7 \times 2 \times 3$ mentally.

Find each product.

2. $(2 \times 4) \times 1 = \blacksquare$
3. $2 \times (1 \times 3) = \blacksquare$
4. $2 \times (4 \times 2) = \blacksquare$
5. $(3 \times 3) \times 2 = \blacksquare$

Practice and Problem Solving

Find each product.

6. $(4 \times 2) \times 5 = \blacksquare$
7. $(3 \times 3) \times 6 = \blacksquare$
8. $8 \times (2 \times 2) = \blacksquare$
9. $(6 \times 1) \times 2 = \blacksquare$
10. $\blacksquare = 5 \times (7 \times 1)$
11. $\blacksquare = 3 \times (3 \times 3)$
12. $\blacksquare = (2 \times 3) \times 5$
13. $\blacksquare = (5 \times 2) \times 7$
14. $(4 \times 2) \times 4 = \blacksquare$

Use the Grouping Property to find the product.

15. $6 \times 1 \times 8 = \blacksquare$
16. $9 \times 2 \times 1 = \blacksquare$
17. $\blacksquare = 7 \times 4 \times 2$
18. $\blacksquare = 9 \times 8 \times 0$
19. $6 \times 5 \times 2 = \blacksquare$
20. $4 \times 2 \times 9 = \blacksquare$

Find the missing factor.

21. $(1 \times \blacksquare) \times 8 = 64$
22. $(2 \times 4) \times \blacksquare = 24$
23. $42 = 7 \times (\blacksquare \times 2)$
24. $2 \times 4 \times \blacksquare = 8$
25. $2 \times 4 \times \blacksquare = 40$
26. $14 = \blacksquare \times 2 \times 7$

27. Ross made 2 pies for each of 3 friends. In each pie he used 3 apples. How many apples did he use?

28. **REASONING** Explain why 18×2 is the same as $9 \times (2 \times 2)$.

29. **REASONING** Explain why $6 \times 1 \times 8$ has the same product as $2 \times 3 \times 8$.

30. **ALGEBRA** Darla had 2 singing lessons a month for 2 months. She learned the same number of songs at each lesson. She learned 12 songs. How many songs did she learn at each lesson?

Mixed Review and Test Prep

Choose $<$, $>$, or $=$ for each ●. (p. 148)

31. 6×8 ● $50 - 2$
32. 7×6 ● 8×5
33. 9×6 ● 6×9
34. Jed has three \$1 bills, 5 quarters, and 2 nickels. How much money does he have? (p. 84)

35. **TEST PREP** Irene bought 100 apples. She made 7 pies. Each pie had 8 apples in it. How many apples were left over? (p. 150)

A 56
B 44
C 34
D 16

EXTRA PRACTICE page H41, Set C

LESSON 4

Problem Solving Skill
Multistep Problems

UNDERSTAND PLAN SOLVE CHECK

Quick Review

1. $(2 \times 3) \times 4 = \blacksquare$
2. $(5 \times 1) \times 7 = \blacksquare$
3. $4 + 16 + 5 = \blacksquare$
4. $7 + 8 + 2 = \blacksquare$
5. $3 \times (4 \times 2) = \blacksquare$

VOCABULARY

multistep problem

KNOW THE SCORE A football team scores 6 points for a touchdown and 3 points for a field goal. The high school team made 1 touchdown and 4 field goals. How many points did they score?

To find how many points in all, you must solve a **multistep problem**, or a problem with more than one step.

Example

STEP 1

Find how many points were scored by touchdowns.

1 touchdown was scored. Each touchdown = 6 points.

$1 \times 6 = 6$ 6 points were scored by touchdowns.

STEP 2

Find how many points were scored by field goals.

4 field goals were scored. Each field goal = 3 points.

$4 \times 3 = 12$ 12 points were scored by field goals.

STEP 3

Find how many points were scored in all.

Add the points scored by touchdowns and field goals.

$6 + 12 = 18$ So, 18 points were scored in all.

Talk About It

- Does it matter if you find the points scored for touchdowns first or the points scored for field goals first? Explain.

Problem Solving Practice

Solve.

1. To raise money for the school, Lucia sold 9 boxes of cards. Ginger sold 7 boxes. Each box cost \$3. How much money did they raise in all?

2. The Wilsons drove 598 miles in 3 days. They drove 230 miles the first day and 175 miles the second day. How far did they go the third day?

Kelsey bought 3 boxes of tacos. Each box had 6 tacos. Then she gave 4 tacos away.

3. Which shows the first step you take to find how many tacos Kelsey had left?
   A $3 + 6 = 9$
   B $6 - 3 = 3$
   C $3 \times 6 = 18$
   D $3 \times 4 = 12$

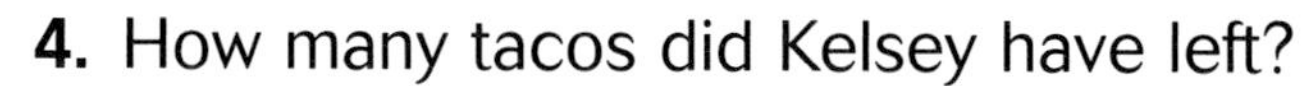

4. How many tacos did Kelsey have left?
   F 12
   G 14
   H 18
   J 22

Mixed Applications

USE DATA For 5–7, use the pictograph.

5. How many students did NOT vote for hot dogs?

6. How many students voted in all?

7. Write a problem about the graph. Exchange with a partner and solve.

8. **What's the Question?** Rob spent \$30 for 4 tickets. He bought 3 children's tickets for \$7 each and 1 adult ticket. The answer is \$9.

FAVORITE HOT LUNCHES

Tacos	
Hot Dogs	
Hamburgers	
Pizza	

Key: Each = 5 votes.

Review/Test

CHECK VOCABULARY AND CONCEPTS

Choose the best term from the box.

factor
Grouping Property
multistep problem

1. A problem with more than one step is a ? . (p. 172)
2. The ? of Multiplication states that when the grouping of factors is changed, the product remains the same. (p. 170)

Solve. (pp. 164–167)

3. Explain how to use $6 \times 10 = 60$ to find 6×9.

CHECK SKILLS

Find the product. (pp. 164–167)

4. $9 \times 7 = ■$
5. $■ = 9 \times 4$
6. $6 \times 9 = ■$
7. $8 \times 9 = ■$
8. $10 \times 5 = ■$
9. $3 \times 10 = ■$
10. $■ = 9 \times 9$
11. $10 \times 7 = ■$

Write a rule for the table. Then copy and complete the table. (pp. 168–169)

12.

Insects	1	2	3	4	5	6	7
Legs	6	12	18	■	■	■	■

Find each product. (pp. 170–171)

13. $(3 \times 1) \times 6 = ■$
14. $■ = 5 \times (2 \times 2)$
15. $(3 \times 3) \times 9 = ■$
16. $4 \times (2 \times 5) = ■$
17. $■ = (2 \times 4) \times 8$
18. $9 \times (4 \times 1) = ■$

CHECK PROBLEM SOLVING

Solve. (pp. 172–173)

19. In March, Mr. Holly's class raised $176. In April, they raised $209. How much do they still need in order to raise $500?
20. Joe bought 5 guppies for $3 each and 8 goldfish for $2 each. How much did he spend?

Standardized Test Prep

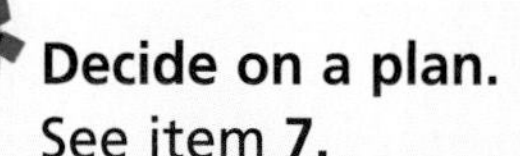

Decide on a plan.
See item **7.**
Look at the table to see the relationship of pints to cups. Choose the rule that describes this relationship.

Also see problem **4,** p. H63.

For 1–7, choose the best answer.

1. There are 9 girls. Each girl has 6 tickets for rides at the fair. How many tickets do the girls have?

A 15 **B** 36 **C** 45 **D** 54

2. The scout leader bought 4 pizzas. Each pizza was cut into 10 pieces. She saved 2 pieces. How many pieces were left for the scouts?

F 16 **G** 20 **H** 38 **J** 42

3. The scout leader spent \$6.78 on drinks. How much change did she get back from \$10.00?

A \$3.22
B \$3.32
C \$3.42
D NOT HERE

4. $(2 \times 5) \times 5 = \blacksquare$

F 12 **G** 15 **H** 25 **J** 50

5. Mona sold 3 boxes of wrapping paper. Jenny sold 4 boxes. They collected \$6 for each box. How much money did they collect altogether?

A \$49 **B** \$42 **C** \$13 **D** \$7

6. Katie has some coins worth 85 cents in her pocket. Which coins could she have?

F 3 quarters
G 2 quarters, 3 dimes, 1 nickel
H 8 dimes, 2 nickels
J 1 quarter, 5 dimes

7. What is the rule for this table?

Pints	1	2	3	4	5	6
Cups	2	4	6	8	10	12

A Multiply the number of pints by 2.
B Add 1 to the number of pints.
C Multiply the number of pints by 4.
D Subtract 2 from the number of cups.

Write What You Know

8. Show two ways to group the factors, and then multiply.

$4 \times 2 \times 5$

Show how you found the products. Write a sentence comparing the products.

9. Copy and complete the table to show the pattern. Write a rule.

Dollars	1	2	3	4	5
Quarters	4	8	12	■	■

Explain how it is used to find the numbers in the table.

PROBLEM SOLVING
MATH DETECTIVE

Follow the Leader

Solve each case. Then copy the riddle. Solve it by writing the letter for each case above the matching answer.

Case 1

Start with 4. **O**
Add 2.
Subtract 3.
Multiply by 5.
Subtract 9.
Multiply by 4.
End Number ■

Case 2

Start with 6. **U**
Multiply by 2.
Subtract 5.
Multiply by 3.
Add 9.
Multiply by 0.
End Number ■

Case 3

Start with 2. **P**
Multiply by 3.
Multiply by 4.
Subtract 4.
Add 7.
Multiply by 1.
End Number ■

Case 4

Start with 5. **Y**
Multiply by 5.
Subtract 15.
Add 5.
Subtract 9.
Multiply by 6.
End Number ■

Case 5

Start with 3. **N**
Add 6.
Multiply by 2.
Subtract 10.
Multiply by 3.
Subtract 4.
End Number ■

Case 6

Start with 7. **H**
Multiply by 5.
Subtract 30.
Multiply by 4.
Add 15.
Subtract 5.
End Number ■

Why doesn't a frog jump when it's sad?

It's too ? ? ? ? ? ? ? .

?	?	?	?	?	?	?
0	**20**	**30**	**24**	**27**	**27**	**36**

Challenge

Multiply with 11 and 12

There are 11 players on a soccer team. In a soccer game, there are two teams on the field. How many players are on the field?

$2 \times 11 = ■$

You can use a multiplication table to find the product.

The product is found where row 2 and column 11 meet.

So, there are 22 players on the field.

×	0	1	2	3	4	5	6	7	8	9	10	11	12
0	0	0	0	0	0	0	0	0	0	0	0	0	0
1	0	1	2	3	4	5	6	7	8	9	10	11	12
2	0	2	4	6	8	10	12	14	16	18	20	22	24
3	0	3	6	9	12	15	18	21	24	27	30	33	36
4	0	4	8	12	16	20	24	28	32	36	40	44	48
5	0	5	10	15	20	25	30	35	40	45	50	55	60
6	0	6	12	18	24	30	36	42	48	54	60	66	72
7	0	7	14	21	28	35	42	49	56	63	70	77	84
8	0	8	16	24	32	40	48	56	64	72	80	88	96
9	0	9	18	27	36	45	54	63	72	81	90	99	108
10	0	10	20	30	40	50	60	70	80	90	100	110	120
11	0	11	22	33	44	55	66	77	88	99	110	121	132
12	0	12	24	36	48	60	72	84	96	108	120	132	144

Talk About It

- Explain how to use a multiplication table to find 6×12.
- What patterns do you notice in the twelves column of the multiplication table?

Try It

Use the multiplication table to solve.

1. $3 \times 11 = ■$

2. $■ = 7 \times 11$

3. $■ = 4 \times 12$

4. $8 \times 11 = ■$

5. $■ = 6 \times 11$

6. $5 \times 12 = ■$

7. $1 \times 12 = ■$

8. $■ = 2 \times 12$

9. $9 \times 12 = ■$

10. 12×7

11. 11×9

12. 11×5

13. 12×3

14. 12×8

Study Guide and Review

VOCABULARY

Choose the best term from the box.

1. The ? means that two numbers can be multiplied in any order. The product is the same. (p. 121)
2. The ? means that when the grouping of factors is changed, the product remains the same. (p. 170)

factors
Grouping Property of Multiplication
Order Property of Multiplication

STUDY AND SOLVE

Chapter 7

Use the Order Property of Multiplication.

Numbers can be multiplied in any order. The product is the same.

$3 \times 4 = 12$

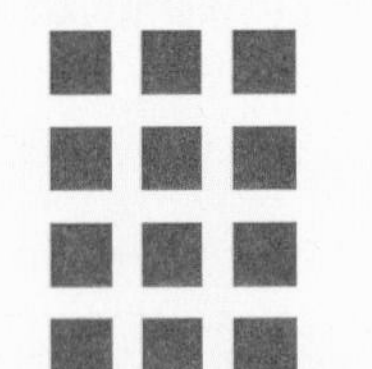

$4 \times 3 = 12$

Find the product. (pp. 120–125)

3. $2 \times 3 = ■$ $\quad 3 \times 2 = ■$
4. $3 \times 7 = ■$ $\quad 7 \times 3 = ■$
5. $4 \times 5 = ■$ $\quad 5 \times 4 = ■$
6. $3 \times 5 = ■$ $\quad 5 \times 3 = ■$
7. $2 \times 7 = ■$ $\quad 7 \times 2 = ■$

Chapter 8

Find missing factors.

$■ \times 7 = 28$

The multiplication table on page 142 can help you. Look down the column for 7 to the product 28. Look across the row from 28. The factor in that row is 4. So, $4 \times 7 = 28$.

Find the missing factor. (pp. 142–143)

8. $■ \times 8 = 16$
9. $5 \times ■ = 30$
10. $9 \times ■ = 45$
11. $3 \times ■ = 3$
12. $4 \times ■ = 0$
13. $■ \times 3 = 15$

Chapter 9

Write multiplication facts with factors 6, 7, and 8.

You can double products of facts you already know to help you find products you don't know.

$6 \times 8 = ■$

Think: $6 \times 4 = 24$

$24 + 24 = 48$, so $6 \times 8 = 48$.

Use the Order Property of Multiplication.

$7 \times 5 = ■$

$5 \times 7 = 35$, so $7 \times 5 = 35$.

Find the product. (pp. 148–153)

14. $6 \times 8 = ■$ **15.** $7 \times 6 = ■$

16. $8 \times 4 = ■$ **17.** $6 \times 6 = ■$

18. $6 \times 3 = ■$ **19.** $4 \times 8 = ■$

20. $8 \times 7 = ■$ **21.** $8 \times 8 = ■$

22. $7 \times 9 = ■$ **23.** $7 \times 7 = ■$

24. $8 \times 5 = ■$ **25.** $7 \times 4 = ■$

Chapter 10

Find a rule for the pattern.

Write a rule for the pattern in the table.

Cars	1	2	3	4	5	6	7	8	9	10
Tires	4	8	12	16	20	■	■	■	■	■

Think: The number of tires is 4 times the number of cars.

Rule: Multiply by 4.

For 26–27, use the table below. (pp. 168–169)

Spiders	1	2	4	5	6	8
Legs	8	16	32	40	■	■

26. Write a rule for the table.

27. Use the rule to complete the table.

PROBLEM SOLVING PRACTICE

Solve. (pp. 126–127, 172–173)

28. Pencils are in packages of 4. Erasers are in packages of 7. Marian bought 16 pencils. How many packages of pencils did she buy? Is there too much or too little information? Explain.

29. A box of cookies costs \$3. A bag of nuts costs \$2. Aimee bought 4 boxes of cookies and 7 bags of nuts for her friends. How much did she spend?

PERFORMANCE ASSESSMENT

TASK A • CLASS PLAY

Materials: square tiles

You need to set up 24 chairs for people to watch the class play. You must put the chairs into rows with an equal number of chairs in each row.

a. Use square tiles to represent the chairs. Make arrays to show two possible ways to set up the chairs. Draw a picture of each array.

b. Write a multiplication sentence for each array.

c. Are the two multiplication sentences you wrote examples of the Order Property of Multiplication? Why or why not?

TASK B • A NEW GAME

You and a friend invent a new game using 2 spinners. You spin the pointer on each spinner. Use the numbers the pointers land on as factors to write a multiplication sentence. If a pointer lands on a shaded section, you can choose any number on that spinner as the factor. The winner of the round is the player with the greater product.

a. Your friend spins a 4 and a 5. You spin a 3 and a shaded section. Which numbers could you choose to make you the winner?

b. Write 3 multiplication sentences for factors you can spin on each spinner that give a product less than 20.

c. You want to change the game so that you can spin factors with products more than 100. What numbers would you put on the spinners?

The Learning Site • Multiplication Mystery

If you know your multiplication facts, it is easier to count groups of coins and other objects.

You can practice your multiplication facts at the Harcourt School Learning Site.

- Go to The Learning Site. **www.harcourtschool.com**
- Click on *Multiplication Mystery.*
- Drag the products into the multiplication table to uncover the picture.
- What picture did you uncover?

Multiplication Mystery

9

This tile is a product of two factors. Drag it to a square where the missing factors meet.

Practice and Problem Solving

Find the product.

1. $2 \times 10 = ■$ **2.** $3 \times 5 = ■$ **3.** $4 \times 2 = ■$ **4.** $6 \times 2 = ■$

5. $7 \times 4 = ■$ **6.** $5 \times 5 = ■$ **7.** $2 \times 7 = ■$ **8.** $3 \times 9 = ■$

Find the missing number.

9. $4 \times ■ = 12$ **10.** $■ \times 8 = 40$ **11.** $■ \times 5 = 35$ **12.** $10 \times ■ = 90$

13. $7 \times ■ = 42$ **14.** $■ \times 7 = 63$ **15.** $8 \times ■ = 56$ **16.** $■ \times 3 = 24$

17. **STRETCH YOUR THINKING** Joanne has 88 stickers. She has 7 sheets of 8 animal stickers. How many sheets of 8 stickers wihtout animals does she have?

18. **Write a Problem** involving two multiplication facts.

Multimedia Math Glossary **www.harcourtschool.com/mathglossary**

19. **Vocabulary** Look up *Order Property of Multiplication* in the Multimedia Math Glossary. Write a problem that uses the examples shown in the glossary and use counters to model it.

Each year some people hike the whole length of the Appalachian Trail. The trail runs 2,167 miles from Maine to Georgia.

in New Jersey and Pennsylvania

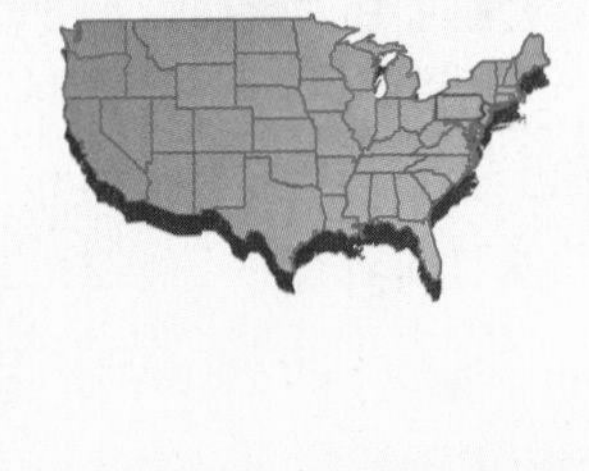

THE APPALACHIAN TRAIL

Each fall many people hike the Appalachian Trail. This is a pretty time of the year, because of the fall colors in the trees.

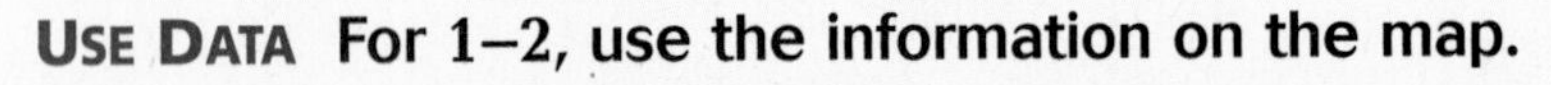

USE DATA For 1–2, use the information on the map.

Suppose you and your family are planning a hike on the Appalachian Trail. The map at the right shows the area you have chosen for your hike. Estimated distances are shown for each trail.

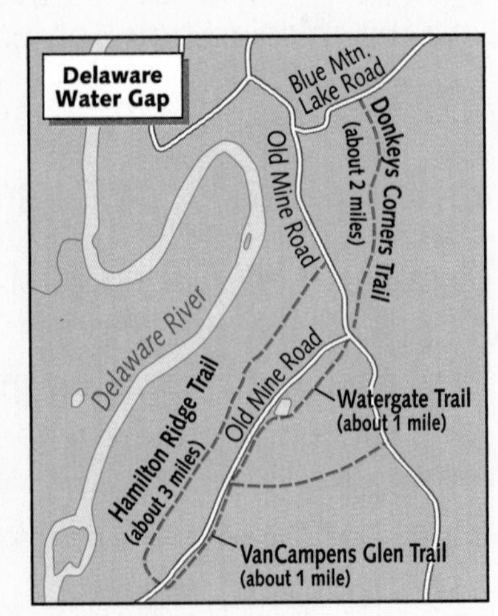

1. If you can hike 2 miles in 1 hour, how far can you hike in 2 hours? in 3 hours?
2. Use the information on the map and your answers to Exercise 1. Plan a hike that will take at least 3 hours.
   - Make a chart to show your route.
   - List each trail you will hike, and show its length.
   - Find the total length of your hike.
   - Find the total time for your hike.

THE DELAWARE WATER GAP

Visitors to the Delaware Water Gap can enjoy hiking, boating, camping, fishing, and watching wildlife.

This table shows distances between points on the Delaware River. The miles listed for each point show its distance from the Depew Access. For example, the distance from Depew to Smithfield Beach is 4 miles. The distance from Poxono to Kittatinny is 8 miles.

Hikers will see many beautiful sights.

DATA LINK

BOAT ACCESS POINTS (miles from Depew Access, NJ)

Depew Access, NJ	Poxono Access, NJ	Smithfield Beach Access, PA	Worthington Access, NJ	Kittatinny Access, NJ
Mile 0	Mile 2	Mile 4	Mile 6	Mile 10

USE DATA For 1–3, use the table.

1. A family canoes down the Delaware River. They travel 2 miles of the river each hour. They start at Depew and travel 3 hours. Where do they stop?
2. A river guide tells campers that they will reach their campsite in 5 hours. They start at Depew, paddling 2 miles each hour. Where will they end up?
3. A fishing boat slowly travels down the river from Poxono. It covers 4 miles in 1 hour. Where will it be in 2 hours?
4. Campsites along a trail are about 7 miles apart. A hiker has passed 6 campsites. Draw an array to show how far the hiker walked.
5. **REASONING** A company rents different kinds of boats. Canoes hold 3 people each, while kayaks hold 2 people, and rafts hold 6 people. One day there are 50 people out in rental boats. Each boat is full. What is a possible group of boats rented this day?

Deer, raccoons, turkey vultures, beavers, and even bears call the Delaware Water Gap home.

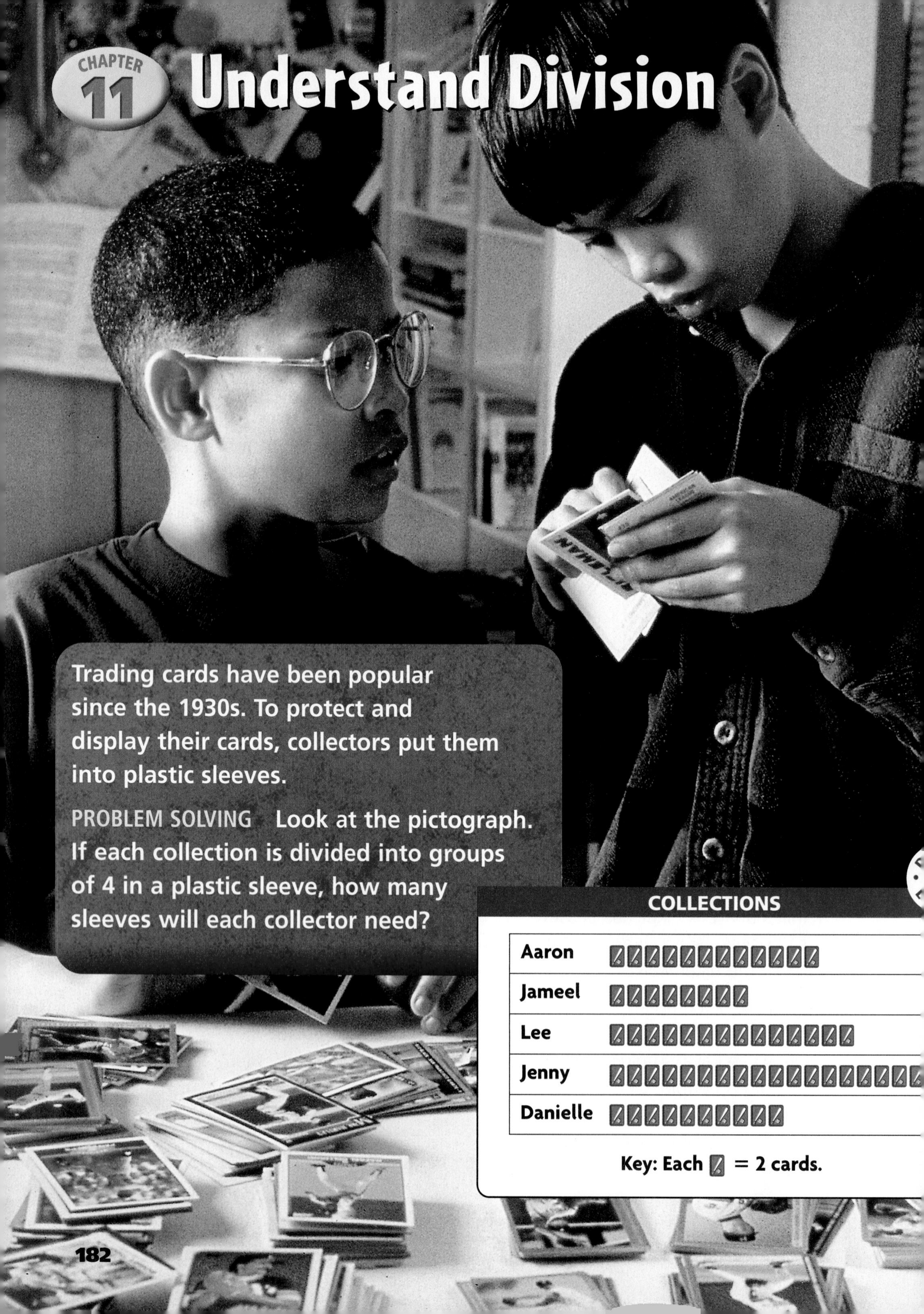

CHAPTER 11 Understand Division

Trading cards have been popular since the 1930s. To protect and display their cards, collectors put them into plastic sleeves.

PROBLEM SOLVING Look at the pictograph. If each collection is divided into groups of 4 in a plastic sleeve, how many sleeves will each collector need?

COLLECTIONS

Aaron	▯▯▯▯▯▯▯▯▯▯▯▯
Jameel	▯▯▯▯▯▯▯▯
Lee	▯▯▯▯▯▯▯▯▯▯▯▯▯▯
Jenny	▯▯▯▯▯▯▯▯▯▯▯▯▯▯▯▯▯▯
Danielle	▯▯▯▯▯▯▯▯▯▯

Key: Each ▯ = 2 cards.

Use this page to help you review and remember important skills needed for Chapter 11.

VOCABULARY

Choose the best term from the box.

factor
product
sum

1. In the multiplication sentence $6 \times 4 = 24$, the number 24 is the __?__.
2. In the multiplication sentence $5 \times 3 = 15$, the number 5 is a __?__.

MEANING OF MULTIPLICATION (For Intervention, see p. H14.)

Copy and complete.

3. 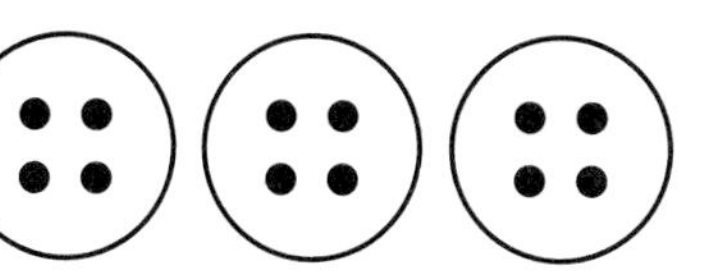

   a. ■ groups of ■
   b. ■ + ■ + ■ = ■
   c. ■ × ■ = ■

4. 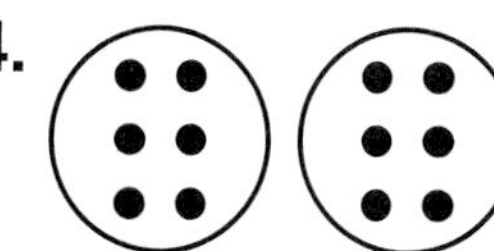

   a. ■ groups of ■
   b. ■ + ■ = ■
   c. ■ × ■ = ■

Write an addition and a multiplication sentence for each.

5.     

6. 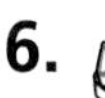   

MULTIPLICATION FACTS THROUGH 10 (For Intervention, see p. H14.)

Find the product.

7. $6 \times 7 =$ ■
8. $3 \times 8 =$ ■
9. ■ $= 8 \times 9$
10. $9 \times 0 =$ ■
11. $5 \times 7 =$ ■
12. ■ $= 10 \times 4$
13. $4 \times 4 =$ ■
14. $2 \times 4 =$ ■
15. ■ $= 9 \times 9$
16. ■ $= 6 \times 8$
17. $8 \times 7 =$ ■
18. $4 \times 3 =$ ■
19. $\begin{array}{r} 1 \\ \times 5 \\ \hline \end{array}$
20. $\begin{array}{r} 2 \\ \times 6 \\ \hline \end{array}$
21. $\begin{array}{r} 3 \\ \times 0 \\ \hline \end{array}$
22. $\begin{array}{r} 5 \\ \times 2 \\ \hline \end{array}$
23. $\begin{array}{r} 0 \\ \times 7 \\ \hline \end{array}$

LESSON 1

The Meaning of Division

Quick Review

1. $6 \times 4 = \blacksquare$
2. $3 \times 5 = \blacksquare$
3. $2 \times 6 = \blacksquare$
4. $7 \times 3 = \blacksquare$
5. $4 \times 3 = \blacksquare$

VOCABULARY
divide

MATERIALS
counters

Explore

When you multiply, you put equal groups together. When you **divide**, you separate into equal groups.

Activity

Divide 14 counters into 2 equal groups. How many counters are in each group?

STEP 1

Use 14 counters. Show 2 groups.

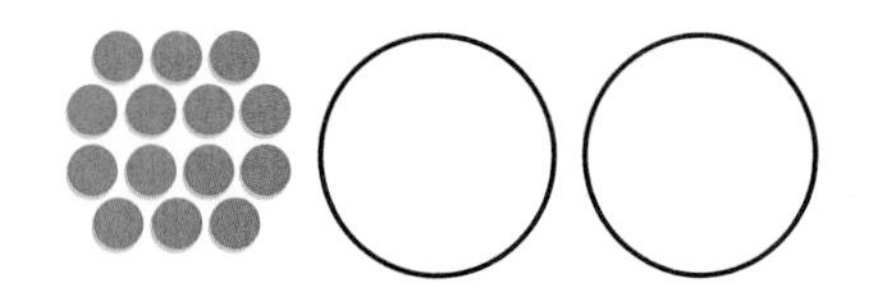

STEP 2

Place a counter in each group.

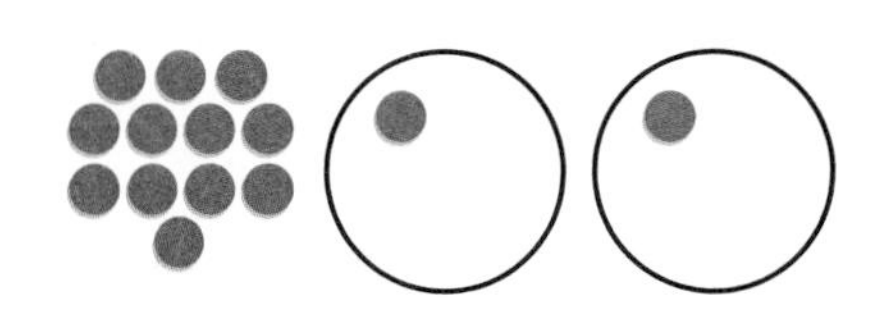

STEP 3

Continue until all counters are used.

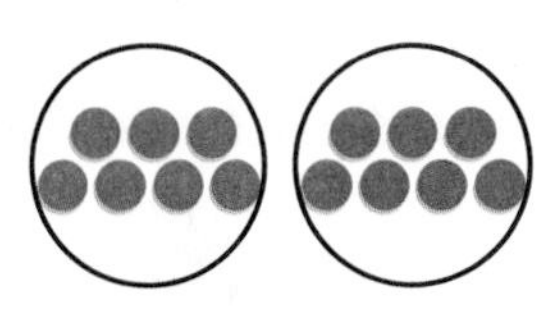

So, there are 7 counters in each group.

Try It

Use counters to make equal groups.

a. Divide 20 counters into 5 equal groups. How many are in each group?

b. Divide 12 counters into equal groups in different ways.

c. How did you find different equal groups using 12 counters?

We are putting 20 counters in 5 equal groups. How many should be in each group?

Connect

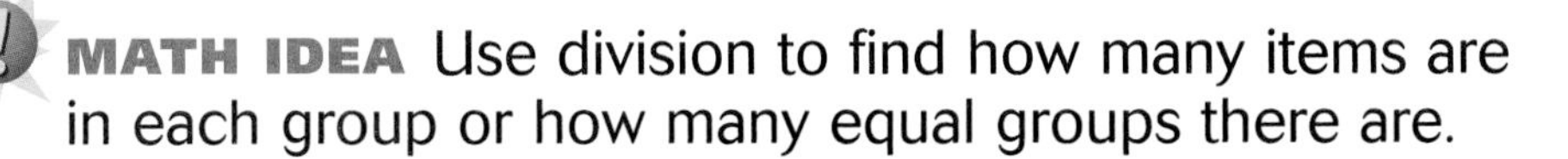

MATH IDEA Use division to find how many items are in each group or how many equal groups there are.

Four friends share 20 marbles equally. How many marbles will each person get?

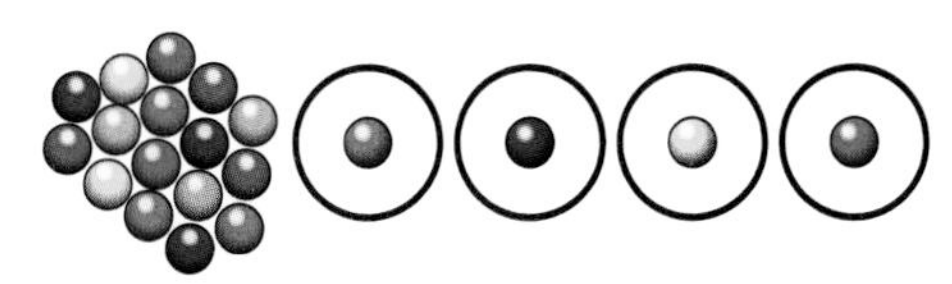

Put one marble in each group until all marbles are used.

Each person will get 5 marbles.

Each person wants 4 marbles. How many people can share 20 marbles?

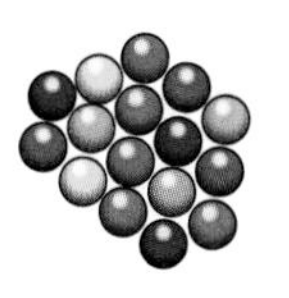

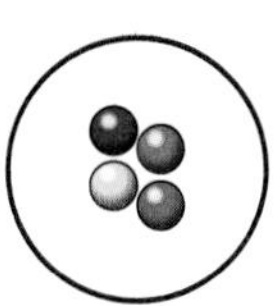

Make equal groups of 4 marbles until all marbles are used.

Five people can share 20 marbles.

Practice and Problem Solving

Copy and complete the table. Use counters to help.

	COUNTERS	NUMBER OF EQUAL GROUPS	NUMBER IN EACH GROUP
1.	15	5	■
2.	21	■	3
3.	24	3	■
4.	28	■	7

For 5–8, use counters.

5. Five friends share 30 stickers equally. How many will each person get?

6. Elijah has 18 books that he wants to put into equal groups. List three different ways that he could do this.

7. **REASONING** Three friends share some grapes equally. If each gets 9 grapes, how many grapes are there all together?

8. **Write About It** Explain how to divide 32 counters into 4 equal groups.

Mixed Review and Test Prep

9. $463 + 297$ (p. 42)

10. $805 - 176$ (p. 58)

11. 2×9 (p. 118)

12. 3×8 (p. 122)

13. **TEST PREP** The Boy Scouts charged $4 to wash each car. How much money did they make for washing 8 cars? (p. 134)

A $2 **B** $4 **C** $12 **D** $32

LESSON

2 Relate Subtraction and Division

Quick Review

How many are there in all?

1. 2 groups of 3
2. 4 groups of 4
3. 5 groups of 2
4. 3 groups of 9
5. 1 group of 8

Learn

GET IN THE GAME Ana has 12 game pieces for a game. Each player gets 4 pieces. How many people can play?

12	÷	4	=	■
↑		↑		↑
number of pieces		number for each player		number of players

Start with 12. Take away groups of 4 until you reach 0. Count the number of times you subtract 4.

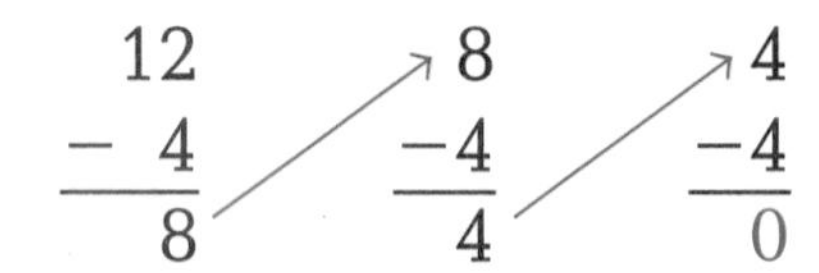

Number of times you subtract 4: **1** **2** **3**

Since you subtract 4 from 12 three times, there are 3 groups of 4 in 12.

So, 3 people can play.

Write: $12 \div 4 = 3$ or $4\overline{)12}$ with quotient 3

Read: Twelve divided by four equals three.

MATH IDEA You can use repeated subtraction to find how many groups when you know how many in all and how many in each group.

- **Discuss** how to skip-count to find $15 \div 5$.

Check

1. **Explain** how to use repeated subtraction to prove that $18 \div 6 = 3$.

Write the division sentence for each.

2. $\begin{array}{r} 12 \\ -\ 6 \\ \hline 6 \end{array} \rightarrow \begin{array}{r} 6 \\ -6 \\ \hline 0 \end{array}$

3. $\begin{array}{r} 8 \\ -2 \\ \hline 6 \end{array} \rightarrow \begin{array}{r} 6 \\ -2 \\ \hline 4 \end{array} \rightarrow \begin{array}{r} 4 \\ -2 \\ \hline 2 \end{array} \rightarrow \begin{array}{r} 2 \\ -2 \\ \hline 0 \end{array}$

Practice and Problem Solving

Write a division sentence for each.

4. $\begin{array}{r} 20 \\ -\ 5 \\ \hline 15 \end{array} \rightarrow \begin{array}{r} 15 \\ -\ 5 \\ \hline 10 \end{array} \rightarrow \begin{array}{r} 10 \\ -\ 5 \\ \hline 5 \end{array} \rightarrow \begin{array}{r} 5 \\ -5 \\ \hline 0 \end{array}$

5. $\begin{array}{r} 24 \\ -\ 8 \\ \hline 16 \end{array} \rightarrow \begin{array}{r} 16 \\ -\ 8 \\ \hline 8 \end{array} \rightarrow \begin{array}{r} 8 \\ -8 \\ \hline 0 \end{array}$

Use subtraction to solve.

6. $15 \div 3 = \blacksquare$
7. $21 \div 7 = \blacksquare$
8. $30 \div 5 = \blacksquare$
9. $36 \div 6 = \blacksquare$
10. $2\overline{)10}$
11. $8\overline{)16}$
12. $7\overline{)35}$
13. $5\overline{)25}$

ALGEBRA **Complete. Write +, −, ×, or ÷ for each ●.**

14. $20 - 5 = 5$ ● 3
15. $24 \div 6 = 18$ ● 14
16. 32 ● $8 = 4 \times 10$
17. 8 ● $2 = 2 + 2$

18. Scott buys 22 baseball cards. He keeps 10 cards and divides the rest equally between 2 friends. How many cards will each friend get?

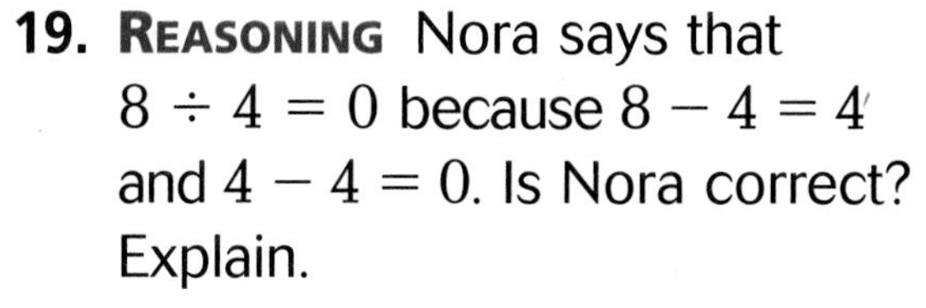

19. **REASONING** Nora says that $8 \div 4 = 0$ because $8 - 4 = 4$ and $4 - 4 = 0$. Is Nora correct? Explain.

20. Explain how to use repeated subtraction to find $100 \div 10$.

Mixed Review and Test Prep

21. $\begin{array}{r} 3 \\ \times 6 \\ \hline \end{array}$ (p. 122)

22. $\begin{array}{r} 4 \\ \times 7 \\ \hline \end{array}$ (p. 150)

23. $\begin{array}{r} \$3.57 \\ +\$7.94 \\ \hline \end{array}$ (p. 88)

24. $\begin{array}{r} \$9.26 \\ -\$5.83 \\ \hline \end{array}$ (p. 88)

25. **TEST PREP** Felipe buys a notebook for $1.39. He pays with a $5 bill. How much change should he get? (p. 86)

A $3.61 **C** $4.61
B $3.71 **D** $6.39

EXTRA PRACTICE page H42, Set A

LESSON 3

Algebra: Relate Multiplication and Division

Quick Review

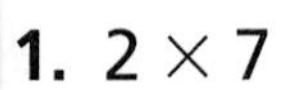

1. 2×7

2. 4×2
3. 2×9
4. 6×3
5. 3×8

VOCABULARY

dividend **divisor**
quotient
inverse operations
variable

Learn

STICK WITH STAMPS Use what you know about arrays and multiplication to understand division.

Mark is putting stamps into his stamp album. Each page holds 18 stamps in 3 equal rows. How many stamps are in each row?

$18 \div 3 = \blacksquare$

18 ↑ number of stamps; 3 ↑ number of rows; ■ ↑ number in each row

Show an array with 18 in 3 equal rows. Find how many are in each row.

Since $3 \times 6 = 18$, then $18 \div 3 = 6$.

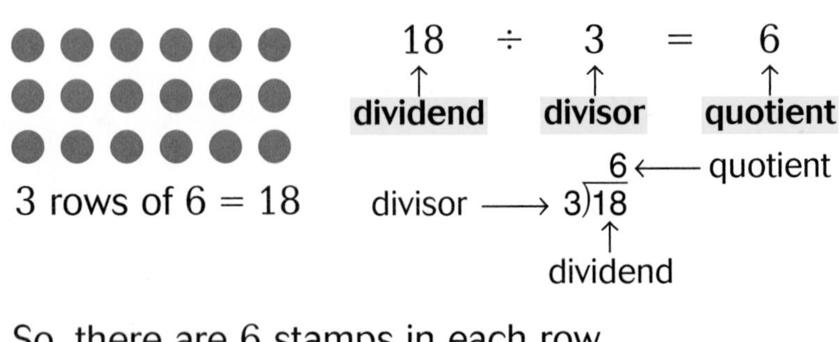

Technology Link

More Practice: Use E-lab, *Exploring Division.*

www.harcourtschool.com/elab2002

So, there are 6 stamps in each row.

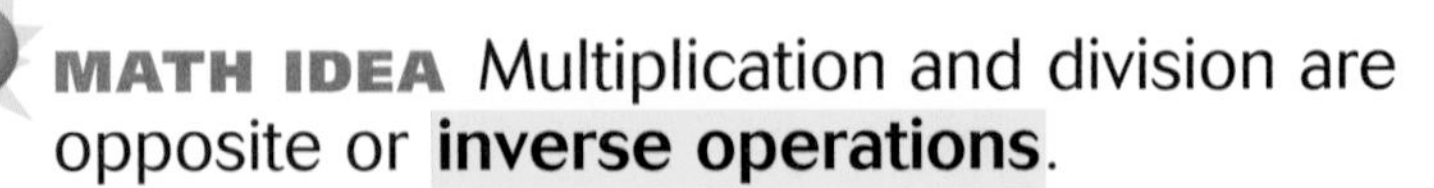

MATH IDEA Multiplication and division are opposite or **inverse operations**.

Examples

A

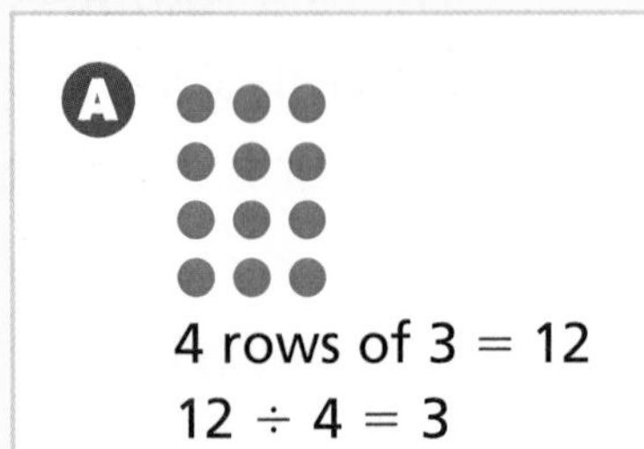

4 rows of 3 = 12
$12 \div 4 = 3$

B

2 rows of 7 = 14
$14 \div 2 = 7$

C

3 rows of 5 = 15
$15 \div 3 = 5$

Use Variables

To divide, think of the related multiplication fact.

$16 \div 2 = ■$

Think: $2 \times ■ = 16$
$2 \times 8 = 16$

2 rows of ■ = 16.

So, $16 \div 2 = 8$.

The box, ■, stands for an unknown number. The box is called a **variable**. Letters can also be used as variables.

$20 \div 5 = a$

Think: $5 \times a = 20$
$5 \times 4 = 20$, so, $a = 4$.

So, $20 \div 5 = 4$.

Examples

A $12 \div 2 = b$

Think: $2 \times b = 12$
$2 \times 6 = 12$, so, $b = 6$.

So, $12 \div 2 = 6$.

B $15 \div 5 = c$

Think: $5 \times c = 15$
$5 \times 3 = 15$, so, $c = 3$.

So, $15 \div 5 = 3$.

MATH IDEA You can use a variable to stand for an unknown number.

Check

1. **Explain** how to use an array to multiply and divide.

Copy and complete.

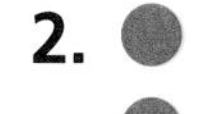

2. 

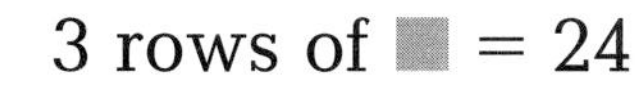

3 rows of ■ = 24

$24 \div 3 = ■$

3. 2 rows of ■ = 18

$18 \div 2 = ■$

4. 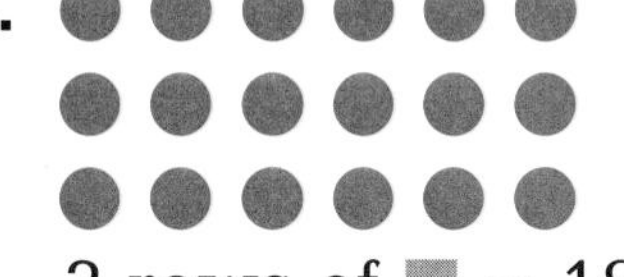

3 rows of ■ = 18

$18 \div 3 = ■$

Find the number that the variable stands for.

5. $12 \div 3 = r$
$r = ?$

6. $16 \div 4 = s$
$s = ?$

7. $24 \div 4 = t$
$t = ?$

LESSON CONTINUES 

Practice and Problem Solving

Copy and complete.

8. 

3 rows of ■ = 21

$21 \div 3 =$ ■

9. 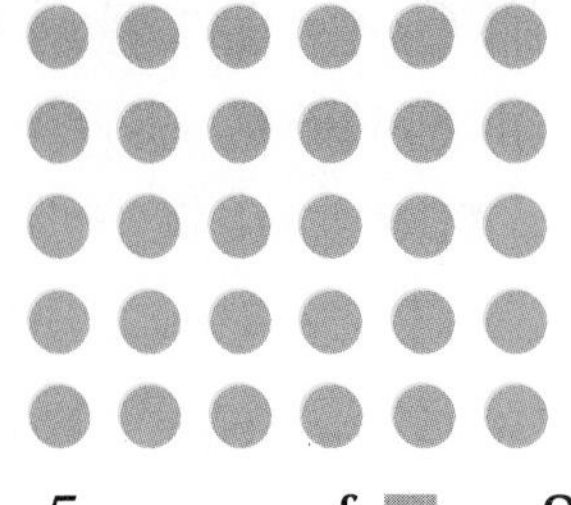

5 rows of ■ = 30

$30 \div 5 =$ ■

10. 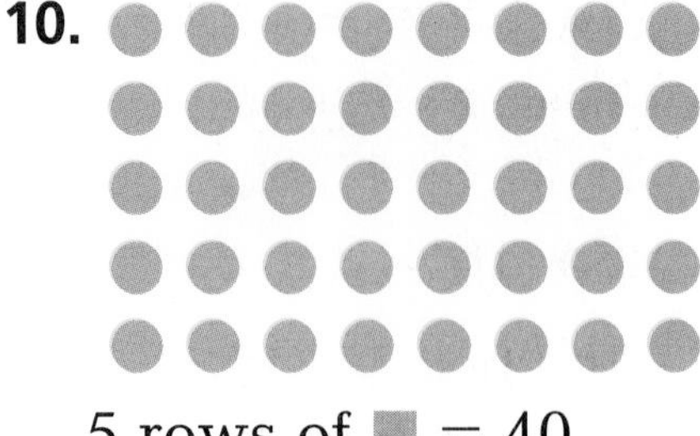

5 rows of ■ = 40

$40 \div 5 =$ ■

Complete each number sentence. Draw an array to help.

11. $3 \times$ ■ $= 18$ $18 \div 3 =$ ■
12. $5 \times$ ■ $= 25$ $25 \div 5 =$ ■
13. $6 \times$ ■ $= 24$ $24 \div 6 =$ ■
14. $3 \times$ ■ $= 24$ $24 \div 3 =$ ■

Find the number that the variable stands for.

15. $12 \div 2 = a$
    $a =$?
16. $15 \div 3 = b$
    $b =$?
17. $18 \div 6 = c$
    $c =$?
18. $14 \div 7 = p$
    $p =$?
19. $25 \div 5 = q$
    $q =$?
20. $24 \div 8 = r$
    $r =$?
21. $5 \times a = 20$
    $a =$?
22. $6 \times b = 18$
    $b =$?
23. $3 \times c = 21$
    $c =$?
24. $p \times 4 = 16$
    $p =$?
25. $q \times 5 = 10$
    $q =$?
26. $r \times 7 = 14$
    $r =$?

ALGEBRA Complete.

27. $4 \times 2 = 24 \div a$
    $a =$?
28. $b \times 3 = 30 \div 5$
    $b =$?
29. $4 \times 1 = c \div 4$
    $c =$?

30. Tory arranged 12 stamps so that 3 were in each row. How many rows did she make?

31. **What's the Question?** Christy puts 36 pennies into 4 equal piles. The answer is 9 pennies.

32. Blake has 21 stamps. He wants to know how many will be in each row if he makes 7 rows. Write a number sentence with a variable that he can use.

33. **REASONING** Mark bakes 14 muffins. He eats 2 muffins and divides the rest equally among 6 friends. What division sentence shows how many muffins each friend gets?

34. Aretha has 18 tadpoles. If she puts 6 tadpoles in each bowl, how many bowls will she need?

35. Colin has 24 toy cars. He puts an equal number of cars into each of 3 boxes. How many cars will be in 2 of the boxes?

Mixed Review and Test Prep

Write + or − for each ●. (p. 186)

36. $20 ● 5 = 5 \times 3$

37. $24 \div 3 = 6 ● 2$

38. $4 \times 9 = 38 ● 2$

39. $7 ● 2 = 36 \div 4$

Find each missing factor. (p. 158)

40. $■ \times 5 = 40$

41. $8 \times ■ = 56$

42. $4 \times ■ = 32$

43. $■ \times 6 = 0$

44. **TEST PREP** Hector puts 1 ice cube in each of 7 cups. How many ice cubes are there in all? (p. 132)

A 1
B 6
C 7
D 8

45. **TEST PREP** These numbers follow a pattern.

14, 21, 28, 35

Which numbers continue the pattern? (p. 136)

F 40, 47, 54
G 42, 49, 56
H 32, 39, 46
J 42, 50, 59

PROBLEM SOLVING LINKUP . . . to Science

Camping Safety

Some spring water is not safe to drink. When you go camping, you should boil the water, use water treatment pills, or use a water filter before you drink any spring water.

1. Andrea's water filter pumps 1 liter of water every 5 minutes. How long will it take to pump 4 liters?

2. Michael's water filter pumps 1 liter of water every 4 minutes. How many liters can he pump in 24 minutes?

LESSON 4

Algebra: Fact Families

Quick Review

1. 3 × ■ = 18
2. ■ × 6 = 18
3. 18 ÷ 6 = ■
4. 5 × ■ = 25
5. 25 ÷ 5 = ■

VOCABULARY

fact family

Learn

FUN FACTS A set of related multiplication and division number sentences is called a **fact family**.

Fact Family for 3, 5, and 15

factor		factor		product	dividend		divisor		quotient
3	×	5	=	15	15	÷	5	=	3
5	×	3	=	15	15	÷	3	=	5

HANDS ON

Activity

Materials: square pieces of paper, scissors

Use this triangle fact card to think of the fact family for 3, 5, and 15.

Make a set of triangle fact cards. Use them to write fact families.

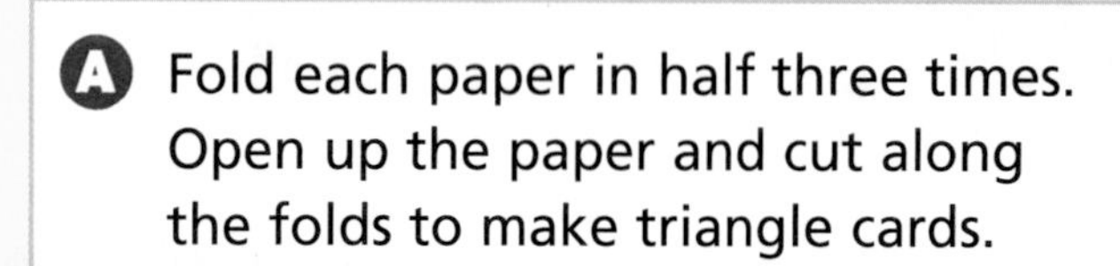

Product 15
Factor 3
5 Factor

A Fold each paper in half three times. Open up the paper and cut along the folds to make triangle cards.

B Make triangle fact cards for each of these products: 12, 15, 18, 20, 24, 25, and 30.

C Write fact families for at least 3 triangle fact cards.

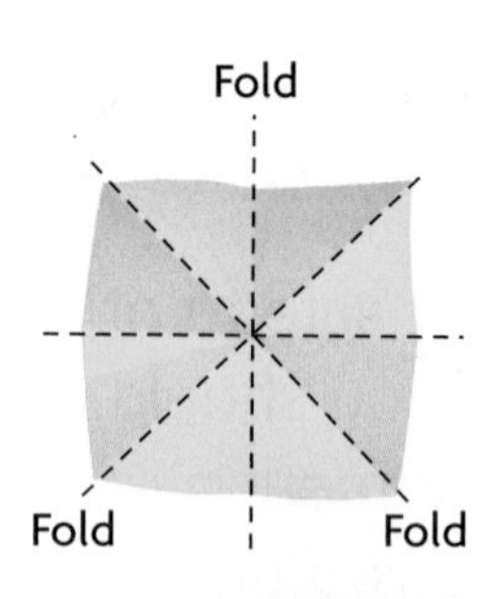

- **REASONING** How many triangle fact cards can you make for the product 12? Explain.

Using a Multiplication Table

MATH IDEA Use related multiplication facts to find quotients or missing divisors in division sentences.

Examples

A Find the quotient.
$12 \div 3 = \blacksquare$

Think: $3 \times \blacksquare = 12$

Find row 3. Look across to find the product 12.
Look up to find the missing factor, 4.
$3 \times 4 = 12$

So, $12 \div 3 = 4$.

B Find the missing divisor.
$30 \div \blacksquare = 5$

Think: $\blacksquare \times 5 = 30$

Find the factor 5 in the top row.
Look down to find the product 30.
Look left to find the missing factor, 6.
$6 \times 5 = 30$

So, $30 \div 6 = 5$.

Remember

3	×	4	=	12
↑		↑		↑
factor		factor		product

×	0	1	2	3	4	5	6
0	0	0	0	0	0	0	0
1	0	1	2	3	4	5	6
2	0	2	4	6	8	10	12
3	0	3	6	9	12	15	18
4	0	4	8	12	16	20	24
5	0	5	10	15	20	25	30
6	0	6	12	18	24	30	36

• **REASONING** How can you use multiplication to check $20 \div 5 = 4$?

Check

1. **Explain** how you can use a multiplication table to show how $5 \times 2 = 10$ and $10 \div 2 = 5$ are related.

Write the missing number for each triangle fact card.

2.

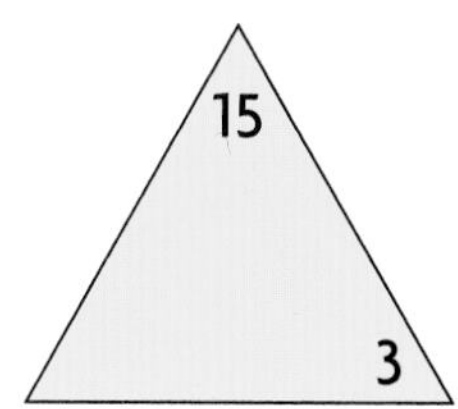

3.

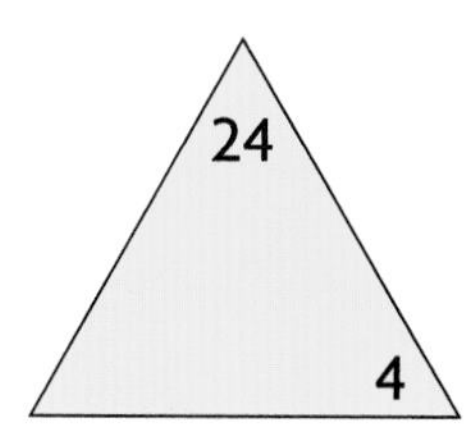

4.

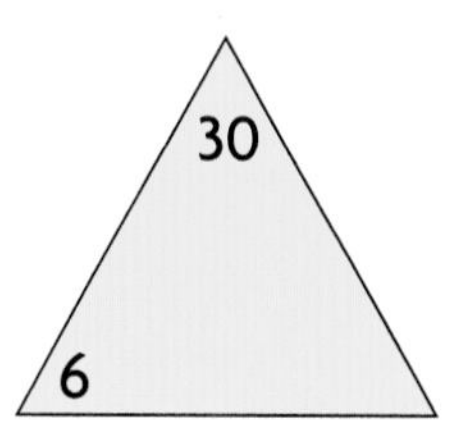

5.

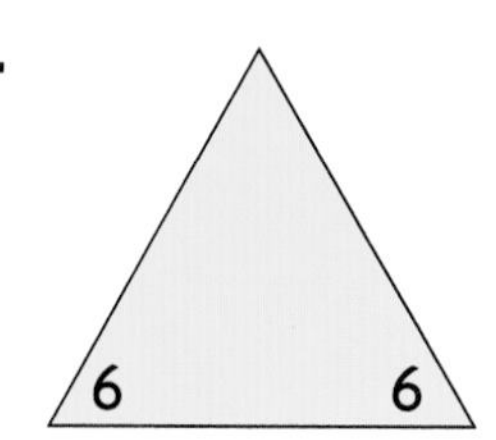

Write the fact family.

6. 3, 6, 18

7. 4, 4, 16

8. 4, 5, 20

9. 3, 7, 21

LESSON CONTINUES

Practice and Problem Solving

Write the missing number for each triangle fact card.

10. 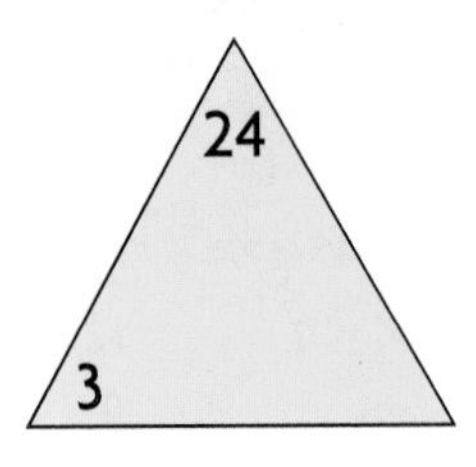

11. 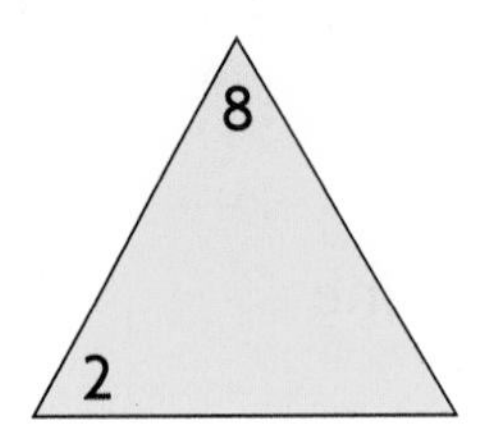

12. 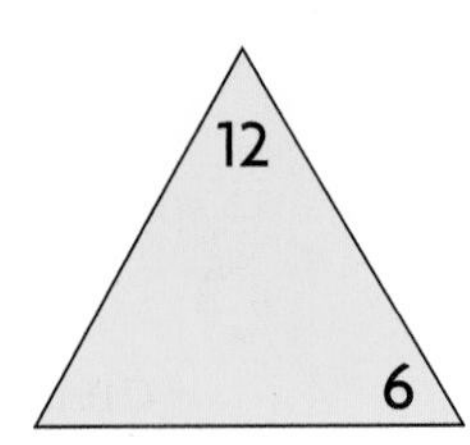

13. 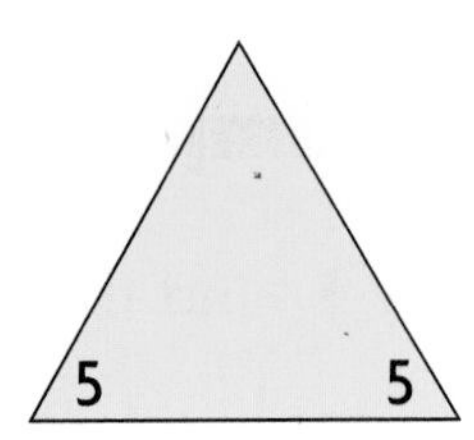

Write the fact family.

14. 5, 6, 30 **15.** 2, 8, 16 **16.** 4, 7, 28 **17.** 5, 5, 25

Find the quotient or product.

18. $3 \times 6 = \blacksquare$ **19.** $6 \times 3 = \blacksquare$ **20.** $18 \div 3 = \blacksquare$ **21.** $18 \div 6 = \blacksquare$

22. $4 \times 9 = \blacksquare$ **23.** $9 \times 4 = \blacksquare$ **24.** $36 \div 4 = \blacksquare$ **25.** $36 \div 9 = \blacksquare$

26. $8 \times 5 = \blacksquare$ **27.** $5 \times 8 = \blacksquare$ **28.** $40 \div 8 = \blacksquare$ **29.** $40 \div 5 = \blacksquare$

Write the other three sentences in the fact family.

30. $3 \times 7 = 21$ **31.** $1 \times 5 = 5$ **32.** $4 \times 3 = 12$

33. $5 \times 3 = 15$ **34.** $6 \times 4 = 24$ **35.** $9 \times 2 = 18$

Find the quotient or the missing divisor.

36. $8 \div 4 = \blacksquare$ **37.** $16 \div 2 = \blacksquare$ **38.** $7 = 21 \div \blacksquare$ **39.** $2 = 12 \div \blacksquare$

40. $24 \div 8 = \blacksquare$ **41.** $10 \div \blacksquare = 2$ **42.** $30 \div \blacksquare = 6$ **43.** $28 \div 4 = \blacksquare$

Complete.

44. $\blacksquare \div 5 = 6 + 3$ **45.** $6 \times \blacksquare = 54 \div 9$ **46.** $42 - 6 = \blacksquare \times 9$

47. What do you notice about the fact family for 6, 6, and 36?

48. REASONING How are $20 \div 5 = 4$ and $20 \div 4 = 5$ alike? How are they different?

49. Geri made 20 bookmarks. She kept 2 and then put an equal number in each of 3 gift boxes. How many bookmarks are in each box?

50. REASONING Kendra says, "There are 3 teaspoons in 1 tablespoon, so there are 15 teaspoons in 5 tablespoons." Do you agree or disagree? Explain.

51. Mr. Tapia has a water bowl and a food bowl for each cat and dog in his pet store. He has 5 dogs and 4 cats. How many bowls does he have?

52. **? What's the Error?** John says that since $4 + 4 = 8$, then $8 \div 4 = 4$. Describe his error and give the correct quotient.

Mixed Review and Test Prep

Find the sum or difference. (p. 88)

53. $\begin{array}{r} \$4.57 \\ +\$0.82 \\ \hline \end{array}$

54. $\begin{array}{r} \$3.19 \\ -\$1.35 \\ \hline \end{array}$

Complete. Write +, −, ×, or ÷ for each ●. (p. 186)

55. $14 \bullet 7 = 7$

56. $9 \bullet 5 = 45$

57. $28 \bullet 4 = 7$

58. $9 \bullet 9 = 18$

59. **TEST PREP** Denise spent 3 hours at math camp each day for 1 week. How many hours did she spend at math camp? (p. 122)

A 3 hours
B 4 hours
C 14 hours
D 21 hours

60. **TEST PREP** Mr. Li picked up Frank from soccer practice 20 minutes before 5:00. What time was it? (p. 94)

F 4:20
G 4:40
H 5:20
J 5:40

PROBLEM SOLVING  Thinker's Corner

Materials: triangle fact cards, paper

Players: 2

a. One player chooses a triangle fact card and holds it so that one number is hidden.

b. The other player names the hidden number and writes the fact family.

c. Players take turns until all cards are used. The player with the most correct number sentences wins.

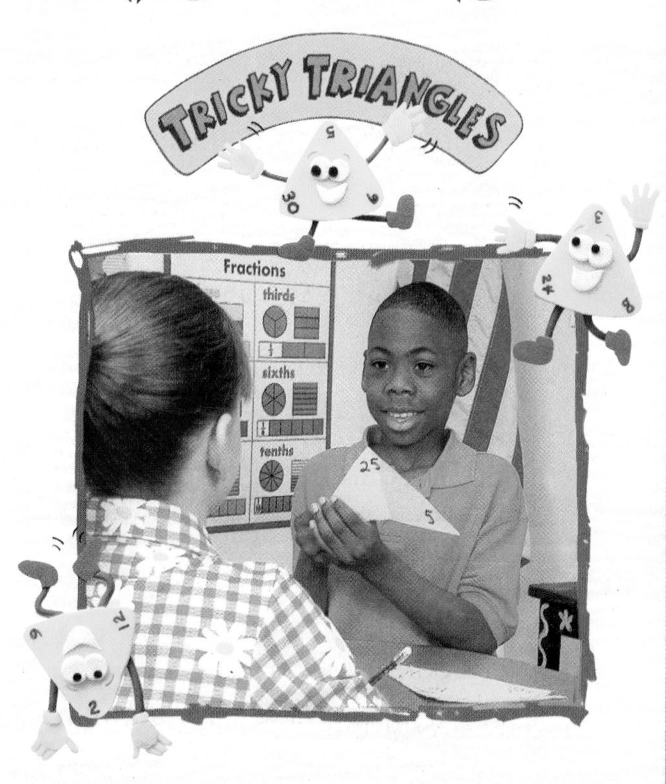

1. How can you make sure your fact families are complete?

2. What card did you hope to choose for your opponent? Explain.

LESSON 5

Problem Solving Strategy
Write a Number Sentence

Quick Review

1. $8 \times 4 = \blacksquare$
2. $30 = 6 \times \blacksquare$
3. $10 \times 5 = \blacksquare$
4. $18 = \blacksquare \times 6$
5. $27 = 9 \times \blacksquare$

PROBLEM Megan puts 36 animal trading cards in her binder. She puts 9 cards on each page. How many pages will Megan need for her cards?

UNDERSTAND

- What are you asked to find?
- What information will you use?
- Is there information you will not use? If so, what?

PLAN

- What strategy can you use?

Write a number sentence to find the number of pages Megan will need.

SOLVE

- How can you use the strategy to solve it?

Write a number sentence and solve.

36	÷	9	=	4
↑		↑		↑
number of trading cards		number on each page		number of pages

So, Megan needs 4 pages for her cards.

CHECK

- How can you decide if your answer is correct?
- What other strategy could you use?

Problem Solving Practice

Write a number sentence to solve.

1. **What if** Megan buys 27 trading cards to add to her collection? How many pages will she need for the new cards?
2. Rosita has 28 cards. She wants to keep 4 cards and divide the rest equally among 4 friends. How many cards will each friend get?

PROBLEM SOLVING STRATEGIES

- Draw a Diagram or Picture
- Make a Model or Act It Out
- Make an Organized List
- Find a Pattern
- Make a Table or Graph
- Predict and Test
- Work Backward
- Solve a Simpler Problem
- **Write a Number Sentence**
- Use Logical Reasoning

Jorge has 45 trading cards in his collection. His binder holds 10 pages. Each page holds 9 trading cards.

3. How many pages will Jorge use for the cards he has?

   A 5 **C** 45
   B 10 **D** 90

4. Which number sentence shows how to find how many trading cards fit in Jorge's binder?

   F $45 + 10 = 55$ **H** $10 + 9 = 19$
   G $9 \times 10 = 90$ **J** $45 \div 9 = 5$

Mixed Strategy Practice

USE DATA For 5–8, use the graph.

5. How many coins in all are in Sebastian's coin collection?
6. Sebastian put the same type of coins on separate pages. The coins from Japan are on page 2. The piasters are just before the coins from France. Which coins are on page 4?
7. Sebastian adds some coins from Mexico to his collection. There are 5 more coins from Japan than from Mexico. How many coins are there from Mexico?

SEBASTIAN'S COIN COLLECTION

francs (France)	
yen (Japan)	
colones (Costa Rica)	
piasters (Egypt)	

Key: Each  **= 5 coins.**

8. **Write a problem** using the data in the pictograph.

CHAPTER 11

Review/Test

CHECK VOCABULARY AND CONCEPTS

Choose the best term from the box.

variable
divide
divisor
quotient
inverse operations

1. Multiplication and division are opposite operations, or ___?___. (p. 188)
2. In $18 \div 6 = 3$, the number 3 is called the ___?___. (p. 188)
3. When you separate into equal groups, you ___?___. (p. 184)
4. A letter that is used to stand for an unknown number is called a ___?___. (p. 189)

Use subtraction to solve. (pp. 186–187)

5. $10 \div 5 = \blacksquare$
6. $27 \div 9 = \blacksquare$
7. $4\overline{)20}$
8. $8\overline{)32}$

CHECK SKILLS

Complete each number sentence.
Draw an array to help. (pp. 188–191)

9. $2 \times \blacksquare = 6$ $\quad 6 \div 2 = \blacksquare$
10. $3 \times \blacksquare = 15$ $\quad 15 \div 3 = \blacksquare$
11. $4 \times \blacksquare = 4$ $\quad 4 \div 4 = \blacksquare$
12. $5 \times \blacksquare = 30$ $\quad 30 \div 5 = \blacksquare$

Find the number that the variable stands for. (pp. 188–191)

13. $14 \div 2 = a$
    $a =$ ___?___
14. $b \times 5 = 20$
    $b =$ ___?___
15. $18 \div 9 = c$
    $c =$ ___?___

Write the fact family. (pp. 192–195)

16. 4, 5, 20
17. 2, 7, 14
18. 3, 8, 24

CHECK PROBLEM SOLVING

Write a number sentence to solve. (pp. 196–197)

19. Ms. Kraft has 20 pencils to divide equally among 5 groups of students. How many pencils does each group get?
20. Fernando has 24 rocks in his collection. If a box holds 6 rocks, how many boxes will Fernando need for his collection?

Standardized Test Prep

TIP! **Decide on a plan.**
See item **7**.

Every 20 minutes means there is a pattern of times. Write this pattern beginning at 8:00.

Also see problem **4,** p. H63.

For 1–7, choose the best answer.

1. Joanne has 40 candy hearts to put into 8 cups. How many hearts will she put in each cup so each cup has the same number?

A 3 **B** 4 **C** 5 **D** 6

2. Kevin gets paid $6 when he mows one lawn. How much money is he paid if he mows 4 lawns?

F $18
G $20
H $24
J NOT HERE

3. Jeremy has 2 pizzas. Each pizza is cut into 8 pieces. He wants to share the pizzas equally among 4 people. How many pieces will each person get?

A 4
B 6
C 10
D 14

4. Which sentence does **not** belong in the fact family for 7, 6, and 42?

F $7 \times 6 = 42$
G $42 - 6 = 36$
H $6 \times 7 = 42$
J $42 \div 7 = 6$

5. Cari took 24 ballet lessons. She took the same number of lessons each week for 8 weeks. Which number sentence tells how many lessons she took each week?

A $24 + 8 = 32$
B $24 - 8 = 16$
C $6 \times 4 = 24$
D $24 \div 8 = 3$

6. $30 \div 5 = ■$

F 5
G 6
H 25
J 30

7. A train leaves the station every 20 minutes starting at 8:00. Marcel gets to the station at 9:10. When will the next train leave?

A 9:20
B 9:30
C 9:35
D 9:45

Write What You Know

8. Write a problem that this number sentence can be used to solve.

$$20 \div 4 = ■$$

Draw a picture to solve the problem.

9. Mr. Remy wants to put 6 pencils on each table. One package has 12 pencils. He says that he needs 3 packages of pencils for 8 tables. Is he correct or incorrect? Explain.

CHAPTER 12 Division Facts Through 5

This farmer is shearing a sheep.

Sheep are an important source of wool for clothing. The white, fluffy fur on a sheep is called *fleece*. An average fleece makes about 48 ounces of yarn.

DATA LINK

AMOUNT OF YARN NEEDED

scarf	🧶🧶🧶🧶
adult sweater	🧶🧶🧶🧶🧶🧶🧶🧶🧶🧶🧶🧶🧶🧶🧶🧶
pair of socks	🧶
dog sweater (medium)	🧶🧶🧶🧶

Key: Each 🧶 = 2 ounces.

PROBLEM SOLVING Look at the pictograph. How many scarves could be made from the fleece of one sheep? Ca 6 dog sweaters be made from the fleece of one sheep?

CHECK WHAT YOU KNOW

Use this page to help you review and remember important skills needed for Chapter 12.

VOCABULARY

Choose the best term from the box.

1. 15 + 26 = 41 is called a ?.

expression
number sentence

SUBTRACTION (For Intervention, see p. H15.)

Find each difference.

2. $14 - 2$
3. $10 - 0$
4. $20 - 4$
5. $18 - 6$
6. $27 - 9$
7. $45 - 9$
8. $21 - 3$
9. $15 - 5$
10. $30 - 10$
11. $16 - 2$

Find each missing number.

12. $24 - ■ = 18$
13. $30 - ■ = 25$
14. $27 - ■ = 24$
15. $32 - ■ = 24$
16. $18 - ■ = 16$
17. $36 = 42 - ■$

MULTIPLICATION FACTS THROUGH 10 (For Intervention, see p. H14.)

Find each product.

18. $6 \times 3 = ■$
19. $3 \times 10 = ■$
20. $■ = 9 \times 2$
21. $7 \times 4 = ■$
22. $■ = 0 \times 7$
23. $■ = 10 \times 1$
24. $5 \times 8 = ■$
25. $3 \times 9 = ■$
26. $1 \times 1 = ■$
27. $6 \times 5 = ■$
28. $4 \times 6 = ■$
29. $■ = 2 \times 10$
30. 9×6
31. 8×9
32. 0×4
33. 7×2

LESSON 1

Divide by 2 and 5

Quick Review

1. $2 \times \blacksquare = 10$
2. $4 \times \blacksquare = 8$
3. $5 \times \blacksquare = 20$
4. $\blacksquare \times 5 = 35$
5. $\blacksquare \times 2 = 18$

Learn

CRAFTY MATH Mrs. Jackson knit 12 hats. She put an equal number of hats on each of 2 shelves in the craft shop. How many hats are on each shelf?

$12 \div 2 = \blacksquare$

Use a related multiplication fact to find the quotient.

Think: $2 \times \blacksquare = 12$

$2 \times 6 = 12$

$12 \div 2 = 6$, or $2\overline{)12}$ with quotient 6

So, there are 6 hats on each shelf.

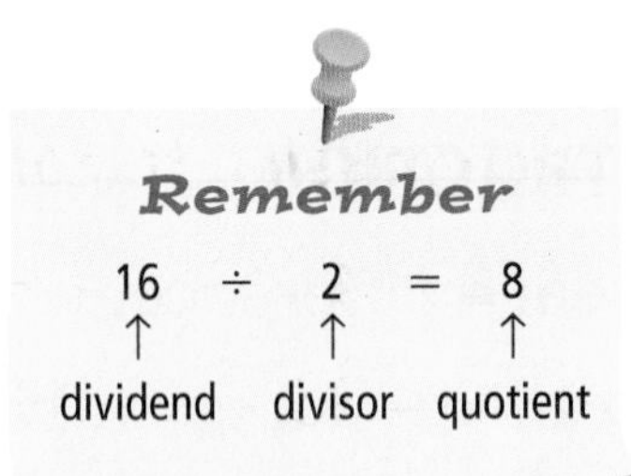

×	0	1	2	3	4	5
0	0	0	0	0	0	0
1	0	1	2	3	4	5
2	0	2	4	6	8	10
3	0	3	6	9	12	15
4	0	4	8	12	16	20
5	0	5	10	15	20	25
6	0	6	12	18	24	30

What if Mrs. Jackson knits 15 hats and puts an equal number of hats on each of 5 shelves? How many hats are on each shelf?

$15 \div 5 = \blacksquare$

Think: $5 \times \blacksquare = 15$ $\quad 5 \times 3 = 15$

$15 \div 5 = 3$, or $5\overline{)15}$ with quotient 3

So, there are 3 hats on each shelf.

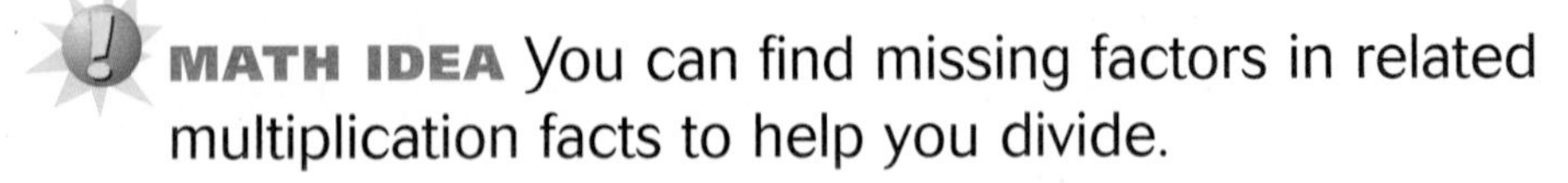

MATH IDEA You can find missing factors in related multiplication facts to help you divide.

- How can you use $2 + 2 + 2 = 6$ to help you find $6 \div 2$?

Check

1. **Explain** how you can use multiplication to check $20 \div 5 = 4$.

Copy and complete each table.

2.

÷	2	4	6	8
2	■	■	■	■

3.

÷	10	15	20	25
5	■	■	■	■

Practice and Problem Solving

Copy and complete each table.

4.

÷	10	12	14	16
2	■	■	■	■

5.

÷	30	35	40	45
5	■	■	■	■

Find each missing factor and quotient.

6. $2 \times \blacksquare = 4$ $\quad 4 \div 2 = \blacksquare$

7. $5 \times \blacksquare = 20$ $\quad 20 \div 5 = \blacksquare$

8. $5 \times \blacksquare = 35$ $\quad 35 \div 5 = \blacksquare$

9. $2 \times \blacksquare = 16$ $\quad 16 \div 2 = \blacksquare$

Find each quotient.

10. $6 \div 2 = \blacksquare$

11. $10 \div 2 = \blacksquare$

12. $\blacksquare = 10 \div 5$

13. $5 \div 5 = \blacksquare$

14. $25 \div 5 = \blacksquare$

15. $\blacksquare = 14 \div 2$

16. $40 \div 5 = \blacksquare$

17. $20 \div 2 = \blacksquare$

18. $2\overline{)2}$

19. $5\overline{)15}$

20. $5\overline{)35}$

21. $2\overline{)16}$

Complete.

22. $10 \div 2 = \blacksquare \times 1$

23. $40 \div 5 = 4 \times \blacksquare$

24. $\blacksquare \div 2 = 3 + 4$

25. **REASONING** What do you notice about the numbers that can be evenly divided by 2?

26. Mrs. Jackson sells hats for $5. She has $15. How many more hats must she sell to have $35 in all?

27. **What's the Error?** Philip used the multiplication fact $2 \times 8 = 16$ to find $8 \div 2 = \blacksquare$. Describe his error. What is the correct quotient?

Mixed Review and Test Prep

28. $\begin{array}{r} 7 \\ \times 8 \\ \hline \end{array}$ (p. 150)

29. $\begin{array}{r} 8 \\ \times 6 \\ \hline \end{array}$ (p. 152)

30. $76 + 67 + 22 = \blacksquare$ (p. 36)

31. $3 \times 2 \times \blacksquare = 48$ (p. 170)

32. **TEST PREP** James made an array with 3 rows of 9 tiles. Choose the number sentence that shows how many tiles are in the array. (p. 120)

A $9 - 3 = 6$

B $3 + 9 = 12$

C $9 \div 3 = 3$

D $3 \times 9 = 27$

EXTRA PRACTICE page H43, Set A

LESSON 2

Divide by 3 and 4

Quick Review

1. ■ × 3 = 18
2. 4 × ■ = 12
3. ■ × 4 = 16
4. 4 × ■ = 32
5. 3 × ■ = 21

Learn

PADDLE POWER The Traveler Scouts want to rent canoes. There are 24 people in the group. A canoe can hold 3 people. How many canoes should the group rent?

24 ÷ 3 = ■

Use a related multiplication fact.

×	0	1	2	3	4	5	6	7	8	9
0	0	0	0	0	0	0	0	0	0	0
1	0	1	2	3	4	5	6	7	8	9
2	0	2	4	6	8	10	12	14	16	18
3	0	3	6	9	12	15	18	21	24	27
4	0	4	8	12	16	20	24	28	32	36
5	0	5	10	15	20	25	30	35	40	45

Think: 3 × ■ = 24
3 × 8 = 24   24 ÷ 3 = 8, or $3\overline{)24}$ with quotient 8

So, the group should rent 8 canoes.

What if the group wants to rent rowboats instead? If each rowboat holds 4 people, how many rowboats should they rent?

24 ÷ 4 = ■

Think: 4 × ■ = 24
4 × 6 = 24   24 ÷ 4 = 6, or $4\overline{)24}$ with quotient 6

So, the group should rent 6 rowboats.

- **REASONING** How can you use 21 ÷ 3 = 7 to find 24 ÷ 3?

Check

1. **Explain** how you can use multiplication to find 12 ÷ 4.

Write the multiplication fact you can use to find the quotient. Then write the quotient.

2. 12 ÷ 3 = ■
3. 8 ÷ 4 = ■
4. 15 ÷ 3 = ■
5. 28 ÷ 4 = ■

Practice and Problem Solving

More Practice: Use Mighty Math Calculating Crew, *Snap Clowns*, Levels **R** and **Y**.

Write the multiplication fact you can use to find the quotient. Then write the quotient.

6. $27 \div 3 = \blacksquare$ **7.** $\blacksquare = 4 \div 4$ **8.** $30 \div 3 = \blacksquare$

9. $16 \div 4 = \blacksquare$ **10.** $18 \div 3 = \blacksquare$ **11.** $\blacksquare = 20 \div 4$

Copy and complete each table.

12.

÷	9	12	15	18
3	■	■	■	■

13.

÷	16	20	24	28
4	■	■	■	■

Find each quotient.

14. $12 \div 4 = \blacksquare$ **15.** $\blacksquare = 6 \div 3$ **16.** $\blacksquare = 14 \div 2$ **17.** $12 \div 2 = \blacksquare$

18. $15 \div 3 = \blacksquare$ **19.** $25 \div 5 = \blacksquare$ **20.** $24 \div 4 = \blacksquare$ **21.** $\blacksquare = 40 \div 4$

22. $\blacksquare = 18 \div 2$ **23.** $32 \div 4 = \blacksquare$ **24.** $\blacksquare = 9 \div 3$ **25.** $30 \div 5 = \blacksquare$

26. $3\overline{)3}$ **27.** $3\overline{)18}$ **28.** $4\overline{)36}$ **29.** $5\overline{)20}$

Complete.

30. $20 \div 4 = 8 - \blacksquare$ **31.** $24 \div 3 = \blacksquare \times 2$ **32.** $36 \div \blacksquare = 18 \div 2$

33. Yusef collected 38 pine cones. He kept 11 pine cones for himself and divided the rest equally among 3 friends. How many pine cones did each friend get?

34. The scouts saw squirrels and birds in the woods. If there were 4 animals and 12 legs, how many squirrels and birds were there?

35. **REASONING** Two numbers have a product of 16 and a quotient of 4. What are they?

36. **Write About It** Explain how to solve $32 \div 4$ in at least 2 different ways.

Mixed Review and Test Prep

37. 3×5 (p. 122)

38. 5×7 (p. 118)

39. 4×7 (p. 134)

40. 9×8 (p. 164)

41. **TEST PREP** Paula had \$1.36. Her aunt gave her \$5.25 for her birthday. How much money does Paula have now? (p. 88)

A \$3.89 **C** \$6.51
B \$5.61 **D** \$6.61

LESSON 3

Divide with 0 and 1

Quick Review

1. $8 \times \blacksquare = 8$
2. $3 \times 0 = \blacksquare$
3. $\blacksquare \times 1 = 4$
4. $0 \times 10 = \blacksquare$
5. $1 \times \blacksquare = 7$

Learn

MOO . . . VE OVER Here are some rules for dividing with 0 and 1.

RULE A

Any number divided by 1 equals that number.

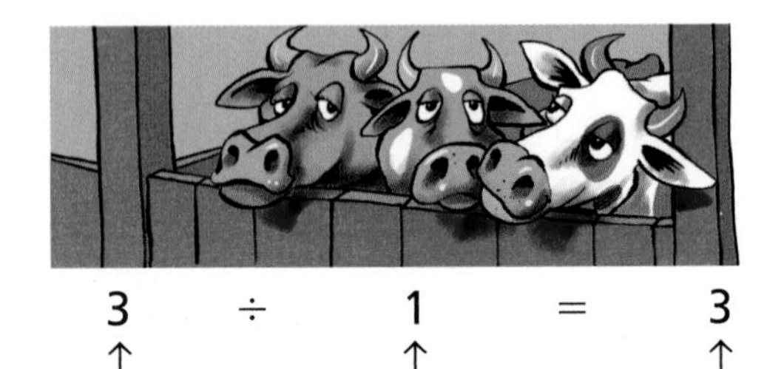

$3 \div 1 = 3$

3 ↑ number of cows; 1 ↑ number of stalls; 3 ↑ number in each stall

If there is only 1 stall, then all of the cows must be in that stall.

RULE B

Any number (except 0) divided by itself equals 1.

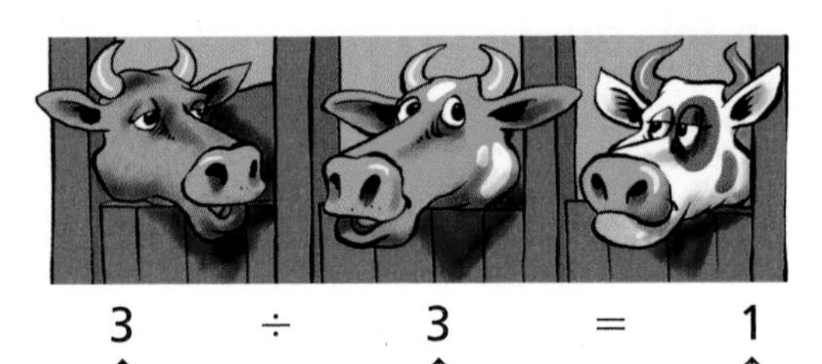

$3 \div 3 = 1$

3 ↑ number of cows; 3 ↑ number of stalls; 1 ↑ number in each stall

If there are the same number of cows and stalls, then one cow goes in each stall.

RULE C

Zero divided by any number (except 0) equals 0.

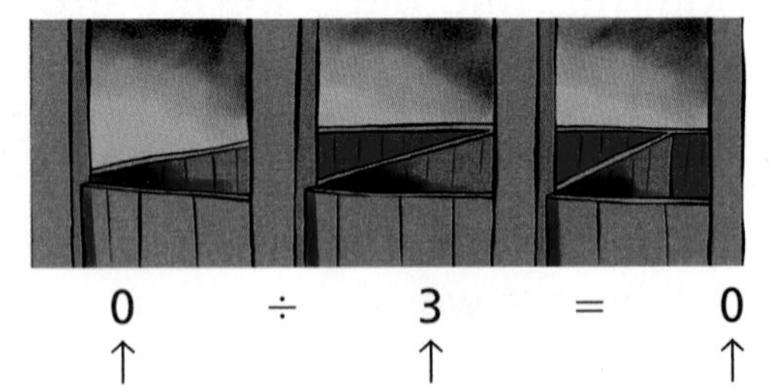

$0 \div 3 = 0$

0 ↑ number of cows; 3 ↑ number of stalls; 0 ↑ number in each stall

If there are no cows, then no matter how many stalls you have, there won't be any cows in the stalls.

RULE D

You cannot divide by 0.

If there are no stalls, then you aren't separating cows into equal groups. So, using division doesn't make sense.

- **REASONING** How can you use multiplication to show that $3 \div 0 = \blacksquare$ doesn't make sense?

Check

1. **Explain** how you can use multiplication to check $0 \div 9 = 0$.

Find each quotient.

2. $3 \div 3 = ■$
3. $■ = 5 \div 1$
4. $0 \div 2 = ■$
5. $■ = 6 \div 6$
6. $7 \div 1 = ■$
7. $0 \div 6 = ■$
8. $■ = 4 \div 4$
9. $10 \div 1 = ■$

Practice and Problem Solving

Find each quotient.

10. $2 \div 1 = ■$
11. $8 \div 8 = ■$
12. $■ = 6 \div 3$
13. $1 \div 1 = ■$
14. $20 \div 5 = ■$
15. $■ = 0 \div 4$
16. $■ = 5 \div 5$
17. $10 \div 2 = ■$
18. $3 \div 1 = ■$
19. $21 \div 3 = ■$
20. $32 \div 4 = ■$
21. $■ = 0 \div 7$
22. $■ = 0 \div 8$
23. $18 \div 2 = ■$
24. $■ = 9 \div 1$
25. $24 \div 4 = ■$
26. Divide 2 by 2.
27. Divide 4 by 1.
28. Divide 0 by 3.
29. Divide 14 by 2.
30. $5\overline{)35}$
31. $9\overline{)9}$
32. $2\overline{)16}$
33. $5\overline{)0}$
34. $2\overline{)14}$
35. $3\overline{)18}$
36. $1\overline{)8}$
37. $4\overline{)36}$
38. $7\overline{)7}$
39. $1\overline{)0}$

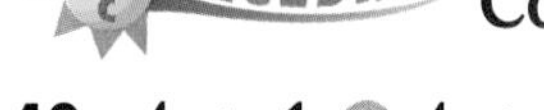

Compare. Write <, >, or = for each ●.

40. $4 \div 1$ ● $4 \div 4$
41. $0 \div 9$ ● $9 \div 1$
42. $6 + 4$ ● $5 \div 1$
43. A farmer has 6 bales of hay. He feeds 2 bales to his cows. He divides the rest equally among 4 stalls. How many bales are in each stall?
44. **REASONING** Chelsea says, "Ask me to divide any number by 1, and I'll give you the quotient." What is her strategy?
45. Use what you know about 0 and 1 to find each quotient.
    a. $398 \div 398 = ■$ b. $971 \div 1 = ■$ c. $0 \div 426 = ■$

Mixed Review and Test Prep

46. $\begin{array}{r} 0 \\ \times 4 \\ \hline \end{array}$ (p. 132)
47. $\begin{array}{r} 5 \\ \times 1 \\ \hline \end{array}$ (p. 132)
48. $387 + 132 + 155 = ■$ (p. 42)
49. $\$10.00 - \$5.69 = ■$ (p. 58)
50. **TEST PREP** Akiko bought 3 sandwiches. Each sandwich cost \$4. How much did she spend for the sandwiches? (p. 122)
    A \$4 **B** \$7 **C** \$12 **D** \$15

EXTRA PRACTICE page H43, Set C

LESSON

Algebra: Write Expressions

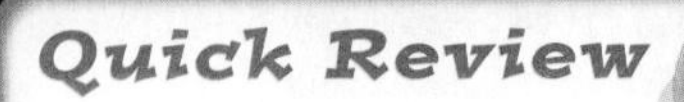

Quick Review

1. $20 \div 4 = \blacksquare$
2. $0 \div 7 = \blacksquare$
3. $10 \div 10 = \blacksquare$
4. $36 \div 4 = \blacksquare$
5. $16 \div 2 = \blacksquare$

Learn

HAPPY CAMPERS The camp counselor divided 21 campers into 3 equal groups. How many campers are in each group?

Write an expression to show how many campers are in each group.

21 campers	divided into	3 groups
↓	↓	↓
21	÷	3

$21 \div 3 = 7$ is a number sentence. 7 is the number of campers in each group.

Remember

An *expression* is part of a number sentence.

Examples:

$3 + 4$	4×2
$25 - 12$	$12 \div 6$

You know a number sentence can be true or false.

$12 \div 3 = 4$ is true.

$12 \times 3 = 4$ is false.

Mr. Gonzales is lining up 4 rows of 5 campers for relay races. There are 20 campers lined up.

$4 \bullet 5 = 20$

Which symbol will make the sentence true?

Try +	$4 + 5 = 20$	*Not* true
Try −	$4 - 5 = 20$	*Not* true
Try ÷	$4 \div 5 = 20$	*Not* true
Try ×	$4 \times 5 = 20$	True

So, the correct operation symbol is ×.

Check

1. **Write** a different expression to describe the relay race problem above.

Write an expression to describe each problem.

2. Nine campers each ate 7 carrot sticks. How many carrot sticks did they eat in all?

3. Jo had 9 carrot sticks. She ate 7 of them. How many carrot sticks does she have left?

Write +, −, ×, or ÷ to make the number sentence true.

4. $9 = 18$ ● 9

5. $6 \times 6 = 4$ ● 9

6. 72 ● $8 = 3 \times 3$

Practice and Problem Solving

Write an expression to describe each problem.

7. Antoine had 24 grapes. His mother gave him 6 more. How many grapes does he have now?

8. Six students share 30 crayons equally. How many crayons does each student get?

9. Beth bought 6 pieces of drawing paper, 8 crayons, and 1 eraser. How many items did she buy?

10. Alison had $0.45. She lost a dime. How much money does she have now?

Write +, −, ×, or ÷ to make the number sentence true.

11. 13 ● $7 = 2 \times 3$

12. $12 + 5 = 9$ ● 8

13. $6 \times 4 = 8$ ● 3

For 14–15, write two different expressions you can use to describe each problem.

14. Carl and 3 of his friends each ate 5 roasted marshmallows. How many marshmallows did Carl and his friends eat in all?

15. Bruce has a photo album with 8 pages. He can fit 6 photos on each page. How many photos can he put in the album?

16. **Write a problem** for each expression.

a. $35 \div 5$ b. 7×3 c. $15 - 3$

Mixed Review and Test Prep

17. $125 + 291 =$ ■ (p. 42)

18. $3{,}538 + 1{,}896 =$ ■ (p. 46)

19. $345 - 77 =$ ■ (p. 58)

20. $622 - 404 =$ ■ (p. 58)

21. **TEST PREP** Anna has a music lesson at 4:30 P.M. The lesson will last 45 minutes. At what time will Anna's lesson end? (p. 98)

A 4:15 P.M.
B 4:45 P.M.
C 5:15 P.M.
D 5:45 P.M.

EXTRA PRACTICE page H43, Set D

LESSON 5

Problem Solving Skill
Choose the Operation

UNDERSTAND | PLAN | SOLVE | CHECK

Quick Review

Choose +, −, ×, or ÷ for each ●.

1. 18 ● 5 = 13
2. 16 ● 8 = 2
3. 3 ● 2 = 6
4. 4 ● 4 = 16
5. 15 − 7 = 4 ● 4

NATURE WALK On a hike, the campers saw 6 chipmunks, 4 deer, and 8 butterflies. How many more butterflies did they see than chipmunks?

This chart can help you decide when to use each operation.

ADD	• Join groups of different sizes.
SUBTRACT	• Take away. • Compare amounts.
MULTIPLY	• Join equal groups.
DIVIDE	• Separate into equal groups. • Find the number in each group.

MATH IDEA Before you solve a problem, decide what operation to use. Write a number sentence to solve the problem.

Since you are comparing amounts, subtract.

$$8 - 6 = 2$$

8 → number of butterflies; 6 → number of chipmunks; 2 → how many more butterflies than chipmunks

So, they saw 2 more butterflies than chipmunks.

- **REASONING** When would you use division to solve a problem?
- Write a number sentence to find how many animals they saw in all.

Problem Solving Practice

Choose the operation. Write a number sentence. Then solve.

1. David collected 8 acorns and 16 wildflowers. He put the same number of wildflowers in each of 8 vases. How many wildflowers were in each vase?

2. The camp counselor put 4 pears, 5 apples, and 7 bananas in a basket. If 3 pieces of fruit were taken, how many pieces of fruit were left?

3. Beth has a scrapbook. Each page can hold 8 small postcards or 6 large postcards. How many small postcards fit on 4 pages?

4. Shawn took 8 photos of birds, 9 photos of wildflowers, and 15 photos of campers. How many photos did he take?

Thirty students went on an amusement park ride. Five students sat in each car. How many cars did they fill?

5. Which number sentence can you use to solve the problem?
   - **A** $30 + 5 = \blacksquare$
   - **B** $30 - 5 = \blacksquare$
   - **C** $30 \div 5 = \blacksquare$
   - **D** $30 \times 5 = \blacksquare$

6. What is the answer to the question?
   - **F** 25 students
   - **G** 6 cars
   - **H** 6 students
   - **J** 1 car

Mixed Applications

Use Data For 7–9, use the graph.

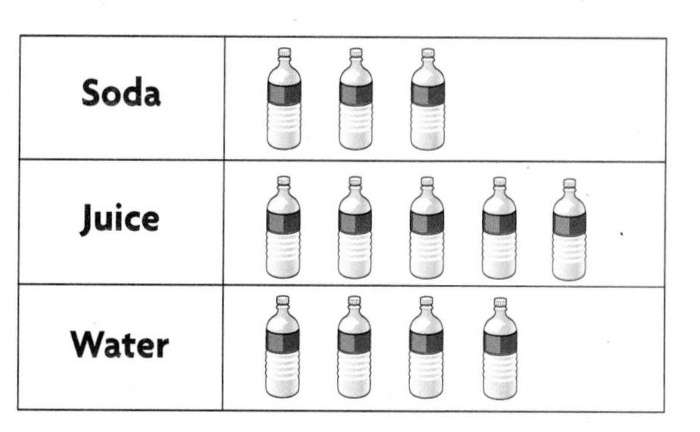

7. Gina took 5 packs of soda to her club meeting. Were there enough bottles of soda for 28 people? Explain.

8. Khar bought 3 packs of water. He gave 4 bottles to friends. How many bottles of water did he have left?

9. **What's the Question?** The answer is 30 bottles of juice.

Review/Test

CHECK VOCABULARY

Choose the best term from the box.

expression
divided
multiplied
zero

1. You cannot divide by _?_. (p. 206)
2. Any number (except 0) _?_ by itself equals 1. (p. 206)
3. A number phrase like $12 \div 2$ or 3×5 is called an _?_. (p. 208)

CHECK SKILLS

Find each quotient. (pp. 202–207)

4. $16 \div 4 = ■$
5. $■ = 21 \div 3$
6. $6 \div 1 = ■$
7. $■ = 25 \div 5$
8. $8 \div 2 = ■$
9. $0 \div 5 = ■$
10. $■ = 9 \div 3$
11. $■ = 8 \div 1$
12. $4\overline{)32}$
13. $1\overline{)10}$
14. $2\overline{)18}$
15. $3\overline{)0}$
16. $5\overline{)15}$
17. $2\overline{)14}$
18. $3\overline{)15}$
19. $5\overline{)0}$
20. $4\overline{)24}$
21. $5\overline{)40}$

Write an expression to describe each problem. (pp. 208–209)

22. Lila made 18 muffins. She put 3 muffins in each bag. How many bags did Lila fill?
23. Kyle had 32 shells. He gave 8 to his friend. How many shells does Kyle have now?

CHECK PROBLEM SOLVING

Choose the operation. Write a number sentence. Then solve. (pp. 210–211)

24. Chiang has 24 trading cards. She puts the cards into piles of 6. How many piles does she make?
25. Casey has 5 packs of stickers. Each pack has 8 stickers. How many stickers does he have?

Standardized Test Prep

TIP! **Check your work.**
See item **4.**
Count by 5 and keep track of the numbers you say until you get to 45.
Also see problem **7**, p. H65.

For 1–8, choose the best answer.

1. On Friday, 25 students will go to the zoo. They will be in 5 equal groups. How many students will be in each group?

A 30 **B** 10 **C** 6 **D** 5

2. Which sentence does **not** belong in the fact family for 2, 8, and 16?

F $16 - 8 = 8$
G $16 \div 2 = 8$
H $8 \times 2 = 16$
J $2 \times 8 = 16$

3. There are 8 cookies for 4 people to share equally. Which number sentence tells how many cookies each person can have?

A $8 + 4 = 12$ **C** $8 \times 4 = 32$
B $8 \div 4 = 2$ **D** $8 - 4 = 4$

4. Farmer Mac has 45 pigs and wants to put 5 pigs in each pen. How many pens will he use?

F 6 **G** 7 **H** 8 **J** 9

5. $\begin{array}{r} 509 \\ +493 \\ \hline \end{array}$

A 1,002 **C** 902
B 996 **D** 196

6. $15 \div 3 = \blacksquare$

F 4 **H** 6
G 5 **J** NOT HERE

7. Mr. Sato bought a television that cost $287. What is that amount rounded to the nearest hundred dollars?

A $200 **C** $300
B $250 **D** $400

8. Samantha bought a sandwich for $1.68, apple chips for $0.79, and a drink for $1.29. How much change did she get back from $5.00?

F $3.76 **H** $1.34
G $2.24 **J** $1.24

Write What You Know

9. Find the quotients. Explain how you found each. Which quotient is greater?

$5 \div 5 = \blacksquare$
$0 \div 5 = \blacksquare$

10. Joey has 27 stamps to put in his stamp book. He wants to put 9 stamps on each page. Write an expression to find how many pages he will need. Explain your expression. Then find the number of pages.

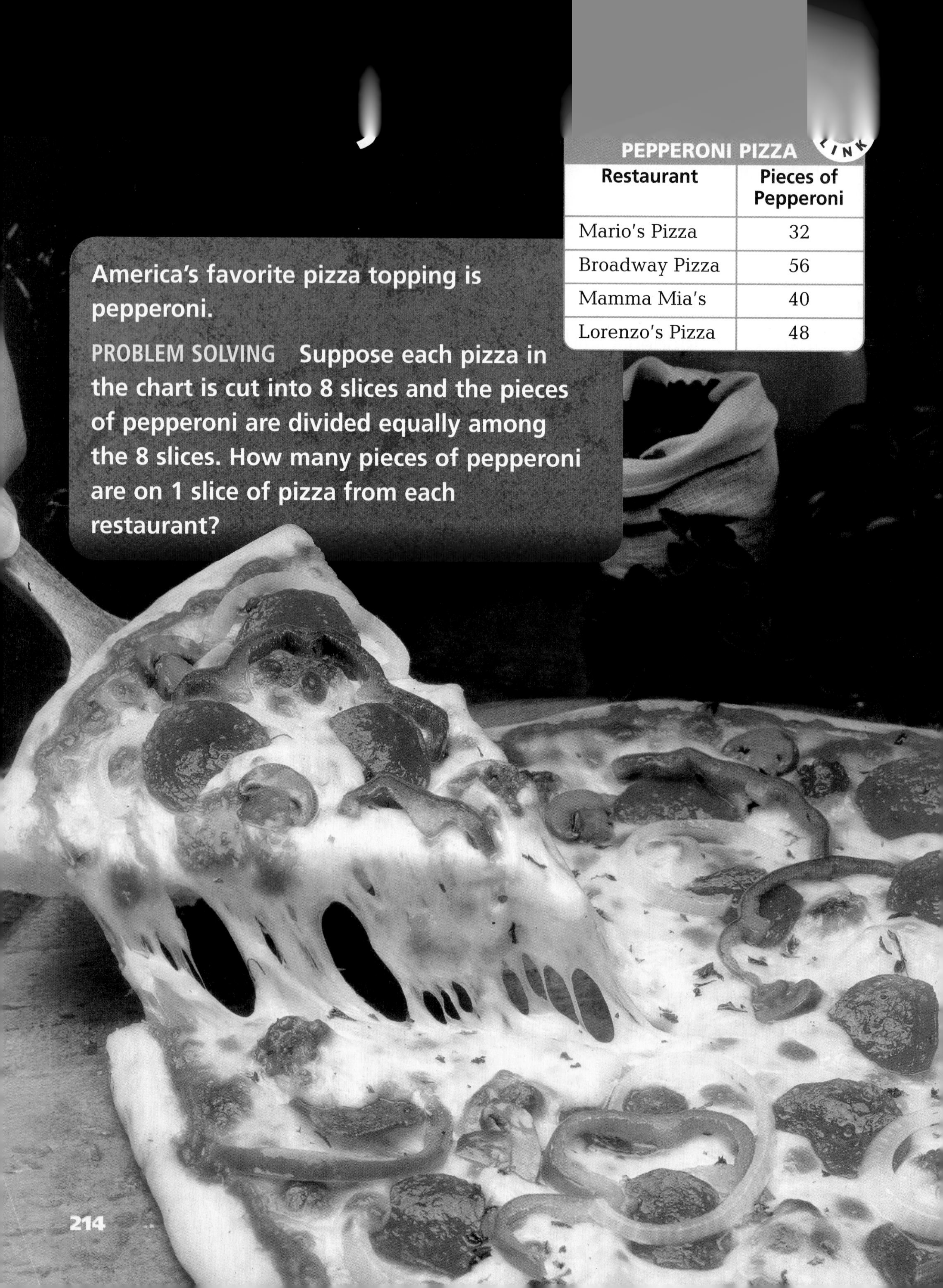

PEPPERONI PIZZA	
Restaurant	**Pieces of Pepperoni**
Mario's Pizza	32
Broadway Pizza	56
Mamma Mia's	40
Lorenzo's Pizza	48

America's favorite pizza topping is pepperoni.

PROBLEM SOLVING Suppose each pizza in the chart is cut into 8 slices and the pieces of pepperoni are divided equally among the 8 slices. How many pieces of pepperoni are on 1 slice of pizza from each restaurant?

CHECK WHAT YOU KNOW

Use this page to help you review and remember important skills needed for Chapter 13.

VOCABULARY

Choose the best term from the box.

Example: $18 \div 3 = 6$

quotient
dividend
divisor

1. In the example, the number 6 is called the _?_.
2. In the example, the number 3 is called the _?_.

DIVISION FACTS THROUGH 5 (For Intervention, see p. H15.)

Find each quotient.

3. $35 \div 5 = ■$ **4.** $8 \div 2 = ■$ **5.** $20 \div 4 = ■$ **6.** $0 \div 5 = ■$

7. $■ = 18 \div 3$ **8.** $12 \div 4 = ■$ **9.** $■ = 9 \div 1$ **10.** $16 \div 2 = ■$

11. $4\overline{)16}$ **12.** $3\overline{)24}$ **13.** $1\overline{)6}$ **14.** $8\overline{)0}$ **15.** $2\overline{)14}$

16. $1\overline{)10}$ **17.** $6\overline{)0}$ **18.** $3\overline{)15}$ **19.** $4\overline{)28}$ **20.** $2\overline{)10}$

ORDER PROPERTY OF MULTIPLICATION (For Intervention, see p. H16.)

Use the Order Property of Multiplication to help you find each product.

21. $3 \times 5 = ■$ $5 \times 3 = ■$ **22.** $7 \times 8 = ■$ $8 \times 7 = ■$

23. $6 \times 8 = ■$ $8 \times 6 = ■$ **24.** $3 \times 6 = ■$ $6 \times 3 = ■$

25. $7 \times 9 = ■$ $9 \times 7 = ■$ **26.** $6 \times 10 = ■$ $10 \times 6 = ■$

MISSING FACTORS (For Intervention, see p. H16.)

Find the missing factor.

27. $3 \times ■ = 27$ **28.** $25 = ■ \times 5$ **29.** $■ \times 6 = 12$ **30.** $45 = 9 \times ■$

31. $6 \times ■ = 48$ **32.** $■ \times 8 = 32$ **33.** $35 = 5 \times ■$ **34.** $20 = ■ \times 4$

35. $■ \times 2 = 18$ **36.** $80 = ■ \times 10$ **37.** $10 \times ■ = 10$ **38.** $56 = 8 \times ■$

LESSON 1

Divide by 6, 7, and 8

Quick Review

1. $6 \times 3 = \blacksquare$
2. $6 \times 7 = \blacksquare$
3. $7 \times 4 = \blacksquare$
4. $7 \times 8 = \blacksquare$
5. $8 \times 4 = \blacksquare$

Learn

IT'S IN THE BAG The Bagel Stop sells bagels in bags of 6. Ramona has 24 fresh bagels to put in bags. How many bags does she need?

$24 \div 6 = \blacksquare$

Use a related multiplication fact to find the quotient.

Think: $6 \times \blacksquare = 24$
$6 \times 4 = 24$

$24 \div 6 = 4$, or $6\overline{)24}$ with quotient 4

So, Ramona needs 4 bags.

Examples

A $63 \div 7 = \blacksquare$
Think: $7 \times \blacksquare = 63$
$7 \times 9 = 63$
$63 \div 7 = 9$, or $7\overline{)63}$ with quotient 9

B $56 \div 8 = \blacksquare$
Think: $8 \times \blacksquare = 56$
$8 \times 7 = 56$
$56 \div 8 = 7$, or $8\overline{)56}$ with quotient 7

MATH IDEA Think of related multiplication facts to help you divide.

- What multiplication fact can you use to find $42 \div 7$? What is the quotient?

Technology Link

More Practice: To learn more about division, watch the Harcourt Math Newsroom Video *Giant Panda Bears.*

Equal Groups

Remember, you can also use equal groups and arrays to help you find a quotient.

Here are two different ways to find 28 ÷ 7.

Jane

I used counters to model equal groups.

28	÷	7	=	■
number of counters		number in each group		number of groups

So, 28 ÷ 7 = 4.

Kevin

I modeled the problem with an array.

28	÷	7	=	■
number in array		number in each row		number of rows

So, 28 ÷ 7 = 4.

- **REASONING** You have used equal groups and arrays to help you find products. Why can you also use them to help you find quotients?

Check

1. **Explain** how you would use a related multiplication fact to find 18 ÷ 6.

Find the missing factor and quotient.

2. 8 × ■ = 16   16 ÷ 8 = ■
3. 7 × ■ = 28   28 ÷ 7 = ■
4. 6 × ■ = 36   36 ÷ 6 = ■
5. 6 × ■ = 30   30 ÷ 6 = ■

Copy and complete each table.

6.

÷	14	21	28	35
7	■	■	■	■

7.

÷	6	12	18	24
6	■	■	■	■

Find the quotient.

8. 21 ÷ 7 = ■
9. 42 ÷ 6 = ■
10. ■ = 56 ÷ 7
11. 32 ÷ 8 = ■
12. $8\overline{)40}$
13. $7\overline{)42}$
14. $8\overline{)24}$
15. $6\overline{)24}$

LESSON CONTINUES

Practice and Problem Solving

Find the missing factor and quotient.

16. $7 \times ■ = 14$ $\quad 14 \div 7 = ■$

17. $6 \times ■ = 60$ $\quad 60 \div 6 = ■$

18. $6 \times ■ = 48$ $\quad 48 \div 6 = ■$

19. $8 \times ■ = 72$ $\quad 72 \div 8 = ■$

20. $7 \times ■ = 42$ $\quad 42 \div 7 = ■$

21. $8 \times ■ = 40$ $\quad 40 \div 8 = ■$

Copy and complete each table.

22.

÷	42	63	56	49
7	■	■	■	■

23.

÷	56	40	48	32
8	■	■	■	■

Find the quotient.

24. $36 \div 6 = ■$

25. $80 \div 8 = ■$

26. $■ = 0 \div 7$

27. $8 \div 1 = ■$

28. $■ = 15 \div 3$

29. $■ = 18 \div 6$

30. $45 \div 5 = ■$

31. $24 \div 8 = ■$

32. $8\overline{)64}$

33. $2\overline{)14}$

34. $7\overline{)28}$

35. $6\overline{)0}$

36. $5\overline{)10}$

37. $7\overline{)7}$

38. $8\overline{)0}$

39. $3\overline{)30}$

40. Divide 42 by 6.

41. Divide 8 by 8.

42. Divide 35 by 5.

Write a division sentence for each.

43.

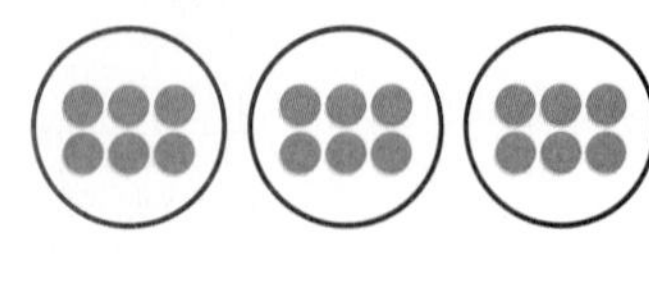

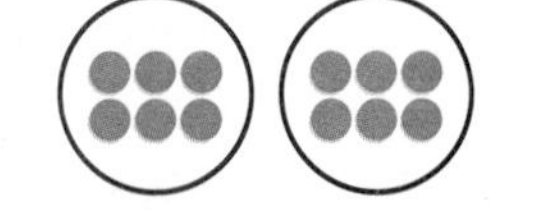

44.

ALGEBRA Complete.

45. $3 + ■ = 49 \div 7$

46. $8 \times 5 = ■ \times 10$

47. $■ - 4 = 24 \div 6$

48. $■ \times 4 = 8 \times 3$

49. $6 \div 6 = 0 + ■$

50. $5 + 3 = 16 \div ■$

51. REASONING Is the quotient $24 \div 6$ greater than or less than the quotient $24 \div 4$? How do you know?

52. ? What's the Question? Hikara bought 35 fruit chews. Fruit chews come in packs of 5. The answer is 7 packs.

53. Jessica has 48 marbles. She puts an equal number of marbles in each of 6 bags. How many marbles are in 2 bags?

54. Asha had 24 pictures of her friends. She put 8 pictures on each page in a photo album. Her album has 20 pages. How many album pages do not have pictures?

Mixed Review and Test Prep

Write the time. (p. 94)

55.

56.

Find the missing factor. (p. 142)

57. $3 \times ■ = 24$

58. $4 \times ■ = 28$

59. $■ \times 8 = 72$

60. $■ \times 1 = 8$

61. TEST PREP Luther read 54 pages of his book on Saturday. He read 39 pages on Sunday. How many pages did he read in all?

A 15
B 83
C 93
D 94

62. TEST PREP Patricia collected 205 stickers. She gave 28 stickers to her sister. How many stickers does Patricia have left? (p. 58)

F 177
G 187
H 233
J 277

PROBLEM SOLVING LINKUP . . . to Reading

STRATEGY • CHOOSE IMPORTANT INFORMATION

Some word problems have more information than you need. Before you solve a problem, find the facts you need to solve the problem.

Mrs. Taylor baked 8 batches of muffins. She had a total of 48 muffins. She also baked 2 cakes. How many muffins were in each batch?

Facts You Need: baked 8 batches of muffins, total of 48 muffins

Fact You Don't Need: baked 2 cakes

$48 \div 8 = 6$

So, there were 6 muffins in each batch.

Write the important facts. Solve the problem.

1. Bonnie made 4 batches of cookies and 3 pies in the morning. She made 3 batches of cookies in the afternoon. She made 63 cookies in all. How many cookies were in each batch?

LESSON

2 Divide by 9 and 10

Quick Review

1. $9 \times 3 = ■$
2. $9 \times 5 = ■$
3. $10 \times 4 = ■$
4. $8 \times 10 = ■$
5. $9 \times 10 = ■$

Learn

PLENTY OF PINS Katie collects different kinds of pins. She has boxes that hold 9 or 10 pins. Help Katie organize her collection.

Examples

A Katie puts her 45 state flag pins in boxes that hold 9 pins each. How many boxes does she need?

$45 \div 9 = ■$

Think: $9 \times ■ = 45$
$9 \times 5 = 45$

$45 \div 9 = 5$, or $9\overline{)45}$ = 5

So, Katie needs 5 boxes for her state flag pins.

B Katie puts her 60 flower pins in boxes that hold 10 pins each. How many boxes does she need?

$60 \div 10 = ■$

Think: $10 \times ■ = 60$
$10 \times 6 = 60$

$60 \div 10 = 6$, or $10\overline{)60}$ = 6

So, Katie needs 6 boxes for her flower pins.

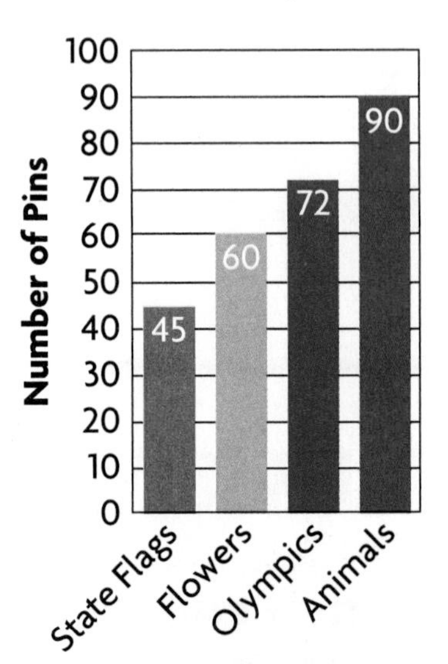

Check

1. **Explain** how to use a related multiplication fact to find $36 \div 9$. What is the quotient?

Copy and complete each table.

2.

÷	9	18	27	36
9	■	■	■	■

3.

÷	20	30	40	50
10	■	■	■	■

Practice and Problem Solving

Copy and complete each table.

4.

÷	54	72	63	81
9	■	■	■	■

5.

÷	70	90	80	100
10	■	■	■	■

Find the quotient.

6. $45 \div 9 =$ ■
7. ■ $= 10 \div 10$
8. ■ $= 0 \div 9$
9. $60 \div 10 =$ ■
10. $9 \div 1 =$ ■
11. $12 \div 6 =$ ■
12. $18 \div 3 =$ ■
13. ■ $= 36 \div 9$
14. ■ $= 50 \div 10$
15. $14 \div 2 =$ ■
16. ■ $= 12 \div 4$
17. $40 \div 5 =$ ■
18. $10\overline{)0}$
19. $6\overline{)42}$
20. $9\overline{)90}$
21. $9\overline{)63}$
22. $8\overline{)64}$
23. $5\overline{)25}$
24. $7\overline{)28}$
25. $3\overline{)24}$
26. Divide 72 by 8.
27. Divide 42 by 7.
28. Divide 70 by 10.

ALGEBRA Write +, −, ×, or ÷ for each ●.

29. 10 ● $10 = 2 - 1$
30. $8 \times 3 = 20$ ● 4
31. 12 ● $7 = 50 \div 10$
32. $3 \times 6 = 2$ ● 9
33. 81 ● $9 = 3 \times 3$
34. $6 \times 7 = 35$ ● 7

35. **REASONING** Ken has 89 patches in his collection. He puts 8 patches on his vest and puts the rest in boxes of 9 patches each. How many boxes does Ken need?

36. **REASONING** Boxes for 9 pins cost \$4 each. Boxes for 10 pins cost \$5 each. Janine has 90 pins. If she wants to spend the least amount of money, what type of boxes should she buy?

37. Trish says, "I can find the quotient of 70 divided by 10 by taking the zero off the end of 70. So, $70 \div 10 = 7$." Do you agree or disagree? Explain.

38. **Write About It** Make a table showing the 9's division facts. Describe any patterns you see in your table.

Mixed Review and Test Prep

39. $8 \times 7 =$ ■ (p. 152)
40. ■ $= 6 \times 9$ (p. 148)
41. $24 \div 3 =$ ■ (p. 204)
42. $28 \div 4 =$ ■ (p. 204)

43. **TEST PREP** Which expression has the same product as $2 \times 3 \times 9$? (p. 170)

A $2 \times 2 \times 8$
B $5 \times 9 \times 1$
C $3 \times 3 \times 7$
D $9 \times 6 \times 1$

EXTRA PRACTICE page H44, Set B

LESSON 3

Practice Division Facts Through 10

Quick Review

1. $7 \times \blacksquare = 56$
2. $\blacksquare \times 3 = 27$
3. $6 \times \blacksquare = 30$
4. $\blacksquare \times 4 = 16$
5. $5 \times \blacksquare = 40$

Learn

BOXED CARS Bobby has 36 toy cars that he wants to put in display boxes. Each display box holds 9 cars. How many display boxes will Bobby need?

$36 \div 9 = \blacksquare$

There are many ways to find the quotient.

A. Use counters.

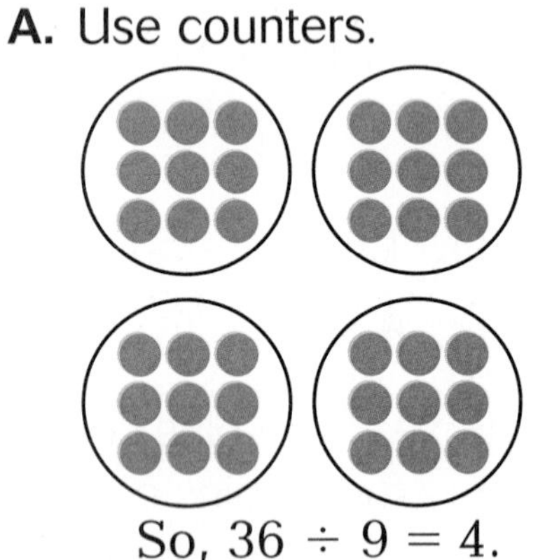

So, $36 \div 9 = 4$.

B. Use an array.

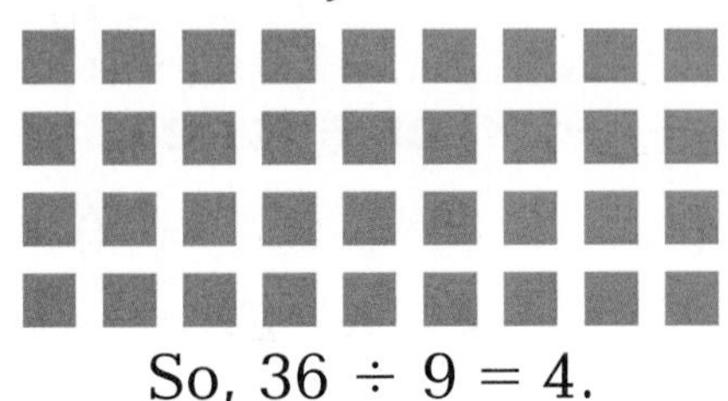

So, $36 \div 9 = 4$.

C. Use repeated subtraction.

	$36 - 9 = 27$	$27 - 9 = 18$	$18 - 9 = 9$	$9 - 9 = 0$
Number of times you subtract 9:	1	2	3	4

So, $36 \div 9 = 4$.

D. Think about fact families.

Fact Family for 4, 9, and 36

factor		factor		product	dividend		divisor		quotient
4	×	9	=	36	36	÷	9	=	4
9	×	4	=	36	36	÷	4	=	9

So, $36 \div 9 = 4$.

Find Missing Factors

E. Use a multiplication table.

Think: $9 \times \blacksquare = 36$

- Find the given factor 9 in the top row.
- Look down the column to find the product, 36.
- Look left across the row to find the missing factor, 4.

$9 \times 4 = 36 \qquad 36 \div 9 = 4$

So, Bobby needs 4 display boxes.

×	0	1	2	3	4	5	6	7	8	9	10
0	0	0	0	0	0	0	0	0	0	0	0
1	0	1	2	3	4	5	6	7	8	9	10
2	0	2	4	6	8	10	12	14	16	18	20
3	0	3	6	9	12	15	18	21	24	27	30
4	0	4	8	12	16	20	24	28	32	36	40
5	0	5	10	15	20	25	30	35	40	45	50
6	0	6	12	18	24	30	36	42	48	54	60
7	0	7	14	21	28	35	42	49	56	63	70
8	0	8	16	24	32	40	48	56	64	72	80
9	0	9	18	27	36	45	54	63	72	81	90
10	0	10	20	30	40	50	60	70	80	90	100

MATH IDEA Use equal groups, arrays, repeated subtraction, fact families, and multiplication tables to help you find quotients.

Check

1. **Explain** how to find $56 \div 8$ in two different ways.

Write a division sentence for each.

2. 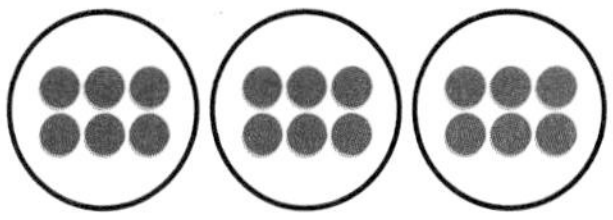

3. 

4. $\begin{array}{r} 21 \\ -\ 7 \\ \hline 14 \end{array} \rightarrow \begin{array}{r} 14 \\ -\ 7 \\ \hline 7 \end{array} \rightarrow \begin{array}{r} 7 \\ -7 \\ \hline 0 \end{array}$

Find the missing factor and quotient.

5. $3 \times \blacksquare = 15 \qquad 15 \div 3 = \blacksquare$
6. $8 \times \blacksquare = 32 \qquad 32 \div 8 = \blacksquare$
7. $4 \times \blacksquare = 40 \qquad 40 \div 4 = \blacksquare$
8. $7 \times \blacksquare = 56 \qquad 56 \div 7 = \blacksquare$

Find the quotient.

9. $10 \div 2 = \blacksquare$
10. $18 \div 9 = \blacksquare$
11. $\blacksquare = 49 \div 7$
12. $80 \div 10 = \blacksquare$
13. $6\overline{)0}$
14. $9\overline{)9}$
15. $8\overline{)40}$
16. $5\overline{)20}$

LESSON CONTINUES

Practice and Problem Solving

Write a division sentence for each.

17. 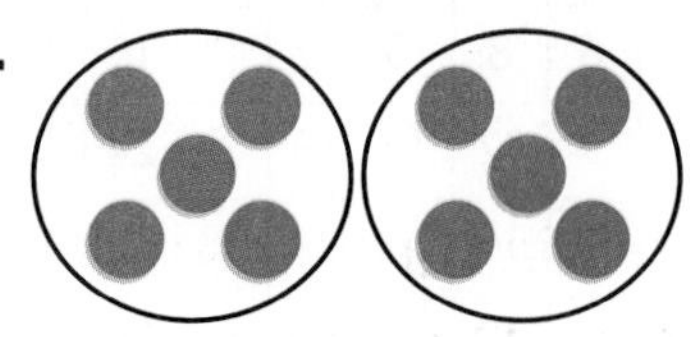

18. 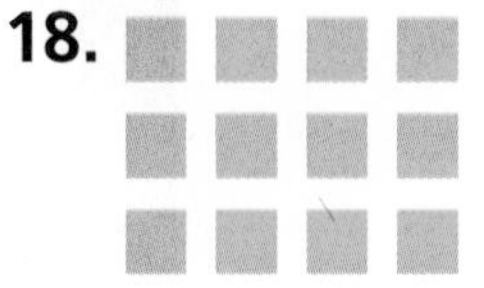

19. $\begin{array}{r} 27 \\ -\ 9 \\ \hline 18 \end{array}$ → $\begin{array}{r} 18 \\ -\ 9 \\ \hline 9 \end{array}$ → $\begin{array}{r} 9 \\ -9 \\ \hline 0 \end{array}$

Find the missing factor and quotient.

20. $10 \times$ ■ $= 90$ $\quad 90 \div 10 =$ ■ **21.** $7 \times$ ■ $= 35$ $\quad 35 \div 7 =$ ■

22. $4 \times$ ■ $= 16$ $\quad 16 \div 4 =$ ■ **23.** $9 \times$ ■ $= 63$ $\quad 63 \div 9 =$ ■

Find the quotient.

24. $10 \div 1 =$ ■ **25.** ■ $= 35 \div 5$ **26.** $50 \div 10 =$ ■ **27.** ■ $= 16 \div 2$

28. ■ $= 81 \div 9$ **29.** ■ $= 20 \div 10$ **30.** $60 \div 6 =$ ■ **31.** $24 \div 3 =$ ■

32. $9\overline{)54}$ **33.** $7\overline{)28}$ **34.** $8\overline{)72}$ **35.** $10\overline{)100}$

36. $8\overline{)24}$ **37.** $2\overline{)0}$ **38.** $4\overline{)4}$ **39.** $3\overline{)21}$

40. Divide 63 by 7. **41.** Divide 30 by 5. **42.** Divide 0 by 9.

Choose the letter of the division sentence that matches each.

a. $42 \div 6 = 7$ **b.** $36 \div 9 = 4$ **c.** $56 \div 7 = 8$ **d.** $27 \div 9 = 3$

43. 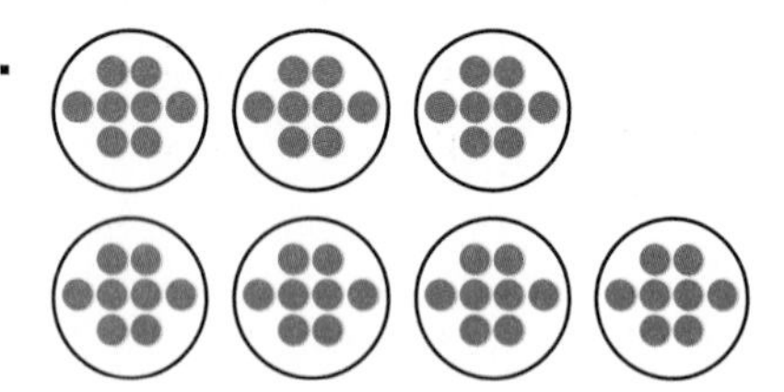

44. 

Compare. Write <, >, or = for each ●.

45. 9×6 ● 9×5 **46.** $24 \div 6$ ● $16 \div 4$ **47.** $4 + 4$ ● $72 \div 8$

48. 8×5 ● 10×4 **49.** $23 - 18$ ● $45 \div 5$ **50.** 3×3 ● $70 \div 10$

51. REASONING Roberta has some boxes with 8 cars in each box. Could Roberta have 20 cars in all? Explain.

52. **Write a problem** about Jonah buying several toy cars that cost $3 each. Use division in your problem.

53. Chi has 2 sheets of stickers. Each sheet has 9 rows of 7 stickers. If Chi uses 2 rows of stickers, how many stickers will Chi have left?

54. Carla put 24 toy animals in 4 boxes. Each box has the same number of animals. How many animals are in 3 boxes?

Mixed Review and Test Prep

55. (p. 36)
$$\begin{array}{r} 35 \\ 22 \\ +15 \\ \hline \end{array}$$

56. (p. 42)
$$\begin{array}{r} 137 \\ 353 \\ +229 \\ \hline \end{array}$$

Compare. Write <, >, or = for each ●. (p. 20)

57. 458 ● 584

58. 1,602 ● 1,062

59. 3,459 ● 3,459

60. 5,891 ● 5,981

61. **TEST PREP** Maria buys pencils for \$1.39, a notebook for \$1.79, and a marker for \$0.85. If she pays with a \$5 bill, how much change will she get? (p. 88)

A \$4.03 C \$1.82
B \$1.97 D \$0.97

62. **TEST PREP** Find the number that the letter *r* stands for. (p. 188)

$$28 \div 7 = r$$

F 3 G 4 H 6 J 7

PROBLEM SOLVING LINKUP ... to Social Studies

Hundreds of years ago, the Miwok Indians lived in the central part of California. Their villages spread from the Pacific Ocean to Yosemite Valley. The Miwoks' main source of food was acorns. Acorns were dried and stored in shelters made from bark, grass, and mud.

1. Acorns were soaked in running streams for 2 to 3 weeks to get rid of the bitter taste. If you began soaking acorns on October 1, on what dates might you check on them?

2. A good acorn harvest might come once every 4 to 5 years. If there was a good harvest of acorns in 1787, in what year might the next good harvest have occurred?

EXTRA PRACTICE page H44, Set C

LESSON 4

Algebra: Find the Cost

Quick Review

1. $2 \times 9 = \blacksquare$
2. $4 \times \blacksquare = 32$
3. $6 \times 2 = \blacksquare$
4. $\blacksquare \times 5 = 45$
5. $10 \times 2 = \blacksquare$

Learn

WHAT'S FOR LUNCH? Mrs. Hugo buys 3 pizzas for her family. How much does Mrs. Hugo spend?

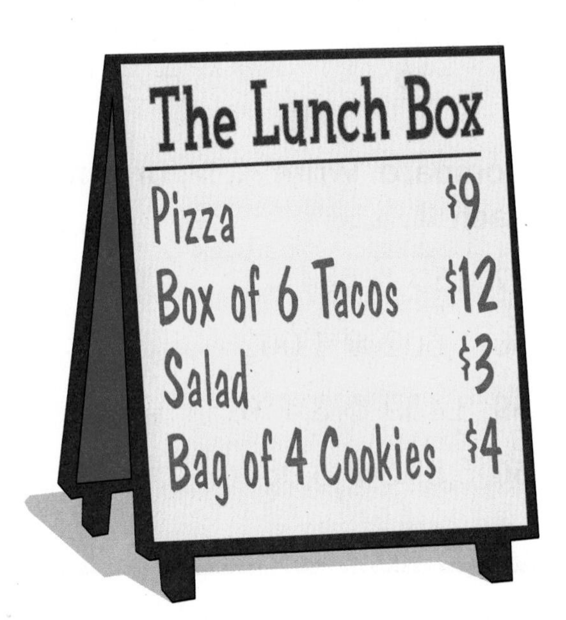

The Lunch Box
Pizza $9
Box of 6 Tacos $12
Salad $3
Bag of 4 Cookies $4

To find the total amount spent, multiply the number of pizzas by the cost of one pizza.

3	×	$9	=	$27
↑		↑		↑
number of pizzas		cost of one		total spent

So, Mrs. Hugo spends $27 for 3 pizzas.

Nicolas buys a box of 6 tacos. How much does each taco cost?

To find the cost of one taco, divide the total amount spent by the number of tacos bought.

$12	÷	6	=	$2
↑		↑		↑
total spent		number of tacos		cost of one

So, each taco costs $2.

MATH IDEA Multiply to find the cost of multiple items or divide to find the cost of one item.

Check

1. **Name** the operation you can use to find the cost of one cookie. Explain.

For 2–3, write a number sentence. Then solve.

2. Alem bought 4 salads. Each salad cost $3. How much did Alem spend?
3. Kim spent $12 on bags of cookies. How many cookies did she buy?

Practice and Problem Solving

For 4–5, write a number sentence. Then solve.

4. Shakira bought 4 hot dogs. Each hot dog cost $4. How much did Shakira spend?

5. Mr. Hess spends $18 for an order of 6 sandwiches. How much does each sandwich cost?

USE DATA For 6–16, use the price list at the right to find the cost of each number of items.

Read-n-Rock

PRICE LIST	
Books	$4 each
Tapes	$7 each
Discs	$9 each

6. 4 discs **7.** 6 tapes **8.** 8 tapes

9. 7 books **10.** 2 tapes **11.** 5 discs

12. 3 books **13.** 9 discs **14.** 5 books

15. 2 discs and 6 books

16. 4 books and 5 tapes

Find the cost of one of each item.

17. 9 markers cost $27. **18.** 6 notepads cost $18. **19.** 3 stamps cost $15.

20. 5 balls cost $30. **21.** 8 pencils cost $8. **22.** 4 games cost $32.

23. 7 toy cars cost $28. **24.** 10 pens cost $20. **25.** 2 T-shirts cost $12.

26. REASONING Ako has $20. She wants to buy rubber stamps that cost $6 each. How many rubber stamps can she buy? Explain.

27. Heidi buys 3 videotapes for $24. She gives the clerk $30. How much does each tape cost? How much change does she get?

Mixed Review and Test Prep

Continue each pattern. (p. 136)

28. 4, 9, 14, 19, ■, ■

29. 44, 40, 36, 32, ■, ■

30. $\begin{array}{r} 2{,}391 \\ +1{,}236 \\ \hline \end{array}$ (p. 46)

31. $\begin{array}{r} 3{,}529 \\ +9{,}382 \\ \hline \end{array}$ (p. 46)

32. TEST PREP Latasha makes 2 sandwiches for each of her 4 friends. She puts 2 slices of ham in each sandwich. How many slices of ham does she use? (p. 172)

A 2 **C** 8
B 4 **D** 16

EXTRA PRACTICE page H44, Set D

LESSON 5

Problem Solving Strategy
Work Backward

PROBLEM Henry baked 3 batches of popovers. The extra batter made 4 more popovers. He made 31 popovers in all. How many popovers does the tin hold?

Quick Review

1. $18 \div 3 = \blacksquare$
2. $12 \div 4 = \blacksquare$
3. $24 \div 6 = \blacksquare$
4. $20 \div 2 = \blacksquare$
5. $10 \div 5 = \blacksquare$

UNDERSTAND

- What are you asked to find?
- What information will you use?

PLAN

- What strategy can you use to solve the problem?

You can *work backward* to find how many popovers the tin holds.

SOLVE

- How can you use the strategy to solve the problem?

Begin with the total number of popovers. Subtract the number of extra popovers from the total.

31	−	4	=	27
↑		↑		↑
total popovers		extra popovers		popovers in 3 batches

Divide to find the number of popovers in each batch.

27	÷	3	=	9
↑		↑		↑
popovers in 3 batches		number of batches		number in each batch

So, Henry's tin holds 9 popovers.

CHECK

- Look back. Does your answer make sense?

Problem Solving Practice

Work backward to solve.

1. **What if** Henry used a different tin to bake 4 batches of popovers? Then he used the extra batter to make 3 more popovers. He made 27 popovers in all. How many popovers does this tin hold?

2. Mr. Jones spent \$30 at the sports shop. He bought a mitt for \$10 and 4 balls. How much did each ball cost?

PROBLEM SOLVING STRATEGIES

- Draw a Diagram or Picture
- Make a Model or Act It Out
- Make an Organized List
- Find a Pattern
- Make a Table or Graph
- Predict and Test
- **Work Backward**
- Solve a Simpler Problem
- Write a Number Sentence
- Use Logical Reasoning

Mr. Lo bought 2 books that cost the same amount. He gave the cashier \$20 and received \$6 in change. How much money did each book cost?

3. Which number sentence shows how to find the total cost of the 2 books?

   A $\$20 + \$6 = \$26$
   B $2 \times \$6 = \$12$
   C $\$20 - \$6 = \$14$
   D $\$20 \div 2 = \$10$

4. How much money did each book cost?

   F \$6
   G \$7
   H \$10
   J \$14

Mixed Strategy Practice

USE DATA For 5–7, use the price list.

5. Zach pays for an apple pie and a bag of cookies with a \$10 bill. He gets his change in quarters and dimes. There are 13 coins in all. How many of each coin does Zach get?

6. **What's the Error?** Lara says a lemon tart and 2 boxes of muffins cost \$11.65. Describe Lara's error and give the correct cost of the items.

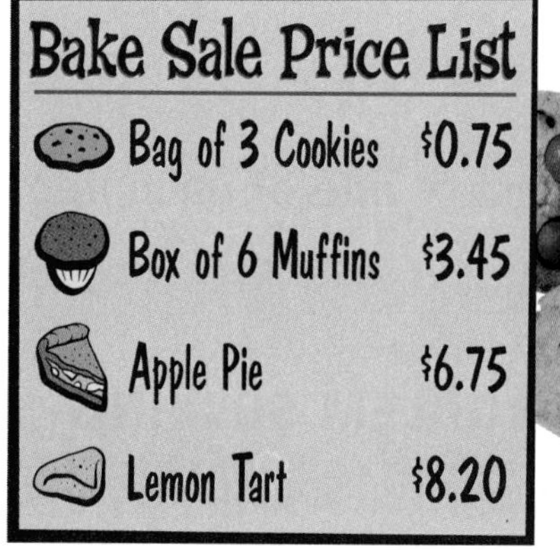

Bake Sale Price List	
Bag of 3 Cookies	\$0.75
Box of 6 Muffins	\$3.45
Apple Pie	\$6.75
Lemon Tart	\$8.20

7. Kenneth has 3 quarters, 2 dimes, 2 nickels, and 5 pennies. List all the ways he can pay for a bag of cookies.

Problem Solving Strategy

Review/Test

CHECK CONCEPTS

Write a division sentence for each. (pp. 222–225)

1. 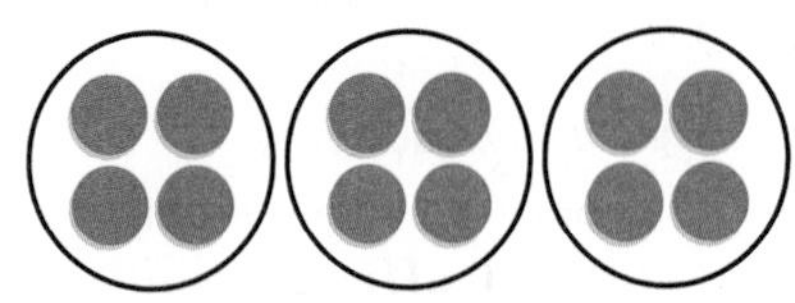

2. 

3.
$$\begin{array}{r} 24 \\ -\ 8 \\ \hline 16 \end{array} \quad \begin{array}{r} 16 \\ -\ 8 \\ \hline 8 \end{array} \quad \begin{array}{r} 8 \\ -8 \\ \hline 0 \end{array}$$

CHECK SKILLS

Find the missing factor and quotient. (pp. 216–219)

4. $8 \times ■ = 40$ $\quad 40 \div 8 = ■$
5. $6 \times ■ = 42$ $\quad 42 \div 6 = ■$
6. $7 \times ■ = 56$ $\quad 56 \div 7 = ■$
7. $8 \times ■ = 32$ $\quad 32 \div 8 = ■$

Find the quotient. (pp. 216–225)

8. $14 \div 7 = ■$
9. $■ = 30 \div 5$
10. $40 \div 4 = ■$
11. $■ = 63 \div 9$
12. $8\overline{)24}$
13. $10\overline{)90}$
14. $2\overline{)14}$
15. $9\overline{)18}$
16. $3\overline{)27}$

For 17–20, use the price list at the right to find the cost of each number of items. (pp. 226–227)

17. 4 balloons
18. 7 noise makers
19. 3 noise makers
20. 9 balloons

PARTY SUPPLIES PRICE LIST	
Balloons	\$3
Noise Makers	\$2

Find the cost of one of each item. (pp. 226–227)

21. 5 notebooks cost \$10.
22. 6 markers cost \$18.
23. 5 caps cost \$35.

CHECK PROBLEM SOLVING

Work backward to solve. (pp. 228–229)

24. Nikki used a tin to make 3 batches of popovers. Then she made 2 extra popovers. She made 26 popovers in all. How many popovers does the tin hold?

25. Roger earned \$35. He made \$15 from a paper route. He also walked 4 dogs after school. How much did Roger charge to walk each dog?

Standardized Test Prep

TIP! **Decide on a plan.**
See item **7.**

Find how many pages Jeremy did not read on the first night. Then you can find the number of pages he read on each of the next 6 nights.

Also see problem **4,** p. H64.

For 1–8, choose the best answer.

1. Mrs. Jenkins made 64 pretzels. She put 8 pretzels on each plate. How many plates did she use?

A 4 **B** 6 **C** 8 **D** 9

2. $42 \div 7 = \blacksquare$

F 9
G 8
H 7
J NOT HERE

3. $54 \div 9 = \blacksquare$

A 5
B 6
C 8
D NOT HERE

4. A box of 8 beanbag animals costs $24. How much does 1 beanbag animal cost?

F $6
G $4
H $3
J $2

5. Carrie bought 7 slices of pizza for her family. Each slice cost $3. How much did Carrie spend on pizza?

A $14 **B** $21 **C** $24 **D** $28

6. The pep club members held a car wash. They charged $4 to wash each car. They washed 8 cars in the first 2 hours. How much did they make?

F $32 **G** $24 **H** $12 **J** $8

7. Jeremy read all 47 pages of his new book. The first night he read 17 pages. On each of the next 6 nights he read an equal number of pages. How many pages did he read on each of these 6 nights?

A 5 **B** 8 **C** 30 **D** 70

8. Leticia put 5 toys on each of 4 shelves. Which number sentence tells how many toys there were?

F $5 + 4 = 9$
G $5 \times 4 = 20$
H $5 - 4 = 1$
J $20 \div 5 = 4$

Write What You Know

9. Mr. Grant needs to put 40 jars into boxes. He wants to put an equal number of jars in each box. Find two ways that he can put the jars into boxes. For each way, tell how many jars are in each box and how many boxes.

10. What number makes this number sentence true?

$$8 \times \blacksquare = 56$$

Write a division sentence that is related to the multiplication sentence. Explain your answer.

PROBLEM SOLVING MATH DETECTIVE

Missing Parts

Put on your thinking cap and use what you know about the relationship between multiplication and division to find the missing factors and products.

Copy and complete each table. Use grid paper to help.

Table 1

×	■	■	4	■
5	15	■	■	■
7	■	56	■	■
■	24	■	■	48
■	■	■	36	■

Table 2

×	■	■	■	■
■	28	■	■	■
■	■	48	24	■
■	63	■	■	18
■	■	■	15	■

STRETCH YOUR THINKING

- Explain how you found the missing factors. Explain how you found the missing products.
- List as many multiplication and division sentences as you can that have a product or a quotient of 6.
- List as many multiplication and division sentences as you can that have a product or a quotient of 24.

Challenge

Divide by 11 and 12

Mr. Samson collected 132 eggs. He wants to put them into egg cartons that hold 12 eggs each. How many egg cartons does Mr. Samson need?

132	÷	12	=	■
↑		↑		↑
number of eggs		number in each carton		number of cartons

×	0	1	2	3	4	5	6	7	8	9	10	11	12
0	0	0	0	0	0	0	0	0	0	0	0	0	0
1	0	1	2	3	4	5	6	7	8	9	10	11	12
2	0	2	4	6	8	10	12	14	16	18	20	22	24
3	0	3	6	9	12	15	18	21	24	27	30	33	36
4	0	4	8	12	16	20	24	28	32	36	40	44	48
5	0	5	10	15	20	25	30	35	40	45	50	55	60
6	0	6	12	18	24	30	36	42	48	54	60	66	72
7	0	7	14	21	28	35	42	49	56	63	70	77	84
8	0	8	16	24	32	40	48	56	64	72	80	88	96
9	0	9	18	27	36	45	54	63	72	81	90	99	108
10	0	10	20	30	40	50	60	70	80	90	100	110	120
11	0	11	22	33	44	55	66	77	88	99	110	121	132
12	0	12	24	36	48	60	72	84	96	108	120	132	144

Use a multiplication table to find the quotient.

Think: $12 \times$ ■ $= 132$

Find the factor 12 in the top row. Look down the column to find the product, 132. Look left along the row to find the missing factor, 11.

$12 \times 11 = 132$

$132 \div 12 = 11$

So, Mr. Samson needs 11 egg cartons.

Talk About It

- Explain how to use repeated subtraction to find $48 \div 12$.
- What patterns do you notice in the elevens column of the multiplication table?

Try It

Use the multiplication table to solve.

1. $99 \div 11 =$ ■ **2.** ■ $= 108 \div 12$ **3.** $110 \div 11 =$ ■ **4.** ■ $= 84 \div 12$

5. $72 \div 12 =$ ■ **6.** $66 \div 11 =$ ■ **7.** ■ $= 144 \div 12$ **8.** $0 \div 11 =$ ■

9. $12\overline{)120}$ **10.** $11\overline{)121}$ **11.** $11\overline{)77}$ **12.** $12\overline{)36}$

Study Guide and Review

VOCABULARY

Choose the best term from the box.

1. A set of related multiplication and division sentences is a ___?___. (p. 192)
2. Any number divided by ___?___ is that number. (p. 206)

one
fact family
quotient
zero

STUDY AND SOLVE

Chapter 11

Use repeated subtraction to divide.

$28 \div 7 = \blacksquare$

$28 - 7 = 21$, $21 - 7 = 14$, $14 - 7 = 7$, $7 - 7 = 0$

You subtracted 7 from 28 four times.
So, $28 \div 7 = 4$.

Write the division sentence shown by the repeated subtraction. (pp. 186–187)

3. $15 - 5 = 10$, $10 - 5 = 5$, $5 - 5 = 0$

4. $32 - 8 = 24$, $24 - 8 = 16$, $16 - 8 = 8$, $8 - 8 = 0$

Use arrays to divide.

$20 \div 4 = \blacksquare$

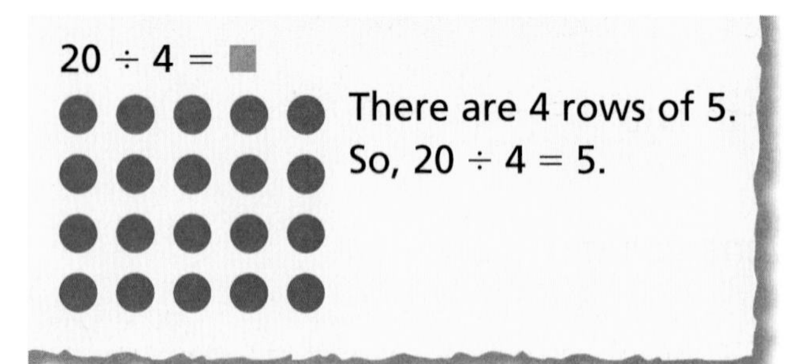

There are 4 rows of 5.
So, $20 \div 4 = 5$.

Use the array to find the quotient. (pp. 188–191)

5. 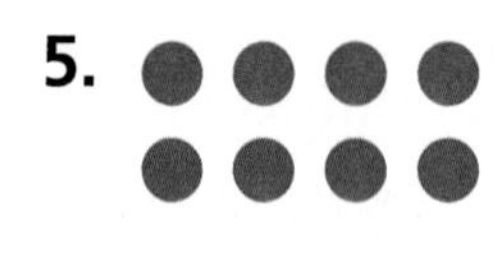

2 rows of $\blacksquare = 8$
$8 \div 2 = \blacksquare$

6.

3 rows of $\blacksquare = 9$
$9 \div 3 = \blacksquare$

Write fact families.

This is the fact family for 3, 4, and 12.
$3 \times 4 = 12$ $12 \div 4 = 3$
$4 \times 3 = 12$ $12 \div 3 = 4$

Write the fact family for each set of numbers. (pp. 192–195)

7. 3, 9, 27
8. 6, 7, 42
9. 5, 8, 40
10. 4, 9, 36

Chapter 12

Use related multiplication facts to find quotients.

$45 \div 5 = \blacksquare$
Think: $5 \times \blacksquare = 45$
$5 \times 9 = 45$
So, $45 \div 5 = 9$, or $5\overline{)45}$ with quotient 9.

Find each quotient. (pp. 202–205)

11. $18 \div 3 = \blacksquare$
12. $30 \div 5 = \blacksquare$
13. $20 \div 4 = \blacksquare$
14. $16 \div 2 = \blacksquare$
15. $40 \div 5 = \blacksquare$
16. $24 \div 4 = \blacksquare$
17. $4\overline{)32}$
18. $3\overline{)9}$
19. $2\overline{)10}$

Chapter 13

Multiply to find the cost of multiple items.

Pens cost $4 each. Find the cost of 7 pens.
7 × $4 = $28
So, 7 pens cost $28.

Beach balls cost $3 each. Find the cost of each number of items. (pp. 226–227)

20. 6 beach balls
21. 2 beach balls
22. 5 beach balls

Divide to find the cost of one item.

8 erasers cost $16.
$16 ÷ 8 = $2
So, each eraser costs $2.

Find the cost of one of each item. (pp. 226–227)

23. 9 tennis balls cost $18.
24. 7 baskets cost $42.
25. 10 notepads cost $10.

PROBLEM SOLVING PRACTICE

Solve. (pp. 196–197, 228-229, 210–211)

26. Marcie bought 5 packs of juice. There are 3 juice boxes in each pack. How many juice boxes did Marcie buy? Write a number sentence and solve.
27. Noah spent 15 minutes eating lunch and then played kickball for 25 minutes. Now it is 12:45 P.M. What time did Noah start eating lunch?
28. Janet had $7.35 and spent $2.50 on a snack. How much money does Janet have left?

PERFORMANCE ASSESSMENT

TASK A • DAISY GARDEN

Materials: counters

Blair has 30 daisy plants. She wants to plant them in her garden so that each row has the same number of plants.

a. Use counters to make a model. Show one way Blair could place the plants in her garden. Draw a picture of your model.

b. Write the multiplication and division sentences that belong to the fact family for the model you drew.

c. How does the model you drew show that multiplication and division are related?

TASK B • AT THE BALL PARK

Kade and Lydia are going to a baseball game. There are special buys at the ball park if you buy more than one of the same item. Kade and Lydia will buy items with friends who want the same thing. Each friend will get one item and will pay an equal part.

Kade has $3 and Lydia has $4.

a. What is one item Kade could buy?

b. Lydia wants to buy a different item. Which item could she buy? Write a division sentence to show how much Lydia's item will cost.

c. If Kade and Lydia combine their money, can they pay for one baseball to share? Explain.

H
H
Specials
4 baseballs for $ 20
3 posters for $ 6
6 caps for $ 18
3 T-shirts for $12
8 mugs for $ 8

Calculator • Find the Unit Cost

Joanne sees a sign that says, "Kites—4 for $15!" How much would Joanne pay for 1 kite?

The unit cost will tell you. The **unit cost** is the cost of one item when several items are sold for a single price.

Find the unit cost. Use a calculator.

STEP 1

Enter the total cost.

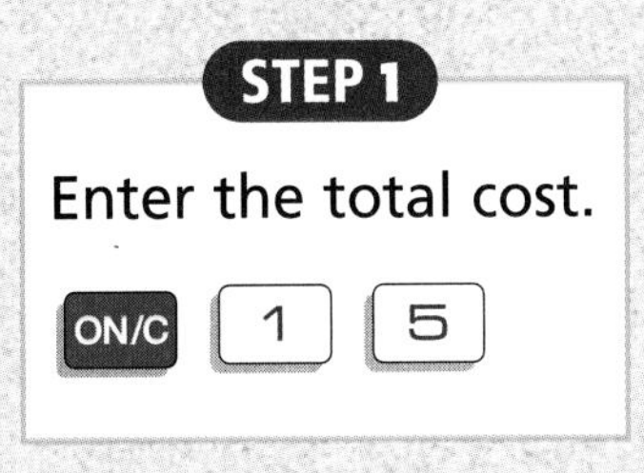

Divide by the number of items.

STEP 3

The quotient is the unit cost.

3.75

So, Joanne would pay $3.75 for 1 kite.

Practice and Problem Solving

Use a calculator to find each unit cost.

1. 4 for $18 **2.** 2 for $7 **3.** 8 for $18

4. 2 for $24 **5.** 5 for $35 **6.** 6 for $39

7. 3 for $27 **8.** 4 for $27 **9.** 4 for $11

Use a calculator to solve.

10. A hobby shop sells 6 puzzles for $12. How much will 2 puzzles cost? Explain.

11. **Stretch Your Thinking** Ricardo bought 5 toy cars for $8.45. How much did each toy car cost?

Multimedia Math Glossary www.harcourtschool.com/mathglossary

12. **Vocabulary** Look up *dividend, divisor,* and *quotient* in the Multimedia Math Glossary. Write a poem that someone could use to help remember these terms.

Many theaters, like the Auditorium Theatre, have two balconies.

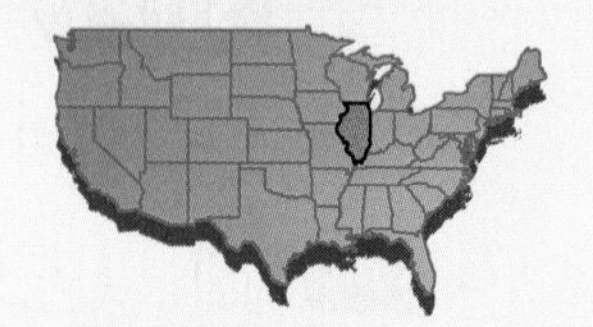

THE AUDITORIUM THEATRE

Many ballet companies and other dance companies perform at the Auditorium Theatre in Chicago.

Center Orchestra Section

STAGE

Row A	□	□	□	□	□	□	□	□
Row B	□	□	□	□	□	□	□	□
Row C	□	□	□	□	□	□	□	□
Row D	□	□	□	□	□	□	□	□

USE DATA **For 1–4, use the diagram.**

1. The diagram above shows 32 seats. Write a number sentence to show how to find the number of seats in each row.
2. If your class went to the Auditorium Theatre, would it fill more than 3 rows, less than 3 rows, or exactly 3 rows? Explain.
3. Each row in the center orchestra section of the Auditorium Theatre has the same number of seats. How many rows are needed for 40 people? Explain.
4. How many rows are needed in the center orchestra section for 72 people? Draw an array and write a number sentence to show your answer.

THE SHUBERT THEATRE

The Shubert Theatre is another theater in Chicago where you can see live performances.

USE DATA For 1–4, use the diagram.

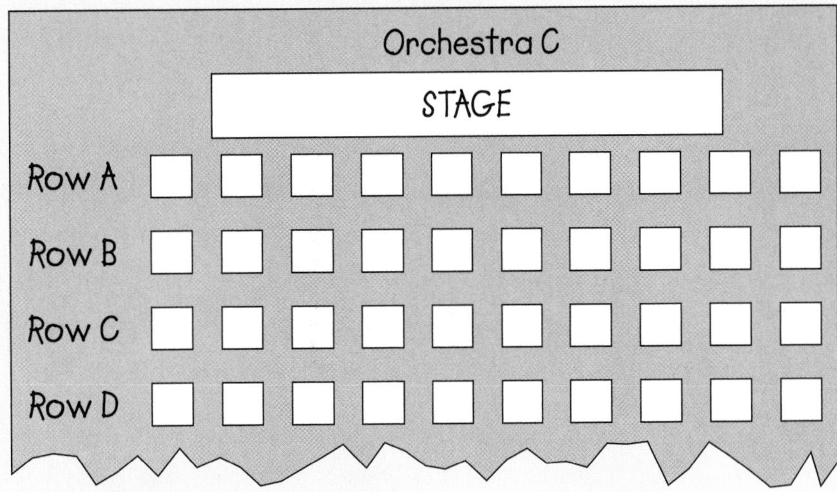

1. The section called Orchestra C is in the center of the first floor. Suppose 50 students want to sit in this section. How many rows of seats will they need? Explain.

2. A group buys 80 seats together in Orchestra C. How many full rows is that? The first row the students sit in is Row C. Then they continue to fill in the rows behind Row C. What is the letter of the last row that they fill in?

3. Suppose a group fills the seats in Rows A, B, C, and D. Write the fact family for this array of seats.

4. **Stretch your thinking** A teacher orders 24 seats in Orchestra C for a field trip. She wants the same number of people to sit in each row. List 3 different ways that they could do this.

In the Shubert Theatre, there are two levels of seats. The seats on the first floor are called orchestra seats. There are 26 rows of seats in the orchestra section. Seats in the upper level are in a section called the loge.

Collect and Record Data

Sue Hendrickson

◀ Sue is the largest, most complete, Tyrannosaurus Rex fossil skeleton ever found.

Dinosaurs lived on Earth many years ago. This Tyrannosaurus Rex skeleton, named Sue, was found in South Dakota in 1990 by Sue Hendrickson. It is 41 feet long.

PROBLEM SOLVING Use the chart. Make a table that shows the dinosaurs in order from smallest to largest.

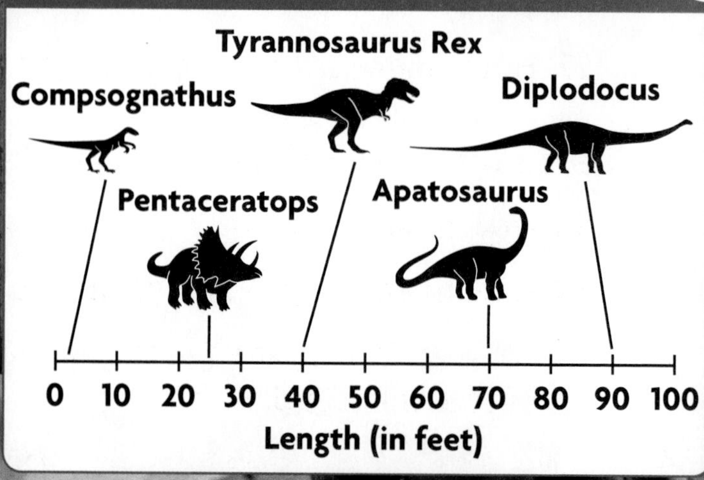

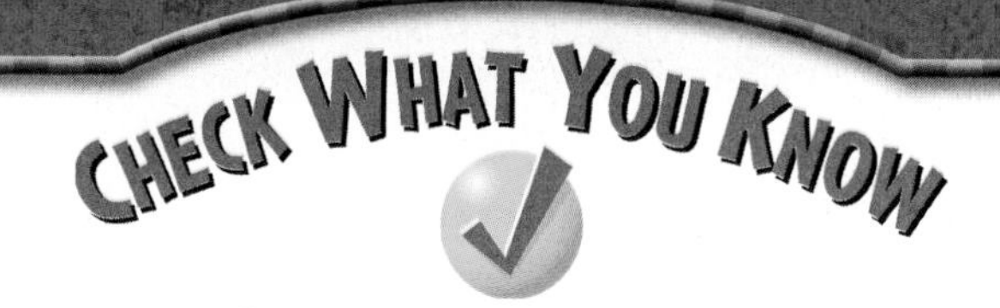

Use this page to help you review and remember important skills needed for Chapter 14.

COLUMN ADDITION (For Intervention, see p. H10.)

Find the sum.

1. $\begin{array}{r} 8 \\ 3 \\ 4 \\ +7 \\ \hline \end{array}$ **2.** $\begin{array}{r} 9 \\ 3 \\ 1 \\ +6 \\ \hline \end{array}$ **3.** $\begin{array}{r} 8 \\ 6 \\ 4 \\ +2 \\ \hline \end{array}$ **4.** $\begin{array}{r} 5 \\ 4 \\ 7 \\ +3 \\ \hline \end{array}$ **5.** $\begin{array}{r} 3 \\ 4 \\ 7 \\ +1 \\ \hline \end{array}$

6. $\begin{array}{r} 4 \\ 3 \\ 8 \\ +5 \\ \hline \end{array}$ **7.** $\begin{array}{r} 2 \\ 4 \\ 7 \\ +4 \\ \hline \end{array}$ **8.** $\begin{array}{r} 5 \\ 5 \\ 5 \\ +6 \\ \hline \end{array}$ **9.** $\begin{array}{r} 7 \\ 3 \\ 6 \\ +1 \\ \hline \end{array}$ **10.** $\begin{array}{r} 9 \\ 5 \\ 9 \\ +3 \\ \hline \end{array}$

11. $5 + 4 + 7 + 2 + 8$ **12.** $1 + 7 + 5 + 9 + 3$ **13.** $6 + 7 + 3 + 4 + 7$

READ A CHART (For Intervention, see p. H17.)

For 14–17, use the information in this chart.

14. What is the title of the chart?

15. How many students like snowy weather best?

16. What kind of weather had the fewest votes?

17. How many students were asked?

OUR FAVORITE WEATHER	
Type	**Students**
Sunny	7
Rainy	4
Snowy	8
Cloudy	3

For 18–20, use the information in this chart.

18. How many students were asked?

19. How many students eat dinner at 6:00?

20. At what time do the greatest number of students eat dinner?

TIME FOR DINNER	
Time	**Students**
5:00	8
5:30	9
6:00	5
6:30	2

LESSON 1

Collect and Organize Data

Quick Review

Write numbers for the tally marks.

1. ||| 2. || 3. 𝍸
4. 𝍸 𝍸 | 5. 𝍸 ||

VOCABULARY

data **tally table** **frequency table**

MATERIALS

pencils, paper

Explore

Information collected about people or things is called **data**.

Kelly's class voted for their favorite dinosaurs and made tables to show the results.

A **tally table** uses tally marks to record data.

FAVORITE DINOSAUR	
Name	**Tally**
Anatosaurus	𝍸 |
Brachiosaurus	𝍸 ||
Tyrannosaurus	𝍸 𝍸 ||
Stegosaurus	|||

A **frequency table** uses numbers to record data.

FAVORITE DINOSAUR	
Name	**Number**
Anatosaurus	6
Brachiosaurus	7
Tyrannosaurus	12
Stegosaurus	3

Collect data about your classmates' favorite dinosaurs. Organize the data in a tally table.

Activity 1

STEP 1

Make a tally table. List four answer choices.

STEP 2

Ask classmates *What's your favorite dinosaur?* Make a tally mark for each answer.

Favorite Dinosaur

Name	Tally

• Why is a tally table good for recording data?

Try It

Decide on a question to ask your classmates.

a. Write four answer choices in a tally table.

b. Ask your classmates the question. Complete the tally table.

Connect

Use the data you collected in the tally table on page 240 to make a frequency table.

Activity 1

STEP 1

Write the title and headings. List the four answer choices.

STEP 2

Count the number of tallies in each row. Write each number in the frequency table.

Favorite Dinosaur	
Name	Number

- Why is a frequency table a good way to show data?

Practice and Problem Solving

1. Make a tally table of three after-school activities. Ask your classmates which activity they like best. Make a tally mark beside the activity for each answer.

2. Use the data from your tally table to make a frequency table. Which after-school activity did the greatest number of classmates choose? the fewest?

3. Which table is better for reading results? Which table is better for collecting choices?

4. Make a frequency table of the data in the tally table. How many students voted for their favorite fruit?

5. **Write a problem** using the information in the Favorite Fruit tally table.

FAVORITE FRUIT	
Name	**Tally**
Grapes	𝍸 𝍸 𝍸 |
Oranges	||||
Apples	𝍸
Bananas	𝍸 |||

Mixed Review and Test Prep

6. $2 \times 10 = ■$ (p. 164)

7. $7 \times 10 = ■$ (p. 164)

8. Which number is greater: 4,545 or 4,454? (p. 20)

9. Round 3,495 to the nearest thousand. (p. 30)

10. **TEST PREP** Ebony buys a sandwich for $3.49. She has a coupon for $0.50 off. She pays with a $5 bill. How much change will she receive? (p. 172)

    A $1.01
    B $1.51
    C $2.01
    D $3.99

LESSON 2

Understand Data

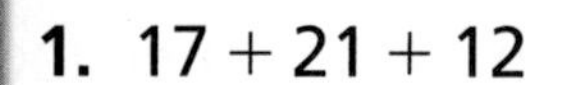

Quick Review

1. 17 + 21 + 12
2. 13 + 9 + 6
3. 24 + 5 + 7
4. 8 + 14 + 7
5. 19 + 13 + 21 + 5

VOCABULARY

survey
results

Learn

SURVEY SAYS . . . A **survey** is a question or set of questions that a group of people are asked. The answers from a survey are called the **results** of the survey.

Jillian and Ted took a survey to find their classmates' favorite snacks. The tally table shows the choices and votes of their classmates.

What are the favorite snacks of their classmates?

Since cookies got the greatest number of votes, 12, cookies are the favorite snack.

- How many students answered Jillian and Ted's survey? How do you know?

FAVORITE SNACK

Snack	Tally
Popcorn	||||
Cookies	𝍸 𝍸 ||
Granola bars	𝍸
Apples	𝍸 ||
Pretzels	||

Check

1. **List** the snacks in order from the most votes to the fewest votes.

For 2–3, use the tally table.

DO YOU HAVE AN OLDER BROTHER OR SISTER?

Answer	Tally
Yes	𝍸 𝍸 ||
No	𝍸 𝍸 𝍸 ||

2. How many people were surveyed?
3. Write a statement that describes the survey results.

Technology Link

More Practice:
Use E-Lab, *Collecting and Organizing Data.*

www.harcourtschool.com/elab2002

Practice and Problem Solving

For 4–7, use the tally table.

4. List the subjects in order from the most votes to the fewest votes.
5. How many students answered the survey?
6. How many more students chose math than chose social studies?
7. How many fewer students chose art than chose reading?

FAVORITE SCHOOL SUBJECT	
Subject	**Tally**
Math	卌 卌 III
Science	卌 IIII
Reading	卌 卌 卌
Social studies	卌 I
Art	卌 II

For 8–11, use the frequency table.

8. How many more students chose basketball than chose baseball?
9. Which sport did the greatest number of students choose?
10. How many students answered this survey?
11. **What if** 5 more students chose basketball? How would that change the results of the survey?

FAVORITE SPORTS	
Sport	**Number**
Basketball	26
Football	15
Baseball	17
Hockey	21
Swimming	30

12. **Write About It** Take a survey to find your classmates' favorite pizza topping. Make a tally table and a frequency table of the data. Explain the results.

13. **Write a problem** using the information from any of the tables in this lesson.

14. **What's the Error?** Lily wrote the following number sentence: $24 \times 0 = 24$. What's her error?

Mixed Review and Test Prep

15. (p. 42) $134 + 159$

16. (p. 42) $365 + 256$

17. (p. 42) $166 + 384$

18. Three friends shared 27 baseball cards equally. How many cards did each friend receive? (p. 204)

19. **TEST PREP** Toshio had 4 packages of 6 bagels. He gave one package away. How many bagels in all does he have now? (p. 122)

A 10 **B** 18 **C** 24 **D** 28

EXTRA PRACTICE page H45, Set A

LESSON 3

Classify Data

Quick Review

Tell if each number is *less than* or *greater than* 345.

1. 298
2. 404
3. 344
4. 354
5. 346

VOCABULARY

classify

Learn

MARBLES IN MOTION You can group, or **classify**, data in many different ways such as by size, color, or shape.

Ms. Vernon gave each pair of students in her class a bag of marbles. She asked students to think about ways that they could group the marbles.

Luis and Joey made a table to show what they did.

Bag of Marbles

	Small	Medium	Large
Blue	2	2	5
Red	3	1	5
Multicolor	3	3	2

- How did Luis and Joey classify their data?

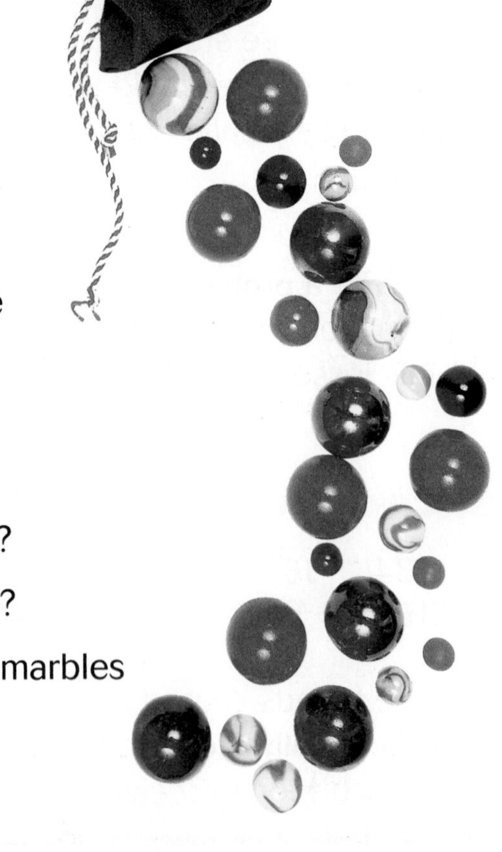

Check

1. **Tell** 3 things that the groupings in the table helped you know about the marbles.
2. How many small blue marbles are there?
3. How many large red marbles are there?
4. How many multicolored marbles are there?
5. How many medium-size marbles are there?
6. How many more large marbles than small marbles are there?

Practice and Problem Solving

For 7–10, use the table.

COLOR OF SHIRTS IN OUR CLASS				
	Green	**Blue**	**Red**	**Yellow**
Girls	1	6	2	5
Boys	3	4	4	0

7. How many girls are wearing red shirts?

8. How many students are wearing blue shirts?

9. How many of the students wearing green shirts are boys?

10. How many students are in the class?

For 11–14, use the table.

PICTURES IN THE ART SHOW				
	Chalk	Crayon	Paint	Pencil
People	3	5	8	1
Animals	4	5	4	3
Plants	6	2	9	5

11. How many pictures are in the art show in all?

12. What is the subject of the greatest number of pictures in the art show: people, animals, or plants?

13. How many more pictures in the art show were made with paint and pencil than with chalk and crayon?

14. There are 12 possible categories of pictures in the art show. Which one was represented the least?

15. Look at the figures at the right. Make a table to classify, or group, the figures.

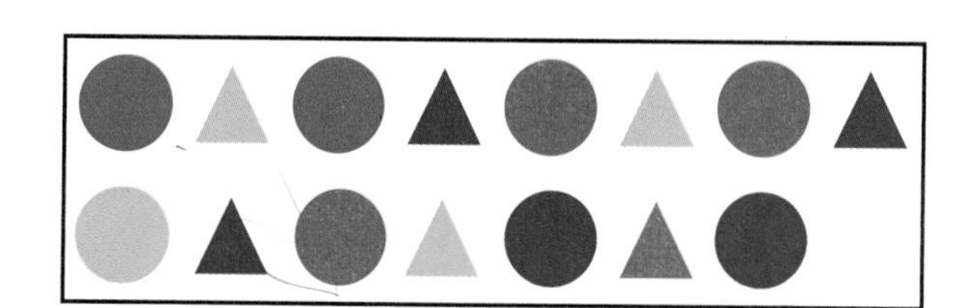

16. There are 11 girls and 8 boys in a music club. Of the girls, 6 are dancers and the rest are singers. Of the boys, 3 are dancers and the rest are singers. Make a table to classify, or group, the students.

Mixed Review and Test Prep

17. $700 - 238 = \blacksquare$ (p. 58)

18. $306 - 67 = \blacksquare$ (p. 58)

19. $5 \times 2 \times 5 = \blacksquare$ (p. 170)

20. If 4 tapes cost $32, how much does 1 tape cost? (p. 226)

21. **TEST PREP** Caroline left her house at 7:15 A.M. She arrived at school 20 minutes later. At what time did she arrive at school? (p. 98)

A 6:55 A.M. **C** 7:30 A.M.
B 7:20 A.M. **D** 7:35 A.M.

EXTRA PRACTICE page H45, Set B

LESSON 4

Problem Solving Strategy
Make a Table

Quick Review

1. 6 + 4 **2.** 3 + 5

3. 8 + 7 **4.** 9 + 6

5. 6 + 7

PROBLEM Leo and Sally did an experiment with the two spinners shown at the right. They spun the pointers and recorded the sum of the two numbers. They spun 20 times. Their results were 2, 4, 6, 5, 4, 5, 4, 4, 2, 4, 4, 3, 4, 5, 6, 4, 6, 4, 3, and 4. Which sum occurred most often?

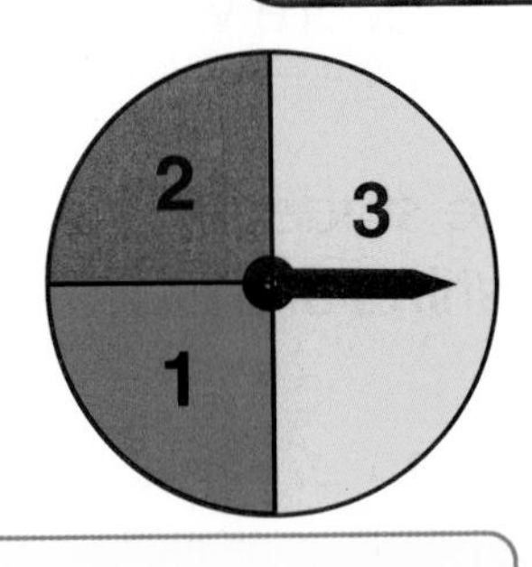

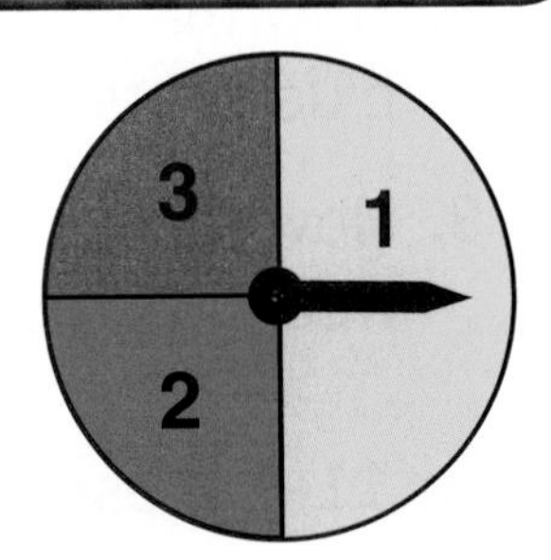

3 + 1 = 4

UNDERSTAND

- What are you asked to find?
- What information will you use?

PLAN

- What strategy can you use to solve the problem?

You can *make a table* to organize the data.

SOLVE

- What should you put in the table?

Leo and Sally recorded the sums in their experiment. So, label one column *Sums*. In this column, list all the sums that are possible. Label another column *Number of Spins*. Use a tally mark to record each spin.

SPINNER EXPERIMENT	
Sums	**Number of Spins**
2	||
3	||
4	~~||||~~ ~~||||~~
5	|||
6	|||

So, the sum 4 occurred most often.

CHECK

- What other strategy could you use to solve the problem?

Problem Solving Practice

Solve.

1. **What if** Leo and Sally spin the pointers 25 times and record the sums in a table? How many tallies should there be?

2. Marta and Dan rolled two number cubes 15 times and recorded the sums. Their results were 9, 9, 5, 6, 9, 3, 7, 7, 3, 5, 12, 12, 10, 8, and 5. Make a table and find the sum rolled most often.

PROBLEM SOLVING STRATEGIES

- Draw a Diagram or Picture
- Make a Model or Act It Out
- Make an Organized List
- Find a Pattern
- **Make a Table or Graph**
- Predict and Test
- Work Backward
- Solve a Simpler Problem
- Write a Number Sentence
- Use Logical Reasoning

Your experiment is to roll two number cubes, numbered 1–6, twenty times to find out what difference you will roll most often.

3. What are all the differences you will list on your table?
   - **A** 0, 2, 4, 6
   - **B** 0, 1, 2, 3, 4, 5, 6
   - **C** 0, 1, 2, 3, 4, 5
   - **D** 1, 2, 3, 4, 5, 6

4. What should the total number of tallies be on your table for this experiment?
   - **F** 10
   - **G** 15
   - **H** 20
   - **J** 25

Mixed Strategy Practice

5. **REASONING** Louise is older than Marsha and Luke. Marsha is younger than Jim. Jim is older than Luke. Al is the youngest of the group. What is the order of the group from youngest to oldest?

6. This was the time when the game ended. If the game lasted 2 hours and 15 minutes, when did the game start?

7. In the library, there are 3 shelves with 9 new books on each shelf. How many new books are on display?

8. Lonnie has 48 dinosaur models. He puts them into 8 bags so that there are the same number in each bag. How many models do 2 bags contain?

9. **What's the Question?** Sylvia spent $7 at the movies. She still had $8 when she got home. The answer is $15.

Review/Test

✓ CHECK VOCABULARY

Choose the best term from the box.

data
survey
classify
tally table
frequency table

1. A question or set of questions that a group of people are asked is a __?__. (p. 242)
2. Information collected about people or things is called __?__. (p. 240)
3. A table that uses tally marks to record data is a __?__. (p. 240)
4. A table that uses numbers to record data is a __?__. (p. 240)

✓ CHECK SKILLS

For 5–6, use the tally table. (pp. 242–243)

5. List the instruments in order from the most votes to the fewest votes.
6. How many people answered the survey?

FAVORITE MUSICAL INSTRUMENT	
Musical Instrument	**Tally**
Guitar	𝍸 𝍸 𝍸
Flute	𝍸 ||||
Drums	𝍸 𝍸 𝍸 |||
Piano	𝍸 𝍸 |||

For 7–8, use the frequency table. (pp. 244–245)

7. How many students have brown eyes?
8. How many girls are in the class?

EYE COLOR OF STUDENTS			
	Blue	**Brown**	**Green**
Girls	4	8	1
Boys	5	7	2

✓ CHECK PROBLEM SOLVING

Solve. (pp. 246–247)

9. There are 12 fourth graders and 13 fifth graders in the Science Club. Five of the fourth graders and 9 of the fifth graders are girls. Make a table to group the students in the Science Club.

10. Sancho rolled two number cubes 10 times and recorded the sums. His results were 7, 6, 11, 2, 8, 9, 6, 8, 11, and 8. Make a table and find the sum rolled most often.

Standardized Test Prep

TIP! Understand the problem.
See item **5**.
How many people voted means the same as the Number of Votes in the frequency table. Decide which operation to use to answer the question.
Also see problem **1**, p. H62.

For 1–7, choose the best answer.

For 1–3, use the tally table.

SHIRT COLOR	
Color	**Tally**
Red	卌 ||
Yellow	|||
Green	卌
Purple	卌 ||||

1. How many shirts were counted?

A 4 **B** 14 **C** 21 **D** 24

2. Which color did the fewest children wear?

F red **G** green **H** yellow **J** purple

3. How many more children wore purple shirts than yellow shirts?

A 4 **B** 5 **C** 6 **D** 7

4. Mrs. Martin has 8 vases in her shop. She wants to put 9 tulips in each vase. She has 55 tulips. How many more tulips does she need?

F 17 **G** 27 **H** 46 **J** 72

For 5–6, use the frequency table.

FAVORITE ICE-CREAM TOPPING	
Topping	**Number of Votes**
Marshmallow	16
Chocolate	32
Nuts	19
Sprinkles	11

5. How many people voted for their favorite topping?

A 32 **B** 68 **C** 76 **D** 78

6. How many votes were there in all for nuts and sprinkles?

F 11 **G** 30 **H** 36 **J** NOT HERE

7. $32 \div 4 = \blacksquare$

A 9 **B** 8 **C** 6 **D** NOT HERE

Write What You Know

8. Bill made a four-section spinner. He used the spinner 20 times and listed the numbers spun.
2, 2, 1, 3, 4, 4, 3, 2, 1, 4, 3, 1, 3, 4, 1, 2, 1, 3, 4, 1
Make a tally table to show the data. Which number was spun most often?

9. Use the table from Exercises 5–6. How many more people voted for chocolate than nuts? Explain how you found your answer.

Analyze and Graph Data

There are about 1,000 giant pandas in the world today. Giant pandas can be 5 feet tall and can weigh more than 300 pounds!

PROBLEM SOLVING Look at the pictograph and make a bar graph of some of the animals at Zoo Atlanta.

ZOO ATLANTA

Animal	Number
Giant panda	🐾🐾
Red panda	🐾🐾
African elephant	🐾🐾🐾
Gorilla	🐾🐾🐾🐾🐾🐾🐾🐾🐾🐾🐾 🐾🐾🐾🐾🐾🐾🐾🐾🐾🐾🐾
Black rhinoceros	🐾🐾
Orangutan	🐾🐾🐾🐾🐾🐾🐾🐾🐾

Key: Each 🐾 = 1 animal.

Giant panda eating bamboo

Use this page to help you review and remember important skills needed for Chapter 15.

TALLY DATA (For Intervention, see p. H17.)

For 1–4, use the tally table.

HOW DO YOU GET TO SCHOOL?	
Car	𝍸 ||||
Walk	𝍸 𝍸 𝍸
Bus	𝍸 𝍸 ||
Bike	𝍸 𝍸 |

1. How many children ride in a car to school?
2. How many children ride their bikes?
3. How many children answered the survey question?
4. How many more children walk than ride the bus?

SKIP-COUNT (For Intervention, see p. H9.)

Find the missing numbers in the pattern.

5. 2, 4, 6, 8, 10, ■, ■, ■
6. 4, 8, 12, 16, 20, ■, ■, ■
7. 45, 40, 35, 30, ■, ■, ■
8. 3, 6, 9, 12, 15, ■, ■, ■
9. 6, 12, 18, 24, 30, ■, ■, ■
10. 80, 70, 60, 50, ■, ■, ■

READ PICTOGRAPHS (For Intervention, see p. H18.)

Use the value of the symbol to find the missing number.

11. If Δ = 2, then

    Δ + Δ + Δ = ■.

12. If □ = 3, then

    □ + □ + □ + □ = ■.

13. If ❀ = 4, then

    ❀ + ❀ + ❀ + ❀ + ❀ + ❀ = ■.

14. If ♥ = 5, then

    ♥ + ♥ + ♥ + ♥ + ♥ + ♥ + ♥ = ■.

15. If ❂ = 2, then

    ❂ + ❂ + ❂ + ❂ + ❂ = ■.

16. If ♦ = 4, then

    ♦ + ♦ + ♦ + ♦ + ♦ = ■.

17. If ☺ = 3, then

    ☺ + ☺ + ☺ + ☺ + ☺ + ☺ = ■.

18. If ☆ = 10, then

    ☆ + ☆ + ☆ + ☆ + ☆ = ■.

LESSON 1

Problem Solving Strategy
Make a Graph

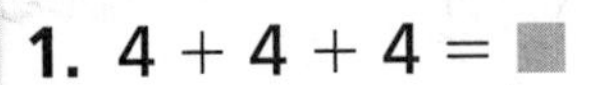

Quick Review

1. 4 + 4 + 4 = ■
2. 3 + 3 + 3 + 3 = ■
3. 2 + 2 + 2 + 2 = ■
4. 5 + 5 + 5 = ■
5. 2 + 2 + 2 = ■

PROBLEM The soccer team sold boxes of greeting cards to raise money. Rafael sold 14 boxes, Joselyn sold 7, Phil sold 24, Ken sold 12, and Felicia sold 10. What is one way the sales could be shown in a graph?

UNDERSTAND

- What are you asked to do?

PLAN

- What strategy can you use to solve the problem?

  You can *make a pictograph*.

SOLVE

- How can you show the data in a pictograph?

  a. Choose a **title** that tells about the graph.

  b. Write a **label** for each row.

  c. Look at the numbers. Choose a **key** to tell how many each picture stands for.

  d. Decide how many **pictures** should be next to each person's name.

GREETING CARDS SOLD

Rafael	✉ ✉ ✉ ✉ ✉ ✉ ✉
Joselyn	✉ ✉ ✉ ½✉
Phil	✉ ✉ ✉ ✉ ✉ ✉ ✉ ✉ ✉ ✉ ✉ ✉
Ken	✉ ✉ ✉ ✉ ✉ ✉
Felicia	✉ ✉ ✉ ✉ ✉

Key: Each ✉ = 2 boxes.

CHECK

- How can you know if the number of pictures in your graph is correct?

Problem Solving Practice

1. **What if** Derek sold 30 boxes of cards, Andy sold 25 boxes, and Kay sold 15 boxes? Make a pictograph to show the information.

2. Barry sold 11 boxes of cards. Using a key of 2, explain how you would show his sales in a pictograph.

PROBLEM SOLVING STRATEGIES

- Draw a Diagram or Picture
- Make a Model or Act It Out
- Make an Organized List
- Find a Pattern
- **Make a Table or Graph**
- Predict and Test
- Work Backward
- Solve a Simpler Problem
- Write a Number Sentence
- Use Logical Reasoning

Torrie and Jeremy made pictographs using the data in the table.

3. Torrie used a key of 3. How many symbols should she draw to show the votes for soccer?

   A 2 **C** 5
   B 3 **D** 6

4. What key did Jeremy use if he drew $7\frac{1}{2}$ symbols to show the votes for football?

   F key of 1 **H** key of 3
   G key of 2 **J** key of 4

FAVORITE SPORTS

Sport	Number of Votes
Soccer	18
Softball	12
Basketball	21
Football	15

Problem Solving Strategy

Mixed Strategy Practice

5. It was 12:05 P.M. when Liza and Nick began eating lunch. Nick finished in 15 minutes. Liza finished 8 minutes later than Nick. At what time did Liza finish lunch?

6. Regina bought a book for $1.80. She gave the cashier a $1 bill and 4 coins. She did not get any change. What coins could she have used?

7. Lloyd spent $3.59, $4.50, and $9.75 for games. He gave the clerk $20.00. How much change should he receive?

8. **REASONING** If December 1 is on a Wednesday, on which day of the week is December 15?

9. Sydney's team scored 89 points, 96 points, 98 points, and 107 points. How many points did the team score in all?

10. **? What's the Question?** Marty bought 2 packs of trading cards. He gave the clerk $10. His change was $4. The answer is $3.

LESSON 2

Read Bar Graphs

Quick Review

1. $4 \times 10 = \blacksquare$
2. $7 \times 7 = \blacksquare$
3. $8 \times 3 = \blacksquare$
4. $\blacksquare \times 6 = 36$
5. $9 \times \blacksquare = 81$

VOCABULARY

bar graph **scale**
horizontal bar graph
vertical bar graph

Learn

YEARS AND YEARS A **bar graph** uses bars to show data. A **scale** of numbers helps you read the number each bar shows.

These bar graphs show the same data.

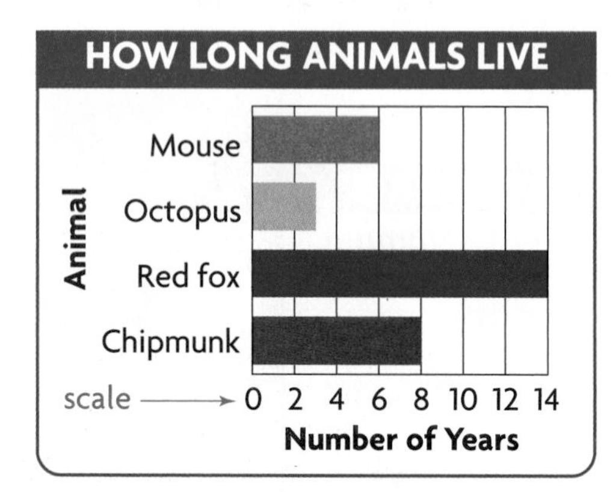

In a **horizontal bar graph**, the bars go across from left to right.

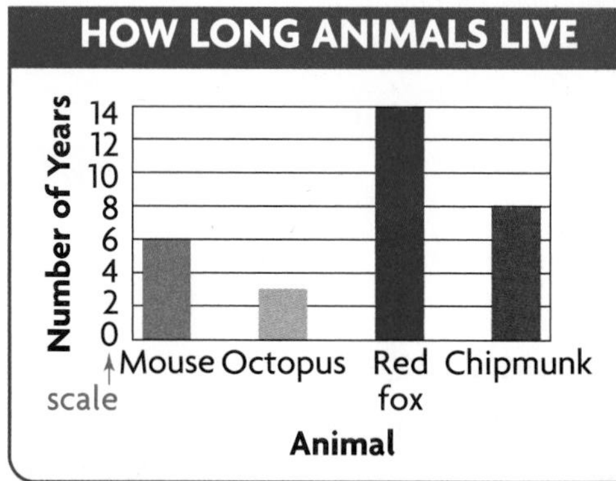

In a **vertical bar graph**, the bars go up.

- What scale is used on the bar graphs? Why is this a good scale?
- How do you read the bar for the octopus, which ends halfway between two lines?

Examples

A

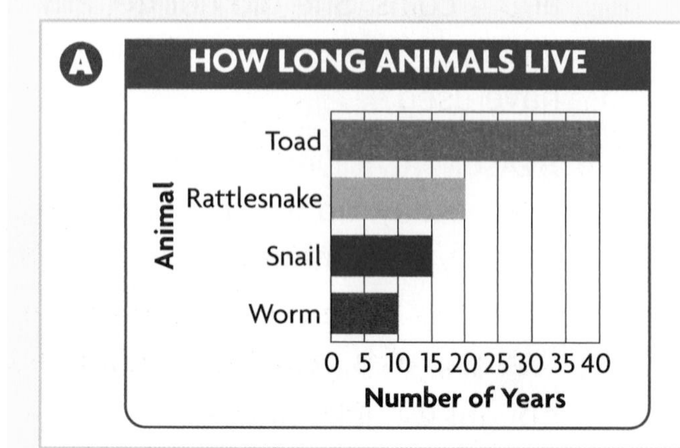

B

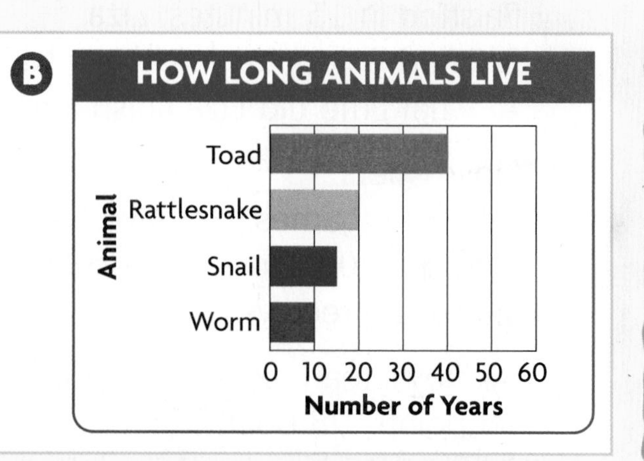

- How are these graphs alike? How are they different?

Check

1. **Explain** how you would use the graph in Example B to tell how long a snail lives.

For 2–3, use the bar graphs in Examples A and B.

2. How long can a toad live?
3. How long can a worm live?

Practice and Problem Solving

For 4–6, use the Sea Animals bar graph.

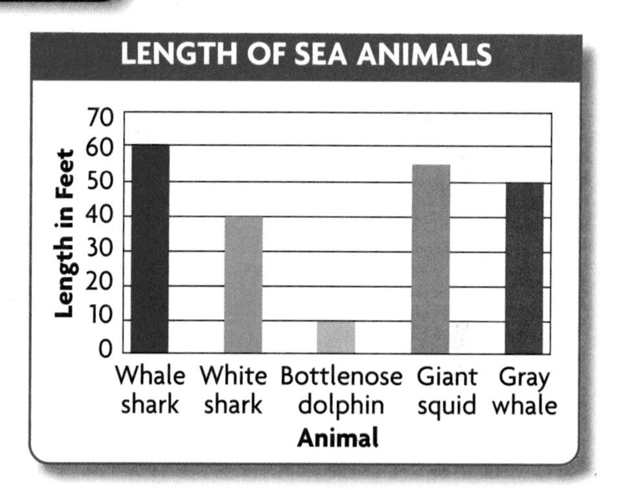

4. Is this a vertical or horizontal bar graph?
5. How long is a giant squid?
6. How much longer is a gray whale than a bottlenose dolphin?

For 7–8, use the Favorite Wild Animals bar graph.

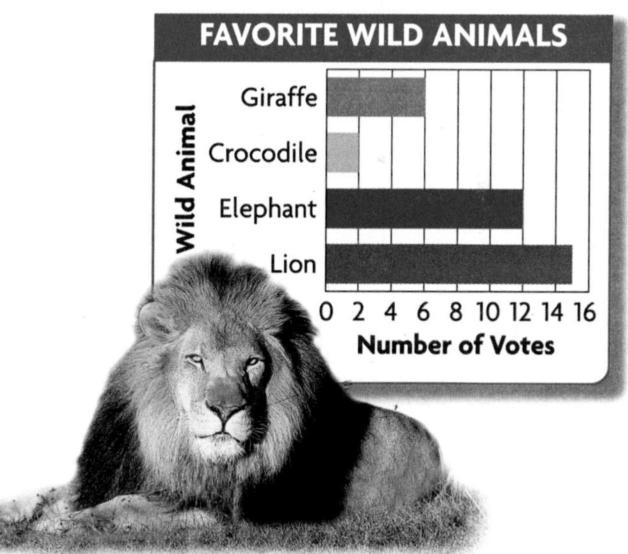

7. Which animal received the most votes?
8. Were there more votes in all for crocodile and lion, or for giraffe and elephant? Explain.
9. **Reasoning** Namiko saw a bar graph showing favorite foods. How could she tell which food got the most votes without looking at the scale?

Mixed Review and Test Prep

10. $6 \div 2 = \blacksquare$ (p. 202)
11. $14 \div 2 = \blacksquare$ (p. 202)
12. $1{,}000 + 300 + 5 = \blacksquare$ (p. 6)
13. Find the product of 7 and 8. (p. 150)
14. **TEST PREP** Meg put 24 counters in 4 equal piles. How many counters are in each pile? (p. 184)
    - **A** 4
    - **B** 6
    - **C** 7
    - **D** 20

LESSON 3

Make Bar Graphs

Quick Review

1. $9\overline{)27}$ 2. $4\overline{)32}$ 3. $6\overline{)36}$
4. $7\overline{)42}$ 5. $7\overline{)63}$

MATERIALS
bar graph pattern, crayons

Explore

You can make a bar graph to show the number of each kind of animal at the St. Louis Zoo. Use the data in the table to make a horizontal bar graph.

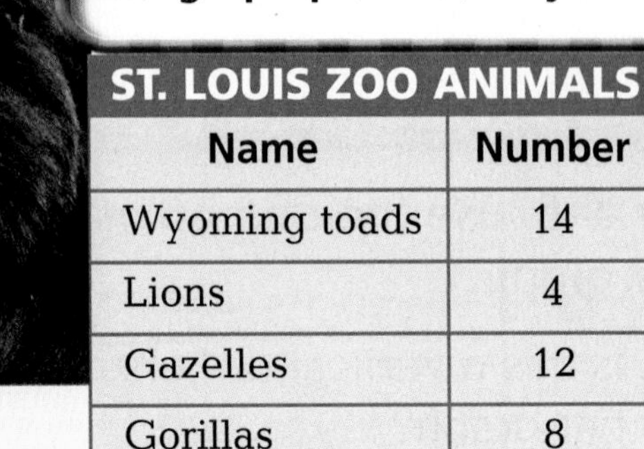

ST. LOUIS ZOO ANIMALS	
Name	Number
Wyoming toads	14
Lions	4
Gazelles	12
Gorillas	8

Activity

STEP 1

Write a title and labels. Decide on the best scale to use, and write the numbers.

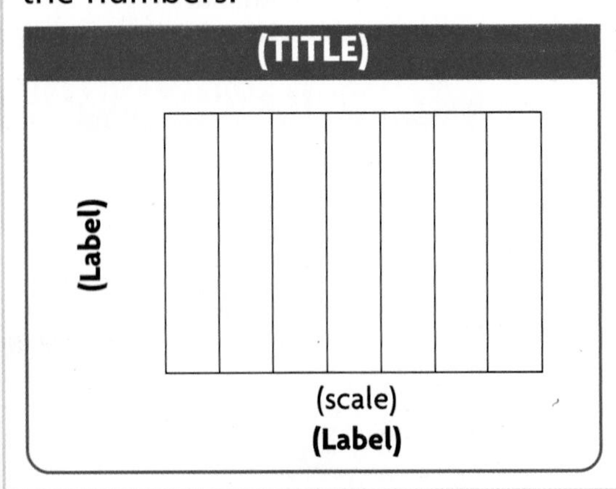

STEP 2

Complete the bar graph. Make the length of each bar equal to the number of each kind of animal.

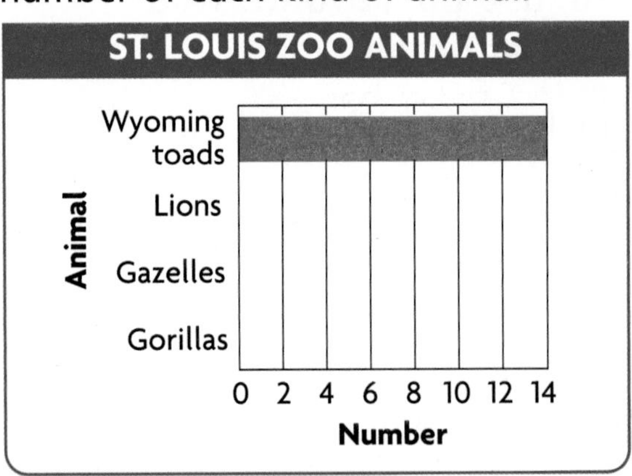

- How did you know where to end the scale?
- **What if** there were only 3 lions at the zoo? Explain how you would draw the bar for lions.

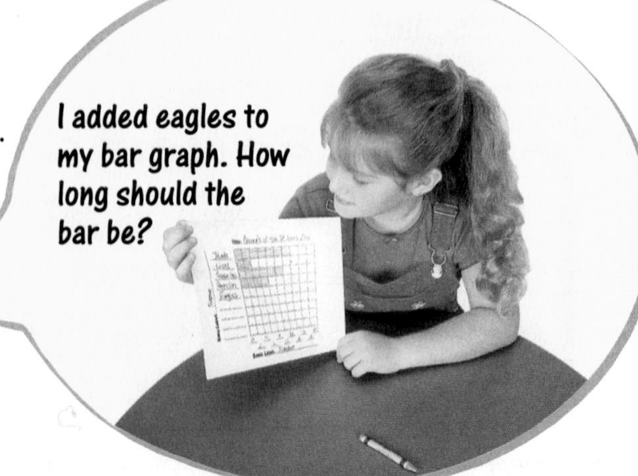

I added eagles to my bar graph. How long should the bar be?

Try It

Use your bar graph.

a. There are 5 eagles at the St. Louis Zoo. Make a bar for the eagles.

b. There are 6 hedgehogs at the St. Louis Zoo. Make a bar for the hedgehogs.

Connect

Decisions about making a bar graph can be made by looking at a frequency table.

a. Decide whether the graph will be horizontal or vertical.

b. Look at the number of different animals to find out how many bars you will need.

c. Look at the number of each animal to help you decide on the scale.

- What decisions were made for the bar graph at the right?

ST. LOUIS ZOO ANIMALS	
Animal	**Number**
Philippine ducks	30
Fennec foxes	10
Black widow spiders	25
Eagles	5

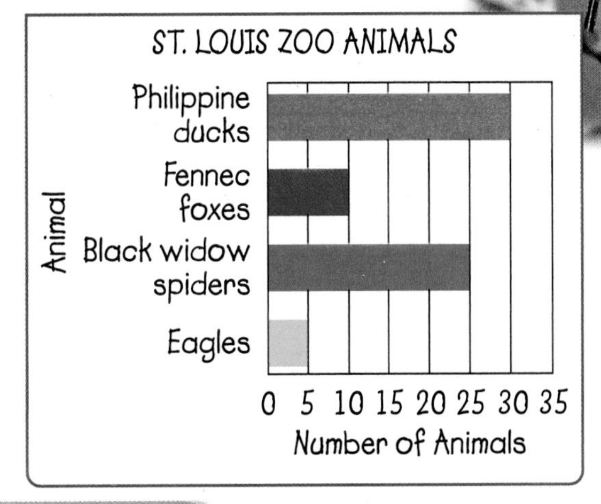

Practice and Problem Solving

1. Copy and complete the Wildlife Center bar graph. Use the data in the table at the right.

WILDLIFE CENTER	
Animal	**Number**
Monkeys	3
Zebras	8
Polar bears	2

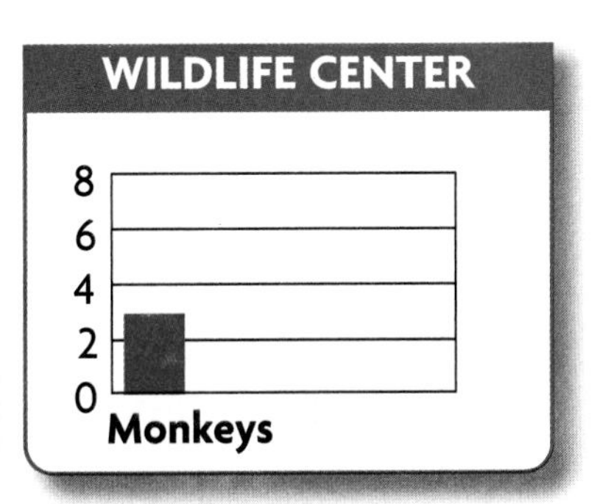

For 2–4, use the Favorite Pets table.

2. Make a bar graph. Why is 5 a good scale to use?

3. For which pet is the bar the longest?

4. Are there more birds and cats, or more dogs and fish? Explain.

FAVORITE PETS	
Animal	**Number**
Birds	30
Fish	15
Dogs	50
Cats	40

Mixed Review and Test Prep

Write <, >, or = for each ●. (p. 156)

5. 10×3 ● 6×5
6. 5×8 ● 7×6
7. 6×6 ● 5×7
8. 4×9 ● 9×5

9. TEST PREP What is the elapsed time from 10:45 A.M. to 1:15 P.M.? (p. 98)

A 2 hr 5 min
B 2 hr 20 min
C 2 hr 30 min
D 2 hr 45 min

LESSON 4

Line Plots

Quick Review

Quick Review

1. $8 \times \blacksquare = 24$
2. $5 \times 4 = \blacksquare$
3. $\blacksquare \times 6 = 48$
4. $\blacksquare \times 5 = 40$
5. $7 \times \blacksquare = 49$

VOCABULARY

line plot
mode
range

Learn

LOTS OF PLOTS All 24 third-grade students measured to find their heights in inches.

HEIGHTS OF THIRD GRADERS	
Height in Inches	**Number of Students**
49	|
50	|||
51	||
52	卌 |
53	卌 ||
54	||||
55	|

5 feet
1 foot

You can make a **line plot** to record each piece of data on a number line.

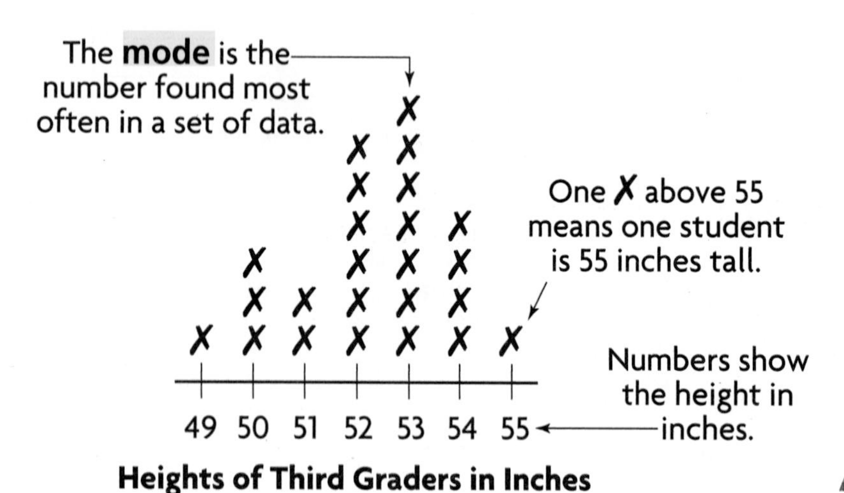

The mode is the number found most often in a set of data.
One X above 55 means one student is 55 inches tall.
Numbers show the height in inches.
49 50 51 52 53 54 55

Heights of Third Graders in Inches

The **range** is the difference between the greatest number and the least number.

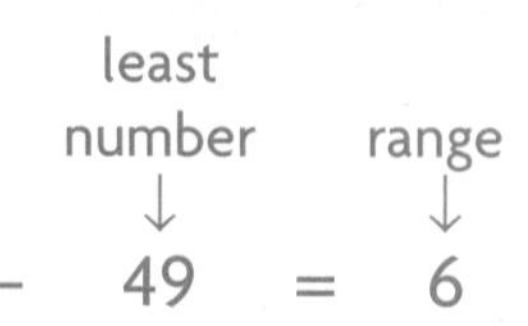

greatest number
least number
range
55 − 49 = 6

- What are the range and mode for this data?
- How many students are 50 inches tall?

Make a Line Plot

Debbie took a survey in her third-grade class to find out how many peanut butter and jelly sandwiches the students ate last week.

She put the survey data in a tally table.

PEANUT BUTTER AND JELLY SANDWICHES	
Number of Sandwiches	Tallies
0	ǀ
1	卌
2	ǀǀǀ
3	卌 ǀ
4	ǀǀǀǀ
5	ǀǀ

Activity

Make a line plot of the data in the table.

Materials: number line

STEP 1

Write a title for the line plot. Label the numbers from 0 to 5.

STEP 2

Draw **X**'s above the number line to show how many students ate each number of sandwiches.

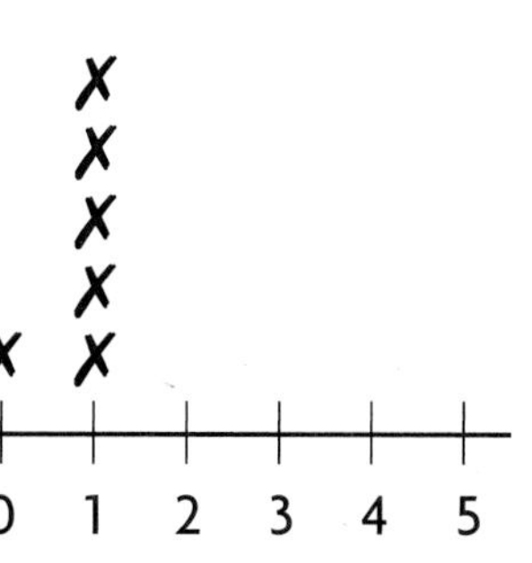

- How are the tally table and the line plot alike? How are they different?

Check

1. **Tell** what the mode is for the data above. Explain.

For 2–3, use your line plot.

2. How many students ate exactly 3 peanut butter and jelly sandwiches last week?

3. How many more students ate 1 or 2 sandwiches than ate 4 or 5 sandwiches?

LESSON CONTINUES 

Practice and Problem Solving

For 4–7, use the line plot at the right.

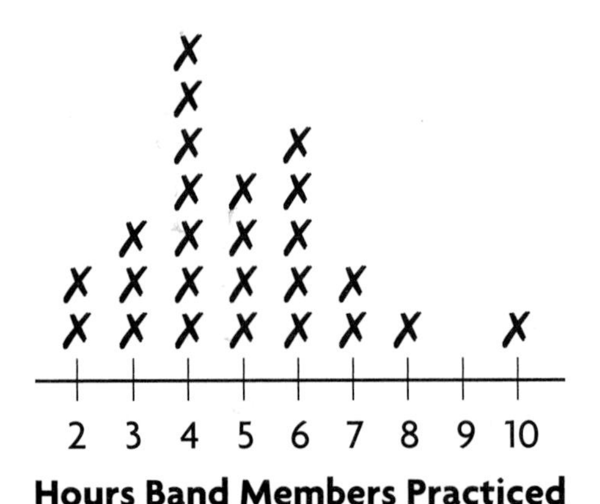

Hours Band Members Practiced

4. The ✗'s on the line plot stand for the band members. What do the numbers stand for?
5. How many band members practiced for just 6 hours?
6. What is the greatest number of hours any student practiced? What is the least number? What is the range for this set of data?
7. Did more band members practice for *less than* 5 hours or for *more than* 5 hours? Explain.

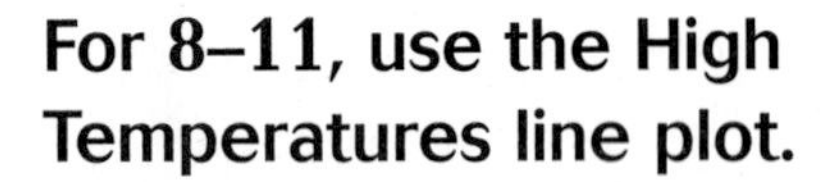

For 8–11, use the High Temperatures line plot.

Mrs. Brown's third-grade class recorded the high temperature each day for the last 3 weeks.

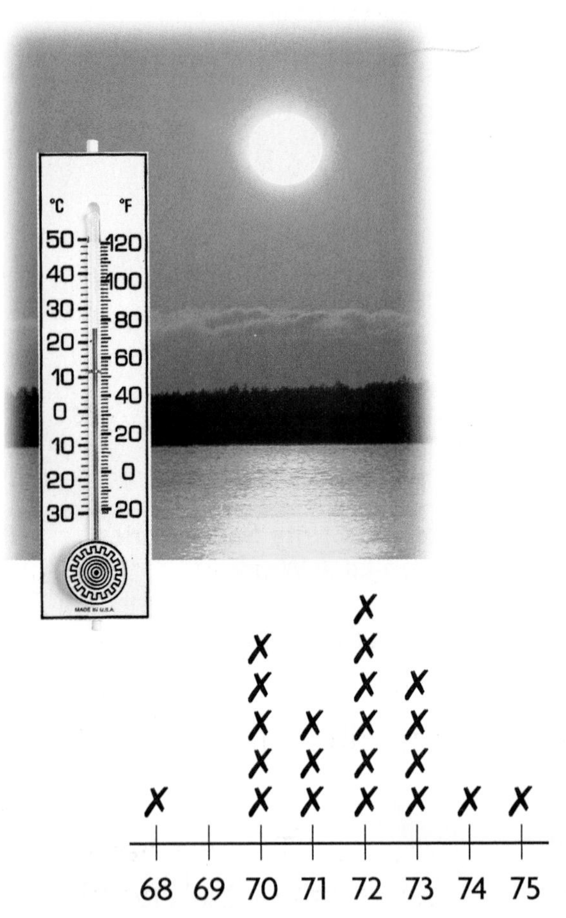

High Temperatures Each Day (in Degrees)

8. What high temperature occurred most often in the last 3 weeks? On how many days?
9. On how many days was the high temperature below 70 degrees?
10. What was the range of high temperatures the last 3 weeks?
11. Predict what you think the high temperature will be this week. Explain.
12. Take a survey to find out how many pets each student has. Show the results in a tally table and a line plot.
13. **Write About It** Explain how you decided which numbers to use in your line plot in Problem 12.

Mixed Review and Test Prep

Find each product. (p. 156)

14. $4 \times 7 = \blacksquare$

15. $3 \times 8 = \blacksquare$

16. $6 \times 6 = \blacksquare$

17. $7 \times 8 = \blacksquare$

Find each quotient. (p. 222)

18. $35 \div 7 = \blacksquare$

19. $27 \div 9 = \blacksquare$

20. $48 \div 8 = \blacksquare$

21. $49 \div 7 = \blacksquare$

22. **TEST PREP** What multiplication fact could you use to solve $8 \div 2$? (p. 188)

A $2 \times 2 = 4$ **C** $8 \times 2 = 16$
B $2 \times 4 = 8$ **D** $4 \times 8 = 32$

23. **TEST PREP** Sonya bought juice for \$1.09 and a sandwich for \$3.45. She paid the cashier with a \$10 bill. How much change should she receive? (p. 88)

F \$5.46 **H** \$6.46
G \$5.56 **J** \$6.56

PROBLEM SOLVING LINKUP . . . to Reading

STRATEGY • USE GRAPHIC AIDS Graphic aids organize information so that it can be compared and analyzed. When you use a graphic aid, you "read" a picture rather than just words.

Brandi took a survey to find how many books her classmates read during the summer. She made a line plot and a bar graph of the data.

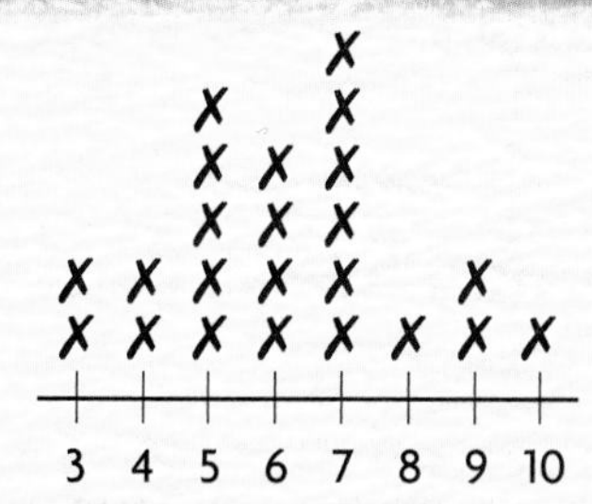

Books Read During the Summer

1. How many classmates did Brandi survey?
2. What are the mode and range of the data?
3. **Write About It** Tell whether you think the bar graph or the line plot is easier to read. Explain the reason for your choice.

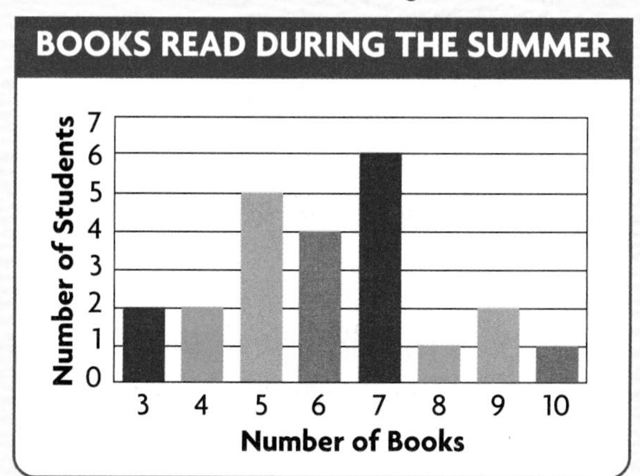

LESSON

5 Locate Points on a Grid

Quick Review

1. 28 ÷ 4 = ■
2. 56 ÷ ■ = 7
3. 35 ÷ 5 = ■
4. 45 ÷ 5 = ■
5. 30 ÷ 3 = ■

VOCABULARY

grid
ordered pair

Learn

GET TO THE POINT Jack is using a map to help him find different animals at the zoo. He wants to see the elephants first. How can this map help Jack find the elephants?

The horizontal and vertical lines on the map make a **grid**.

Start at 0. Move 2 spaces to the right.
Then, move 3 spaces up.
The elephants are located at (2,3).

(2,3)

The first number tells how many spaces to move to the right.

The second number tells how many spaces to move up.

GRID

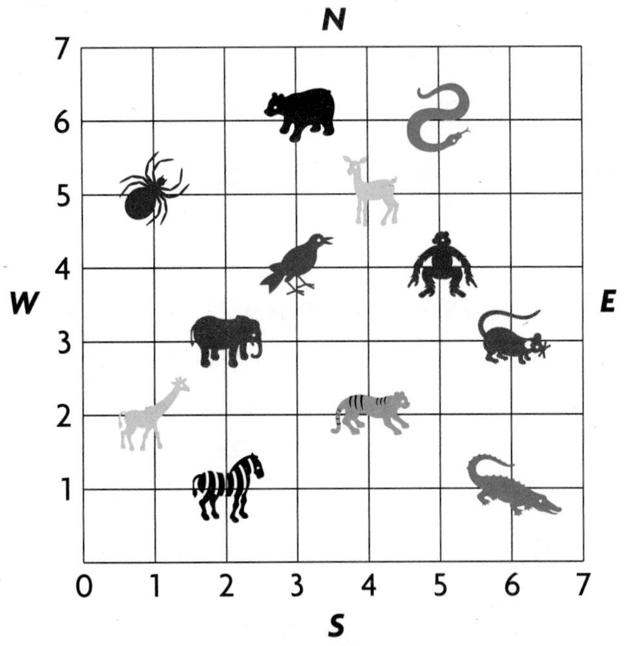

MATH IDEA An **ordered pair** of numbers within parentheses, like (2,3), names a point on a grid.

Example

Which animal is found at (5,4) on the grid?

Start at 0.
Move 5 spaces to the right.
Then move 4 spaces up.
So, the monkey is found at (5,4).

- Tell how to find the ordered pair for the spider on the grid.

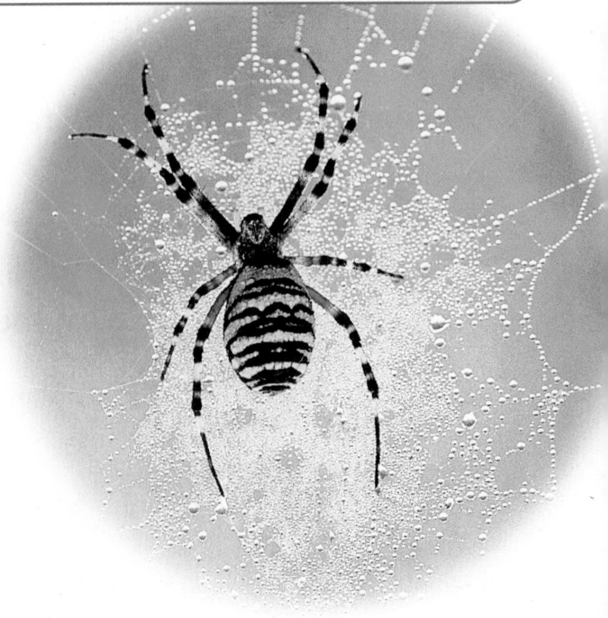

Wasp spider ▲

Check

1. **Explain** how you would find (11,12) on a 12-by-12 grid.

For 2–5, use the zoo grid on page 262.

2. Does (4,3) show the same point on the grid as (3,4)? Explain.

Write the ordered pair for each animal.

3. tiger
4. bear
5. giraffe

▲ Emerald tree boa

Practice and Problem Solving

For 6–11, use the zoo grid on page 262. Write the ordered pair for each animal.

6. bird
7. snake
8. zebra
9. mouse
10. deer
11. alligator

For 12–17, use the grid at the right. Write the letter of the point named by the ordered pair.

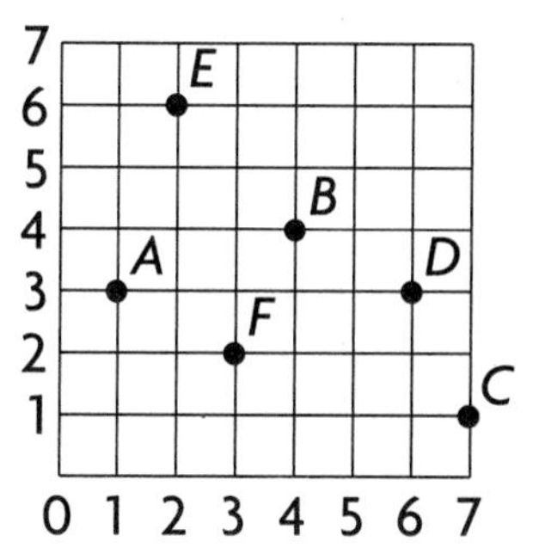

12. (6,3)
13. (1,3)
14. (4,4)
15. (2,6)
16. (7,1)
17. (3,2)

18. **REASONING** Rachel saw the zebra at (2,1) on the zoo grid. Mary saw the elephant at (2,3). What ordered pair names the point between the zebra and the elephant?

19. **What's the Error?** Jerry said, "To find the point (2,3), start at 0, move 3 spaces to the right and 2 spaces up." What error did Jerry make?

20. Walt has 45 pennies. Arthur has 11 nickels. Which boy has more money? How much more does he have?

Mixed Review and Test Prep

21. $5 \times 9 = \blacksquare$ (p. 118)
22. $7 \times 6 = \blacksquare$ (p. 150)
23. $6 \times 8 = \blacksquare$ (p. 148)
24. $4 \times 7 = \blacksquare$ (p. 134)

25. **TEST PREP** Find the difference of 567 and 288. (p. 58)
    A 279 **C** 379
    B 289 **D** 855

EXTRA PRACTICE page H46, Set C

LESSON

6 Read Line Graphs

Quick Review

1. 5 × ■ = 0
2. ■ × 9 = 63
3. 30 = ■ × 6
4. 9 × 4 = ■
5. 7 × ■ = 7

VOCABULARY

line graph

Learn

LINE UP A **line graph** is a graph that uses a line to show how something changes over time. What was the normal, or average, temperature in Des Moines, Iowa, in March?

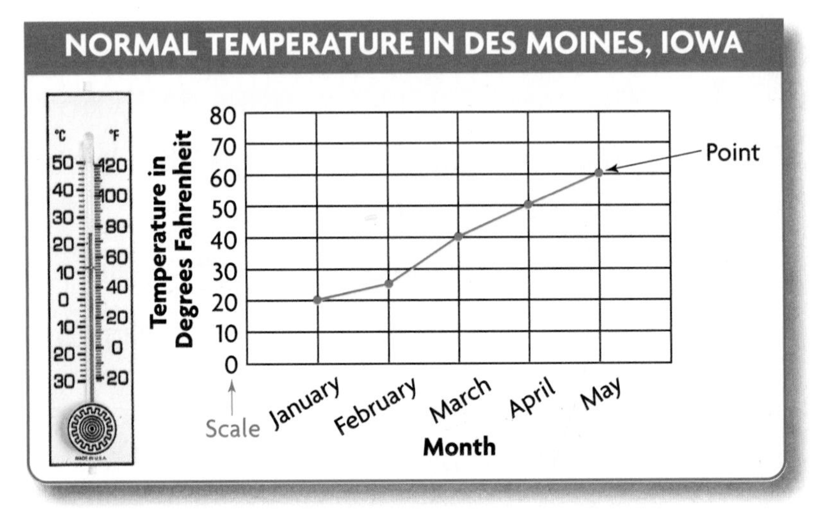

A line graph is like a grid.

a. From 0 find the vertical line for March. Move up to the point.

b. Follow the horizontal line left to the scale.

c. The point for March is at 40 degrees.

So, the normal temperature in Des Moines in March is 40 degrees.

Technology Link

To learn more about data and graphs, watch the **Harcourt Math Newsroom Video,** ***Fish Census.***

Check

1. **Tell** what you notice about the temperature in Des Moines from January to May.

For 2–3, use the line graph above.

2. What is the normal temperature in January?
3. In what month is the normal temperature 60 degrees?

Practice and Problem Solving

For 4–6, use the Alaska line graph.

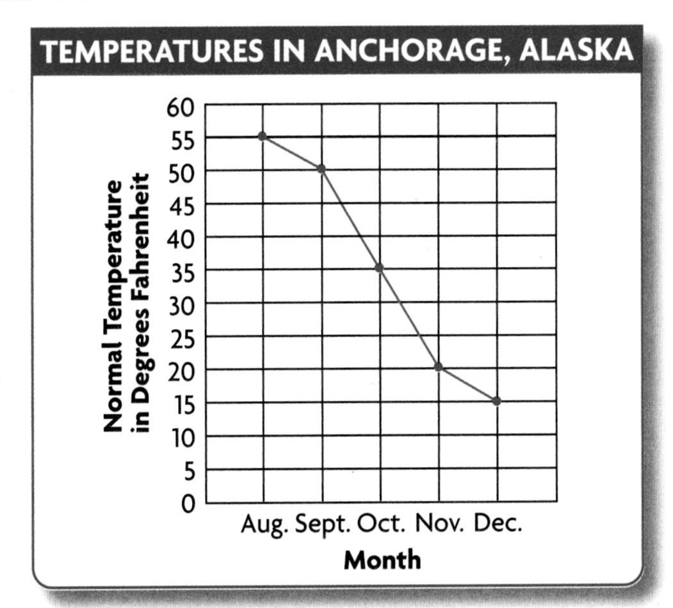

4. What is the normal temperature in December?

5. In what month is the normal temperature 50 degrees?

6. How many degrees higher is the normal temperature in August than in November?

For 7–9, use the Pumpkin line graph.

7. On what day were the most pumpkins sold? the fewest?

8. On which days were more than 8 pumpkins sold?

9. On which two days were the same number of pumpkins sold? How many pumpkins?

10. **Write a problem** using one of the line graphs above.

11. Sachio decorated 17 cookies and Linda decorated 12 cookies. They gave 20 cookies to their neighbors. How many cookies are left?

12. The total cost of a watch and a calculator is $50. The watch costs $20 more than the calculator. Find the cost of the calculator.

Mixed Review and Test Prep

Find the quotient.

13. $3\overline{)12}$ (p. 204) **14.** $4\overline{)24}$ (p. 204)

15. $5\overline{)15}$ (p. 202) **16.** $2\overline{)18}$ (p. 202)

17. **TEST PREP** In a multiplication sentence, the product is 56. One factor is 8. What is the other factor? (p. 142)

A 6 **B** 7 **C** 8 **D** 9

Review/Test

CHECK VOCABULARY AND CONCEPTS

For 1–3, choose the best term from the box.

line plot
line graph
mode
range

1. The number that occurs most often in a set of data is the __?__. (p. 258)
2. A number line that is used to record each piece of data is a __?__. (p. 258)
3. A graph that shows change over time is a __?__. (p. 264)
4. How would you find (4,5) on a grid? (p. 262)

For 5–7, choose the letter that names each. (pp. 252–255; 258–261; 264–265)

A. bar graph **B.** pictograph **C.** line plot **D.** line graph

5.

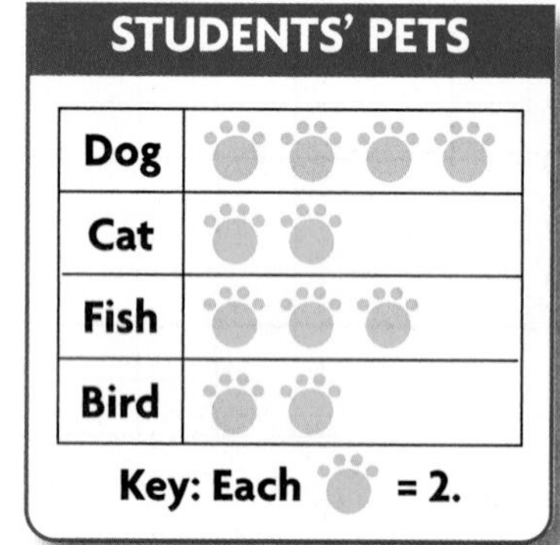

6.

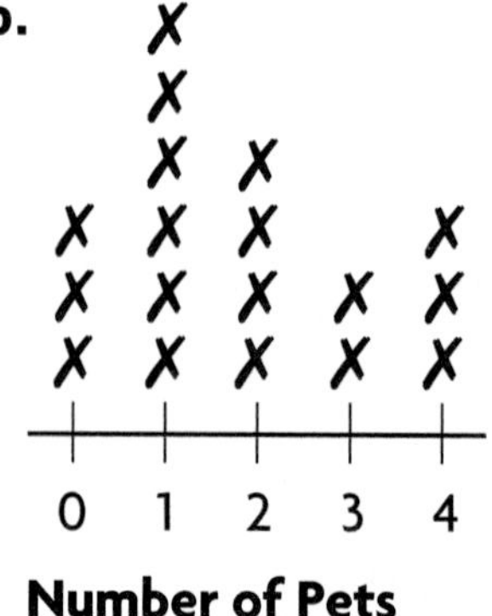

7.

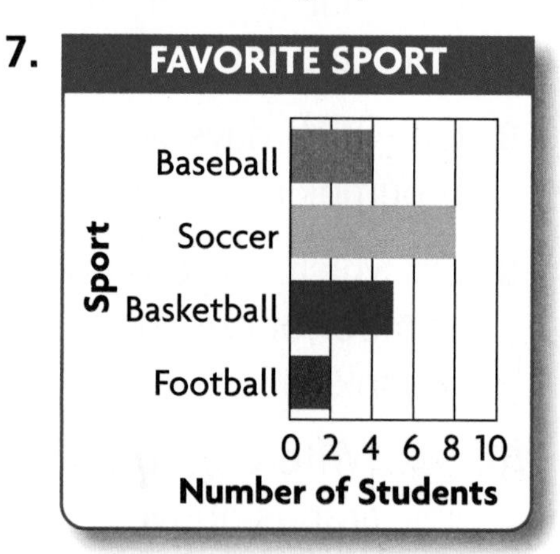

CHECK SKILLS

For 8–9, use the bar graph above. (pp. 254–257)

8. What scale is used on this graph?
9. How many students chose soccer?

For 10–11, use the line plot above. (pp. 258–261)

10. What is the mode for this data?
11. What is the range?

CHECK PROBLEM SOLVING

For 12–15, use the pictograph above. (pp. 252–253)

12. What key is used on this graph?
13. How many pets does a [paw print] equal?
14. How many more dogs than cats do the students have?
15. How many pets do the students have in all?

Standardized Test Prep

TIP! **Get the information you need.** See item **3.**

You know that each symbol means 3 votes. You need to find the number of votes for raisins in the data for Favorite Dried Fruit.

Also see problem **3,** p. H63.

For 1–5, choose the best answer.

1. 4,892 − 304

A 4,498 **C** 4,592
B 4,588 **D** NOT HERE

2. How many more students chose spring than chose fall?

F 30 **H** 16
G 26 **J** 4

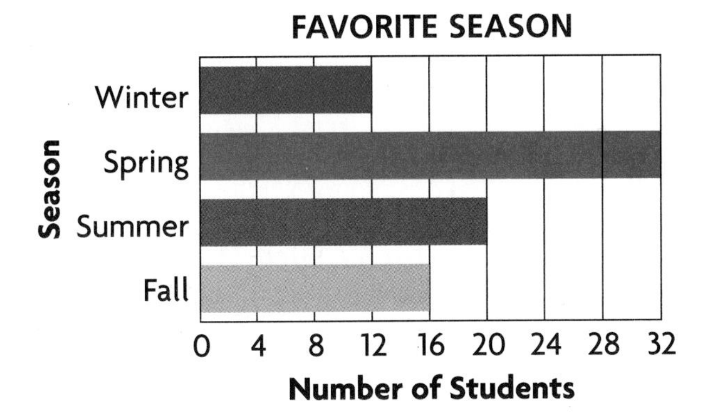

Kelly made a pictograph using this data.

Favorite Dried Fruit	
Raisins	24
Bananas	12
Cherries	18

3. If each symbol represents 3 votes, how many symbols will show the votes for raisins?

A 3 **B** 6 **C** 8 **D** 12

4. What is the ordered pair for the star on the grid?

F (1,3) **H** (2,2)
G (4,5) **J** (5,4)

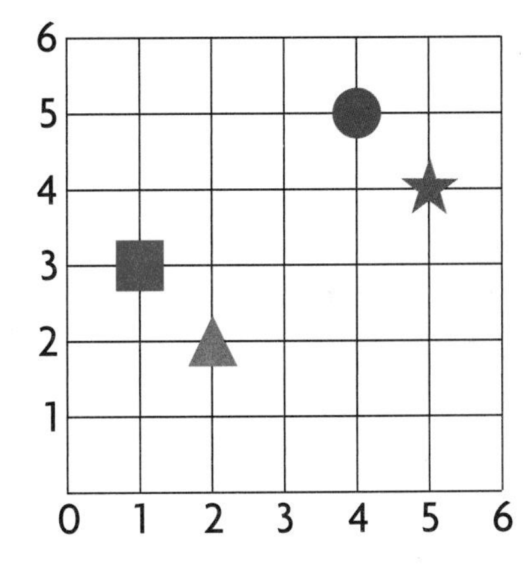

5. What is the value of the 3 in 16,035?

A 3 **B** 30 **C** 300 **D** 3,000

Write What You Know

6. Make a pictograph of the data for Favorite Dried Fruit. Include a key for your pictograph, and explain how you decided how many pictures to use for each fruit.

7. Look at the Favorite Season bar graph. How do you know how many students voted for each season? How many students voted in all? Show how you found the answer.

CHAPTER 16 Probability

Joshua trees, like this one, grow in the desert.

Temperatures differ from place to place. In McMurdo, Antarctica, the temperature is very cold all year. In Nairobi, Kenya, it is always warm. In Aspen, Colorado, the temperature changes from season to season.

AVERAGE TEMPERATURES (°F) FOR FIVE DAYS IN JANUARY

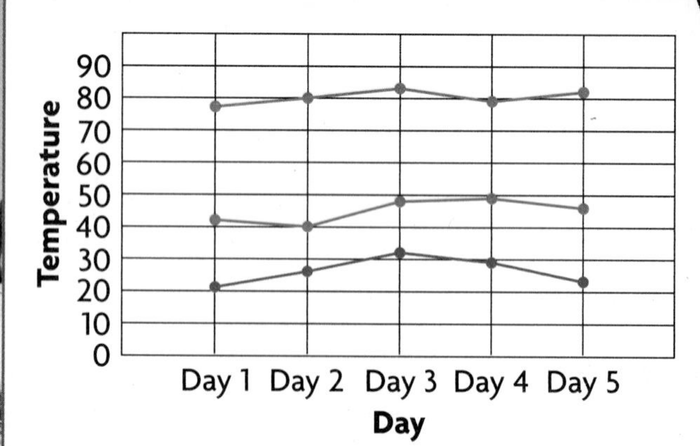

PROBLEM SOLVING Use the line graph to predict what the temperature will be in each place the next day.

CHECK WHAT YOU KNOW

Use this page to help you review and remember important skills needed for Chapter 16.

IDENTIFY PARTS OF A WHOLE (For Intervention, see p. H18.)

For 1–3, write the fraction that names the red part of the spinner.

1. 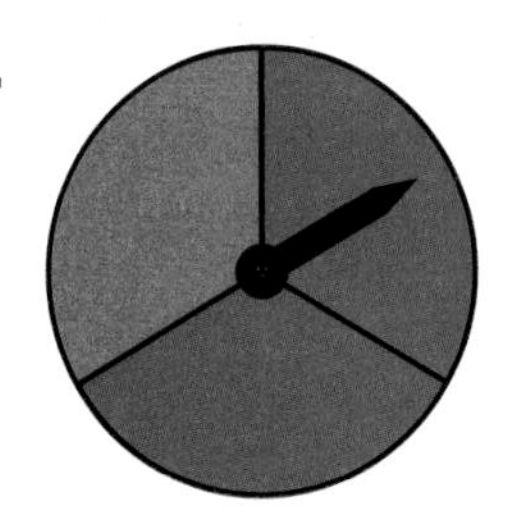

2. 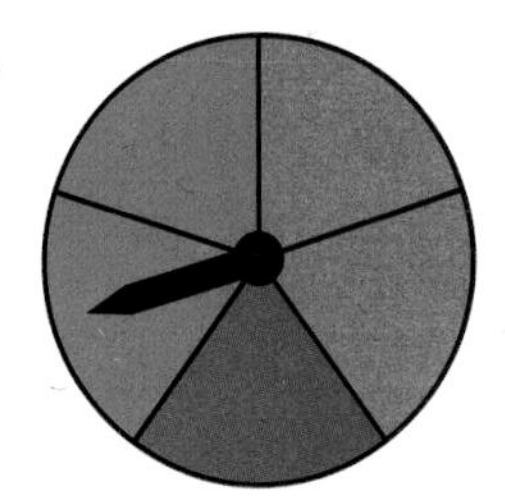

3. 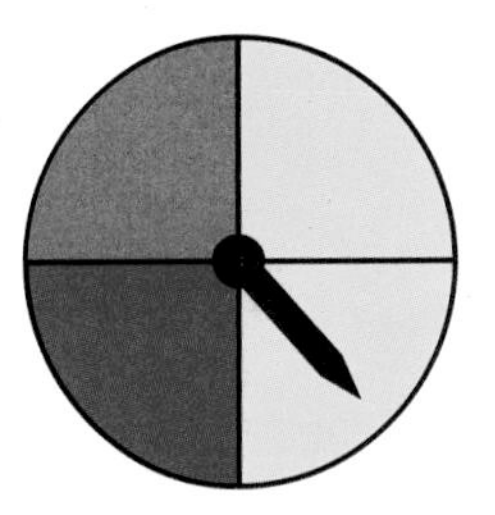

COMPARE PARTS OF A WHOLE (For Intervention, see p. H19.)

For 4–7, write the color shown by the largest part of each spinner.

4. 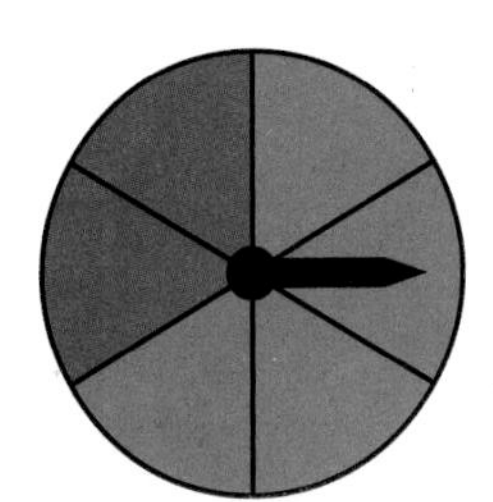

5. 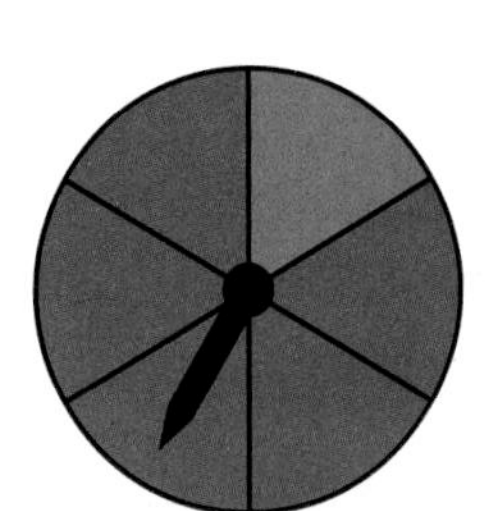

6. 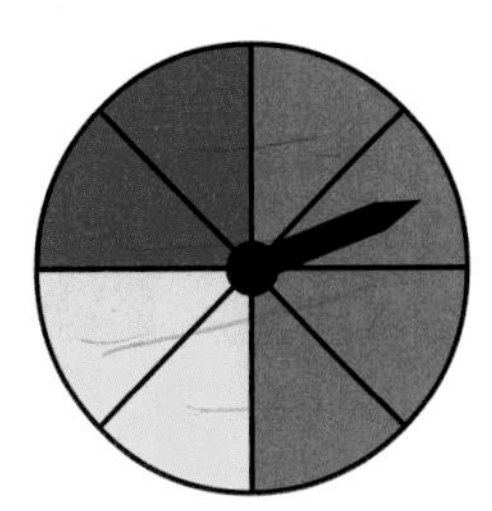

7. 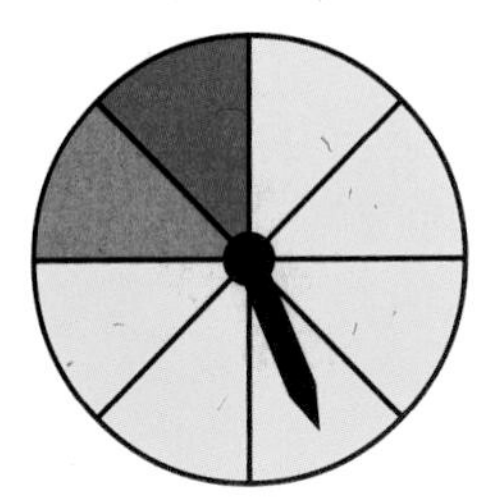

USE A TALLY TABLE (For Intervention, see p. H19.)

For 8–12, use the tally table.

8. What color was landed on the most times?
9. What color was landed on the fewest times?
10. How many more times was purple landed on than blue?
11. How many fewer times was red landed on than green?
12. How many times was the spinner used in all?

SPINNER RESULTS

Color	Tallies
yellow	||
green	||||
red	||
blue	|
purple	~~||||~~

LESSON 1

Certain and Impossible

Learn

WAY, NO WAY An **event** is something that happens. An event is **certain** if it will always happen. An event is **impossible** if it will never happen.

CERTAIN

A. You get wet if you jump into a swimming pool full of water.

B. Ice cubes feel cold.

IMPOSSIBLE

A. A rock will turn into a piece of cheese.

B. A mouse will talk to you today.

- Name some events that are certain. Name some that are impossible.

Quick Review

Complete. Use <, >, or = for each ●.

1. $5 + 23$ ● 28

2. 54 ● $95 - 40$

3. $436 - 25$ ● 409

4. $87 + 6$ ● 95

5. $32 \div 4$ ● $64 \div 8$

VOCABULARY

event
certain
impossible

Alonso drew this spinner for a game he made. Is it *impossible* for him to spin orange on his spinner?

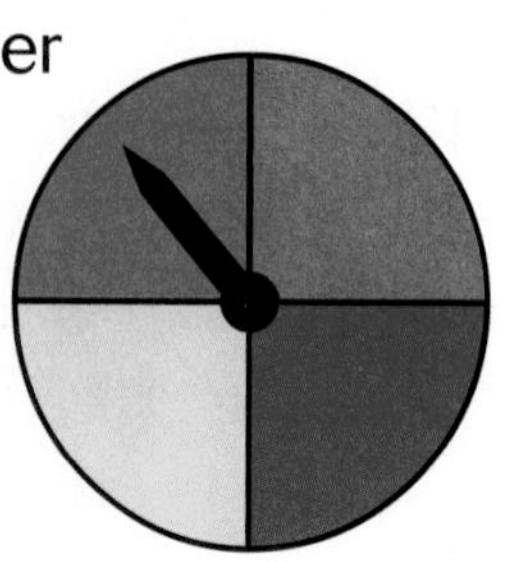

- How do you know?

Leslie put marbles in a bag for an experiment. Is it *certain* that she will pull a blue or a yellow marble?

- How do you know?

Check

1. Decide if it is certain or impossible that you will spin yellow or red on this spinner. Is it certain or impossible that you will spin green?

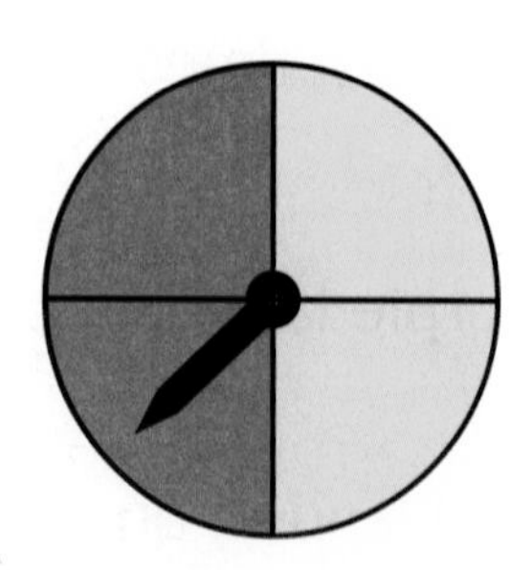

2. Is a spin of orange certain or impossible? Why?

Technology Link

More Practice: Use Mighty Math Number Heroes, *Probability,* Level A.

Tell whether each event is *certain* or *impossible*.

3. You could pull a red marble from this bag.

4. You could spin green or yellow on this spinner.

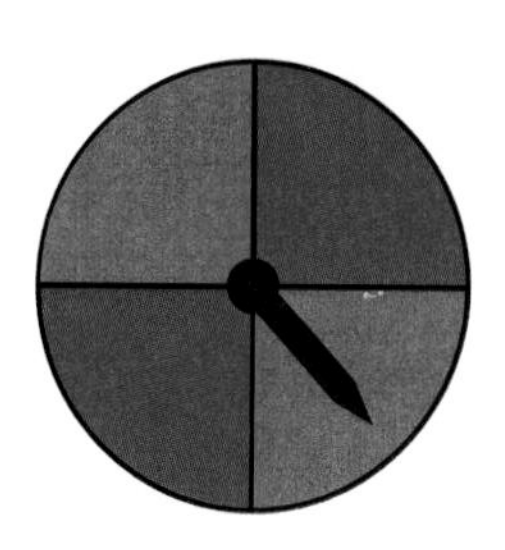

Practice and Problem Solving

Tell whether each event is *certain* or *impossible*.

5. If you add any two 1-digit numbers, the sum will be 19.

6. A box contains 100 pennies. You choose a coin from the box. It is a penny.

For 7–10, look at the bag of marbles at the right. Tell whether each pull is *certain* or *impossible*.

7. a marble

8. a yellow marble

9. two blue marbles

10. a red, blue, or green marble

11. Kaitlin rolled two cubes each numbered 1 through 6. If she adds the numbers, is it certain or impossible that she will get a sum less than 2?

12. Jeffrey has 21 marbles. If he gives 9 marbles away, he will have twice as many marbles as Miguel. How many marbles does Miguel have?

13. **REASONING** Elena rolls a cube numbered 1 through 6. She rolls the cube 4 times and adds the numbers. Their sum is 18. Two of the numbers she rolls are sixes. What numbers could Elena have rolled?

Mixed Review and Test Prep

14. $246 + 323$ (p. 42)

15. $768 - 515$ (p. 58)

16. $4{,}819 + 5{,}073$ (p. 46)

17. $6{,}342 - 5{,}107$ (p. 62)

18. **TEST PREP** Verna's math class begins at 11:30 A.M. and lasts for 1 hour 15 minutes. At what time is the class over? (p. 98)

A 11:45 A.M.
B 12:00 P.M.
C 12:30 P.M.
D 12:45 P.M.

EXTRA PRACTICE page H47, Set A

LESSON

2 Likely and Unlikely

Quick Review

What time is 20 minutes earlier?

1. 11:25 **2.** 10:40

3. 3:05 **4.** 12:17

5. 9:00

VOCABULARY

likely unlikely outcome

Learn

MAYBE, MAYBE NOT An event is **likely** to happen if it has a good chance of happening.

An event is **unlikely** if it does not have a good chance of happening.

An **outcome** is a possible result of an experiment.

Naomi and Ellis are playing a game. They use a spinner 25 times each.

- Look at the spinner. Which outcome is most likely? Which outcome is most unlikely? Explain.

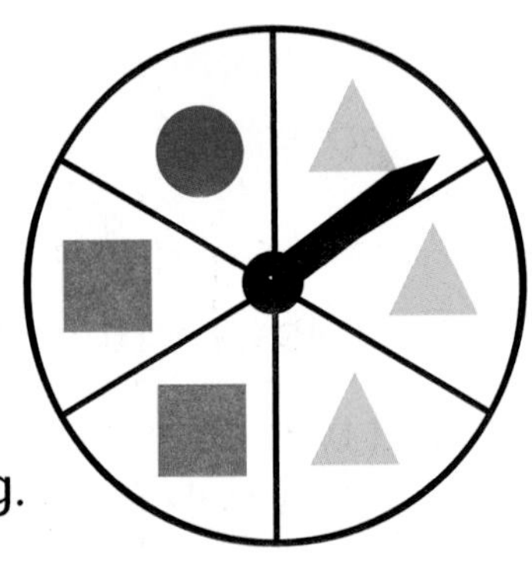

MATH IDEA An event is sometimes likely or unlikely depending on its chance of happening.

Check

1. **Name** the color marble you are most likely to pull from this bag of marbles. Explain your choice.
2. **Name** the color marble you are most unlikely to pull. Explain.

For 3–4, look at the spinner.

3. Name the color you are most likely to spin.
4. Name the color you are most unlikely to spin.

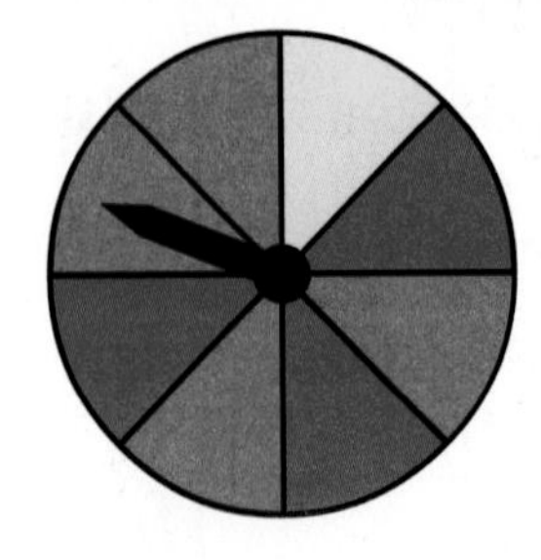

Problem Solving Practice

Name the color:

5. that is most likely to be pulled.

6. that you are most likely to spin.

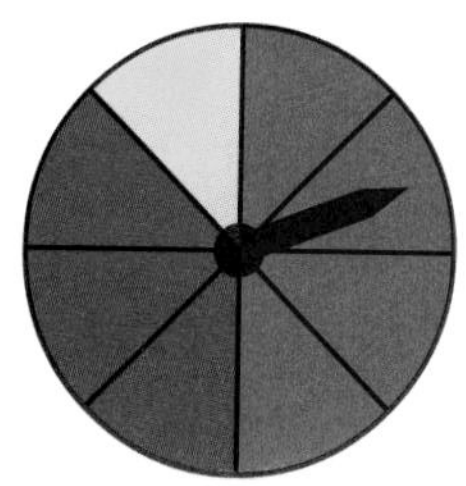

7. that is most unlikely to be pulled.

8. that you are most unlikely to spin.

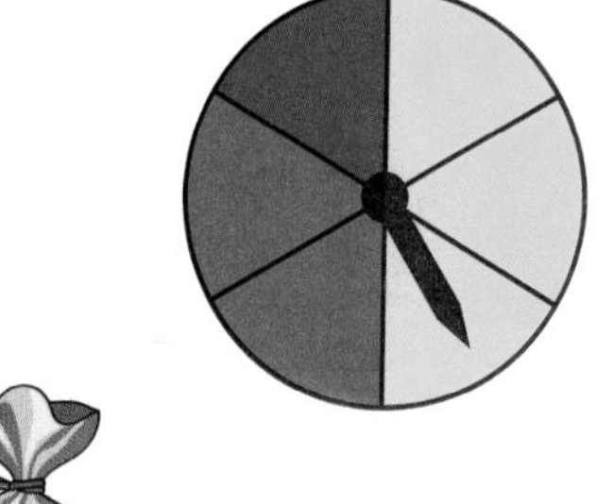

For 9–12, look at the bag of marbles at the right. Write *impossible, unlikely, likely,* or *certain* to match each event.

9. Pull a marble.

10. Pull a yellow marble.

11. Pull a black marble.

12. Pull a blue marble.

13. REASONING What is the difference between *likely* and *certain*?

14. REASONING What is the difference between *unlikely* and *impossible*?

15. Chris and Hope play a pattern game. Chris says 3, and Hope says 12. Chris says 4, and Hope says 16. Chris says 5, and Hope says 20. What does Hope say when Chris says 7? Explain.

16. **? What's the Question?** Della bought five items at the store. She was given $5.27 change from a $10 bill. The answer is $4.73.

Mixed Review and Test Prep

17. $8\overline{)48}$ (p. 216)

18. $6\overline{)42}$ (p. 216)

19. $\begin{array}{r} 7 \\ \times 4 \\ \hline \end{array}$ (p. 150)

20. $\begin{array}{r} 5 \\ \times 6 \\ \hline \end{array}$ (p. 118)

21. TEST PREP Jiro's mother baked 18 cupcakes. She gave an equal number to each of 6 people. How many cupcakes did each person receive? (p. 216)

A 1 **B** 2 **C** 3 **D** 4

EXTRA PRACTICE page H47, Set B

LESSON 3

Possible Outcomes

Quick Review

1. 5×6 2. 3×9
3. $56 = 8 \times$ ■
4. $4 \times$ ■ $= 16$ 5. 7×8

VOCABULARY

possible outcome
equally likely
predict

MATERIALS

coins

Explore

A **possible outcome** is something that has a chance of happening.

Two outcomes are **equally likely** if they have the same chance of happening.

Activity

Record the outcomes of tossing a coin.

STEP 1

Look at a coin. Decide on the possible outcomes.

The possible outcomes are heads and tails.

STEP 2

Toss the coin 20 times. Record the results in a tally table.

COIN TOSS	
Outcome	Tallies
Heads	
Tails	

- What did you notice about your results?

Since they have an equal chance of happening, the outcomes *heads* and *tails* are equally likely.

When you do an experiment, you can **predict**, or tell what you think will happen.

We are tossing a coin 50 times. How many times do you think the coin will show heads?

Try It

a. **What if** you tossed the coin 50 times? Predict how many times you would expect to toss heads.

b. Now toss 2 coins. What are the possible outcomes? Toss the coins 25 times. Record the results.

Connect

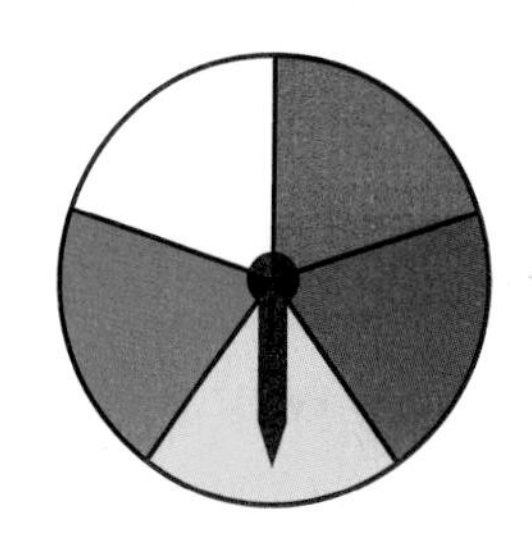

Look at this spinner. There are 5 possible outcomes. The pointer can land on red, blue, yellow, green or white.

Each space on the spinner is the same size, so the chance is *equally likely* that you will spin any color.

The chance is *1 out of 5* that you will spin any color.

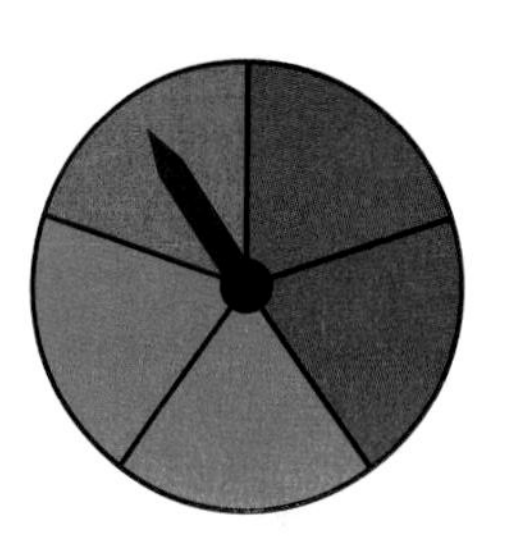

- What are the possible outcomes for this spinner? Which outcomes are equally likely? Explain.
- What is the chance that you will spin blue? that you will spin green? that you will spin red?

Practice and Problem Solving

For 1–4, list the possible outcomes of each event.

1. rolling a cube numbered 1–6

2. pulling blocks from this bag

3. using this spinner

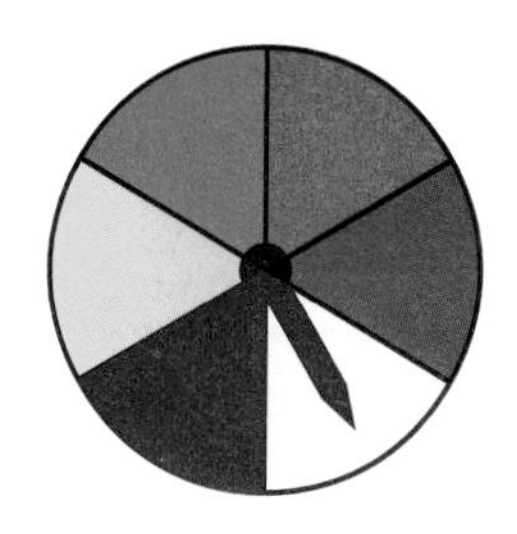

4. pulling blocks from this bag

5. Maggie used the spinner at the right. The pointer landed 6 times on red, 3 times on blue, and 1 time on yellow. Predict the color it will land on next. What is the chance that she will spin yellow?

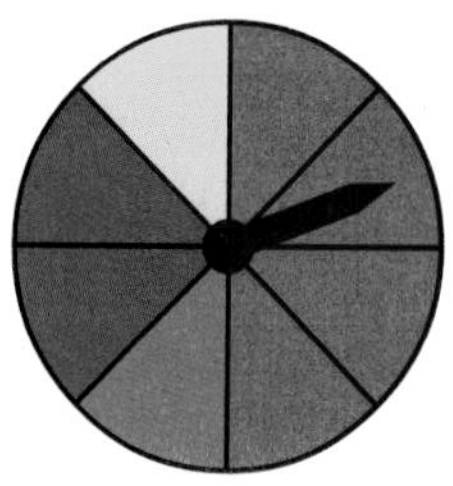

Mixed Review and Test Prep

6. 9×3 (p. 164)

7. 3×6 (p. 122)

8. $8 \times ■ = 64$ (p. 142)

9. $12 \div 2$ (p. 202)

10. **TEST PREP** Gary has $2.78 in his pocket and $3.14 in his bank. He wants to buy a book for $6.50. How much more does he need? (p. 88)

A $1.58 **B** $1.08 **C** $0.58 **D** $0.08

LESSON 4

Experiments

Quick Review

1. ■ ÷ 8 = 8
2. ■ ÷ 4 = 9
3. 49 ÷ 7
4. 72 ÷ ■ = 9
5. 24 ÷ 6

Learn

TRY IT One way to find out how likely it is for outcomes to occur is to conduct experiments.

HANDS ON

Activity 1

Materials: 4-part spinner pattern

STEP 1

Make a spinner that has four equal parts. Color the parts red, green, blue, and yellow.

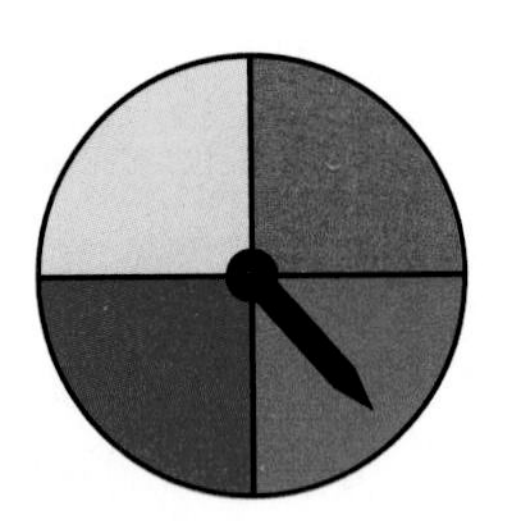

STEP 2

Make a tally table. List all the possible outcomes. Spin 20 times. Record the outcomes in the table.

OUTCOMES	
Color	Tallies
red	
blue	
green	
yellow	

STEP 3

Make a bar graph of your data to show the results of your experiment.

SPINNER EXPERIMENT

Color: Red, Blue, Green, Yellow

0 1 2 3 4 5 6 7 8 9 10

Number of Spins

- Compare the graphs from all the experiments in the class. What do the graphs show about the outcomes?
- How would the outcome of your experiment change if 2 parts of the spinner were red?

Activity 2

Materials: color tiles; paper bag

STEP 1

Put 1 red tile, 3 green tiles, and 6 blue tiles in a paper bag.

STEP 2

Make a tally table. List all the possible outcomes.

STEP 3

Pull one tile, record the color, and put it back in the bag. Make 40 pulls.

STEP 4

Make a bar graph of the data to show the results of your experiment.

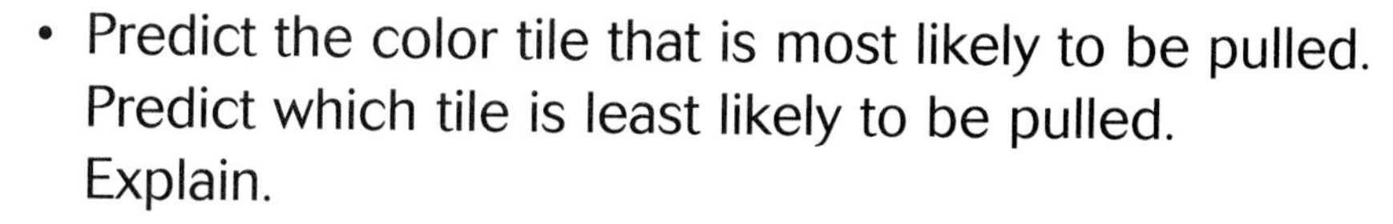

- Predict the color tile that is most likely to be pulled. Predict which tile is least likely to be pulled. Explain.
- What is the chance that you will pull either a red or a blue tile?

Check

For 1–2, use the spinners.

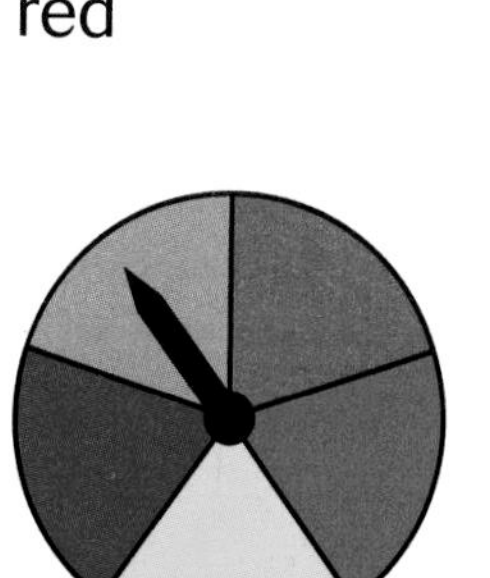

A

1. **Choose** which spinner has equally likely outcomes. Explain.
2. Name an unlikely outcome for Spinner B.
3. Draw a spinner on which all of the outcomes are equally likely.
4. Name all the possible outcomes for this bag of color tiles. Which are unlikely? Which are equally likely? Which is most likely?

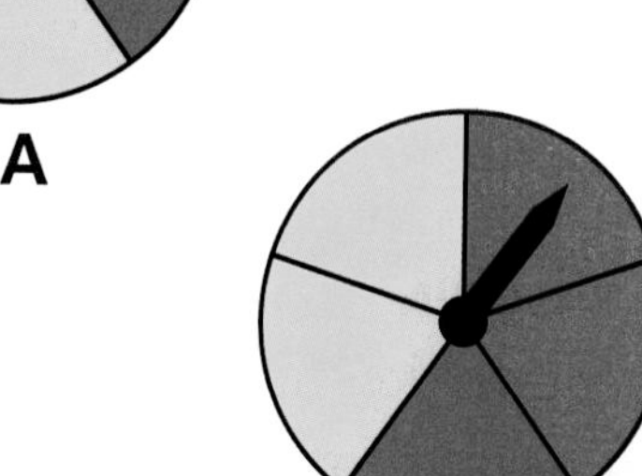

B

Practice and Problem Solving

5. Name all the possible outcomes for this spinner. Which are unlikely? Which are equally likely? Which is most likely?

6. Keshawn pulled color tiles from this bag. What is the chance that he will pull a red tile?

7. **USE DATA** Pamela drew the bar graph at the right to show part of the results of an experiment. Predict which color marble she will pull most often.

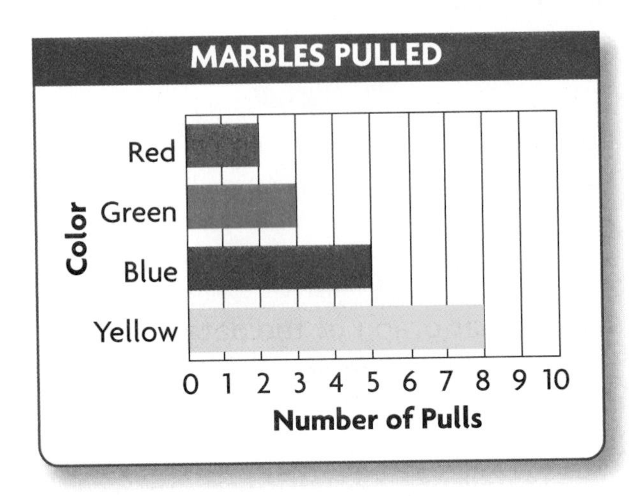

8. Toss a coin 10 times. Record each toss in order. Then repeat the experiment. Is there a pattern in the results? What outcome do you notice?

9. Draw a spinner. Color it so that the chance of spinning red is 2 out of 4, of spinning green is 1 out of 4, and of spinning blue is 1 out of 4.

10. Spencer rolled a cube numbered 1, 2, 3, 4, 4, and 4. He rolled the cube 25 times. Which number do you think he rolled most often? Explain.

11. Andi started making a spinner at 11:45 A.M. She finished making it 25 minutes later. At what time did Andi finish making her spinner?

12. **Write About It** Explain what a spinner should look like if the three outcomes are all equally likely.

13. **REASONING** Gustavo had 15 cookies. He gave 3 cookies to each of his friends. He kept 3 cookies for himself. With how many friends did he share his cookies?

14. Heather went to the mall and bought 3 tapes. Each tape cost $9. Then she bought lunch for $6. She has $4 left. How much money did Heather start with?

Mixed Review and Test Prep

Add. (p. 42)

15. $\begin{array}{r} 127 \\ +418 \\ \hline \end{array}$ **16.** $\begin{array}{r} 246 \\ +309 \\ \hline \end{array}$ **17.** $\begin{array}{r} 485 \\ +438 \\ \hline \end{array}$ **18.** $\begin{array}{r} 314 \\ +255 \\ \hline \end{array}$ **19.** $\begin{array}{r} 380 \\ +163 \\ \hline \end{array}$

Find the missing addend. (p. 68)

20. $15 + \blacksquare = 54$ **21.** $\blacksquare + 212 = 245$ **22.** $138 + \blacksquare = 250$

23. **TEST PREP** Hank bought a new comb for $1.59. He paid with a $5 bill. How much change should he get? (p. 88)

A $2.41 **C** $3.41
B $2.71 **D** $4.51

24. **TEST PREP** Katya bought a card for $1.85 and ribbon for $0.98. She paid with a $10 bill. How much change should she get? (p. 88)

F $5.71 **H** $7.07
G $6.77 **J** $7.17

PROBLEM SOLVING LINKUP . . . to Science

Scientists know that hurricanes most often form in September, October, and November. Hurricanes always have very high-speed winds that travel in circles. They form over warm, tropical ocean waters and then usually move west or northwest. Scientists use this information to predict hurricanes.

1. Which storm is most likely to become a hurricane? Explain.
2. Which storm is most unlikely to become a hurricane? Explain.
3. Which storms are equally likely to become hurricanes?

SUMMER 2000 TROPICAL STORMS

Storm	Circular winds	Warm ocean waters	High-speed winds	September to November	Moving west/ northwest
Alberto	✓	✓	✓		✓
Beryl	✓	✓	✓		
Debby	✓	✓	✓		✓
Ernesto	✓	✓	✓	✓	✓
Gordon	✓	✓	✓	✓	

EXTRA PRACTICE page H47, Set C

LESSON 5

Predict Outcomes

Learn

WEATHER OR NOT You can use the results of an experiment to predict what will happen in the future.

Meteorologists are scientists who predict weather. They study weather patterns from around the world to help them.

Quick Review

1. 25¢ = ■ pennies
2. 9 nickels = ■¢
3. ■ dimes = $1.00
4. 4 quarters = ■ pennies
5. 50¢ = ■ nickels

HANDS ON

Activity

STEP 1

Take the outdoor temperature every day at the same time for two weeks.

STEP 2

Record each temperature on a line plot. Use a scale:

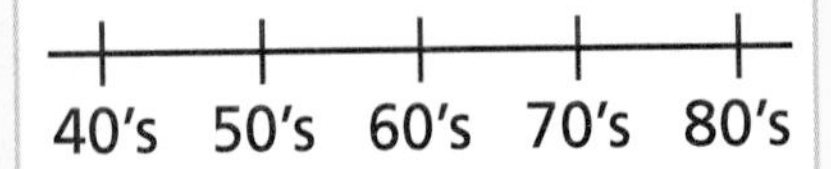

STEP 3

Use the data to predict the temperature for the next day.

Take the temperature the next day to check your prediction.

- How did you decide on your prediction? Explain.
- What was the temperature the next day? How did it compare with your prediction?

Check

1. Look at the line plot. Tell what you would predict for the next day's temperature. Explain.

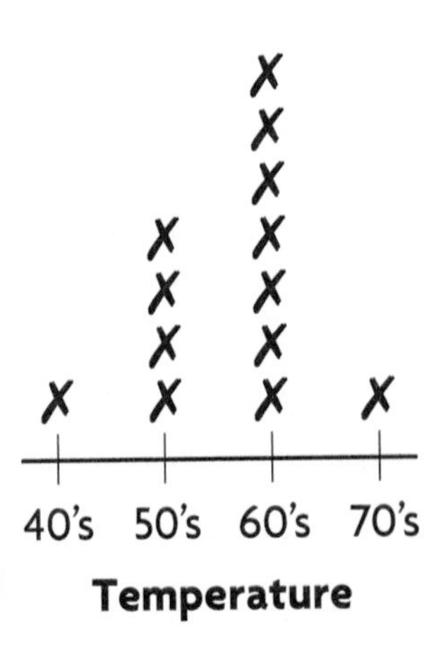

2. **What if** the temperatures for seven days in a row were 46, 48, 52, 55, 46, 48, and 47? What temperature would you predict for the next day? Explain.

Practice and Problem Solving

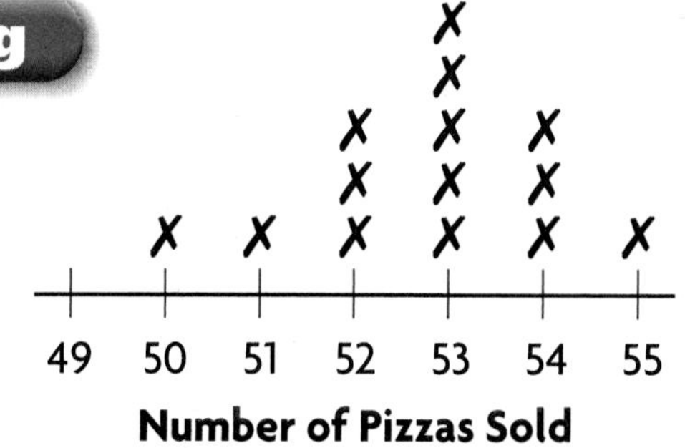

3. The line plot at the right shows the number of pizzas sold at Annie's Pizza Shop for the last two weeks. How many pizzas do you predict will be sold tomorrow?

4. This tally table shows the pulls from a bag of tiles. Predict which color is least likely to be pulled next.

OUTCOMES	
Color	Tallies
red	卌
blue	卌 卌 卌 ||
green	卌 ||
yellow	卌 ||||

5. The line plot below shows the results of rolling a number cube. Predict which number you would most likely roll.

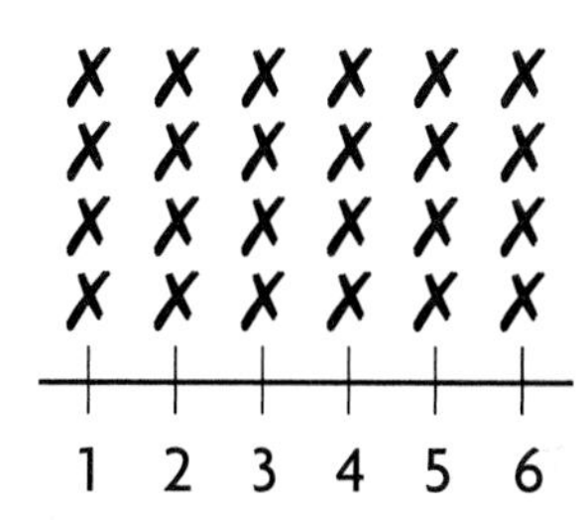

6. **Reasoning** Martha rolls a number cube 30 times for an experiment. The number cube is labeled with these numbers: 1, 2, 3, 4, 5, 6. Predict about how many times she will roll an even number.

7. The library has 392 fiction books and 514 nonfiction books. Estimate how many of these books the library has in all.

8. Jared has a collection of 42 baseball cards. He trades 2 of his cards for 6 other cards. How many cards does he have now?

9. **Reasoning** You toss a coin in the air one time. Predict how it will land. Explain.

10. **? What's the Error?** Todd says he will most likely pull a red cube from the bag. What's his error? Explain.

Mixed Review and Test Prep

Tell the value of the blue digit. (p. 6)

11. 7,593

12. 2,904

13. $\begin{array}{r} 8 \\ \times 7 \\ \hline \end{array}$ (p. 152)

14. $5\overline{)40}$ (p. 202)

15. **TEST PREP** Each of 4 students bought 8 packs of cards. How many packs of cards did they buy in all? (p. 134)

A 18 **B** 22 **C** 24 **D** 32

EXTRA PRACTICE page H47, Set D

LESSON 6

Problem Solving Skill
Draw Conclusions

UNDERSTAND PLAN SOLVE CHECK

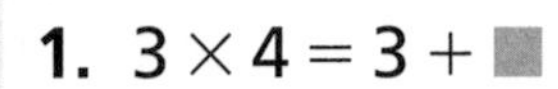

1. $3 \times 4 = 3 + ■$
2. $20 - 2 = 3 \times ■$
3. $3 \times 4 = ■ \times 2$
4. $6 \times ■ = 3 \times 8$
5. $6 \times 6 = ■ \times 4$

VOCABULARY
fair

IT'S NOT FAIR All of the students in Omar's class made math games. Omar made a spinner game in which four cars move around a track. Each time a color is landed on, the car of that color moves forward a space. Which spinner would make Omar's game fair?

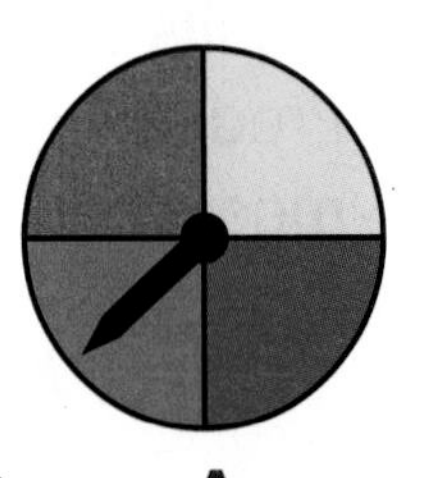
A

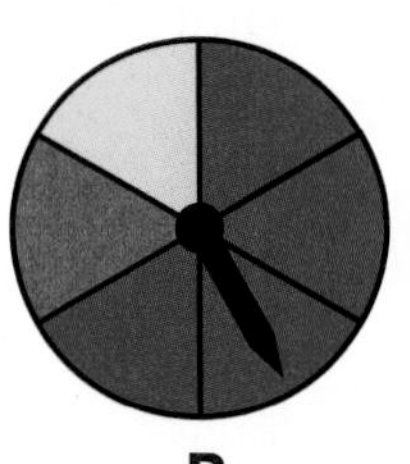
B

Example

STEP 1

Find out what the problem asks.

Which spinner would make Omar's game fair?

STEP 2

Find out what a fair game is.

A game is **fair** if every player has an equal chance to win.

STEP 3

Look at the spinners to see if every player would have an equal chance to win.

Spinner A has 4 equal parts: 1 red, 1 green, 1 yellow, and 1 blue.

Spinner B has 3 parts, but the parts are not equal. The blue part is larger than the red or yellow parts.

STEP 4

Draw a conclusion. Decide which spinner would make Omar's game fair.

Every player has an equal chance to win with Spinner A.

So, Spinner A would make Omar's game fair.

Talk About It

- In Marcie's game, two players take turns pulling a marble from a bag to move their pieces. Which bag of marbles will make Marcie's game fair?
- **REASONING** In a fair game, will each player always win the same number of times? Explain.

A

B

Practice and Problem Solving

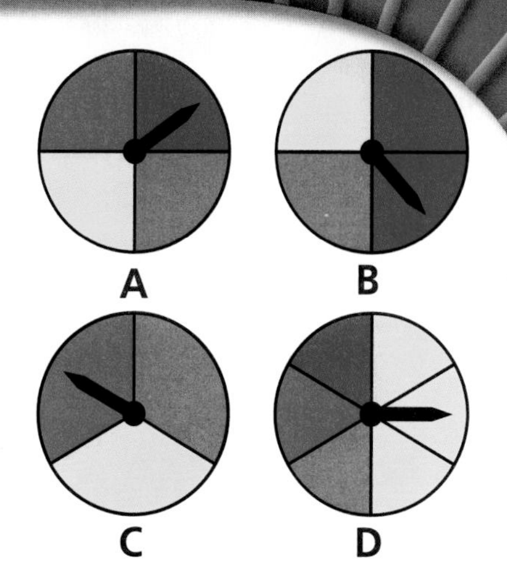

For 1–3, use the spinners.

1. Which of the Spinners A–D are fair? Explain.
2. Which of the Spinners A–D are unfair? Explain.
3. For each unfair spinner, what is the most likely outcome?

For 4–5, use these bags of marbles.

A

B

C

D

4. Which statement is true?
   A All bags have an equal number of marbles.
   B Every player has an equal chance to win with Bag D.
   C Every player has an equal chance to win with Bag C.
   D Bag A is fair.

5. Which of the bags of marbles are fair?
   F Bags A and B
   G Bags B, C, D
   H Bags B and C
   J Bags A and D

Mixed Applications

6. Pat is thinking of two numbers whose sum is 49 and whose difference is 1. What numbers is she thinking of?

7. Aaron has 8 nickels in one hand and 45¢ in nickels in the other hand. How many nickels does he have in all?

8. **REASONING** Anita and Cathy have \$20 together. If Cathy gives Anita \$3, they will each have the same amount of money. How much money does Anita have?

9. **REASONING** I am a number between 60 and 70. If you keep subtracting tens from me, you reach 2. What number am I?

10. **Write About It** Choose an unfair spinner or bag of marbles from above. Tell how to make it fair.

CHAPTER 16

Review/Test

CHECK VOCABULARY AND CONCEPTS

Choose the best term from the box.

1. An event that will always happen is ___?___. (p. 270)

2. An outcome that probably won't happen is ___?___. (p. 272)

certain
impossible
likely
unlikely

Tell whether each event is *certain* or *impossible*.

3. Next year, December 13 will come before December 12. (pp. 270–271)

4. Snow will melt when the temperature is above freezing. (pp. 270–271)

CHECK SKILLS

5. What are the possible outcomes of using this spinner? (pp. 274–275)

6. What outcomes are equally likely when you use this spinner? (pp. 274–275)

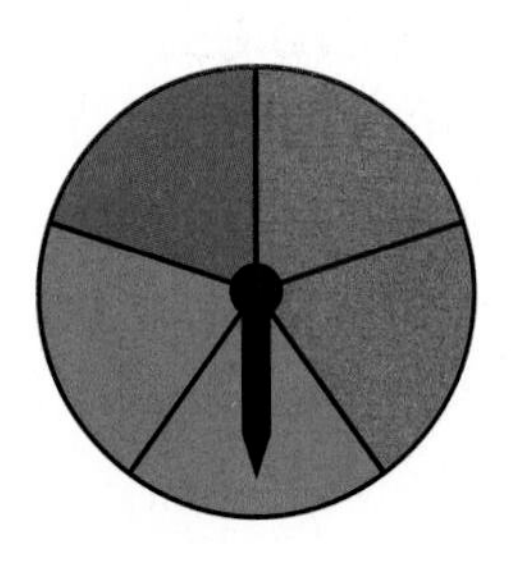

CHECK PROBLEM SOLVING

Choose the spinner that is fair. Choose the bag of marbles that is fair. Write *A* or *B*. (pp. 282–283)

7.

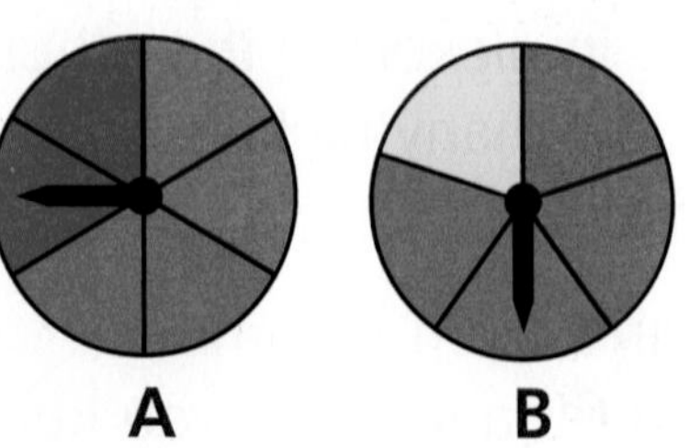

A B

8.

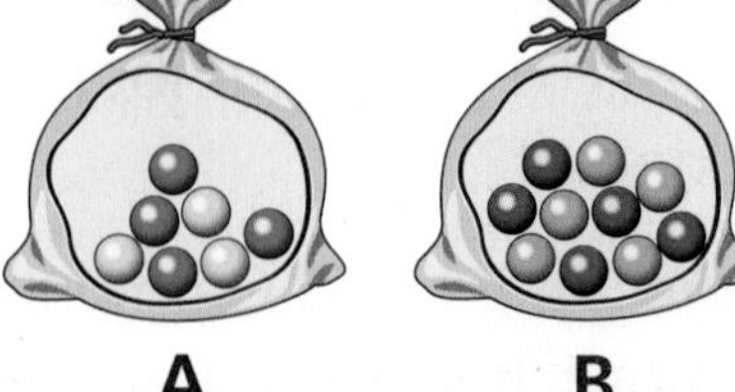

A B

For 9–10, use the spinner at the right. (pp. 282–283)

9. Is this spinner fair? Explain.

10. Are you likely or unlikely to spin yellow? Explain.

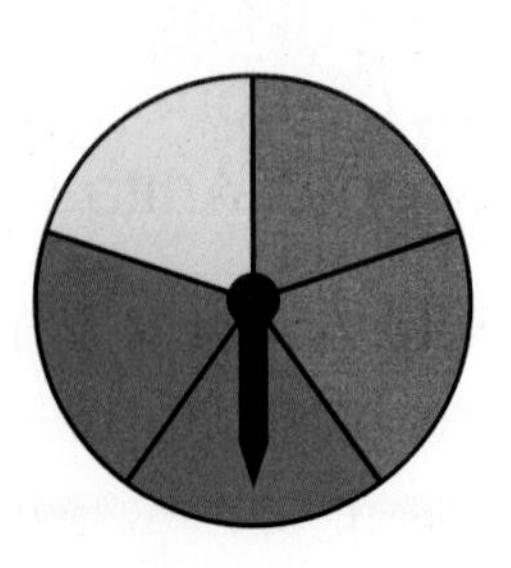

Standardized Test Prep

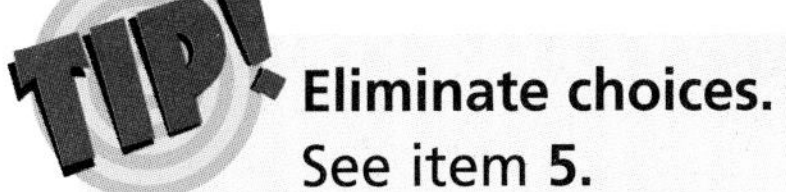

Eliminate choices.
See item **5**.

Relate each statement to the spinner until you find the one that is true.

Also see problem **5**, p. H64.

For 1–6, choose the best answer.

1. A spinner has four parts of equal size with the colors yellow, orange, purple, and green. Which event is **impossible**?

A The pointer lands on green.
B The pointer lands on orange.
C The pointer lands on red.
D The pointer lands on purple.

2. Jack will pull a marble out of the bag. Which color of marble is he most likely to get?

F yellow
G red
H blue
J green

3. $8 \times 7 = \blacksquare$

A 48
B 56
C 64
D NOT HERE

4. What is the chance that you will spin blue on this spinner?

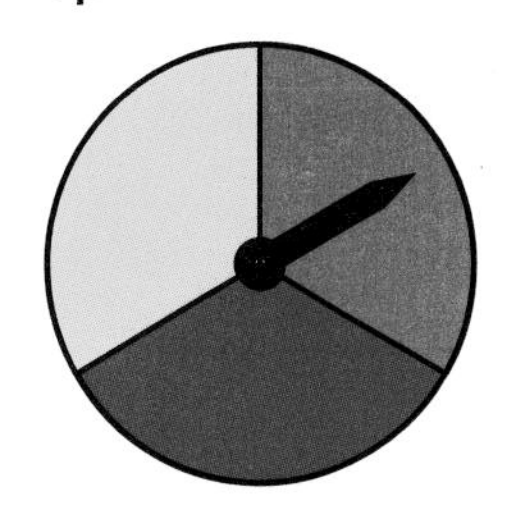

F 1 out of 3
G 2 out of 3
H 3 out of 4
J 3 out of 3

5. Which statement is true about the spinner?

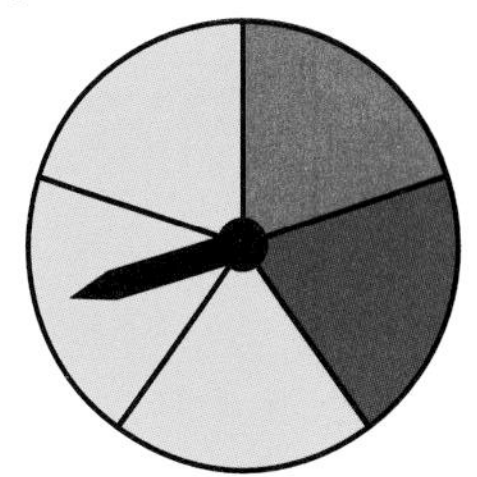

A A player has an equal chance of spinning red, blue, or yellow.
B A player is most likely to spin red.
C A player is most likely to spin yellow.
D A player is most likely to spin blue.

6. $356 + 298 = \blacksquare$

F 654 **G** 658 **H** 664 **J** 678

Write What You Know

7. Draw a spinner on which the numbers 4, 5, and 6 are equally likely to be spun. Explain your answer.

8. Describe how you can draw a spinner on which the number 1 is most unlikely to be spun.

PROBLEM SOLVING
MATH DETECTIVE

The Spinning Wheel

Put on your thinking cap. Use the clues to help you draw a spinner for each situation.

1. The chance of spinning red or blue is equally likely.

2. The chance of spinning blue is certain.

3. The chance of spinning red is impossible.

4. The chance of spinning red is slightly greater than the chance of spinning blue.

5. The chance of spinning red or yellow is equally likely, but the chance of spinning blue is greater than either red or yellow.

6. There is only a slight chance of spinning blue.

7. The chance of spinning either red, yellow, blue, or green is equally likely.

8. The chance of spinning blue or yellow is not very likely and the chance of spinning red is highly likely.

9. The chance of spinning red, blue, and yellow is equally likely.

STRETCH YOUR THINKING

If a spinner is divided into 12 equal sections, how would you label the spinner to have a slightly better chance of spinning red than blue?

Challenge

Find Mean and Median

Cari counted the number of robins she saw on her way to school. She recorded the data in a tally table.

Robins I Saw	
Monday	|||
Tuesday	𝍸 |||
Wednesday		
Thursday	𝍸	
Friday	||

Suppose Cari had seen the same total number of robins, but she had seen an equal number each day. How many robins would she have seen each day?

A **mean** is a number which can be used to represent all the numbers in a set of data.

MATERIALS: unit cubes
Make stacks of cubes to model the number of robins counted each day.

Mon. Tues. Wed. Thurs. Fri.

To find the mean, rearrange the stacks so that each stack has the same number of cubes. Count the number in each stack.

Mon. Tues. Wed. Thurs. Fri.

So, Cari would have seen 4 robins each day.
The mean is 4.

The **median** is the middle number in an ordered series of numbers.

Make stacks of cubes to model the number of robins counted each day. Place the stacks in order from least to greatest.

To find the median, count the number in the middle stack.

median

So, the median of the set of data is 3 robins.

Try It

- Use cubes. Find the mean and median of 2, 8, and 5.

Study Guide and Review

VOCABULARY

Choose the best term from the box.

data
bar graph
possible outcome
line graph

1. Something that has a chance of happening is a __?__. (p. 274)
2. A graph that uses a line to show how something changes over time is a __?__. (p. 264)

STUDY AND SOLVE

Chapter 14

Interpret data from a survey.

FAVORITE JUICE			
	Apple	Orange	Grape
Boys	6	8	3
Girls	5	4	6

How many boys voted in all?
Think: The boys are listed in the first row. Find the sum of these numbers.
6 + 8 + 3 = 17
So, 17 boys voted in all.

For 3–5, use the table. (pp. 242–245)

3. How many more boys than girls voted in the survey?
4. What is the most popular juice among the girls?
5. What is the most popular juice overall? Explain.

Chapter 15

Read a bar graph.

How many students play soccer?
Step 1: Find the bar for soccer.
Step 2: Compare the height of the bar with the scale to find the number of students who play soccer.
So, 8 of the students play soccer.

For 6–7, use the bar graph. (pp. 254–255)

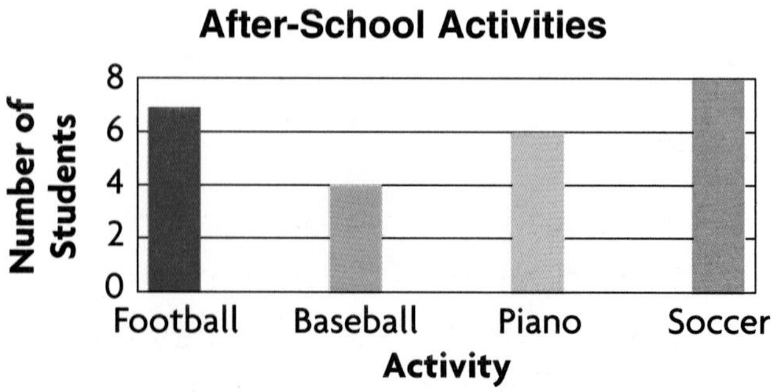

6. How many more students play soccer than play baseball?
7. How many students play football?

Find points on a grid.

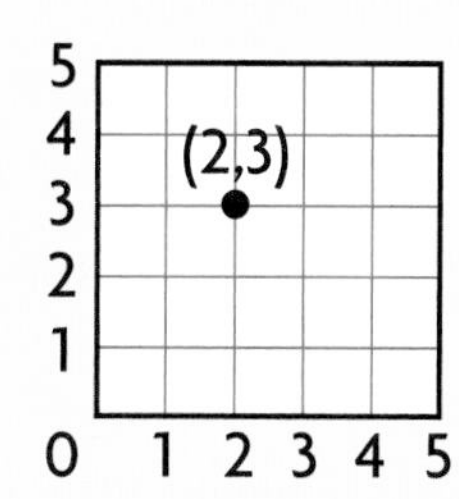

The **ordered pair** (2,3) tells you to move 2 spaces to the right of 0 and 3 spaces up.

Use the grid. (pp. 262–263)

8. Explain how to find (3,4) on the grid.

9. Explain how to find (4,3) on the grid.

Chapter 16

Predict outcomes of experiments.

Margo pulled a paper clip from a bag, recorded the color, and put it back. After 10 times she had pulled 1 orange, 3 red, and 6 blue paper clips.

What color paper clip do you predict she will pull next?

Think: Since she pulled a blue paper clip more than half the time, it is most likely that she will pull a blue paper clip next.

Use the tally table. (pp. 280–281)

This tally table shows the pulls from a bag of marbles.

OUTCOMES	
Color	**Tallies**
Yellow	||||
Green	𝍸 |||
Purple	𝍸 |

10. Name the color marble which is least likely to be pulled.

11. Name the color marble which is most likely to be pulled.

PROBLEM SOLVING PRACTICE

Solve. (pp. 252–253, 282–283)

12. Mr. Lind surveyed his customers to find out what to add to the menu. There were 20 votes for chili, 15 votes for ice cream, and 10 votes for pizza. Make a bar graph that shows the results of the survey.

13. Jill is making a game with a bag of marbles. She has 20 blue marbles and 12 orange marbles. How many blue marbles and orange marbles should she put in a bag to make the game fair?

PERFORMANCE ASSESSMENT

TASK A • OUTDOOR FUN

You decide to write a report about your classmates' favorite outdoor game.

a. Take a survey to find your classmates' favorite outdoor game. Copy the tally table and the frequency table and record their choices on both tables.

b. Decide which type of graph would be best to display the data. Then make the graph.

c. Write a question your classmates could answer by looking at your graph.

OUTDOOR GAMES	
Name	Tally

OUTDOOR GAMES	
Name	Number

TASK B • TRY YOUR LUCK

Materials: 8 different markers

You want to make a spinner for a new game. Each section of the spinner will have a different solid color. Two players each choose a color and spin the pointer. If the pointer lands on the chosen color, the player gets one point. The player with more points after 10 spins wins the game.

a. Make a spinner with 8 equal sections. Color each section a different color.

b. List all the outcomes of using the spinner 10 times.

c. Explain why this spinner gives each player the same chance of winning.

Mighty Math Number Heroes • Probability

You can use Mighty Math Number Heroes to solve probability problems.

- Click on *Handsome Chance.*
- Click [button]. Choose Level A. Complete the activity by finding *certain, possible,* or *impossible* outcomes.
- Click [button]. Choose Level G. Complete the activity by finding *likely* and *unlikely* outcomes.

Practice and Problem Solving

1. Click [button]. Choose Level I. Complete the activity by finding *certain, possible,* or *impossible* outcomes.
2. Click [button]. Choose Level J. Complete the activity by comparing probabilities.

Use Mighty Math Number Heroes to solve.

3. **WRITE A PROBLEM** about a spinner that you make. Draw a picture of the spinner. Name a certain outcome, an impossible outcome, and a likely outcome.
4. **REASONING** Leslie tosses a coin 49 times. The coin lands on heads 30 times and tails 19 times. If Leslie tosses the coin again, is it likely to land on heads or tails? Explain.

Multimedia Math Glossary **www.harcourtschool.com/mathglossary**

5. **Vocabulary** Visit the Multimedia Math Glossary to review *survey* and *results.* Conduct your own survey and display the results in a *frequency table* or a *tally table.*

in Iowa

At the Effigy Mounds National Monument, some mounds are shaped like bears and others are shaped like birds.

THE EFFIGY MOUNDS NATIONAL MONUMENT

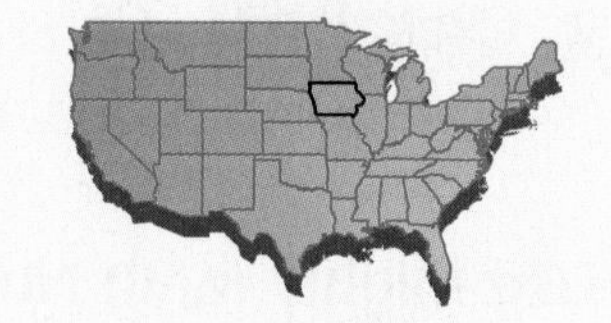

Long ago, Eastern Woodland Indians living on the plains built thousands of mounds. At the National Monument, you can see 200 of these mounds. Some of the mounds are effigies, or mounds shaped like animals.

Suppose you have made up a game to sell in the park's gift shop. Your game has 20 cards that match some of the park's mounds. Fifteen cards show bear mounds, and 5 show bird mounds. The first player puts the cards in a bag and pulls one out without looking.

1. Is it *certain* or *impossible* that the player will pull a card showing a deer?
2. What are the possible outcomes?
3. Is it *more likely* that the player will pull a bear card or a bird card?
4. Describe an outcome that is *certain*.
5. What is the chance that the player will pull a card showing a bear? Explain how you know.
6. **REASONING** How could you change the game so that it is *equally likely* that you will pull each kind of card?

WILDLIFE

Many kinds of birds live at Effigy Mounds National Monument. You might see hawks, eagles, herons, egrets, falcons, and owls. Each fall Effigy Mounds National Monument holds a hawk watch. Hawks and other migrating birds can be seen flying along the Mississippi River.

Eagles, falcons, owls, and hawks are all raptors. Raptors are birds that hunt for their food.

Red shouldered hawk

1. Take a survey. Find out which of the four kinds of raptors your classmates would most like to see. Make a tally table and a frequency table of the data.

2. What kind of graph would you use to display your data? Explain why you made that choice.

3. Make a graph to display your data.

4. **WRITE A PROBLEM** that your classmates could answer by looking at your graph. Exchange with a partner and solve.

Animal shaped mounds at Effigy Mounds National Monument.

CHAPTER 17 Solid and Plane Figures

Collecting model trains is a popular hobby. The largest model railway in the world is in New Jersey. It has 8 miles of track and 135 model trains.

PROBLEM SOLVING Look carefully at the photo. Name the different shapes that you see.

CHECK WHAT YOU KNOW

Use this page to help you review and remember important skills needed for Chapter 17.

IDENTIFY SOLID FIGURES (For Intervention, see p. H23.)

Choose the best term from the box.

SOLID FIGURES
- cone
- cube
- cylinder
- square pyramid
- rectangular prism
- sphere

1. 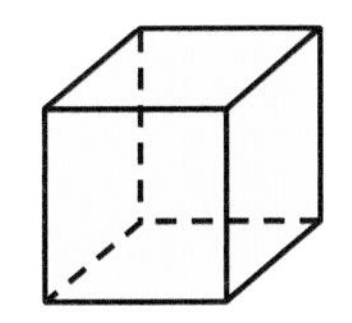
2. 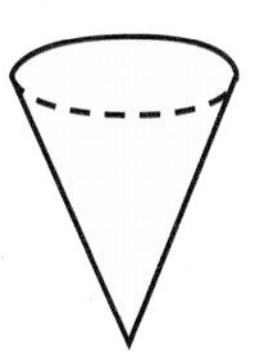
3. 
4. 
5. 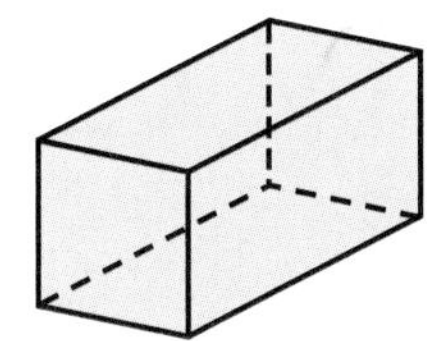
6. 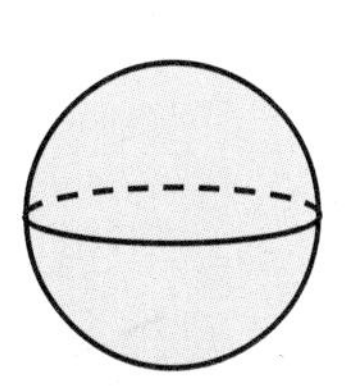
7. 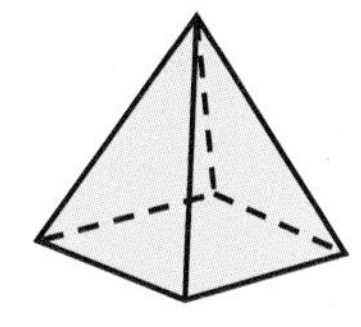
8. 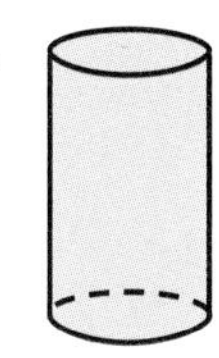
9. 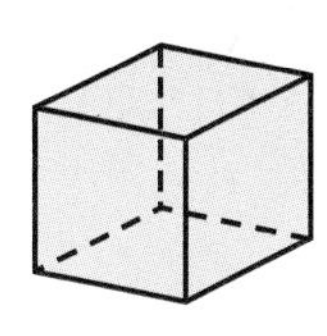

IDENTIFY PLANE FIGURES (For Intervention, see p. H23.)

Choose the best term from the box.

PLANE FIGURES
- circle
- rectangle
- square
- triangle

10. 
11. 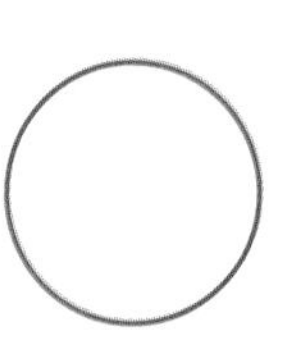
12. 
13. 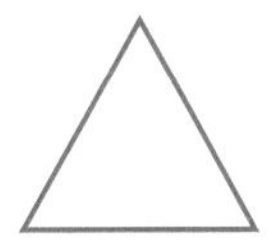
14. 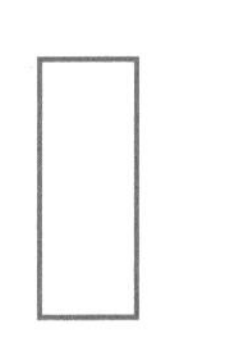
15. 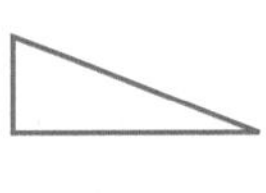
16. 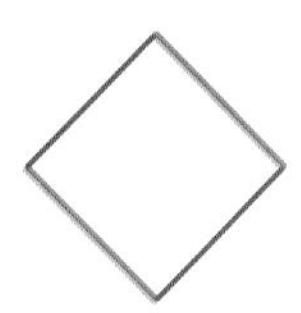
17. 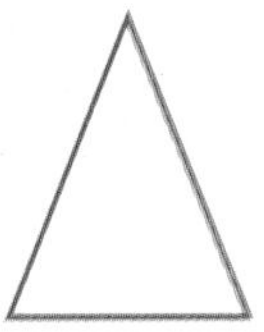
18. 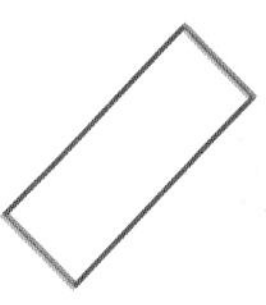

LESSON 1

Solid Figures

Quick Review

Name each plane figure.

1. 

2. 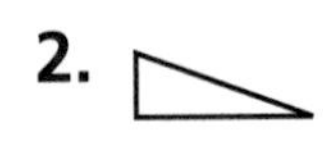

3. 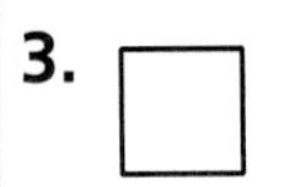

4. 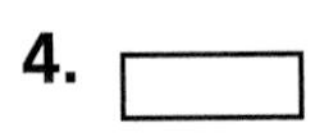

5.  

VOCABULARY

face edge vertex

Learn

FIGURE IT OUT Use names of solid figures to describe objects around you.

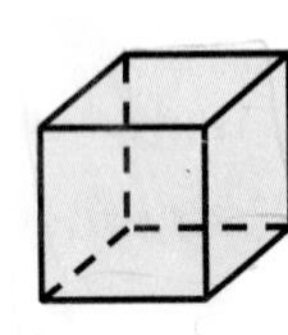

cube

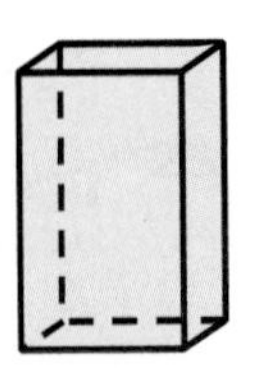

rectangular prism

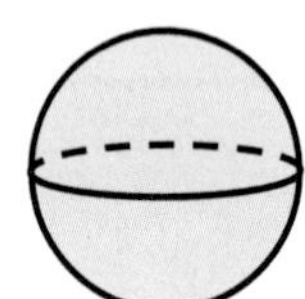

sphere

cylinder

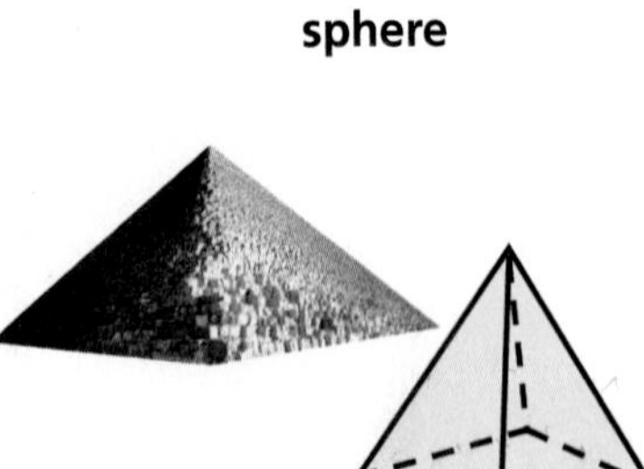

square pyramid

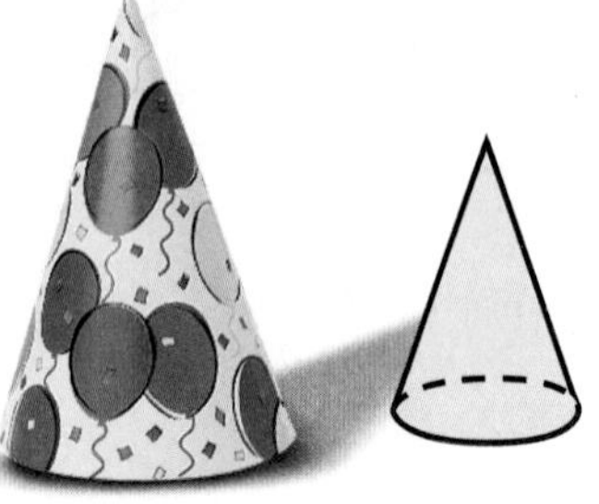

cone

A **face** is a flat surface of a solid figure.

An **edge** is the line segment formed where two faces meet.

A **vertex** is a corner where three or more edges meet. Two or more corners are called vertices.

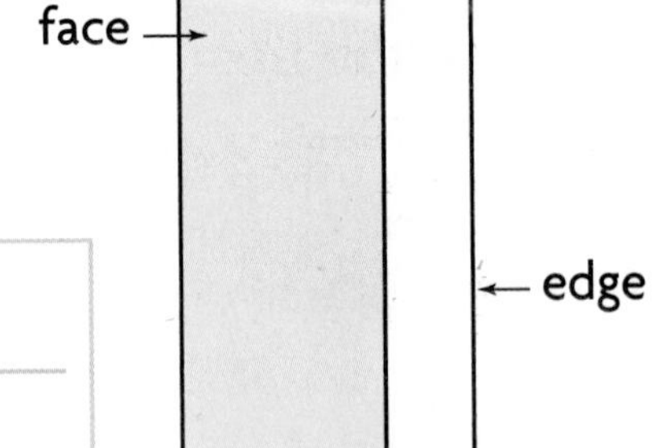

A rectangular prism has 6 faces, 12 edges, and 8 vertices.

- How many edges does a cube have? a sphere?
- **REASONING** Which solid figures will roll? Explain how you know.

Tracing Faces

Use names of plane figures to describe the faces of solid figures.

Activity

Materials: solid figures (square pyramid, rectangular prism, cube), paper, crayons

Trace the faces of several solid figures. Then name the faces that make up each solid figure.

STEP 1

On a large sheet of paper, make a chart like the one below. Trace the faces of each solid figure.

Name of Figure	Faces	Names and Number of Faces
Square pyramid	□ △△ △△	

STEP 2

Record the names and number of faces for each solid figure.

Name of Figure	Faces	Names and Number of Faces
Square pyramid	□ △△ △△	1 square 4 triangles

- **REASONING** Use the words *all, some,* or *none* to describe the faces of the solid figures you traced.

MATH IDEA Some solid figures have faces, edges, and vertices. Faces of solid figures are plane figures such as squares, rectangles, and triangles.

Technology Link

More Practice:
Use E-Lab, *Tracing and Naming Faces.*
www.harcourtschool.com/elab2002

Check

1. **Explain** how you know a sphere has no faces.

Name the solid figure that each object looks like.

2. 

3. 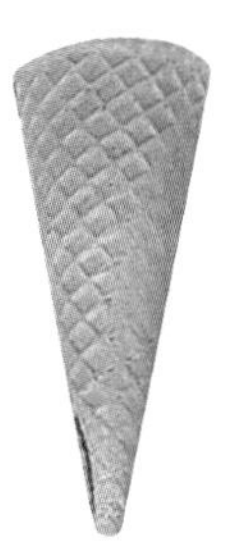

4. 

5. 

LESSON CONTINUES 

Practice and Problem Solving

Name the solid figure that each object looks like.

6. 

7. 

8. 

9. 

Which solid figure has the faces shown? Write *a, b,* or *c*.

10. 

11. 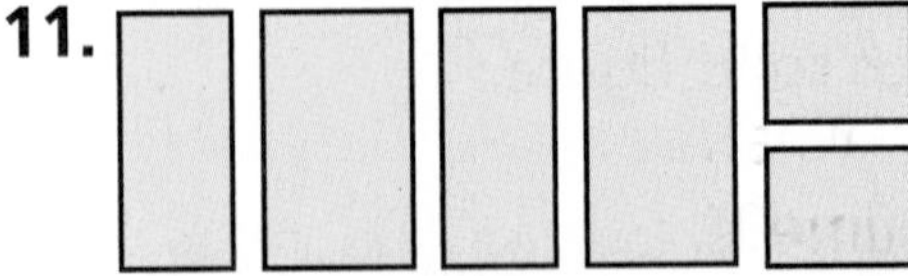

12. 

a. **rectangular prism**
b. **square pyramid**
c. **cube**

Copy and complete the table.

	FIGURE	FACES	EDGES	VERTICES
13.	Rectangular prism	■	■	■
14.	Sphere	■	■	■
15.	Cube	■	■	■
16.	Square pyramid	■	■	■

17. REASONING An analogy is a comparison of similar features of objects. For example, *day* is to *light* as *night* is to *darkness*. Complete each analogy.

a. A cereal *box* is to a *rectangular prism* as a *ball* is to a __?__.

b. A *square* is to a *cube* as a *rectangle* is to a __?__.

18. **Write About It** List objects you might find at a grocery store that look like each of the following solid figures. Think of at least two objects for each figure.

a. sphere

b. rectangular prism

c. cylinder

19. Josh painted a box shaped like a rectangular prism. Each face was a different color. How many colors did Josh use?

20. How are a cube and a rectangular prism alike? How are they different?

21. **Write a problem** about a solid figure. Give clues about the figure. Exchange with a classmate and decide what the figure is.

Mixed Review and Test Prep

22. $7 \times 4 = \blacksquare$ (p. 134)

23. $8 \times 0 = \blacksquare$ (p. 132)

24. $3 \times \blacksquare = 15$ (p. 142)

25. $4 \times \blacksquare = 8$ (p. 142)

Find each sum or difference. (p. 88)

26. $\$10.48 + \$\ 6.97$

27. $\$8.36 + \$4.52$

28. $\$9.41 - \$3.73$

29. $\$10.00 - \$\ 1.25$

30. **TEST PREP** How much is one $5 bill, 8 quarters, 2 dimes, 2 nickels, and 1 penny? (p. 84)

A $8.31 C $8.06
B $8.26 D $7.31

31. **TEST PREP** Wendy says, "The movie starts at ten minutes after six." At what time does the movie start? (p. 94)

F 5:50 H 6:10
G 6:01 J 10:06

PROBLEM SOLVING Thinker's Corner

VISUAL THINKING You can use what you know about the faces of solid figures to model solid figures.

Materials: paper, pencil, scissors, tape

a. Trace the shape at the right onto a piece of paper. Be sure to trace the dotted lines.

b. Cut out the shape along the solid lines.

c. Fold each of the 4 points up on the dotted lines.

d. Tape the edges of the figure together.

1. Name the figure you made.
2. Name the faces of the figure.

LESSON 2

Combine Solid Figures

Quick Review

Name each solid figure.

1. 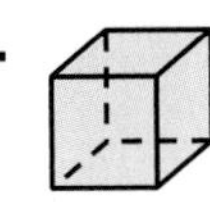
2. 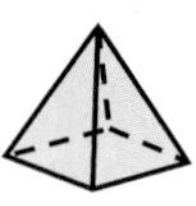
3. 
4. 
5. 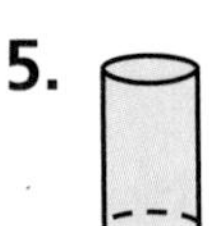

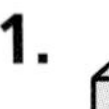

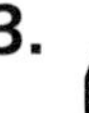

Learn

PUT IT ALL TOGETHER Some objects are made up of two or more solid figures put together. Look at the house on the right. What solid figures make up the shape of the house?

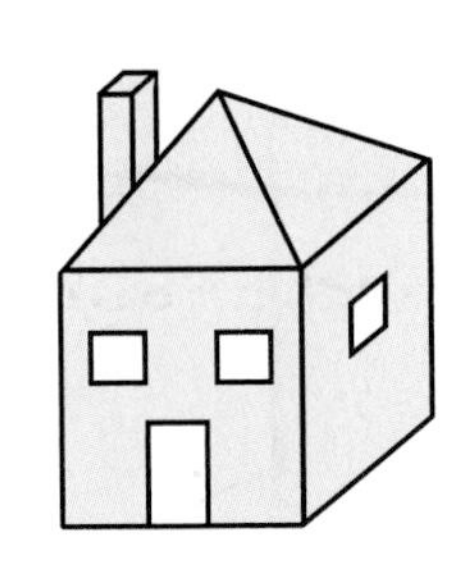

Look at each part of the house separately. Think about the solid figures you know.

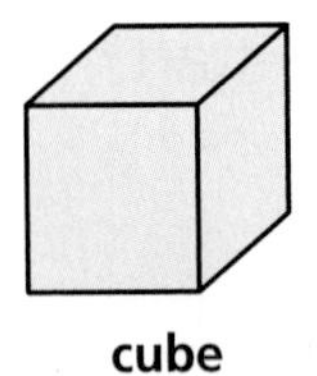
cube

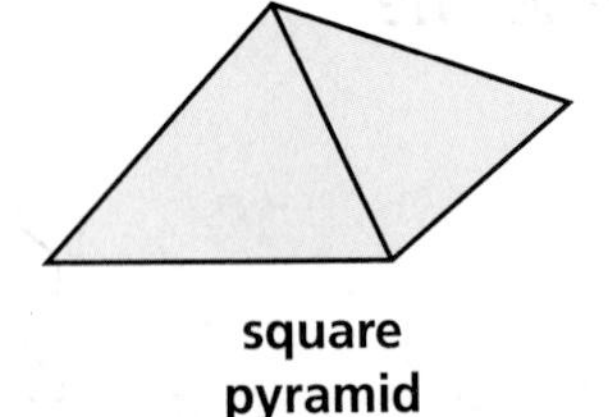
square pyramid

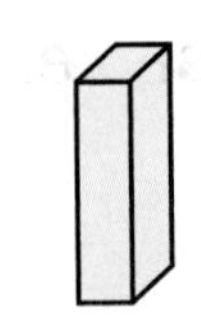
rectangular prism

So, the house is made up of a cube, a square pyramid, and a rectangular prism.

MATH IDEA Solid figures can be combined to make different solid objects.

Examples

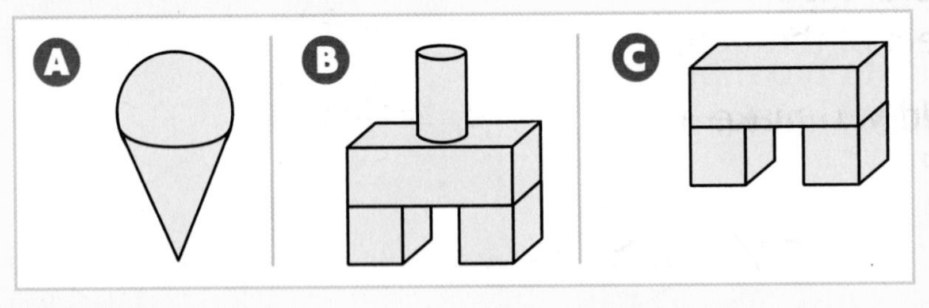

- What solid figures are used to make Object A?
- What solid figures are used to make Object B?

Check

1. **Explain** how you can make Object B look like Object C.

Name the solid figures used to make each object.

2. 

3. 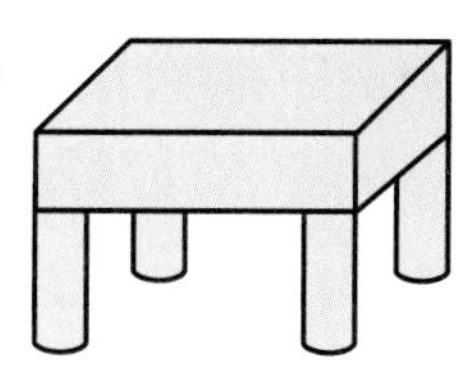

4. 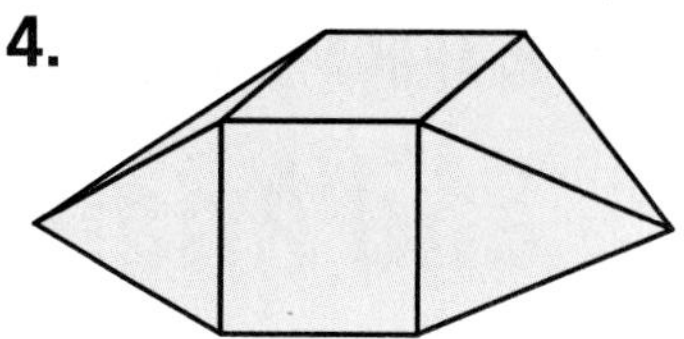

Practice and Problem Solving

Name the solid figures used to make each object.

5. 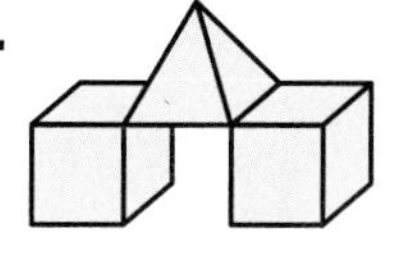

6. 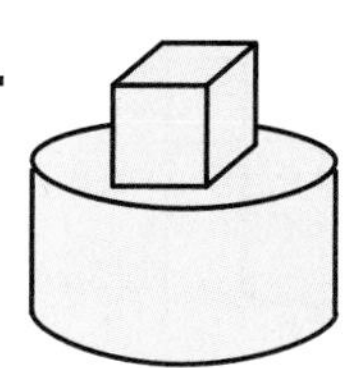

7. 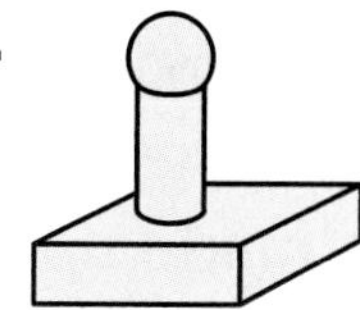

Each pair of objects should be the same. Name the solid figure that is missing.

8. 

9. 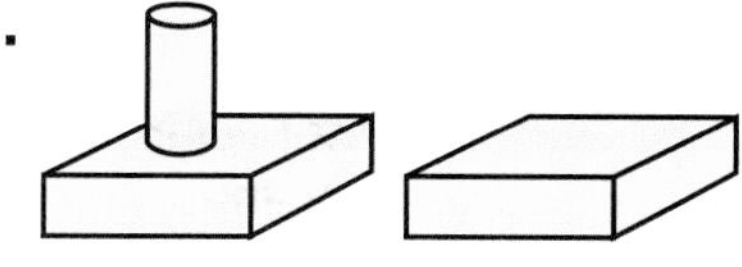

10. 

11. 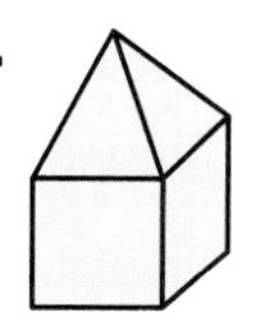

12. 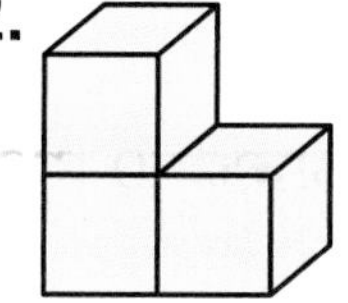

13. 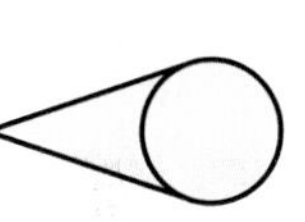 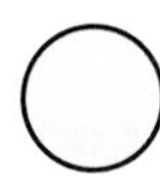

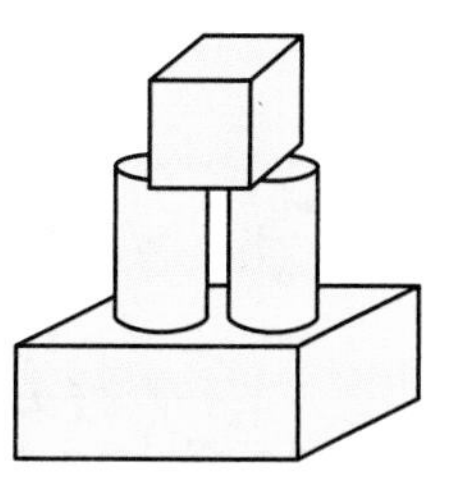

14. Name at least two different faces that make up the solid object at the right.

15. What solid figures could you make if you cut a cube in half?

Mixed Review and Test Prep

16. Round 765 to the nearest hundred. (p. 28)

17. Round 1,080 to the nearest thousand. (p. 30)

18. $\begin{array}{r} 343 \\ +239 \\ \hline \end{array}$ (p. 42)

19. $\begin{array}{r} 751 \\ -390 \\ \hline \end{array}$ (p. 58)

20. **TEST PREP** Four friends shared 32 crackers equally. How many crackers did each friend get? (p. 204)

A 4
B 7
C 8
D 28

EXTRA PRACTICE page H48, Set B

LESSON 3

Line Segments and Angles

Quick Review

Write the number of sides each figure has.

1. ▭ 2. □ 3. △

4. ○ 5. ◺

VOCABULARY

line, point, line segment, ray, angle, right angle

Learn

POINT TO POINT Victor drew this plane figure by connecting points on grid paper. The sides of his figure are made up of line segments.

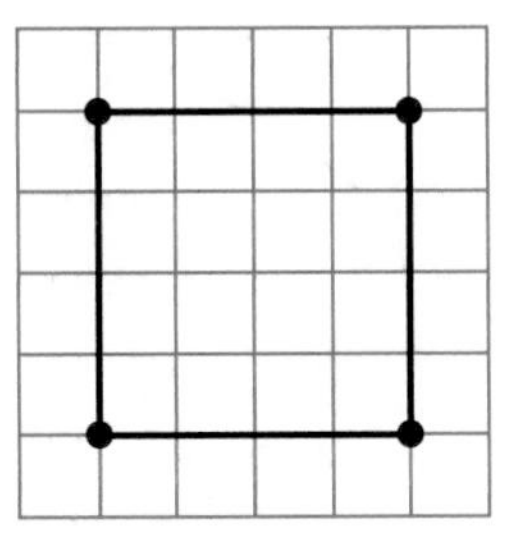

A **line** is straight. It continues in both directions. It does not end.	⟷
A **point** is an exact position or location.	point
A **line segment** is straight. It is the part of a line between two points, called endpoints.	•—•

- What plane figure did Victor draw?
- How many sides does the figure have? How many line segments?
- **REASONING** How are lines and line segments alike? How are they different?

Identifying Angles

A **ray** is a part of a line. It has one endpoint. It is straight and continues in one direction.

An **angle** is formed by two rays with the same endpoint.

A **right angle** is a special angle. It forms a square corner.

Technology Link

To learn more about lines, watch the **Harcourt Math Newsroom Video,** ***Georgia Dome*****.**

Some angles are *less than* a right angle.

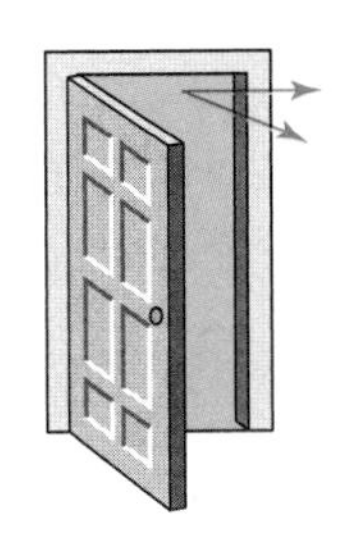

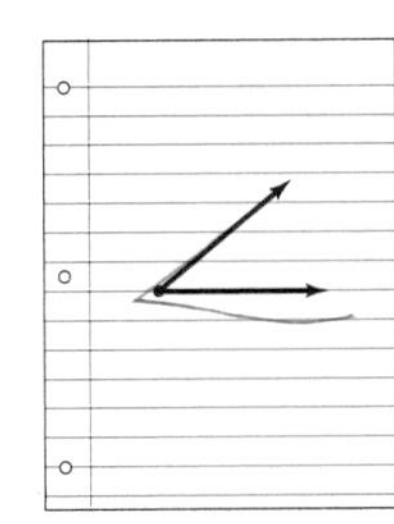

Some angles are *greater than* a right angle.

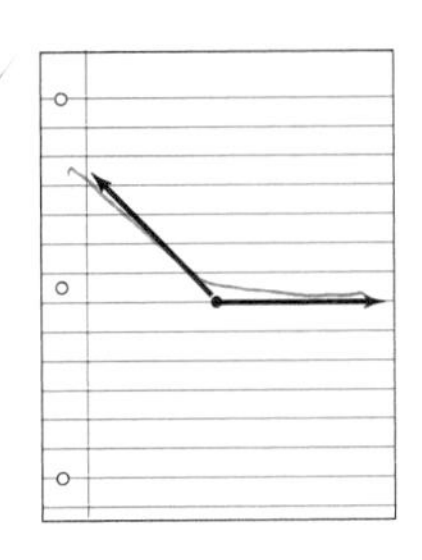

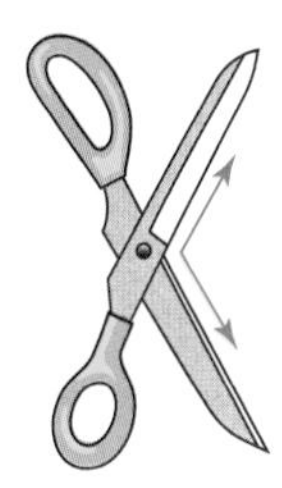

Look at the photo of Victor.

- What plane figure did Victor draw?
- How many angles does the figure have? How many right angles?

MATH IDEA You can identify line segments and angles in plane figures. You can tell if an angle is a right angle or if it is greater than or less than a right angle.

Check

1. **Draw** a picture of an angle that is greater than a right angle. Explain how you know it is greater than a right angle.

Name each figure.

2. 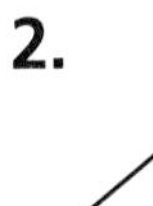

3. 

4. 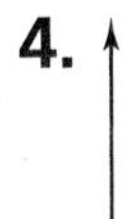

5. •

LESSON CONTINUES

Practice and Problem Solving

Name each figure.

6. 7. 8. 9.

Use a corner of a piece of paper to tell whether each angle is a *right angle, greater than* a right angle, or *less than* a right angle.

10. 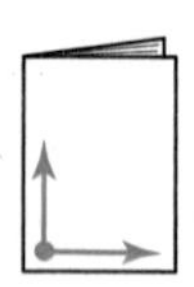

11. 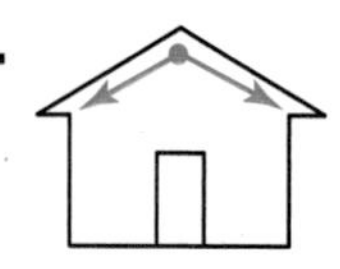

12. 

13. 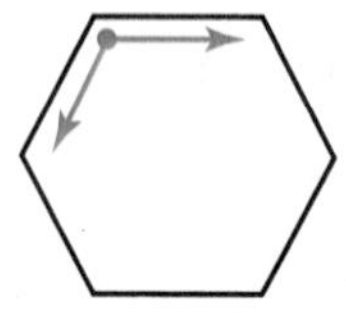

14. 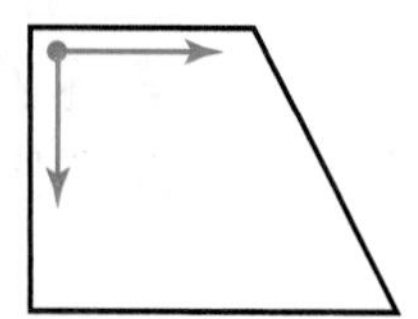

15. 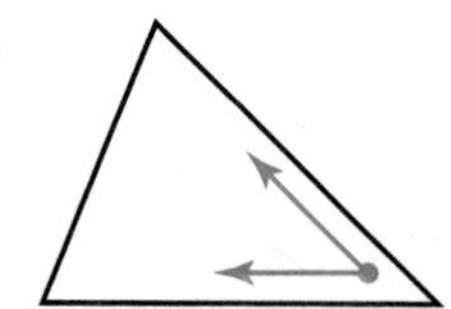

Copy and complete the table.

	Figure	Number of Line Segments	Number of Angles	Number of Right Angles
16.		■	■	■
17.		■	■	■
18.		■	■	■
19.		■	■	■
20.		■	■	■
21.		■	■	■

22. Compare the number of line segments and the number of angles for each figure in the table. What pattern do you notice?

23. **Write About It** Draw a right angle and a triangle that has a right angle. Describe the parts of each figure.

24. **REASONING** Blanca says that a figure with 2 right angles can be a triangle. Do you agree or disagree? Draw a picture to help.

25. List at least 3 objects in the room that contain right angles. How can you be sure the angles are right angles?

Mixed Review and Test Prep

Find each quotient. (p. 222)

26. $16 \div 2 = ■$
27. $36 \div 4 = ■$
28. $25 \div 5 = ■$
29. $21 \div 3 = ■$

Find each product. (p. 31)

30. $5 \times 416 = ■$
31. $3 \times 178 = ■$
32. $8 \times 409 = ■$
33. $7 \times 380 = ■$

34. **TEST PREP** What is the value of the blue digit in 3,495? (p. 6)
A 4,000
B 400
C 40
D 4

35. **TEST PREP** Will bought a magazine for \$2.79 and 3 pieces of candy for 9¢ each. How much money did he spend? (p. 172)
F \$2.88
G \$2.96
H \$3.06
J \$5.49

PROBLEM SOLVING

Thinker's Corner

CLOCKS AND ANGLES Use what you learned in this lesson to describe the angles made by the hands of a clock.

Write whether each angle is a *right angle, greater than* a right angle, or *less than* a right angle.

1. 
7:45

2. 
11:15

3. 
9:00

4. 
3:30

5. **REASONING** The hour hand is between 3 and 4. The minute hand is pointing to the 5. What time is it? Describe the angle made by the hands.

6. Joy left for school at 7:50 A.M. It took her 15 minutes to walk to school. What time is it? Describe the angle made by the hands.

LESSON

4 Types of Lines

Quick Review

Write *right angle*, *greater than* a right angle, or *less than* a right angle.

1. 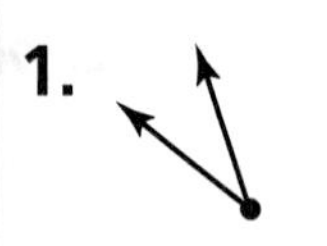

2. 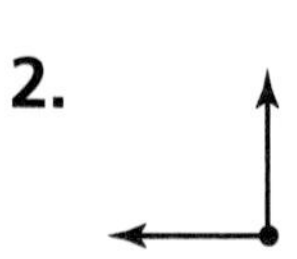

3. 

4. 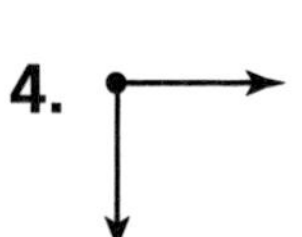

5. 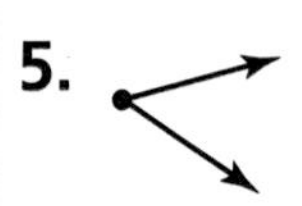

VOCABULARY

intersecting lines
parallel lines

Learn

GET IN LINE Here are some ways to describe the relationships between lines.

Lines that cross are **intersecting lines**. Intersecting lines form angles.

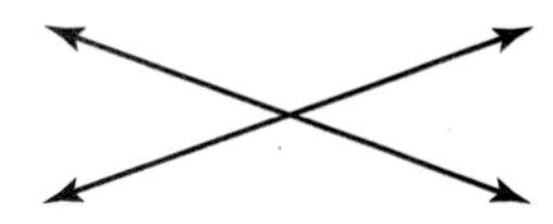

Some intersecting lines cross to form right angles.

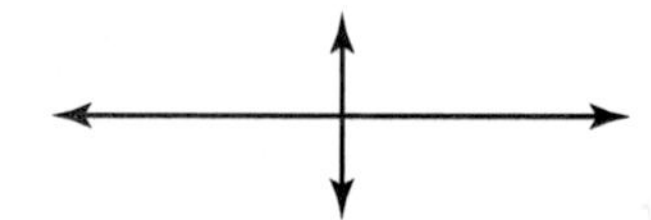

Lines that never cross are **parallel lines**. Since parallel lines never cross, they do not form angles.

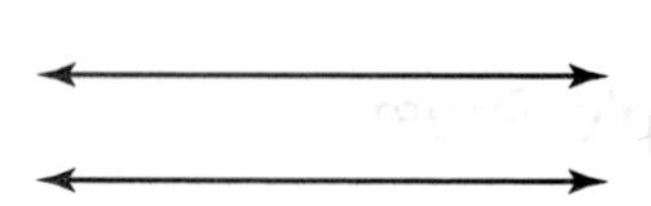

You may see intersecting and parallel lines in the world around you.

REASONING Are the angles in the climbing dome right angles? How can you check?

MATH IDEA You can classify lines as intersecting or parallel. Some intersecting lines cross to form right angles.

Check

1. **Explain** how to tell whether the swing chains are intersecting or parallel.

Describe the lines. Write *parallel* or *intersecting*.

2.

3.

4.

Practice and Problem Solving

Describe the lines. Write *parallel* or *intersecting*.

5. 

6. 

7. 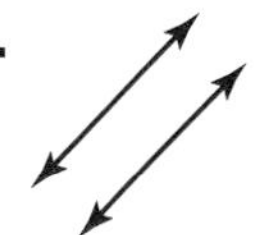

Describe the intersecting lines. Write *form right angles* or *do not form right angles*.

8. 

9. 

10. 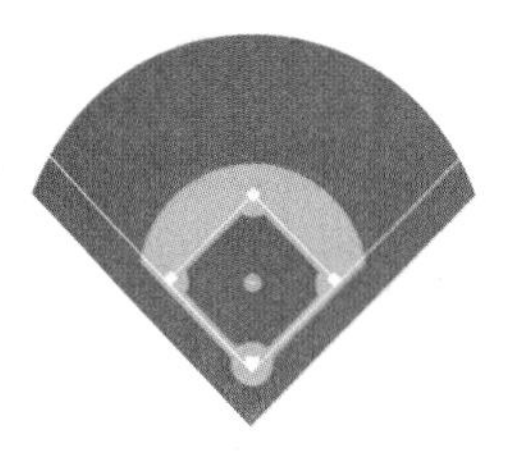

USE DATA For 11–13, use the map at the right.

11. Which street does not cross Oak Street?

12. Describe the angle the bicycle trail makes at the intersection of Oak Street and Pine Street.

13. **What's the Question?** The answer is Maple Street.

14. **REASONING** Use what you know about line relationships to describe the sides of a rectangle.

15. **Write About It** Draw and label sets of intersecting and parallel lines. Use a ruler to help. Include lines that cross to form right angles. Describe the angles in each of your drawings.

Mixed Review and Test Prep

Find the quotient. (p. 222)

16. $8\overline{)40}$

17. $7\overline{)49}$

18. $9\overline{)72}$

19. $10\overline{)30}$

20. **TEST PREP** Talia is paid $4 to baby-sit for 1 hour. How much will she earn in 5 hours? (p. 118)

A $9

B $20

C $25

D $30

EXTRA PRACTICE page H48, Set D

LESSON 5

Circles

Quick Review

Name each figure.

1. •———•
2. •——→
3. ←——→
4. 
5. •

VOCABULARY

center
diameter
radius

MATERIALS large and small paper clips, two pencils, ruler

Explore

Draw two different circles. Follow Steps 1–4, using large and small paper clips.

Activity

STEP 1

Draw a point. •

STEP 2

Draw a circle by placing pencils in the ends of the paper clip. The pencil on the point should not move.

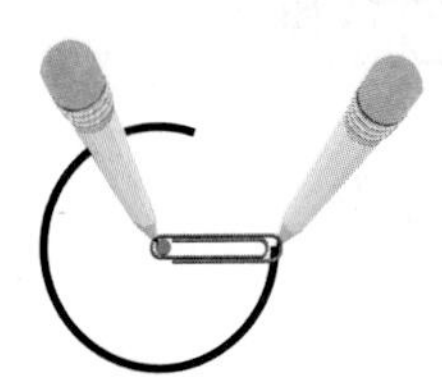

STEP 3

Use a ruler to draw a line segment from the point to anywhere on the circle.

STEP 4

Use a ruler to draw a line across the circle and through the point.

These are the circles I drew. How are my circles alike?

Talk About It

- How are the circles you drew alike? How are they different?

Connect

The **center** of a circle is a point in the middle of a circle. It is the same distance from anywhere on the circle.

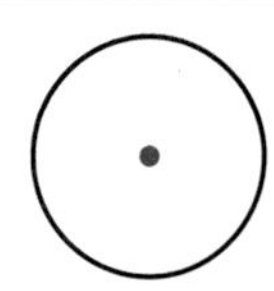

A **radius** of a circle is a line segment. Its endpoints are the center of the circle and any point on the circle.

A **diameter** of a circle is a line segment. It passes through the center. Its endpoints are points on the circle.

Practice and Problem Solving

For 1–4, use the circle at the right.

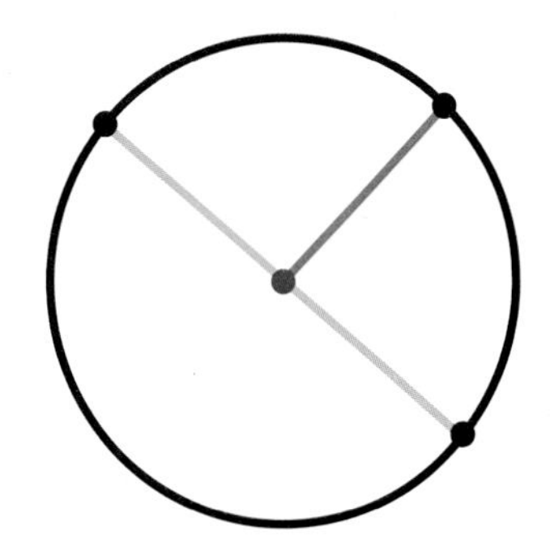

1. What part of the circle is red?
2. What part of the circle is blue?
3. What part of the circle is yellow?
4. Describe the relationship between the radius and the diameter.
5. **What's the Error?** Ty says that a circle with a diameter of 5 inches is larger than a circle with a radius of 3 inches. Describe his error. Draw a picture to help.
6. **REASONING** Josh makes 3 cuts in a pie. Each cut is a different diameter of the pie. How many pieces does Josh have?

Mixed Review and Test Prep

7. $3 \times 7 = \blacksquare$ (p. 122)
8. $9 \times 8 = \blacksquare$ (p. 164)
9. $4{,}623 + 3{,}957$ (p. 46)
10. $8{,}017 - 6{,}459$ (p. 62)
11. **TEST PREP** A sandwich costs \$3.55. Don pays with a \$5 bill. How much change should Don get? (p. 88)

    A \$1.45 **C** \$2.45
    B \$1.55 **D** \$3.45

LESSON

6 Problem Solving Strategy
Solve a Simpler Problem

Quick Review

Write the number of faces each figure has.

1. 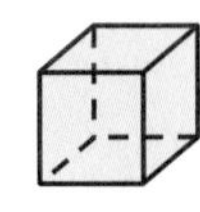
2.
3. 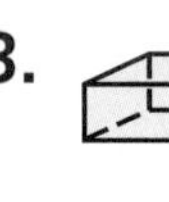
4. 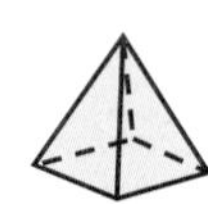
5. 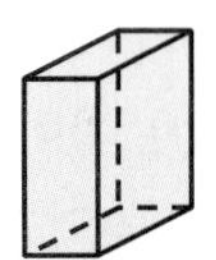

PROBLEM The patterns on the faces of the Soma Cube puzzle contain squares. How many squares are on all faces of the Soma Cube?

The Soma Cube puzzle was discovered in 1936 by the Danish poet and scientist Piet Hein.

UNDERSTAND

- What are you asked to find?
- What information will you use?
- Is there information you will not use? If so, what?

PLAN

- What strategy can you use to solve the problem?

You can solve a simpler problem.

SOLVE

- How can you use the strategy to solve the problem?

Count the number of squares on one face of the Soma Cube.

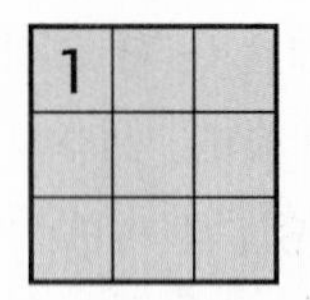

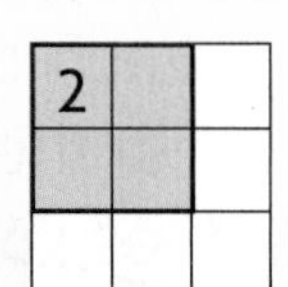

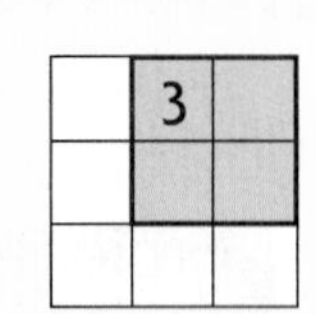

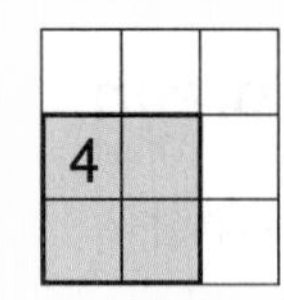

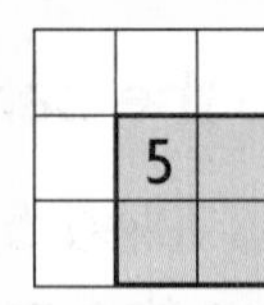

6	7	8
9	10	11
12	13	14

Add to find the number of squares on all 6 faces of the Soma Cube.

$14 + 14 + 14 + 14 + 14 + 14 = 84$

So, there are 84 squares on all faces of the Soma Cube.

CHECK

- How can you decide whether your answer is correct?

Problem Solving Practice

- Draw a Diagram or Picture
- Make a Model or Act It Out
- Make an Organized List
- Find a Pattern
- Make a Table or Graph
- Predict and Test
- Work Backward
- **Solve a Simpler Problem**
- Write a Number Sentence
- Use Logical Reasoning

Solve a simpler problem.

1. **What if** you count only the smallest squares on the Soma Cube? How many small squares are on all faces of the Soma Cube?
2. The faces of the Soma Cube also contain rectangles like the one below. How many of these rectangles are on all faces of the Soma Cube?

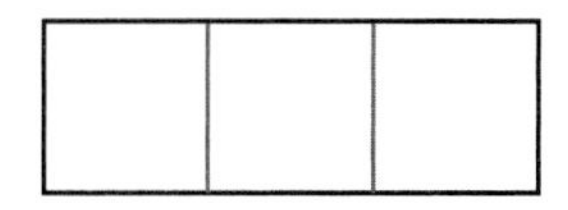

This square pyramid is at the Louvre Museum in Paris, France. The patterns on the faces contain triangles.

The pyramids were designed by the Chinese American architect Ieoh Ming Pei. The glass lets light into underground halls and rooms of the museum.

3. How many triangles are on one face of the pyramid? Hint: The triangles are not all the same size.

   A 1 triangle **C** 5 triangles
   B 4 triangles **D** 10 triangles

4. Which number sentence shows how to find the number of triangles on 4 faces of the pyramid?

   F $10 \times 5 = 50$ **H** $4 \times 4 = 16$
   G $10 + 10 + 10 + 10 = 40$ **J** $1 + 1 + 1 + 1 = 4$

Mixed Strategy Practice

USE DATA For 5–7, use the bar graph.

5. Ms. Colmery's class filled 5 rows and 2 extra chairs in the museum auditorium. How many chairs were in each row?
6. **REASONING** Describe at least two ways chairs can be arranged in equal rows for Mr. Leong's class.
7. How many students visited the museum in all? Write a number sentence and solve.

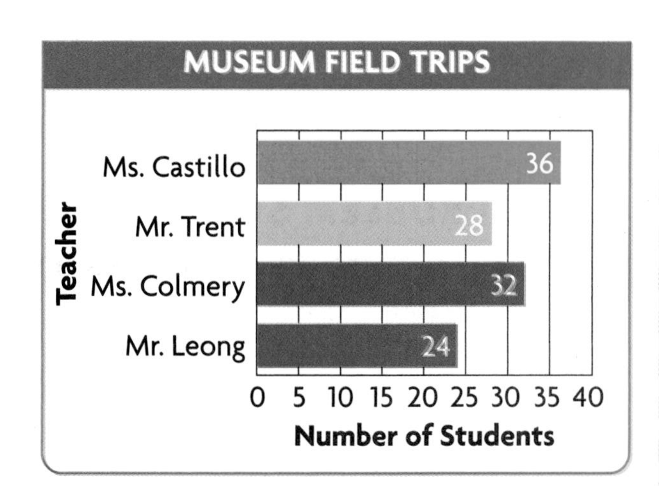

CHAPTER 17

Review/Test

CHECK VOCABULARY AND CONCEPTS

For 1–2, choose the best term from the box.

face
edge
radius

1. A flat surface of a solid figure is a __?__. (p. 294)
2. A line segment whose endpoints are the center of a circle and any point on the circle is a __?__. (p. 306)
3. A solid figure has 6 square faces. What is it? (pp. 294–297)
4. A solid figure has 6 rectangular faces. What is it? (pp. 294–297)

CHECK SKILLS

5. Name the solid figures used to make this object. (pp. 298–299)

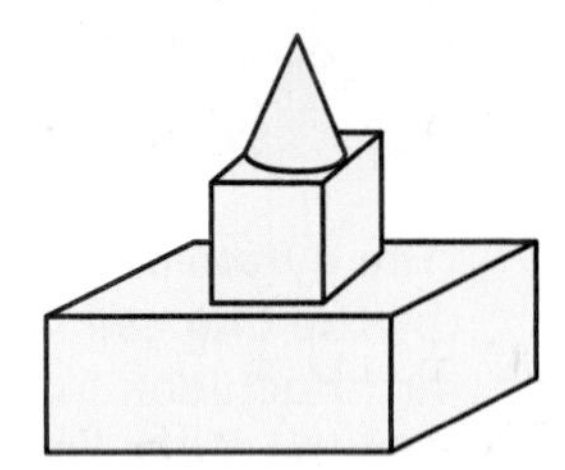

6. Write if each angle is a *right angle, greater than* a right angle, or *less than* a right angle. (pp. 300–303)

a.

b. 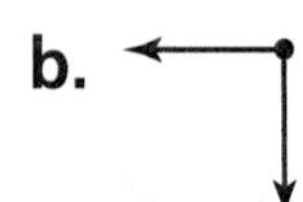

Name each figure. (pp. 300–303)

7. **8.** **9.** **10.**

Describe the lines. Write *intersecting* or *parallel*. (pp. 304–305)

11.

12. 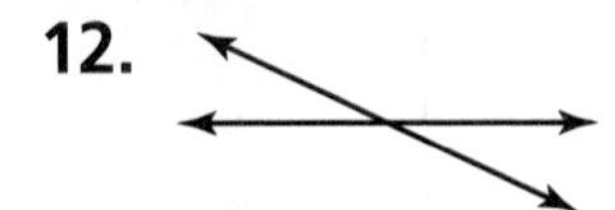

13. 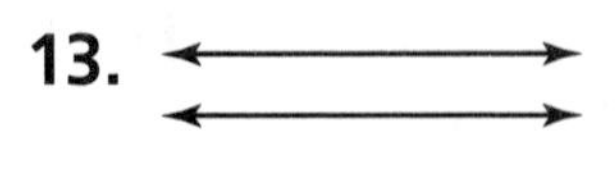

CHECK PROBLEM SOLVING

For 14–15, use the figure at the right. (pp. 308–309)

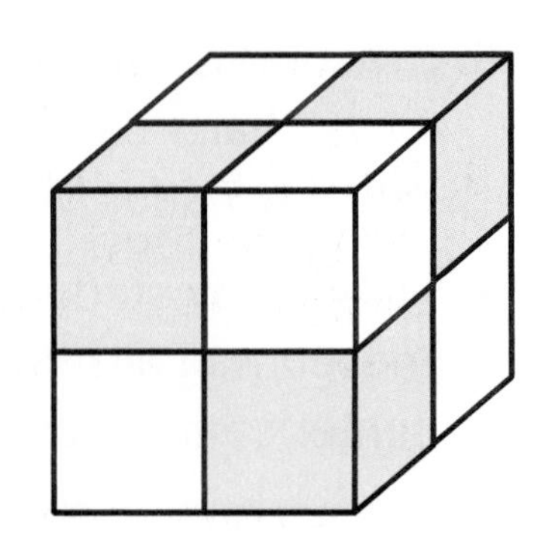

14. Each face of this cube has a checkerboard pattern of blue and white squares. How many blue squares are on all faces of the cube?
15. How many blue and white squares are on all faces of the cube? Explain how you know.

Standardized Test Prep

Check your work.
See item **4**.

Draw a picture to check your work. Be sure you draw exactly what is described.
Also see problem **7**, p. H65.

For 1–8, choose the best answer.

1. Which solid figures could describe a cereal box and a soup can?

   A cone and cube
   B cone and sphere
   C rectangular prism and cylinder
   D sphere and cylinder

2. Which solid figure has no faces, no edges, and no vertices?

   F cylinder
   G sphere
   H cone
   J cube

3. Which is a name for a plane figure that has exactly 3 line segments?

   A triangle
   B square
   C cube
   D cylinder

4. $48 \div 6 = \blacksquare$

   F 6
   G 7
   H 9
   J NOT HERE

5. Which number sentence shows one way to find the number of faces on 2 cubes?

   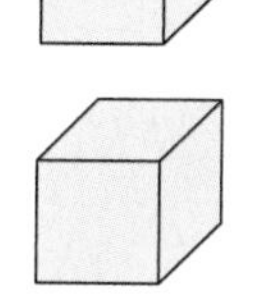

   A $8 \times 2 = 16$
   B $6 + 8 + 6 + 8 = 28$
   C $6 + 6 = 12$
   D $2 + 4 + 8 = 14$

6. George and Joey saved money to buy a gift for their mother. Together they have \$7.42. George saved \$3.58. How much did Joey save?

   F \$11.00
   G \$4.84
   H \$3.84
   J NOT HERE

7. How is 5,016 written in expanded form?

   A $5{,}000 + 10 + 6$
   B $5{,}000 + 100 + 6$
   C $5{,}000 + 160$
   D $500 + 10 + 6$

8. Darren drew a circle. Then he drew a line segment from the center of the circle to a point on the circle. Which word best describes his line segment?

   F center
   G radius
   H diameter
   J side

Write What You Know

9. Describe a rectangular prism by using the terms *faces, edges,* and *vertices*. How many of each are there? Describe the shapes of the faces.

10. Draw a figure that has at least one angle that is greater than a right angle. Draw a ring around that angle. Explain how you know the angle is greater than a right angle.

CHAPTER 18

Polygons

Traffic signs help drivers know about the roads they are driving on. Different shapes are used for different signs. The stop sign is the only sign in the shape of an octagon. Signs for school zones are in the shape of a pentagon. Yield signs are triangles.

PROBLEM SOLVING Make a list of the different signs you see on your way to school. Tell what geometric shapes are used for the signs.

Rockland County, New York

CHECK WHAT YOU KNOW

Use this page to help you review and remember important skills needed for Chapter 18.

SIDES AND CORNERS (For Intervention, see p. H24.)

Tell the number of sides and corners in each figure.

1. 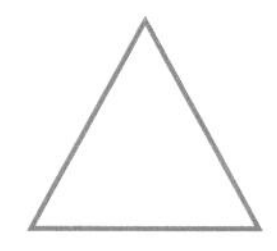
2. 
3. 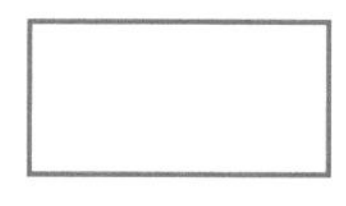
4. 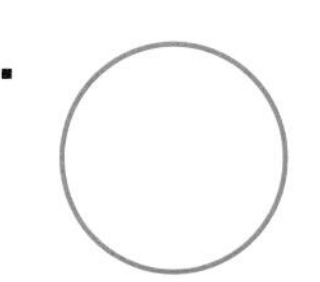
5. 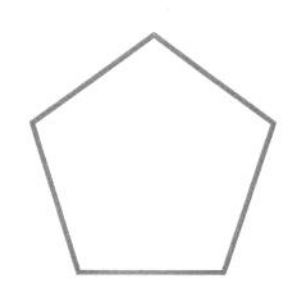
6. 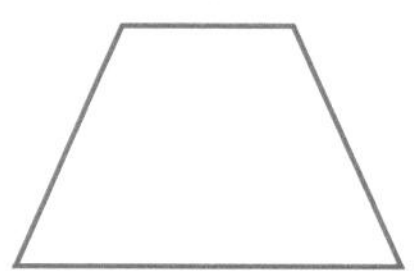
7. 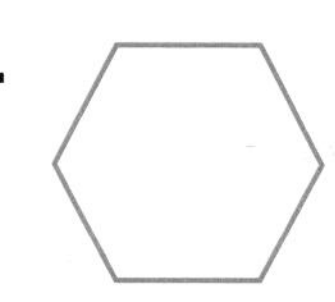
8. 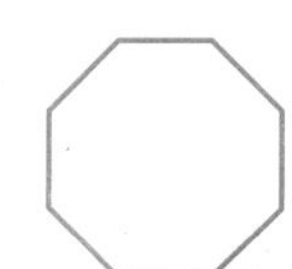
9. 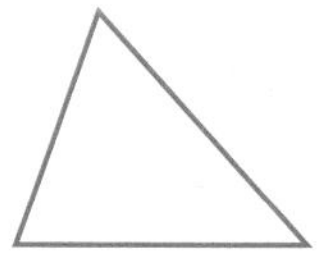

GEOMETRIC PATTERNS (For Intervention, see p. H24.)

Name the missing figure in each pattern. Describe the pattern.

10.  ___?___

11. 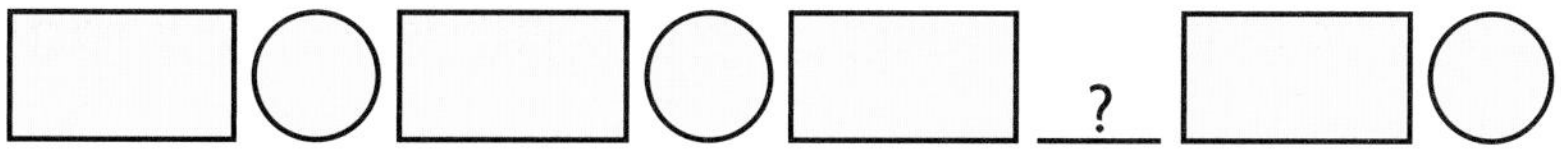 ___?___

12. 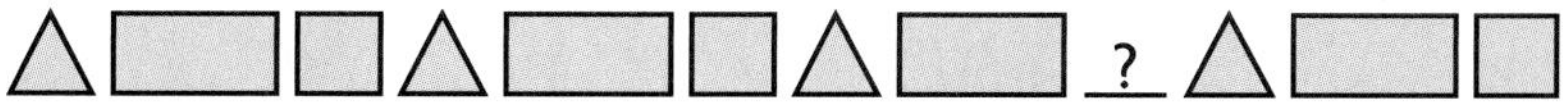 ___?___

13. 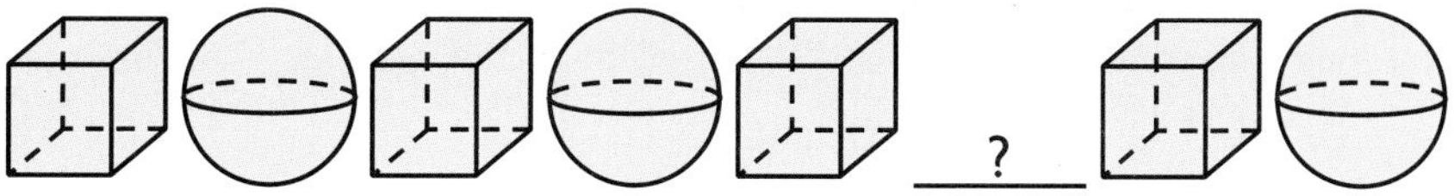 ___?___

14. 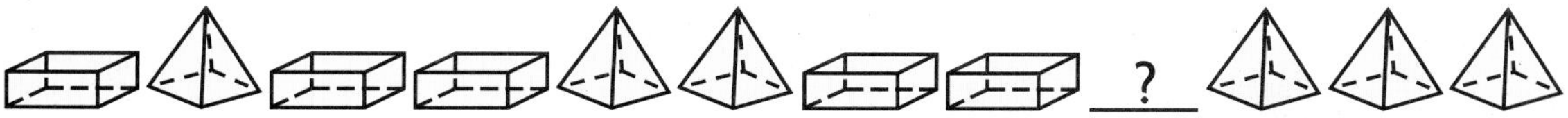 ___?___

LESSON 1

Polygons

Quick Review

Write the number of sides each figure has.

1. 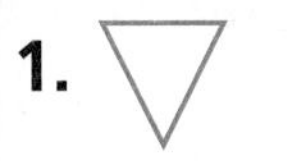
2. 
3. 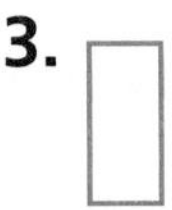
4. ◇
5. ○

VOCABULARY

polygon **quadrilateral** **pentagon** **hexagon** **octagon**

Learn

SHAPE UP! A closed figure begins and ends at the same point. An open figure has ends that do not meet. A **polygon** is a closed plane figure with straight sides. The sides of a polygon are line segments.

polygons

A 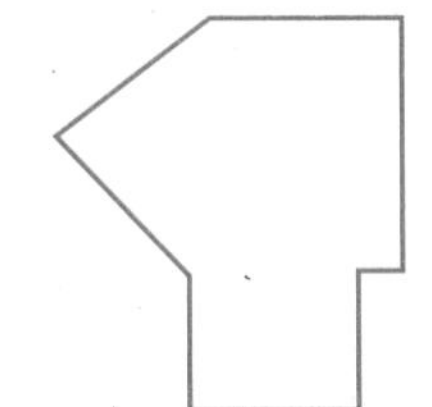

B 

not polygons

C 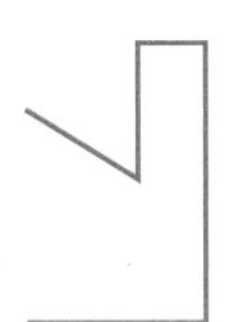

D 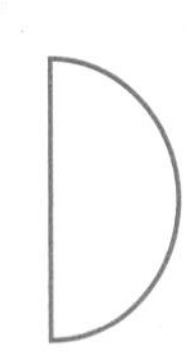

- Explain why figures C and D are *not* polygons.

MATH IDEA You can name and sort polygons by the number of *sides* or *angles* they have.

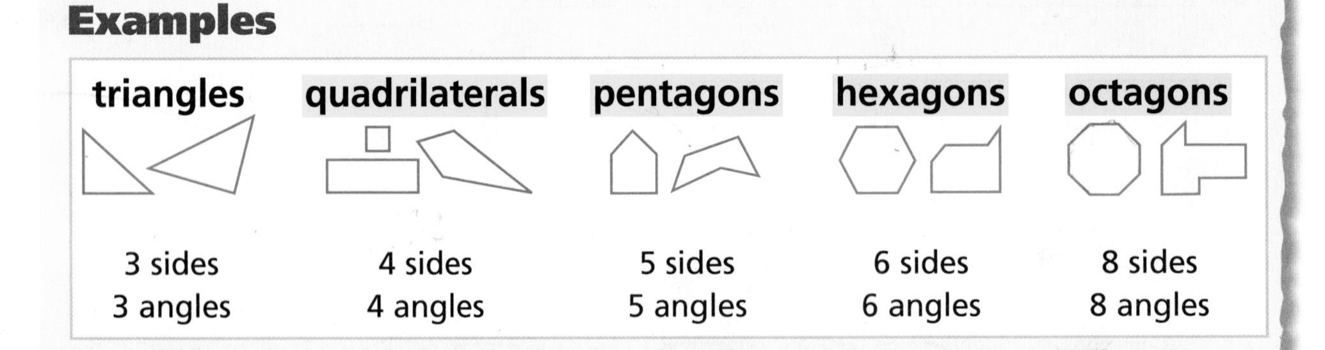

Examples

triangles	quadrilaterals	pentagons	hexagons	octagons
3 sides 3 angles	4 sides 4 angles	5 sides 5 angles	6 sides 6 angles	8 sides 8 angles

- What do you notice about the number of sides and the number of angles in polygons?

Check

1. **Explain** why a circle is *not* a polygon.

Tell if each figure is a polygon. Write *yes* or *no*.

2. 
3. 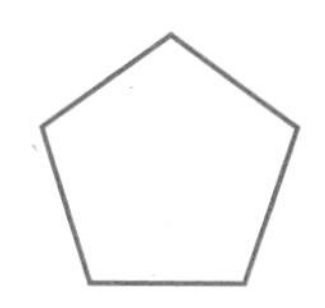
4. 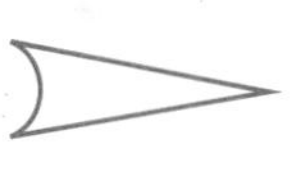
5. 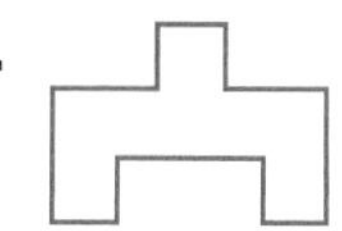
6. 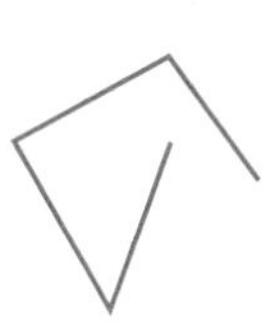

▶ Practice and Problem Solving

Tell if each figure is a polygon. Write *yes* or *no*.

7. 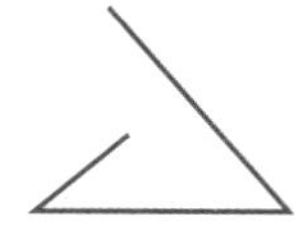
8. 
9. 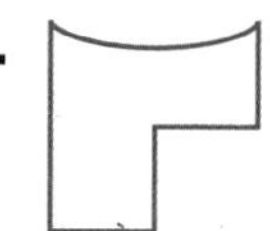
10. 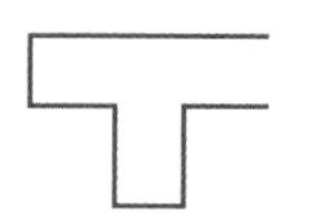
11. 

Write the number of sides and angles each polygon has. Then name the polygon.

12. 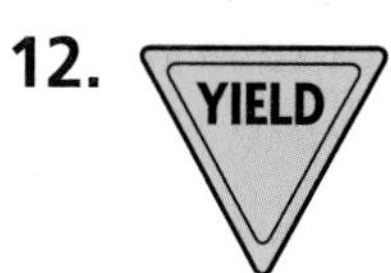

13. 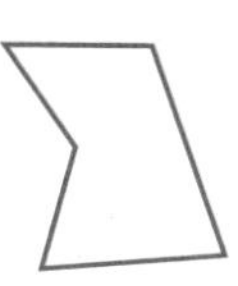
14. 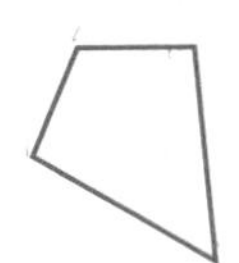
15. 

16. 

For 17–19, write the letters of the figures that answer the questions.

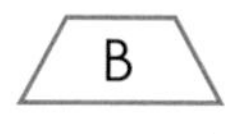

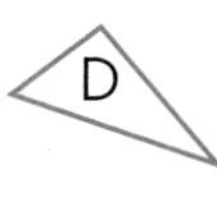

17. Which are polygons?

18. Which is a quadrilateral?

19. Which figures have some angles that are less than a right angle?

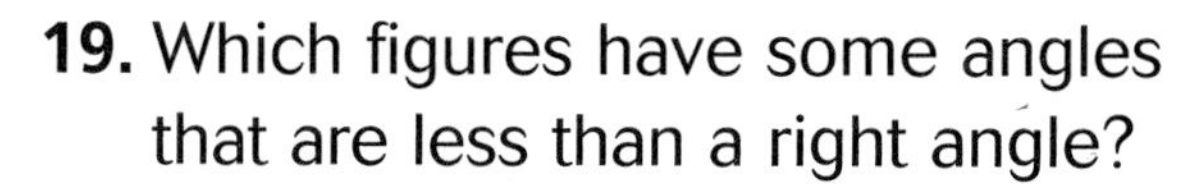

20. **REASONING** What is another name for a rectangle? Explain how you know.

21. **Write About It** Draw polygons with 3, 4, 5, 6, and 8 sides. Then label each polygon.

Mixed Review and Test Prep

22. 466 (p. 42)
+527

23. 684 (p. 42)
+359

24. 700 (p. 58)
−293

25. 900 (p. 58)
−564

26. **TEST PREP** Dawson got to the soccer field at 3:35 P.M. The game started 40 minutes later. What time did the game start? (p. 98)

A 3:55 P.M. C 4:15 A.M.
B 4:00 P.M. D 4:15 P.M.

EXTRA PRACTICE page H49, Set A

LESSON
2 Triangles

Quick Review

Write if each angle is a *right angle, greater than* a right angle, or *less than* a right angle.

1.
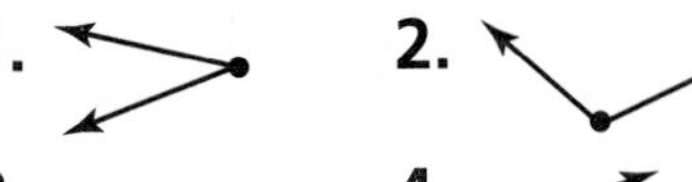
2.

3.
4.
5.
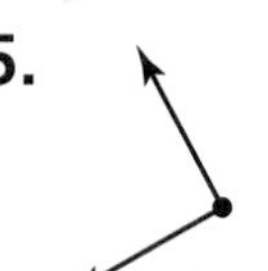

VOCABULARY

equilateral **isosceles** **scalene** **right triangle** **obtuse triangle** **acute triangle**

Learn

TIME FOR TRIANGLES Beverly and Armando sorted these triangles in different ways.

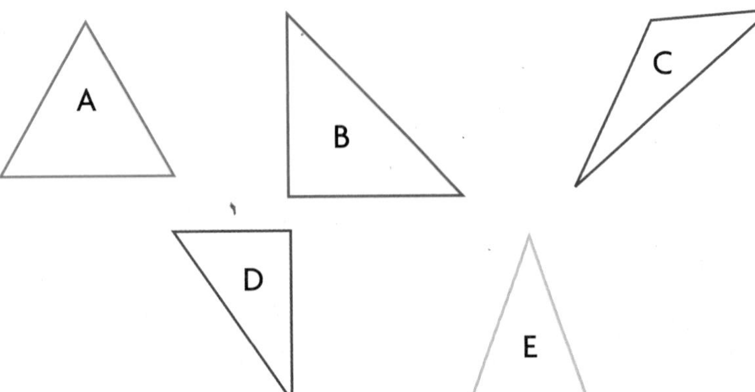

This is how Beverly sorted the triangles.

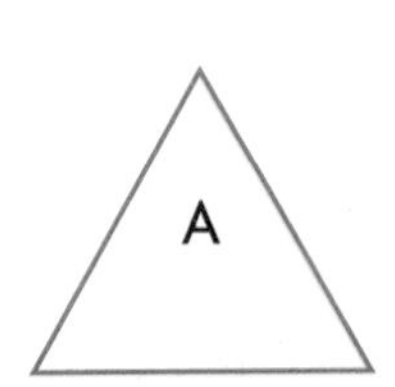

All sides are equal.

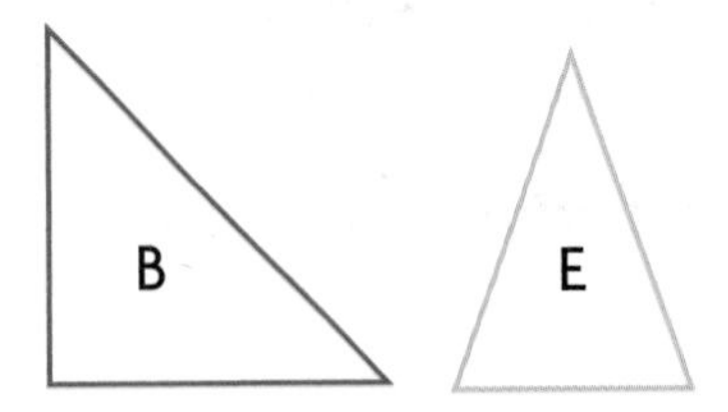

Two sides are equal.

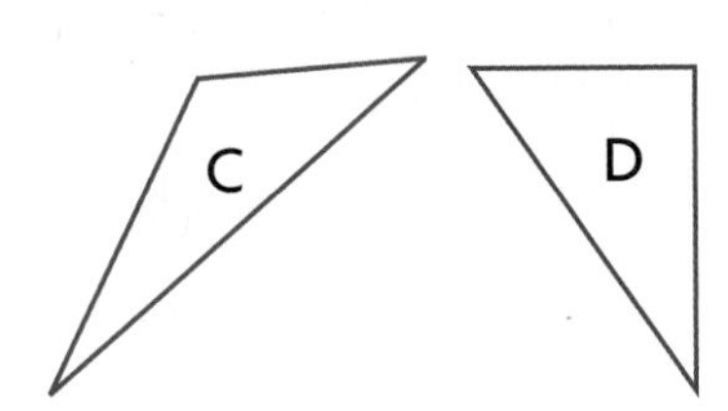

No sides are equal.

This is how Armando sorted the triangles.

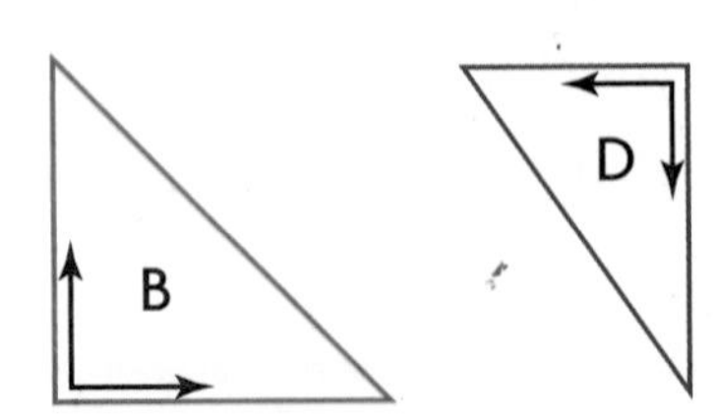

One angle is a right angle.

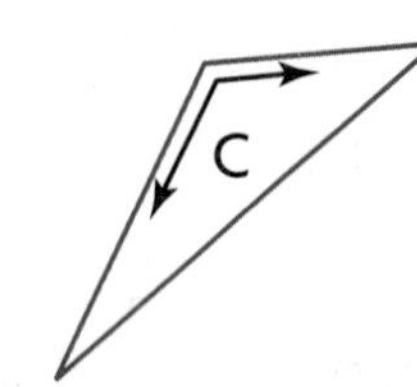

One angle is greater than a right angle.

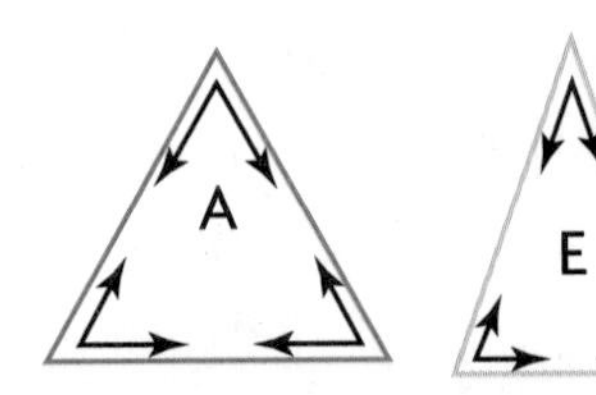

All angles are less than a right angle.

- How did Beverly sort the triangles? How did Armando sort the triangles?
- How can you check if an angle is greater than or less than a right angle?

Name Triangles

You can name triangles by their equal sides.

equilateral triangle

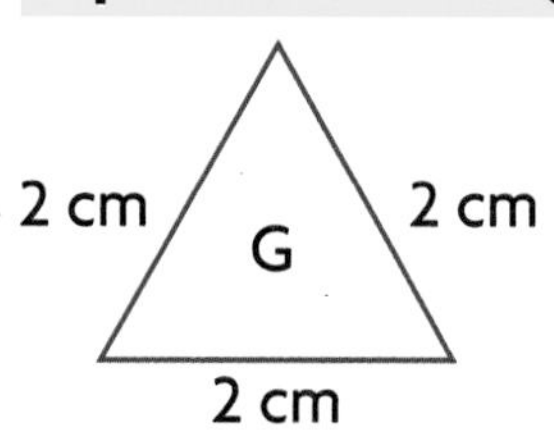

3 equal sides

isosceles triangle

3 cm
3 cm
H
2 cm

2 equal sides

scalene triangle

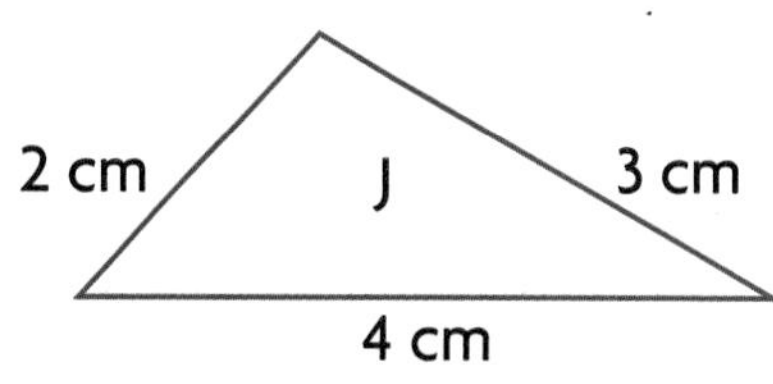

0 equal sides

You can name triangles by their angles.

right triangle

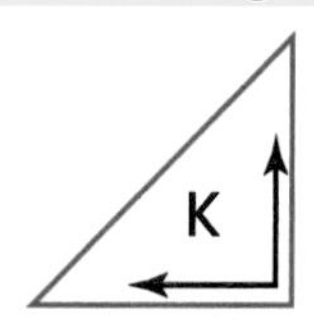

1 right angle

obtuse triangle

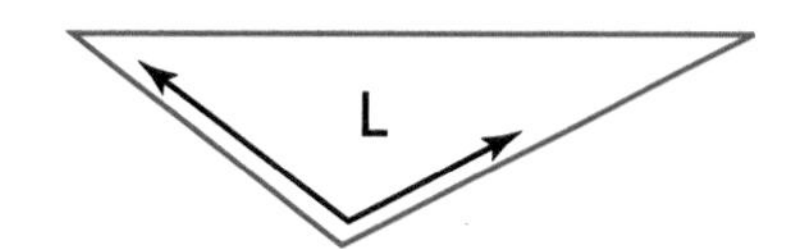

1 angle greater than a right angle

acute triangle

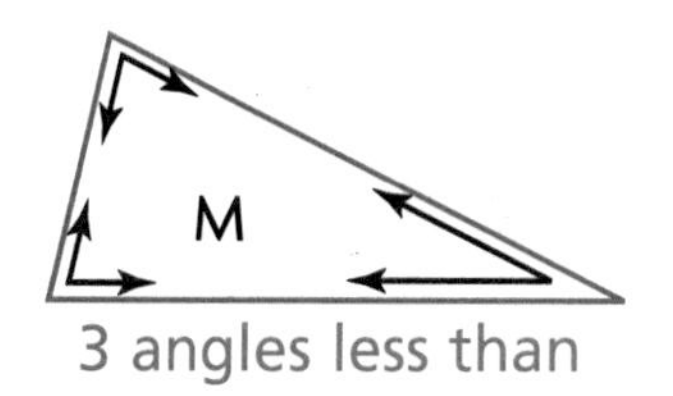

3 angles less than a right angle

- How are triangles J and M alike? How are they different?

MATH IDEA You can name and sort triangles by their sides or their angles.

Technology Link

More Practice: Use Mighty Math Number Heroes, *Geoboard,* Level Q.

Check

1. **Describe** triangle N by its sides. Then describe it by its angles.

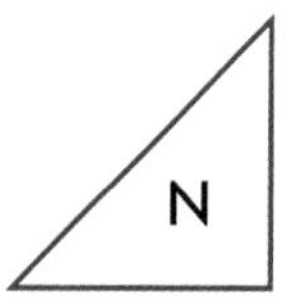

For 2–5, use the triangles at the right. Write *O, P, Q,* or *R*.

2. Which triangles have 0 equal sides?
3. Which triangle is isosceles?
4. Which triangles have 3 angles less than a right angle?
5. Which triangle is obtuse?

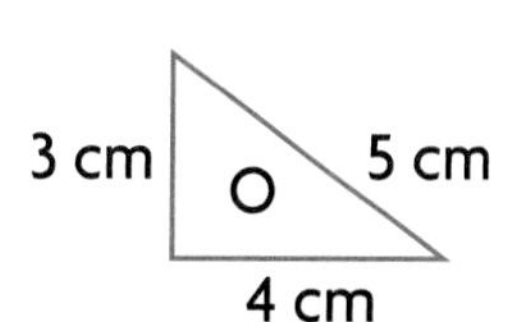

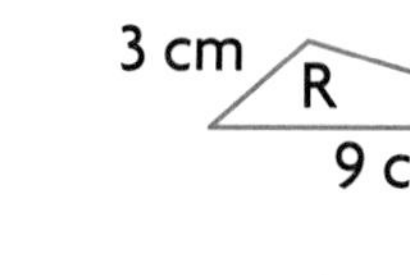

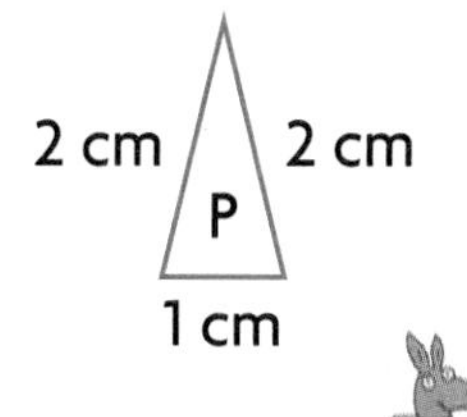

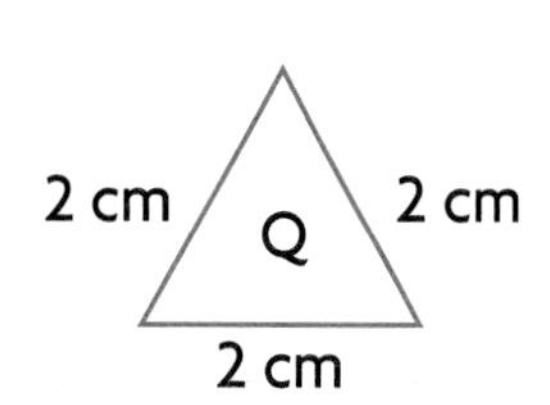

LESSON CONTINUES 

Practice and Problem Solving

For 6–8, use the triangles at the right. Write *A, B,* or *C*.

6. Which triangle is scalene?

7. Which triangles have at least 2 equal sides?

8. Which triangle has 1 angle that is greater than a right angle?

For 9–12, write one letter from each box to describe each triangle.

a. equilateral triangle
b. isosceles triangle
c. scalene triangle

d. right triangle
e. obtuse triangle
f. acute triangle

9. 2 cm, 2 cm, 2 cm

10.

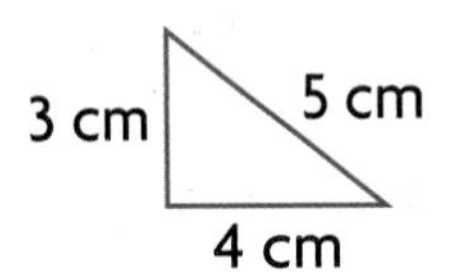

11.

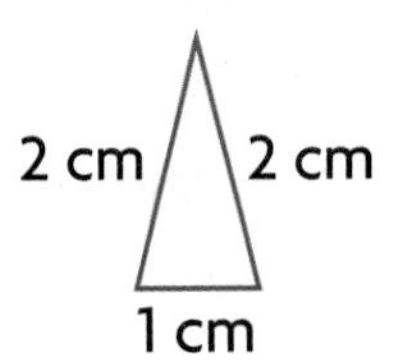

12. 3 cm, 2 cm, 4 cm

For 13–15, name each triangle. Write *equilateral, isosceles,* or *scalene*.

13.

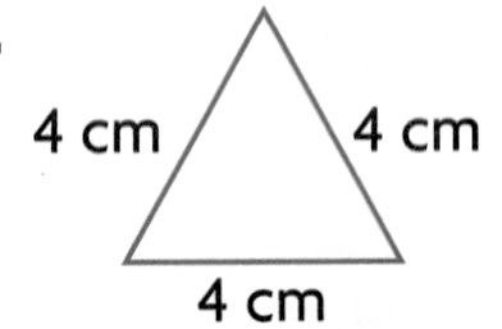

14. 4 cm, 6 cm, 8 cm

15. 2 cm, 2 cm, 3 cm

For 16–18, name each triangle. Write *right, obtuse,* or *acute*.

16.

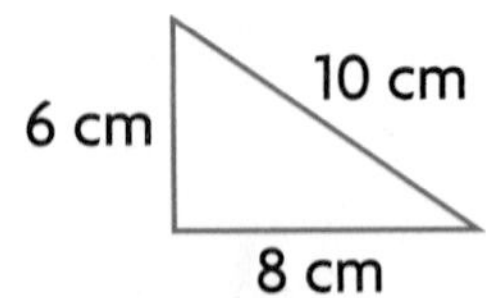

17.

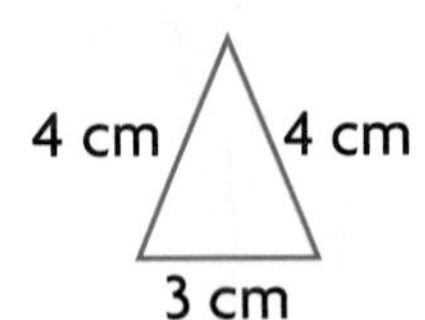

18.

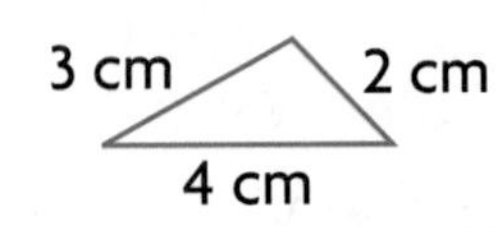

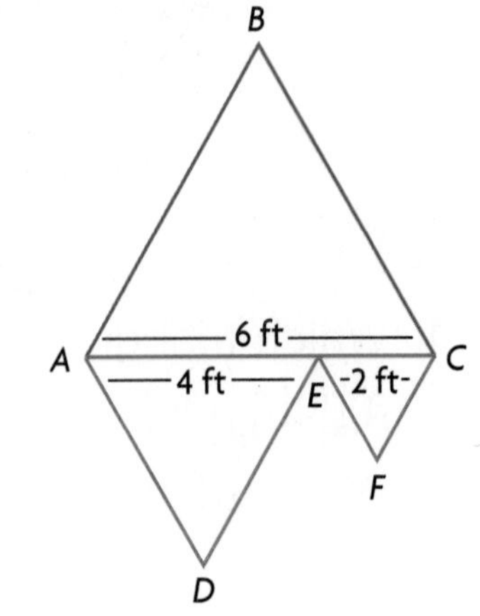

USE DATA For 19–20, use the diagram at the right.

19. Mrs. Liu has a garden with paths that are equilateral triangles. The shortest path from *A* to *C* is the green path. Which path is longer: the blue path or the red path? Explain.

20. **What's the Question?** The answer is 4 feet longer.

21. A right triangle has sides that are 3 inches, 4 inches, and 5 inches. Is this triangle equilateral, isosceles, or scalene? Explain.

22. Draw a triangle with 2 equal sides and one right angle. You may use grid paper to help. Name the triangle.

Mixed Review and Test Prep

23. $\begin{array}{r} 6 \\ \times 8 \\ \hline \end{array}$ (p. 148)

24. $\begin{array}{r} 9 \\ \times 7 \\ \hline \end{array}$ (p. 164)

25. $\begin{array}{r} 3 \\ \times 9 \\ \hline \end{array}$ (p. 122)

26. $5\overline{)35}$ (p. 202)

27. $8\overline{)48}$ (p. 216)

28. $7\overline{)49}$ (p. 216)

For 29–30, use the bar graph. (p. 254)

29. **TEST PREP** How many more people went to the Summer Concert and Storytelling Festival than to the Heritage Parade?

A 20 **C** 110
B 70 **D** 130

30. **TEST PREP** About 40 more people are expected to attend the Summer Concert next year. Choose the best number of seats to set up for the concert.

F 50 **H** 120
G 80 **J** 130

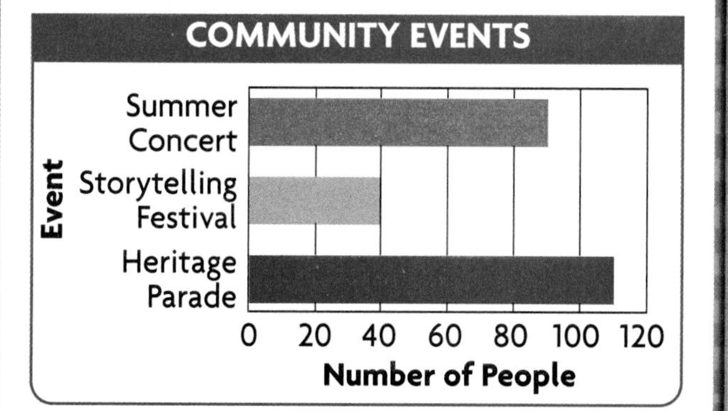

PROBLEM SOLVING LINKUP . . . to Art

Many artists use triangles in their works. The Native American blanket at the right uses different kinds of triangles. Find and name as many triangles on the blanket as you can.

Materials: grid paper, pencil, crayons

1. Draw and color a blanket design on grid paper. Use different kinds of triangles in your design.
2. Trade designs with a classmate and describe the triangles you see.

EXTRA PRACTICE page H49, Set B

LESSON

3 Quadrilaterals

Quick Review

Describe the lines. Write *parallel* or *intersecting.*

1. 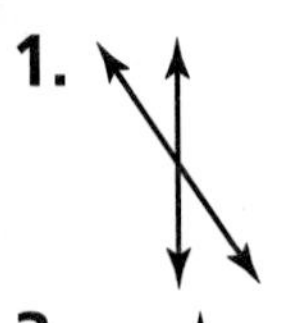

2. 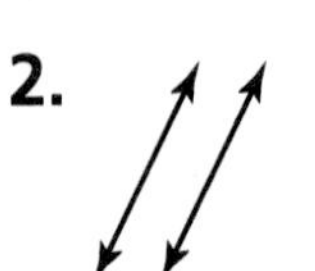

3. 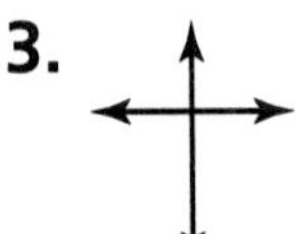

4. 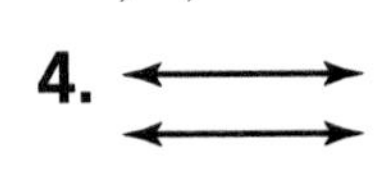

5. 

VOCABULARY

parallelogram
rhombus

LIMIT OF FOUR Polygons with 4 sides and 4 angles are quadrilaterals.

quadrilaterals | **not quadrilaterals**

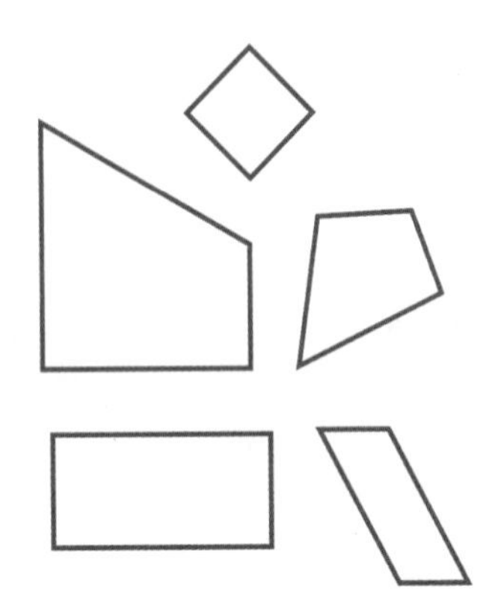

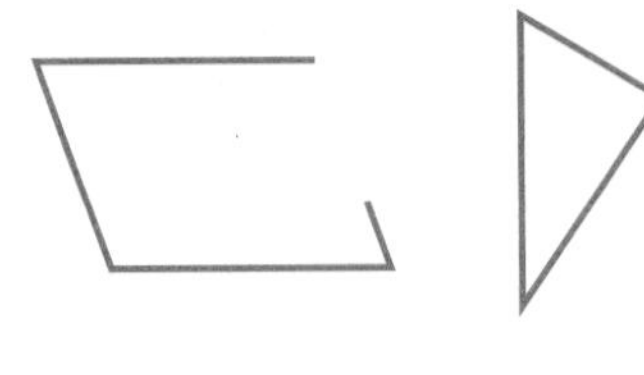

An angle in a quadrilateral can be a right angle or greater than or less than a right angle. The sides of a quadrilateral can be parallel.

Examples

A

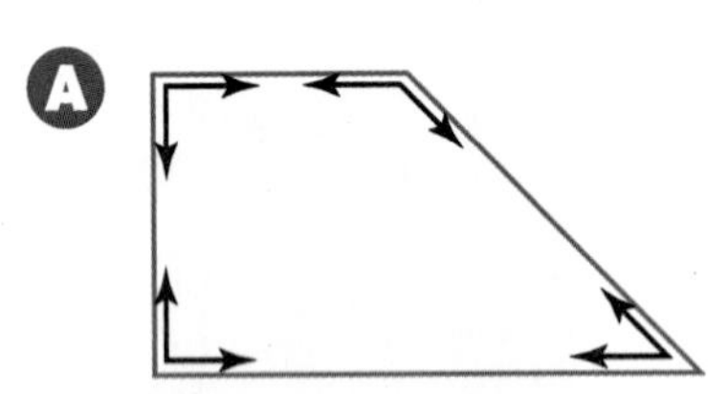

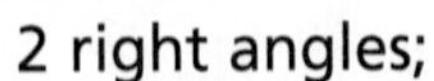

2 right angles;
1 angle is less than a right angle;
1 angle is greater than a right angle;
1 pair of parallel sides.

B

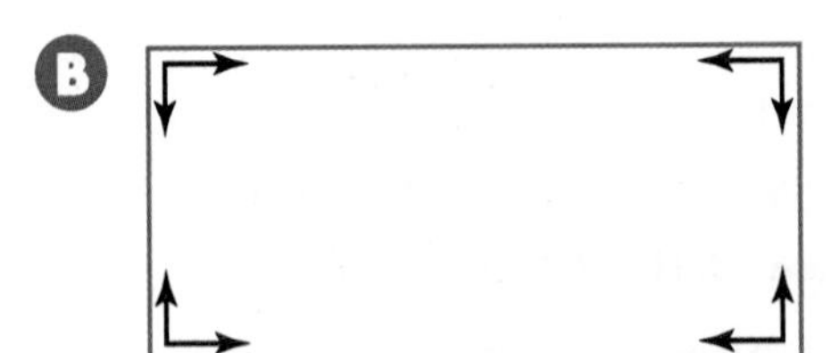

4 right angles;
2 pairs of parallel sides.

- How can you check if two sides of a quadrilateral are parallel?

Name Quadrilaterals

Here are some names of quadrilaterals with parallel sides.

parallelograms

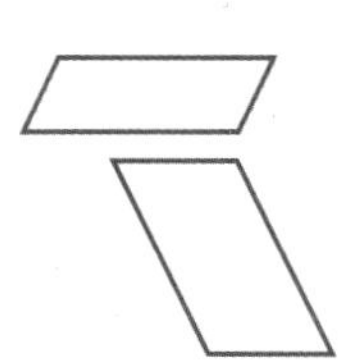

2 pairs of parallel sides
2 pairs of equal sides

rhombuses

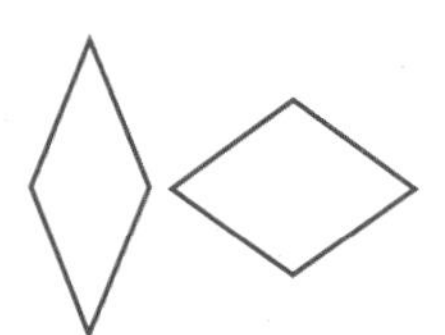

2 pairs of parallel sides
4 equal sides

rectangles

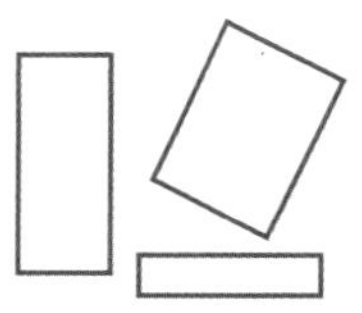

2 pairs of parallel sides
2 pairs of equal sides
4 right angles

squares

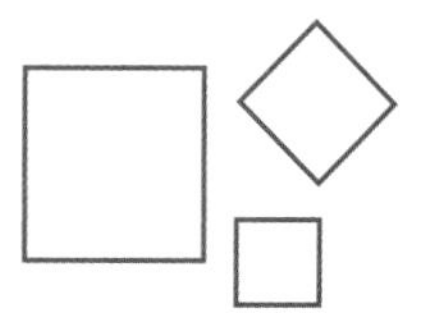

2 pairs of parallel sides
4 equal sides
4 right angles

- How are a rectangle and a square alike? How are they different?

MATH IDEA You can name and sort quadrilaterals by looking at their sides and angles.

Check

1. **Describe** the sides and angles of this quadrilateral. What is another name for it?

For 2–4, use the quadrilaterals at the right.

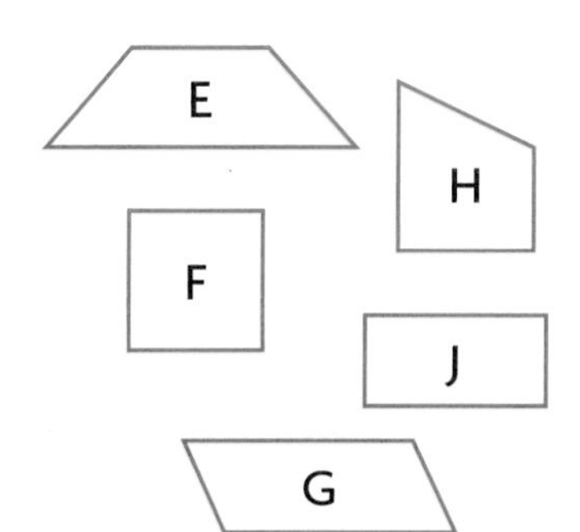

2. Which quadrilaterals have 2 pairs of parallel sides?
3. Which quadrilaterals have 2 or more right angles?
4. How are quadrilateral E and quadrilateral G alike? How are they different?

Write as many names for each quadrilateral as you can.

5\.

6\.

7\.

8\. 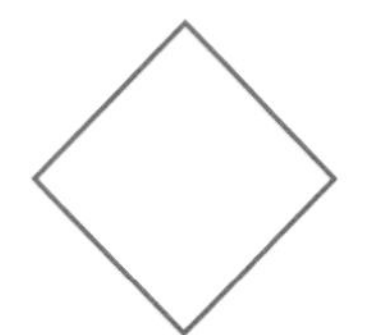

9\. 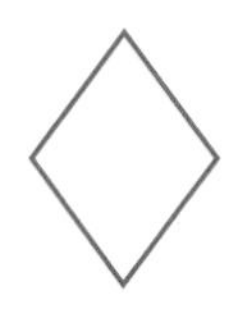

10\. 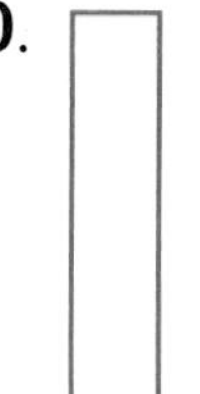

LESSON CONTINUES

Practice and Problem Solving

For 11–13, use the quadrilaterals at the right. Write *A, B, C, D,* and *E.*

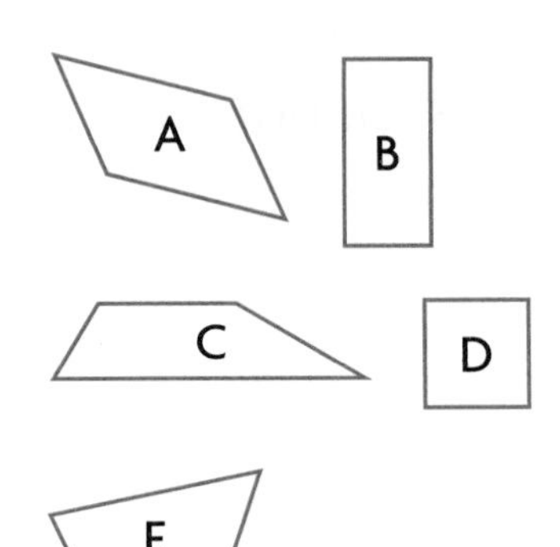

11. Which quadrilaterals have 2 pairs of equal sides?

12. Which quadrilaterals have no right angles?

13. How are quadrilateral A and quadrilateral D alike? How are they different?

For 14–19, write as many names for each quadrilateral as you can.

14.

15. 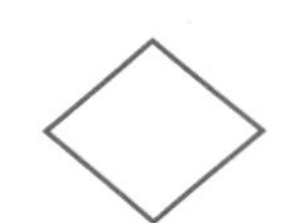

16.

17.

18. 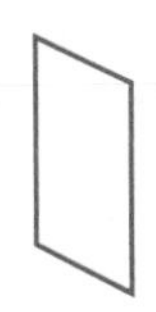

19. 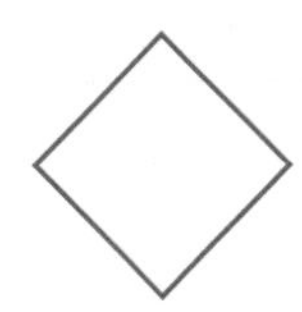

For 20–23, write *all* the letters that describe each quadrilateral.

20.

21.

22.

23.

a. It has 4 equal sides.
b. It has 2 pairs of parallel sides.
c. It has 4 right angles.
d. It has 2 pairs of equal sides.

24. REASONING How is the blue figure like the figures to its right?

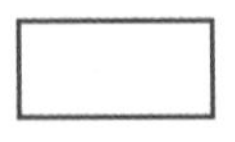

25. I have 4 equal sides and 4 right angles. What am I?

26. I have 5 sides and 5 angles. What am I?

27. REASONING Can a quadrilateral have no equal sides and no parallel sides? Draw a picture to help you explain.

28. **Write About It** Draw and label 4 different quadrilaterals on grid paper. Explain how each is different.

29. Akemi sees a tile with 4 right angles. She says it must be a square. Do you agree or disagree? Explain.

30. Dante drew a quadrilateral with 4 right angles and 2 pairs of parallel sides. What could he have drawn?

Mixed Review and Test Prep

Which number is greater? (p. 20)

31. 9,362 or 9,529

32. 5,108 or 5,018

Which number is less? (p. 20)

33. 2,814 or 2,148 **34.** 8,730 or 998

35. Round 6,398 to the nearest thousand. (p. 30)

36. Maurice will meet his mother in 30 minutes. What time will it be? (p. 94)

For 37–38, use the grid. (p. 262)

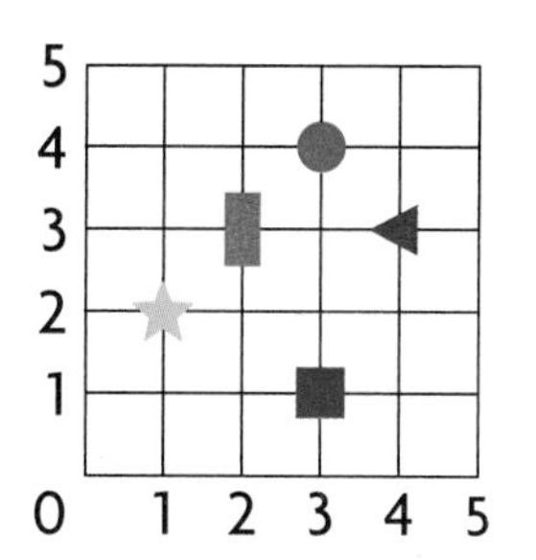

37. TEST PREP Which is the ordered pair for the triangle?

A (1,2) **B** (2,3) **C** (3,4) **D** (4,3)

38. TEST PREP Which shape is found at (3,4) on the grid?

F square
G circle
H rectangle
J triangle

PROBLEM SOLVING LINKUP . . . to Reading

STRATEGY • USE GRAPHIC AIDS Graphic aids such as charts, diagrams, and maps help display information visually. Sometimes the information needed to solve a problem is given in a graphic aid. The diagram at the right shows that all squares are quadrilaterals.

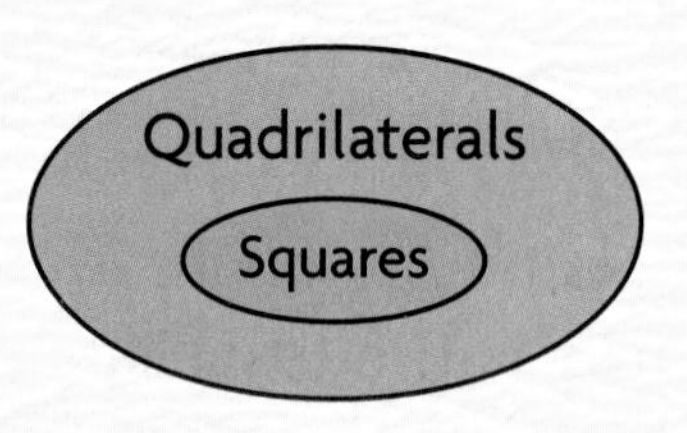

1. What does this diagram show?

Quadrilaterals
Rectangles

2. What does this diagram show?

Rectangles
Squares

3. Erin says a square is always a rectangle but a rectangle is not always a square. Do you agree or disagree? Explain.

4. Make a diagram to show the relationship among polygons, quadrilaterals, and parallelograms.

EXTRA PRACTICE page H49, Set C

LESSON 4

Combine Plane Figures

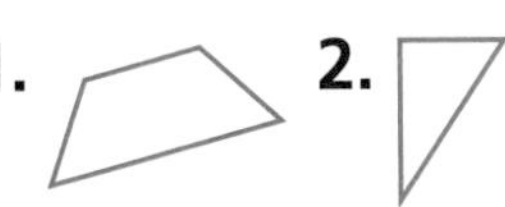

Quick Review

Name each figure.

1.  2. 

3. 4. 5.

VOCABULARY

tessellate
tessellation

Learn

TWIST AND TURN M. C. Escher, a Dutch artist, created many works of art by combining figures in a special way.

When plane figures combine to cover a surface without overlapping or leaving any space between them, those figures **tessellate**. The repeating pattern formed by the figures is called a **tessellation**.

These figures tessellate.

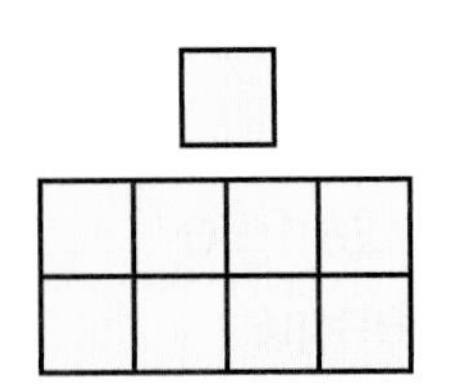

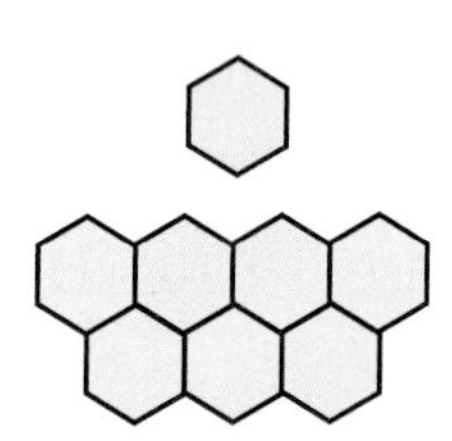

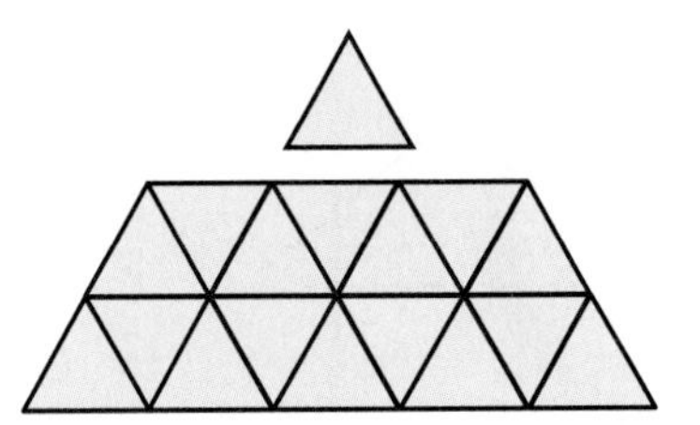

These figures do not tessellate.

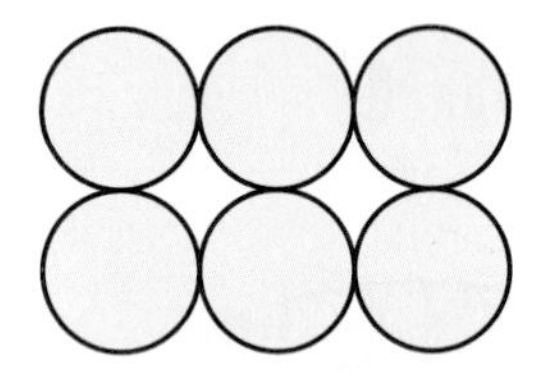

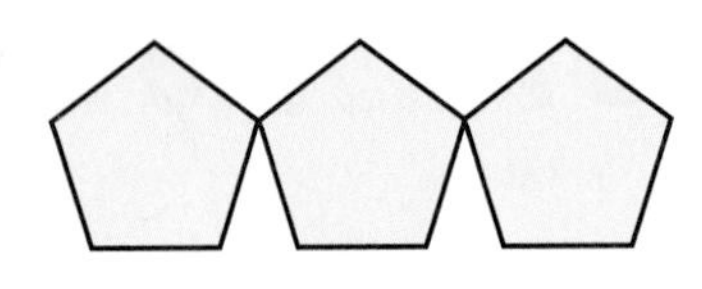

- Name the figure in the art by M. C. Escher. Does the figure tessellate? How do you know?

Symmetry Drawing E21 by M. C. Escher.

MATH IDEA Some plane figures can be combined to form tessellations.

Check

1. **Explain** how you know that circles do not tessellate.

Tell if each figure will tessellate. Write *yes* or *no*.

2. 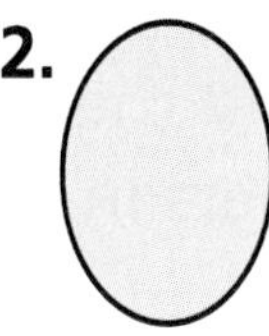

3. 

4. 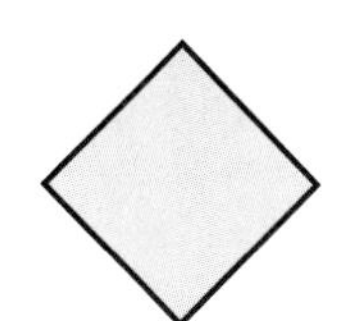

5. 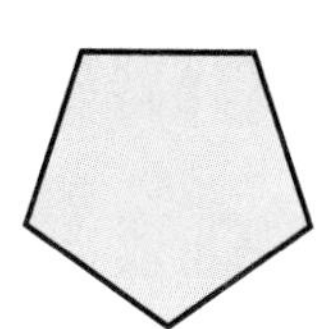

Practice and Problem Solving

Tell if each figure will tessellate. Write *yes* or *no*.

6. 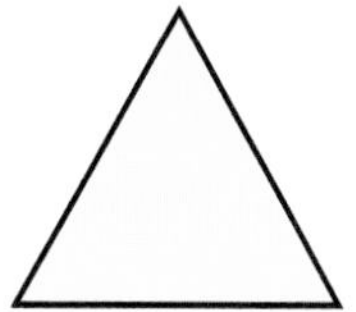

7. 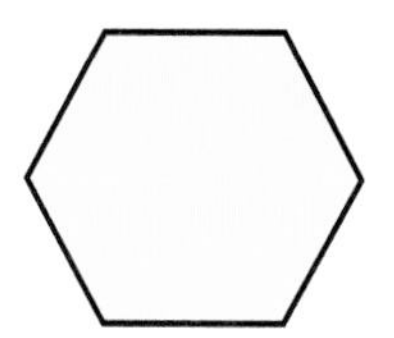

8. 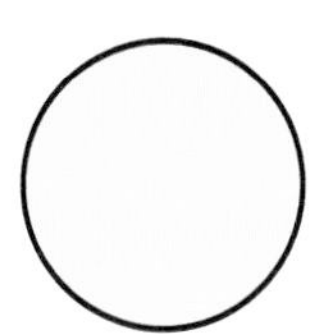

9. 

Trace and cut out each figure. Use each figure to make a tessellation. You may color your design.

10. 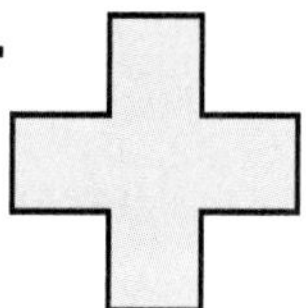

11. 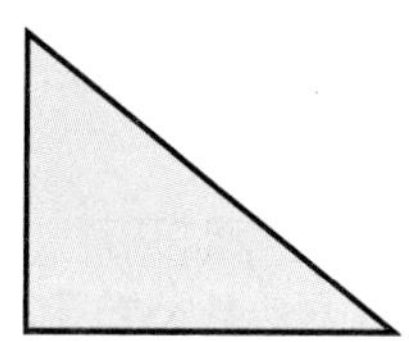

12. 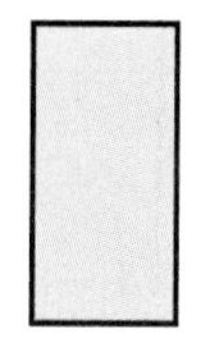

13. Is this a tessellation? Explain why or why not.

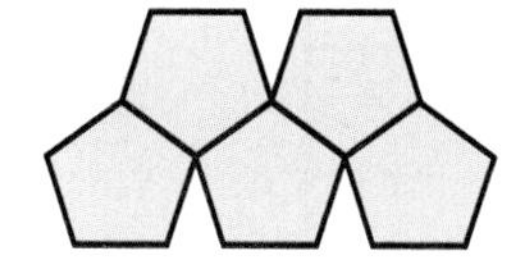

14. **Write About It** Explain how you know that this figure will not tessellate.

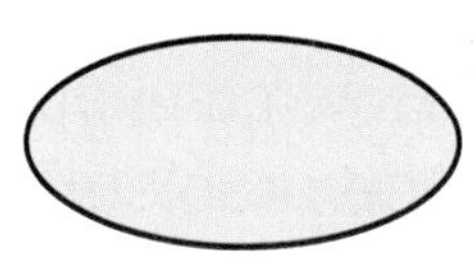

15. Pilar wants to buy 6 pencils. Store A sells 6 pencils for \$0.49. Store B's price is \$0.09 each. Where should Pilar buy her pencils in order to spend the least amount of money? Explain.

Mixed Review and Test Prep

For 16–19, find each sum. (p. 36)

16.	17.	18.
43	32	15
11	28	59
+27	+16	+15

19. $27 + 24 + 38 = ■$

20. **TEST PREP** What is this number in standard form? (p. 10)

$30{,}000 + 500 + 2$

A 352 **C** 30,502

B 3,502 **D** 35,002

EXTRA PRACTICE page H49, Set D

LESSON 5

Problem Solving Strategy
Find a Pattern

PROBLEM Karl is making a design with pattern blocks. What will be the next 4 pattern blocks in his design?

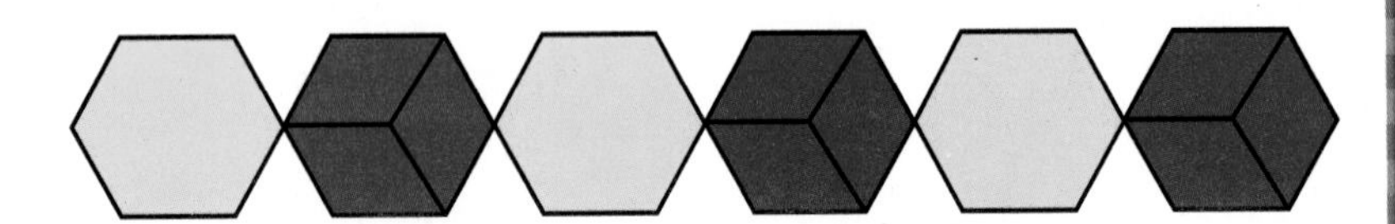

Quick Review

Write the rule and the next three numbers in each pattern.

1. 12, 19, 26, 33, ■, ■, ■
2. 4, 8, 12, 16, 20, ■, ■, ■
3. 54, 48, 42, 36, ■, ■, ■
4. 26, 23, 20, 17, ■, ■, ■
5. 9, 18, 27, 36, ■, ■, ■

UNDERSTAND

- What are you asked to find?
- What information will you use?

Pattern Blocks

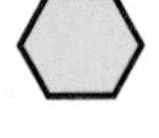
hexagon

rhombus

trapezoid

square

triangle

PLAN

- What strategy can you use to solve the problem?

  You can *find a pattern.*

SOLVE

- How can you use the strategy to solve the problem?

  Use the order of the pattern blocks in the design to identify a pattern.

  Think: hexagon, 3 rhombuses, hexagon, 3 rhombuses, hexagon, 3 rhombuses.

So, the next 4 pattern blocks in the design are a hexagon and 3 rhombuses.

CHECK

- How can you decide if your answer is right?

Problem Solving Practice

Find a pattern to solve.

1. **What if** Karl makes this design with pattern blocks? What will be the next 3 blocks in his design?

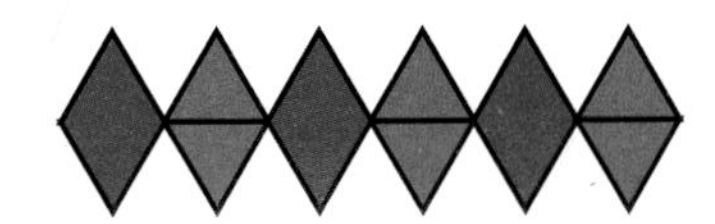

2. Valerie is drawing this pattern around the border of a picture frame. What will be the next three shapes in her pattern?

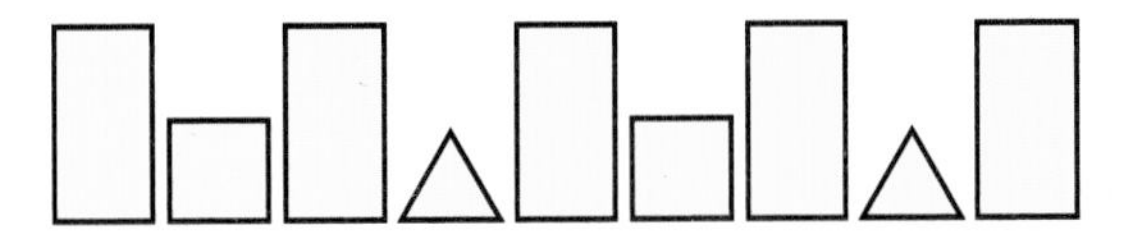

3. Draw a design using pattern block shapes. Ask a classmate to name the next 3 shapes.

PROBLEM SOLVING STRATEGIES

- Draw a Diagram or Picture
- Make a Model or Act It Out
- Make an Organized List
- **Find a Pattern**
- Make a Table or Graph
- Predict and Test
- Work Backward
- Solve a Simpler Problem
- Write a Number Sentence
- Use Logical Reasoning

Julio writes this number pattern: 4, 12, 20, 28, 36.

4. Which describes Julio's number pattern?

   A Multiply by 3.
   B Subtract 8.
   C Add 8.
   D Divide by 3.

5. What are the next three numbers in Julio's pattern?

   F 46, 38, 30
   G 38, 46, 54
   H 40, 48, 55
   J 44, 52, 60

Mixed Strategy Practice

USE DATA For 6–9, use the graph.

6. Make a bar graph to show the information in the graph.

7. Andrea wants to know how far a garden snail can move in 1 hour. Write a number sentence and solve.

8. **What's the Question?** The answer is 5 times as fast.

9. How long would it take the sloth to move 40 feet?

TIME TO MOVE 20 FEET

Giant Tortoise	
Garden Snail	
Sloth	

Key: Each (clock) = 2 minutes.

Review/Test

CHECK VOCABULARY

Choose the best term from the box.

equilateral
isosceles
quadrilateral
hexagon
tessellation

1. A triangle with 3 equal sides is ? . (p. 317)
2. A repeating pattern formed by figures that cover a surface without overlapping or leaving any space between them is a ? . (p. 324)
3. Any polygon that has 4 sides and 4 angles is a ? . (p. 314)

CHECK SKILLS

Write the number of sides and angles each polygon has. Then name the polygon. (pp. 314–315)

4. 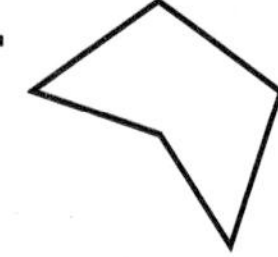
5. 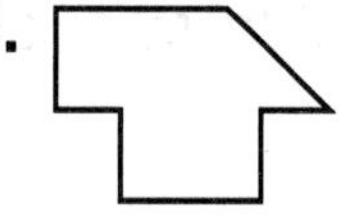
6. 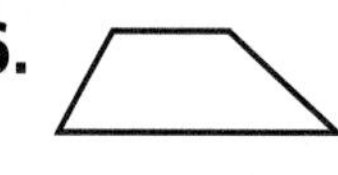
7. 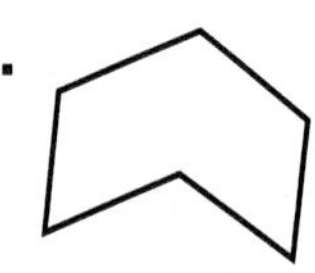

Name each triangle. Write *equilateral, isosceles,* or *scalene.* (pp. 316–319)

8. 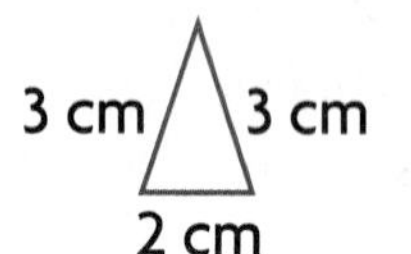

9. 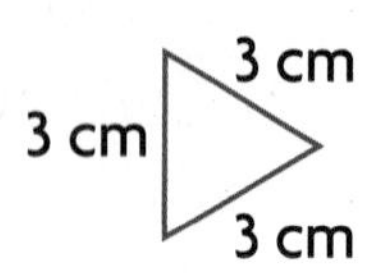

10. 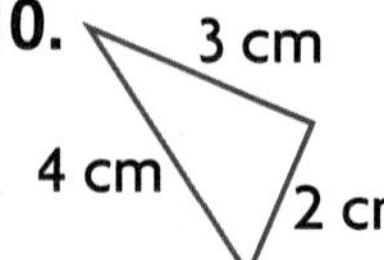

11. 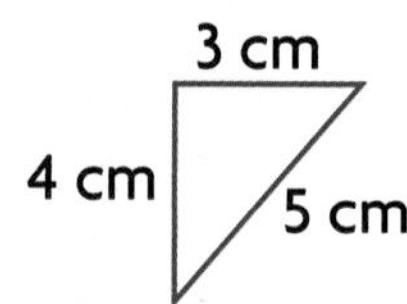

Write as many names for each quadrilateral as you can. (pp. 320–323)

12. 
13. 
14. 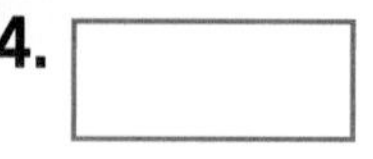
15. 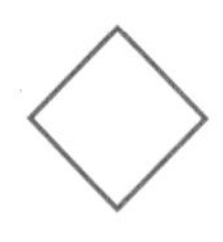

Tell if each figure will tessellate. Write *yes* or *no.* (pp. 324–325)

16. 
17. 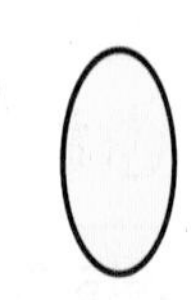
18. 
19. 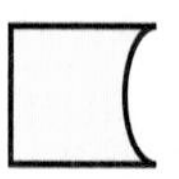

CHECK PROBLEM SOLVING

Find a pattern to solve. (pp. 326–327)

20. Stan made this pattern. What will be the next 3 shapes in his pattern? Describe his pattern.

Standardized Test Prep

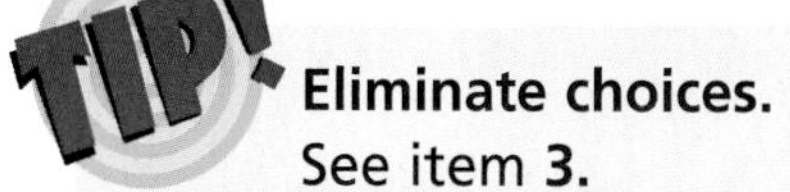

Eliminate choices.
See item **3**.

Eliminate any figures that do not have 4 sides. Then eliminate any figure that does not always have all sides equal.

Also see problem **5**, p. H64.

For 1–7, choose the best answer.

1. Which figure is **not** a polygon?

A

B 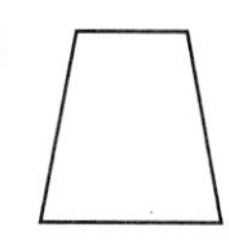

C

D

2. Margaret ate breakfast at 7:15 A.M. She left the house at 7:40 A.M. and arrived at school 20 minutes later. What time did Margaret get to school?

F 7:35 A.M.
G 7:55 A.M.
H 8:00 A.M.
J 8:05 A.M.

3. Which figure always has 4 equal sides?

A rectangle
B rhombus
C triangle
D parallelogram

4. Which figure always has 2 pairs of parallel sides?

F triangle
G circle
H quadrilateral
J square

5. Which numbers would be next in this pattern?

25, 21, 17, 13

A 12, 11, 10
B 11, 9, 7
C 9, 7, 5
D 9, 5, 1

6. Pablo was at summer camp every day for 6 weeks. How many days did Pablo spend at camp?

F 60
G 42
H 30
J NOT HERE

7. Jamie spent \$17 on a book bag. He still had \$9 when he got home. How much money did Jamie have before he bought the book bag?

A \$36
B \$26
C \$8
D NOT HERE

Write What You Know

8. Draw a group of rectangles. Draw a group of parallelograms. Describe the angles and sides of rectangles. Describe the angles and sides of parallelograms.

9. Draw a triangle with two equal sides and one right angle. Name the triangle in two ways and explain your answer.

Congruence and Symmetry

Many geometric figures have symmetry. Many plants and animals also have symmetry. Some figures have more than one line of symmetry.

PROBLEM SOLVING Look at the shell. Is the dashed line a line of symmetry? How do you know?

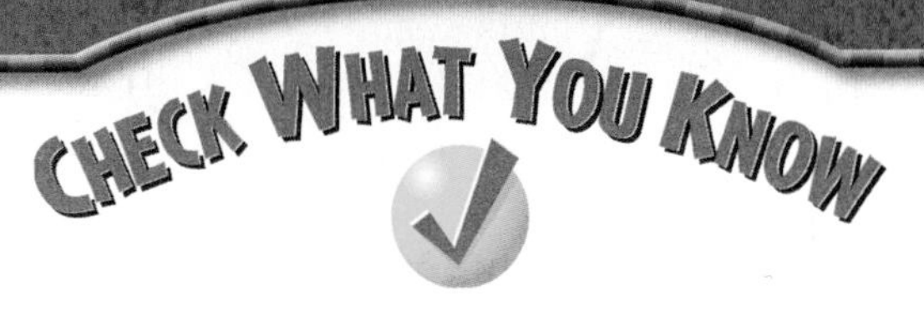

Use this page to help you review and remember important skills needed for Chapter 19.

VOCABULARY

Choose the best term from the box.

1. A closed plane figure with straight sides is a ? .
2. A polygon with 4 sides and 4 angles is a ? .
3. Lines that never cross are ? .

intersecting lines
parallel lines
polygon
quadrilateral

SAME SIZE SAME SHAPE (For Intervention, see p. H25.)

Tell whether the figures are the same size and shape. Write *yes* or *no*.

4.

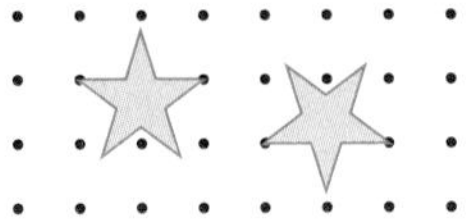

5.

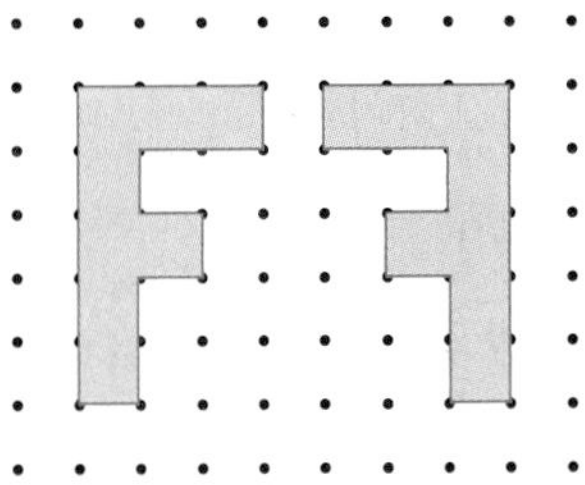

6.

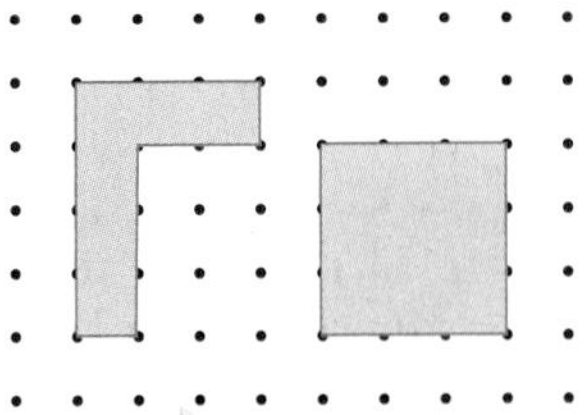

7.

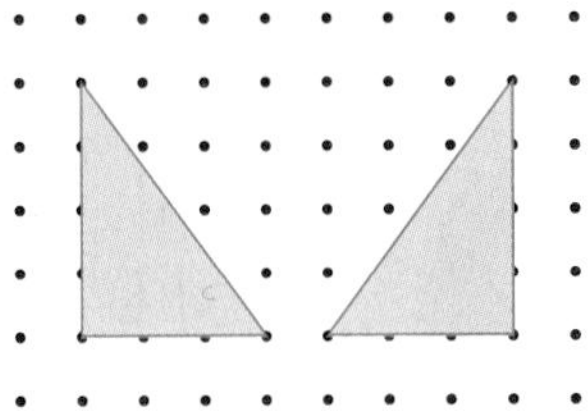

8.

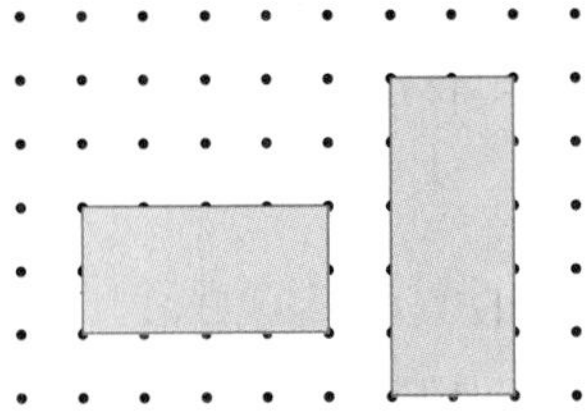

9.

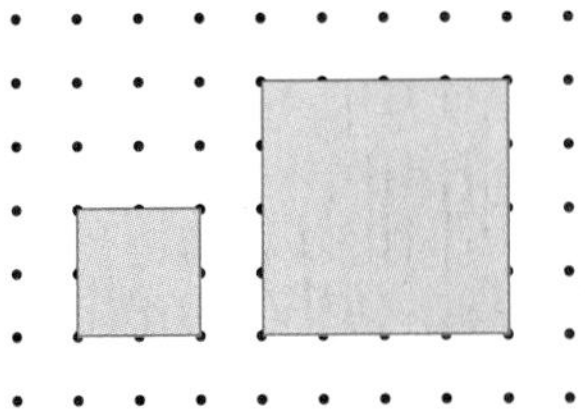

10.

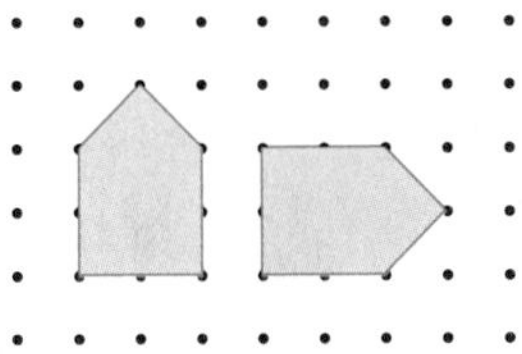

11.

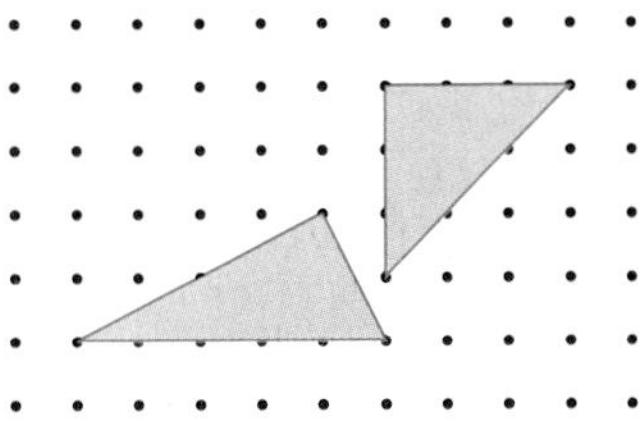

12.

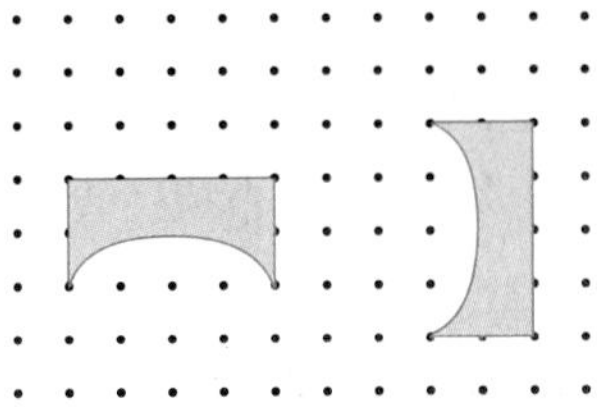

LESSON 1

Congruent Figures

ALL THE SAME **Congruent** figures have the same *size* and *shape*. Figures can be in different positions and still be congruent.

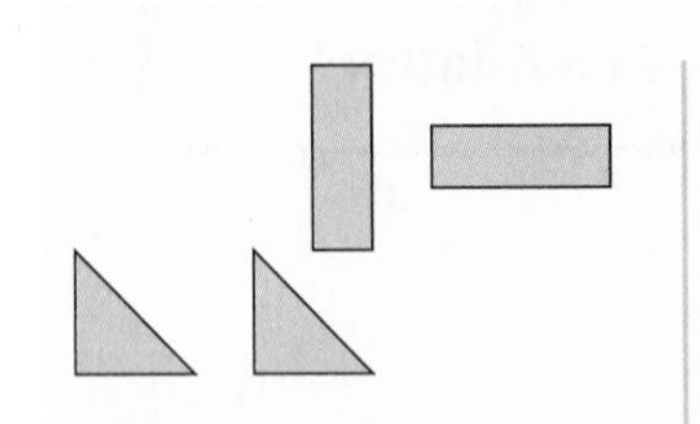

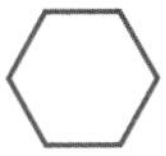

These pairs of figures are congruent.

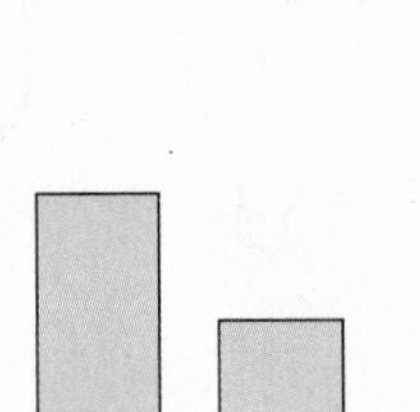

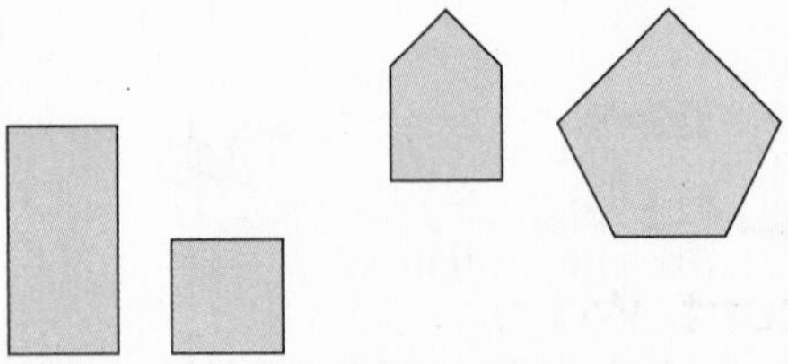

These pairs of figures are not congruent.

Quick Review

Write the number of sides and angles each polygon has.

1. (triangle)
2. (rectangle)
3. (hexagon)
4. (square)
5. (pentagon)

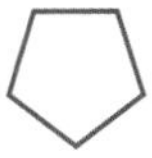

VOCABULARY
congruent

MATERIALS
pattern blocks, triangle dot paper, crayons

Use pattern blocks to find and build congruent figures.

- Sort a group of pattern blocks. Look for blocks that are the same size and shape. Put congruent pieces together.
- Use only small green triangles to make a figure that is congruent to the yellow hexagon. On triangle dot paper, draw the figure you made.
- Use any pattern blocks to make a different figure that is congruent to the yellow hexagon. Draw the figure you made.

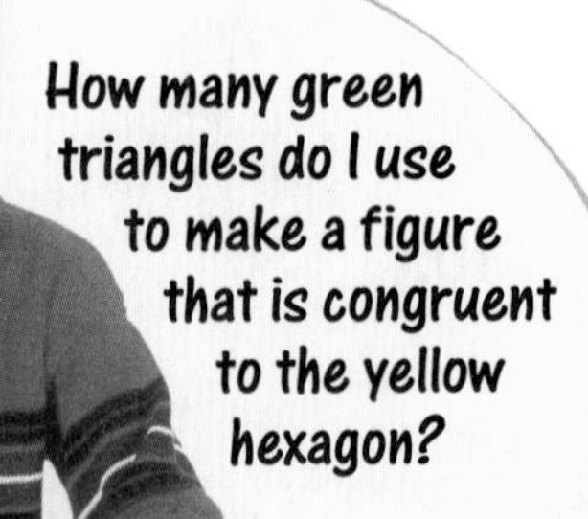

How many green triangles do I use to make a figure that is congruent to the yellow hexagon?

Talk About It

- How do you know that the figures you made are congruent?

Connect

More Practice:
Use E-Lab, *Using Congruent Figures.*
www.harcourtschool.com/elab2002

You can tell if two figures are congruent by tracing one figure and placing it over the second figure. If they are exactly the same, they are congruent.

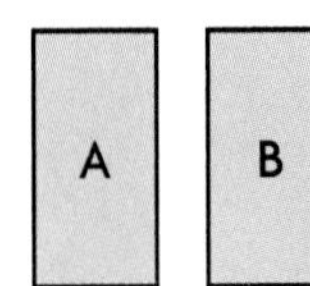

Trace and cut out rectangle A. Place it over rectangle B.

Are rectangles A and B congruent?

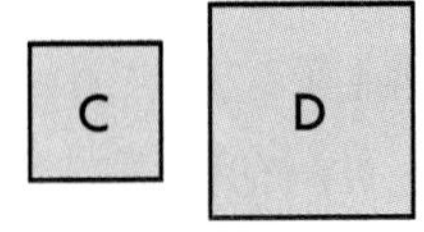

Trace and cut out square C. Place it over square D.

Are squares C and D congruent?

Practice and Problem Solving

Tell if the figures are congruent. Write *yes* or *no*.

1. 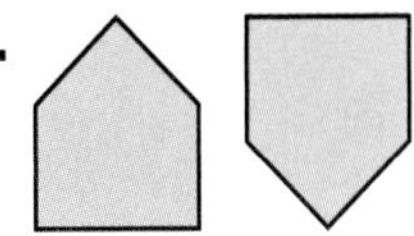

2. 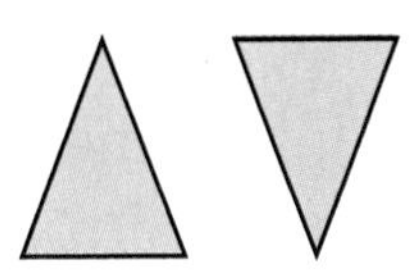

3. 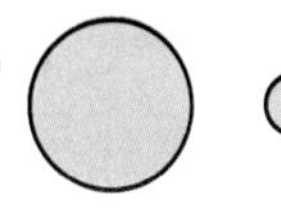

4. 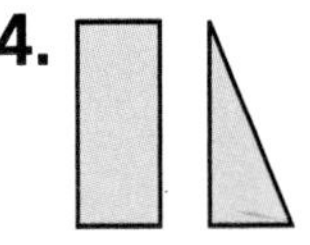

5. 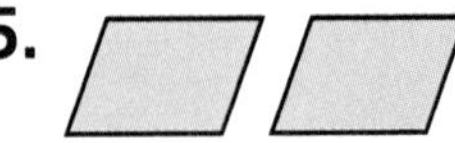

6. 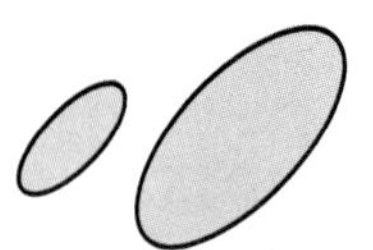

7. **REASONING** Joe says that all squares are congruent. Do you agree or disagree? Explain.

8. **Write About It** Copy this figure on grid paper. Draw a congruent figure. Then explain how you know the figures are congruent.

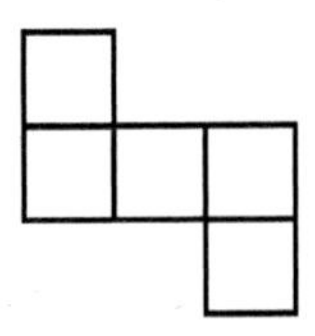

Mixed Review and Test Prep

9. $328 + 846$ (p. 42)

10. $574 + 659$ (p. 42)

11. $728 - 485$ (p. 58)

12. $932 - 267$ (p. 58)

13. **TEST PREP** Karson's play began at 6:25 P.M. It lasted 50 minutes. What time did the play end? (p. 98)

A 5:35 P.M. **B** 6:50 P.M. **C** 7:15 P.M. **D** 7:50 P.M.

Symmetry

Quick Review

Tell if each figure is congruent to the figure at the right. Write *yes* or *no*.

1. 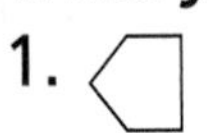 2. 3.

4. 5.

VOCABULARY

line of symmetry

MATERIALS

paper, scissors

Learn

HALF AND HALF An imaginary line that divides a figure in half is a **line of symmetry**. If you fold a figure along a line of symmetry, both sides match.

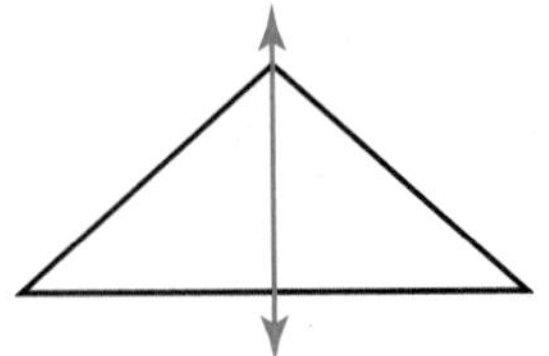

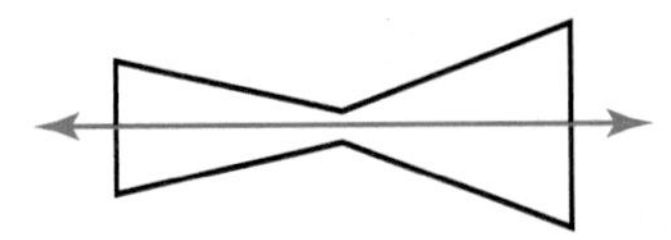

Activity

STEP 1

Fold a piece of paper in half. Draw a figure along one side of the fold. Cut out the figure.

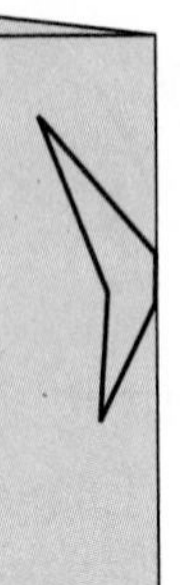

STEP 2

Unfold the figure. One side is like a mirror image of the other side.

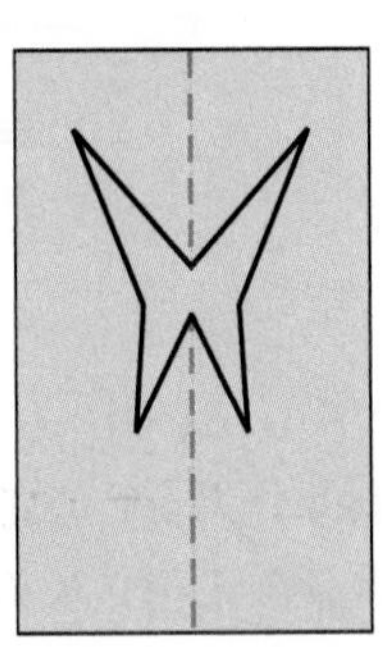

- Where is the line of symmetry on your figure?
- Is the left half of your figure congruent with the right half? Explain how you know.

MATH IDEA Some figures have one or more lines of symmetry. Some figures have no lines of symmetry.

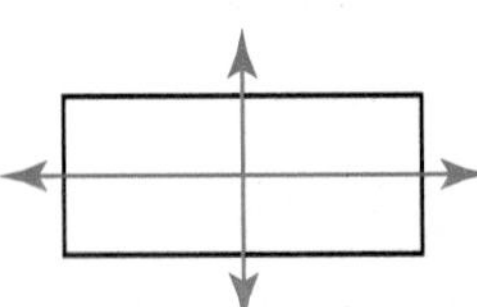

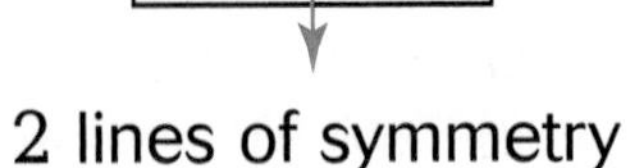

2 lines of symmetry

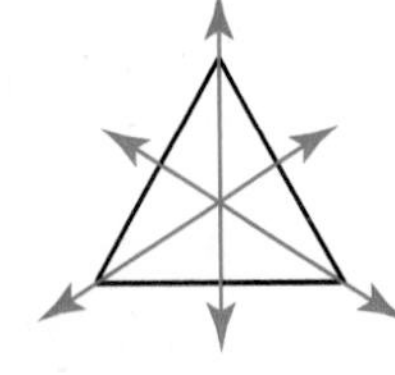

3 lines of symmetry

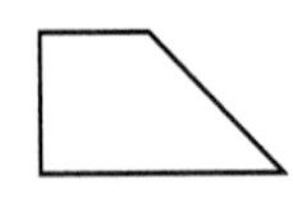

0 lines of symmetry

Check

1. **Explain** how you know the blue line in the figure at the right is a line of symmetry.

Tell if the blue line is a line of symmetry. Write *yes* or *no*.

2. 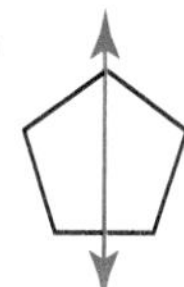

3. 

4. 

Practice and Problem Solving

Tell if the blue line is a line of symmetry. Write *yes* or *no*.

5. 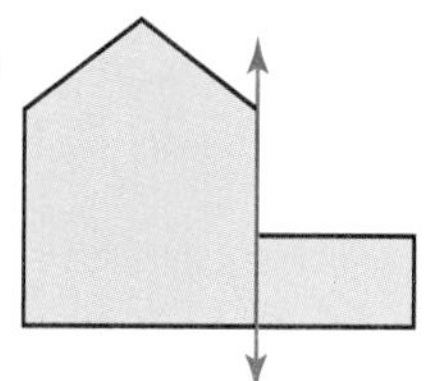

6. 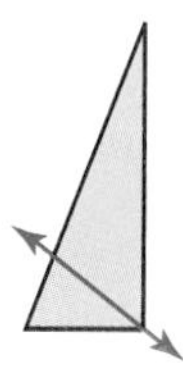

7. 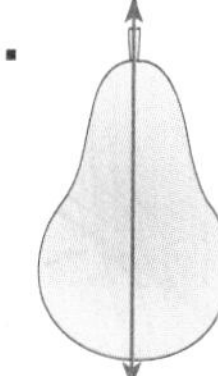

Write how many lines of symmetry each figure has.

8. 

9. 

10. 

11. **What's the Error?** Evan drew this figure and line of symmetry. Describe Evan's error and explain how Evan can check his line of symmetry.

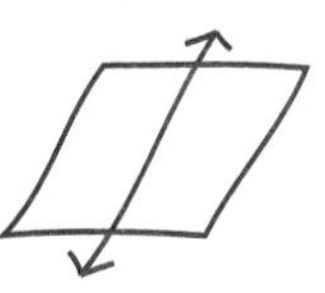

12. Jody says that a circle has too many lines of symmetry to count. Do you agree or disagree? Explain.

13. **Write About It** Draw a plane figure with more than one line of symmetry. Describe each line of symmetry.

Mixed Review and Test Prep

Find the product. (p. 170)

14. $(4 \times 2) \times 5$

15. $7 \times (3 \times 0)$

Write + or − to make the sentence true. (p. 68)

16. 32 ● 26 = 6

17. 48 ● 24 = 72

18. **TEST PREP** Larry bought a book for \$6.75. He paid with a \$10 bill. How much change did he get? (p. 88)

A \$2.25
B \$3.25
C \$6.75
D \$16.75

EXTRA PRACTICE page H50, Set A

LESSON 3

Similar Figures

Learn

SIZE WISE Figures that have the same shape but may have different sizes are called **similar** figures.

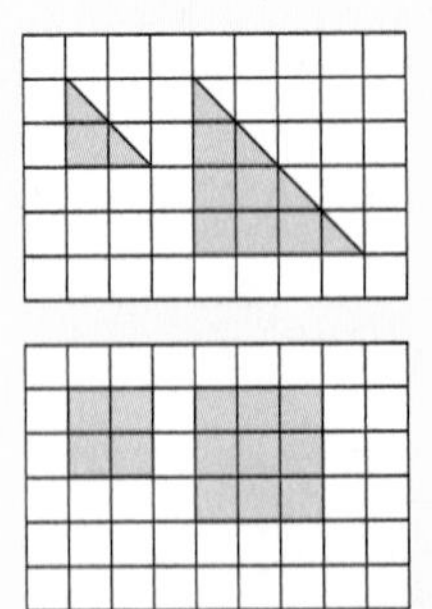

These pairs of figures are similar.

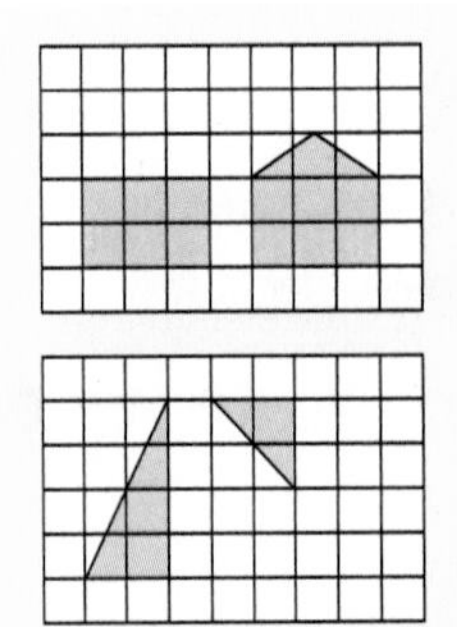

These pairs of figures are not similar.

When you enlarge or reduce the size of a figure, the figure you make is similar to the first one.

Quick Review

Tell if the blue line is a line of symmetry. Write *yes* or *no*.

1.
2.
3.
4.
5.

VOCABULARY

similar

MATERIALS

1-inch grid paper, 1-centimeter grid paper

Activity

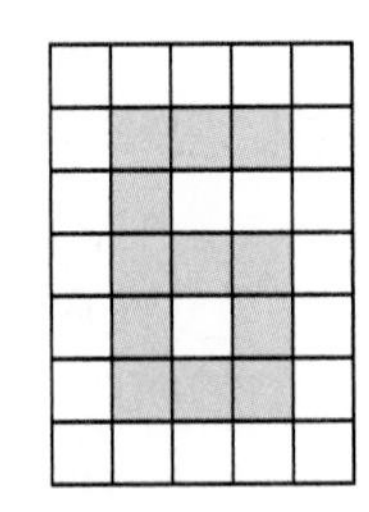

- Copy the figure at the right on 1-inch grid paper. Copy one square at a time.
- Is the figure you drew similar to the figure at the right? Is it congruent? Explain how you know.
- Draw a figure on 1-centimeter grid paper. Copy the figure you drew one square at a time on 1-inch grid paper.
- Compare the two figures you drew. Tell what you know about the figures.

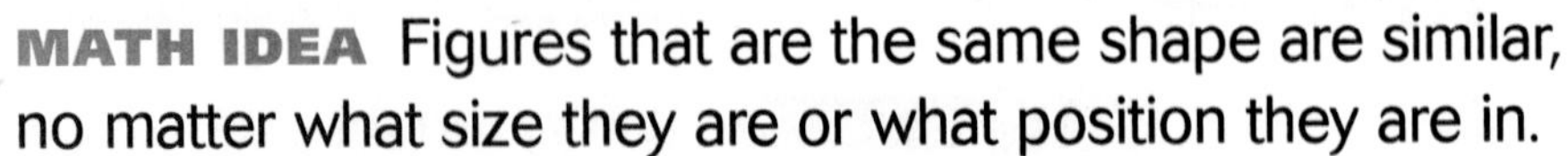

MATH IDEA Figures that are the same shape are similar, no matter what size they are or what position they are in.

Check

1. **Explain** whether or not two figures can be both similar and congruent.

Tell if the figures are similar. Write *yes* or *no*.

2. 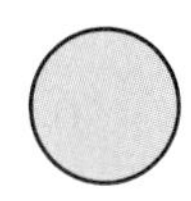

3. 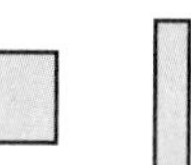

4. 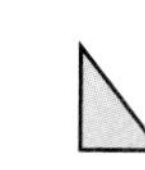

Practice and Problem Solving

Tell if the figures are similar. Write *yes* or *no*.

5. 

6. 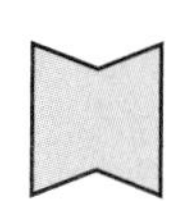

7. 

Draw a similar figure for each. Use 1-inch grid paper.

8. 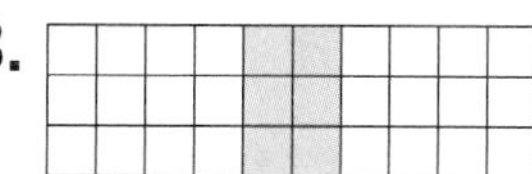

9. 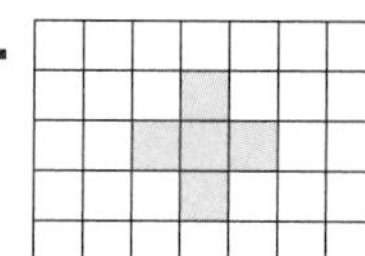

10. 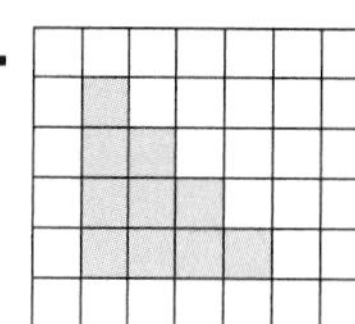

Draw a similar figure for each. Use 1-centimeter grid paper.

11. 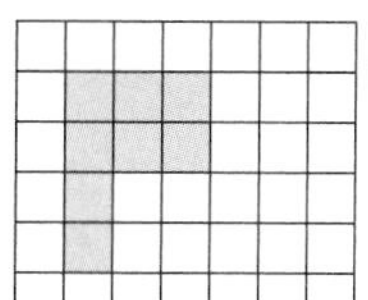

12. 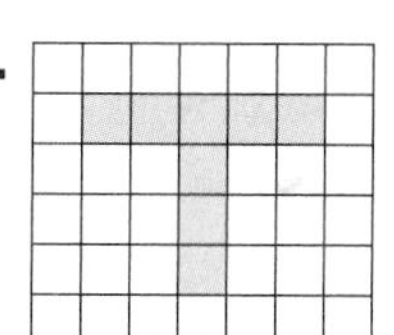

13. 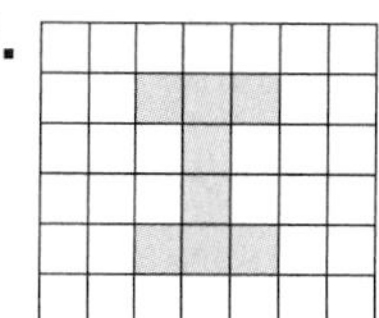

14. **REASONING** Do figures have to be in the same position for them to be similar? Explain.

15. Devon says that congruent figures are also similar figures. Do you agree or disagree? Explain.

16. **Write About It** Draw a design on 1-inch grid paper. Have a classmate draw the design on 1-centimeter grid paper. Are the designs similar? How can you tell?

Mixed Review and Test Prep

17. $4 \times 6 = \blacksquare$ (p. 156)

18. $8 \times 10 = \blacksquare$ (p. 164)

19. $28 \div 7 = \blacksquare$ (p. 216)

20. $32 \div 4 = \blacksquare$ (p. 204)

21. **TEST PREP** Choose a rule for the table. (p. 168)

bicycles	1	2	3	4
wheels	2	4	6	8

A Multiply bicycles by 2.

B Multiply wheels by 2.

C Add 2 to bicycles.

D Subtract 2 from wheels.

LESSON 4

Slides, Flips, and Turns

Quick Review

Tell if each figure is congruent to the figure at the right. Write *yes* or *no*.

1.
2.
3.
4.
5.

VOCABULARY
slide flip turn

MATERIALS
pattern blocks, triangle dot paper

Explore

A plane figure can be moved in different ways.

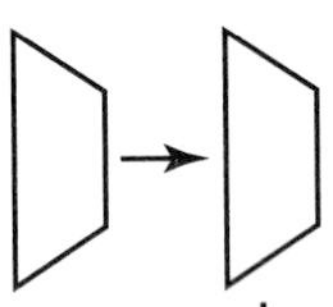
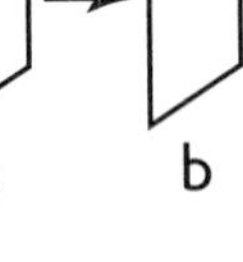

slide

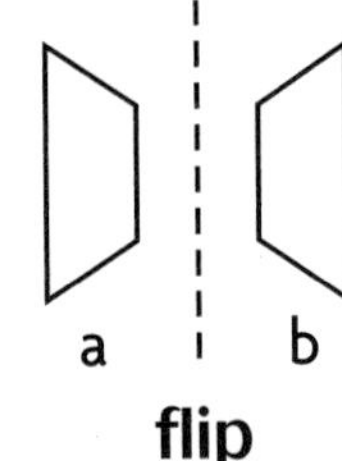

flip

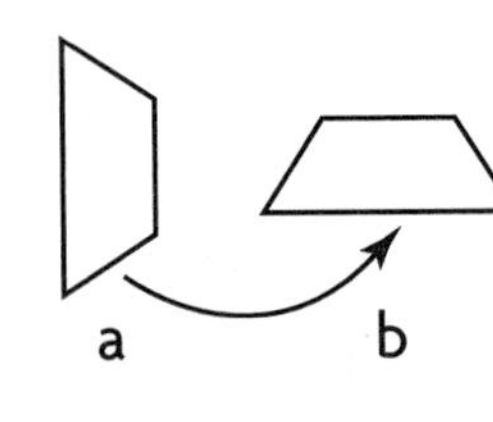

turn

Activity

Use pattern blocks to slide, flip, and turn figures.

- Choose a pattern block. Show different ways to move the block.

STEP 1

Trace the block on triangle dot paper.

STEP 2

Slide the block and trace it. Label the drawing "slide."

slide

STEP 3

Flip the block and trace it. Label the drawing "flip."

flip

STEP 4

Turn the block and trace it. Label the drawing "turn."

turn

- Choose a different pattern block. Repeat the steps above.

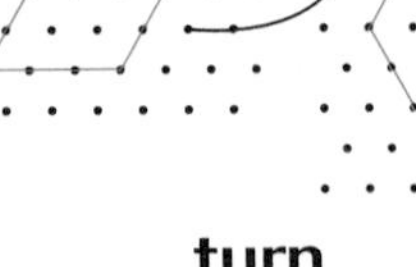

More Practice: Use Mighty Math Number Heroes, *Geoboard,* Levels F and N.

Talk About It

- Look at your drawings. Does a slide ever look like a flip? Does a flip ever look like a turn?
- **REASONING** Does the size or shape of the block change when you slide, flip, or turn it? Explain.

Connect

Think about how you use slides, flips, and turns to describe motions of real-world objects.

You can slide a book across the table.

You can flip a pancake.

You can turn a puzzle piece.

MATH IDEA You can describe a motion used to move a plane figure as a slide, a flip, or a turn.

Practice and Problem Solving

Tell what kind of motion was used to move each plane figure. Write *slide, flip,* or *turn*.

1. 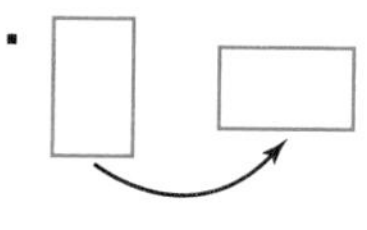

2. 

3. 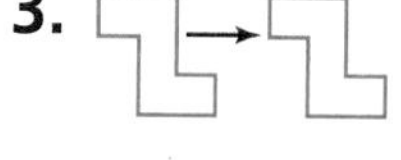

4. 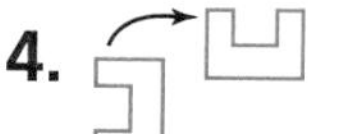

5. 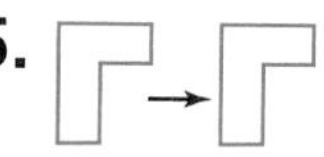

6. 

7. Which figure shows what Figure A would look like after a flip. Write X, Y, or Z.

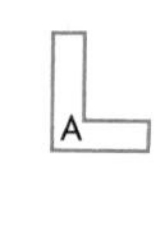

X Y Z

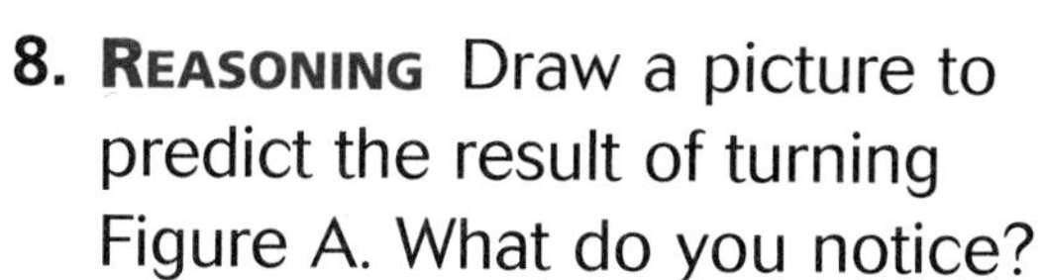

8. **REASONING** Draw a picture to predict the result of turning Figure A. What do you notice?

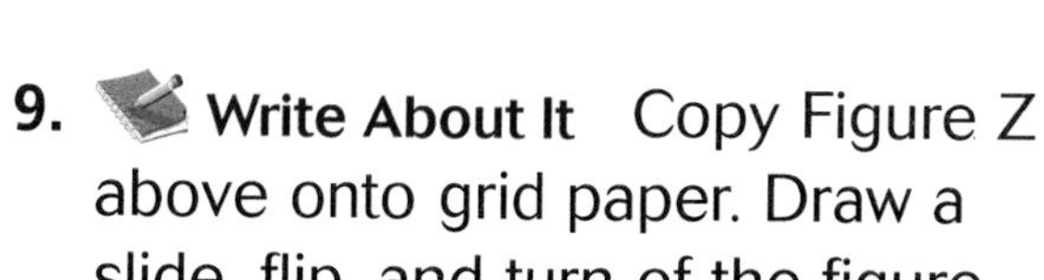

9. **Write About It** Copy Figure Z above onto grid paper. Draw a slide, flip, and turn of the figure. Be sure to label each drawing.

Mixed Review and Test Prep

10. $9 \times 3 = \blacksquare$ (p. 122)

11. $6 \times 5 = \blacksquare$ (p. 118)

Write a rule for each pattern and the next three numbers. (p. 136)

12. 67, 63, 59, 55, ■, ■, ■

13. 28, 36, 44, 52, ■, ■, ■

14. **TEST PREP** Six toy cars cost $54. How much does one cost? (p. 226)

A $6 **C** $8

B $7 **D** $9

LESSON 5

Problem Solving Strategy
Make a List

PROBLEM Chantal and Miguel each made a solid figure with blocks. Are the two solid figures congruent?

Quick Review

Tell if the figures below are similar to the figure at the right. Write *yes* or *no*.

1. 
2. 
3. 
4.
5.

UNDERSTAND

- What are you asked to find?
- What information will you use?
- Is there information you will not use? If so, what?

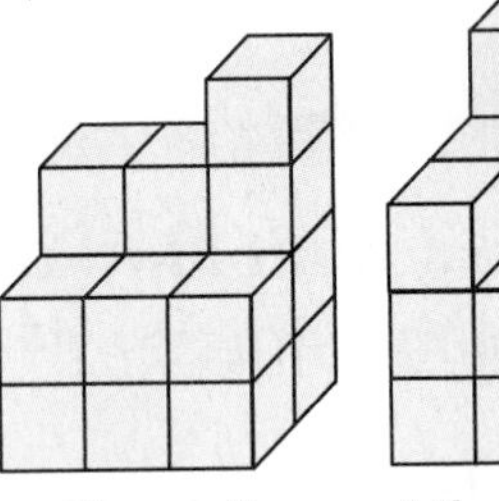

Chantal's figure Miguel's figure

PLAN

- What strategy can you use to solve the problem?

  You can *make a list* of the number of cubes in each layer. Then check if the figures match.

SOLVE

- How can you use the strategy to solve the problem?

  You can make a list of the cubes in each layer.

Chantal's figure	Miguel's figure
Layer 1 — 6 cubes	Layer 1 — 6 cubes
Layer 2 — 6 cubes	Layer 2 — 6 cubes
Layer 3 — 3 cubes	Layer 3 — 3 cubes
Layer 4 — 1 cube	Layer 4 — 1 cube
$6 + 6 + 3 + 1 = 16$ cubes	$6 + 6 + 3 + 1 = 16$ cubes

- So, the two figures are congruent because they have the same number of cubes and have the same shape.

CHECK

- How else could you solve this problem?

Problem Solving Practice

Make a list to solve.

1. Kendal built a solid figure with connecting cubes. There are 6 cubes in the first layer, 1 cube in the second layer, and 1 cube in the third layer. Could the solid figure at the right be the one Kendal made? Explain.

PROBLEM SOLVING STRATEGIES

- Draw a Diagram or Picture
- Make a Model or Act It Out
- **Make an Organized List**
- Find a Pattern
- Make a Table or Graph
- Predict and Test
- Work Backward
- Solve a Simpler Problem
- Write a Number Sentence
- Use Logical Reasoning

For 2–3, use the figures below.

2. How many cubes were used to build Figure A?

   A 7 **C** 9
   B 8 **D** 10

3. How many cubes should be added to Figure B to be congruent to Figure A?

   F 1 **H** 3
   G 2 **J** 4

Mixed Strategy Practice

USE DATA For 4–5, use the line plot at the right. (p. 258)

4. The X's on the line plot stand for students. How many students saw more than 4 movies last year?

5. What is the range for this set of data? What is the mode?

6. There were 438 people on a train. At the station, 113 people got off and 256 people got on. How many people are on the train now? (p. 58)

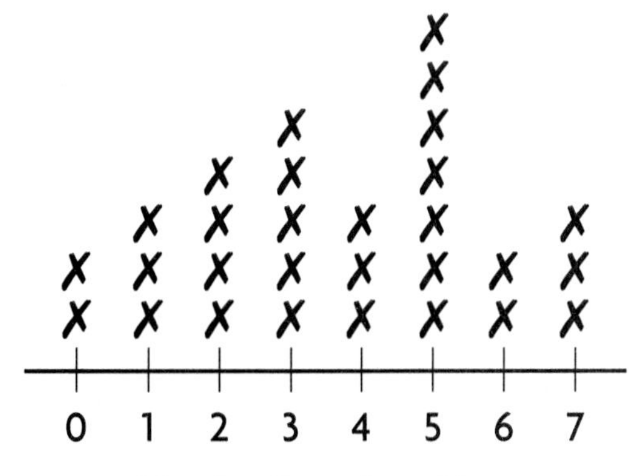

7. Mimi bought a sandwich for \$2.75 and a carton of milk for \$1.25. She paid with a \$5 bill. How much change should she get? (p. 88)

Review/Test

CHECK VOCABULARY AND CONCEPTS

Complete. Choose the best word(s) from the box.

line of symmetry
similar
congruent

1. Figures that are the same size and shape are _?_. (p. 332)
2. An imaginary line that divides a figure in half is a _?_. (p. 334)

Tell what kind of motion was used to move each plane figure. Write *slide, flip,* or *turn.* (pp. 338–339)

3. 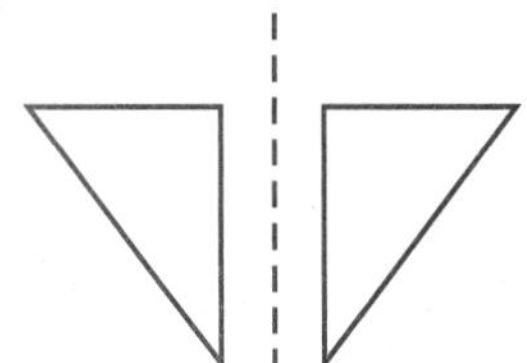

4. 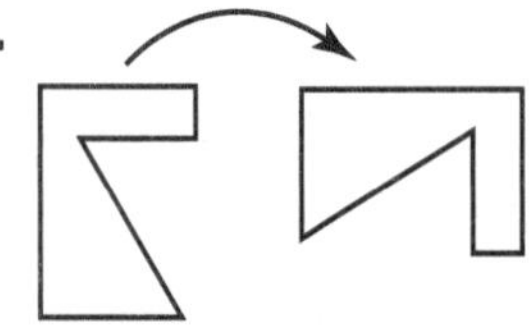

5. 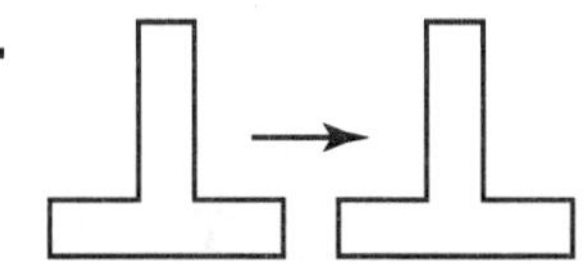

CHECK SKILLS

Tell if the blue line is a line of symmetry. Write *yes* or *no.* (pp. 334–335)

6. 

7. 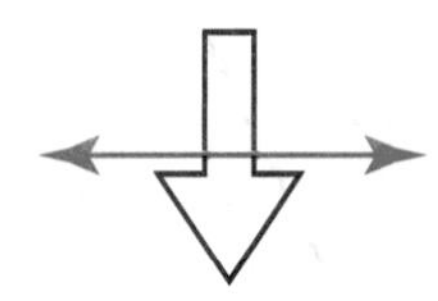

8. 

Tell if the figures are similar. Write *yes* or *no.* (pp. 336–337)

9. 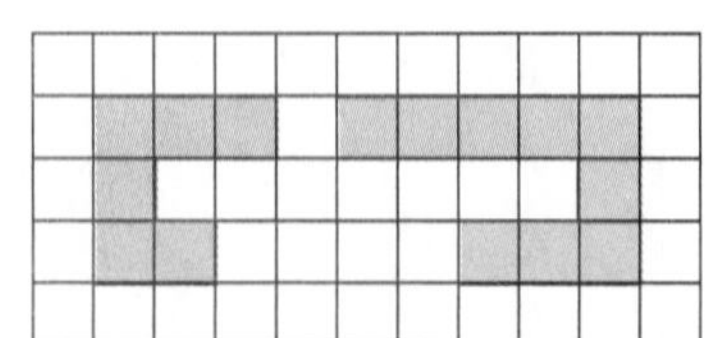

10. 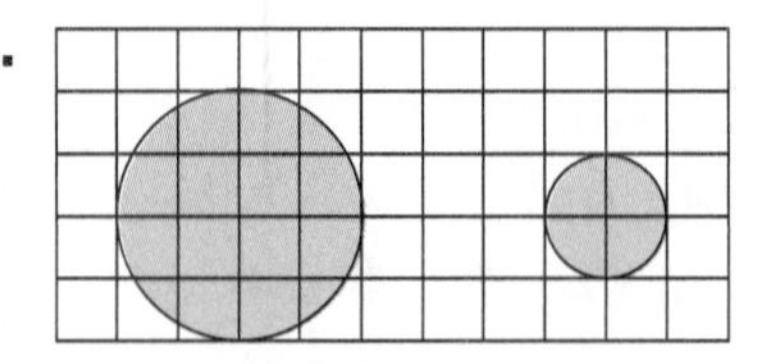

11. 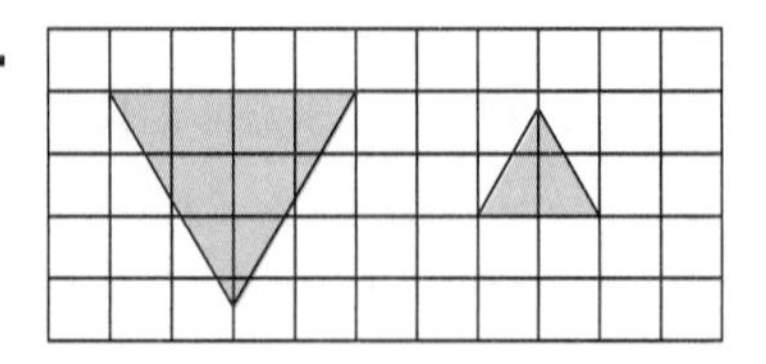

CHECK PROBLEM SOLVING

Solve. (pp. 340–341)

12. Make a list of the number of cubes in each layer of Figure A and Figure B. Are the figures congruent? Explain.

Figure A

Figure B

Standardized Test Prep

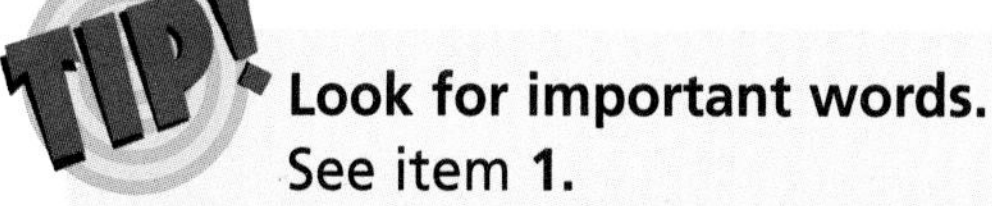

Look for important words.
See item **1.**

Never is an important word. It means the opposite of *always.* Quadrilaterals always have 4 sides and 4 angles, so you need to look for the figure that never has 4 sides and 4 angles.

Also see problem **2,** p. H62.

For 1–7, choose the best answer.

1. Which figure could never be named a quadrilateral?

A triangle **C** rectangle
B parallelogram **D** square

2. Which letter does **not** have a line of symmetry?

F P **H** O
G M **J** V

3. In which figure is each angle less than a right angle?

A **C**
B **D**

4. Which figure is congruent to this figure?

F **H**
G **J**

5. What is five thousand, fifteen written in standard form?

A 5,015 **C** 15,000
B 5,050 **D** NOT HERE

6. Stanley is going to the basketball game. He leaves his house at 6:45 P.M. The game starts at 7:30 P.M. How much time will Stanley have to get to the game?

F 35 minutes **H** 45 minutes
G 40 minutes **J** 1 hour

7. The Redbrook School had a book fair. The sales were \$43.76 the first day, \$21.98 the second day, and \$56.32 the last day. What was the total amount of sales?

A \$122.96 **C** \$121.06
B \$122.06 **D** NOT HERE

Write What You Know

8. How many lines of symmetry does this figure have?

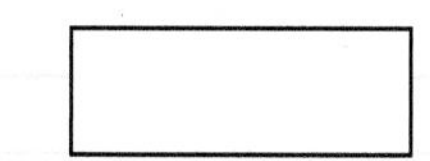

Copy this figure and draw the lines of symmetry. Explain why the lines you drew are lines of symmetry.

9. Are the two figures below congruent? How can you tell?

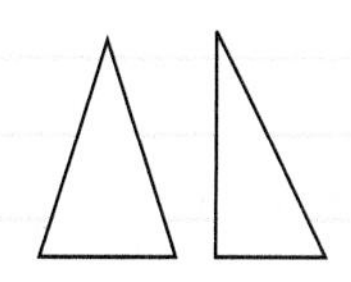

PROBLEM SOLVING
MATH DETECTIVE

On the Edge

These line segments cost 5¢, 10¢, and 25¢ as shown.

5¢ ——   10¢ ————   25¢ ————————————

Use your detective skills to find the cost of each shape.

1.

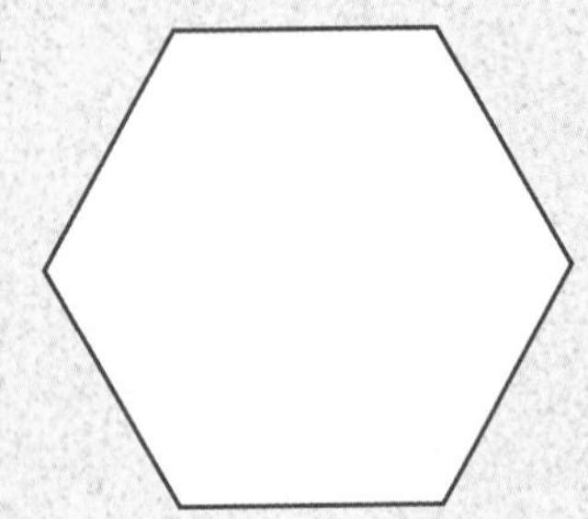

2.

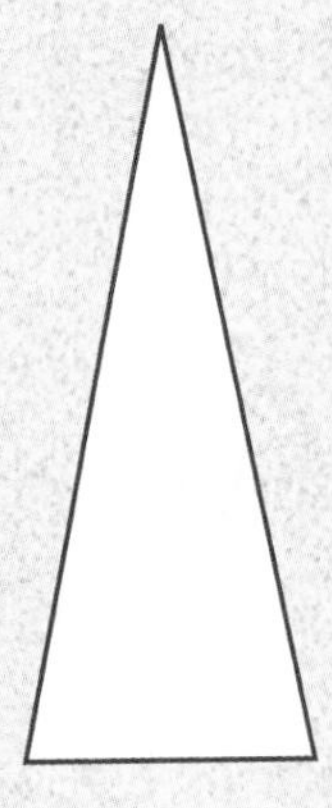

3.

4.

5.

6.

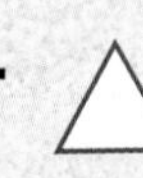

7.

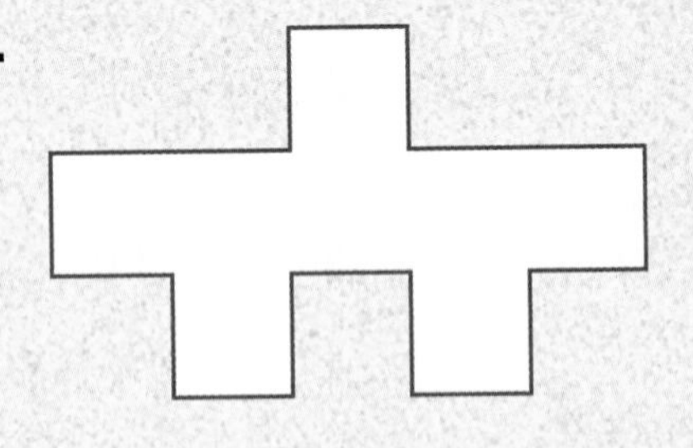

8.

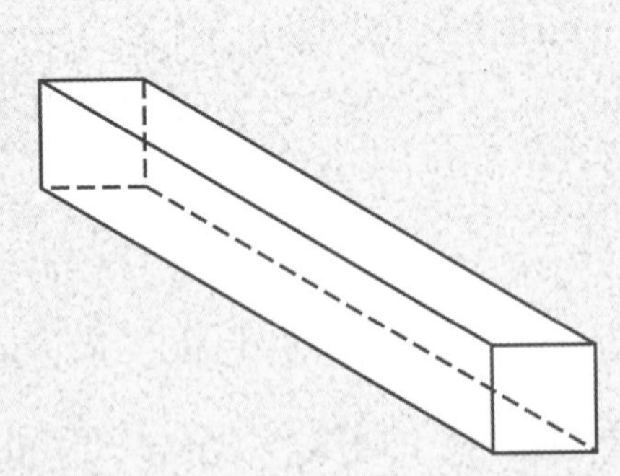

STRETCH YOUR THINKING

9. Draw a shape that costs 35¢.

10. Draw a shape that costs $1.20.

Challenge

Patterns with Plane Figures

hexagon rhombus

trapezoid triangle

Tom and Anna used pattern blocks to make shape patterns.

Tom made this pattern. What are the next three blocks in his pattern?

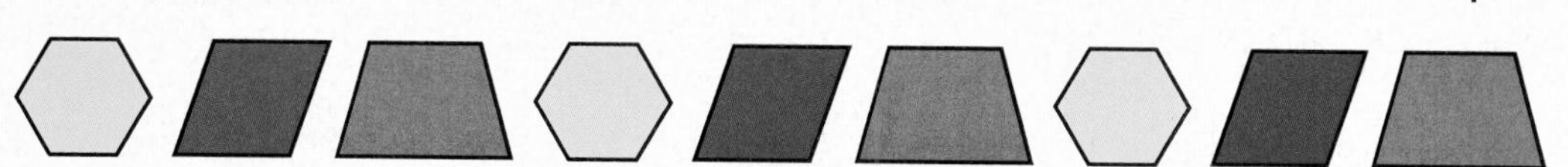

Find where the pattern repeats. Then continue the pattern.

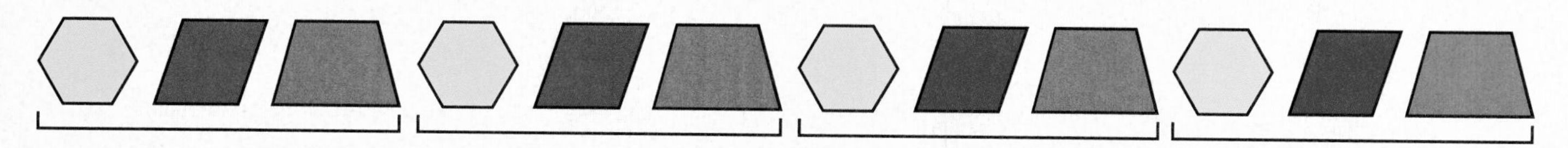

So, the next three blocks are a hexagon, a rhombus, and a trapezoid.

Anna made this pattern. Then she took two of the pieces away.

- What shapes are missing from Anna's shape pattern? Explain how you know.

Try It

1. Tell what shapes are missing in the pattern.

2. Draw the next two shapes in the pattern.

3. Draw a pattern. Ask a classmate to tell what shapes continue the pattern.

Study Guide and Review

VOCABULARY

Choose the best term from the box.

face
polygon
vertex

1. A closed plane figure with straight sides is a __?__. (p. 314)
2. A corner of a solid figure where three or more edges meet is called a __?__. (p. 294)

STUDY AND SOLVE

Chapter 17

Describe solid figures.

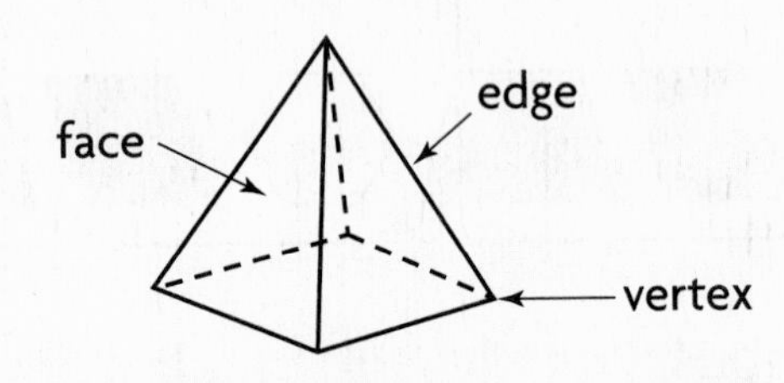

This square pyramid has 5 faces, 8 edges, and 5 vertices. One face is a square, and the other four faces are triangles.

For 3–5, use the figure. (pp. 294–297)

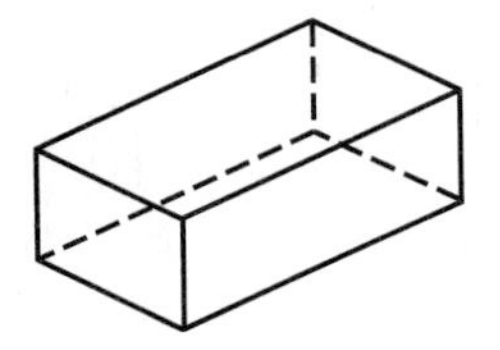

3. How many faces, edges, and vertices does the figure have?
4. What shape are the faces of this figure?
5. What is this figure called?

Tell if an angle is a right angle, greater than a right angle, or less than a right angle.

This angle is a **right angle**. It forms a square corner.

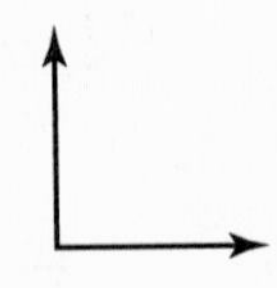

Describe the angles below.

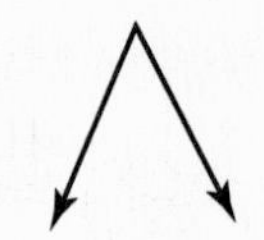

This angle is *less than* a right angle.

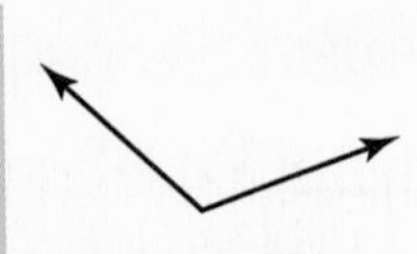

This angle is *greater than* a right angle.

Write whether the angle is a *right angle, greater than* a right angle, or *less than* a right angle. (pp. 300–303)

6. 

7. 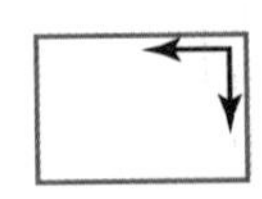

8. 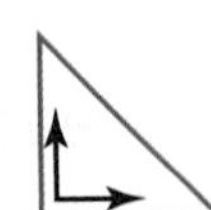

9. 

Chapter 18

Classify triangles and quadrilaterals.

Which figure is an equilateral triangle?

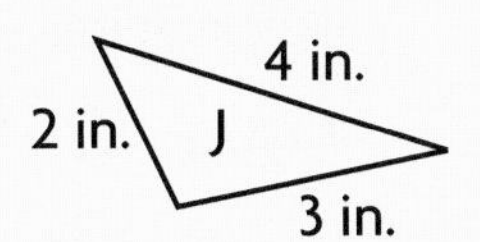

2 in. K 2 in.
2 in.

Figure K is an equilateral triangle. It has 3 equal sides.

Which figure is a parallelogram?

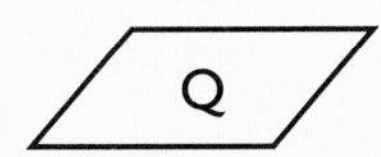

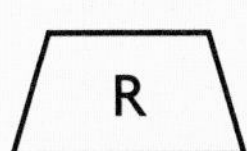

Figure Q is a parallelogram. It has 2 pairs of equal sides and 2 pairs of parallel sides.

For 10–13, use the figures. (pp. 316–323)

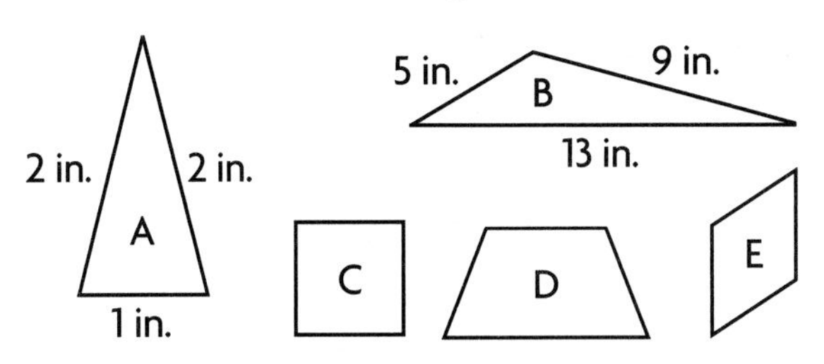

10. Which is a scalene triangle?

11. Which is an isosceles triangle?

12. Which quadrilaterals have 2 pairs of parallel sides?

13. How many right angles does figure C have? figure D?

Chapter 19

Identify lines of symmetry.

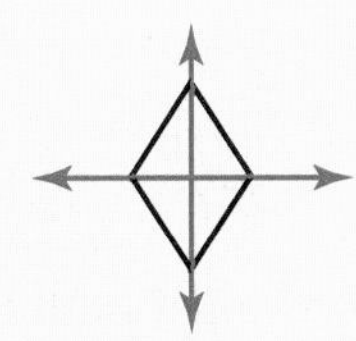

Some figures have more than 1 line of symmetry.

For 14–15, tell how many lines of symmetry each object has. (pp. 334–335)

14.

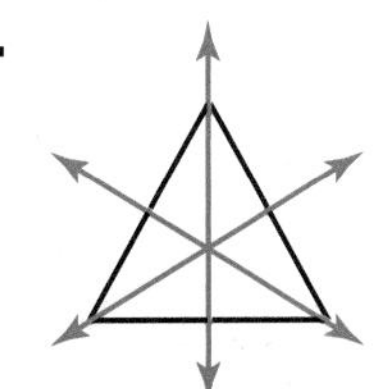

15.

PROBLEM SOLVING PRACTICE

Solve. (pp. 308–309, 326–327)

16. Jeff put a different letter on each small square of this cube. How many letters did he use?

17. Dimitri wrote this number pattern. Describe the pattern. Then write the next 3 numbers in his pattern.
72, 64, 56, 48, ■, ■, ■

PERFORMANCE ASSESSMENT

TASK A • ART CLASS

Matthew is learning how to draw these solid figures in art class.

triangle	point
square	line segment
polygon	angle
rectangle	perpendicular
quadrilateral	right angle
parallelogram	parallel

a. Choose one of the solid figures. Tell how many faces, edges, and vertices it has.

b. Draw each plane figure that is a face of the solid figure you chose. Label each plane figure with its name.

c. Write at least three sentences to describe one of the faces. Use as many of the terms from the box at the top of the page as you can.

TASK B • MATH T-SHIRTS

Materials: pattern blocks, ruler

The math club members want to design special T-shirts to wear on meeting days. These are the rules for the design.

Design Rules

1. The design must be made of pattern-block shapes.
2. The design must be in the shape of a triangle.
3. One side of the triangle must be at least 6 inches long.
4. Congruent shapes must be the same color.

a. Follow all the design rules listed in the box. First, use pattern blocks to make a design for the T-shirt. Then, draw your design.

b. Explain whether your design has any lines of symmetry.

c. Explain how you would enlarge the design for a poster.

E-Lab • Symmetry

A line of symmetry divides a plane figure into 2 congruent parts. You can fold a figure along its line of symmetry, and both sides will match.

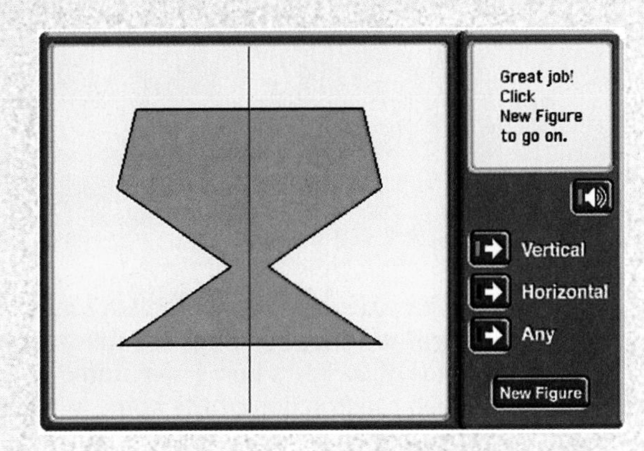

You can use E-Lab to practice drawing lines of symmetry.

- Click *Symmetry*.
- Click *Vertical*. Then click *New Figure* to begin.
- Drag the line of symmetry to the right so that both halves match.
- Draw a picture of the figure and its line of symmetry.

Is any straight line through a figure a line of symmetry? Explain.

Practice and Problem Solving

Use E-Lab to solve.

1. Click *Horizontal*. Then click *New Figure*. Drag the line of symmetry to the correct position. Draw a picture of the figure and its line of symmetry.
2. Click *Any*. Then click *New Figure*. Drag a line of symmetry to the correct position. Draw a picture of the figure and its line of symmetry.

Solve.

3. **What's the Error?** Erika drew the figure at the right and its line of symmetry. What's wrong with Erika's drawing?

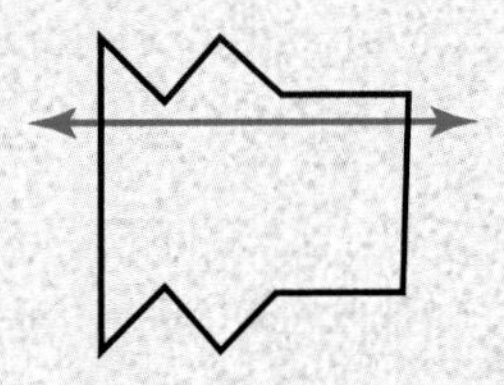

Multimedia Math Glossary **www.harcourtschool.com/mathglossary**

4. **Vocabulary** Visit the Multimedia Math Glossary to learn more about *lines of symmetry* and *congruent figures*. Draw 3 figures, each with a different number of lines of symmetry. You may color your figures.

More than 400 of Wright's buildings are still standing today. These are photos of Wright's home and studio in Oak Park, Illinois. He began construction of the home in 1889. The studio was completed in 1898.

PROBLEM SOLVING ON LOCATION

with Frank Lloyd Wright

IN OAK PARK, ILLINOIS

Frank Lloyd Wright lived from 1867 to 1959. During that time he designed and built more than 500 homes, museums, and office buildings.

1. Look at the different views of the house. Find a polygon that has a line of symmetry. Draw the polygon and show a line of symmetry.
2. Describe a pair of congruent figures in one of the photos. Tell how you know the figures are congruent.
3. Look for an example of a right angle, an angle less than a right angle, and an angle greater than a right angle. Trace the angles from the photos and label each angle.
4. Look at the photos and find one example of parallel lines and one example of intersecting lines. Draw the lines.

IN MILL RUN, PENNSYLVANIA

Frank Lloyd Wright was an architect perhaps best known for houses he designed. Fallingwater is the weekend retreat he built for the Edgar J. Kaufmann family in 1936. It is now open for public tours mid-March through November.

For 1–3, use the picture.

1. Name a solid figure in the photo. Tell what part of the photo shows this figure.
2. How many faces, vertices, and edges does the solid figure you named in Exercise 1 have?
3. Write three sentences to describe the plane figure in the photo. Use as many of the words from the box below as you can.

triangle	parallelogram	line segment	right angle	square
polygon	quadrilateral	rectangle	circle	intersecting

Frank Lloyd Wright designed houses and buildings that seemed to be part of the landscape. In Fallingwater, the stream flows parallel to and partially under the house. The horizontal levels appear to be floating amid the rock ledges.

20 Customary Units

On July 21, 1969, Neil Armstrong became the first person to set foot on the moon's surface. On Earth, objects weigh 6 times as much as they weigh on the moon.

PROBLEM SOLVING Look at the table. What would the moon rocks weigh on Earth?

DATA LINK

WEIGHTS OF OBJECTS ON THE MOON

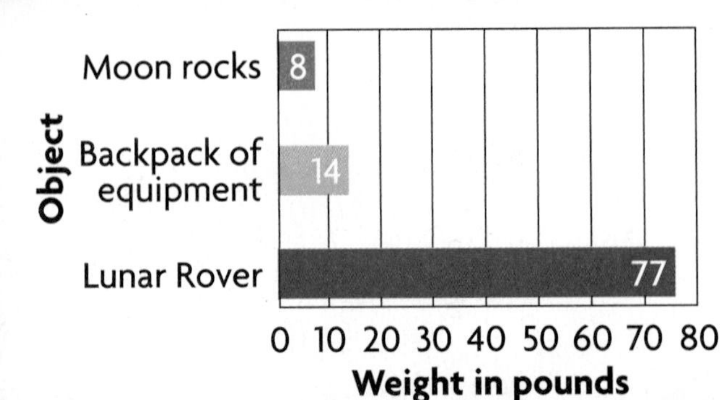

Use this page to help you review and remember important skills needed for Chapter 20.

inch
sum

VOCABULARY

Choose the correct term from the box.

1. An __?__ is a customary unit used to measure length.

USE A CUSTOMARY RULER (For Intervention, see p. H26.)

For 2–3, use the picture at the right.

2. Is nail A less than or greater than 2 inches long?
3. Which nail is about 3 inches long?

A B

MEASURE TO THE NEAREST INCH (For Intervention, see p. H26.)

Write the length to the nearest inch.

4. 

5.

inches 1 2 3 4 5

FIND A RULE (For Intervention, see p. H27.)

Write a rule for the table. Then copy and complete the table.

6.

gloves	1	2	3	4	5	6
fingers	5	10	15	■	■	■

7.

rings	1	2	3	4	5	6
cost	\$3	\$6	\$9	■	■	■

8.

butterflies	1	5	3	4	7	9
wings	2	10	6	■	■	■

9.

packs	3	2	7	5	1	8
crackers	21	14	49	■	■	■

LESSON 1

Length

Quick Review

1. 7 × ■ = 42
2. ■ × 5 = 45
3. ■ × 8 = 32
4. 6 × ■ = 48
5. 9 × ■ = 63

Learn

HOW LONG? An estimate is an answer that is close to the actual answer. You can estimate length by using an item close to 1 inch, like a small paper clip or your knuckle.

Remember

To use a ruler:

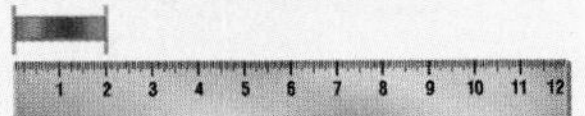

- line up one end of the object with the end of the ruler.
- find the inch mark closest to the object's other end.

The ribbon is 2 inches long to the nearest inch.

HANDS ON

Activity

MATERIALS: paper clips, ruler

STEP 1

Copy the table.

LENGTHS OF RIBBONS		
Color	Estimate	Measure
green	about 1 inch	1 inch
blue		
purple		
orange		
red		

STEP 2

Use paper clips to estimate the length of the blue ribbon. Record your estimate in the table.

STEP 3

Use a ruler. Measure the length of the blue ribbon to the nearest inch. Record your measurement in the table.

STEP 4

Repeat Steps 2 and 3 for the purple, orange, and red ribbons.

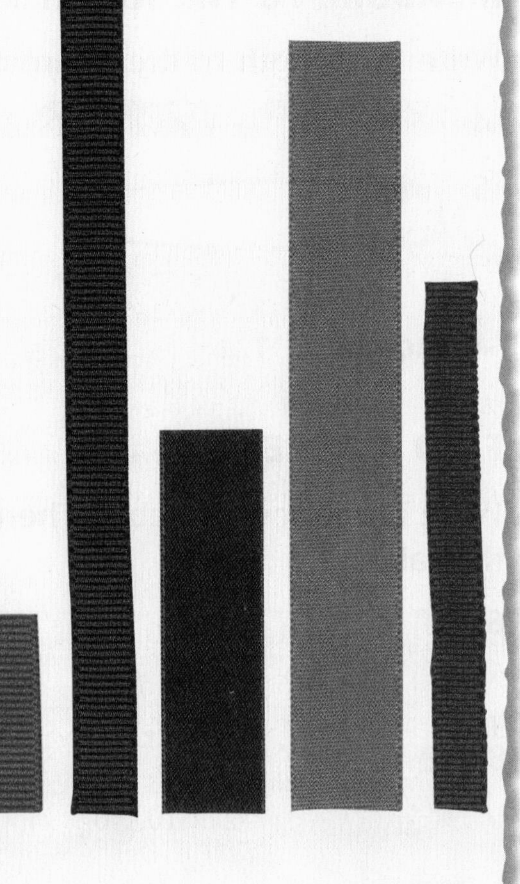

- Why does it make sense to use a small paper clip to estimate inches?

Measuring

You can also measure to the nearest half inch.

Example

What is the length of this crayon to the nearest half inch?

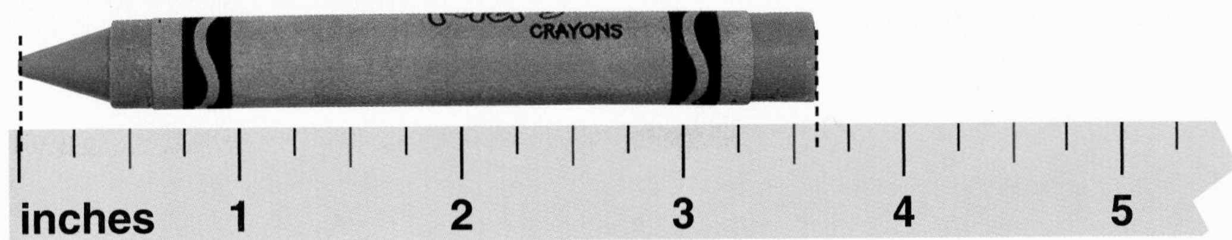

To measure to the nearest half inch:

STEP 1

Line up one end of the crayon with the left side of the ruler.

STEP 2

Find the $\frac{1}{2}$-inch mark that is closest to the other end of the crayon.

So, the length of the crayon to the nearest half inch is $3\frac{1}{2}$ inches.

Technology Link

To learn more about customary measurement, watch the Harcourt Math Newsroom Video *Blue Whale Tracking.*

Check

1. **Describe** where you find $\frac{1}{2}$-inch marks on a ruler.

Estimate the length in inches. Then use a ruler to measure to the nearest inch.

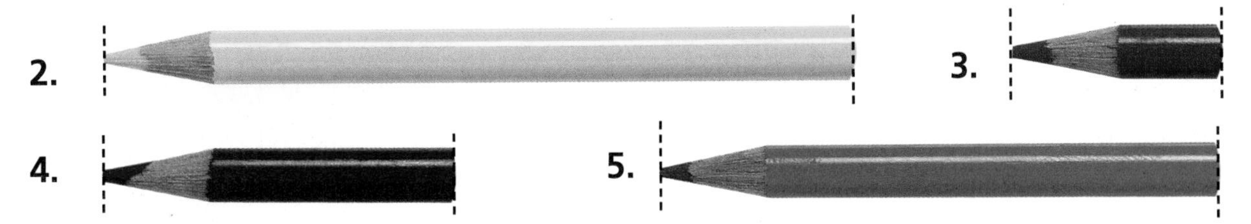

2.
3.
4.
5.

Measure the length to the nearest half inch.

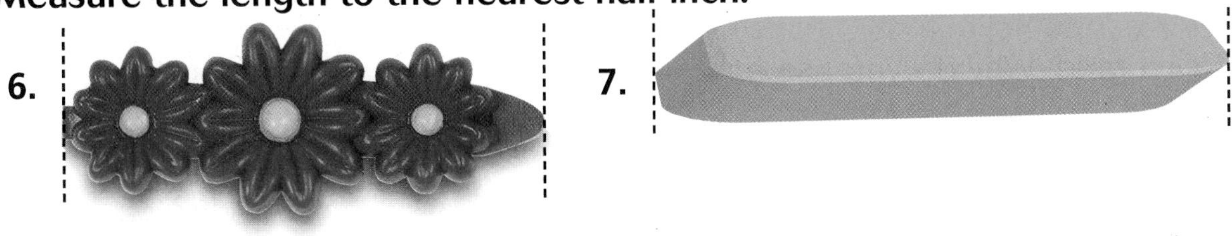

6.
7.
8.

LESSON CONTINUES

Practice and Problem Solving

Estimate the length in inches. Then use a ruler to measure to the nearest inch.

9.

10.

11.

Measure the length to the nearest inch.

12. 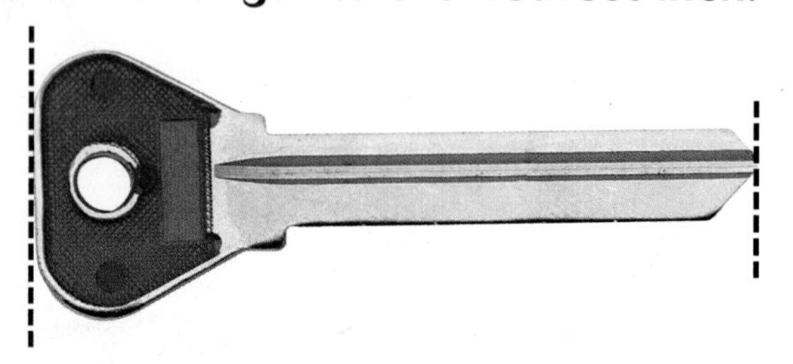

13. 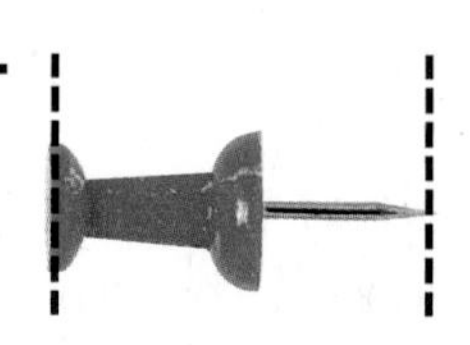

Measure the length to the nearest half inch.

14. 

15. 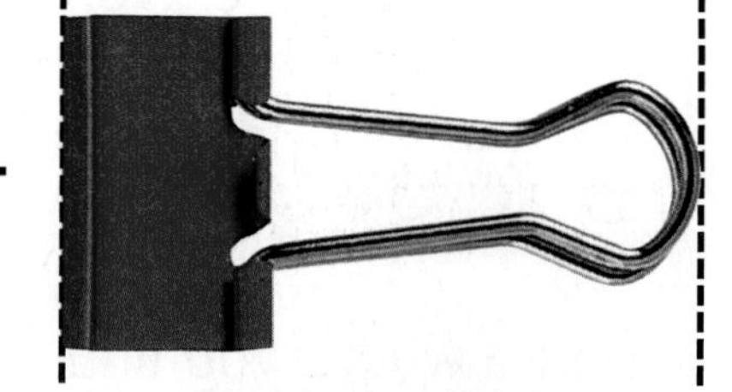

16.

17. 

18.

Use a ruler. Draw a line for each length.

19. 1 inch **20.** $2\frac{1}{2}$ inches **21.** 4 inches **22.** $5\frac{1}{2}$ inches

23. **REASONING** Why do you use a ruler instead of your knuckle to measure inches?

24.  **What's the Question?** Lori had 2 rolls of ribbon. Each roll had 12 inches of ribbon. After she used some ribbon for a costume, she had 18 inches left. The answer is 6 inches.

25. Joyce used 72 beads to make 9 necklaces with an equal number of beads on each. How many beads were on 2 necklaces?

26. A brush measures $6\frac{1}{2}$ inches. Between which two inch marks does the end of the brush lie?

27. Without using a ruler, draw a line about 8 inches long. Measure the line you drew to the nearest inch. How close was your line to 8 inches?

28. Suppose you need at least 5 inches of yarn for an art project. Is this yellow piece of yarn long enough? Explain.

Mixed Review and Test Prep

29. How many faces does the solid figure have? (p. 294)

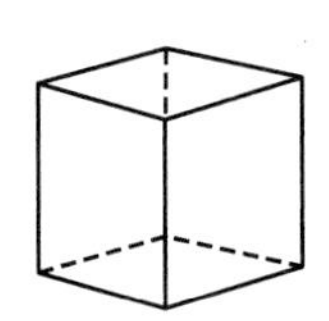

30. TEST PREP Noah had 8 rows of 6 toy cars. He gave 5 cars to his brother. How many cars does Noah have left? (p. 172)

A 58 **B** 53 **C** 43 **D** 19

31. $3 \times 1 \times 4 = ■$ (p.170)

32. $5 \times 2 \times 4 = ■$ (p.170)

33. TEST PREP The movie begins at 11:15 A.M. It is 2 hours and 10 minutes long. What time will it end? (p. 98)

F 11:25 A.M. **H** 1:15 P.M.
G 12:25 P.M. **J** 1:25 P.M.

PROBLEM SOLVING

Thinker's Corner

Measure the pieces of yarn and break the code! To find the correct letter for each blank, match the measurement and the color of the yarn.

W
T
T
H

S
E
C

R
A
E

What did the mother bird call the baby bird?

?	?	?	?	?	?	?	?	?	?
$1\frac{1}{2}$ inches	2 inches	1 inch	1 inch	$1\frac{1}{2}$ inches	2 inches	1 inch	$\frac{1}{2}$ inch	$\frac{1}{2}$ inch	$1\frac{1}{2}$ inches

EXTRA PRACTICE page H51, Set A

LESSON 2

Inch, Foot, Yard, and Mile

Quick Review

1. 41 + 52
2. 38 + 39
3. 12 + 34
4. 27 + 63
5. 12 + 12 + 12

VOCABULARY

foot (ft)
yard (yd)
mile (mi)

Learn

CHOOSE AND USE You know that an inch (in.) is used to measure length and distance. Other customary units used to measure length and distance are the **foot (ft)**, **yard (yd)**, and **mile (mi)**.

Example

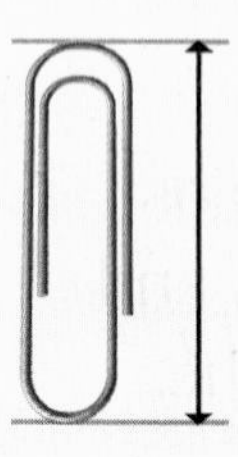

A paper clip is about 1 inch long.

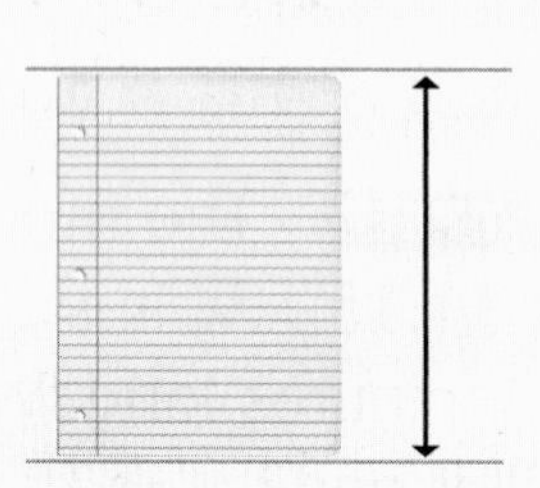

A piece of notebook paper is about 1 foot long.

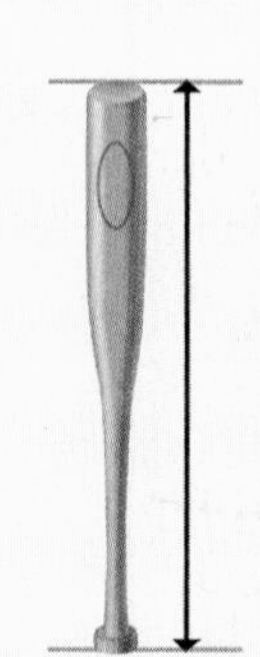

A baseball bat is about 1 yard long.

You can walk 1 mile in about 20 minutes.

TABLE OF MEASURES
1 foot = 12 inches
1 yard = 3 feet = 36 inches
1 mile = 5,280 feet

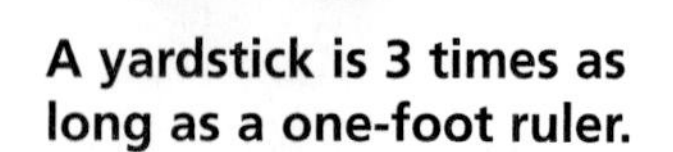

A yardstick is 3 times as long as a one-foot ruler.

• Explain which unit you would use to measure the length of your hand.

Check

1. **Write** the names of two objects in your classroom you would measure using feet.

Choose the unit you would use to measure each. Write *inch, foot, yard,* or *mile*.

2. the length of a pencil
3. the length of a butterfly
4. the length of a parking lot
5. the distance a train goes in 30 minutes

Practice and Problem Solving

Choose the unit you would use to measure each.
Write *inch, foot, yard,* or *mile*.

6. the height of a refrigerator
7. the length of your sneaker
8. the distance between your classroom and the cafeteria
9. the distance you could walk in two hours
10. the length of the school gym
11. the length of a spoon

Choose the best unit of measure.
Write *inches, foot* or *feet* , *yards*, or *miles*.

12. Sal rides the bus 3 _?_ to school.
13. A football is about 1 _?_ long.
14. The distance between the floor and the doorknob is about 3 _?_ .
15. Sarah's math book is 11 _?_ long.
16. Angie thinks this grasshopper is about 4 inches long. Do you agree with her estimate? Measure to check and record the length.

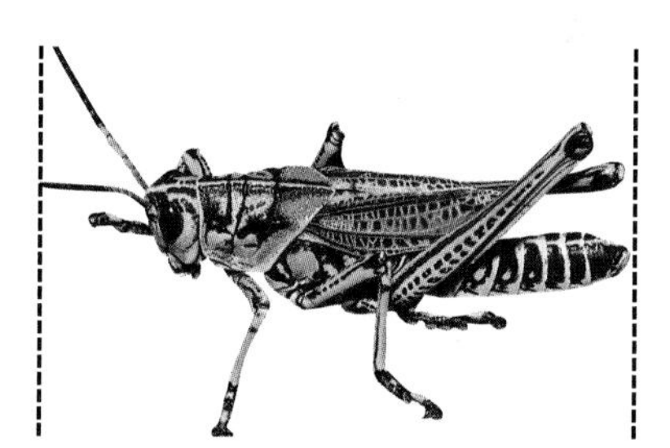

17. Mitchell got 5 stickers from each of 6 friends. He bought 13 more stickers. How many stickers does Mitchell have in all?
18. **Write About It** Estimate the distance from your desk to the classroom door in yards. Then measure the distance. Record your estimate and the actual measurement.

Mixed Review and Test Prep

For 19–22, use the graph. (p. 254)

19. How many votes are for red?
20. Which color has the most votes?
21. What is the total number of votes?
22. What kind of graph is this?

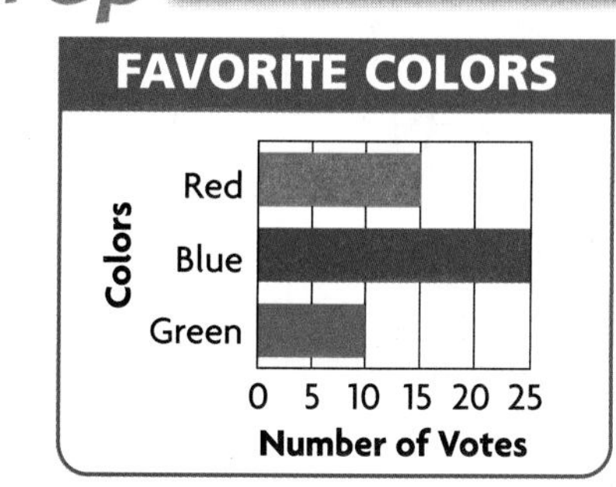

23. **TEST PREP** 4,623 – 2,307 (p. 62)

A 2,314 **C** 2,916
B 2,316 **D** 6,930

LESSON 3

Capacity

Quick Review

1. $2\overline{)8}$
2. $4\overline{)16}$
3. $4\overline{)8}$
4. $4\overline{)32}$
5. $2\overline{)16}$

VOCABULARY
capacity
cup (c)
pint (pt)
quart (qt)
gallon (gal)

MATERIALS
cup, pint, quart, and gallon containers; water, rice, or beans

Explore

Capacity is the amount a container can hold. **Cup (c)**, **pint (pt)**, **quart (qt)**, and **gallon (gal)** are customary units for measuring capacity.

cup (c)

pint (pt)

quart (qt)

gallon (gal)

Activity

Copy the table to help find how many cups are in a pint, a quart, and a gallon.

NUMBER OF CUPS		
	Estimate	Measure
Cups in a pint?		
Cups in a quart?		
Cups in a gallon?		

STEP 1 Estimate the number of cups it will take to fill the pint container. Record your estimate.

STEP 2 Fill a cup and pour it into the pint container. Repeat until the pint container is full.

STEP 3 Record the actual number of cups it took to fill the pint container.

STEP 4 Repeat Steps 1–3 for the quart and the gallon containers.

Try It

a. How many pints does it take to fill a quart?

b. How many pints does it take to fill a gallon?

We are using pints to fill a quart. How many pints will it take?

Connect

How are cups, pints, quarts, and gallons related?

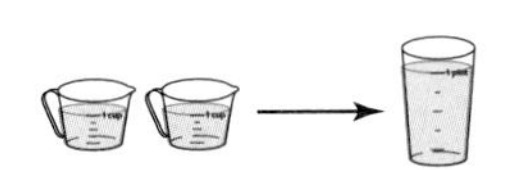

2 cups in 1 pint

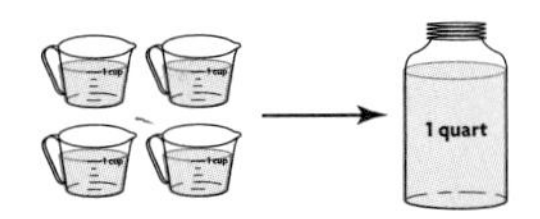

4 cups in 1 quart

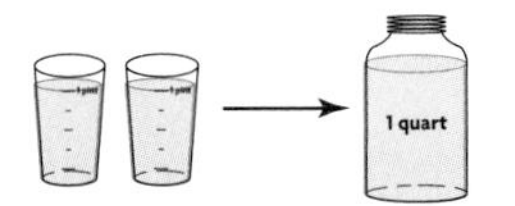

2 pints in 1 quart

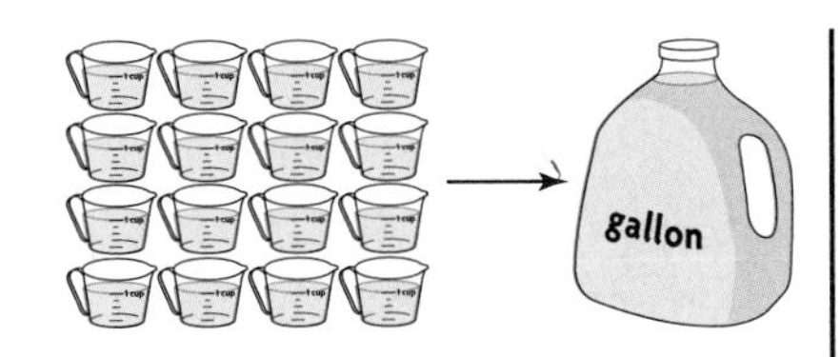

16 cups in 1 gallon

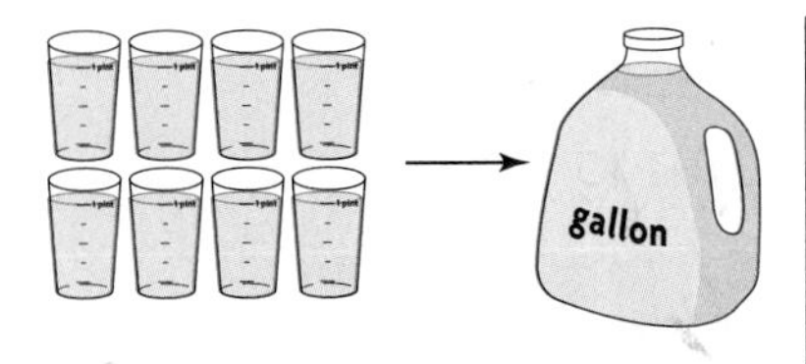

8 pints in 1 gallon

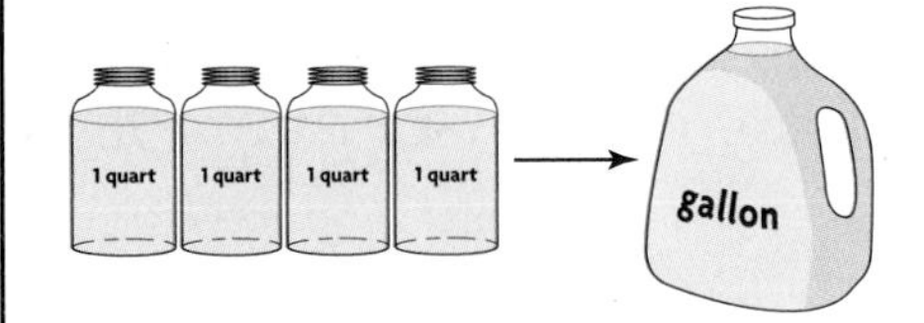

4 quarts in 1 gallon

Practice and Problem Solving

Choose the better estimate.

1.

1 cup or
1gallon?

2.

3 cups or
3 quarts?

3.

2 cups or
2 quarts?

4.

10 pints or
10 gallons?

Compare. Write <, >, or = for each ●.

5. 1 cup ● 3 pints

6. 1 gallon ● 3 quarts

7. 4 pints ● 1 quart

8. **ESTIMATION** Leticia estimates that she needs about 2 quarts of juice to serve 7 cups. Do you agree? Explain.

9. Write these units of capacity in order from least to greatest: *pint, cup, gallon, quart.*

Mixed Review and Test Prep

Multiply. (p. 148)

10. $6 \times 5 = \blacksquare$

11. $6 \times 9 = \blacksquare$

Divide. (p. 216)

12. $56 \div 7 = \blacksquare$

13. $49 \div 7 = \blacksquare$

14. **TEST PREP** Donato gave away 23 marbles. He bought 16 more. Then he had 361 marbles. How many did he have to begin with? (p. 42)

A 368 **B** 345 **C** 322 **D** 300

LESSON 4

Weight

Quick Review

Compare. Use <, >, or = for each ●.

1. $16 \div 8$ ● 2×2
2. 10 ● $24 \div 3$
3. $81 \div 9$ ● 9
4. $18 \div 3$ ● $10 \div 2$
5. 13 ● 3×4

VOCABULARY
ounce (oz) pound (lb)

MATERIALS
simple balance
pennies

Explore

An **ounce (oz)** and a **pound (lb)** are customary units for measuring weight.

9 pennies weigh about 1 ounce.

144 pennies weigh about 1 pound.

You can estimate and then weigh objects to decide if they weigh about 1 ounce or about 1 pound.

Activity

STEP 1

Place 9 pennies on one side of a balance to show 1 ounce. Find two objects that you think weigh about 1 ounce each. Weigh them to check.

STEP 2

Record what your objects are and whether they weigh more than, less than, or the same as 1 ounce.

STEP 3

Place 144 pennies on one side of the balance to show 1 pound. Find two objects that you think weigh about 1 pound each. Weigh them to check.

STEP 4

Record what your objects are and whether they weigh more than, less than, or the same as 1 pound each.

- Give an example of a small object that is heavier than a large object.

45 pennies weigh about 5 oz. What things in your classroom weigh about 5 oz?

Try It

Name an object that weighs about each amount.

a. 5 ounces **b.** 3 pounds

Connect

How are pounds and ounces related?

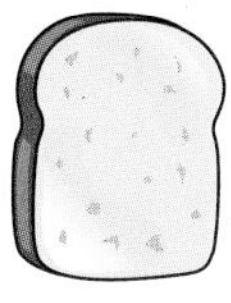 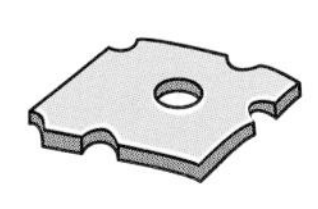 

These things each weigh about 1 ounce.

A loaf of bread weighs about 1 pound.

16 ounces = 1 pound

Practice and Problem Solving

Choose the unit you would use to weigh each. Write *ounce* or *pound*.

1. 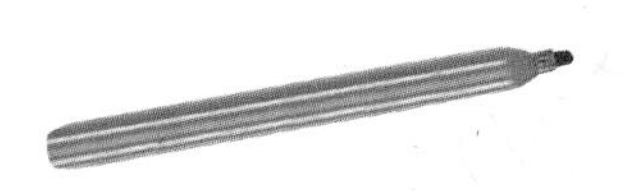

2. 

3. 

4. 

Choose the better estimate.

5. 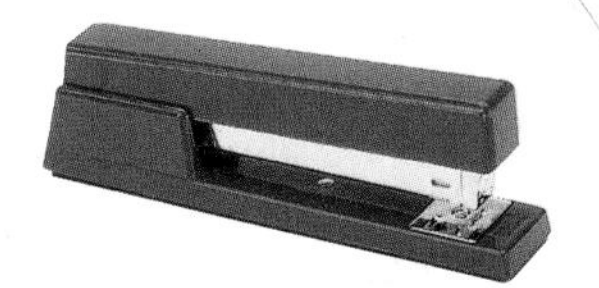

1 ounce or 1 pound?

6. 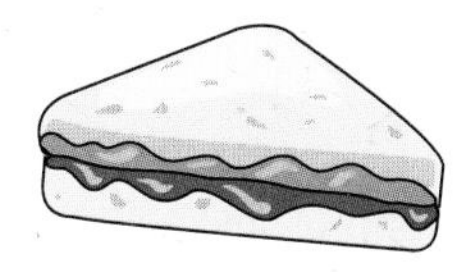

3 ounces or 3 pounds?

7. 

2 ounces or 2 pounds?

8. 

2 ounces or 2 pounds?

9. **REASONING** Bill has 24 cookies divided equally into 6 bags. How many are in 3 bags?

10. Mr. Reynolds cooked three 6-ounce hamburgers. Did he use more than one pound of ground meat? Explain.

11. **Write a problem** about items from your home. Use pounds and ounces in your problem.

12. $1 \times 5 \times 7 = ■$ (p. 170)

13. $3 \times 2 \times 8 = ■$ (p. 170)

14. $2 \times 3 \times 3 = ■$ (p. 170)

15. $12 + 16 + 8 = ■$ (p. 36)

16. **TEST PREP** Which shape is shown? (p. 314)

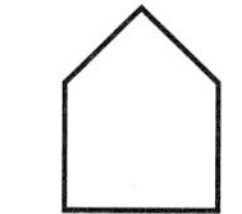

A square
B pentagon
C circle
D rectangle

5 Ways to Change Units

Quick Review

Choose the smaller measurement in each.

1. pound, ounce
2. cup, pint
3. quart, cup
4. yard, foot
5. pint, quart

Learn

CHANGE IT The students in Mrs. Lopez's class need 32 cups of juice for a picnic. How many quarts of juice should they buy?

To change cups into quarts, they must know how these units are related.

- A quart is larger than a cup.
- 4 cups = 1 quart

cup (c)

quart (qt)

Remember

Table of Measures

Length
12 inches = 1 foot
3 feet = 1 yard

Capacity
2 cups = 1 pint
4 cups = 1 quart
2 pints = 1 quart
8 pints = 1 gallon
4 quarts = 1 gallon

Jake and Theresa used different ways to change cups into quarts.

Jake
I'll draw 32 cups. I'll circle groups of 4 to show quarts.

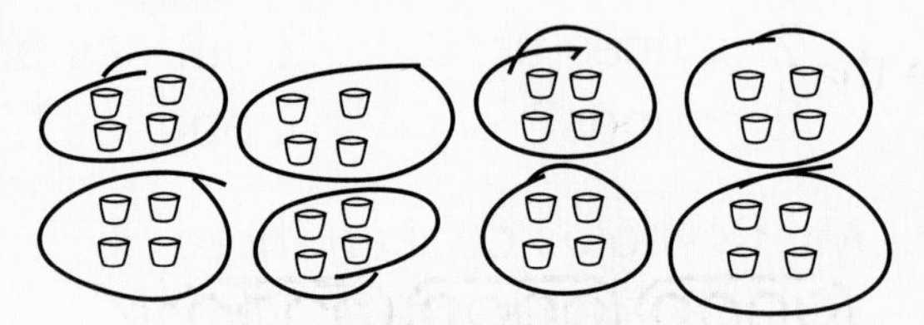

There are 8 groups of 4 cups. So, 32 cups equals 8 quarts.

Theresa
I'll make a table.

Quarts	1	2	3	4	5	6	7	8
Cups	4	8	12	16	20	24	28	32

The table shows that 8 quarts equals 32 cups.

So, they should buy 8 quarts of juice.

MATH IDEA To change one unit into another, first decide how the units are related.

- Explain why Jake drew circles around groups of 4 cups.

Check

1. **Write** how many pints are in 1 quart. Use the Table of Measures on page 362 to help.

Copy and complete. Use the Table of Measures to help.

2. Change gallons to pints.
   larger unit ___?___
   1 gallon = ■ pints

3. Change feet to inches.
   larger unit ___?___
   1 foot = ■ inches

Practice and Problem Solving

Copy and complete. Use the Table of Measures to help.

4. Change yards to feet.
   larger unit ___?___
   1 yard = ■ feet

5. Change quarts to gallons.
   larger unit ___?___
   ■ quarts = 1 gallon

Change the units. Use the Table of Measures to help.

6. ■ cups = 1 pint

   12 cups = ■ pints

7. ■ inches = 1 foot

feet	1
inches	12

   ■ inches = 3 feet

8. Callie has 23 inches of yarn. Is this more than or less than 2 feet? Explain.

9. **What's the Error?** Dylan drew this model to find how many gallons equal 24 pints. Describe his error. Draw the correct model.

(OOOO) (OOOO) (OOOO)
(OOOO) (OOOO) (OOOO)

Mixed Review and Test Prep

Multiply. (p. 164)

10. $9 \times 4 =$ ■
11. $9 \times 7 =$ ■

Divide. (p. 204)

12. $12 \div 4 =$ ■
13. $28 \div 4 =$ ■

14. **TEST PREP** Yolanda had \$2.19. She lost a quarter. She wants to buy a book for \$2.15. How much more does she need? (p. 88)

A \$0.24
B \$0.21
C \$0.11
D \$0.04

EXTRA PRACTICE page H51, Set C

LESSON 6

Algebra: Rules for Changing Units

Quick Review

1. $4 \times 3 = ■$
2. $6 \times 2 = ■$
3. $2 \times 8 = ■$
4. $5 \times 4 = ■$
5. $4 \times 6 = ■$

Learn

WHAT'S THE RULE? You can use rules to help you change units. To make a rule for changing units, you must first decide how the units are related.

Examples

A Use a rule to find the number of quarts in 5 gallons.

Remember: 4 quarts = 1 gallon

To change a larger unit into a smaller unit, multiply.

gallons	1	2	3	4	5
quarts	4	8	12	16	20

Rule:
Multiply the number of gallons by 4.

■ quarts = 5 gallons
$■ = 5 \times 4$
$20 = 5 \times 4$
So, there are 20 quarts in 5 gallons.

B Use a rule to find the number of gallons in 12 quarts.

Remember: 4 quarts = 1 gallon

To change a smaller unit into a larger unit, divide.

quarts	4	8	12
gallons	1	2	3

Rule:
Divide the number of quarts by 4.

■ gallons = 12 quarts
$■ = 12 \div 4$
$3 = 12 \div 4$
So, there are 3 gallons in 12 quarts.

Check

1. **Discuss** why you multiply to change a larger unit into a smaller unit.

Use the rules to change the units. (2 pints = 1 quart)

2. How many pints are in 5 quarts?

Rule: Multiply the number of quarts by 2.

■ $= 5 \times 2$

■ pints = 5 quarts

3. How many quarts are in 8 pints?

Rule: Divide the number of pints by 2.

■ $= 8 \div 2$

■ quarts = 8 pints

Practice and Problem Solving

Use the rules to change the units. (4 cups = 1 quart)

4. How many quarts are in 24 cups?

Rule: Divide the number of cups by 4.

■ $= 24 \div 4$

■ quarts = 24 cups

5. How many cups are in 7 quarts?

Rule: Multiply the number of quarts by 4.

■ $= 7 \times 4$

■ cups = 7 quarts

Write a rule and change the units. You may make a table to help.

6. How many feet are in 3 yards?

■ feet = 3 yards

7. How many yards are in 12 feet?

■ yards = 12 feet

8. How many pints are in 16 cups?

■ pints = 16 cups

9. How many cups are in 6 pints?

■ cups = 6 pints

10. **REASONING** Gail is 5 feet tall. How many inches tall is she? Copy and complete the table at the right.

feet	1	2	3	4	5
inches	12				

Mixed Review and Test Prep

Divide. (p. 206)

11. $3 \div 1 =$ ■

12. $4 \div 4 =$ ■

13. $0 \div 1 =$ ■

14. $1 \div 1 =$ ■

15. **TEST PREP** What is the sum of 416 and 295? (p. 42)

A 121

B 611

C 701

D 711

EXTRA PRACTICE page H51, Set D

LESSON 7

Problem Solving Skill
Use a Graph

UNDERSTAND | PLAN | SOLVE | CHECK

Quick Review

1. $3 \times 4 = \blacksquare$
2. $6 \times 4 = \blacksquare$
3. $4 \times 2 = \blacksquare$
4. $3 \times 2 = \blacksquare$
5. $5 \times 4 = \blacksquare$

GRAPHING GALLONS This graph shows how quarts and gallons are related. Stacy and Ethan used the graph to decide how many quarts are in 6 gallons.

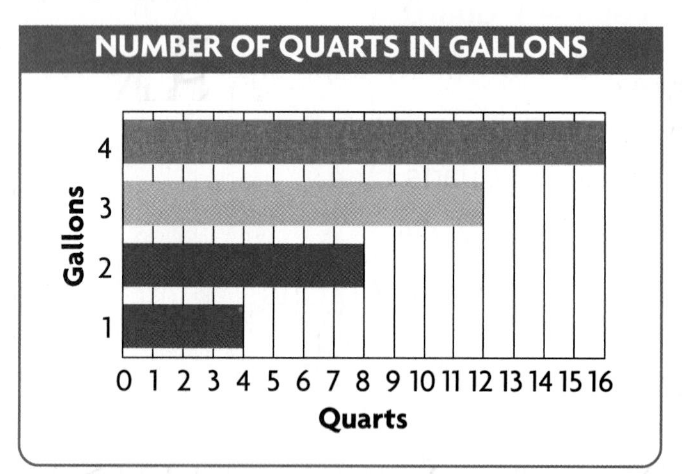

Stacy
By looking at the purple bar, I can see that 4 quarts = 1 gallon. I'll write a rule to change gallons to quarts.

Rule:
Multiply the number of gallons by 4.
■ quarts = 6 gallons x 4
■ quarts = 6 x 4
24 = 6 x 4

So, there are 24 quarts in 6 gallons.

Ethan
I'll look for a pattern in the lengths of the bars.

1 gallon = 4 quarts
2 gallons = 8 quarts
3 gallons = 12 quarts
4 gallons = 16 quarts

Each bar increases by 4.
5 gallons is 16 + 4 = 20.
6 gallons is 20 + 4 = 24.

So, there are 24 quarts in 6 gallons.

Talk About It

- How could Stacy tell that there are 4 quarts in 1 gallon?

Problem Solving Practice

USE DATA For 1–5, use the graph.

1. Explain how you can use the graph to find how many cups are in 1 pint.
2. How many cups are in 4 pints?
3. How many pints are in 6 cups?

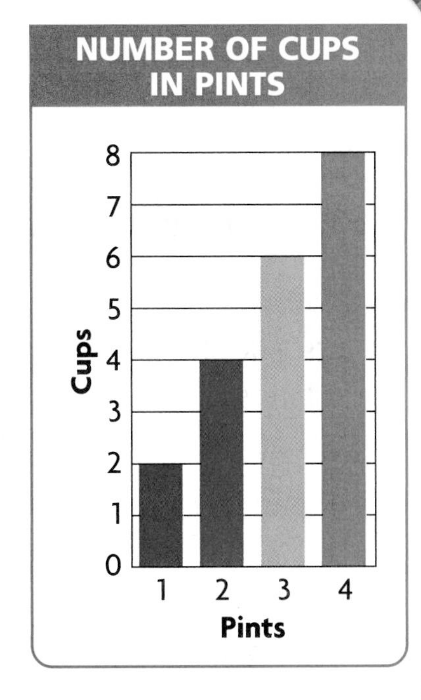

Choose the letter of the correct answer.

4. Which rule could you use to help find how many cups are in 5 pints?
   - **A** Multiply the number of pints by 4.
   - **B** Multiply the number of pints by 2.
   - **C** Multiply the number of gallons by 4.
   - **D** Multiply the number of cups by 3.
5. How many cups are in 5 pints?
   - **F** 5 cups
   - **G** 10 cups
   - **H** 15 cups
   - **J** 20 cups

Mixed Applications

Solve.

6. Alan had 6 boxes of cookies. Each box had 8 cookies. He sold 2 boxes. How many cookies did he have left?
7. A yard of fabric costs $3. Faye bought 4 yards of fabric and 11 inches of ribbon. How much did she pay for the fabric?

8. Mr. Castillo delivered 3 cases of pasta to each of 3 stores. Each case had 8 boxes of pasta. How many boxes of pasta did he deliver?
9. Toni went fishing every day for 2 weeks. During the third week she went fishing on Monday and Friday. How many days did Toni go fishing in the three weeks?

Review/Test

CHECK CONCEPTS

Choose the better estimate. (pp. 358–361)

1. 

1 ounce or
1 pound?

2. 

2 pints or
2 gallons?

3. 

30 ounces or
30 pounds?

CHECK SKILLS

Measure the length to the nearest inch. (pp. 352–355)

4. 

5. 

Choose the unit you would use to measure each.
Write *inch, foot, yard,* or *mile*. (pp. 352–355)

6. the length of a marker

7. the height of a wall

Write the rule and change the units.
You may make a table to help. (pp. 356–357)

8. How many feet are in 2 yards?

■ feet = 2 yards

9. How many pints are in 8 cups?

■ pints = 8 cups

CHECK PROBLEM SOLVING

USE DATA For 10–12, use the graph. (pp. 366–367)

10. How many pints are in 1 quart?

11. How many pints are in 4 quarts?

12. How many quarts are in 4 pints?

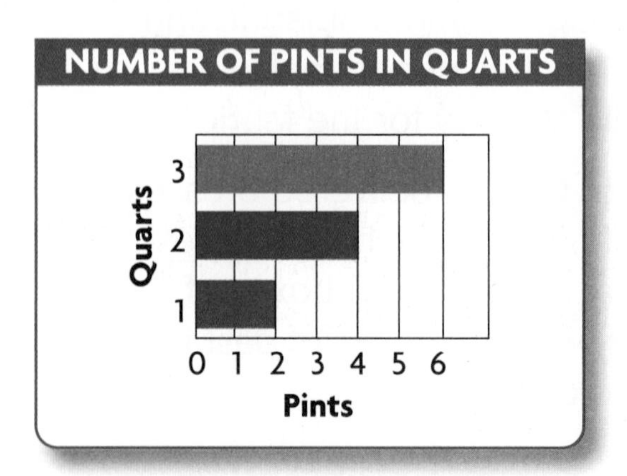

Standardized Test Prep

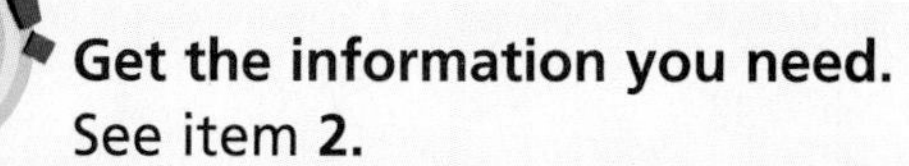

Get the information you need. See item **2.**

You are expected to know which units are used to measure weight. Think about a soccer ball you have held then find an amount that is reasonable.

Also see problem **3,** p. H63.

For 1–8, choose the best answer.

1. Measure the length of the string to the nearest inch.

A 1 inch
B 2 inches
C 3 inches
D 4 inches

2. Which is the most reasonable estimate for the weight of a soccer ball?

F 1 pound
G 1 ounce
H 1 cup
J 1 inch

3. How many inches are in 4 feet?

feet	1	2	3	4
inches	12	24	36	■

A 4 inches
B 12 inches
C 48 inches
D 52 inches

4. 4,872 − 3,469 = ■

F 1,417
G 1,413
H 1,403
J NOT HERE

5. Which statement is true?

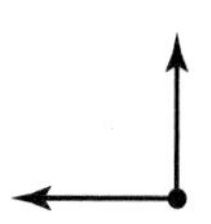

A This angle is less than a right angle.
B This angle is a right angle.
C This angle is greater than a right angle.
D NOT HERE

6. There are 8 pints in 1 gallon. How many pints are in 3 gallons?

F 24 pints
G 16 pints
H 8 pints
J NOT HERE

7. How many feet are in 3 yards?

A 3 ft
B 6 ft
C 9 ft
D 12 ft

8. Which statement is **not** true?

F $2,125 < 2,373$
G $578 > 475$
H $5,300 + 42 = 5,342$
J $9,123 > 9,124$

Write What You Know

9. Patsy wants to measure the length of her pet mouse. What unit should she use? Explain.

10. Copy and complete the table to show how many quarts there are in each number of gallons. What rule did you follow?

Gallons	1	2	3	4
Quarts	4	■	■	■

CHAPTER 21

Metric Units

During the summer months, one of the hottest places in the United States is Death Valley.

PROBLEM SOLVING The table shows temperatures during July in four cities and in Death Valley. How much cooler is the high temperature for Key West, FL, than Death Valley?

DATA LINK

JULY TEMPERATURES

Place	Temperature	
	High	Low
Death Valley, CA	46°C	31°C
Salt Lake City, UT	33°C	18°C
Columbus, OH	29°C	17°C
Philadelphia, PA	30°C	19°C
Key West, FL	33°C	28°C

Death Valley

Use this page to help you review and remember important skills needed for Chapter 21.

VOCABULARY

Choose the correct term from the box.

centimeter
inch

1. A metric unit used to measure length is a ? .

USE A METRIC RULER (For Intervention, see p. H27.)

For 2–4, use the yarn shown at the right.

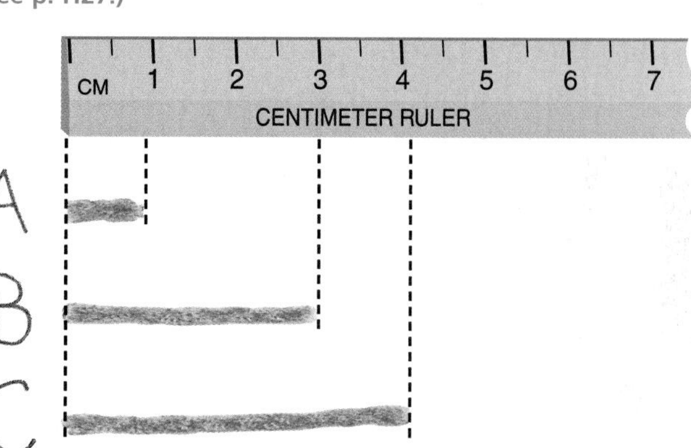

2. Which piece of yarn is 3 centimeters long?

3. Which piece of yarn is about 1 centimeter long?

4. Which piece of yarn is about 1 centimeter longer than yarn B?

MEASURE TO THE NEAREST CENTIMETER (For Intervention, see p. H28.)

Write the length to the nearest centimeter.

5.

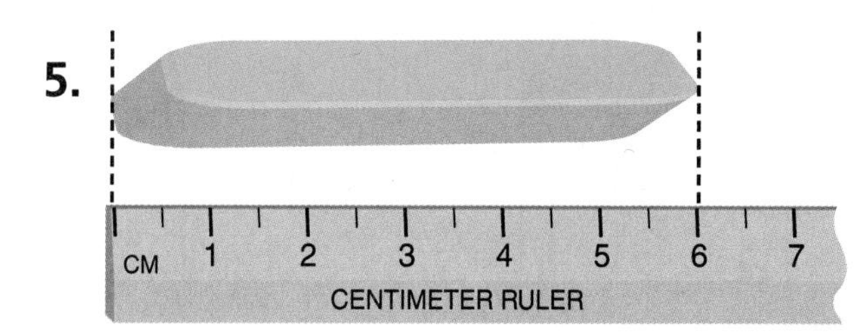

6.

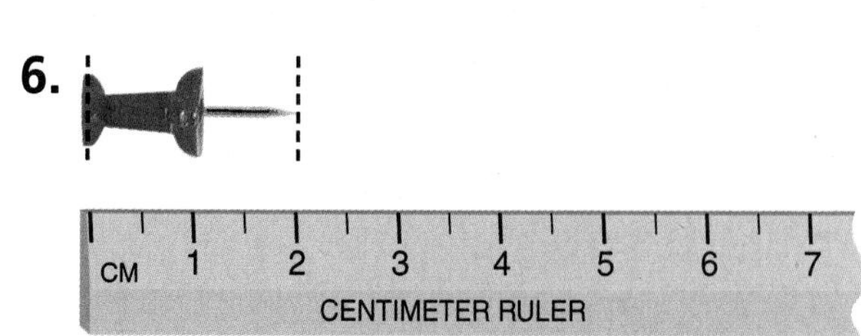

7.

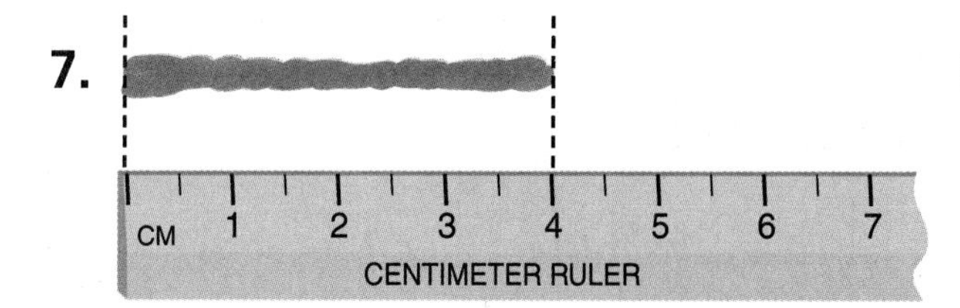

8.

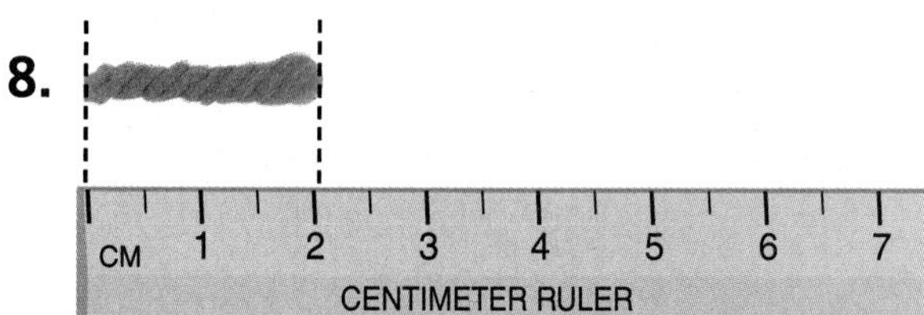

LESSON 1

Length

Quick Review

1. 5×10
2. 10×7
3. 10×10
4. 10×1
5. 9×10

VOCABULARY

centimeter (cm)
decimeter (dm)
meter (m)
kilometer (km)

Learn

MAKE IT METRIC In the metric system, **centimeter (cm)**, **decimeter (dm)**, **meter (m)**, and **kilometer (km)** are units used to measure length and distance.

A *centimeter* is about the width of your index finger.

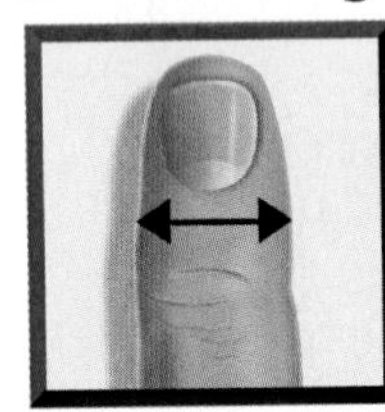

A *decimeter* is about the width of an adult's hand.

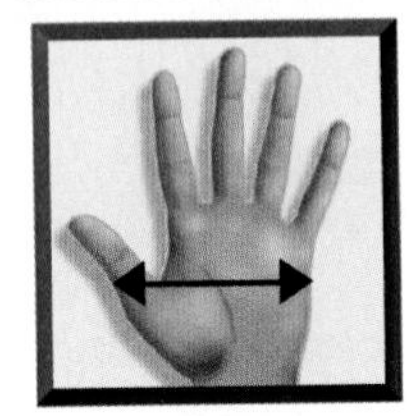

Your armspan is about 1 *meter* long.

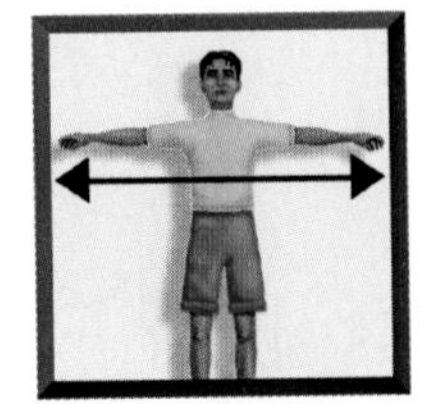

A *kilometer* is a little more than half a mile.

Activity

Materials: centimeter ruler

STEP 1

Copy the table.

LENGTH IN CENTIMETERS		
Object	Estimate	Measure

STEP 2

Write four objects that you would measure using centimeters.

STEP 3

Estimate the length of each object in centimeters. Record your estimates.

STEP 4

Use a centimeter ruler. Measure the length of each object to the nearest centimeter. Record your measurements.

Relating Units

How are centimeters, decimeters, meters, and kilometers related?

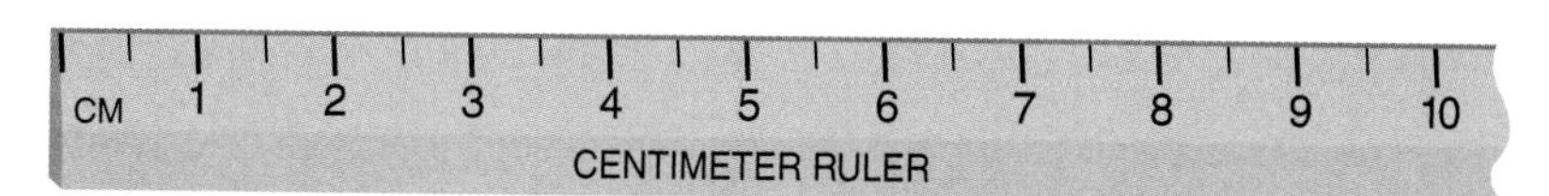

10 centimeters = 1 decimeter

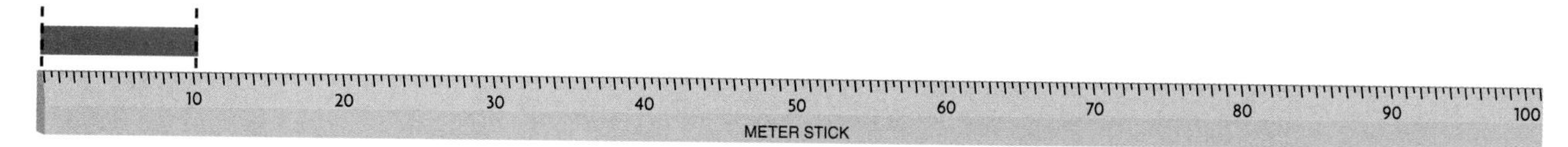

100 centimeters = 1 meter

10 decimeters = 1 meter

1,000 meters = 1 kilometer

- How could you multiply to find how many centimeters are in 3 decimeters? (HINT: 10 centimeters = 1 decimeter.)

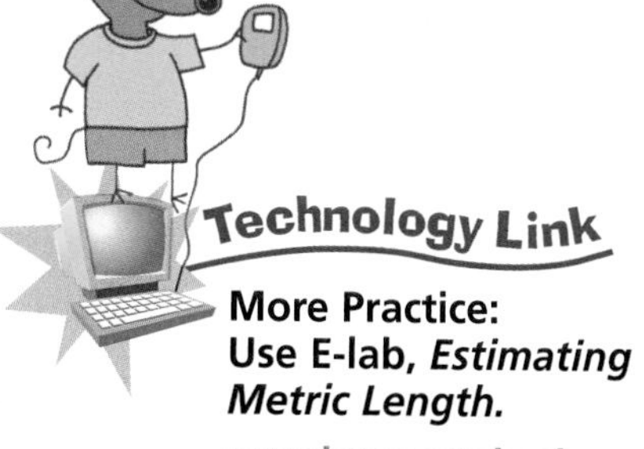

Technology Link

More Practice:
Use E-lab, *Estimating Metric Length.*

www.harcourtschool.com/elab2002

Check

1. **Explain** how you would find the number of centimeters for objects that do not line up exactly with a centimeter mark.

Estimate the length in centimeters. Then use a ruler to measure to the nearest centimeter.

2.

3.

4.

5.

Choose the unit you would use to measure each. Write *cm, m,* or *km*.

6. the length of a crayon
7. the length of a chalkboard
8. the length of a carrot
9. the length of a playground
10. the distance you could walk in 1 hour
11. the length of a butterfly

LESSON CONTINUES

Practice and Problem Solving

Estimate the length in centimeters. Then use a ruler to measure to the nearest centimeter.

12. 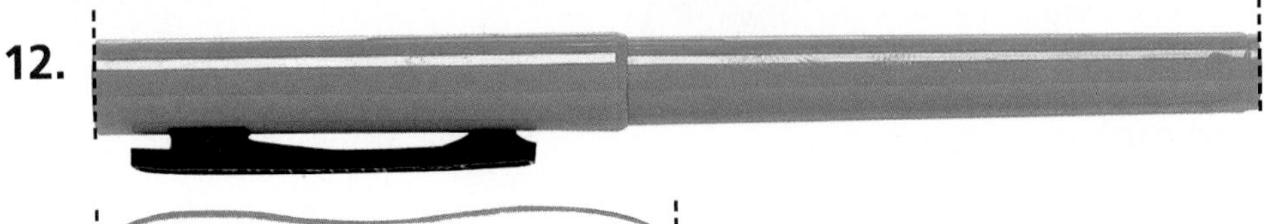

13.

14. 

15. 

16. 

17.

Choose the unit you would use to measure each. Write *cm, m,* or *km*.

18. the distance you can ride a bike in 30 minutes

19. the length of a crayon

20. the distance from your classroom to the playground

21. the length of your classroom

Choose the better estimate.

22. Alicia biked 2 __?__ to get to the store.
    A kilometers B decimeters

23. Carole's ponytail is 1 __?__ long.
    A decimeter B kilometer

24. The wall is 3 __?__ high.
    A centimeters B meters

25. The paper clip is 3 __?__ long.
    A decimeters B centimeters

Use a ruler. Draw a line for each length.

26. 2 centimeters

27. 1 decimeter

28. 14 centimeters

29. **Write About It** Choose 3 objects inside your classroom and 3 outside. Estimate and measure the lengths. Record the results. Tell what tool you used to measure each.

30. **REASONING** Chad says 23 centimeters is the same as 2 decimeters plus 3 centimeters. Do you agree? Explain.

31. **What's the Error?** Nick said that this line measures about 2 cm. Describe his error. Give the correct measure.

32. Sarah drew this poster for her science project. What is the length of the side of her poster in decimeters?

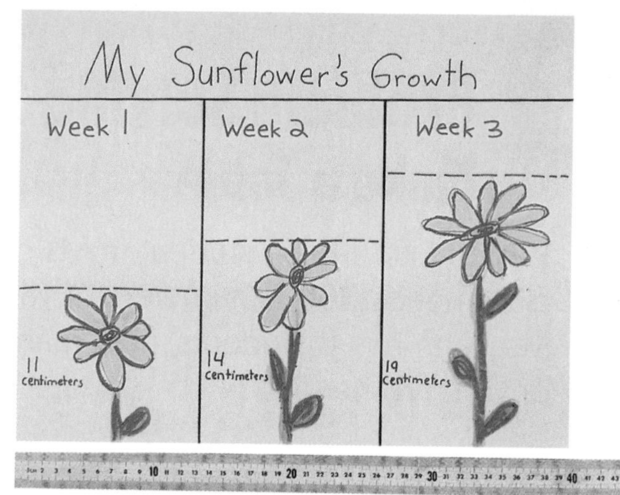

Mixed Review and Test Prep

Find the product. (p. 164)

33. $9 \times 6 = \blacksquare$ 34. $4 \times 9 = \blacksquare$ 35. $7 \times 9 = \blacksquare$ 36. $10 \times 9 = \blacksquare$

37. **TEST PREP** Which angle is less than a right angle? (p. 300)

A 
B
C
D 

38. **TEST PREP** Which angle is greater than a right angle? (p. 300)

F 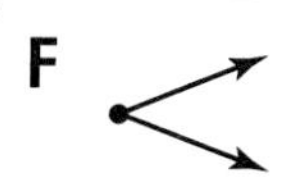
G 
H 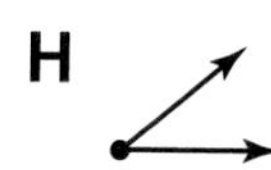
J 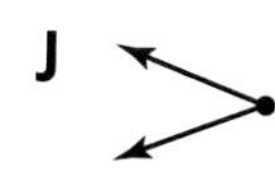

PROBLEM SOLVING  Thinker's Corner

MATERIALS: number cube labeled 3–8; centimeter ruler; two different colored pencils

A. Play with a partner. On a sheet of paper, draw 6 circles that are not near each other.

B. Roll the number cube. Put the point of your pencil on a circle and use the ruler to draw a line as many centimeters long as the number you rolled. The goal is to connect all the circles.

LESSON 2

Problem Solving Strategy
Make a Table

Quick Review

1. 10×2
2. $200 + 200$
3. $100 + 100$
4. $400 + 400$
5. $1,000 + 1,000$

PROBLEM Each student in Ms. Tahn's art class needs 200 centimeters of yarn. If Ms. Tahn has 8 students, how many meters of yarn are needed?

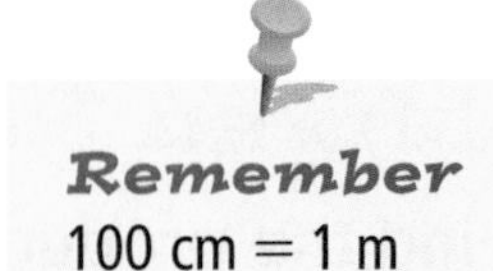

Remember
100 cm = 1 m

UNDERSTAND

- What are you asked to find?
- What information will you use?
- Is there any information you will not use? If so, what?

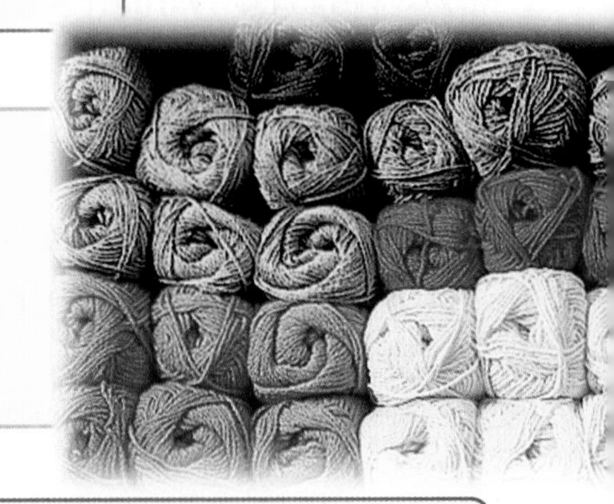

PLAN

- What strategy can you use to solve the problem?

You can *make a table* to show how to change centimeters into meters.

SOLVE

- How can you use the strategy to solve the problem?

Think: 100 cm = 1 m.

Students	1	2	3	4	5	6	7	8
Centimeters	200	400	600	800	1,000	1,200	1,400	1,600
Meters	2	4	6	8	10	12	14	16

The 8 students need 1,600 centimeters of yarn.
1,600 centimeters = 16 meters

So, 16 meters of yarn are needed.

CHECK

- How can you decide if your answer is correct?

Problem Solving Practice

Use *make a table* to solve.

1. **What if** Ms. Tahn had 10 students? Use the table on page 376 to help decide how many meters of yarn are needed.
2. Patty needs 500 centimeters of ribbon for a costume. How many meters of ribbon does she need?
3. Ruby buys 3 meters of blue ribbon and 1 meter of red ribbon. How many centimeters of ribbon does she have?

PROBLEM SOLVING STRATEGIES

- Draw a Diagram or Picture
- Make a Model or Act It Out
- Make an Organized List
- Find a Pattern
- **Make a Table or Graph**
- Predict and Test
- Work Backward
- Solve a Simpler Problem
- Write a Number Sentence
- Use Logical Reasoning

Barbara jogged 3,000 meters. How many kilometers did she jog? (HINT: 1 km = 1,000 m)

4. Which table helps solve the problem?

A

Kilometers	1	2	3
Meters	1,000	2,000	3,000

B

Centimeters	100	200	300
Meters	1	2	3

C

Meters	1	2	3
Decimeters	10	20	30

D

Decimeters	1	2	3
Centimeters	10	20	30

5. Which answers the question?

F 3,000 km
G 300 km
H 30 km
J 3 km

Mixed Strategy Practice

6. Reece had some walnuts. He ate 3 walnuts. Then he gave 4 friends 5 walnuts each. He had 1 walnut left. How many walnuts did he start with?
7. Betsy is 8 years old. She has 4 red beads and 8 blue beads. If she gives an equal number of beads to 3 friends, how many beads will each friend get?
8. Adam was second in line. Susan stood behind Adam and in front of Jean. Tim was first in line. Who was fourth in line?
9. Spencer arranged his photos in 4 rows of 4. What is another way he could arrange his photos in equal rows?

LESSON 3

Capacity: Liters and Milliliters

Quick Review

1. 300 + 400 = ■
2. 800 + 100 = ■
3. 2,000 + 2,000 = ■
4. 250 + 250 = ■
5. 900 + 200 = ■

VOCABULARY
milliliter (mL)
liter (L)

MATERIALS plastic glass that holds about 250 mL, liter container, water

Explore

Capacity can be measured by using metric units such as the **milliliter (mL)** and **liter (L)**.

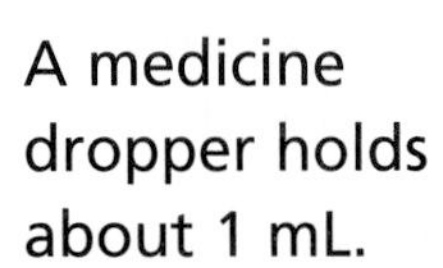

A medicine dropper holds about 1 mL.

A water glass holds about 250 mL.

A water bottle holds about 1 L.

Activity

How many milliliters are in 1 liter?

STEP 1

Use the plastic water glass. Pour 250 mL of water into the liter container. Record the number of milliliters you have poured.

STEP 2

Repeat until the liter container is full. Keep track of how many glasses you have poured, and how many milliliters you have poured.

I poured one glass of water into the liter container. How many more glasses will it take to fill it?

Try It

a. How many glasses did you pour to fill the liter container?

b. How many milliliters did it take to fill the liter container?

Connect

How are liters and milliliters related?

1,000 milliliters = 1 liter

Liters	1	2	3	4
Milliliters	1,000	2,000	3,000	4,000

1,000 mL, or 1 L

Practice and Problem Solving

Choose the better estimate.

1. 

3 mL or
3 L?

2. 

400 mL or
400 L?

3. 

2 mL or
2 L?

4. 

200 mL or
200 L?

Choose the unit you would use to measure the capacity of each. Write *mL* or *L*.

5. 

6. 

7. 

8. 

9. REASONING Rashad has a liter of juice to pour into his thermos. If his thermos holds 500 mL, how many times can he fill his thermos?

10. **What's the Question?** Jamal had some stamps. He gave away 2 stamps and then arranged the rest in 4 rows of 3. The answer is 14 stamps.

Mixed Review and Test Prep

Compare. Use <, >, or = for each ●.

11. 4×8 ● 36 (p. 134)

12. $56 \div 7$ ● 7 (p. 216)

13. 3×7 ● 7×3 (p. 122)

14. $18 \div 2$ ● $27 \div 3$ (p. 222)

15. TEST PREP Jackie put 8 beads on each of 4 bracelets, and 20 beads on a necklace. How many beads did she use in all? (p. 172)

A 12   **C** 40
B 32   **D** 52

LESSON 4

Mass: Grams and Kilograms

Quick Review

1. 10 + 40 = ■
2. 65 + 10 = ■
3. 1,000 + 7,000 = ■
4. 200 + 600 = ■
5. 6,000 + 5,000 = ■

VOCABULARY
gram (g)
kilogram (kg)

MATERIALS classroom objects, small paper clips, simple balance, book with a mass of about 1 kilogram

Explore

The **gram (g)** and the **kilogram (kg)** are metric units for measuring mass.

A paper clip has a mass of about 1 gram.

A large book has a mass of about 1 kilogram.

Activity

Find the mass of objects in your classroom.

STEP 1

Place 10 paper clips on one side of the simple balance to show 10 g.

STEP 2

Find an object that you think might equal 10 g. Use the balance to check.

STEP 3

Repeat Steps 1 and 2 for 25 g and 1 kg. Use the book to show 1 kg.

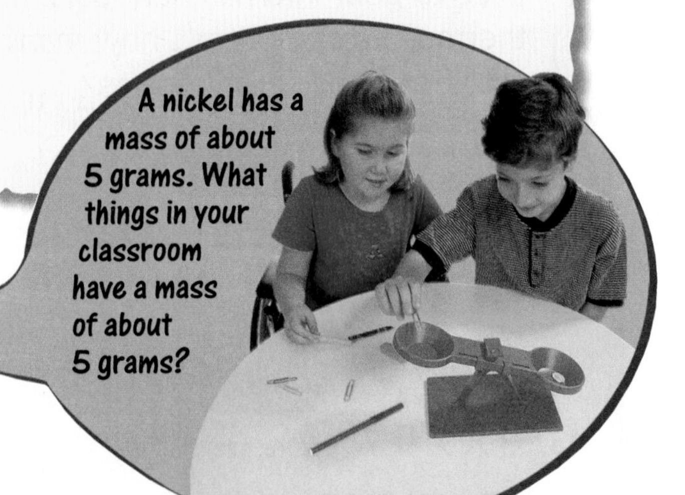

A nickel has a mass of about 5 grams. What things in your classroom have a mass of about 5 grams?

Try It

Name an object that has a mass of each amount.

a. 5 grams **b.** 2 kilograms

Connect

This cat has a mass of 5 kg. How many grams is that?

1,000 grams = 1 kilogram

Kilograms	1	2	3	4	5
Grams	1,000	2,000	3,000	4,000	5,000

So, the cat has a mass of 5,000 g.

Practice and Problem Solving

Choose the better estimate.

1.

200 g or 200 kg?

2.

18 g or 18 kg?

3.

6 g or 6 kg?

Choose the tool and unit to measure each.

4. length of a pencil **5.** mass of a grape

6. capacity of a bucket

Tools	Units	
ruler	cm	g
liter container	kg	mL
simple balance	L	m

7. **Write About It** Do objects of about the same size always have about the same mass? Give an example.

8. Kim's kitten has a mass of 2 kg. How many grams is that?

9. Sue had 3 red pens and 9 blue pens. She put the same number of pens into 2 cups. How many pens were in each cup?

Mixed Review and Test Prep

For 10–13, subtract. (p. 58)

10. 441 − 78 **11.** 703 − 264

12. 157 − 139 **13.** 500 − 167

14. **TEST PREP** Lino had 15 photos. He put an equal number of photos on each of 3 album pages. How many photos are on each page? (p. 204)

A 45 **B** 10 **C** 5 **D** 3

LESSON 5

HANDS ON

Measure Temperature

Quick Review

Skip-count. Write the missing number.

1. 10, 15, 20, 25, ■
2. 80, 75, ■, 65, 60
3. 110, 115, 120, ■, 130
4. 95, ■, 105, 110, 115
5. 90, ■, 86, 84, 82

VOCABULARY
degrees Fahrenheit (°F)
degrees Celsius (°C)

MATERIALS Celsius and Fahrenheit thermometers

Explore

Degrees Fahrenheit (°F) are customary units of temperature, and **degrees Celsius (°C)** are metric units of temperature.

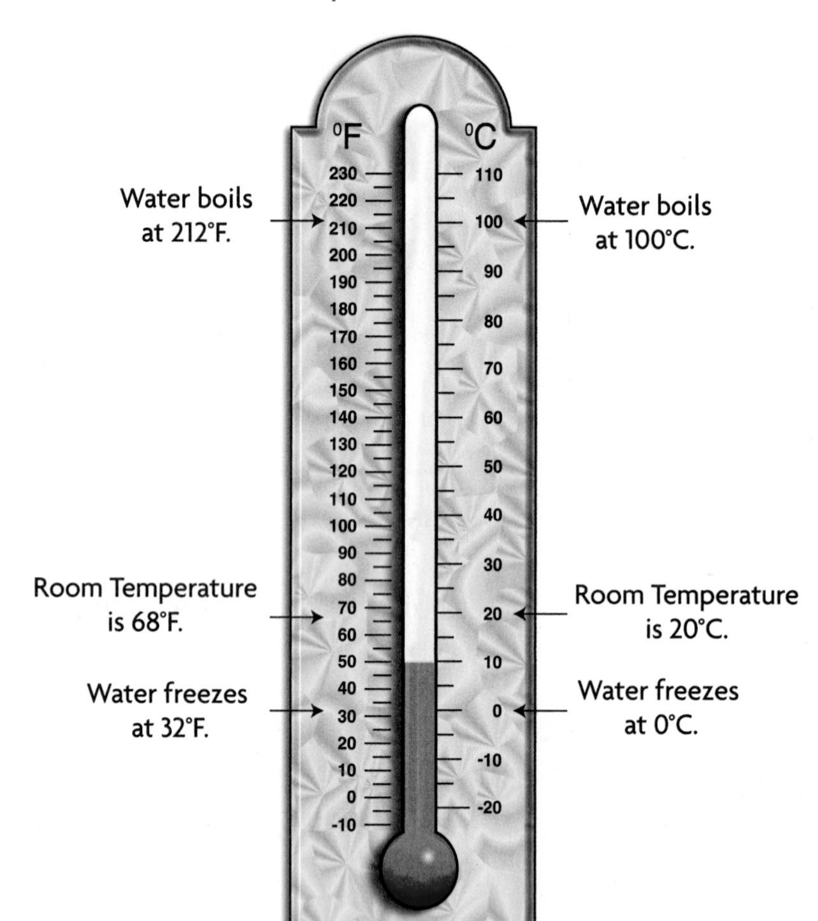

Activity

STEP 1

Estimate the temperature outside the classroom in degrees Celsius and in degrees Fahrenheit. Record your estimates.

STEP 2

Measure the temperature outside the classroom using thermometers. Record the differences between your estimates and the actual measurements.

Try It

What is the temperature now?

a. The temperature was 71°F. It dropped 20°.

b. The temperature was 25°C. It dropped 3°.

Connect

To read a thermometer, find the number at the top of the red bar. The thermometer on page 382 shows that the temperature is

50°F **Read:** fifty degrees Fahrenheit

10°C **Read:** ten degrees Celsius

Practice and Problem Solving

Write each temperature in °F.

1.

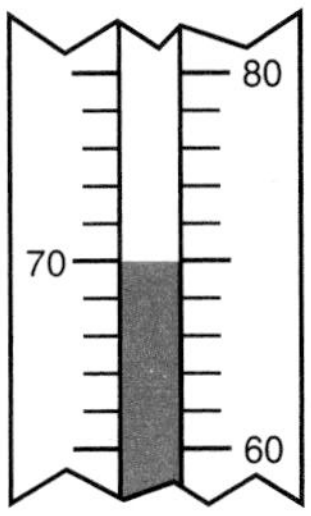

2.

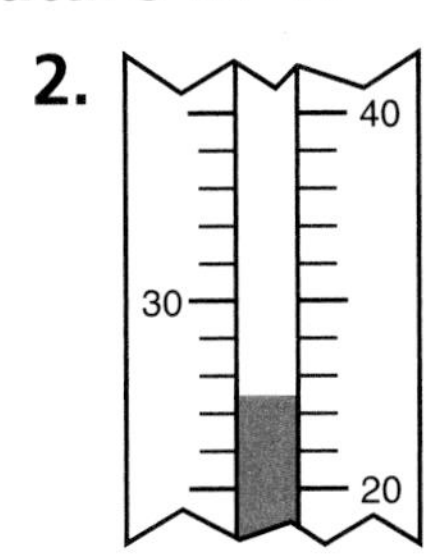

Write each temperature in °C.

3.

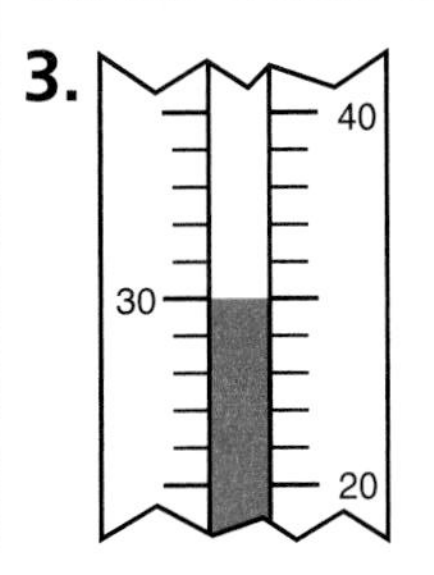

4.

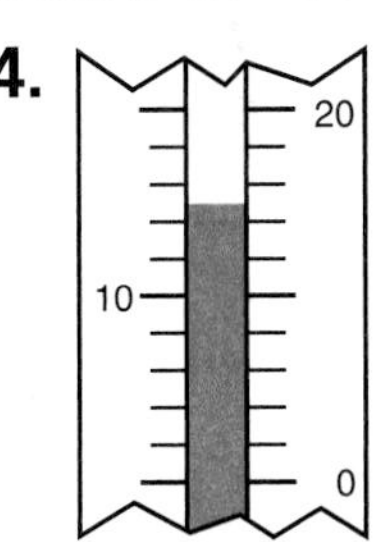

Choose the better estimate.

5.

25°F or 95°F?

6.

56°F or 98°F?

Choose the better estimate.

7.

8°C or 25°C?

8.

0°C or 22°C?

9. The temperature was 45°F at 9:00 A.M. It was 62°F at noon. How many degrees did the temperature rise?

Mixed Review and Test Prep

Find the number that the variable stands for. (p. 188)

10. $8 \times a = 72$ **11.** $9 \times b = 81$

12. $c \times 9 = 36$ **13.** $9 \times d = 54$

14. TEST PREP Tom put 42 dimes into 6 equal piles. How many dimes were in 2 piles? (p. 172)

A 14 **B** 10 **C** 8 **D** 6

Review/Test

CHECK VOCABULARY AND CONCEPTS

Choose the best term from the box.

degrees Celsius
degrees Fahrenheit
gram
liter
meter

1. Capacity can be measured by using metric units such as the __?__. (p. 378)
2. A metric unit for measuring mass is the __?__. (p. 380)
3. Customary units of temperature are __?__. (p. 382)
4. Metric units of temperature are __?__. (p. 382)

Choose the better estimate. (pp. 378–383)

5. 

2 mL or 2 L?

6. 

2 g or 2 kg?

7. 

30°F or 60°F?

CHECK SKILLS

Estimate the length in centimeters. Then use a ruler to measure to the nearest centimeter. (pp. 372–375)

8.

9. 

Choose the unit you would use to measure each. Write *cm, m,* or *km*. (pp. 372–375)

10. the length of a pencil
11. the distance to the moon
12. the length of a bee
13. the distance you run in 10 seconds

CHECK PROBLEM SOLVING

Use *make a table* to solve. (pp. 376–377)

14. Ashley has 2 meters of yarn. How many centimeters of yarn does she have?
15. Ray had 1 meter of string. Then he bought 400 centimeters more. How many meters does he have in all?

Standardized Test Prep

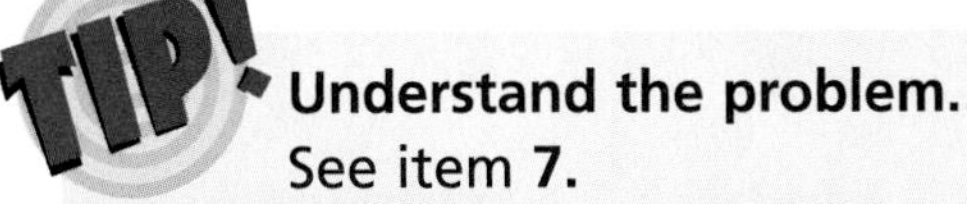

Understand the problem.
See item **7.**

You need to understand that *most likely* means it will probably happen more often than any of the other choices.

Also see problem **1,** p. H62.

For 1–7, choose the best answer.

1. Which temperature does the thermometer show?

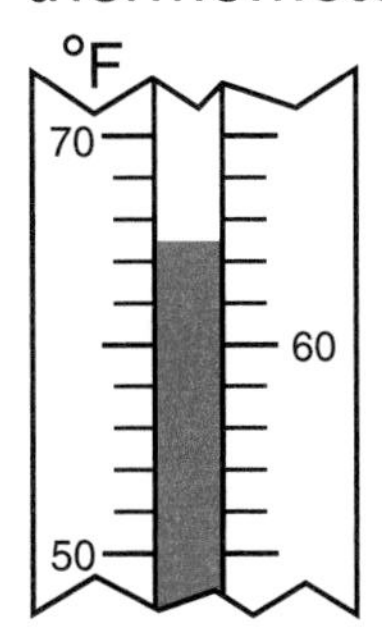

A 60°F
B 65°F
C 70°F
D 75°F

---

2. Which is the best estimate for the amount of water in a fish tank?

F 1 mL
G 10 mL
H 1 L
J 10 L

---

3. Which is a unit for measuring the *mass* of a tennis raquet?

A meter
B liter
C gram
D inch

---

4. Four tapes cost a total of $32. How much would one tape cost?

F $8
G $9
H $32
J $36

---

5. Which is the best estimate of the outdoor temperature?

A 0°C
B 25°C
C 32°C
D 100°C

---

6. A solid figure has 6 faces. All the faces are squares. Which solid figure is described?

F square pyramid
G sphere
H cone
J cube

---

7. On what color is the pointer *most likely* to land?

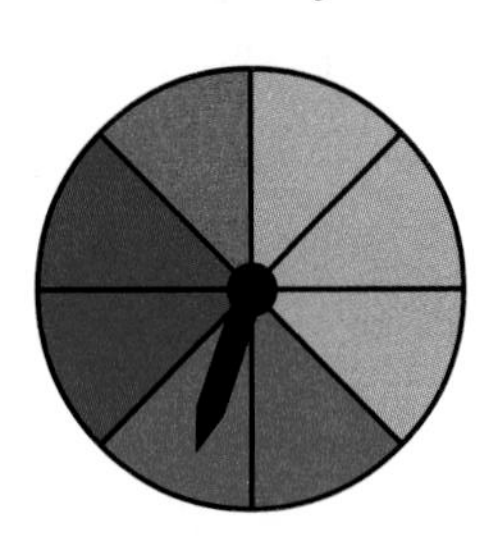

A orange
B green
C blue
D red

Write What You Know

8. Describe how a parallelogram and a rectangle are alike. How are they different?

9. For each metric unit below, give two examples of what could be measured using those units.
liter, meter, gram

CHAPTER 22

Perimeter, Area, and Volume

People have grown flowers for thousands of years. In the 1600s, tulips in Holland were so valuable that the bulbs were used as money!

PROBLEM SOLVING Look at the garden diagram below. How can you find how many feet of fencing you would need to go around that garden?

CHECK WHAT YOU KNOW

Use this page to help you review and remember important skills needed for Chapter 22.

COLUMN ADDITION (For Intervention, see p. H10.)

Find each sum.

1. $\begin{array}{r} 3 \\ 7 \\ +5 \\ \hline \end{array}$ **2.** $\begin{array}{r} 2 \\ 4 \\ +8 \\ \hline \end{array}$ **3.** $\begin{array}{r} 1 \\ 3 \\ +9 \\ \hline \end{array}$ **4.** $\begin{array}{r} 36 \\ 12 \\ +\ 9 \\ \hline \end{array}$

5. $\begin{array}{r} 12 \\ 7 \\ +23 \\ \hline \end{array}$ **6.** $\begin{array}{r} 44 \\ 9 \\ +\ 6 \\ \hline \end{array}$ **7.** $\begin{array}{r} 245 \\ 65 \\ +\ 72 \\ \hline \end{array}$ **8.** $\begin{array}{r} 99 \\ 201 \\ +\ 31 \\ \hline \end{array}$

9. $\begin{array}{r} 102 \\ 40 \\ +\ 14 \\ \hline \end{array}$ **10.** $\begin{array}{r} 57 \\ 400 \\ +\ \ 3 \\ \hline \end{array}$ **11.** $\begin{array}{r} 142 \\ 231 \\ +105 \\ \hline \end{array}$ **12.** $\begin{array}{r} 65 \\ 19 \\ +15 \\ \hline \end{array}$

MULTIPLICATION FACTS (For Intervention, see p. H11.)

Find each product.

13. $7 \times 4 = \blacksquare$ **14.** $3 \times 6 = \blacksquare$ **15.** $2 \times 5 = \blacksquare$

16. $4 \times 9 = \blacksquare$ **17.** $7 \times 2 = \blacksquare$ **18.** $6 \times 6 = \blacksquare$

19. $9 \times 8 = \blacksquare$ **20.** $10 \times 3 = \blacksquare$ **21.** $4 \times 5 = \blacksquare$

22. $\begin{array}{r} 3 \\ \times 7 \\ \hline \end{array}$ **23.** $\begin{array}{r} 7 \\ \times 8 \\ \hline \end{array}$ **24.** $\begin{array}{r} 9 \\ \times 6 \\ \hline \end{array}$ **25.** $\begin{array}{r} 5 \\ \times 5 \\ \hline \end{array}$ **26.** $\begin{array}{r} 2 \\ \times 8 \\ \hline \end{array}$ **27.** $\begin{array}{r} 7 \\ \times 6 \\ \hline \end{array}$

28. $\begin{array}{r} 3 \\ \times 3 \\ \hline \end{array}$ **29.** $\begin{array}{r} 9 \\ \times 5 \\ \hline \end{array}$ **30.** $\begin{array}{r} 10 \\ \times\ 2 \\ \hline \end{array}$ **31.** $\begin{array}{r} 7 \\ \times 9 \\ \hline \end{array}$ **32.** $\begin{array}{r} 1 \\ \times 6 \\ \hline \end{array}$ **33.** $\begin{array}{r} 4 \\ \times 4 \\ \hline \end{array}$

34. $\begin{array}{r} 5 \\ \times 6 \\ \hline \end{array}$ **35.** $\begin{array}{r} 2 \\ \times 9 \\ \hline \end{array}$ **36.** $\begin{array}{r} 3 \\ \times 5 \\ \hline \end{array}$ **37.** $\begin{array}{r} 8 \\ \times 6 \\ \hline \end{array}$ **38.** $\begin{array}{r} 5 \\ \times 8 \\ \hline \end{array}$ **39.** $\begin{array}{r} 2 \\ \times 6 \\ \hline \end{array}$

LESSON 1

Perimeter

Quick Review

1. $5 + 2 + 4 = ■$
2. $6 + 7 + 2 = ■$
3. $1 + 8 + 9 = ■$
4. $6 + 6 + 3 = ■$
5. $4 + 4 + 4 = ■$

VOCABULARY
perimeter

MATERIALS
square tiles

Explore

The distance around a figure is called its **perimeter**.

You can build a rectangle by making an array with square tiles.

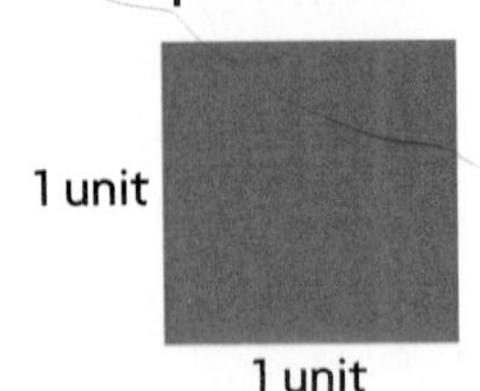

Each side of a square tile has a length of 1 unit.

Then you can find the perimeter of the rectangle by counting the units.

Activity

STEP 1

Make a rectangle by placing 3 rows with 5 square tiles in each row.

	16	15	14	13	12	
1						11
2						10
3						9
	4	5	6	7	8	

STEP 2

Count the number of units around the outside of the rectangle.

STEP 3

Record the number of units you counted. This is the perimeter of the rectangle in units.

MATH IDEA You can find the perimeter of a figure by counting the number of units around the outside of the figure.

How can I use square tiles to find the perimeter of my math book?

MATH

Try It

Use square tiles to find the perimeter of each.

a. your math book **b.** your desktop

Connect

You can count units to find the perimeters of these plane figures.

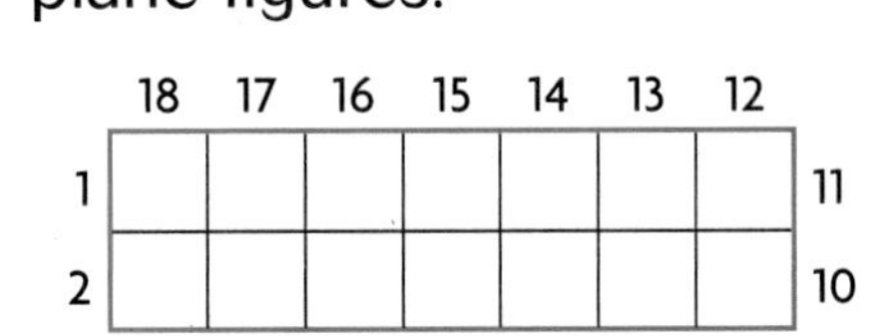

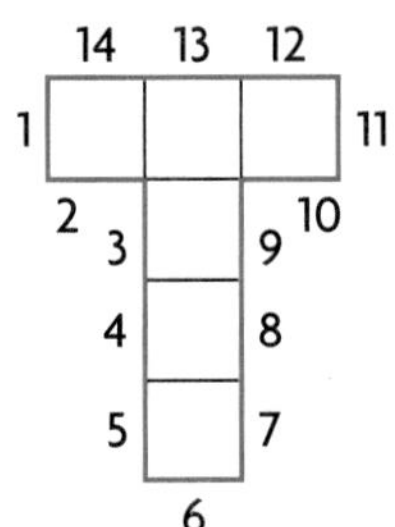

Perimeter = 18 units

Perimeter = 14 units

Practice and Problem Solving

Find the perimeter of each figure.

1. 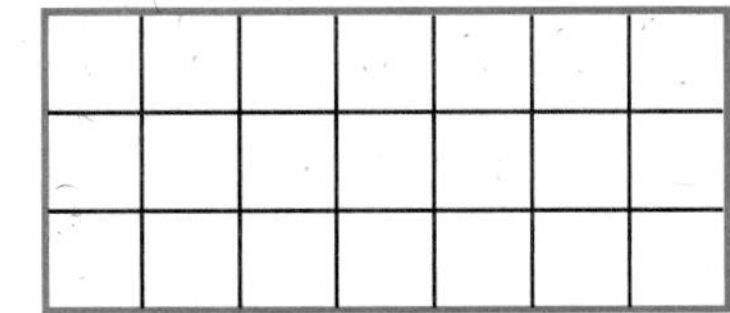

2. 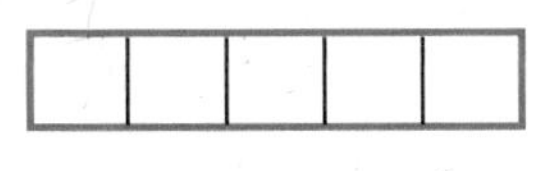

3. 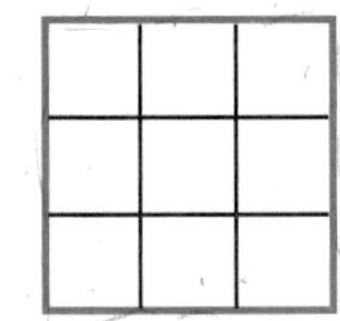

4. 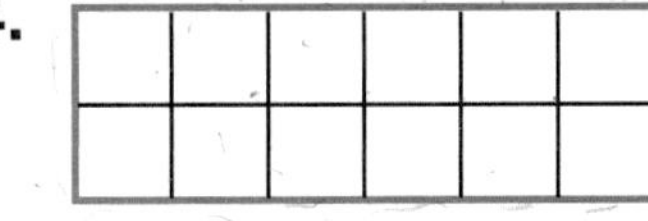

5. 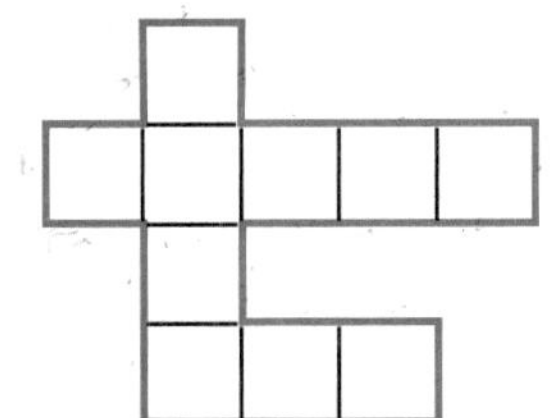

6. 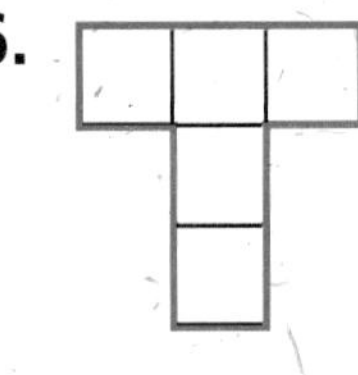

7. **What's the Error?** Elise wrote that the perimeter of this rectangle is 6 units. Describe her error. Write the correct perimeter.

8. Marita made a rectangle with 5 rows of 6 tiles. What is the perimeter of her rectangle?

Mixed Review and Test Prep

9. $27 \div 9 = \blacksquare$ (p. 220)

10. $72 \div 8 = \blacksquare$ (p. 216)

11. $404 - 139 = \blacksquare$ (p. 58)

12. Find the missing factor. (p. 142)
$7 \times \blacksquare = 42$

13. TEST PREP Lin has 2 bags of sandwiches. Each bag holds 2 sandwiches. Each sandwich is cut into 4 pieces. How many pieces of sandwich does Lin have in all? (p. 170)

A 4 **B** 8 **C** 10 **D** 16

LESSON 2

Estimate and Find Perimeter

Quick Review

1. $2 + 3 + 3 = ■$
2. $5 + 6 + 7 = ■$
3. $7 + 3 + 6 = ■$
4. $4 + 2 + 9 = ■$
5. $10 + 4 + 4 + 2 = ■$

Learn

AROUND AND AROUND You can estimate the perimeter of your math book.

Activity

MATERIALS: toothpicks, paper clips

STEP 1

Copy the table. Estimate the perimeter of your math book in paper clips and in toothpicks. Record your estimates.

PERIMETER OF MY MATH BOOK

	Estimate	Measurement
Number of paper clips		
Number of toothpicks		

STEP 2

Use paper clips. Record how many paper clips it takes to go around all the edges of your math book.

STEP 3

Use toothpicks. Record how many toothpicks it takes to go around all the edges of your math book.

- How does your estimate compare with your actual measurement?
- Did it take more paper clips or more toothpicks to measure the perimeter of your math book? Explain.

Use a Ruler

You can use a ruler to find the perimeter.

Example 1

Find the perimeter of the bookmark in centimeters.

Use a ruler to find the lengths of the sides.
Add the lengths of the sides.

10 cm + 2 cm + 10 cm + 2 cm = 24 cm

The perimeter is 24 cm.

You can add the lengths of the sides to find the perimeter.

Example 2

Find the perimeter of the figure.

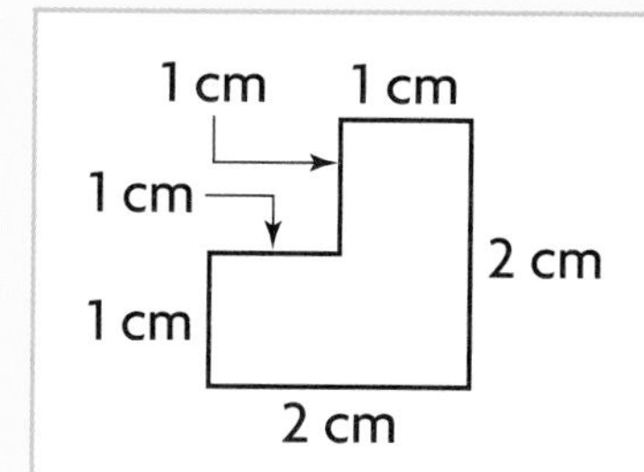

Add the lengths of the sides:
1 cm + 2 cm + 2 cm + 1 cm + 1 cm + 1 cm = 8 cm

The perimeter is 8 cm.

- Why do you need to measure only 2 sides of the bookmark to find its perimeter?

Technology Link
More Practice: Use Mighty Math Carnival Countdown, *Pattern Block Roundup*, Level N.

Check

1. **Discuss** whether you could measure only 2 sides to find the perimeter in Example 2.

Find the perimeter.

2.

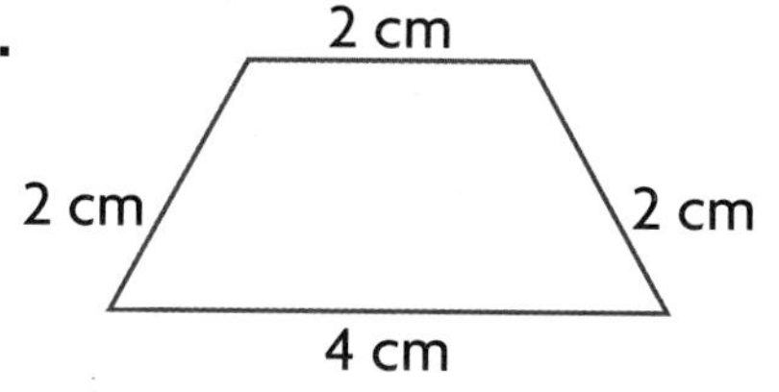

3.

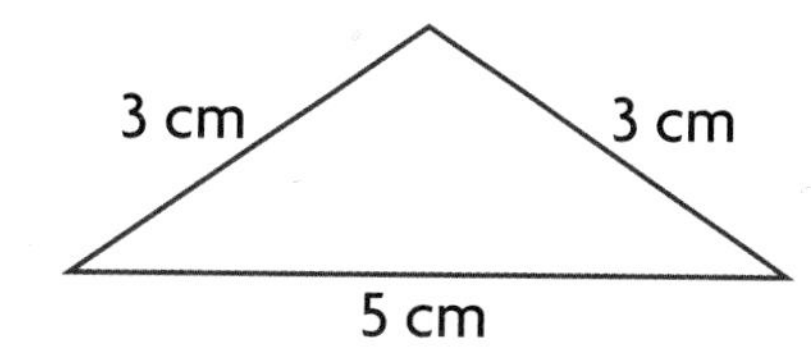

LESSON CONTINUES

Practice and Problem Solving

Find the perimeter.

4.

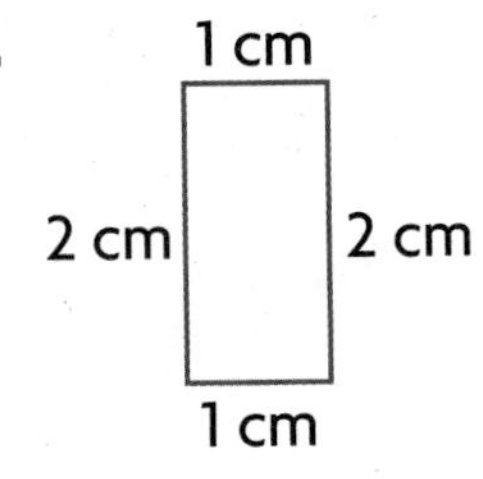

5.

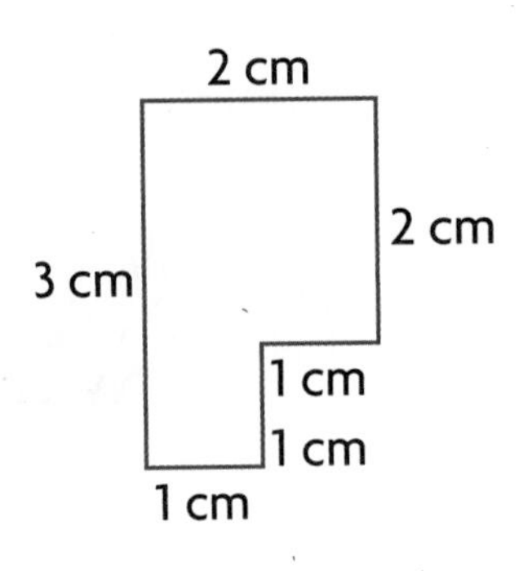

6.

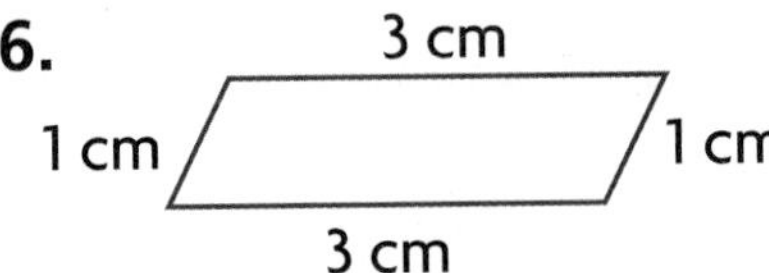

7.

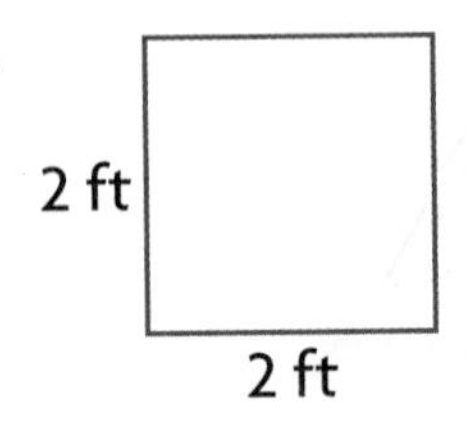

8.

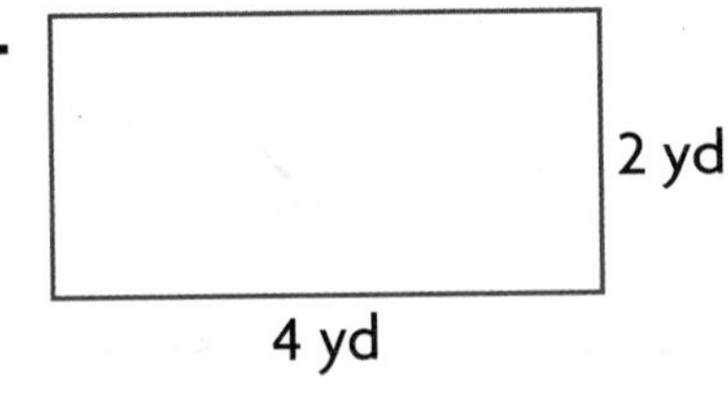

9.

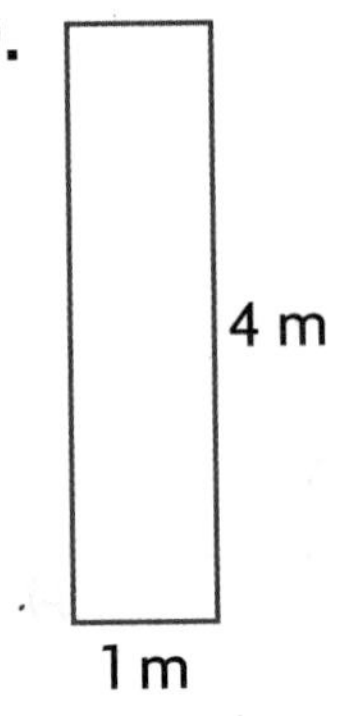

Use your centimeter ruler to find the perimeter.

10.

11.

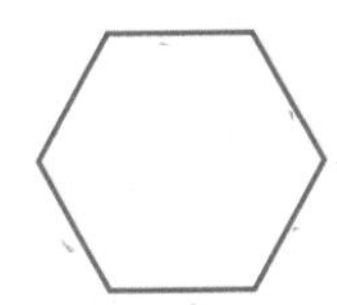

12. **Reasoning** The length of one side of a square is 6 cm. What is the perimeter of the square?

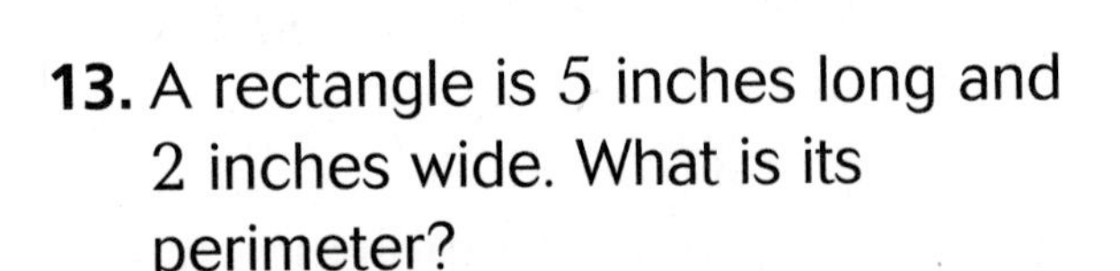

13. A rectangle is 5 inches long and 2 inches wide. What is its perimeter?

14. Use graph paper. Draw a rectangle with a perimeter of 12 units.

15. Jana's beach towel is 5 feet long and 3 feet wide. What is its perimeter?

16. ALGEBRA This triangle has a perimeter of 8 cm. How long is Side C?

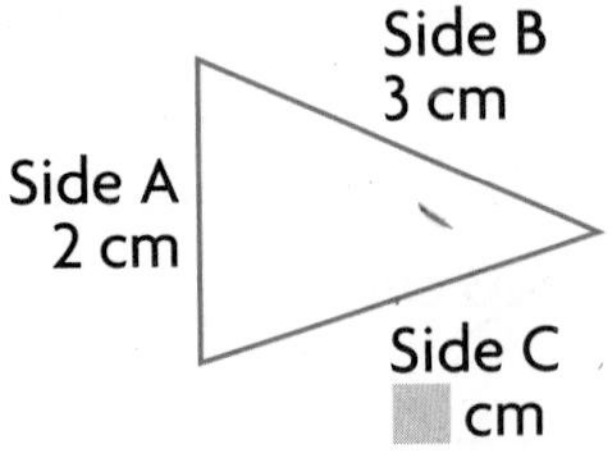

17. **Write About It** Find an object with a perimeter you could measure by using inches. Explain how to estimate and measure its perimeter. Then measure the perimeter with a ruler.

Mixed Review and Test Prep

USE DATA **For 18–22, use the graph.** (p. 252)

18. How many students voted for cheese?

19. How many more students voted for cheese than for vegetables?

20. How many students voted for crackers and trail mix?

21. **TEST PREP** How many students did NOT vote for popcorn?

A 21
B 24
C 27
D 31

22. **TEST PREP** How many students voted in all?

F 12
G 32
H 40
J 50

PROBLEM SOLVING LINKUP . . . to Social Studies

In 1806, Thomas Jefferson built a house in the Blue Ridge Mountains of Virginia. Jefferson built a special room that is a perfect cube at the center of the house. This room is 20 feet long, 20 feet wide, and 20 feet tall. Jefferson's granddaughter used to draw in this room because a large window called a skylight was in the ceiling.

1. There are 32 panes of glass in the center room skylight. The panes are in 2 rows of 16. How else might Jefferson have arranged the panes in equal rows?

2. The floor of the center room is a square with each side measuring 20 feet. What is the perimeter of this floor?

EXTRA PRACTICE page H53, Set A

LESSON 3

Area of Plane Figures

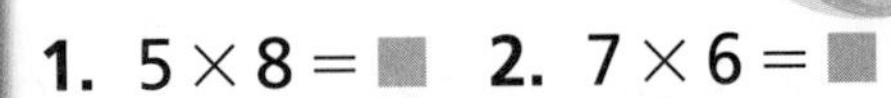

Quick Review

1. $5 \times 8 = ■$
2. $7 \times 6 = ■$
3. $3 \times 3 = ■$
4. $2 \times 8 = ■$
5. $6 \times 4 = ■$

VOCABULARY
square unit
area

MATERIALS
square tiles
grid paper

Explore

A **square unit** is a square with a side length of 1 unit. You use square units to measure area. **Area** is the number of square units needed to cover a flat surface.

1 square unit

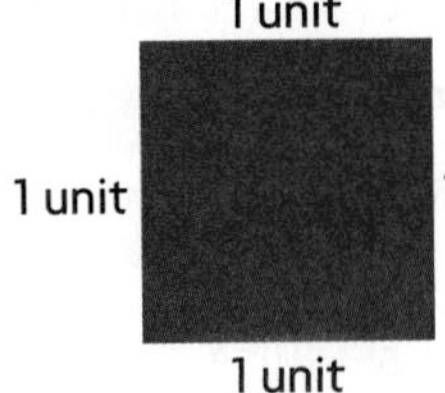

Activity

Use square tiles to find the area of your math book cover.

STEP 1

Estimate how many squares will cover your math book. Then place square tiles in rows on the front of your math book. Cover the whole surface.

STEP 2

Use grid paper. Draw a picture to show how you covered the math book.

STEP 3

Count and record the number of square tiles you used. This number is the book cover's area in square units.

MATH IDEA You can find the area of a surface by counting the number of square units needed to cover the surface.

- Look at the picture you made. How could you use multiplication to find the area?

How many rows of tiles do I need to cover an index card?

Try It

Use square tiles to find the area of each.

a. an index card
b. a sheet of notebook paper

Connect

To find the area of a rectangle, multiply the number of rows times the number in each row.

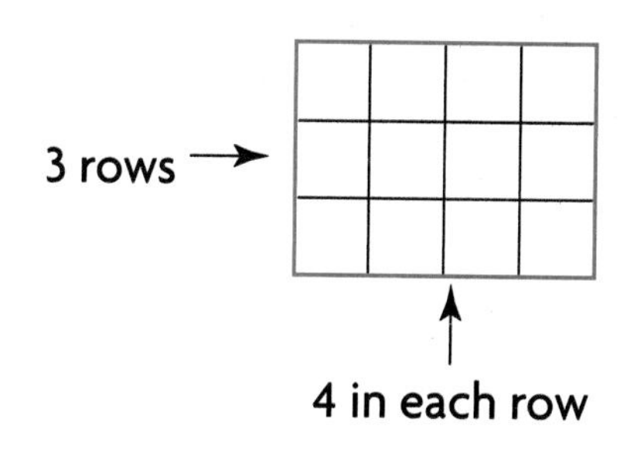

number of rows		number in each row		area
↓		↓		↓
3	×	4	=	12 square units

Technology Link

More Practice: Use E-Lab, *Finding Area.*

www.harcourtschool.com/elab2002

Practice and Problem Solving

Find the area of each figure. Write the area in square units.

1. 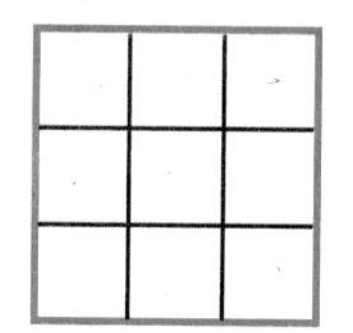

2. 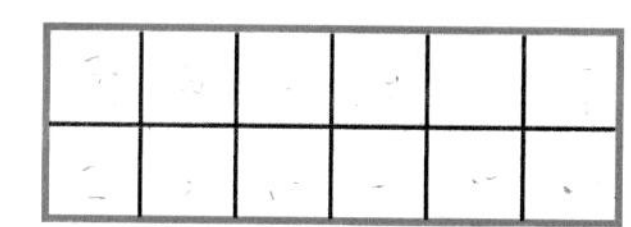

3. 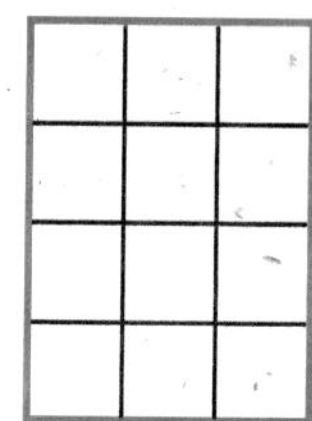

4. 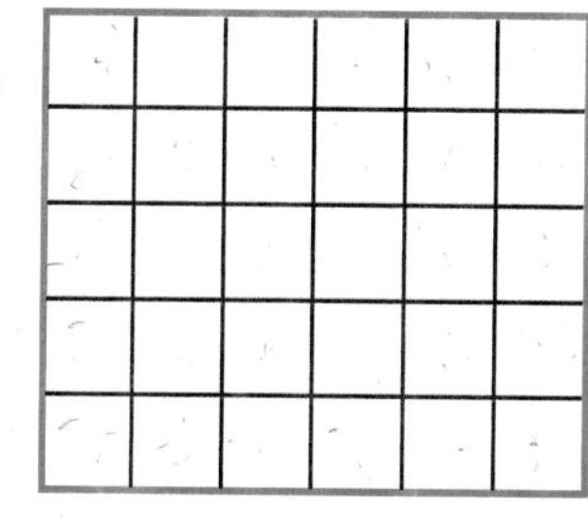

5. 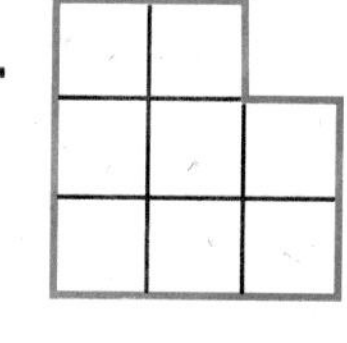

6. 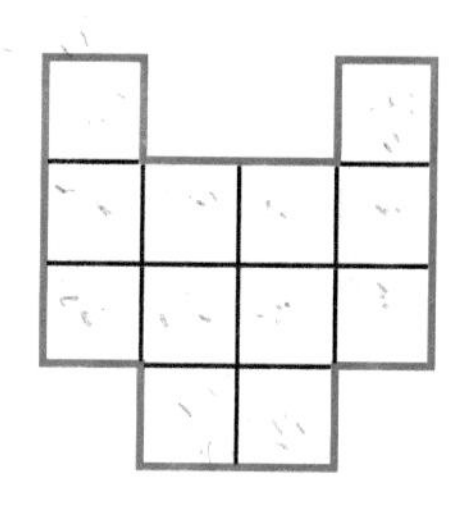

7. REASONING Look at 2 and 3. What do you notice about the area and perimeter of these figures?

8. **What's the Question?** Rachel's blanket is 6 feet wide and 4 feet long. The answer is 24 square feet.

Mixed Review and Test Prep

Find each missing addend. (p. 68)

9. $5 + ■ = 12$

10. $6 + ■ = 14$

11. $■ + 9 = 15$

12. $■ + 4 = 11$

13. TEST PREP Erma had 45 beads. She put 9 beads on each key chain. How many key chains did she make? (p. 220)

A 10 **B** 9 **C** 5 **D** 3

LESSON 4

Problem Solving Skill
Make Generalizations

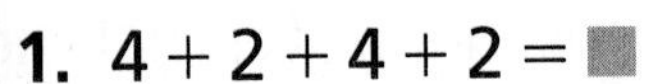

Quick Review

1. $4 + 2 + 4 + 2 = ■$
2. $8 + 3 + 8 + 3 = ■$
3. $3 \times 4 = ■$
4. $7 \times 3 = ■$
5. $8 \times 4 = ■$

UNDERSTAND PLAN SOLVE CHECK

DON'T FENCE ME IN Maura plans to plant a flower garden and put a fence around it. She has 12 feet of fencing to make a square or rectangular garden. If she wants to have the most area possible, should her fence be a square or a rectangle?

Maura draws a picture to show all the square and rectangular gardens she can make.

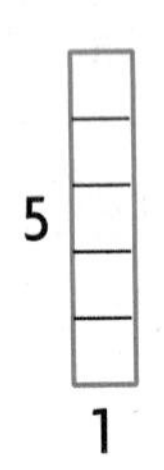

Perimeter:
$1 + 5 + 1 + 5 = 12$ feet
Area: $5 \times 1 = 5$ square feet

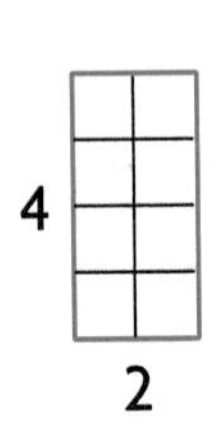

Perimeter:
$2 + 4 + 2 + 4 = 12$ feet
Area: $4 \times 2 = 8$ square feet

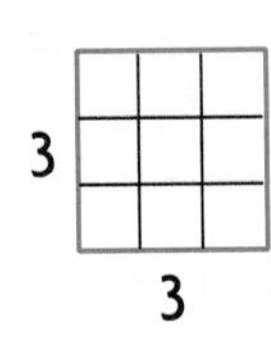

Perimeter:
$3 + 3 + 3 + 3 = 12$ feet
Area: $3 \times 3 = 9$ square feet

Order the areas: $9 > 8 > 5$

9 square feet is the greatest area.

So, Maura should build a square fence.

Talk About It

- **Describe** how the area changes when rectangles with the same perimeter change from long and thin to square.
- **What if** Maura had 20 feet of fencing? To have the greatest area, should her fence be a square or a rectangle?

Problem Solving Practice

1. **What if** Maura had 8 feet of fencing? She wants to make a rectangle or square with the greatest possible area. How long should it be? How wide should it be?

2. Kyle has 8 feet of fencing to make a play yard for his rabbit. What rectangle could he fence in that would have the least possible area?

Jane used 16 inches of ribbon to make a rectangular frame.

3. Which drawing shows the frame Jane made?

A

B

C

D

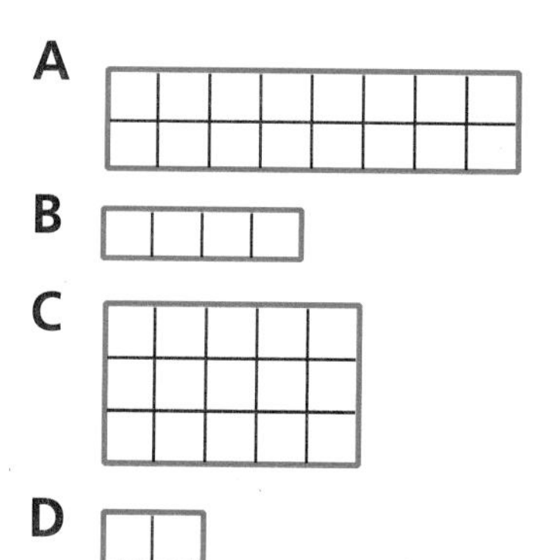

4. Which shows how Jane could increase the area of the frame?

F

G

H

J

Mixed Applications

5. Abe bought 3 muffins for $1 each and 2 cartons of milk for $0.50 each. How much did he spend?

6. Ted eats 1 sandwich and drinks 2 glasses of milk each day. How many glasses of milk does he drink in a week?

7. **REASONING** I am a 2-digit number less than 20. I can be divided evenly into groups of 4. I cannot be divided evenly into groups of 3. What number am I?

8. **Write a problem** about finding the perimeter of a pen for a guinea pig. Make sure you include how long and wide the pen is. Trade with a partner and solve.

LESSON

5 Estimate and Find Volume

Quick Review

1. $1 \times 4 \times 2 = ■$
2. $2 \times 3 \times 2 = ■$
3. $5 \times 1 \times 2 = ■$
4. $3 \times 2 \times 8 = ■$
5. $4 \times 2 \times 3 = ■$

VOCABULARY

volume
cubic unit

Learn

FILL IT UP **Volume** is the amount of space a solid figure takes up.

A **cubic unit** is used to measure volume. A cubic unit is a cube with a side length of 1 unit. You can use color cubes to show cubic units.

1 cubic unit

HANDS ON

Activity

Use color cubes to find the volume of a box.

MATERIALS: color cubes, small box

STEP 1

Estimate how many cubes it will take to fill the box. Record your estimate.

STEP 2

Count the cubes you use. Place the cubes in rows along the bottom of the box. Then continue to make layers of cubes until the box is full.

STEP 3

Record how many cubes it took to fill the box. This is the volume of the box in cubic units.

- How does your estimate compare with the actual volume?

MATH IDEA To measure the volume of a solid, find the number of cubic units needed to fill the solid.

Find the Volume

When you cannot count each cube, you can think about layers to find the volume.

Example

Find the volume of each solid.

Since you cannot see each cube, look at the top layer of cubes. For a rectangular prism, number of layers $\times$ cubes in each layer $=$ Volume.

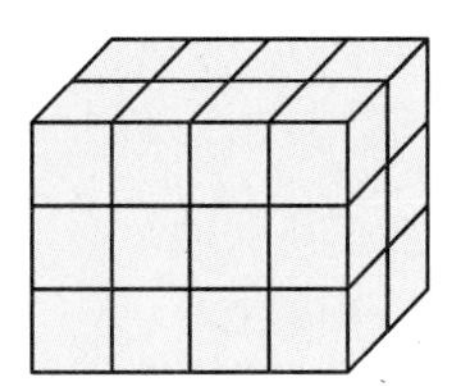

3 layers $\times$ 8 cubes per layer $=$ 24 cubic units
So, the volume is 24 cubic units.

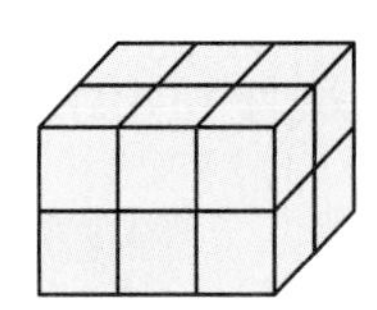

2 layers $\times$ 6 cubes per layer $=$ 12 cubic units
So, the volume is 12 cubic units.

Check

1. **Describe** how you could build another solid with 24 cubic units. You may use color cubes to help.

Use cubes to make each solid. Then write the volume in cubic units.

2. 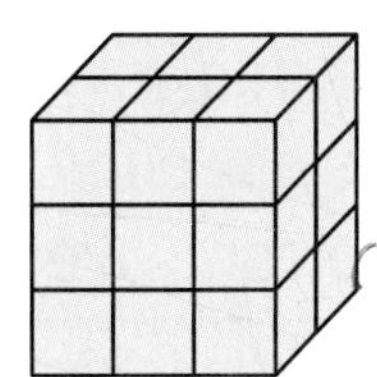

3. 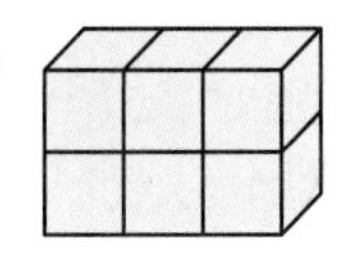

4. 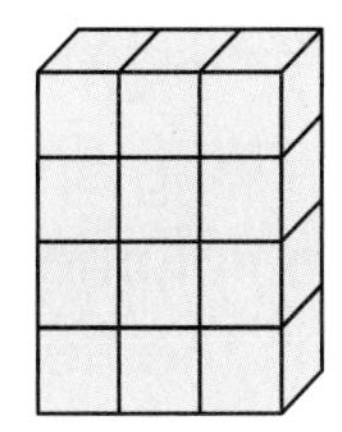

5. 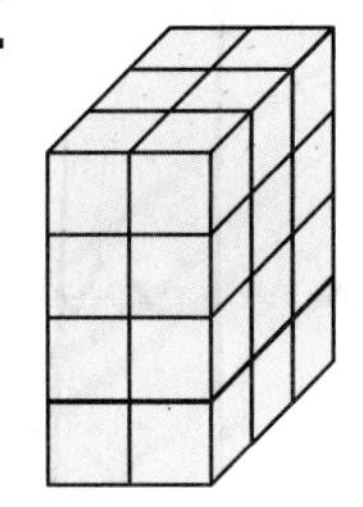

LESSON CONTINUES 

Practice and Problem Solving

Use cubes to make each solid. Then write the volume in cubic units.

6. 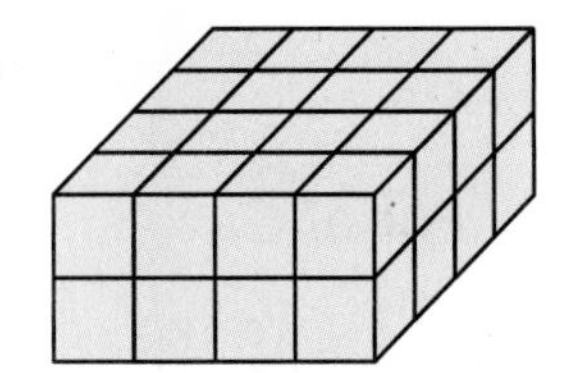

7. 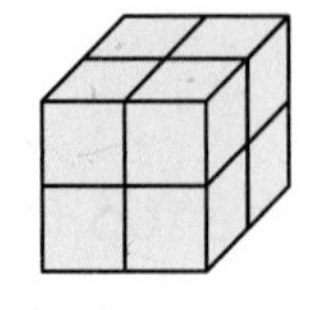

8. 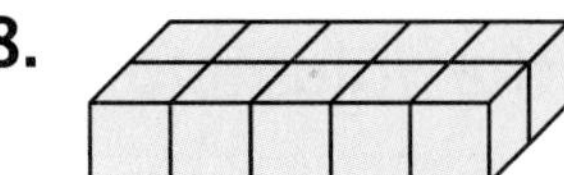

Find the volume of each solid. Write the volume in cubic units.

9. 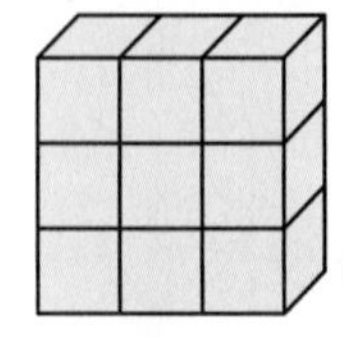

10. 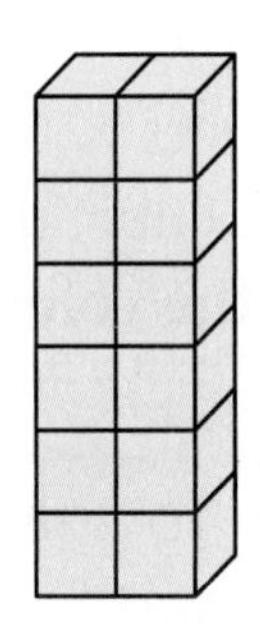

11. 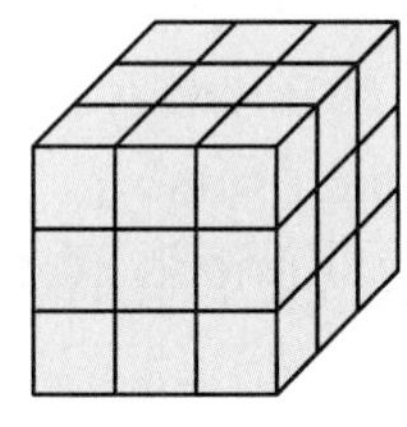

12. 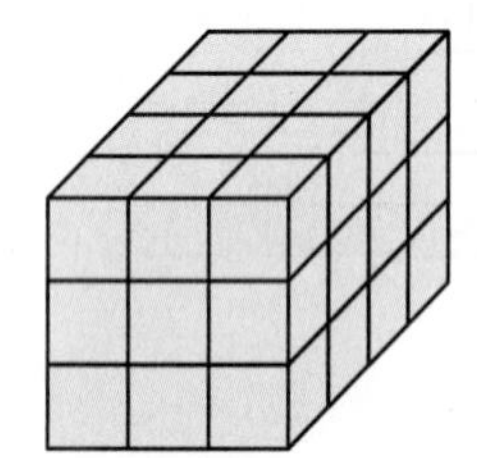

13. 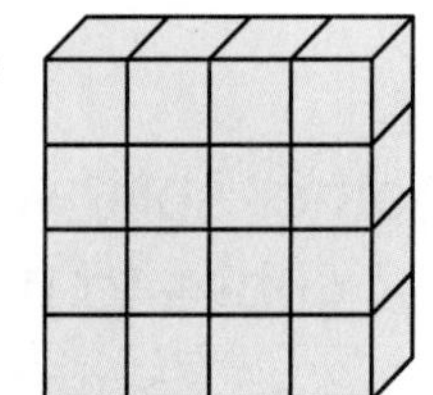

14. 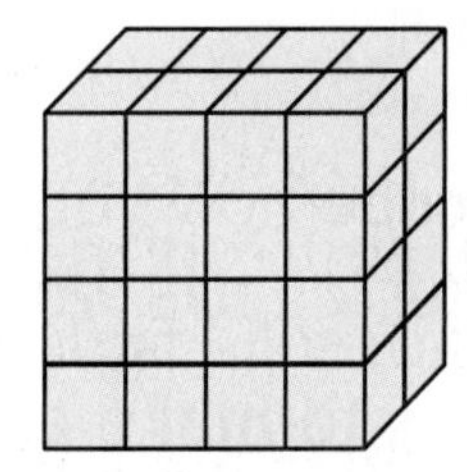

15. **ALGEBRA** Each layer of Andrew's prism is 6 cubic units. Its volume is 12 cubic units. How many layers are in the prism? You may use cubes to help.

16. **What's the Error?** Justin found the volume of this solid. He said the volume was 16 cubic units. Describe his error. Give the correct volume.

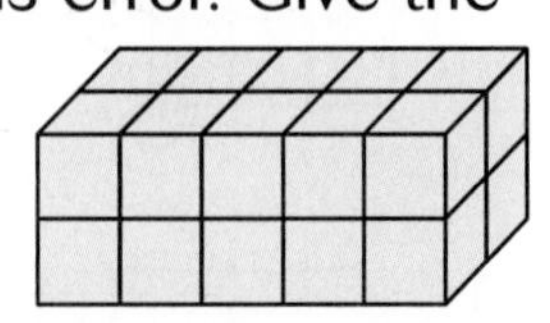

17. Sam's box is 2 cubes long, 2 cubes wide, and 3 cubes high. What is the volume of his box?

18. Todd's box is 4 cubes long and 4 cubes wide. It has a volume of 32 cubic units. What is the height of the box?

19. **Write About It** How is finding the area of a figure different from finding the volume of a solid?

Mixed Review and Test Prep

Find each sum or missing addend. (p. 68)

20. $45 + 12 + 5 = ■$

21. $35 + 15 + 20 = ■$

22. $102 + ■ + 15 = 120$

23. $■ + 46 + 10 = 61$

24. $13 + 22 + ■ = 42$

25. $200 + ■ = 302$

26. **TEST PREP** Jane had 205 stickers. She gave 28 stickers to her sister. How many stickers does Jane have left? (p. 58)

A 175 C 180
B 177 D 182

27. **TEST PREP** Which is NOT true? (p. 132)

F $0 \times 4 = 0$ H $1 \times 3 = 3$
G $0 \times 0 = 1$ J $7 \times 1 = 7$

PROBLEM SOLVING LINKUP ... to Reading

STRATEGY • ANALYZE INFORMATION To solve some problems, you need to *analyze,* or look carefully at, each part.

Analyze these drawings to identify the solid figure. Notice that the drawings of the top, side, and front views are plane shapes, not solid figures.

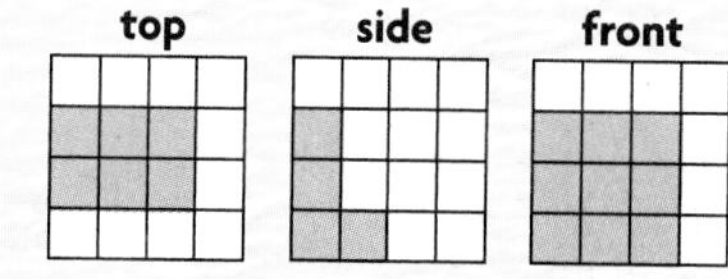

So, the solid figure looks like this.

Choose the solid figure that each set of drawings shows.

1.
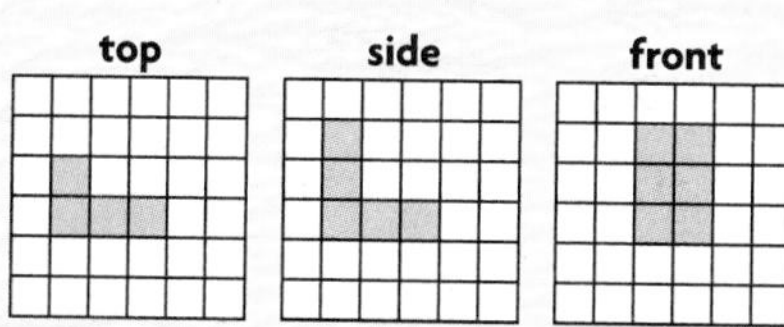

2.
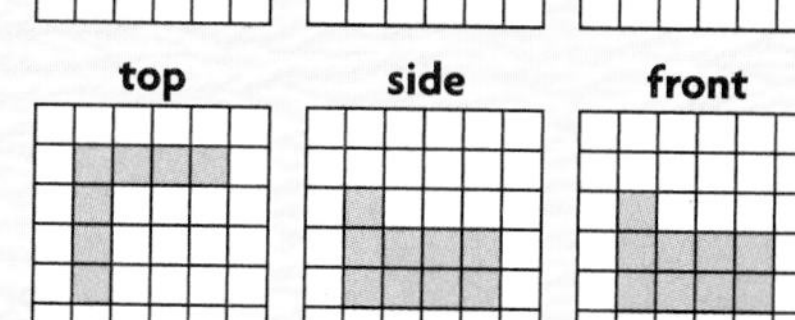

3.
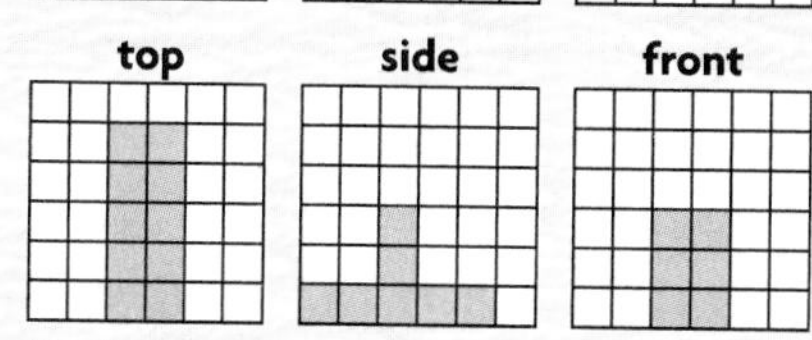

a.
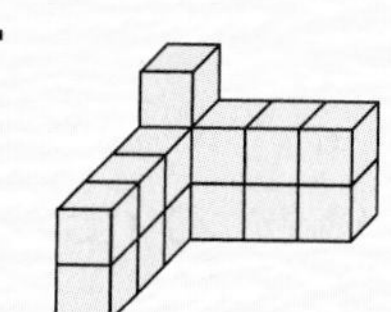

b.
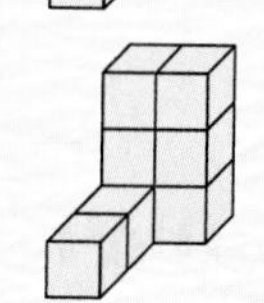

c.
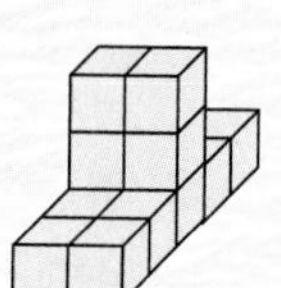

4. Build a figure with connecting cubes. Then use grid paper and draw its top, side, and front views.

EXTRA PRACTICE page H53, Set B

Review/Test

CHECK VOCABULARY AND CONCEPTS

Choose the best term from the box.

1. To measure volume, you use ? . (p. 398)
2. The distance around a figure is called its ? . (p. 388)

area
cubic units
perimeter

Find the perimeter of each figure. (pp. 388–389)

3. 

4.

5. 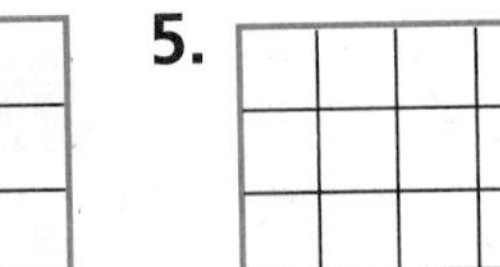

Write the area in square units. (pp. 394–395)

6. 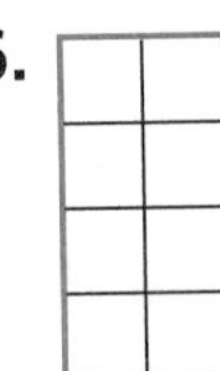

7. 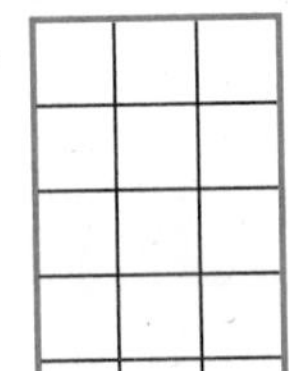

8.

CHECK SKILLS

Find the perimeter. (pp. 390–393)

9. 1 cm 1 cm 1 cm

10. 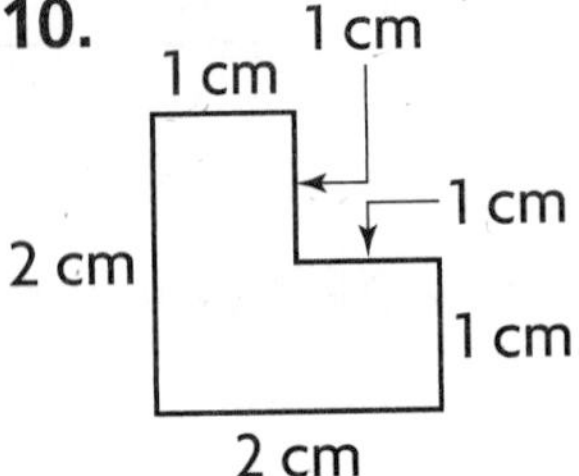

11. 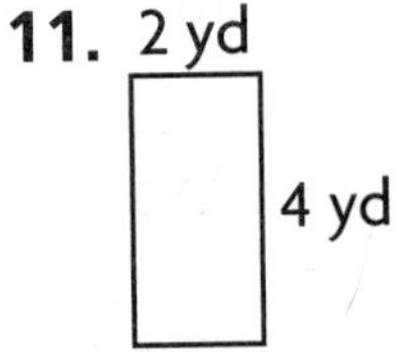

12. 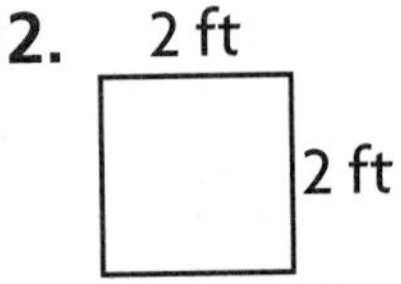

Write the volume in cubic units. (pp. 398–401)

13. 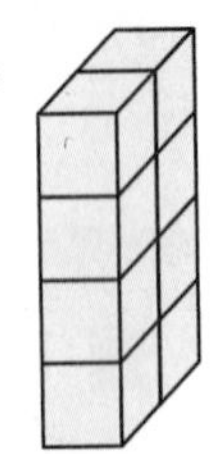

14. 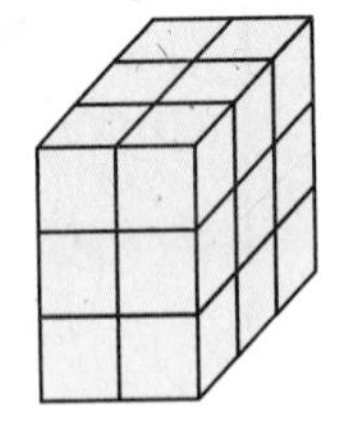

15. 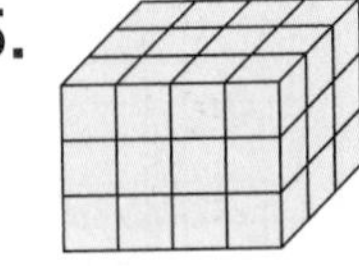

CHECK PROBLEM SOLVING

Solve. (pp. 396–397)

16. Pedro has 20 inches of ribbon. He wants to make a rectangle or square with the greatest possible area. How wide should it be? How long should it be?

17. Nora has 14 feet of fencing. She wants to make a rectangle with an area less than 10 square feet. How long should it be? How wide should it be?

Standardized Test Prep

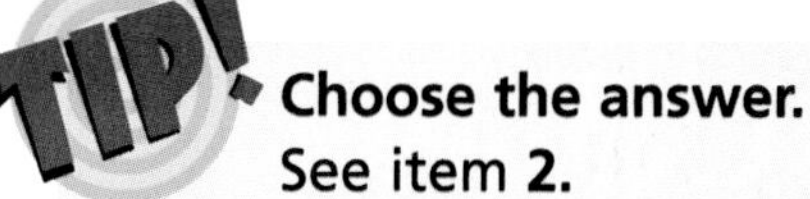

Choose the answer.
See item **2.**
Find the area for each answer choice. Compare and order the areas to find the greatest one.

Also see problem **6,** p. H65.

For 1–5, choose the best answer.

1. What is the perimeter?

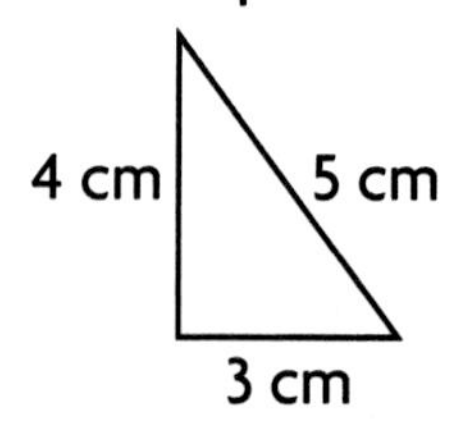

A 5 cm
B 12 cm
C 7 cm
D 60 cm

2. Debbie has 24 feet of fencing. She wants to build a pen for her puppies. Which measurements would have the greatest area?

F 3 feet wide, 9 feet long
G 4 feet wide, 8 feet long
H 5 feet wide, 7 feet long
J 6 feet wide, 6 feet long

3. What is the volume?

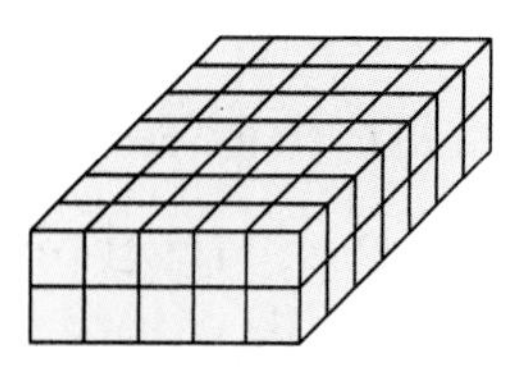

A 70 cubic units
B 35 cubic units
C 14 cubic units
D 10 cubic units

For 4–5, use the graph.

FAVORITE CAR COLOR

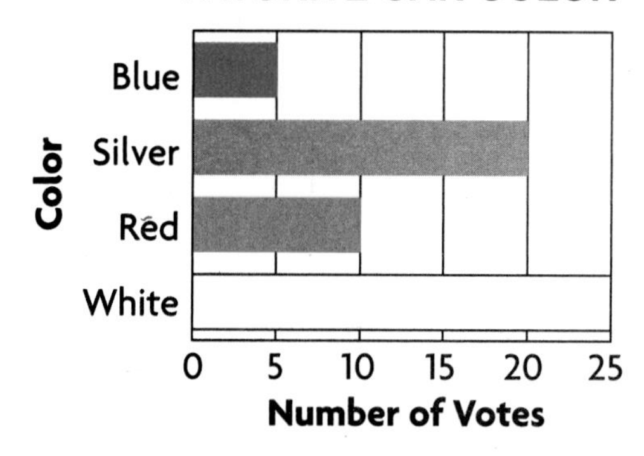

4. Which color received 20 votes?

F red
G white
H blue
J silver

5. How many people voted for a favorite car color?

A 60
B 50
C 40
D NOT HERE

Write What You Know

6. Explain how you can find the area of the figure below. What is the area?

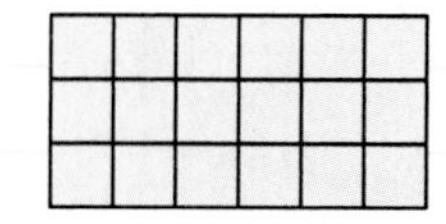

7. Explain how to find the perimeter of the figure below. What is the perimeter?

5 ft
2 ft

PROBLEM SOLVING MATH DETECTIVE

Around and Around You Go

REASONING Use grid paper to copy the array. Write the letters in each square as shown. Then read each clue and follow directions to solve each case!

Clue 1: What is the perimeter of the rectangle?

A	B	C	D
E	F	G	H
I	J	K	L

Clue 2: Use the same sheet of grid paper. Draw the figure with squares A, D, I, and L missing. What is the perimeter of this figure?

Clue 3: How is the perimeter of the second figure related to the perimeter of the first figure?

Clue 4: Look at the figure you drew in Clue 2. Can you remove squares from this figure and keep the same perimeter? If so, draw the new figure.

Clue 5: Look at the figure you drew in Clue 1. How could you remove a square or squares to change the perimeter to 16 units? Draw the new figure.

STRETCH YOUR THINKING Use grid paper. Draw an array with a perimeter of 12 units. Then copy that array, but remove squares to make a different figure with a perimeter of 12 units.

Challenge

Find Circumference of a Circle

Suppose you want to make a bank out of an empty can. How could you find the distance around the can?

The distance around a circle is its **circumference.**

Remember
The distance around an object or plane figure is called its perimeter.

Activity

MATERIALS: can, string, centimeter ruler

STEP 1

Wrap the string around the can.

STEP 2

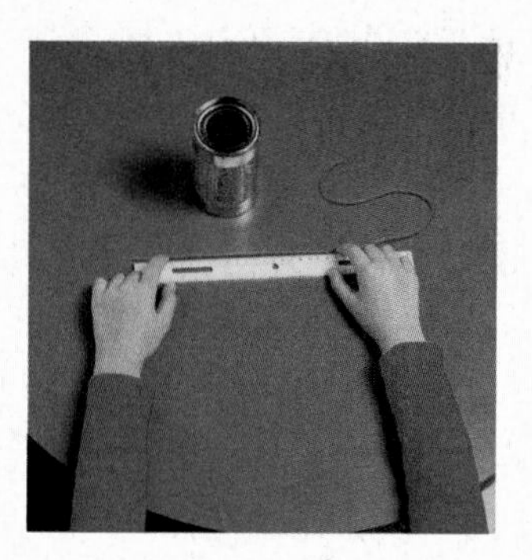

Use a ruler to measure the length of the string.

STEP 3

Copy the table. Write the circumference of the can to the nearest centimeter.

Object	Circumference
Can	■
Cup	■
Quarter	■
Doorknob	■

- How is finding the circumference of a circle like finding the perimeter of a rectangle? How is it different?

Try It

Find the circumference of each object. Record the measurement in your table.

1. drinking cup

2. quarter

3. doorknob

UNIT 7

Study Guide and Review

VOCABULARY

Choose the best term from the box.

square unit
perimeter

1. A square with a side length of 1 unit is a ___?___. (p. 394)

STUDY AND SOLVE

Chapter 20

Measure to the nearest inch.

The pin is 3 inches long to the nearest inch.

Measure to the nearest inch. (pp. 352–355)

2.

3.

Change units.

Use a rule to find the number of quarts in 2 gallons.
Remember: 4 quarts = 1 gallon
Rule: Multiply the number of gallons by 4.
■ quarts = 2 gallons
■ = 2 × 4

8 = 2 × 4
So, there are 8 quarts in 2 gallons.

Use the rule to change the units (8 pints = 1 gallon). (pp. 362–365)

4. How many pints are in 5 gallons?
Rule: Multiply the number of gallons by 8.
■ = 5 × 8
■ pints = 5 gallons

Chapter 21

Measure to the nearest centimeter.

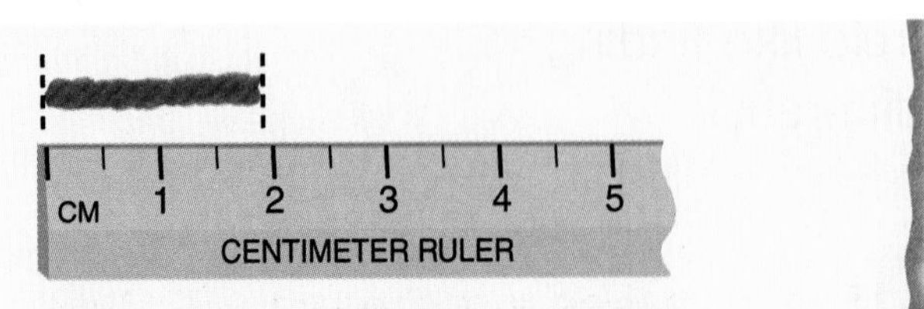

The string is 2 cm long to the nearest centimeter.

Measure to the nearest centimeter. (pp. 372–375)

5.

6.

7.

8.

Estimate length and distance.

Choose the better estimate.

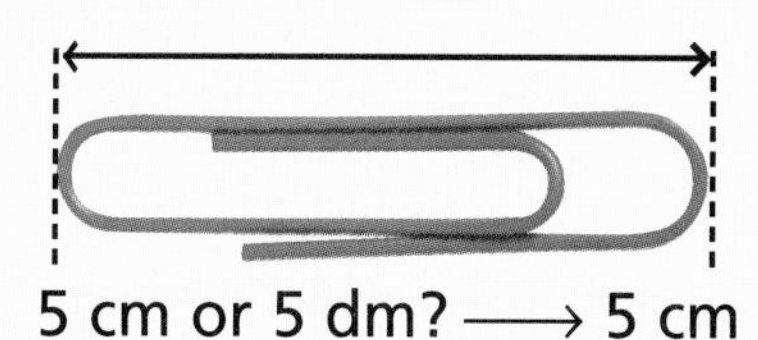

5 cm or 5 dm? ⟶ 5 cm

Choose the better estimate.
(pp. 372–375)

9. Your thumb: 5 dm or 5 cm?
10. A lamp post: 3 dm or 3 m?
11. A pen: 15 cm or 15 dm?

Chapter 22

Find perimeter and area.

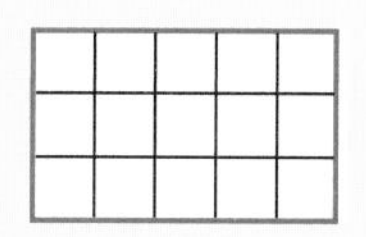

Add the lengths of the sides to find the perimeter.
$3 + 5 + 3 + 5 = 16$
So, the perimeter is 16 units.
Count the number of square units to find the area.
So, the area is 15 square units.

Find the perimeter. (pp. 388–393)

12. 3 cm, 1 cm, 1 cm, 3 cm

Write the area in square units.
(pp. 394–395)

13.

14.

Find volume.

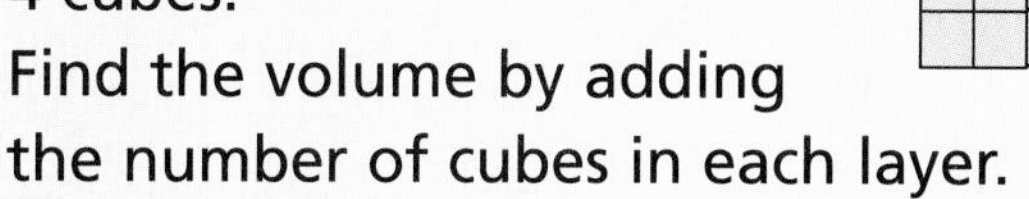

This figure has 2 layers of 4 cubes.
Find the volume by adding the number of cubes in each layer.
$4 + 4 = 8$
So, the volume is 8 cubic units.

Write the volume in cubic units.
(pp. 398–401)

15. 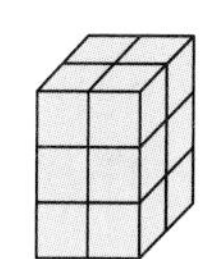

16. 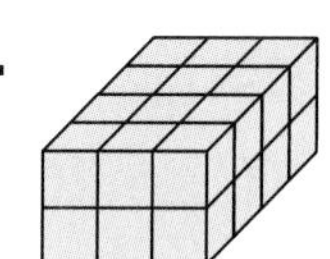

PROBLEM SOLVING PRACTICE

Solve. (pp. 376–377, 396–397)

17. Linda's room is 400 cm wide. How many meters wide is it?

18. Jed makes a rectangle with the greatest possible area using 16 inches of string. What are the length and width of his rectangle?

PERFORMANCE ASSESSMENT

TASK A • YARN DOLLS

Materials: ruler

Natalie is making yarn dolls. These pictures show the lengths of some of the pieces of yarn Natalie needs.

1.

2.

3.

YARN MEASUREMENT	
Estimate	**Actual**
1.	
2.	
3.	

a. Copy the table. Estimate the length of each piece of yarn to the nearest inch. Record your estimates in the table.

b. Use a ruler to measure each piece of yarn to the nearest half inch. Record your measurements next to the estimates.

c. Compare each estimate to the measured length. Write *greater than, less than,* or *equal.*

TASK B • HOME FOR A PET

Tracy and her dad are making a pen for Tracy's pet rabbit. They will use 20 feet of fencing. The pen will have a rectangular shape.

a. On a grid, draw two possible rectangular-shaped pens that they can make. Label the length and width of each rectangle. Write the area of each pen inside the rectangle.

b. Which of the pens has the greater area? Is this the greatest area possible for a pen with a perimeter of 20 feet? Explain how you know.

Mighty Math Number Heroes • Perimeter and Area

You can use Mighty Math Number Heroes to find perimeter and area.

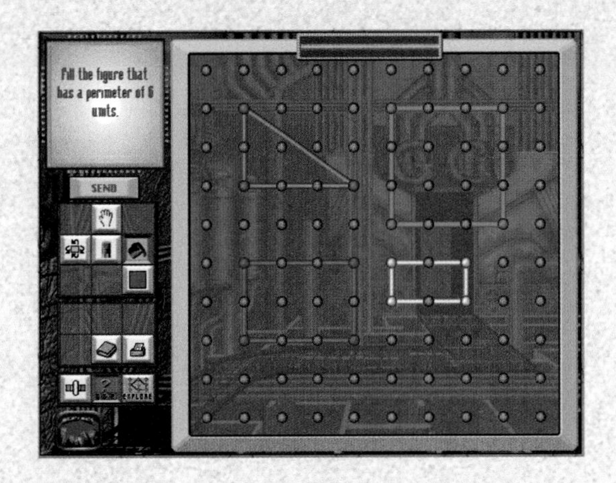

- Click on *Geoboard*.
- Click [icon]. Choose Level G, O, or W.
- Answer at least 5 questions.
- Draw pictures to record your work.

Practice and Problem Solving

Click [EXPLORE icon]. Use the Geoboard to draw each figure. Use grid paper to record each figure you draw.

1. Draw a rectangle that has a perimeter of 10 units.
2. Draw a square that has a perimeter of 12 units.
3. Draw a polygon that has a perimeter of 14 units.
4. Draw a polygon that has a perimeter of 18 units.
5. Draw a rectangle that has an area of 14 square units.
6. Draw a square that has an area of 25 square units.
7. Jessica draws a rectangle on the Geoboard. The width is 4 units and the length is 5 units. What are the perimeter and area of the rectangle?
8. **Write a problem** about the perimeter and area of a playground. Use the Geoboard to draw a picture for your problem.

Multimedia Math Glossary www.harcourtschool.com/mathglossary

9. **Vocabulary** Review the customary units used to measure capacity. Draw a picture to show how to change *pints* to *cups* and *gallons* to *quarts*.

When Cornelius, the Cobasaurus, was built, it was the largest maze in the world.

PROBLEM SOLVING ON LOCATION

at the Farm

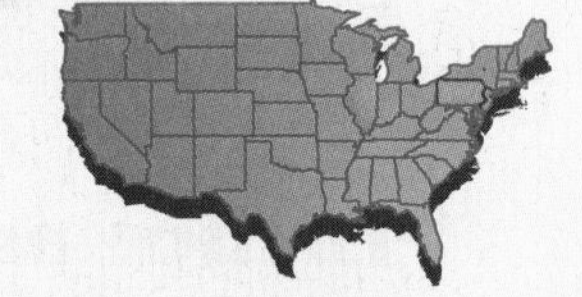

ANNVILLE, PENNSYLVANIA

You use a pencil to find the way out of a maze in a puzzle. Since the maze in this photo is in a cornfield, people walk through it to find the way out!

CORNELIUS, THE COBASAURUS	
Estimated Length	693 feet
Estimated Width	206 feet
Total Length of Paths	2 miles

For 1–5, use the information in the table.

1. Suppose you walk the perimeter of this maze. About how far would you walk?

2. A classmate says that this maze is 600 feet longer than it is wide. Is this true? Explain.

3. Which is longer, a walk around the perimeter of the maze or a walk around the paths? (HINT: One mile = 5,280 feet.) How much longer? Explain.

4. Suppose you want to measure the perimeter of a maze like Cornelius in metric units. What unit would you use?

5. **REASONING** Suppose another maze is built. It has the same perimeter as Cornelius. Its length and width are not the same measurements as the length and width of Cornelius. What could the length and width of the new maze be? Tell how you found your answer.

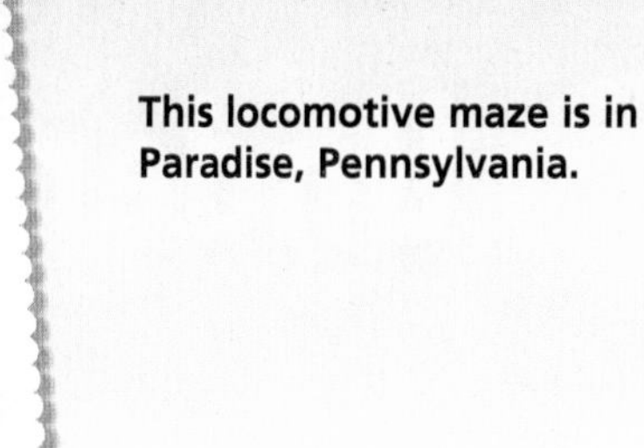

This locomotive maze is in Paradise, Pennsylvania.

It would probably take about 2 hours to walk through this maze of Albert Einstein.

VERNON, NEW JERSEY

Albert Einstein was a famous scientist. He lived in New Jersey for much of his life. This maze was made to look like him.

Draw your own maze. You will need centimeter grid paper and a pencil.

1. Draw a square with sides that are each 15 centimeters long. Label the length and width. What is the perimeter?
2. Shade in some of the centimeter squares to form a maze. Then draw the winning path through your maze. Include some dead-end paths in your maze.
3. Find the length of the winning path. Count the centimeters on the grid paper. What is the length of your winning path?

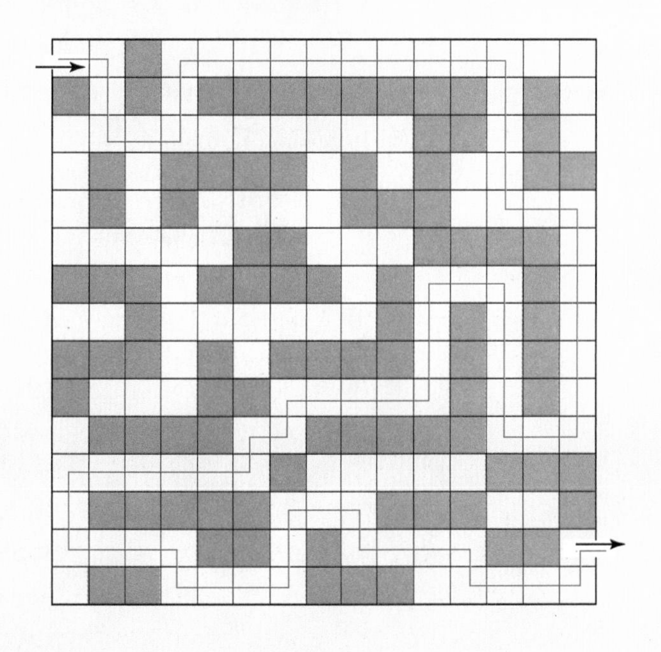

Understand Fractions

There are thousands of different kinds of insects in the world. Some are very large. Others are so small that we can't even see them.

PROBLEM SOLVING The table shows some insects and their sizes. Which insect is the smallest? Which is the largest?

SIZES OF INSECTS

Insect	Length
Ant	$\frac{1}{4}$ inch
Mosquito	$\frac{1}{8}$ inch
Grasshopper	2 inches
Goliath beetle	$4\frac{1}{2}$ inches

African Goliath beetle

Use this page to help you review and remember important skills needed for Chapter 23.

MODEL PARTS OF A WHOLE (For Intervention, see p. H28.)

Write how many equal parts make up the whole figure. Then write how many parts are *not* shaded.

1. 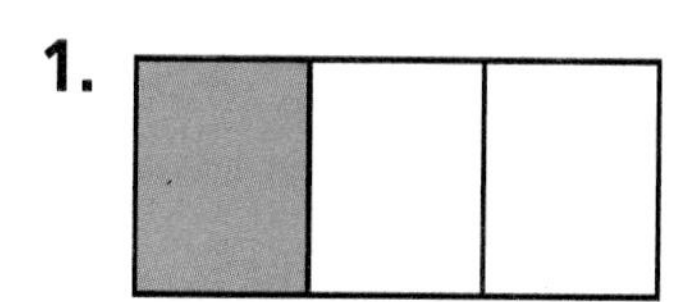

2. 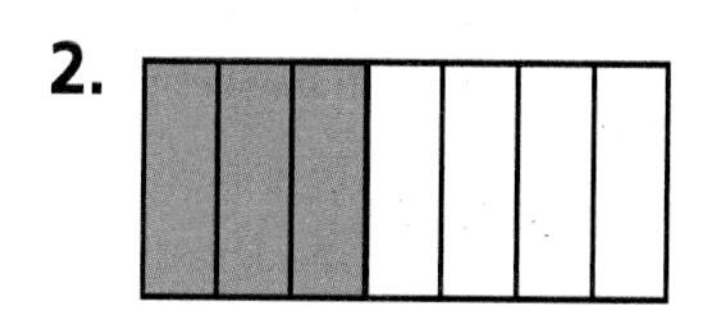

3. 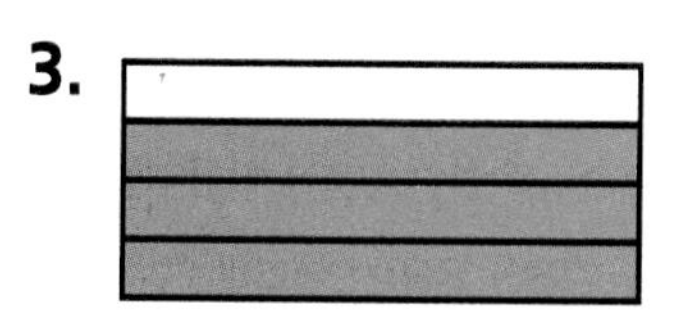

4. 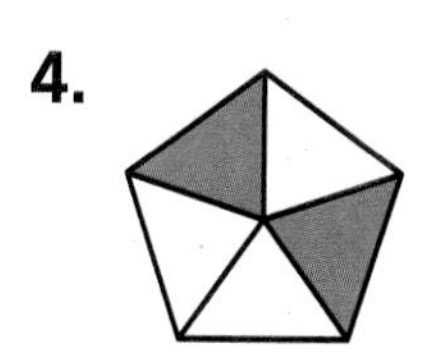

5. 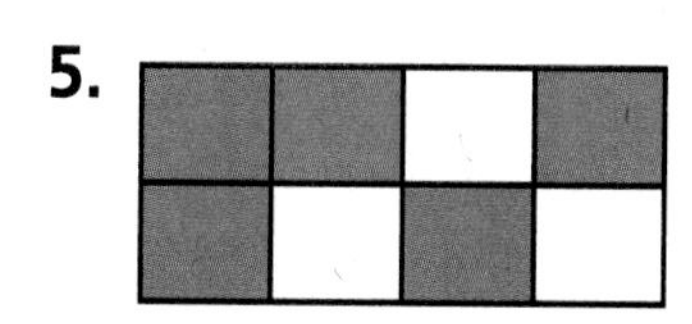

6. 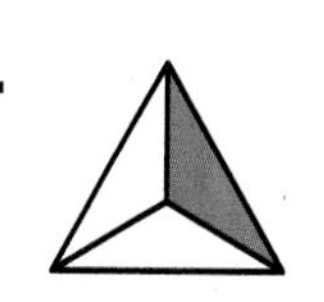

Write how many equal parts make up the whole figure. Then write how many parts *are* shaded.

7. 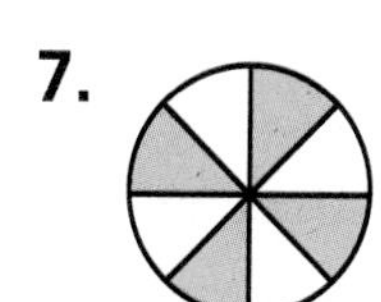

8. 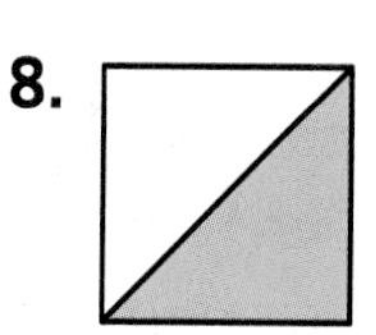

9. 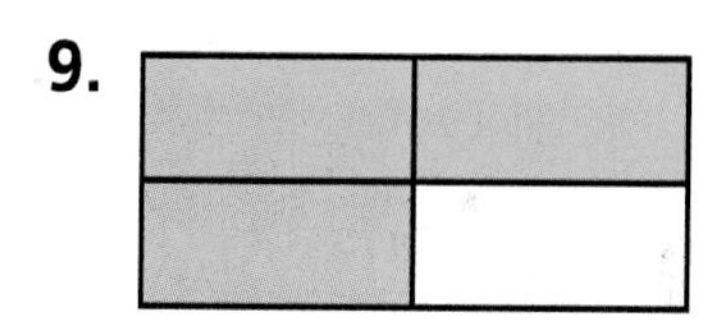

MODEL PARTS OF A GROUP (For Intervention, see p. H29.)

Write the number in each group. Then write the number in each group that is *not* green.

10. ★☆☆☆

11. 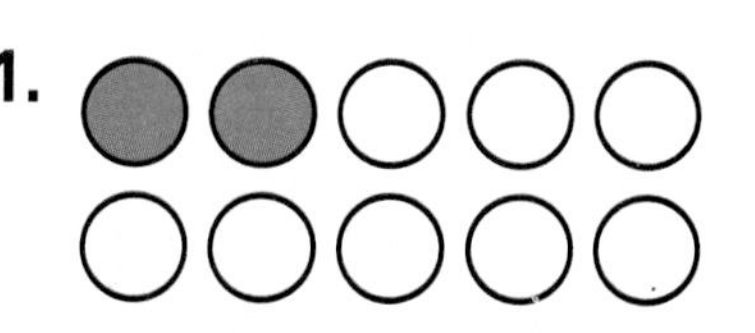

12. 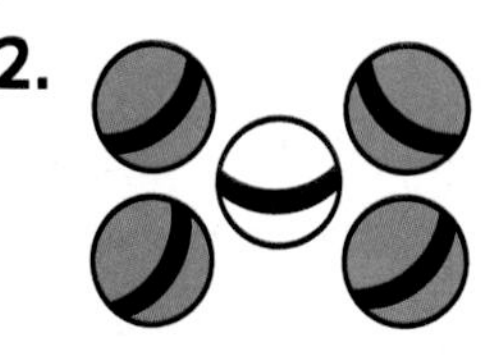

Write the number in each group. Then write the number in each group that *is* green.

13. ▲▲▲△△△ △△△△△△

14. 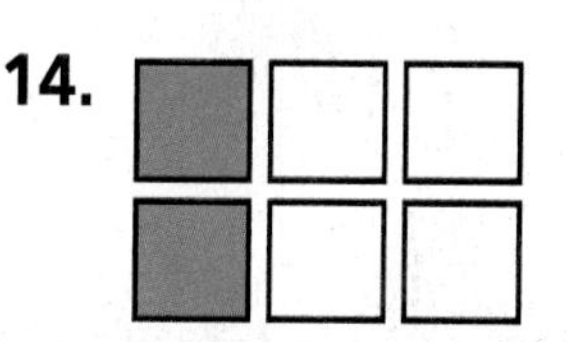

15. 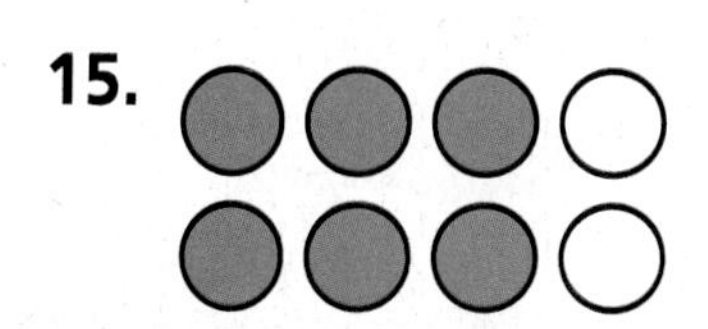

LESSON

Count Parts of a Whole

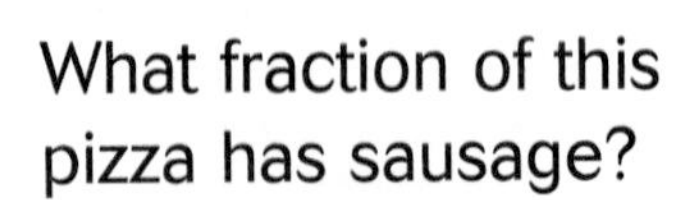

Learn

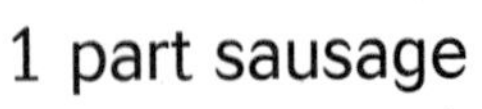

ALL TOGETHER

A number that names part of a whole or part of a set is called a **fraction**.

What fraction of this pizza has sausage?

1 part sausage → 1 ← numerator
6 equal parts in all → 6 ← denominator

Read: one sixth **Write:** $\frac{1}{6}$

one part out of six parts
1 divided by 6

So, $\frac{1}{6}$ of the pizza has sausage.

The **numerator** tells how many parts are being counted.

The **denominator** tells how many equal parts are in the whole.

- What fraction of the pizza does *not* have sausage? Explain how you know.

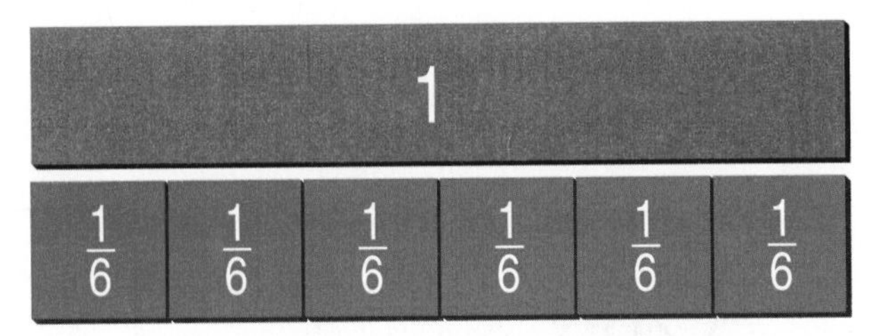

These fraction bars show how a whole can be divided into sixths, or six equal parts.

Quick Review

Find the missing number in the pattern.

1. 2, 4, 6, 8, ■
2. 12, 11, 10, 9, ■
3. 3, 6, 9, 12, ■
4. 11, 9, 7, 5, ■
5. 4, 8, 12, 16, ■

VOCABULARY

fraction
numerator
denominator

Examples A whole can be divided into equal parts.

A

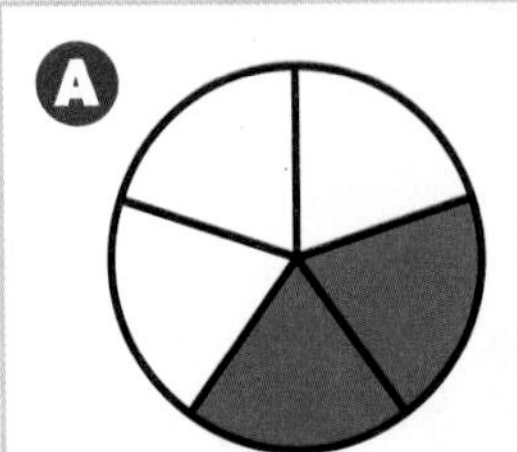

$\frac{2}{5}$
two fifths

B

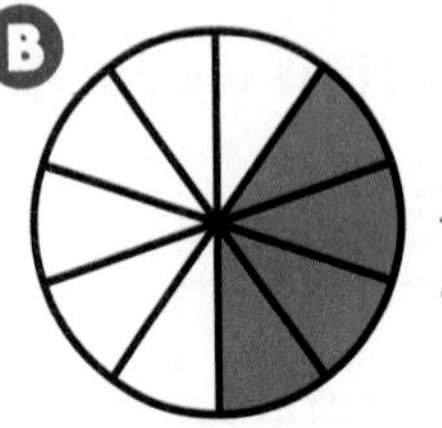

$\frac{4}{10}$
four tenths

C

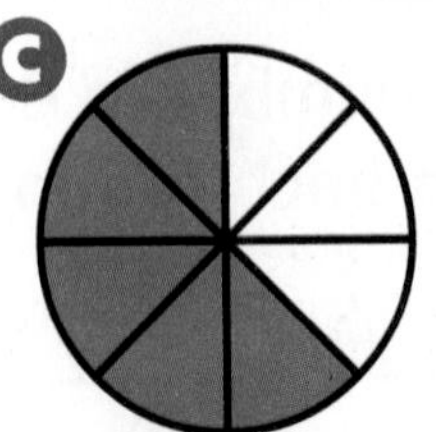

$\frac{5}{8}$
five eighths

Counting Equal Parts

You can count equal parts, such as sixths, to make one whole.

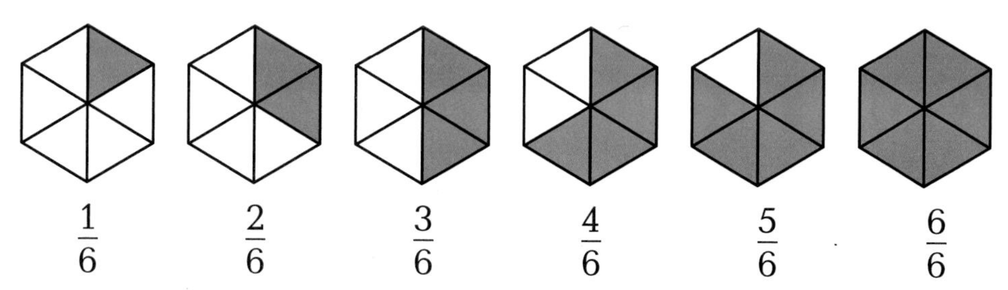

$\frac{1}{6}$ $\frac{2}{6}$ $\frac{3}{6}$ $\frac{4}{6}$ $\frac{5}{6}$ $\frac{6}{6}$

$\frac{6}{6}$ = one whole

A number line can show parts of one whole.

The part from 0 to 1 on these number lines shows one whole. The line can be divided into any number of equal parts.

This number line is divided into sixths.

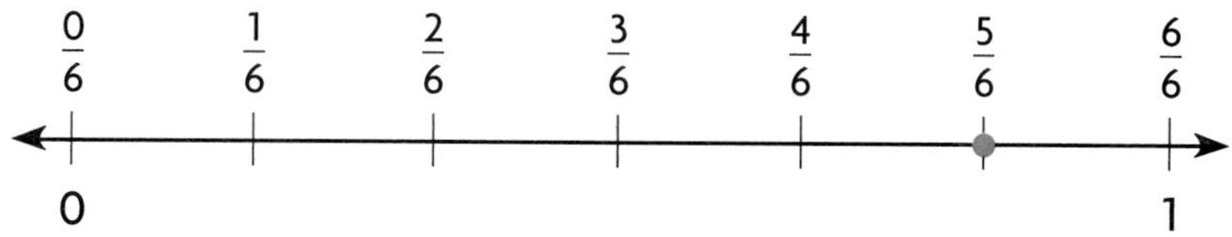

The point shows the location of $\frac{5}{6}$.

Examples

A This number line is divided into thirds.

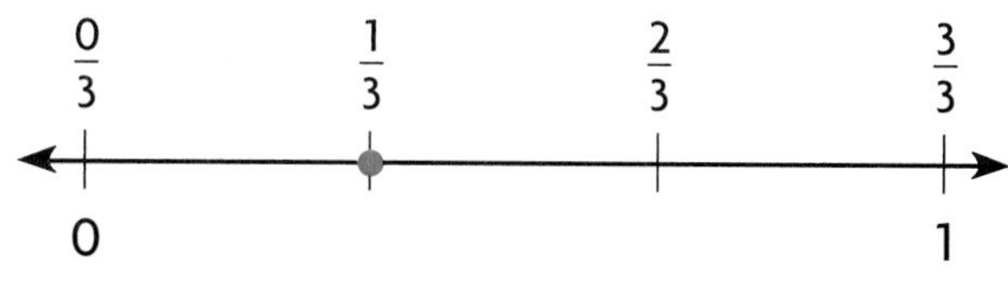

The point shows the location of $\frac{1}{3}$.

B This number line is divided into fourths.

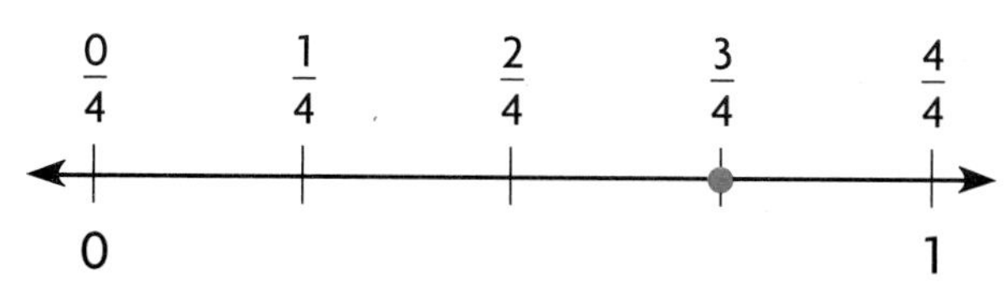

The point shows the location of $\frac{3}{4}$.

Check

1. **Write** how to count by eighths to make one whole.

Write a fraction in numbers and words that names the shaded part.

2. 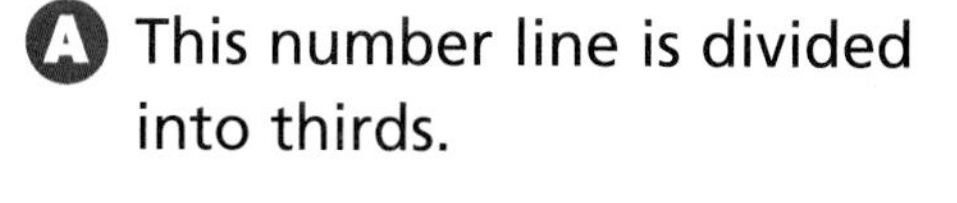

3. 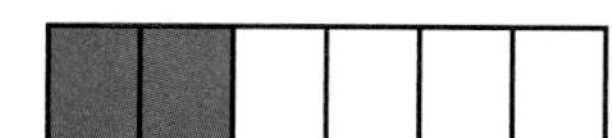

4. 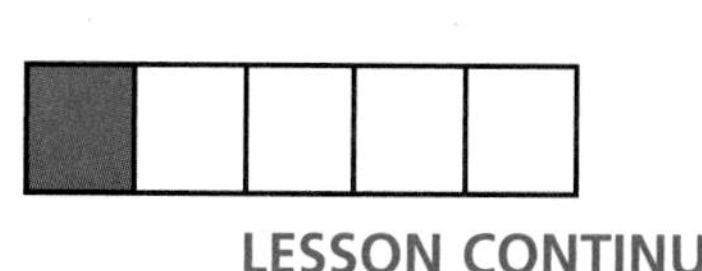

LESSON CONTINUES

Practice and Problem Solving

Write a fraction in numbers and words that names the shaded part.

5. 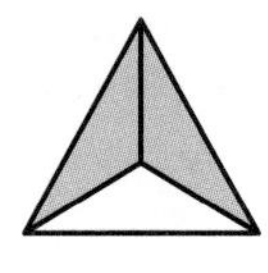

6. 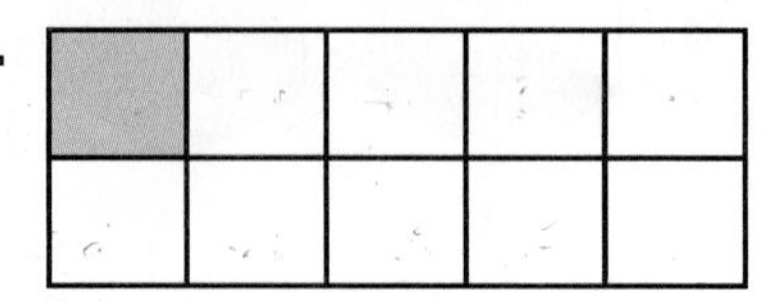

7. 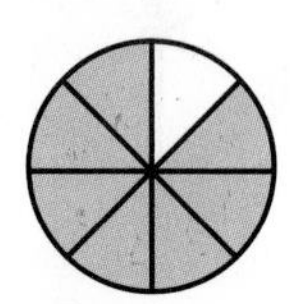

8. 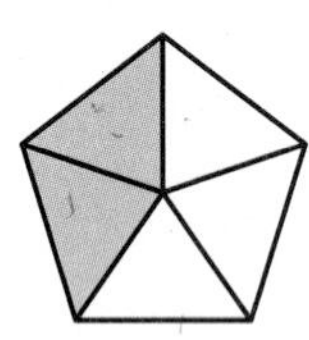

9. 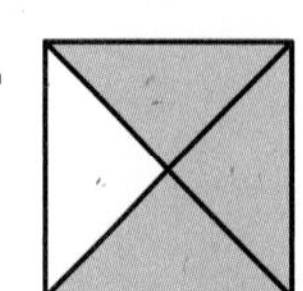

10. 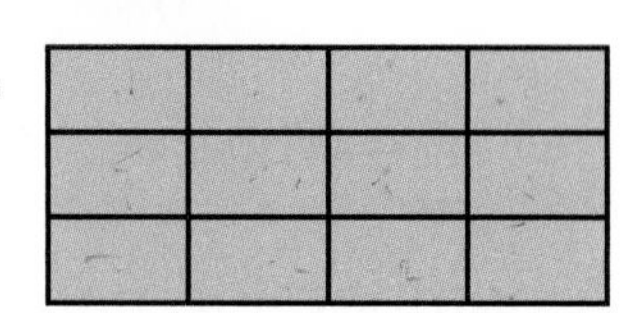

Write the fraction, using numbers.

11. one fourth

12. four out of nine

13. six sevenths

14. two divided by three

15. three fifths

16. five out of twelve

Write a fraction to describe each shaded part.

17. 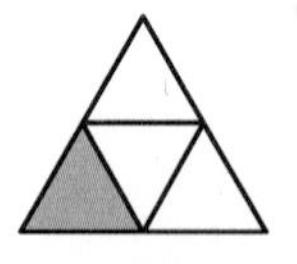 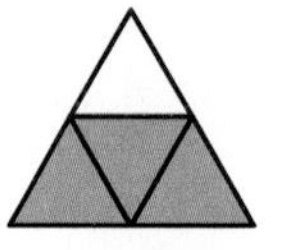 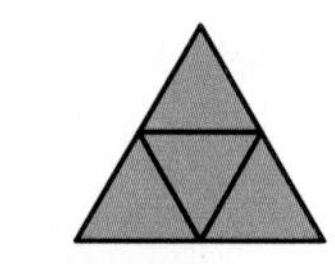

18. 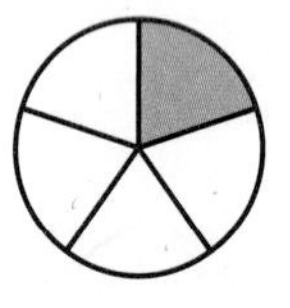 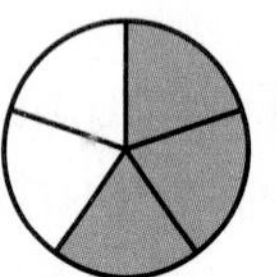 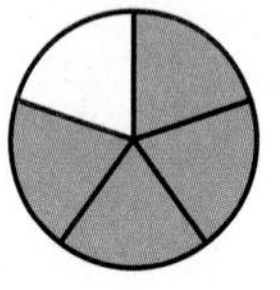 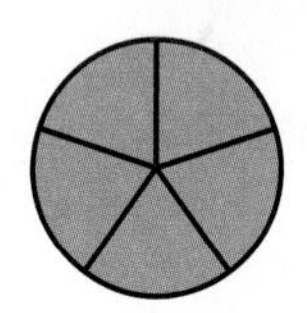

ALGEBRA **Write a fraction that names the point of each letter on the number line.**

19. 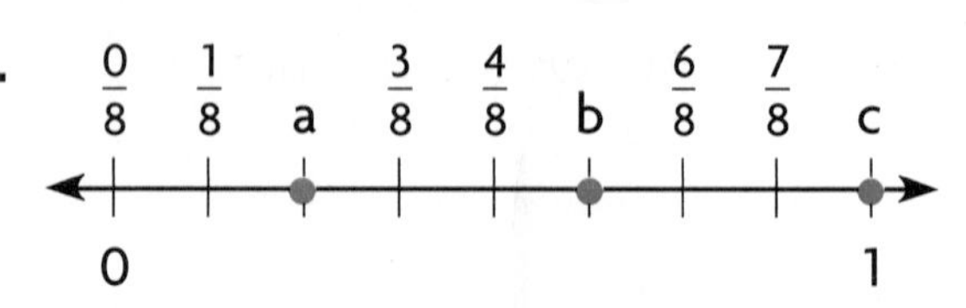

20. 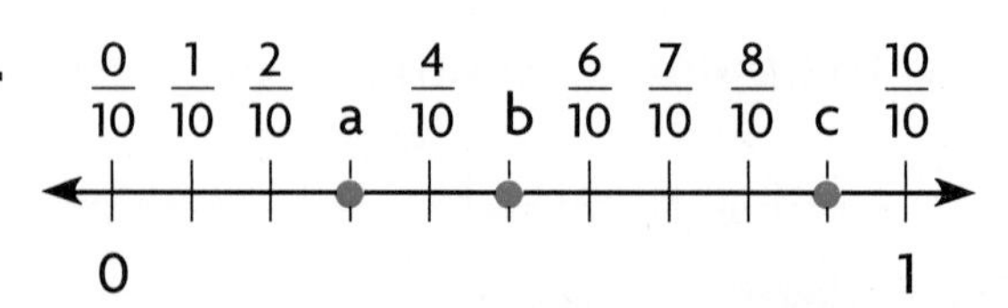

21. **REASONING** There are two pizzas the same size. One is cut into 6 equal pieces. The other is cut into 8 equal pieces. Which pizza has smaller pieces?

22. **What's the Error?** Lydia said the fraction that names this drawing is $\frac{6}{4}$. Explain Lydia's mistake.

23. Suppose you and 4 friends share equal pieces of a pie. What fraction describes your part?

24. Use crayons to draw a picture that shows each fraction.

a. $\frac{1}{2}$ purple b. $\frac{1}{4}$ green

Mixed Review and Test Prep

Find each difference. (p. 58)

25. $245 - 52 = \blacksquare$

26. $736 - 379 = \blacksquare$

27. $400 - 118 = \blacksquare$

Find each product.

28. $4 \times 8 = \blacksquare$ (p. 134)

29. $6 \times 6 = \blacksquare$ (p. 148)

30. $7 \times 9 = \blacksquare$ (p. 150)

31. $7 \times 0 = \blacksquare$ (p. 132)

32. $9 \times 9 = \blacksquare$ (p. 164)

33. $5 \times 1 = \blacksquare$ (p. 132)

34. **TEST PREP** Which polygon has 3 sides and 3 angles? (p. 314)

A triangle
B quadrilateral
C pentagon
D hexagon

35. **TEST PREP** Theo measured the distance from his house to the hobby store across town. Which unit did he use to measure? (p. 356)

F inch
G foot
H yard
J mile

PROBLEM SOLVING LINKUP . . . to Social Studies

These alphabet flags are used on ships to send messages in code. For example, if a ship flies the flag for the letter P, that ship is about to sail out of the harbor. Ships carry books that explain the codes in nine different languages.

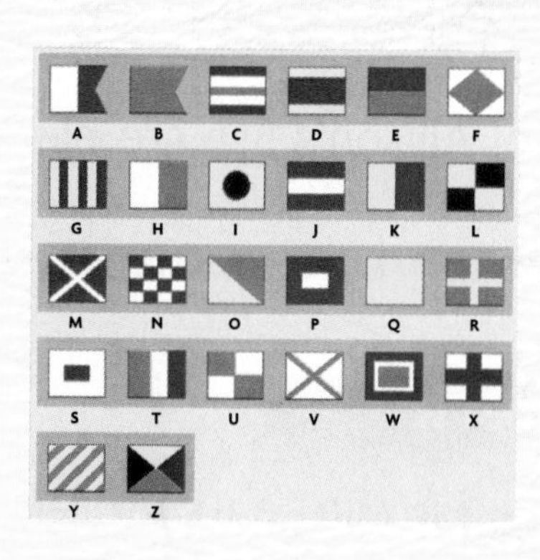

USE DATA For 1–4, use the flags.

1. Look at the flag for the letter G. What fraction names the part of the flag that is yellow?

2. Look at all of the flags. Which of them are divided into four equal parts, or fourths?

3. Look at the flag for the letter N. Into how many equal parts is the flag divided?

4. Write a fraction that names one part of each of these flags: L, O, and T.

LESSON 2

Count Parts of a Group

Quick Review

Write the fraction that comes next in the counting pattern.

1. $\frac{0}{3}, \frac{1}{3}, \frac{2}{3},$ ■
2. $\frac{1}{5}, \frac{2}{5}, \frac{3}{5},$ ■
3. $\frac{2}{6}, \frac{3}{6}, \frac{4}{6},$ ■
4. $\frac{3}{8}, \frac{4}{8}, \frac{5}{8},$ ■
5. $\frac{3}{10}, \frac{4}{10}, \frac{5}{10},$ ■

Learn

GO FISH Allison and her dad went to Pal's Pet Store. Allison chose 8 fish and her dad chose 8 fish. What fraction of the fish are black mollies?

Allison's Fish

Dad's Fish

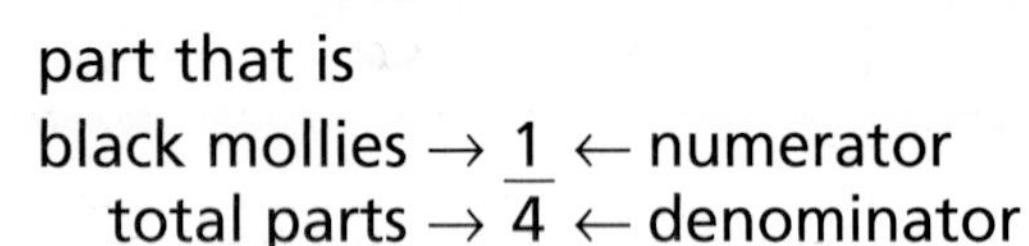

part that is
black mollies → $\underline{1}$ ← numerator
total parts → 4 ← denominator

Read: one fourth, or one out of four

Write: $\frac{1}{4}$

So, $\frac{1}{4}$ of Allison's fish are black mollies.

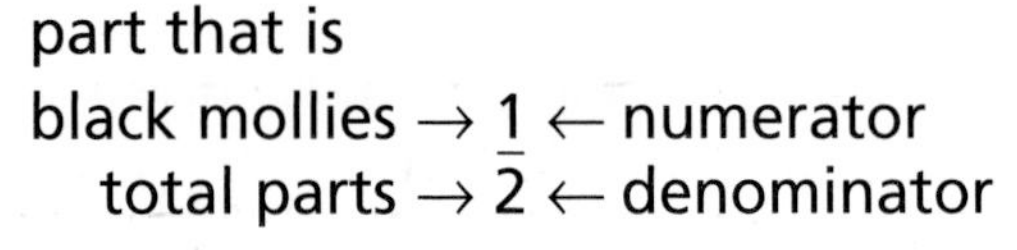

part that is
black mollies → $\underline{1}$ ← numerator
total parts → 2 ← denominator

Read: one half, or one out of two

Write: $\frac{1}{2}$

So, $\frac{1}{2}$ of Dad's fish are black mollies.

MATH IDEA Use fractions to show parts of a group.

Check

1. **What if** Allison had 2 fish in each of 6 bags? What fraction names the fish in one bag?

Write the fraction that names the circled part of each group.

2.

3.

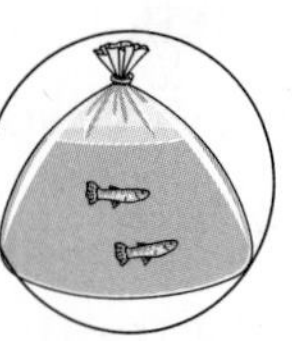

Practice and Problem Solving

For 4–7, write the fraction that names the part of each group that is circled.

4.

5. ◉ ● ● ○ ●

6.

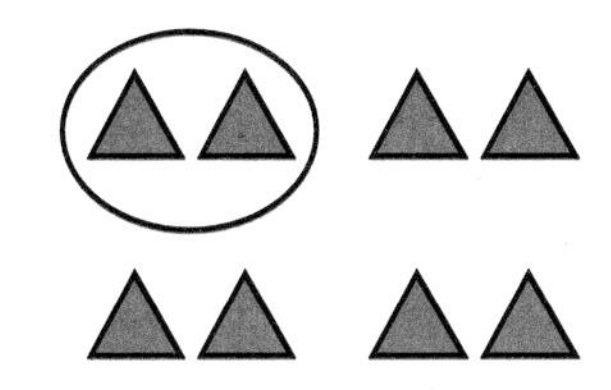

7.

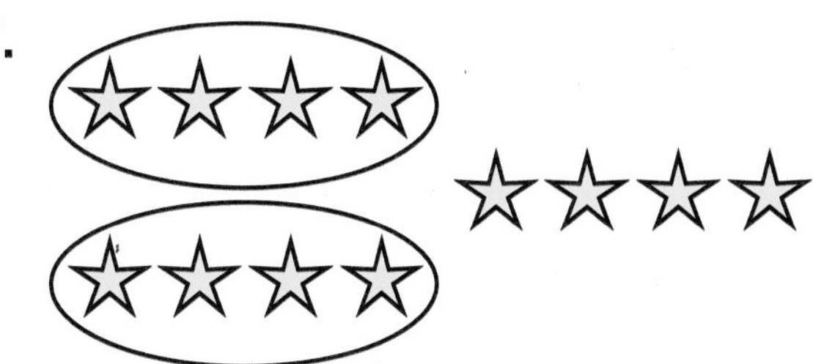

8. Draw 9 circles. Circle $\frac{1}{3}$ of them.

9. Draw 6 triangles. Circle $\frac{1}{2}$ of them.

10. Draw 16 stars. Circle $\frac{3}{4}$ of them.

Use a pattern to complete the table.

11.	Model	○○○○○	●○○○○	■	●●●○○	●●●●○	●●●●●
12.	Total number of parts	5	■	5	5	5	5
13.	Number of green parts	0	1	2	■	4	5
14.	Fraction of green parts	$\frac{0}{5}$	$\frac{1}{5}$	$\frac{2}{5}$	$\frac{3}{5}$	$\frac{4}{5}$	■

15. Esther has 12 ribbons. Of those ribbons, $\frac{1}{12}$ are red and $\frac{2}{12}$ are blue. The rest are yellow. How many yellow ribbons does she have?

16. **What's the Question?** Jonas has 4 blue tiles, 3 green tiles, and 1 yellow tile. The answer is $\frac{7}{8}$.

17. **Write a problem** in which a fraction is used to name part of a group. Tell what the numerator and denominator mean.

Mixed Review and Test Prep

Find each quotient. (p. 206)

18. $4 \div 4 = $ ■

19. $9 \div 1 = $ ■

20. $0 \div 7 = $ ■

21. $8 \div 8 = $ ■

22. **TEST PREP** Find the number that the variable stands for.
$n \times 7 = 21$ (p. 188)

A 2
B 3
C 4
D 14

LESSON

3 Equivalent Fractions

Quick Review

Name the fraction for the shaded part.

1. 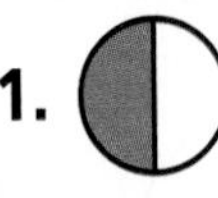
2. 
3. 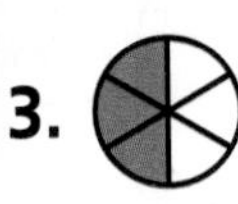
4. 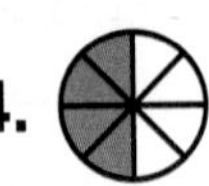
5. 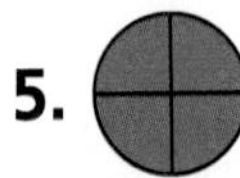

VOCABULARY

equivalent fractions

Learn

EQUAL PARTS Two or more fractions that name the same amount are called **equivalent fractions**.

What other fractions name $\frac{1}{2}$?

Activity 1

MATERIALS: sheet of paper

STEP 1

Fold a sheet of paper in half. Shade one half of the paper blue.

1 out of 2 equal parts is blue.

$\frac{1}{2}$ of the paper is blue.

STEP 2

Fold the paper in half again.

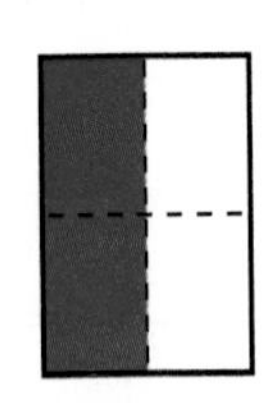

2 out of 4 equal parts are blue.

$\frac{2}{4}$ of the paper is blue.

STEP 3

Fold the paper in half a third time.

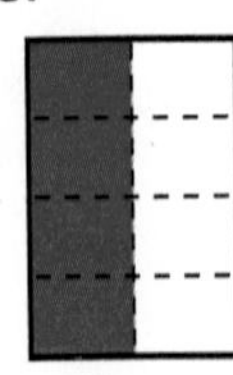

4 out of 8 equal parts are blue.

$\frac{4}{8}$ of the paper is blue.

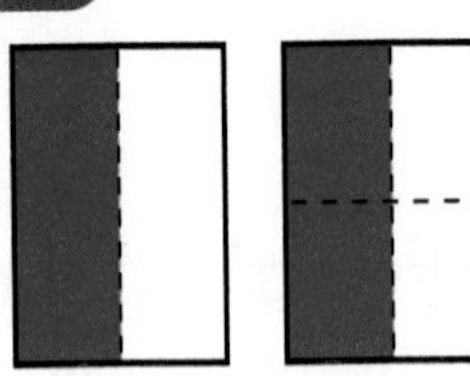

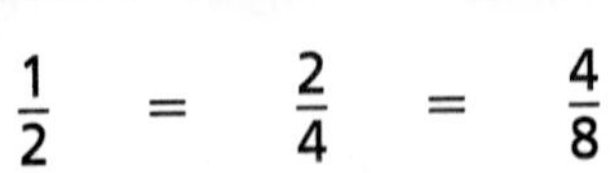

$\frac{1}{2} = \frac{2}{4} = \frac{4}{8}$

So, $\frac{1}{2}$, $\frac{2}{4}$, and $\frac{4}{8}$ are all names for $\frac{1}{2}$.

They are equivalent fractions.

Technology Link

More Practice:
Use E-lab, *Equivalent Fractions.*

www.harcourtschool.com/elab2002

Activity 2

Use fraction bars to find equivalent fractions.
What fraction is equivalent to $\frac{2}{3}$?

Start with the bar for 1 whole. Line up two $\frac{1}{3}$ bars for $\frac{2}{3}$.

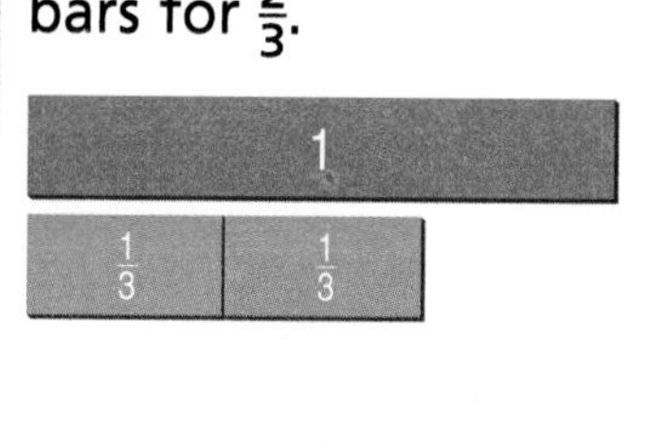

STEP 2

Use $\frac{1}{6}$ bars to match the length of the bars for $\frac{2}{3}$.

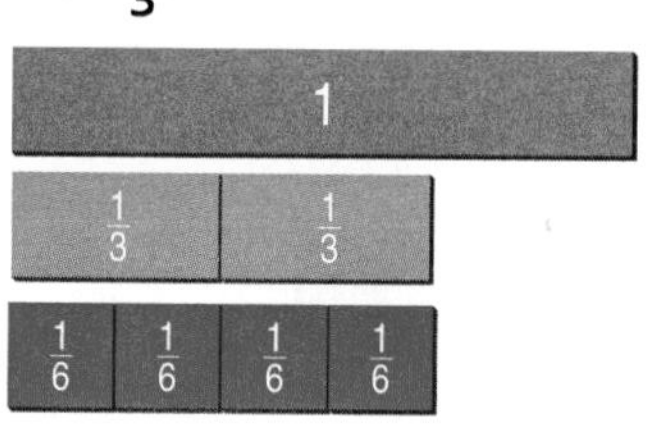

STEP 3

Count the number of $\frac{1}{6}$ bars that make up $\frac{2}{3}$. Write the equivalent fraction.

Count: $\frac{1}{6}$ $\frac{2}{6}$ $\frac{3}{6}$ $\frac{4}{6}$

Write: $\frac{2}{3} = \frac{4}{6}$

- How can you tell that the fraction bars show equivalent fractions?
- Use sixths, eighths, and tenths fraction bars. Find fractions that are equivalent to $\frac{1}{2}$.

Examples

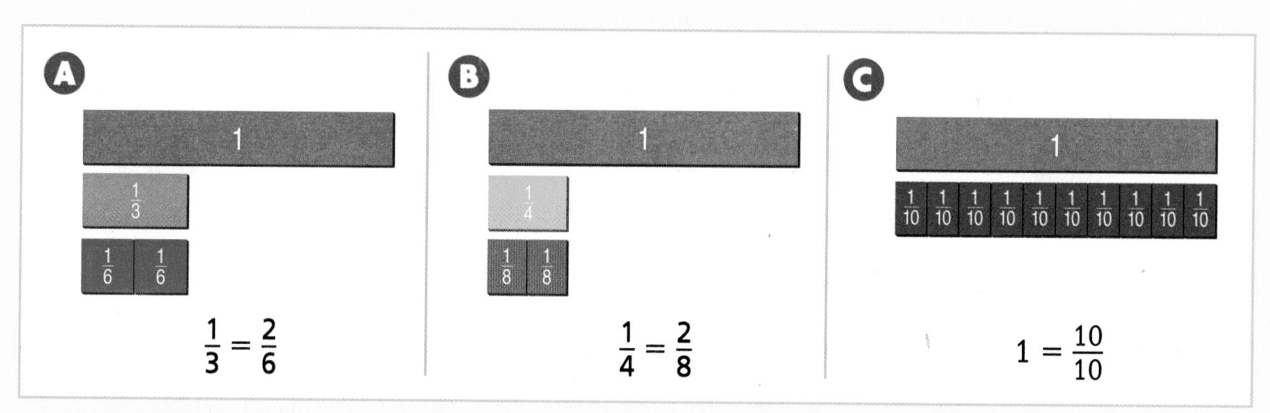

Check

1. **Explain** how to use fraction bars to decide if $\frac{2}{4}$ and $\frac{2}{3}$ are equivalent.

Find an equivalent fraction. Use fraction bars.

2.

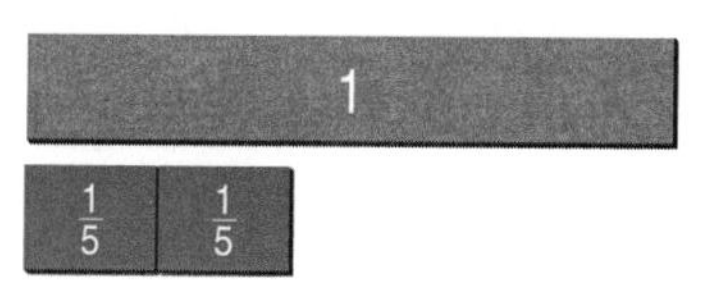

3.

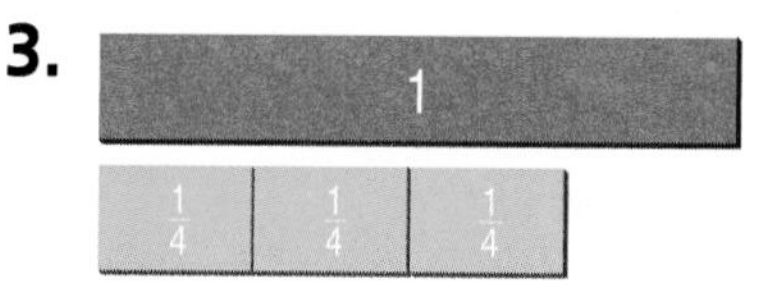

LESSON CONTINUES

Practice and Problem Solving

Find an equivalent fraction. Use fraction bars.

4. 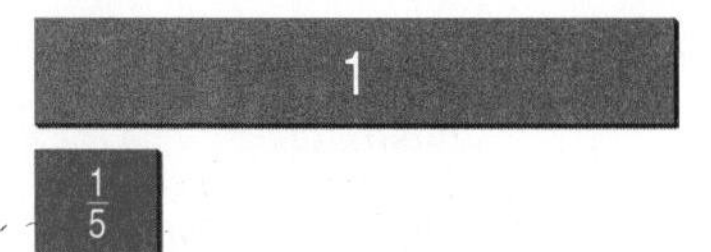

5. 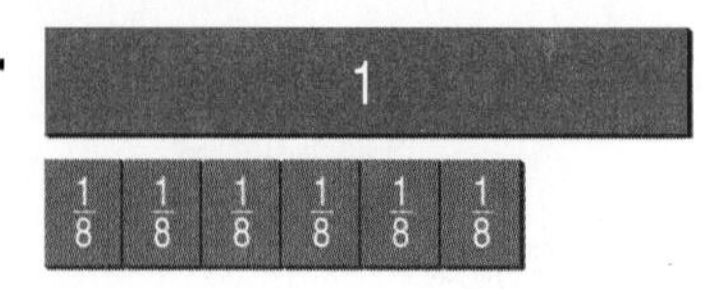

6. 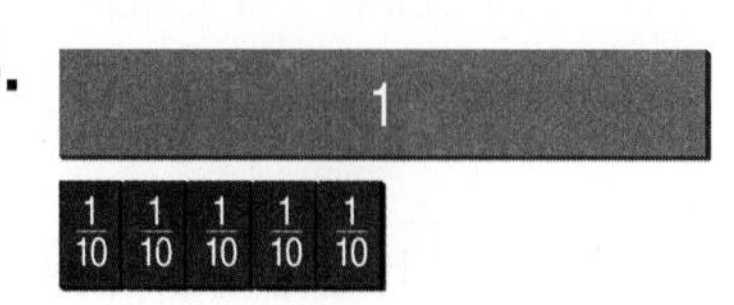

Find the missing numerator. Use fraction bars.

7. 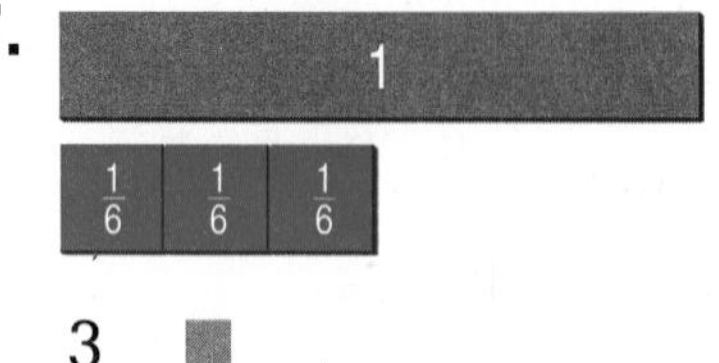

$\frac{3}{6} = \frac{■}{2}$

8. 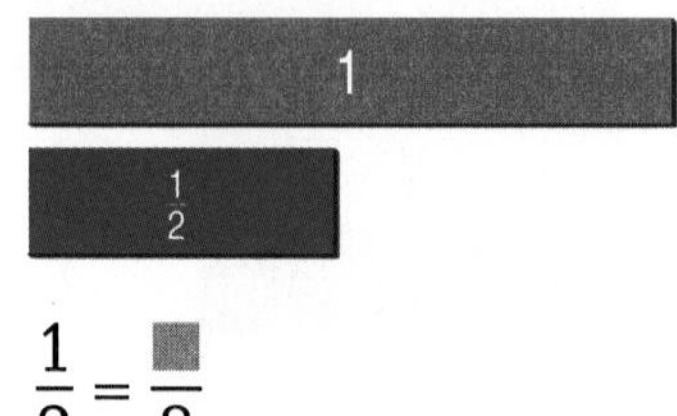

$\frac{1}{2} = \frac{■}{8}$

9. 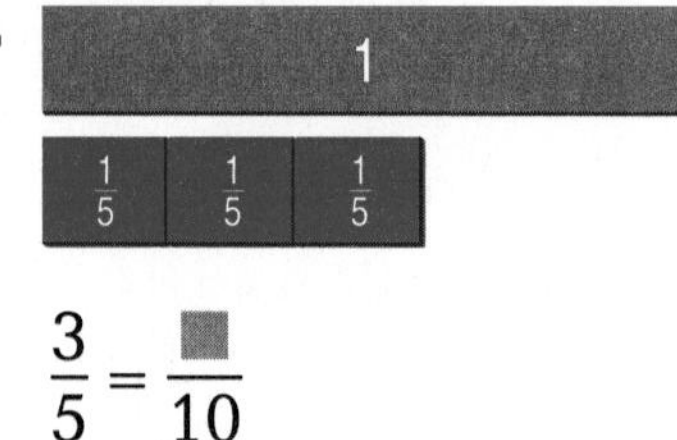

$\frac{3}{5} = \frac{■}{10}$

10. $\frac{3}{9} = \frac{■}{3}$

11. $\frac{1}{2} = \frac{■}{10}$

12. $\frac{2}{4} = \frac{■}{12}$

13. $\frac{1}{4} = \frac{■}{8}$

14. $\frac{2}{6} = \frac{■}{3}$

15. $\frac{3}{5} = \frac{■}{10}$

16. Write the fraction that names the shaded part of each. Then tell which fractions are equivalent.

a.

b. 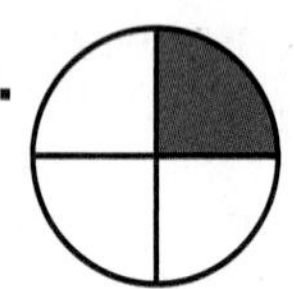

c. 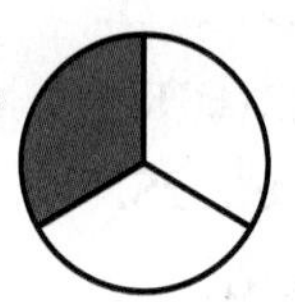

d. 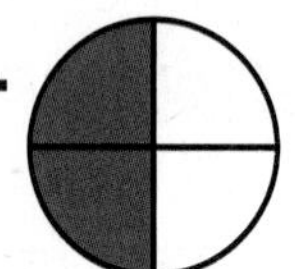

e. 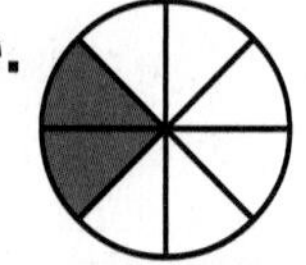

17. Molly's age is an even number between 20 and 29. Her age can be evenly divided by 3. How old is Molly?

18. Pablo likes oranges and plums, but not apples or pears. What fraction of the fruit does Pablo like?

19. ALGEBRA In his yard, Mr. York has a 42 foot tall pine tree. The pine tree is 15 feet taller than his oak tree. How tall is his oak tree?

20. **Write About It** Explain how to use fraction bars to find equivalent fractions for $\frac{4}{8}$.

21. Jason ate $\frac{2}{6}$ of the pie, Wesley ate $\frac{1}{3}$ of it, and Joe ate the rest. Who ate the most pie?

22. **REASONING** Sumi used 6 of one kind of fraction bar to show $\frac{1}{2}$. What kind of fraction bars did she use?

Mixed Review and Test Prep

Round to the nearest thousand. (p. 30)

23. 833 **24.** 2,497 **25.** 5,555 **26.** 39,670

Find the quotient. (p. 222)

27. $90 \div 9$ **28.** $56 \div 8$ **29.** $49 \div 7$ **30.** $72 \div 9$

31. **TEST PREP** Emma left at 11:15 A.M. and returned home at 12:37 P.M. How long was she gone? (p. 98)

A 1 hour 12 minutes
B 1 hour 22 minutes
C 1 hour 42 minutes
D 1 hour 52 minutes

32. **TEST PREP** Which set is equivalent to 3 quarters? (p. 80)

F 7 nickels, 25 pennies
G 6 dimes, 5 pennies
H 1 quarter, 9 nickels
J 5 dimes, 5 nickels

PROBLEM SOLVING LINKUP ... to Science

Did you know that there may be as many as 14,000 different kinds of ants in the world?

Not all ants are alike. Different species of ants live in different places, eat different foods, and are different colors and sizes. Here are some interesting facts about three kinds of ants.

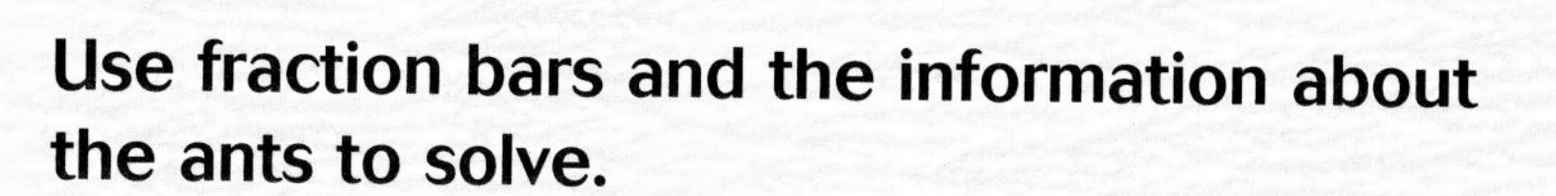

Use fraction bars and the information about the ants to solve.

1. Find an equivalent fraction for the length of
   a. the fire ant **b.** the bulldog ant.
2. Find how many odorous house ants it would take to equal the length of one bulldog ant. Write the equivalent fractions.

Fire ants
- are red in color
- will sting
- are about $\frac{1}{4}$ inch long

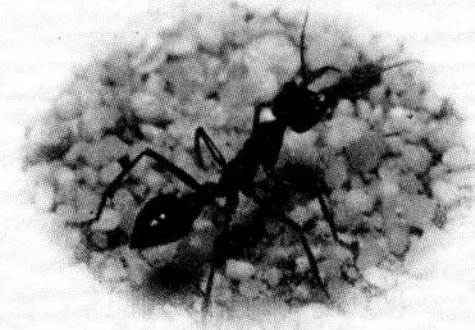

Bulldog ants
- eat meat
- are found in Australia
- are about $\frac{4}{5}$ inch long

Odorous house ants
- are brown or black in color
- smell bad when crushed
- are about $\frac{1}{10}$ inch long

EXTRA PRACTICE page H54, Set C

LESSON 4

Compare and Order Fractions

Quick Review

Compare. Write $<$, $>$, or $=$ for each ●.

1. 27 ● 37
2. 398 ● 399
3. 241 ● 214
4. 4,012 ● 4,012
5. 5,431 ● 5,341

Learn

SIZE IT UP Fraction bars can help you compare parts of a whole.

Examples

A Compare $\frac{1}{4}$ and $\frac{2}{4}$.

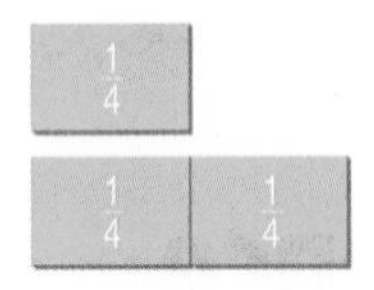

The bar for $\frac{1}{4}$ is shorter than the bars for $\frac{2}{4}$.

So, $\frac{1}{4} < \frac{2}{4}$, or $\frac{2}{4} > \frac{1}{4}$.

B Compare $\frac{1}{3}$ and $\frac{1}{4}$.

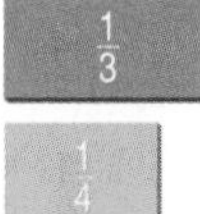

The bar for $\frac{1}{4}$ is shorter than the bar for $\frac{1}{3}$.

So, $\frac{1}{4} < \frac{1}{3}$, or $\frac{1}{3} > \frac{1}{4}$.

- When the denominator is greater, is the fraction bar longer or shorter? Why?

Tiles can help you compare parts of a group.

Examples

A Compare $\frac{2}{5}$ and $\frac{3}{5}$.

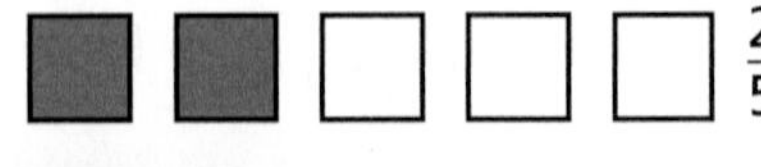

$\frac{2}{5}$

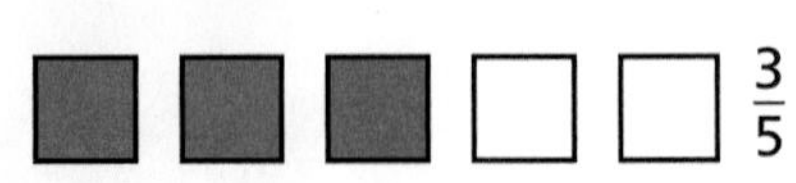

$\frac{3}{5}$

3 tiles are more than 2 tiles.

So, $\frac{3}{5} > \frac{2}{5}$, or $\frac{2}{5} < \frac{3}{5}$.

B Compare $\frac{4}{6}$ and $\frac{2}{3}$.

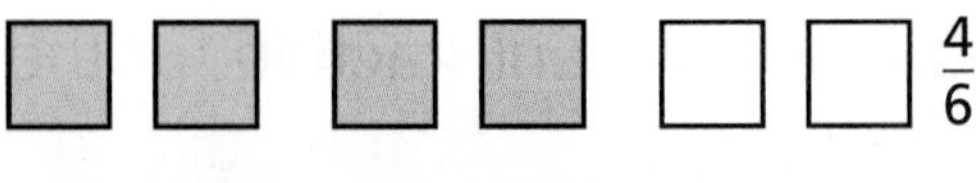

$\frac{4}{6}$

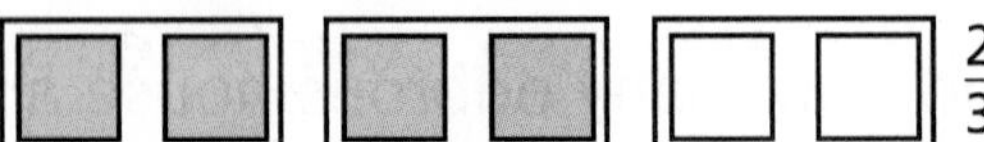

$\frac{2}{3}$

4 tiles is the same as 4 tiles.

So, $\frac{4}{6} = \frac{2}{3}$.

- How do you compare fractions when the denominators are the same? When the denominators are different?

Ordering Fractions

You can order three or more fractions from least to greatest or from greatest to least.

Cassie needs $\frac{3}{8}$ cup raisins, $\frac{1}{4}$ cup chocolate chips, and $\frac{2}{3}$ cup peanuts to make a trail mix. She wants to know which ingredient she needs the most of. Use fraction bars to order the fractions from greatest to least.

Examples

STEP 1

Compare the fractions.

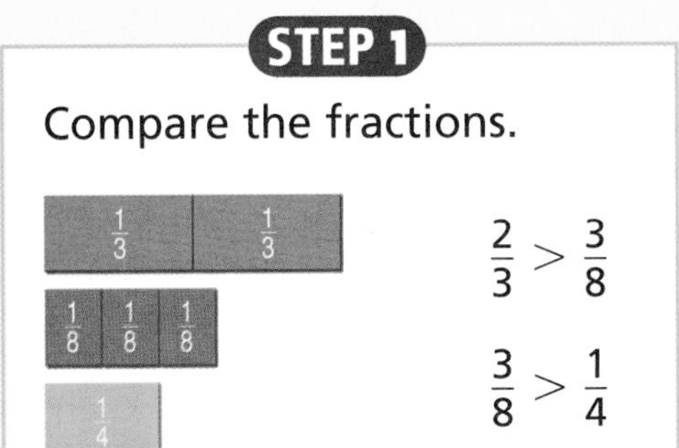

$\frac{2}{3} > \frac{3}{8}$

$\frac{3}{8} > \frac{1}{4}$

STEP 2

Order the fractions from greatest to least.

Think: $\frac{2}{3} > \frac{3}{8} > \frac{1}{4}$

Write: $\frac{2}{3}, \frac{3}{8}, \frac{1}{4}$

So, the fractions in order from greatest to least are $\frac{2}{3}$, $\frac{3}{8}$, and $\frac{1}{4}$.

- Order the fractions from least to greatest.

Check

1. **Describe** what happens to the size of fraction bars when the denominators become greater as in $\frac{1}{2}$, $\frac{1}{3}$, and $\frac{1}{4}$.
2. **Describe** what happens to the size of fraction bars when the denominators become smaller, as in $\frac{1}{8}$, $\frac{1}{6}$, and $\frac{1}{4}$.

Compare. Write <, >, or = for each ●.

3.

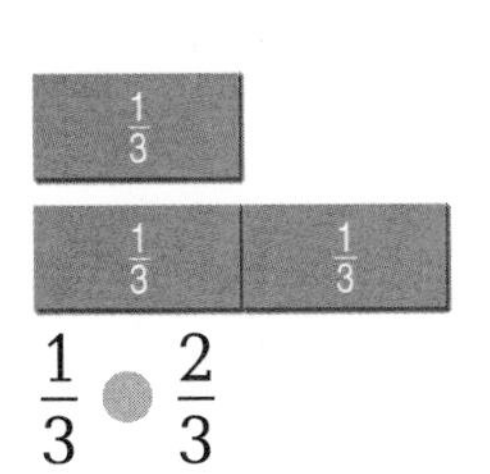

$\frac{1}{3}$ ● $\frac{2}{3}$

4.

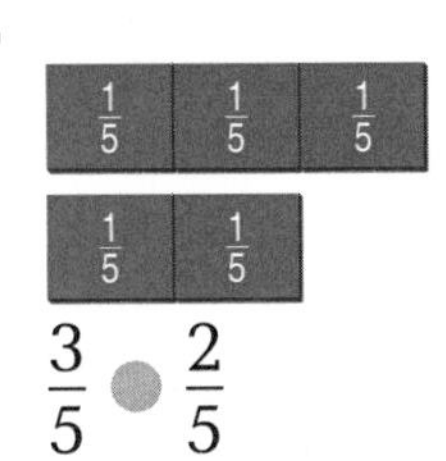

$\frac{3}{5}$ ● $\frac{2}{5}$

5.

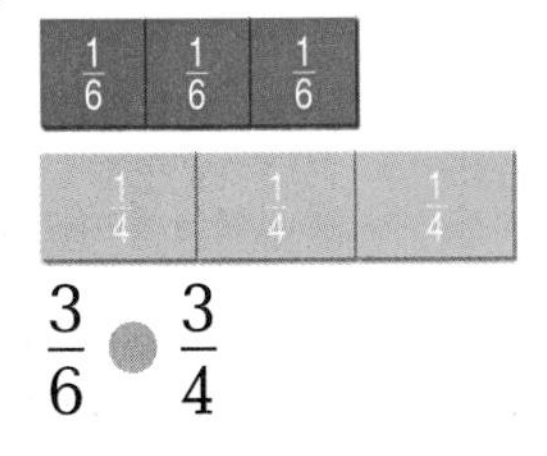

$\frac{3}{6}$ ● $\frac{3}{4}$

LESSON CONTINUES

Practice and Problem Solving

Compare. Write <, >, or = for each ●.

6. 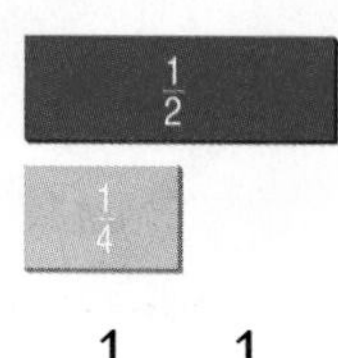

$\frac{1}{2}$ ● $\frac{1}{4}$

7. 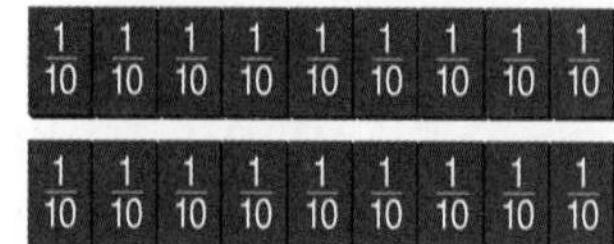

$\frac{9}{10}$ ● $\frac{9}{10}$

8. 

$\frac{3}{12}$ ● $\frac{5}{8}$

9. 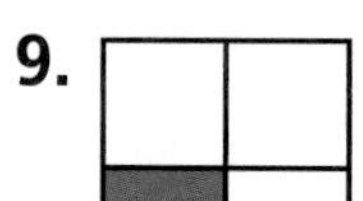

$\frac{1}{4}$ ● $\frac{2}{4}$

10. 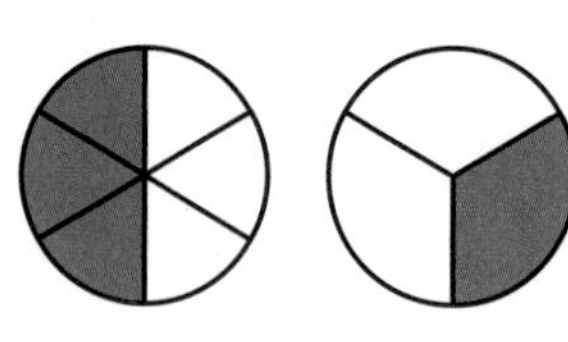

$\frac{3}{6}$ ● $\frac{1}{3}$

11. 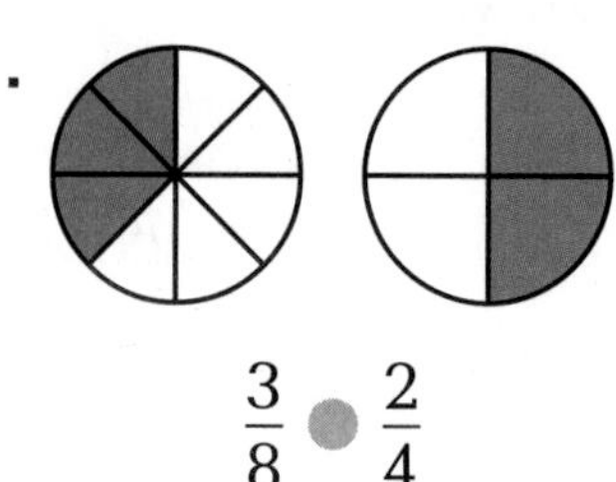

$\frac{3}{8}$ ● $\frac{2}{4}$

Compare the part of each group that is green.
Write <, >, or = for each ●.

12. 

$\frac{2}{7}$ ● $\frac{3}{7}$

13. 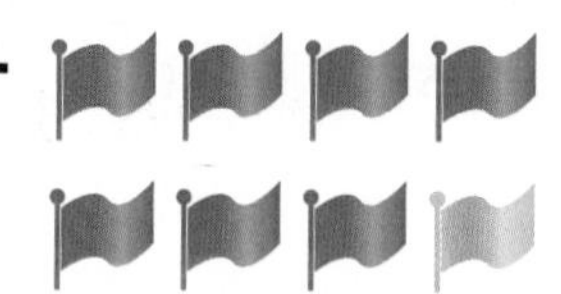

$\frac{4}{4}$ ● $\frac{3}{4}$

14. 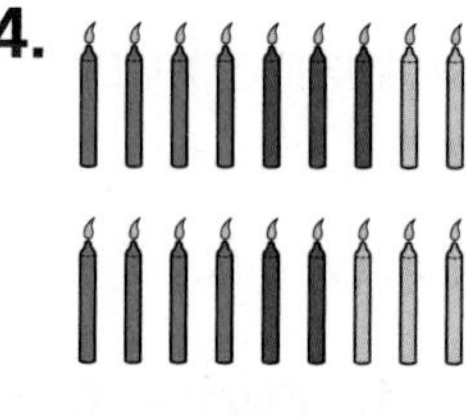

$\frac{4}{9}$ ● $\frac{4}{9}$

15. Order $\frac{1}{3}$, $\frac{1}{6}$, and $\frac{4}{6}$ from greatest to least.

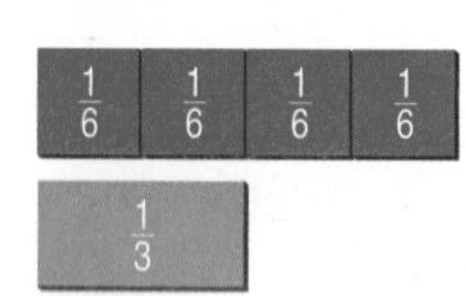

16. Order $\frac{1}{2}$, $\frac{3}{4}$, and $\frac{2}{5}$ from least to greatest.

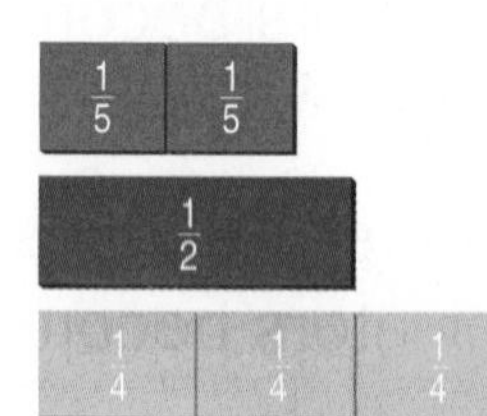

USE DATA For 17–19, use the bar graph.

17. How many leaves did Kevin collect in all?

18. What fraction of the leaves are red? yellow? orange?

19. Order the fractions of leaves from the greatest to the least amount.

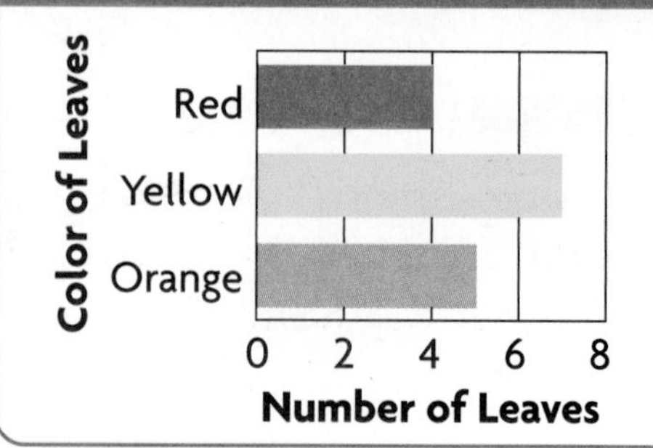

20. **REASONING** Terry says that because 10 is greater than 6, $\frac{5}{10}$ is greater than $\frac{5}{6}$. Do you agree or disagree? Explain.

21. Write the next two fractions in the pattern. Explain the rule.
$\frac{1}{8}, \frac{2}{8}, \frac{3}{8}, \frac{4}{8}, \frac{5}{8}, ■, ■$

Mixed Review and Test Prep

Order each set of numbers from least to greatest. (p. 24)

22. 45, 54, 50
23. 110, 111, 101
24. 199, 89, 98
25. 455, 555, 545

Find the number that the variable stands for. (p. 188)

26. $4 \times s = 36$ $s = \underline{?}$
27. $9 \times r = 45$ $r = \underline{?}$
28. $56 \div 7 = q$ $q = \underline{?}$
29. $81 \div 9 = p$ $p = \underline{?}$

30. **TEST PREP** Cheryl bought two markers for $0.55 each and a notebook for $1.79. How much did she spend in all? (p. 88)

A $1.89
B $2.34
C $2.79
D $2.89

31. **TEST PREP** How many feet are in 4 yards? (p. 362)

F 20 feet
G 12 feet
H 9 feet
J 6 feet

PROBLEM SOLVING LINKUP . . . to Social Studies

At the end of December, some African-Americans celebrate Kwanzaa, a 7-day festival that ends with a feast, or "karumu," of healthful foods on December 31. They put a straw mat, called a "mkeka," in the middle of the table. Fresh fruit on the mat represents African harvest festivals.

1. Of the seven Kwanzaa candles, 3 are red, 1 is black, and the rest are green. What fraction names the part that is green?

2. Clarence takes $\frac{3}{4}$ cup grapes, $\frac{1}{2}$ cup strawberries, and $\frac{1}{4}$ cup blueberries from the mkeka. Use fraction bars to put the fruits in order from least to greatest.

LESSON 5

Problem Solving Strategy
Make a Model

PROBLEM Three classmates ran a relay in the Spring Sports Day race. George ran $\frac{2}{3}$ mile, Rosa ran $\frac{7}{8}$ mile, and Ben ran $\frac{5}{6}$ mile. Who ran the farthest?

Quick Review

Compare. Write <, >, or = for each ●.

1. $\frac{2}{3}$ ● $\frac{2}{3}$
2. $\frac{4}{6}$ ● $\frac{3}{6}$
3. $\frac{1}{4}$ ● $\frac{1}{2}$
4. $\frac{4}{8}$ ● $\frac{5}{10}$
5. $\frac{2}{3}$ ● $\frac{1}{6}$

UNDERSTAND

- What are you asked to find?
- What information will you use?
- Is there information you will not use? If so, what?

PLAN

- What strategy can you use to solve the problem?

Make a model to show what part of a mile each person ran.

SOLVE

- How can you use the strategy to solve the problem?

Model the problem by using fraction bars.

Line up the fraction bars for $\frac{2}{3}$, $\frac{7}{8}$, and $\frac{5}{6}$.

Compare the lengths of the fraction bars.

Since $\frac{7}{8} > \frac{5}{6} > \frac{2}{3}$, Rosa ran the farthest.

George

Rosa

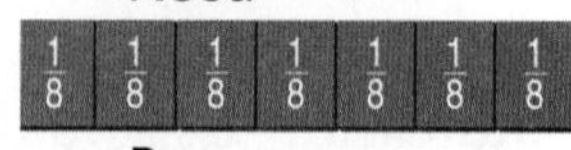

Ben

CHECK

- What other strategy could you use to solve the problem?

Problem Solving Practice

PROBLEM SOLVING STRATEGIES

- Draw a Diagram or Picture
- **Make a Model or Act It Out**
- Make an Organized List
- Find a Pattern
- Make a Table or Graph
- Predict and Test
- Work Backward
- Solve a Simpler Problem
- Write a Number Sentence
- Use Logical Reasoning

Use *make a model* to solve.

1. **What if** George had run $\frac{9}{10}$ mile? Who would have run the farthest?
2. Tina used $\frac{2}{3}$ cup of milk, $\frac{1}{4}$ cup of sugar, and $\frac{1}{2}$ cup of nuts in a recipe. Of which ingredient did she use the most?

Mr. Collins asked his science class to read a book on insects. By Friday, Alfredo had read $\frac{1}{2}$ of the book. Courtney had read $\frac{2}{5}$ of the book. Sandy had read $\frac{3}{8}$ of the book.

3. Which statement is true?

   A $\frac{3}{8} > \frac{2}{5} > \frac{1}{2}$

   B $\frac{1}{2} > \frac{2}{5} > \frac{3}{8}$

   C $\frac{1}{2} > \frac{3}{8} > \frac{2}{5}$

   D $\frac{2}{5} > \frac{3}{8} > \frac{1}{2}$

4. Who had read the greatest part of the book?

   F Alfredo
   G Courtney
   H Sandy

Mixed Strategy Practice

5. **REASONING** Mohammed folded a sheet of notebook paper in half and then folded it in half again. He unfolded the paper and shaded one of the equal parts red. What fraction shows how many parts are red?
6. Eric had 74 baseball cards. He traded 26 of his cards for 12 of Paul's cards. How many cards does he have now?
7. Arlo has basketball practice from 3:30 to 5 o'clock. It takes him a half hour to get home and one hour to do his homework before dinner. At what time does he eat dinner?
8. **REASONING** Kathy is cutting an apple into 8 equal slices. Julie is cutting an apple into 6 equal slices. Kathy says her slices will be larger because 8 is greater than 6. Do you agree? Explain.

LESSON

Mixed Numbers

Quick Review

Compare. Write <, >, or = for each ●.

1. $\frac{1}{3}$ ● $\frac{2}{3}$
2. $\frac{2}{4}$ ● $\frac{1}{2}$
3. $\frac{5}{6}$ ● $\frac{3}{6}$
4. $\frac{3}{8}$ ● $\frac{5}{8}$
5. $\frac{5}{5}$ ● $\frac{3}{3}$

VOCABULARY

mixed number

Learn

APPLE DAY Kevin's mother made a plate of apple pieces for an after-school snack. Each piece was $\frac{1}{6}$ of an apple. How many apples did she use?

Kevin counted the pieces.

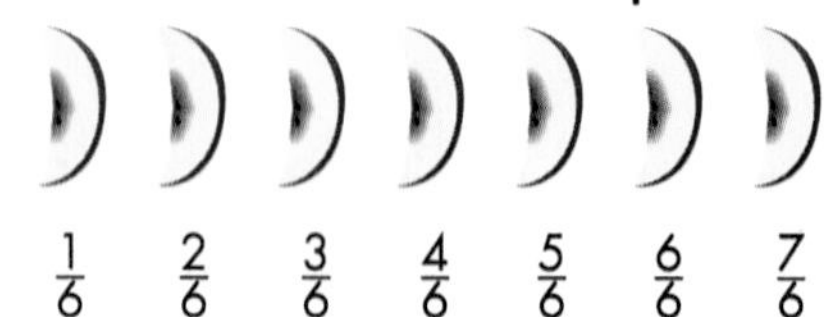

$\frac{1}{6}$ $\frac{2}{6}$ $\frac{3}{6}$ $\frac{4}{6}$ $\frac{5}{6}$ $\frac{6}{6}$ $\frac{7}{6}$

$\frac{6}{6} = 1$ whole apple So, $\frac{7}{6} = 1 + \frac{1}{6}$ or $1\frac{1}{6}$

Read: one and one sixth

So, Kevin's mother used $1\frac{1}{6}$ apples.

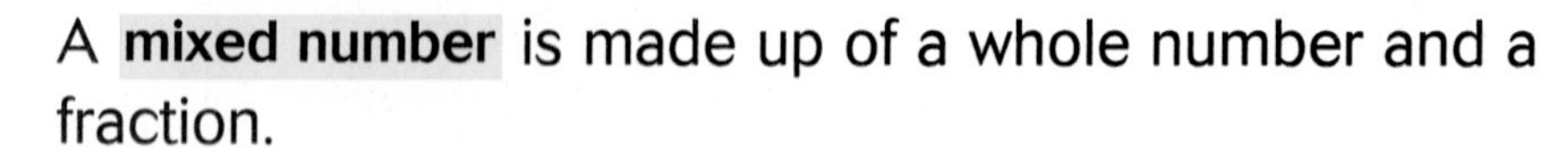

A **mixed number** is made up of a whole number and a fraction.

To show $1\frac{1}{6}$ on the number line, first draw a jump of $\frac{6}{6}$ or 1. Then draw a jump of $\frac{1}{6}$.

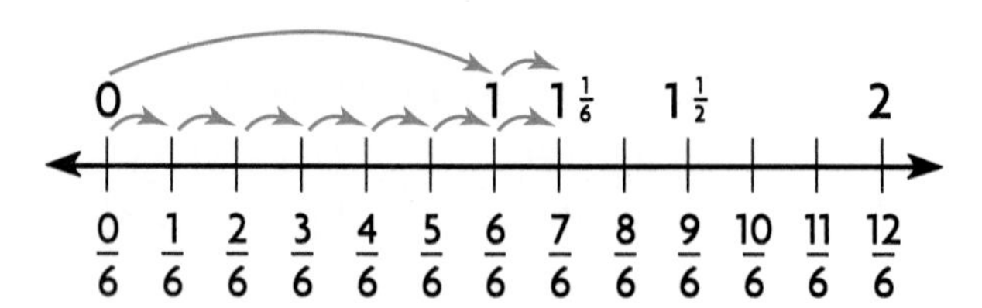

• What is the mixed number for $\frac{11}{6}$?

Check

Write a mixed number for the parts that are shaded.

1. 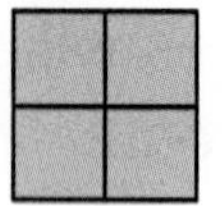 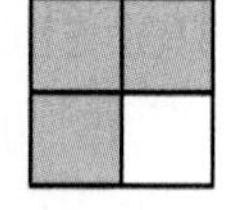

2. 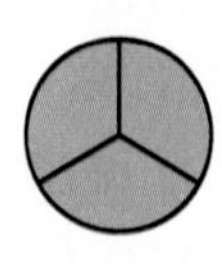 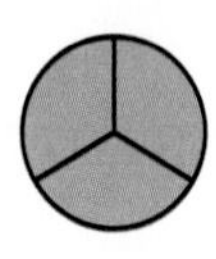 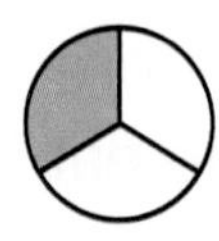

3. 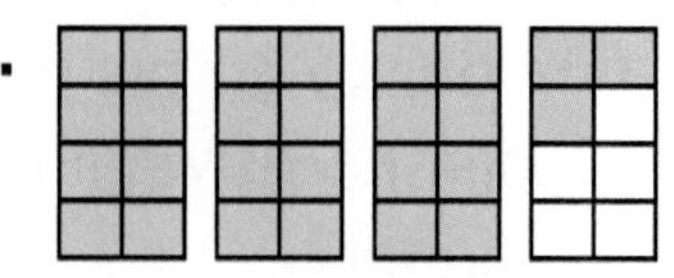

Problem Solving Practice

Write a mixed number for the parts that are shaded.

4. 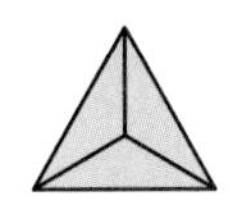 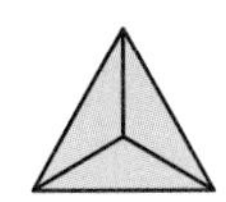 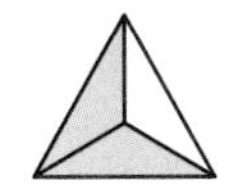

5. 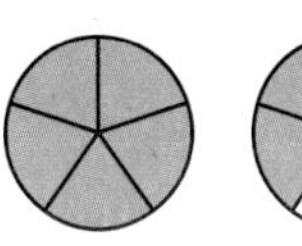

6.

For 7–11, use the number line to write the mixed number.

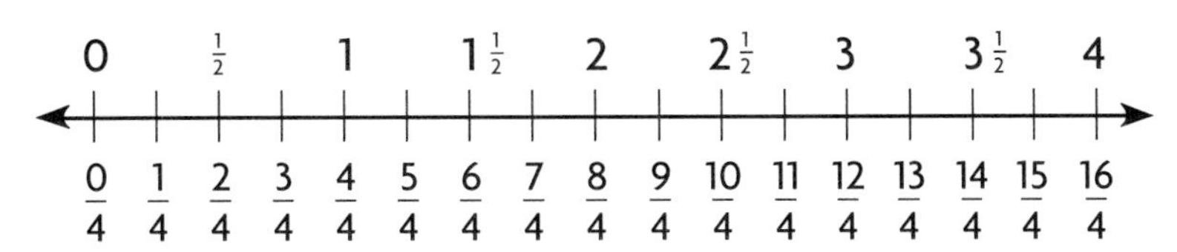

7. $\frac{15}{4}$

8. $\frac{7}{4}$

9. $\frac{11}{4}$

10. $\frac{5}{4}$

11. $\frac{9}{4}$

12. Benita needs $\frac{1}{4}$ apple for each muffin. How many apples will she need for 12 muffins?

13. **What's the Error?** Benita ate $1\frac{1}{3}$ muffins. Her brother Felipe ate $1\frac{2}{6}$ muffins. Felipe said that he ate more. Explain Felipe's mistake.

14. Oren said that there were $2\frac{1}{2}$ pizzas. If all of the pieces were eighths, how many pieces of pizza were there?

15. **Write a problem** in which a fraction can be written as a mixed number.

Mixed Review and Test Prep

Find the number the variable stands for. (p. 188)

16. $7 \times r = 35$
    $r = ?$

17. $s \times 4 = 36$
    $s = ?$

18. $24 \div t = 6$
    $t = ?$

19. $u \div 8 = 2$
    $u = ?$

20. **TEST PREP** Which is NOT true? (p. 422)

A $\frac{3}{4} = \frac{6}{8}$

B $\frac{3}{8} < \frac{4}{8}$

C $\frac{2}{5} > \frac{4}{5}$

D $\frac{3}{6} = \frac{1}{2}$

EXTRA PRACTICE page H54, Set E

Review/Test

CHECK VOCABULARY AND CONCEPTS

Choose the best term from the box.

fraction
numerator
denominator
mixed number

1. In the fraction $\frac{3}{8}$, the 3 is called the __?__. (p. 412)
2. A number that names part of a whole or part of a set is a __?__. (p. 412)
3. A number that is made up of a whole number and a fraction is called a __?__. (p. 428)

Find an equivalent fraction. Use fraction bars. (pp. 418–421)

4. 

5. 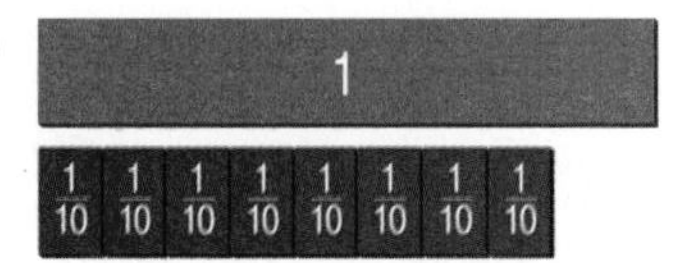

6. 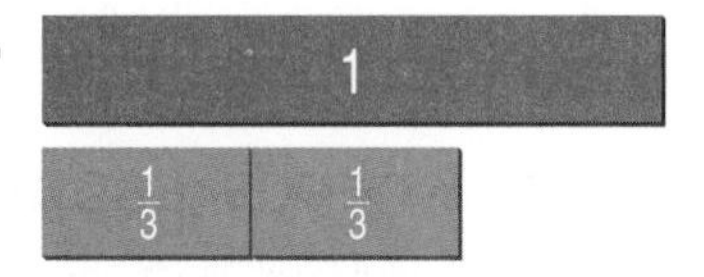

CHECK SKILLS

Write a fraction in numbers and in words that names the shaded part. (pp. 412–415, 428–429)

7. 

8. 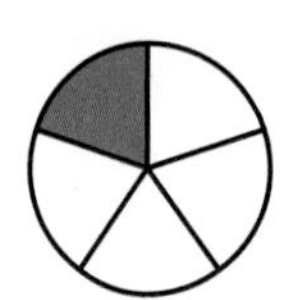

9. 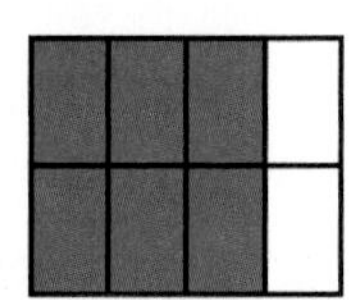

10. 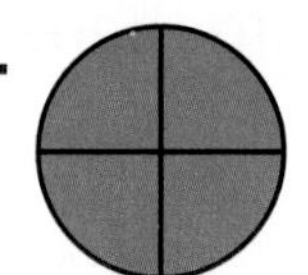 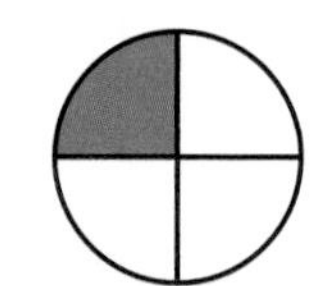

Write a mixed number for the parts that are shaded. (pp. 428–429)

11. 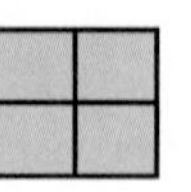 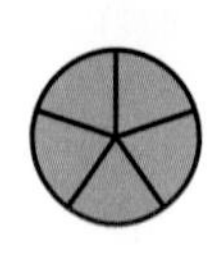

12. 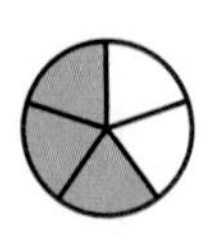 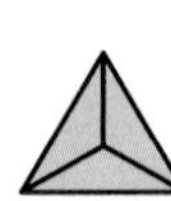 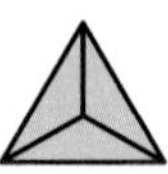

13. 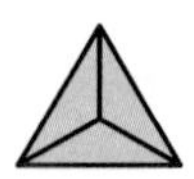 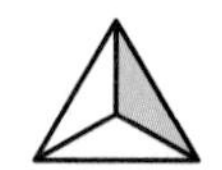

CHECK PROBLEM SOLVING

Use *make a model* to solve. (pp. 426–427)

14. Luke walks $\frac{1}{2}$ mile, Karen walks $\frac{4}{5}$ mile, and Darius walks $\frac{1}{3}$ mile. Who walks the farthest?

15. Kumiko has 3 yellow pins, 4 blue pins, and 2 red pins. What fraction of the pins are red?

Standardized Test Prep

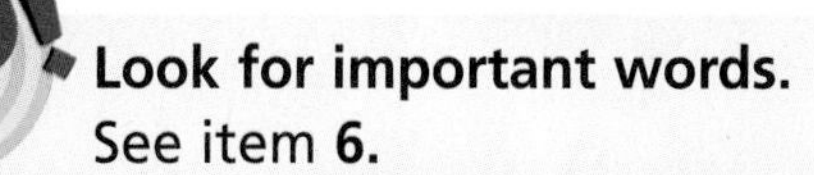

TIP! Look for important words.
See item **6.**

Important words are *order from least to greatest.* Compare and order the fractions in each answer choice. Be sure you start with the least.

Also see problem **2,** p. H62.

For 1–6, choose the best answer.

1. Which fraction names the shaded part of the group?

A $\frac{1}{9}$ B $\frac{1}{5}$ C $\frac{3}{9}$ D $\frac{5}{9}$

2. Which fraction names the shaded part?

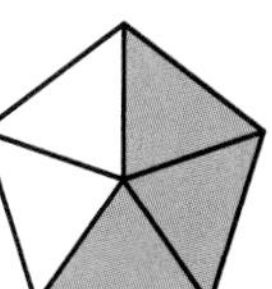

F $\frac{1}{5}$ H $\frac{3}{5}$

G $\frac{2}{5}$ J $\frac{2}{3}$

3. A quadilateral has __?__ sides and __?__ angles.

A 3 C 6

B 4 D NOT HERE

4. Which symbol makes this a true number sentence?

$\frac{1}{2} \bullet \frac{2}{3}$

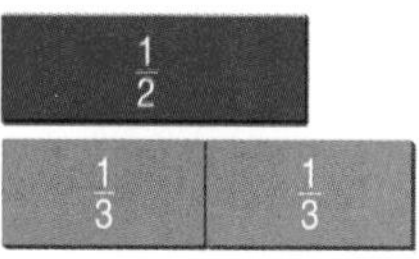

F $>$ H $<$

G $=$ J $\times$

5. Which number makes this number sentence true?

$$213 + \blacksquare = 701$$

A 488 C 588

B 498 D 914

6. Which set of fractions is in order from least to greatest?

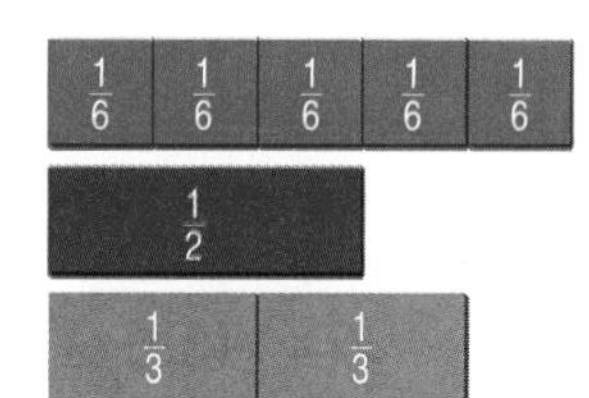

F $\frac{5}{6}, \frac{1}{2}, \frac{2}{3}$ H $\frac{1}{2}, \frac{5}{6}, \frac{2}{3}$

G $\frac{1}{2}, \frac{2}{3}, \frac{5}{6}$ J $\frac{2}{3}, \frac{1}{2}, \frac{5}{6}$

Write What You Know

7. Ray and 7 friends want to share a pie. Show how to cut the pie into equal parts. Shade the part to show the fraction that each person will get. Write the fraction.

8. Explain what equivalent fractions are. Use equivalent fractions to write the number that makes the sentence true.

$\frac{1}{3} = \frac{\blacksquare}{6}$

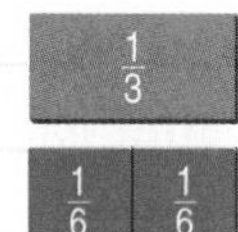

CHAPTER 24

Add and Subtract Like Fractions

Corn bread was a favorite food of early pioneers. It is usually served warm with butter.

PROBLEM SOLVING This recipe serves 4 people. How can you change the recipe to serve 8 people? Rewrite the recipe.

Corn Bread

1 cup cornmeal

$\frac{1}{4}$ cup flour

$\frac{1}{2}$ teaspoon baking powder

$\frac{1}{8}$ teaspoon baking soda

$\frac{3}{4}$ teaspoon salt

1 egg

1 tablespoon corn oil

$\frac{3}{4}$ cup buttermilk

Mix dry ingredients. Add egg, oil, and buttermilk. Mix well. Pour batter into a hot iron skillet. Bake at 425° for 20 to 25 minutes.

Use this page to help you review and remember important skills needed for Chapter 24.

VOCABULARY

Choose the best term from the box.

numerators
equivalent fractions

1. Two or more fractions that name the same amount, such as $\frac{2}{4}$ and $\frac{1}{2}$, are called __?__ .

NAME THE FRACTION (For Intervention, see p. H29.)

Name the fraction for the part that is shaded.

2. 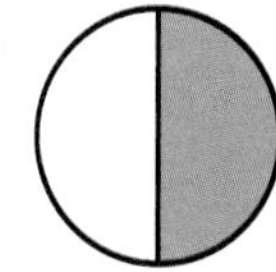

3. 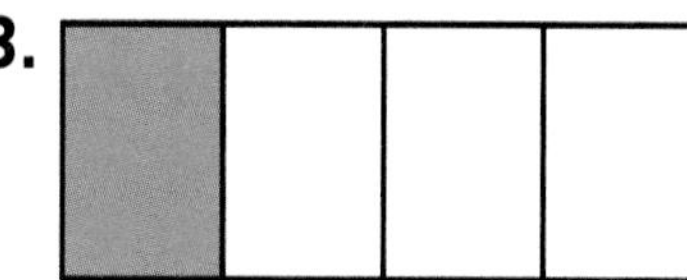

4. 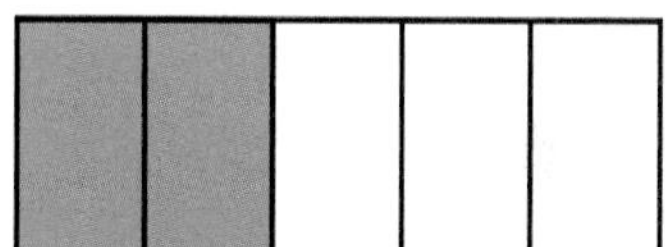

5. 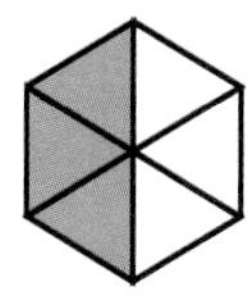

6. 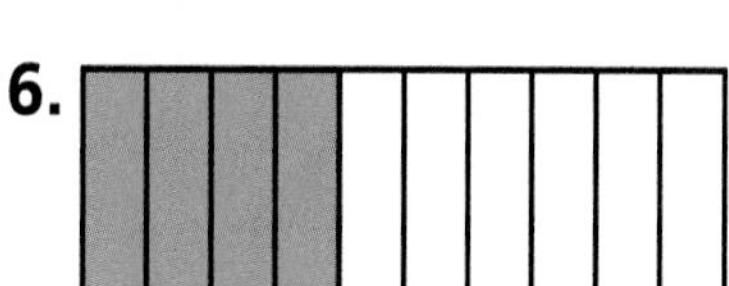

7. 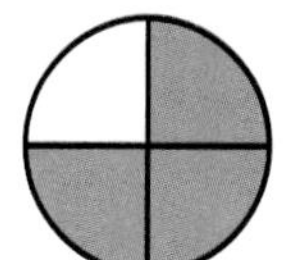

Write the fraction for the part that is green.

8. 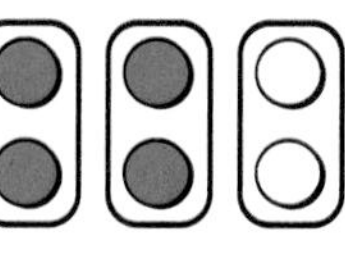

9. 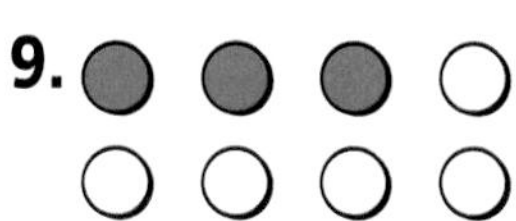

10. 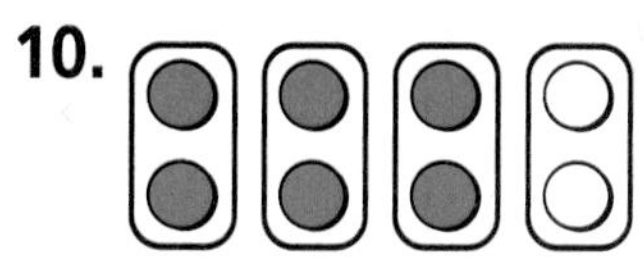 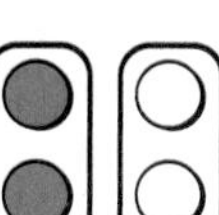

COMPARE FRACTIONS (For Intervention, see p. H30.)

Compare. Write <, >, or = for each ●.

11. 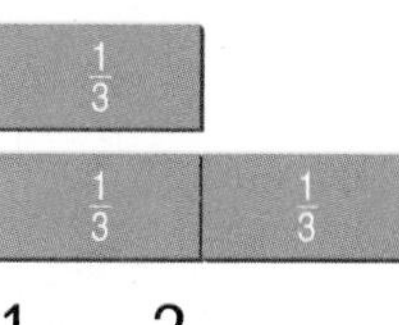

$\frac{1}{3}$ ● $\frac{2}{3}$

12. 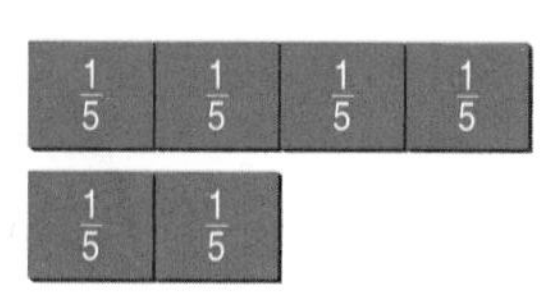

$\frac{4}{5}$ ● $\frac{2}{5}$

13. 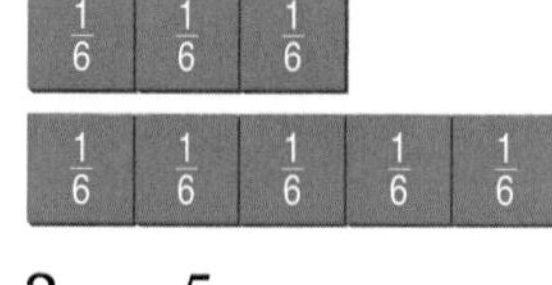

$\frac{3}{6}$ ● $\frac{5}{6}$

14. 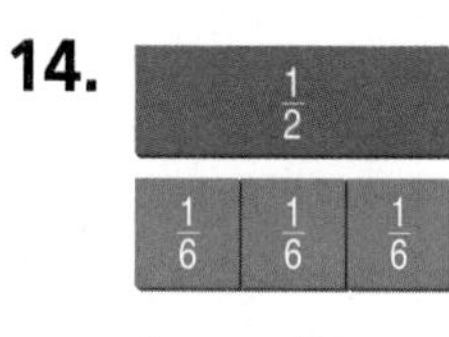

$\frac{1}{2}$ ● $\frac{3}{6}$

15. 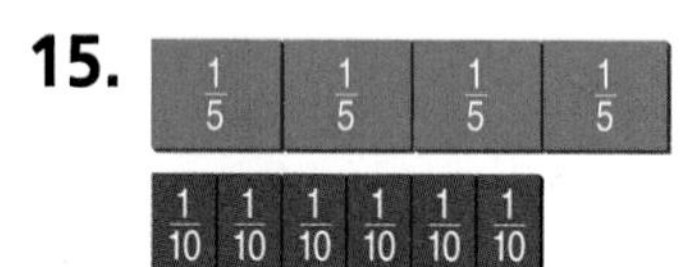

$\frac{4}{5}$ ● $\frac{6}{10}$

16. 1/4 1/4
1/8 1/8 1/8 1/8
$\frac{2}{4}$ ● $\frac{4}{8}$

LESSON 1

Add Fractions

Quick Review

Name an equivalent fraction.

1. $\frac{4}{6}$ 2. $\frac{2}{8}$ 3. $\frac{2}{4}$
4. $\frac{2}{10}$ 5. $\frac{1}{2}$

VOCABULARY
like fractions

MATERIALS
fraction bars

Explore

Fractions that have the same denominator are called **like fractions**.

Ryan and his mother are making fruit punch. The recipe says to add $\frac{2}{4}$ cup orange juice and $\frac{1}{4}$ cup pineapple juice. How much juice is needed altogether?

Remember
$\frac{1}{2}$ → numerator
→ denominator

Activity

Use fraction bars to find $\frac{2}{4} + \frac{1}{4}$.

STEP 1

Model $\frac{2}{4}$ with fraction bars.

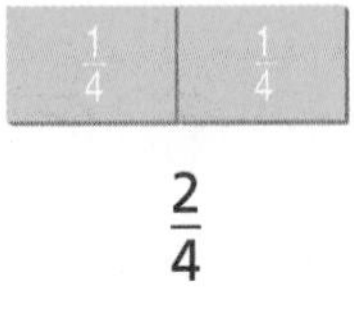

$\frac{2}{4}$

STEP 2

Add one more $\frac{1}{4}$ fraction bar.

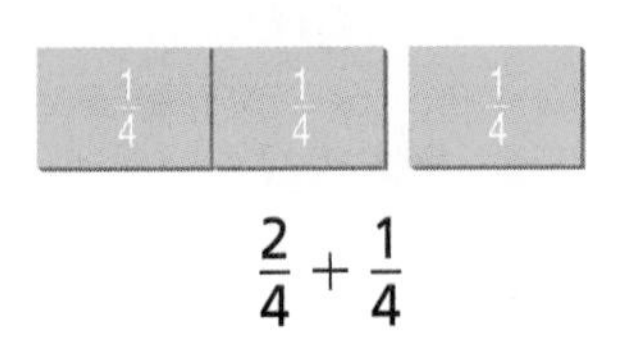

$\frac{2}{4} + \frac{1}{4}$

STEP 3

Count the number of $\frac{1}{4}$ fraction bars.

$\frac{1}{4}, \frac{2}{4}, \frac{3}{4}$ or $\frac{2}{4} + \frac{1}{4} = \frac{3}{4}$

So, the recipe calls for $\frac{3}{4}$ cup juice altogether.

- Why doesn't the denominator change when you find the sum?
- Explain how you could use fraction bars to find $\frac{1}{3} + \frac{1}{3}$.

I'm counting the fraction bars to find the sum: 1/6, 2/6, 3/6...
What comes next?

Try It

Use fraction bars to find the sum.

a. $\frac{3}{6} + \frac{1}{6}$ b. $\frac{3}{5} + \frac{2}{5}$

Connect

Add like fractions by adding the numerators.

Example

Tom peeled an orange. It had 8 wedges. He ate 1 wedge, or $\frac{1}{8}$ of it. Then he ate 4 more wedges, or $\frac{4}{8}$ of it. What fraction did he eat?

Model

Add the number of $\frac{1}{8}$ wedges that Tom ate.

$\frac{1}{8} + \frac{4}{8}$

Record

1 wedge + 4 wedges = 5 wedges

$\frac{1}{8} + \frac{4}{8} = \frac{5}{8}$

So, Tom ate $\frac{5}{8}$ of the orange.

MATH IDEA When you add like fractions, you add the numerators, and the denominators stay the same.

Practice and Problem Solving

Find the sum.

1. 

$\frac{1}{5} + \frac{2}{5} = ■$

2. 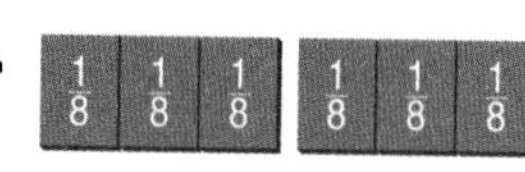

$\frac{3}{8} + \frac{3}{8} = ■$

3. 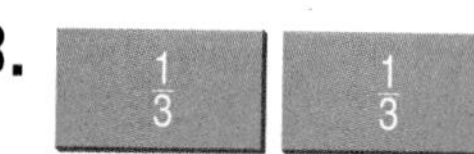

$\frac{1}{3} + \frac{1}{3} = ■$

Use fraction bars to find the sum.

4. $\frac{2}{5} + \frac{2}{5} = ■$
5. $\frac{1}{6} + \frac{4}{6} = ■$
6. $\frac{4}{10} + \frac{3}{10} = ■$
7. $\frac{3}{8} + \frac{2}{8} = ■$
8. Kris has 6 plums and 2 apples. She buys 2 more apples. What fraction of the fruit are apples?
9. A recipe calls for $\frac{1}{4}$ cup sugar. Celia wants to double the recipe. How much sugar will she need?

Mixed Review and Test Prep

10. $7 \times 9 = ■$ (p. 150)
11. $8 \times 6 = ■$ (p. 152)
12. $72 \div 9 = ■$ (p. 220)
13. 1 foot = ■ inches (p. 356)
14. **TEST PREP** Which figure has four right angles? (p. 320)

A triangle
B rectangle
C circle
D ray

LESSON

2 Add Fractions

Quick Review

Compare. Write <, >, or = for each ●.

1. $\frac{5}{8}$ ● $\frac{7}{8}$
2. $\frac{3}{4}$ ● $\frac{1}{4}$
3. $\frac{3}{6}$ ● $\frac{1}{2}$
4. $\frac{6}{10}$ ● $\frac{2}{5}$
5. $\frac{5}{10}$ ● $\frac{4}{8}$

VOCABULARY

simplest form

Learn

PIECES PLUS When you add fractions, you can show the sum in simplest form. A fraction is in **simplest form** when it uses the largest fraction bars possible.

Abby and Chris are sharing a sub sandwich for lunch. The whole sandwich has 8 pieces. Abby ate 2 pieces. Chris ate 2 pieces. How much of the sandwich did they eat?

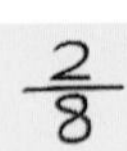

$\frac{2}{8}$ = ?

Find $\frac{2}{8} + \frac{2}{8}$ in simplest form.

Activity

Materials: fraction bars

STEP 1

Model $\frac{2}{8}$ with fraction bars.

$\frac{2}{8}$

STEP 2

Add $\frac{2}{8}$.

$\frac{2}{8} + \frac{2}{8} = \frac{4}{8}$

STEP 3

Find the largest fraction bar that is the same length.

$\frac{2}{8} + \frac{2}{8} = \frac{4}{8}$

$\frac{4}{8}$ in simplest form is $\frac{1}{2}$.

So, Abby and Chris ate $\frac{1}{2}$ of the sub sandwich.

• How do you know if a fraction is in simplest form?

33. A basketball game is divided into four quarters. Why is the break period after the first two quarters called *halftime*?

34. James bought a model car for \$9.49. He paid with a \$20 bill. How much change did he receive?

Mixed Review and Test Prep

Find the difference.

35. $724 - 182$ (p. 58)

36. $400 - 63$ (p. 58)

37. $888 - 491$ (p. 58)

38. $1,355 - 628$ (p. 62)

39. $3,742 - 1,281$ (p. 62)

Find the missing addend. (p. 42)

40. $35 + \blacksquare = 74$

41. $\blacksquare + 658 = 845$

42. $198 + \blacksquare = 451$

43. TEST PREP Which fraction is equivalent to $\frac{4}{6}$? (p. 418)

A $\frac{3}{4}$ **B** $\frac{2}{3}$ **C** $\frac{6}{12}$ **D** $\frac{1}{2}$

44. TEST PREP A box is 4 cubes long, 3 cubes wide, and 2 cubes high. What is the volume of this box? (p. 398)

F 9 cubic units **H** 20 cubic units

G 16 cubic units **J** 24 cubic units

PROBLEM SOLVING

Thinker's Corner

Solve the riddle!

On which side does a chicken have more feathers?

Find the sum. Write the answer in simplest form. Then find the fraction below that matches. Record the letter from that box.

$\frac{3}{9} + \frac{1}{9}$	T	$\frac{2}{6} + \frac{2}{6}$	O	$\frac{5}{12} + \frac{5}{12}$	S	$\frac{1}{8} + \frac{1}{8}$	D
$\frac{5}{8} + \frac{2}{8}$	U	$\frac{2}{5} + \frac{2}{5}$	E	$\frac{5}{12} + \frac{6}{12}$	N	$\frac{3}{8} + \frac{3}{8}$	E
$\frac{6}{12} + \frac{1}{12}$	O	$\frac{5}{10} + \frac{1}{10}$	I	$\frac{1}{6} + \frac{2}{6}$	T	$\frac{1}{10} + \frac{2}{10}$	H

? ?   ? ? ?   ? ? ? ? ? ? ?!

$\frac{7}{12}$ $\frac{11}{12}$ $\frac{4}{9}$ $\frac{3}{10}$ $\frac{4}{5}$ $\frac{2}{3}$ $\frac{7}{8}$ $\frac{1}{2}$ $\frac{5}{6}$ $\frac{3}{5}$ $\frac{1}{4}$ $\frac{3}{4}$

EXTRA PRACTICE page H55, Set A

LESSON 3

Subtract Fractions

Quick Review

Find the difference.

1. 12 – 8 **2.** 15 – 6

3. 18 – 5 **4.** 21 – 8

5. 25 – 11

MATERIALS
fraction bars

Explore

Rebecca's dad has $\frac{7}{8}$ of his pan of corn bread to share with the family. If Rebecca eats $\frac{2}{8}$ of the corn bread, how much of the corn bread is left?

Activity

Use fraction bars to find $\frac{7}{8} - \frac{2}{8}$.

STEP 1

Model $\frac{7}{8}$ with fraction bars.

$\frac{1}{8}$ $\frac{1}{8}$ $\frac{1}{8}$ $\frac{1}{8}$ $\frac{1}{8}$ $\frac{1}{8}$ $\frac{1}{8}$

STEP 2

Take away 2 of the fraction bars, or $\frac{2}{8}$.

STEP 3

Count the number of $\frac{1}{8}$ fraction bars left.

There are five $\frac{1}{8}$ fraction bars left.

So, $\frac{5}{8}$ of the corn bread is left.

- Why doesn't the denominator change when you find the difference?

Try It

Use fraction bars to find the difference.

a. $\frac{3}{6} - \frac{1}{6} = \blacksquare$

b. $\frac{5}{8} - \frac{2}{8} = \blacksquare$

If I take away one of the $\frac{1}{6}$ bars, how many sixths will be left?

Connect

You subtract like fractions by subtracting the numerators.

Example

$\frac{7}{8} - \frac{2}{8} = ■$

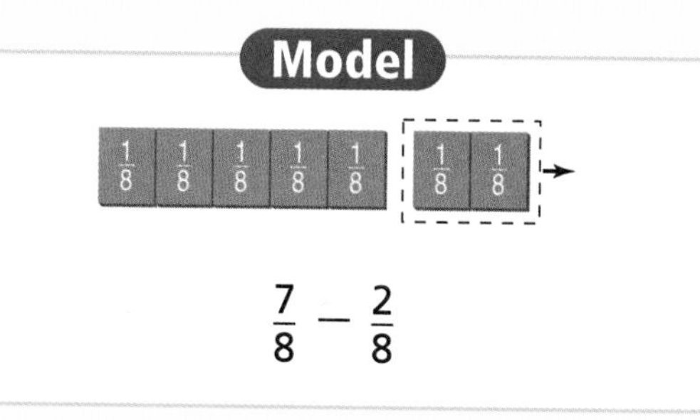

Record

bars you start with		bars you take away		bars that are left
↓		↓		↓
$\frac{7}{8}$	$-$	$\frac{2}{8}$	$=$	$\frac{5}{8}$

MATH IDEA When you subtract like fractions, you subtract the numerators, and the denominators stay the same.

Practice and Problem Solving

Find the difference.

1. 1/8 1/8 1/8 1/8 1/8 1/8 1/8

$\frac{7}{8} - \frac{4}{8} = ■$

2. 1/10 1/10 1/10 1/10 1/10 1/10 1/10 1/10

$\frac{8}{10} - \frac{3}{10} = ■$

3. 1/12 1/12 1/12 1/12 1/12 1/12

$\frac{6}{12} - \frac{2}{12} = ■$

Use fraction bars to find the difference.

4. $\frac{4}{5} - \frac{1}{5} = ■$ **5.** $\frac{6}{8} - \frac{3}{8} = ■$ **6.** $\frac{9}{10} - \frac{7}{10} = ■$ **7.** $\frac{2}{3} - \frac{1}{3} = ■$

8. $\frac{4}{6} - \frac{2}{6} = ■$ **9.** $\frac{11}{12} - \frac{6}{12} = ■$ **10.** $\frac{7}{8} - \frac{4}{8} = ■$ **11.** $\frac{8}{10} - \frac{7}{10} = ■$

12. Malcolm practiced guitar from 11:40 A.M. to 12:55 P.M. How long did he practice?

Mixed Review and Test Prep

Find the sum. (p. 36)

13. 73 + 86 + 95

14. 45 + 19 + 71

15. 67 + 24 + 36

16. Write the value of the 2 in 56,299. (p. 10)

17. **TEST PREP** Wanda earned \$30 mowing lawns, \$25 raking leaves, and \$24 baby-sitting. How much more does Wanda need to have a total of \$100? (p. 172)

A \$100 more **C** \$45 more

B \$79 more **D** \$21 more

LESSON

4 Subtract Fractions

Quick Review

Find the sum. Write the answer in simplest form.

1. $\frac{3}{12} + \frac{1}{12} = \blacksquare$
2. $\frac{1}{4} + \frac{2}{4} = \blacksquare$
3. $\frac{5}{8} + \frac{1}{8} = \blacksquare$
4. $\frac{3}{10} + \frac{6}{10} = \blacksquare$
5. $\frac{2}{6} + \frac{3}{6} = \blacksquare$

Learn

MANY MORE Kara and Eli shared a pack of 10 graham crackers. Kara ate 5 crackers, or $\frac{5}{10}$ of them. Eli ate 3 crackers, or $\frac{3}{10}$ of them. What fraction tells how many more of the crackers Kara ate than Eli?

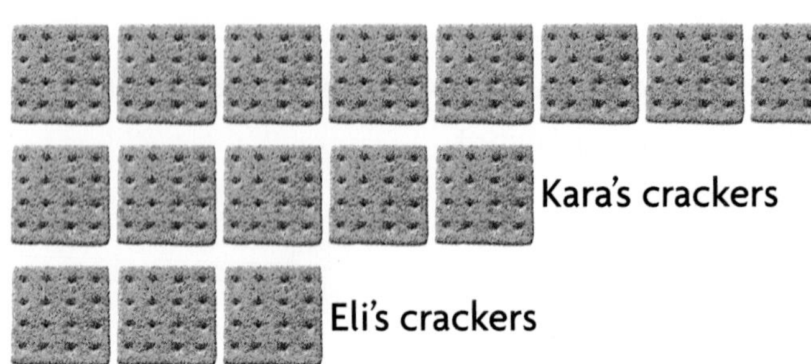

Activity

Use fraction bars to find $\frac{5}{10} - \frac{3}{10}$ in simplest form.

Materials: fraction bars

STEP 1

Model $\frac{5}{10}$ and $\frac{3}{10}$ with fraction bars.

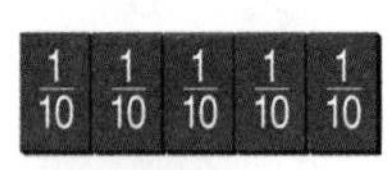

1/10 1/10 1/10

STEP 2

Compare the bars to find the difference.

1/10 1/10 1/10 1/10 1/10

1/10 1/10 1/10 ?

1/10 1/10

STEP 3

Find the largest fraction bar that is the same length.

$\frac{2}{10}$ in simplest form is $\frac{1}{5}$.

$\frac{5}{10} - \frac{3}{10} = \frac{2}{10}$, or $\frac{1}{5}$

So, Kara ate $\frac{1}{5}$ more of the graham crackers than Eli.

- How much of the pack did Kara and Eli eat altogether? How much is left?

Subtracting Fractions

What if Kara and Eli made 8 peanut butter crackers and ate 4 of the crackers. What fraction of the crackers are left?

Use fraction bars to find $\frac{8}{8} - \frac{4}{8}$ in simplest form.

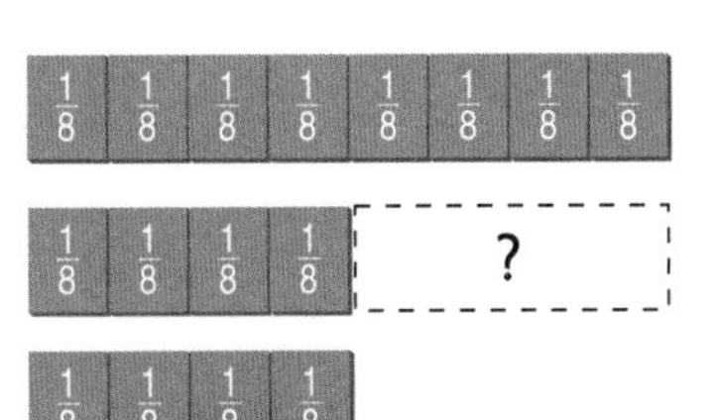

$\frac{8}{8}$ = the total number of crackers

$\frac{4}{8}$ = the 4 crackers they ate

$\frac{4}{8}$ = the crackers that are left

$\frac{4}{8}$ in simplest form is $\frac{1}{2}$.

$\frac{8}{8} - \frac{4}{8} = \frac{4}{8}$, or $\frac{1}{2}$

So, $\frac{1}{2}$ of the crackers are left.

MATH IDEA You can compare fraction bars to subtract fractions and to find the simplest form.

Check

1. **Explain** how you can use fraction bars to find $\frac{3}{4} - \frac{1}{4}$ in simplest form.

Compare. Find the difference. Write the answer in simplest form.

2. 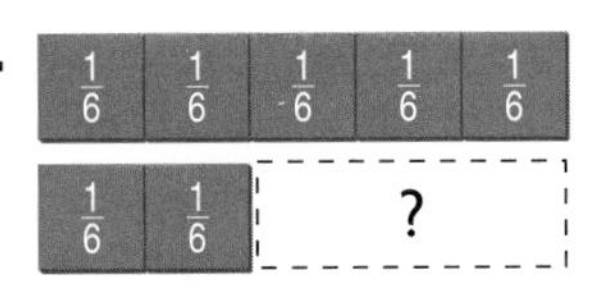

$\frac{5}{6} - \frac{2}{6} = \blacksquare$

3. 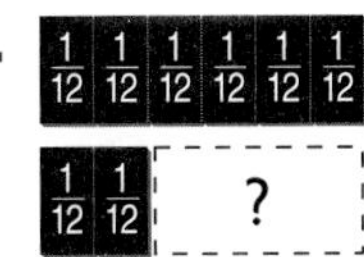

$\frac{6}{12} - \frac{2}{12} = \blacksquare$

4. 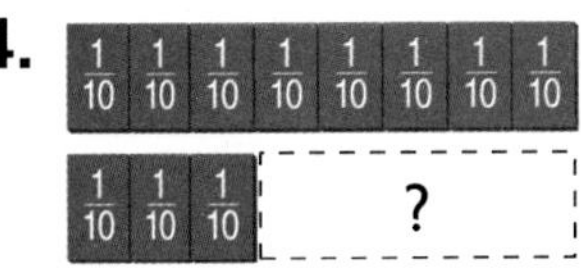

$\frac{8}{10} - \frac{3}{10} = \blacksquare$

5. 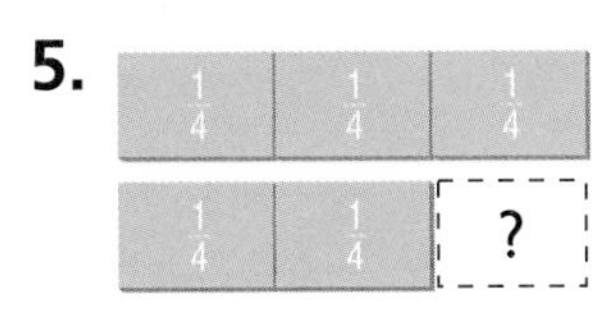

$\frac{3}{4} - \frac{2}{4} = \blacksquare$

6. 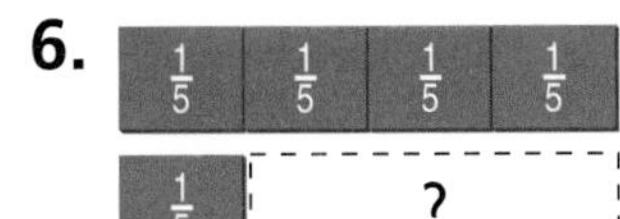

$\frac{4}{5} - \frac{1}{5} = \blacksquare$

7. 

$\frac{5}{8} - \frac{4}{8} = \blacksquare$

LESSON CONTINUES

Practice and Problem Solving

Compare. Find the difference. Write the answer in simplest form.

8. 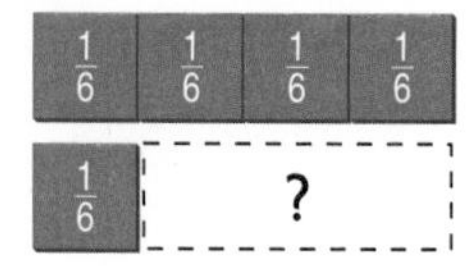

$\frac{4}{6} - \frac{1}{6} = \blacksquare$

9. 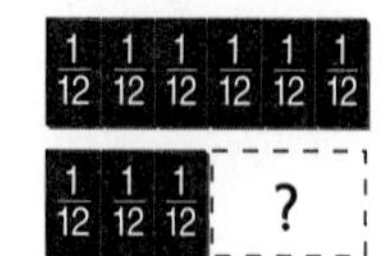

$\frac{6}{12} - \frac{3}{12} = \blacksquare$

10. 

$\frac{3}{8} - \frac{2}{8} = \blacksquare$

11. 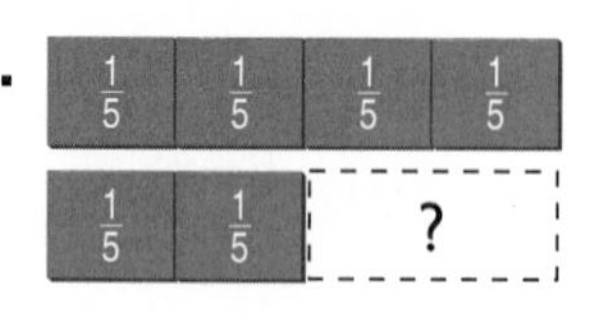

$\frac{4}{5} - \frac{2}{5} = \blacksquare$

12. 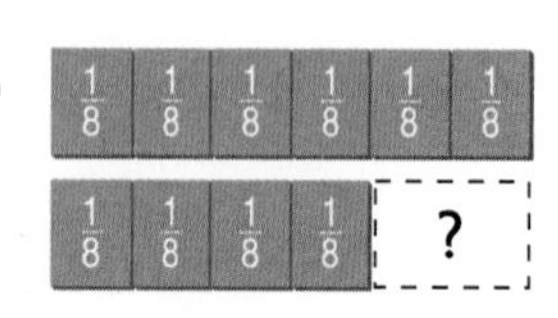

$\frac{6}{8} - \frac{4}{8} = \blacksquare$

13. 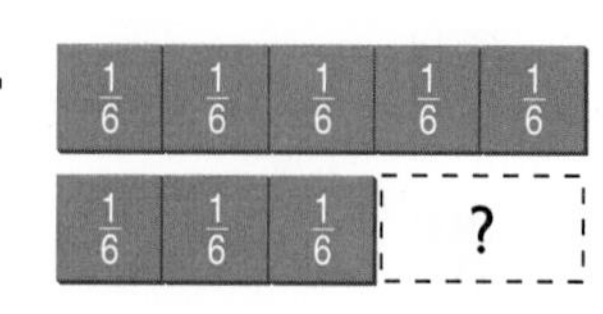

$\frac{5}{6} - \frac{3}{6} = \blacksquare$

Find the difference. Write the answer in simplest form. Use fraction bars.

14. $\frac{2}{3} - \frac{1}{3} = \blacksquare$

15. $\frac{4}{5} - \frac{1}{5} = \blacksquare$

16. $\frac{4}{6} - \frac{2}{6} = \blacksquare$

17. $\frac{5}{6} - \frac{1}{6} = \blacksquare$

18. $\frac{7}{10} - \frac{2}{10} = \blacksquare$

19. $\frac{7}{8} - \frac{5}{8} = \blacksquare$

20. $\frac{10}{12} - \frac{7}{12} = \blacksquare$

21. $\frac{6}{8} - \frac{1}{8} = \blacksquare$

22. $\frac{11}{12} - \frac{9}{12} = \blacksquare$

23. $\frac{8}{10} - \frac{5}{10} = \blacksquare$

24. $\frac{5}{6} - \frac{3}{6} = \blacksquare$

25. $\frac{6}{8} - \frac{2}{8} = \blacksquare$

26. **REASONING** There are 10 letters in the word *California*. The letter *a* is $\frac{2}{10}$ of the word, and the letter *C* is $\frac{1}{10}$ of the word. What fraction of the word is the letter *i*? What fraction of the word are the vowels?

27. Dana cut an apple pie into 8 equal pieces. She shared the pie with 4 of her friends. Dana and each of her friends ate 1 piece of pie. What fraction of the pie is left?

28. **MEASUREMENT** A pancake recipe calls for $\frac{2}{3}$ cup pancake mix and $\frac{1}{3}$ cup milk. How much more mix than milk is needed?

29. **What's the Error?** Haley wrote $\frac{5}{12} - \frac{3}{12} = 2$. What was her error?

30. Viola has $\frac{7}{8}$ of a cake to share with friends. If she and her friends eat $\frac{4}{8}$ of the cake, how much of the cake will Viola have left?

31. **REASONING** An apple was cut into equal-size slices. Kinji ate 3 of them. If 6 slices of apple are left, what fraction of the apple did Kinji eat?

Mixed Review and Test Prep

Find the quotient. (p. 220)

32. $36 \div 9$
33. $40 \div 10$
34. $63 \div 9$
35. $100 \div 10$
36. $70 \div 10$
37. $45 \div 9$
38. $72 \div 9$
39. $10 \div 10$
40. $20 \div 10$
41. $54 \div 9$
42. $81 \div 9$
43. $60 \div 10$

44. **TEST PREP** How many centimeters are in 3 meters? (p. 372)

A 3
B 30
C 300
D 3,000

45. **TEST PREP** Roller-coaster tickets cost \$1.50 each. Samantha bought 2 tickets. How much change should she receive from \$5.00? (p. 88)

F \$2.00
G \$3.00
H \$3.50
J \$4.00

PROBLEM SOLVING LINKUP . . . to Geography

You can use what you know about adding and subtracting fractions to find distances on a map. The scale tells you how distances are measured. On this map each unit represents $\frac{1}{10}$ of a mile.

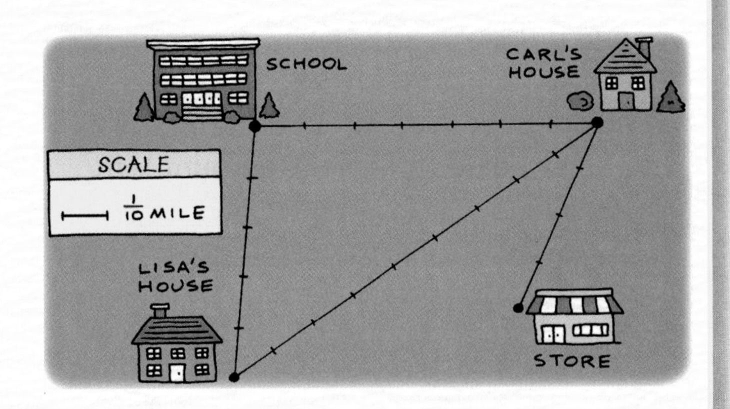

What if you are visiting your friend Carl at his house after school? Use the map to answer the questions.

1. How much farther is the school than the store from Carl's house?

2. How far is Lisa's house from the school?

3. Which is farther from Carl's house, Lisa's house or the school? How much farther?

4. **REASONING** Carl says the store is $\frac{1}{2}$ mile from his house. Is he right? Explain your reasoning.

LESSON 5

Problem Solving Skill
Reasonable Answers

UNDERSTAND › PLAN › SOLVE › CHECK

Quick Review

Find the difference.

1. $\frac{8}{12} - \frac{3}{12} = \blacksquare$
2. $\frac{9}{10} - \frac{2}{10} = \blacksquare$
3. $\frac{5}{6} - \frac{4}{6} = \blacksquare$
4. $\frac{7}{8} - \frac{4}{8} = \blacksquare$
5. $\frac{2}{3} - \frac{1}{3} = \blacksquare$

SOUNDS GOOD Whenever you solve a problem, always check to see that your answer is reasonable and makes sense.

Tanya's mother made a very large chocolate chip cookie. Tanya and Allie each ate $\frac{3}{8}$ of the cookie. What fraction of the cookie was left?

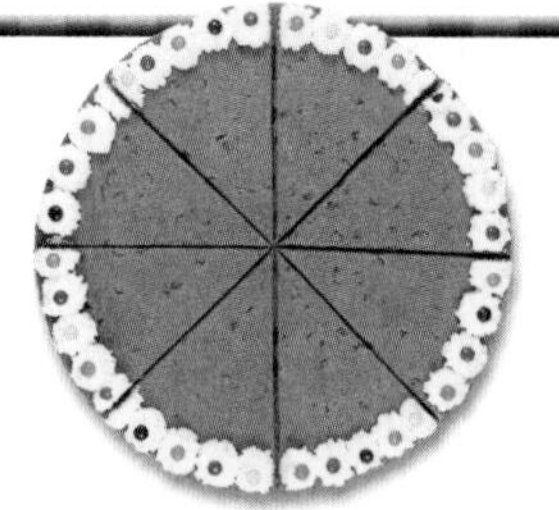

1 whole cookie, or $\frac{8}{8}$

Example

STEP 1

Find out what the problem asks.

How much of the cookie was left after Tanya and Allie each ate $\frac{3}{8}$ of it?

STEP 2

Add to find how much of the cookie was eaten.

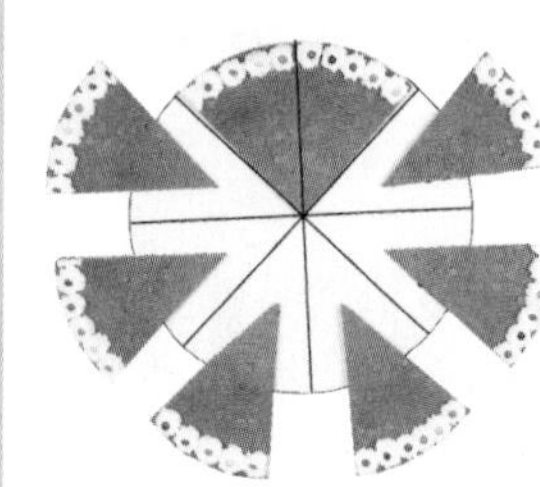

$\frac{3}{8} + \frac{3}{8} = \frac{6}{8}$

STEP 3

Subtract the amount eaten from the whole cookie to find the amount left.

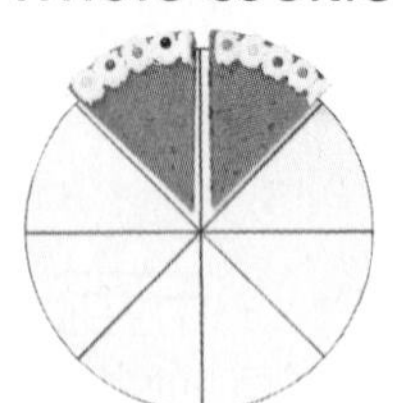

$\frac{8}{8} - \frac{6}{8} = \frac{2}{8}$

So, $\frac{2}{8}$, or $\frac{1}{4}$, of the cookie was left.

STEP 4

Check to see that your answer is reasonable and makes sense.

Think: The girls ate more than $\frac{1}{2}$ of the cookie. So, less than $\frac{1}{2}$ will be left. Since $\frac{2}{8}$ is less than $\frac{1}{2}$, it is a reasonable answer.

Talk About It

- Why is $\frac{8}{8}$ used for the whole cookie?
- Why wouldn't it be reasonable to decide that $\frac{5}{8}$ of the cookie was left?

Problem Solving Practice

Solve. Tell how you know your answer is reasonable.

1. Gil hiked $\frac{2}{5}$ of the mountain trail and rested. Then he hiked another $\frac{1}{5}$ of the trail. How much of the trail does he have left to hike?

2. Clyde opened a new box of cereal. He ate $\frac{1}{3}$ of the cereal in the box. How much cereal was left?

Amy planted $\frac{1}{4}$ of her garden on Monday and $\frac{1}{4}$ of her garden on Tuesday. She planted the rest on Wednesday. How much of the garden did she plant on Wednesday?

3. How can you solve the problem?
   - **A** Compare $\frac{1}{4}$ and $\frac{1}{4}$.
   - **B** Find $\frac{1}{4} + \frac{1}{4}$.
   - **C** Find $\frac{1}{4} + \frac{1}{4}$, and then subtract from $\frac{4}{4}$.
   - **D** Find $\frac{1}{4} - \frac{1}{4}$.

4. What is the answer to the question?
   F $\frac{1}{4}$ **G** $\frac{1}{2}$ **H** $\frac{3}{4}$ **J** $\frac{4}{4}$

Mixed Applications

5. **REASONING** Selma cut a pan of brownies into 3 equal pieces. Then she cut each of those pieces in half. She ate 2 of the pieces. What fraction of the brownies were left? Explain.

6. **ALGEBRA** Satoko had a total of 12 apples and pears. She traded each pear for 2 apples. Then she had 15 apples in all. How many of each fruit did Satoko have to begin with?

7. **Write About It** Describe the pattern below. What will the fourteenth shape in the pattern be?

Review/Test

CHECK VOCABULARY AND CONCEPTS

Choose the best term from the box.

like fractions
simplest form
denominator

1. Fractions that have the same denominator are __?__. (p. 434)
2. When a fraction uses the largest fraction bar or bars possible, it is in __?__. (p. 436)

CHECK SKILLS

Find the sum. Write the answer in simplest form. Use fraction bars. (p. 436–439)

3. $\frac{2}{4} + \frac{1}{4} = ■$
4. $\frac{3}{5} + \frac{1}{5} = ■$
5. $\frac{2}{8} + \frac{4}{8} = ■$
6. $\frac{3}{12} + \frac{2}{12} = ■$
7. $\frac{2}{6} + \frac{1}{6} = ■$
8. $\frac{3}{8} + \frac{2}{8} = ■$
9. $\frac{2}{10} + \frac{7}{10} = ■$
10. $\frac{1}{3} + \frac{2}{3} = ■$

Find the difference. Write the answer in simplest form. Use fraction bars. (pp. 442–445)

11. $\frac{4}{6} - \frac{2}{6} = ■$
12. $\frac{7}{10} - \frac{3}{10} = ■$
13. $\frac{6}{8} - \frac{4}{8} = ■$
14. $\frac{5}{5} - \frac{3}{5} = ■$
15. $\frac{10}{12} - \frac{7}{12} = ■$
16. $\frac{2}{4} - \frac{1}{4} = ■$
17. $\frac{8}{10} - \frac{5}{10} = ■$
18. $\frac{8}{12} - \frac{4}{12} = ■$

CHECK PROBLEM SOLVING

Solve. Tell how you know your answer is reasonable.
(pp. 446–447)

19. Joe gave $\frac{1}{8}$ of his football cards to Pete and $\frac{3}{8}$ of his cards to Ron. What fraction of his cards did he keep for himself?

20. On Monday Raquel read $\frac{1}{10}$ of her library book. On Tuesday she read $\frac{1}{10}$ of her book. What fraction of her book does she have left to read?

Standardized Test Prep

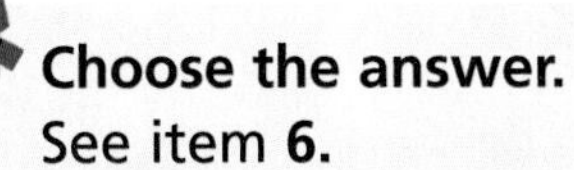

Choose the answer.
See item **6.**

Subtract and look for an answer that matches yours. If your answer doesn't match, look for another form of the number, such as simplest form.

Also see problem **6,** p. H64.

For 1–8, choose the best answer.

1. $\frac{7}{10} + \frac{2}{10} = \blacksquare$

A $\frac{9}{20}$ **C** $\frac{9}{10}$

B $\frac{5}{20}$ **D** $\frac{5}{10}$

2. Trisha had $\frac{6}{8}$ of a pizza to share with her friend. She and her friend ate $\frac{3}{8}$ of the pizza. What fraction was left?

F $\frac{9}{8}$ **G** $\frac{3}{8}$ **H** $\frac{9}{16}$ **J** $\frac{3}{4}$

3. $\frac{2}{6} + \frac{3}{6} = \blacksquare$

A $\frac{5}{6}$ **C** $\frac{5}{12}$

B $\frac{2}{6}$ **D** 1

4. Which number is even?

F 2,227 **H** 2,685

G 2,243 **J** 3,330

5. $7 \times (2 \times 4) = \blacksquare$

A 13 **C** 56

B 14 **D** NOT HERE

6. $\frac{5}{6} - \frac{3}{6} = \blacksquare$

F $\frac{1}{6}$ **H** $\frac{2}{3}$

G $\frac{1}{3}$ **J** $\frac{8}{6}$

7. The third graders want to measure the length of a hallway. Which unit would be best to measure the length of a hall?

A cups **C** degrees

B pounds **D** yards

8. Rudy read $\frac{2}{5}$ of his new book last night. How much is left to read?

F $\frac{2}{5}$ **H** $\frac{4}{5}$

G $\frac{3}{5}$ **J** $\frac{5}{2}$

Write What You Know

9. A pie is cut into 8 equal slices. Derek's family eats 2 slices on Monday and 3 slices on Tuesday. What fraction is left? Draw a picture to solve.

10. Find the difference. Explain how you found the answer.

$$\frac{9}{10} - \frac{6}{10} = \blacksquare$$

Decimals and Fractions

Swimming is a sport in which amounts of time are used to keep score. The times have to be very exact, so decimals are used to show 100 parts of one second. The table shows some of the world's fastest freestyle swimmers and their recorded times.

PROBLEM SOLVING Use the table to find which swimmer had the fastest time.

DATA LINK

FINAL TIMES IN 100 M FREESTYLE

Swimmer	Time
Alexander Popov	48.21 seconds
Matt Biondi	48.42 seconds
Michael Klim	48.98 seconds
Fernando Scherer	48.69 seconds

CHECK WHAT YOU KNOW

Use this page to help you review and remember important skills needed for Chapter 25.

NAME THE FRACTION (For Intervention, see p. H29.)

Write a fraction for the shaded part.

1.

2.

3. 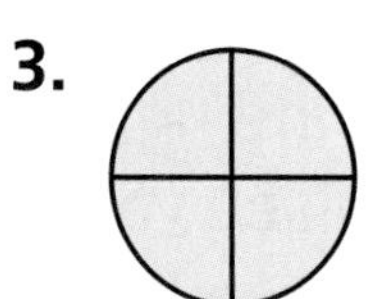

Write a fraction that names the part of each group that is shaded.

4. 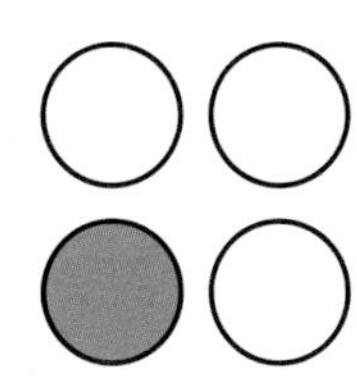

5. 

6. 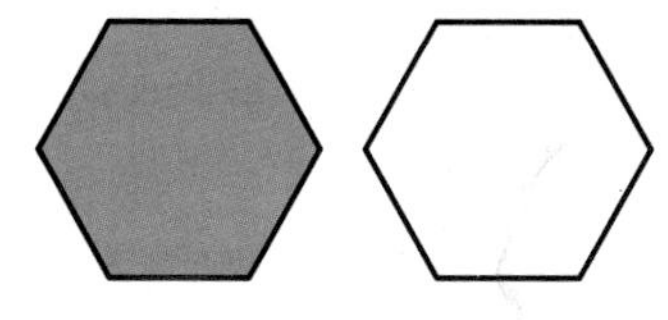

COMPARE FRACTIONS (For Intervention, see p. H30.)

Compare. Write <, >, or = for each ●.

7. 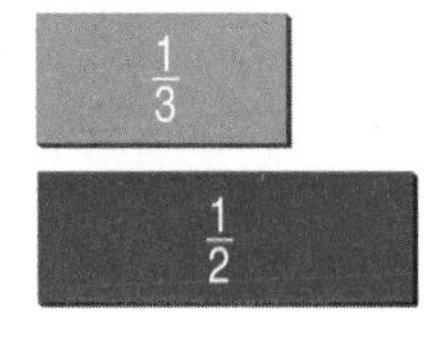

$\frac{1}{3}$ ● $\frac{1}{2}$

8. 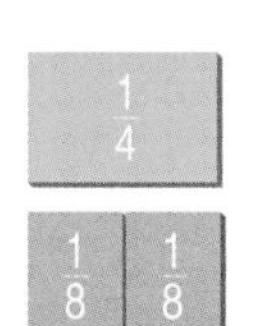

$\frac{1}{4}$ ● $\frac{2}{8}$

9. 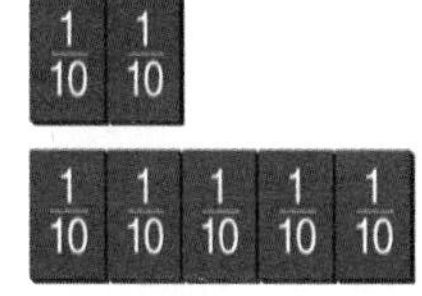

$\frac{2}{10}$ ● $\frac{5}{10}$

Compare the part of each group that is shaded. Write <, >, or = for each ●.

10. 
$\frac{5}{5}$ ● $\frac{2}{5}$

11. 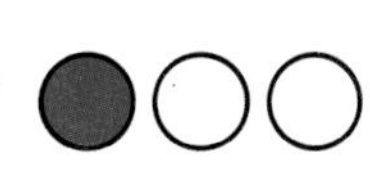
$\frac{1}{3}$ ● $\frac{2}{3}$

12. 
$\frac{3}{4}$ ● $\frac{0}{4}$

13. $\frac{3}{6}$ ● $\frac{4}{6}$

14. $\frac{4}{4}$ ● $\frac{4}{4}$

15. 
$\frac{6}{7}$ ● $\frac{5}{7}$

LESSON 1

Relate Fractions and Decimals

Quick Review

Write each fraction in words.

1. $\frac{1}{3}$
2. $\frac{2}{5}$
3. $\frac{2}{10}$
4. $\frac{3}{8}$
5. $\frac{1}{2}$

VOCABULARY

decimal
tenth

Learn

FAIR SHARE A **decimal** is a number with one or more digits to the right of the decimal point. A decimal uses place value to show values less than one, such as tenths.

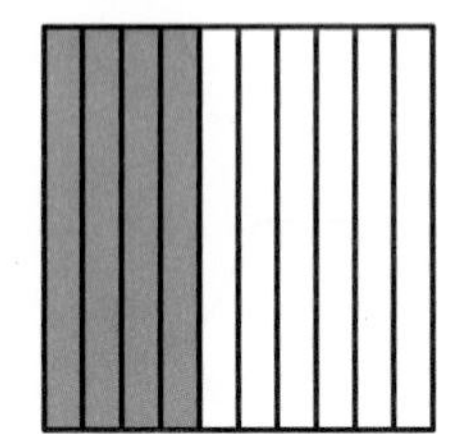

This square has 10 equal parts. Each equal part is one **tenth**.

Technology Link

More Practice: Use Mighty Math Number Heroes, *Fraction Fireworks,* Levels Q and Y.

Fraction	Decimal
Write: $\frac{4}{10}$	**Write:** 0.4 (decimal point)
Read: four tenths	**Read:** four tenths

The fraction $\frac{4}{10}$ and the decimal 0.4 name the same amount.

MATH IDEA You can use a fraction or a decimal to show values in tenths.

Examples

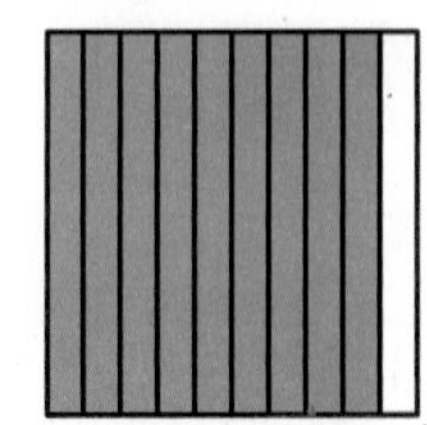

Fraction: $\frac{9}{10}$
Decimal: 0.9
Read: nine tenths

B

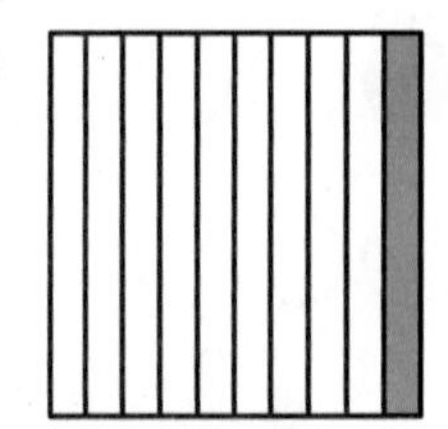

Fraction: $\frac{1}{10}$
Decimal: 0.1
Read: one tenth

C

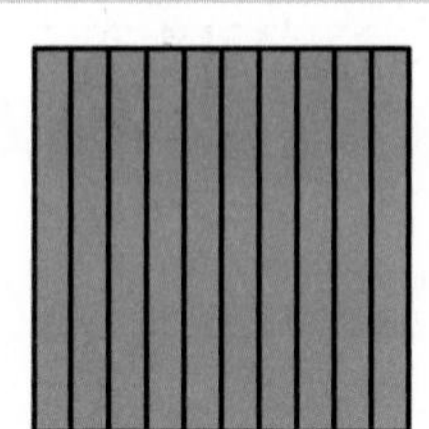

Fraction: $\frac{10}{10}$
Decimal: 1.0
Read: ten tenths or one

- How many parts on the square would you shade to show 0.3?

Check

1. **Write** a fraction that shows the amount of crispy treats that are left. Then write the same amount as a decimal.

Write the fraction and decimal for the shaded part.

2. 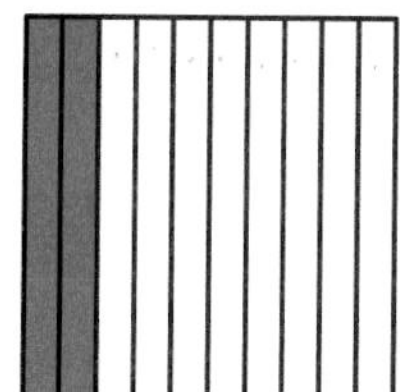

3. 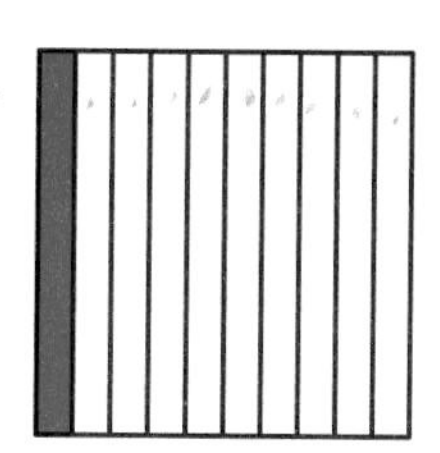

4. 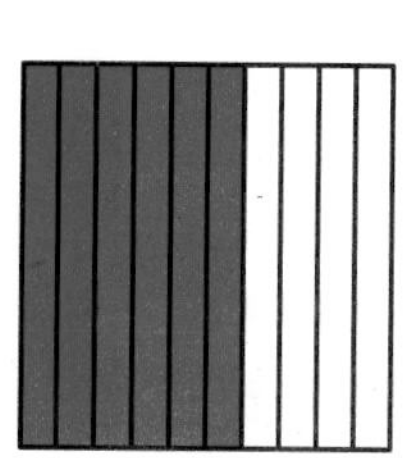

Practice and Problem Solving

Write the fraction and decimal for the shaded part.

5. 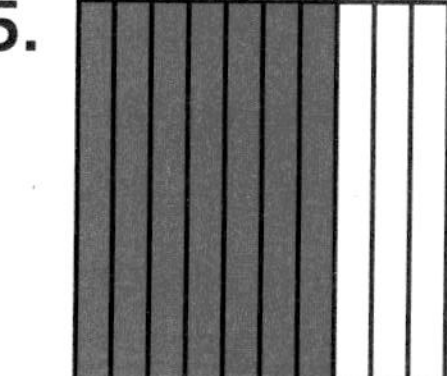

6. 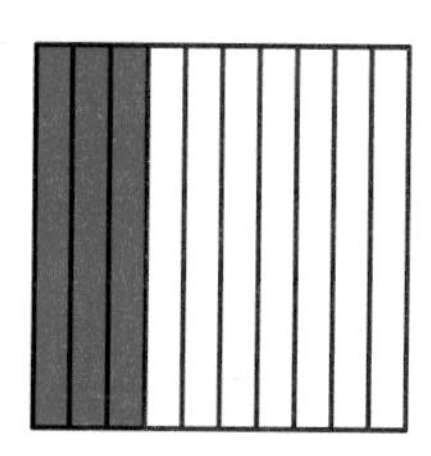

7. 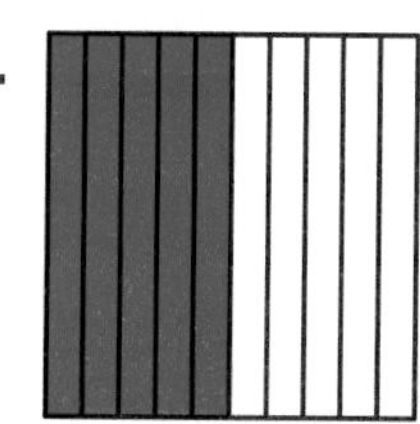

8. 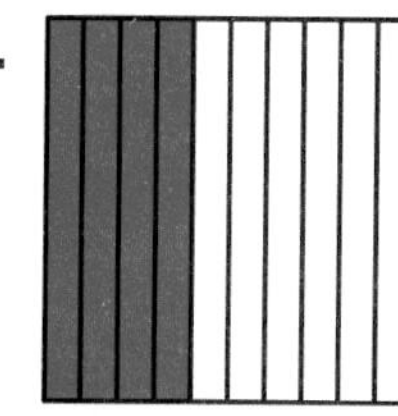

9. 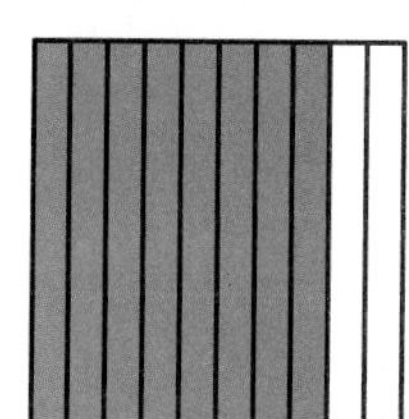

10. 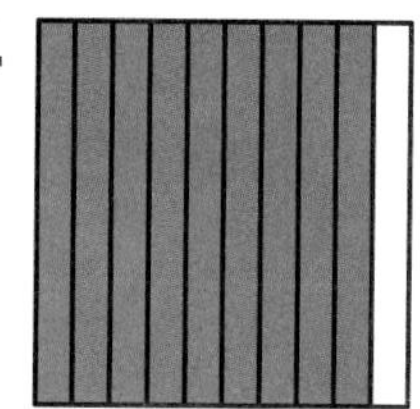

11. **REASONING** Look at Exercise 8. How could you use subtraction to find the decimal amount that is NOT shaded?

12. **Write About It** Look at the fraction bars at the right. Explain why 0.5 is the same as $\frac{1}{2}$.

$\frac{1}{2}$				
$\frac{1}{10}$	$\frac{1}{10}$	$\frac{1}{10}$	$\frac{1}{10}$	$\frac{1}{10}$

Mixed Review and Test Prep

Find the quotient. (p. 216)

13. $49 \div 7 = \blacksquare$

14. $63 \div 7 = \blacksquare$

15. $42 \div 6 = \blacksquare$

16. $32 \div 8 = \blacksquare$

17. **TEST PREP** Frieda puts 5 beads on each of 9 necklaces. How many beads does she use? (p. 118)

A 45 **B** 54 **C** 72 **D** 81

EXTRA PRACTICE page H56, Set A

LESSON 2

Tenths

Quick Review

1. $\frac{1}{10} + \frac{2}{10} = ■$
2. $\frac{2}{10} + \frac{5}{10} = ■$
3. $\frac{2}{10} - \frac{1}{10} = ■$
4. $\frac{4}{10} - \frac{4}{10} = ■$
5. $\frac{7}{10} + \frac{2}{10} = ■$

MATERIALS
decimal models
markers

Remember

A mixed number is made up of a whole number and a fraction.

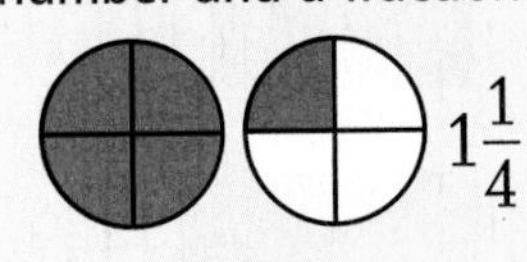

$1\frac{1}{4}$

Explore

This decimal model shows six tenths.

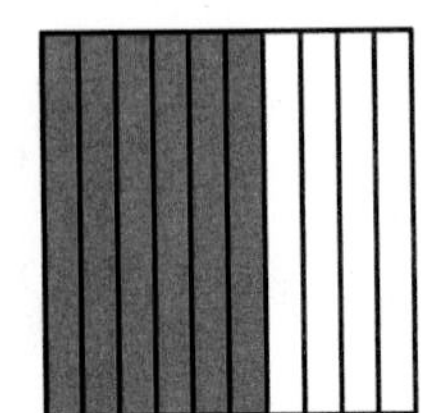

Write: $\frac{6}{10}$ or 0.6

Activity

Use a decimal model to show 0.2, or $\frac{2}{10}$.

STEP 1

Shade the decimal model to show two tenths.

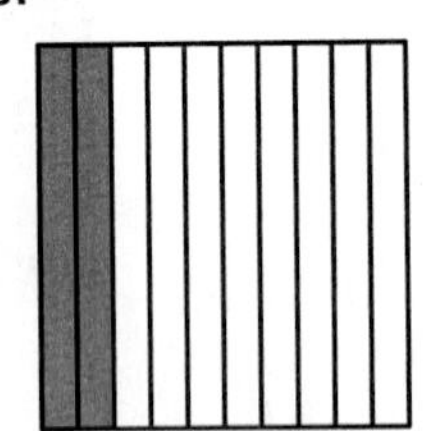

STEP 2

Below your decimal model, write the fraction and decimal amount you have shown.

- Use a new decimal model. Shade and label the decimal model to show 0.8, or $\frac{8}{10}$.

Try It

Shade and label decimal models to show each amount.

a. $\frac{9}{10}$, or 0.9 **b.** $\frac{5}{10}$, or 0.5

c. $\frac{7}{10}$, or 0.7 **d.** $\frac{1}{10}$, or 0.1

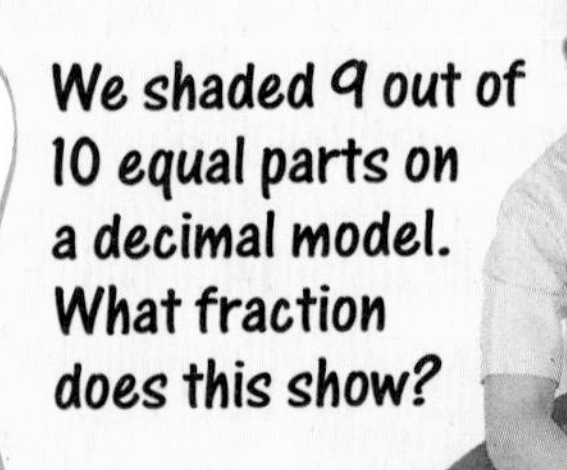

We shaded 9 out of 10 equal parts on a decimal model. What fraction does this show?

Connect

You can write a fraction or a decimal to show tenths.

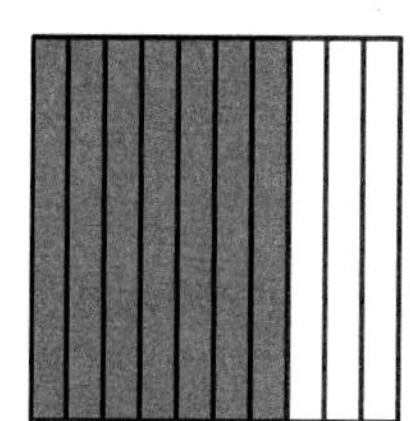

Write: $\frac{7}{10}$ or 0.7

Read: seven tenths

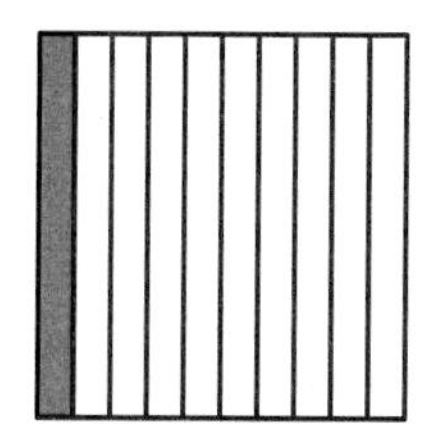

Write: $\frac{1}{10}$ or 0.1

Read: one tenth

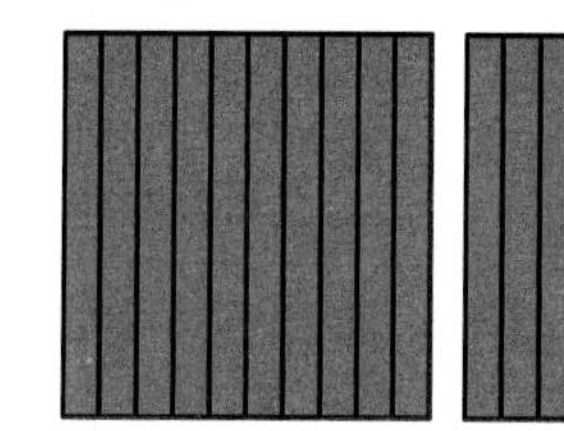

Write: $1\frac{5}{10}$ or 1.5

Read: one and five tenths

Practice and Problem Solving

Use decimal models to show each amount. Then write the decimal.

1. $\frac{3}{10}$ **2.** $\frac{8}{10}$ **3.** $\frac{4}{10}$ **4.** $1\frac{6}{10}$

Write each fraction or mixed number as a decimal.

5. $\frac{2}{10}$ **6.** $\frac{9}{10}$ **7.** $1\frac{1}{10}$ **8.** $\frac{7}{10}$

Write each decimal as a fraction or mixed number.

9. 0.5 **10.** 1.4 **11.** 0.3 **12.** 0.1

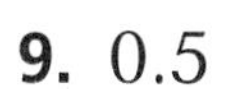

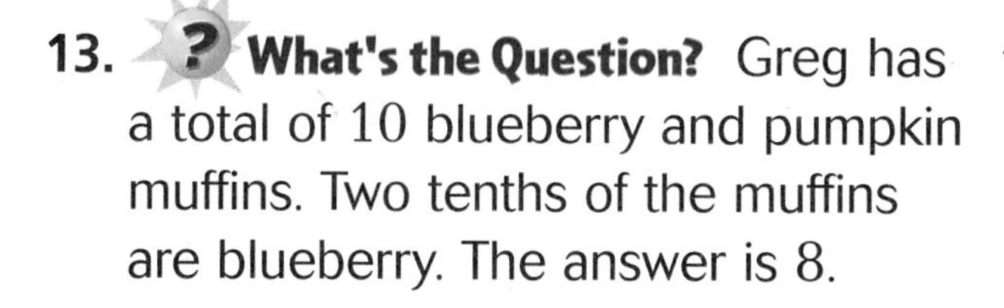

13. **What's the Question?** Greg has a total of 10 blueberry and pumpkin muffins. Two tenths of the muffins are blueberry. The answer is 8.

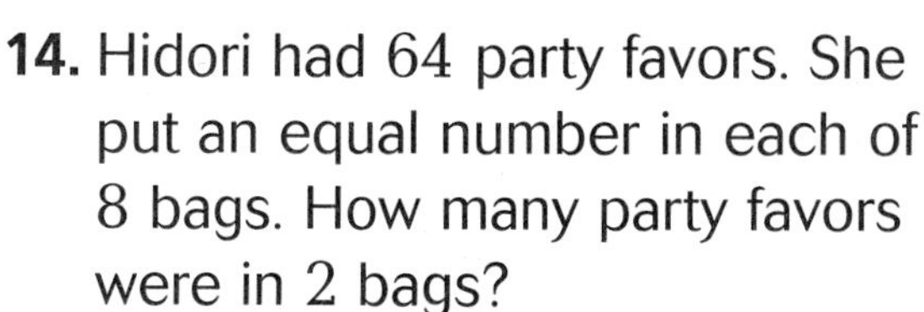

14. Hidori had 64 party favors. She put an equal number in each of 8 bags. How many party favors were in 2 bags?

Mixed Review and Test Prep

Write <, >, or = for each ●. (p. 138)

15. 4×9 ● 36

16. 9×7 ● 8×8

17. 3×6 ● 9×2

18. 8×7 ● 9×6

19. **TEST PREP** Sara had \$4.08. She bought a sticker for \$0.29. How much money did she have left? (p. 88)

A \$3.89 **C** \$3.79

B \$3.81 **D** \$3.70

LESSON 3

Hundredths

Explore

Each of these decimal models has 100 equal parts. Each equal part is one **hundredth**.

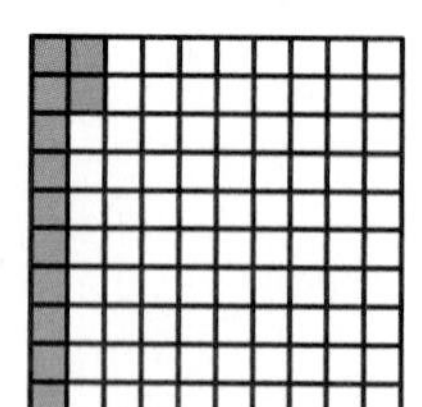

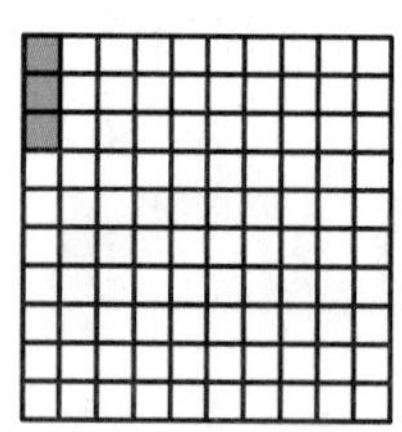

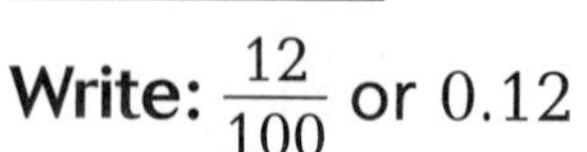

Write: $\frac{12}{100}$ or 0.12
Read: twelve hundredths

Write: $\frac{3}{100}$ or 0.03
Read: three hundredths

Quick Review

Write each fraction as a decimal.

1. $\frac{1}{10}$
2. $\frac{3}{10}$
3. $\frac{6}{10}$
4. $\frac{9}{10}$
5. $\frac{2}{10}$

VOCABULARY
hundredth

MATERIALS
decimal models
markers

Activity

Use a decimal model to show 0.05, or $\frac{5}{100}$.

STEP 1

Shade the decimal model to show five hundredths.

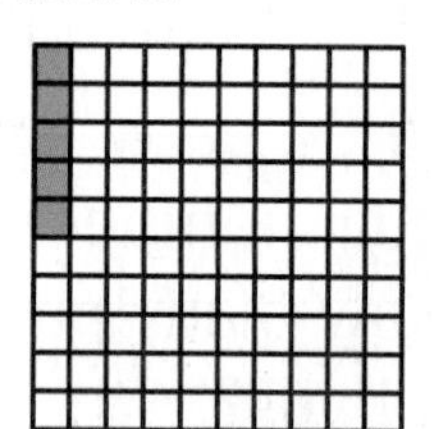

STEP 2

Below your model, write the fraction and decimal you have shown.

How many equal parts should I shade in all to show 0.06?

- Use a new decimal model. Shade and label the decimal model to show 0.13, or $\frac{13}{100}$.

Try It

Use decimal models to show:

a. 0.06

b. 0.60

Connect

You can write a fraction or a decimal to show hundredths.

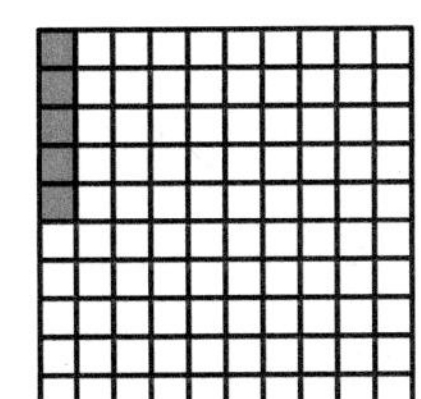

Write: $\frac{5}{100}$ or 0.05
Read: five hundredths

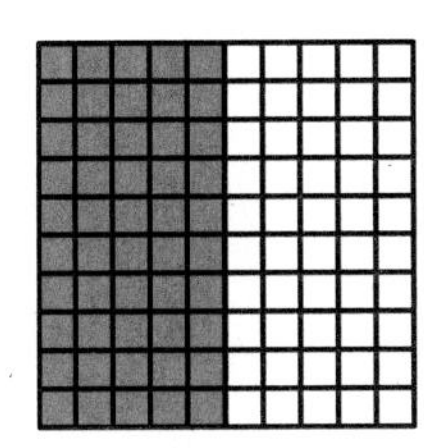

Write: $\frac{50}{100}$ or 0.50
Read: fifty hundredths

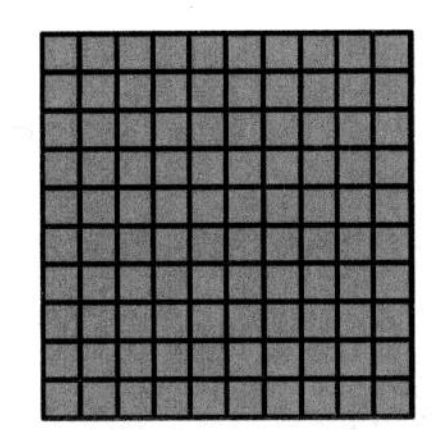
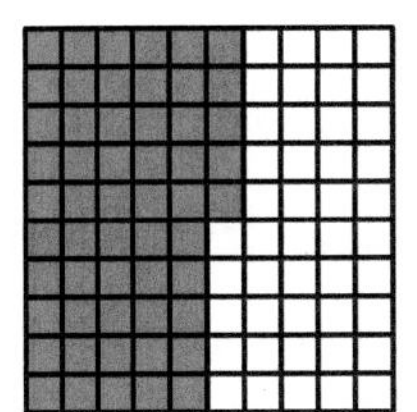

Write: $1\frac{55}{100}$ or 1.55
Read: one and fifty-five hundredths

Practice and Problem Solving

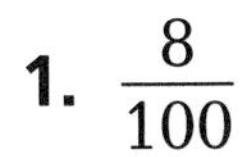

Technology Link
More Practice: Use E-Lab, ***Hundredths.***
www.harcourtschool.com/elab2002

Use decimal models to show each amount. Then write the decimal.

1. $\frac{8}{100}$ **2.** $\frac{10}{100}$ **3.** $\frac{24}{100}$ **4.** $1\frac{22}{100}$

Write each fraction or mixed number as a decimal.

5. $\frac{6}{100}$ **6.** $\frac{15}{100}$ **7.** $\frac{70}{100}$ **8.** $1\frac{57}{100}$

Write each decimal as a fraction or mixed number.

9. 0.01 **10.** 0.56 **11.** 1.20 **12.** 0.02 **13.** 1.05

14. **What's the Error?** Sean said that this model shows 0.90. Describe his error. Write the correct decimal.

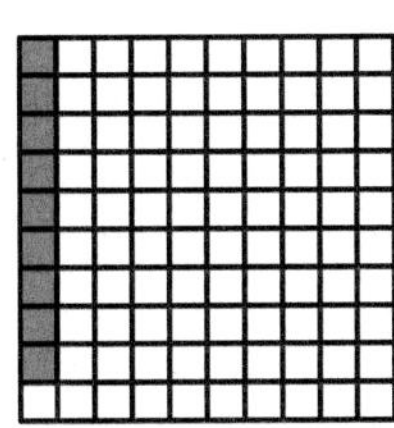

15. Janet and Kim are sharing a pizza. If Janet eats $\frac{3}{8}$ of the pizza, and Kim eats $\frac{2}{8}$, what fraction of the pizza will be left?

Mixed Review and Test Prep

16. 3,921 + 765 = ■ (p. 46)

17. 593 + 1,421 = ■ (p. 46)

18. 843 – 258 = ■ (p. 58)

19. 706 – 55 = ■ (p. 58)

20. TEST PREP Jacob's photo is 5 in. wide and 7 in. long. What is its perimeter? (p. 390)

A 12 in. **C** 26 in.
B 24 in. **D** 35 in.

LESSON

Read and Write Decimals

Quick Review

Write each number in standard form.

1. 400 + 50 + 3
2. 100 + 70 + 6
3. 300 + 0 + 3
4. 900 + 20 + 1
5. 200 + 40 + 9

Learn

CUT THE CAKE Richard went to a party. There was a large cake, cut into 100 equal pieces. If 35 pieces were eaten, what decimal shows how much of the cake was eaten?

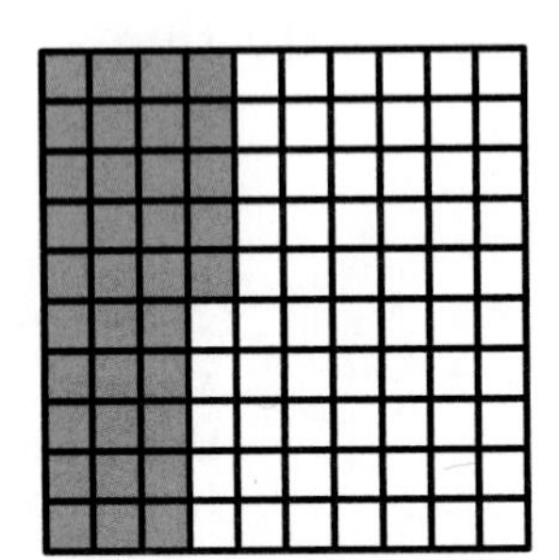

ONES	•	TENTHS	HUNDREDTHS
0	•	3	5

Write: 0.35 **Read:** thirty-five hundredths

So, 0.35 of the cake was eaten.

You can also show a decimal in expanded form.

expanded form: 0.3 + 0.05

0.35 is the same as 3 tenths 5 hundredths.

Example

ONES	•	TENTHS	HUNDREDTHS
2	•	4	6

standard form: 2.46

word form: two and forty-six hundredths

expanded form: 2.0 + 0.4 + 0.06

2.46 is the same as 2 ones 4 tenths 6 hundredths.

- What is the standard form for 0.5 + 0.06, or 5 tenths 6 hundredths?

Check

1. Write the expanded form for 0.32.

Write the word form and expanded form for each decimal.

2.

ONES	•	TENTHS	HUNDREDTHS
0	•	6	2

3.

ONES	•	TENTHS	HUNDREDTHS
3	•	5	5

Practice and Problem Solving

Write the word form and expanded form for each decimal.

4.

ONES	•	TENTHS	HUNDREDTHS
0	•	7	4

5.

ONES	•	TENTHS	HUNDREDTHS
0	•	1	3

6.

ONES	•	TENTHS	HUNDREDTHS
4	•	5	1

7.

ONES	•	TENTHS	HUNDREDTHS
0	•	2	9

Write *tenths* or *hundredths*.

8. 0.16 = 1 tenth 6 _?_

9. 0.20 = 2 _?_ 0 hundredths

Write the missing number.

10. 0.90 = ■ tenths 0 hundredths

11. 0.09 = ■ tenths 9 hundredths

12. Mrs. Lightfoot displayed her students' drawings. She had 4 rows of 5 drawings on one wall, and 3 rows of 7 drawings on another wall. How many were displayed in all?

13. Raul says that 4 tenths is the same as 40 hundredths. Do you agree? Explain. Use this decimal square to help.

Mixed Review and Test Prep

Find the number that the variable stands for. (p. 188)

14. $a \times 4 = 32$

15. $7 \times b = 21$

16. $8 \times c = 40$

17. $d \times 5 = 30$

18. **TEST PREP** Pilar put 6 olives on each pizza. If she made 8 pizzas, how many olives did she use? (p. 148)

A 14 **B** 24 **C** 40 **D** 48

EXTRA PRACTICE page H56, Set B

LESSON 5

Compare and Order Decimals

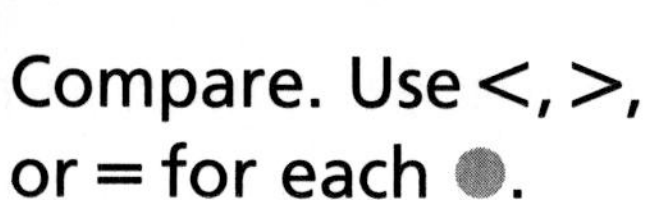

Quick Review

Compare. Use <, >, or = for each ●.

1. 35 ● 53
2. 901 ● 1,093
3. 1,243 ● 1,423
4. 30 + 5 ● 25 + 10
5. 45 + 6 ● 55 − 5

Learn

TAKE YOUR PLACE A place-value chart can help you compare tenths.

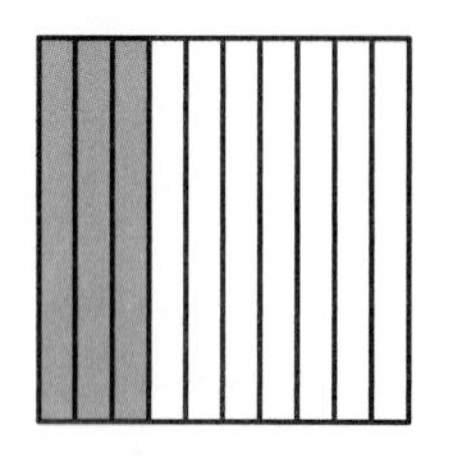

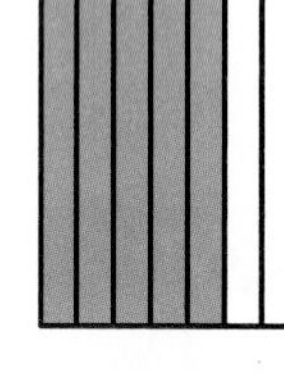

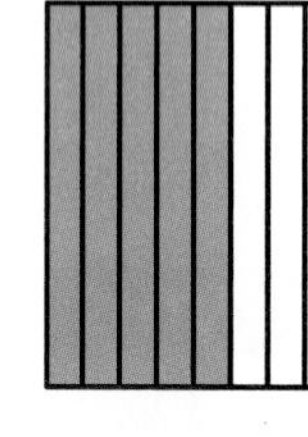

0.3 < 0.5

ONES	•	TENTHS
0	•	3
0	•	5

- Compare ones.
- Compare tenths.
  0.3 is less than 0.5.

You can also compare decimals with hundredths.

Swimmers and other athletes compare their times written in hundredths of seconds.

ONES	•	TENTHS	HUNDREDTHS
4	•	7	9
4	•	5	1

4.79 > 4.51

- Begin with the digit in the greatest place value.
- Compare digits in each place.
- 7 tenths > 5 tenths.
  So, 4.79 seconds is greater than 4.51 seconds.

You can use a number line to order decimals.

Example

Write 0.7, 0.3, and 0.9 in order from least to greatest.

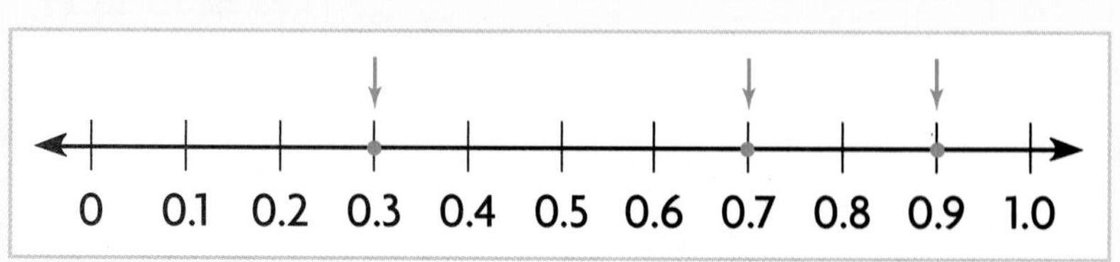

The numbers from least to greatest are 0.3, 0.7, 0.9.

- How could you use the number line to compare 0.5 and 0.8?

Check

1. **Explain** how to compare 3.54 and 3.52.

Compare. Write $<$ or $>$ for each ●.

2.

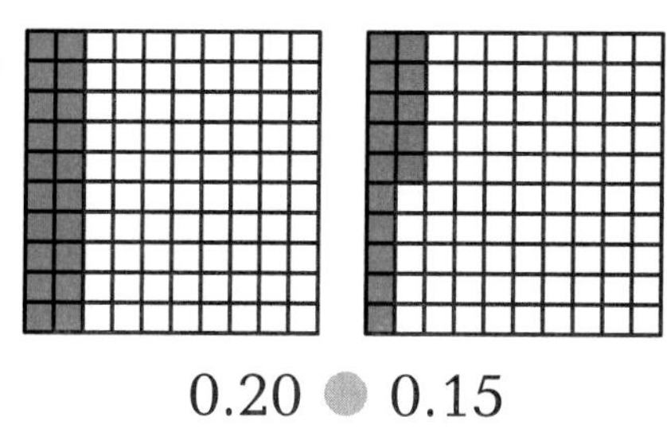

0.20 ● 0.15

3.

ONES	•	TENTHS
4	•	1
4	•	6

4.1 ● 4.6

Practice and Problem Solving

Compare. Write $<$ or $>$ for each ●.

4.

ONES	•	TENTHS
1	•	5
1	•	6

1.5 ● 1.6

5.

ONES	•	TENTHS	HUNDREDTHS
2	•	5	1
2	•	3	9

2.51 ● 2.39

6. 2.4 ● 2.1
7. 3.5 ● 2.6
8. 0.22 ● 0.25
9. 1.67 ● 1.76

For 10–14, order the decimals from least to greatest. For 10–12, use the number line.

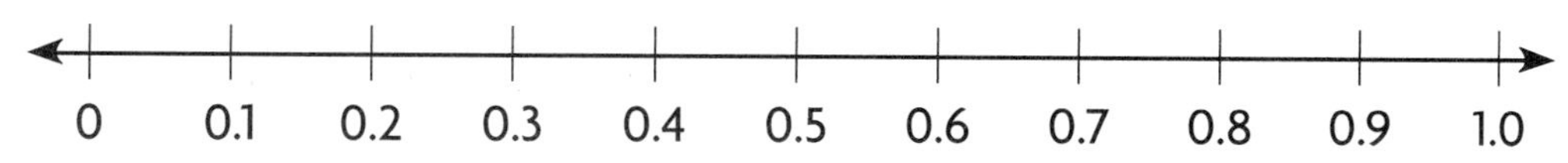

10. 0.5, 0.2, 0.1
11. 0.6, 0.5, 0.8
12. 0.1, 0.3, 0.2
13. 0.13, 0.09, 0.2
14. 0.11, 0.3, 0.04

15. **Write a problem** that compares decimals with ones, tenths, and hundredths. Trade with a partner and solve.

16. Silvia had 47 stickers. She gave the same number of stickers to each of 4 friends, and kept 11. How many did each friend get?

Mixed Review and Test Prep

Find the sum. (p. 42)

17. 132 + 65
18. 209 + 371
19. 587 + 439
20. 76 + 366

21. **TEST PREP** $72 \div 9 = ■$ (p. 220)

A 8 **B** 7 **C** 6 **D** 5

EXTRA PRACTICE page H56, Set C

LESSON

6 Problem Solving Skill
Reasonable Answers

UNDERSTAND PLAN SOLVE CHECK

Quick Review

Write each fraction as a decimal.

1. $\frac{1}{10}$
2. $\frac{50}{100}$
3. $\frac{4}{10}$
4. $\frac{5}{10}$
5. $\frac{71}{100}$

LIGHT WEIGHT Brody weighed a grapefruit at the grocery store. The grapefruit weighed 0.5 pound. He took 2 grapefruit to the checkout counter. The checker said that he bought 3.3 pounds of grapefruit. Is that amount reasonable?

MATH IDEA When you check if an answer is reasonable, decide if the answer makes sense compared with the facts in the problem.

Read the problem again. What information do you have?

One grapefruit weighs 0.5 pound.
Brody bought 2 grapefruit.

Find about how much 2 grapefruit weigh.

Think: 0.5 is the same as $\frac{5}{10}$.

$\frac{5}{10}$ is the same as $\frac{1}{2}$.

Each grapefruit weighs about half a pound. That means that two grapefruit weigh about 1 pound.

Compare 1 pound and 3.3 pounds. Is 3.3 pounds reasonable?

No, 3.3 pounds is not reasonable.

Problem Solving Practice

Solve.

1. Brody found that a plum weighs $\frac{1}{4}$ pound. He took 2 plums to the grocery clerk. The clerk said that 2 plums weigh $\frac{1}{2}$ pound. Is the clerk's total reasonable? Explain.

2. Zach had 2.5 pounds of ground meat to cook for dinner. After his family ate dinner, he said he had about 3 pounds of ground meat left. Is his estimate reasonable? Explain.

Nieta spent $4.15 on lunch. She paid with a $5 bill.

3. Which is a reasonable answer for how much change Nieta got?
   - **A** No change
   - **B** $0.85
   - **C** $1.85
   - **D** $3.85

4. **What if** Nieta gave the clerk a $5 bill to pay for her lunch, which cost $3.15? Which is a reasonable answer for how much change she got?
   - **F** No change
   - **G** $0.85
   - **H** $1.85
   - **J** $3.85

Mixed Applications

5. Ms. Cortez built a fence around her garden. The length of the garden is 4 feet and the width is 3 feet. What is the perimeter of her fence?

6. Mia baked 30 cookies. She gave 6 cookies to each of 4 teachers. Then she gave half of the rest of the cookies to her sister. How many cookies did Mia keep?

7. I am a number less than 40 that can be divided equally into groups of 7. The sum of my digits is 8. What number am I?

8. Neil bought 2 sandwiches. He paid with a $10 bill. He received $4 in change. How much did each sandwich cost?

USE DATA For 9-10, use the graph.

9. How many more students voted for otters and seals than voted for whales and seals?

10. How many students did NOT vote for dolphins?

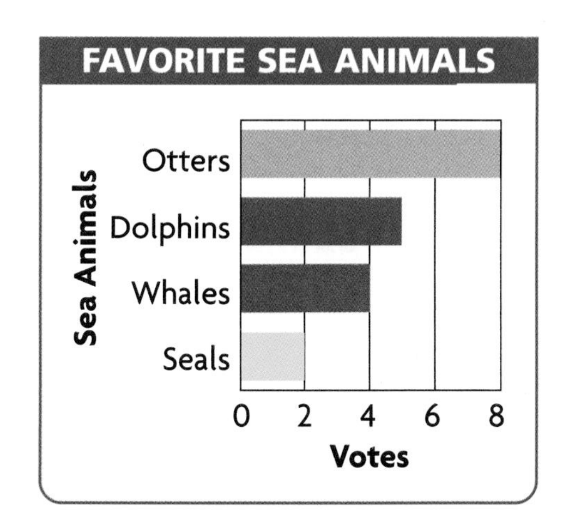

Problem Solving Skill

Review/Test

CHECK VOCABULARY AND CONCEPTS

Choose the best term from the box.

decimal
product

1. A number with one or more digits to the right of the decimal point is a ___?___ . (p. 452)

Write each fraction or mixed number as a decimal. (pp. 454–457)

2. $\frac{3}{10}$ 3. $\frac{9}{10}$ 4. $1\frac{2}{10}$ 5. $\frac{3}{100}$ 6. $\frac{80}{100}$

CHECK SKILLS

Write the fraction and decimal for the shaded part. (pp. 452–453)

7. 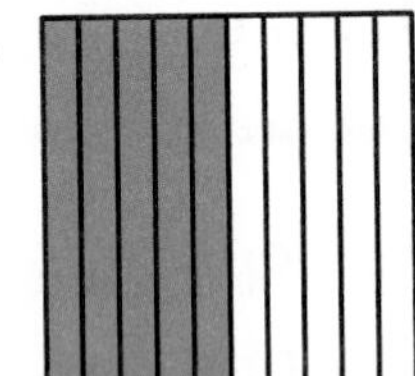

8. 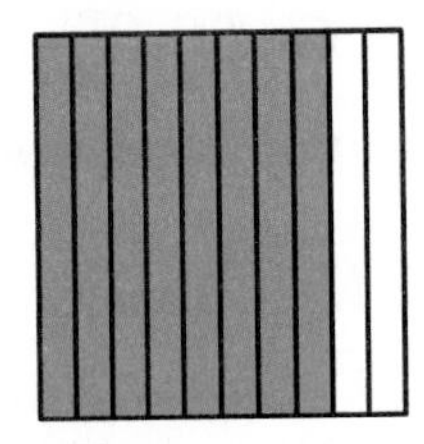

9. 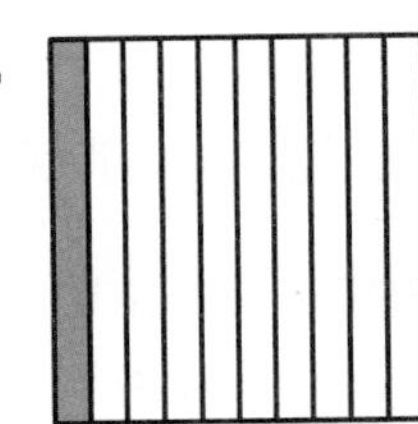

10. 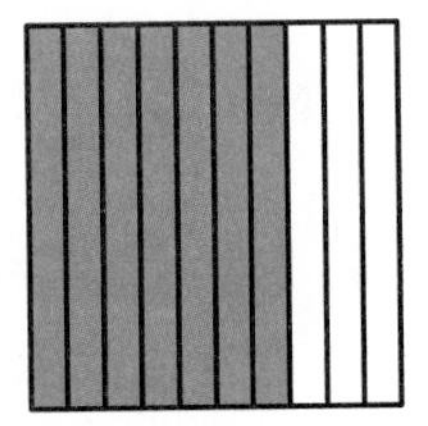

Write the missing number. (pp. 458–459)

11. $0.11 = ■$ tenth 1 hundredth

12. $0.07 = 0$ tenths ■ hundredths

13. $0.02 = ■$ tenths 2 hundredths

14. $0.35 = ■$ tenths 5 hundredths

Compare. Write $<$ or $>$ for each ●. (pp. 460–461)

15. 1.9 ● 1.1 16. 3.42 ● 3.41 17. 0.13 ● 0.31

Order from least to greatest. Use the number line to help. (pp. 460–461)

18. 0.9, 0.3, 0.4 19. 0.1, 0.8, 0.7

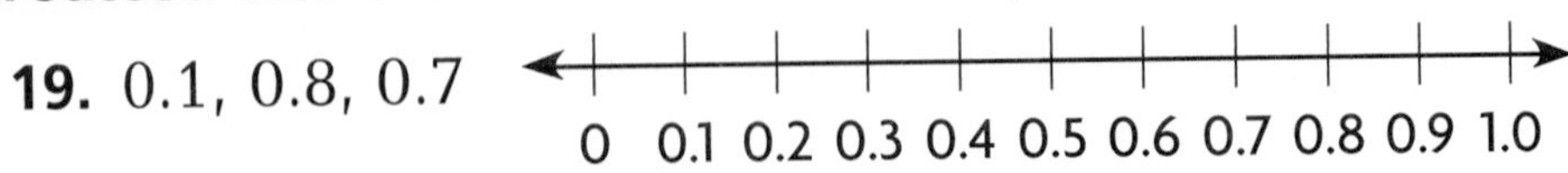

CHECK PROBLEM SOLVING

Solve. (pp. 462–463)

20. Quinn is being weighed. He asks that 2 pounds be subtracted from his weight because he has his shoes on. Each shoe weighs 0.4 pound. Is his request reasonable? Explain.

Standardized Test Prep

TIP! **Look for important words.**
See item **8.**

An important word is *reasonable.* All answer choices include the word *about* so an exact amount is not needed.

Also see problem **2,** p. H62.

For 1–8, choose the best answer.

1. Which decimal names the shaded part?

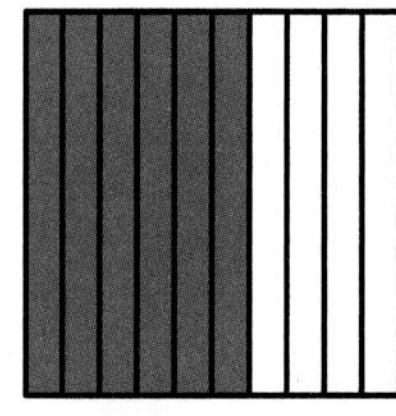

A 0.5 **B** 0.6 **C** 0.7 **D** 6.0

2. What time does the clock show?

F 7:17 **H** 4:37
G 6:17 **J** 3:37

3. Which is another name for $1\frac{5}{100}$?

A 0.05 **C** 1.05
B 0.15 **D** 1.50

4. Which symbol makes this true?

0.3 ● 0.07

F $<$ **G** $>$ **H** $=$

5. $\begin{array}{r} 6,147 \\ -3,218 \\ \hline \end{array}$

A 2,929 **C** 2,819
B 2,829 **D** NOT HERE

6. Which is another name for $\frac{33}{100}$?

F 33 **H** 0.33
G 3.3 **J** 0.033

7. Which point on the number line represents 0.5?

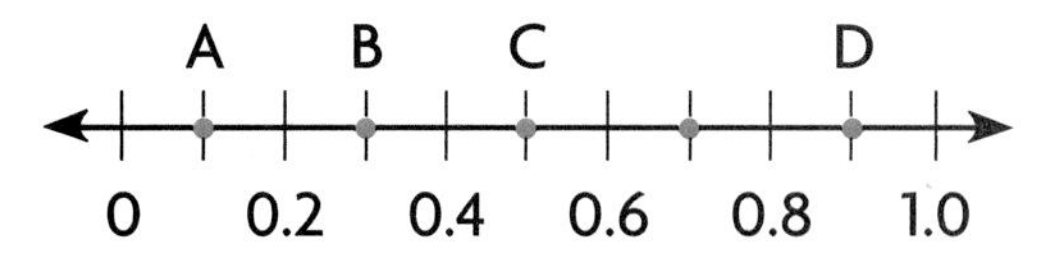

A point A **C** point C
B point B **D** point D

8. Jan spent $5.29 on school supplies. She paid with a $10 bill. Which is a reasonable amount for Jan's change?

F about $0.70 **H** about $6.00
G about $5.00 **J** about $7.00

Write What You Know

9. Name a decimal that is greater than 0.02. Explain your answer.

10. Make a spinner. Name two outcomes that are equally likely. Name an outcome that would be unlikely for your spinner.

CHAPTER 26 Decimals and Money

In the United States, 1 dollar is equal to 100 pennies. In India, 1 rupee is equal to 100 paise. In both countries, money amounts can be written as decimals.

PROBLEM SOLVING Use the table. List the fewest possible dimes and pennies that make each amount.

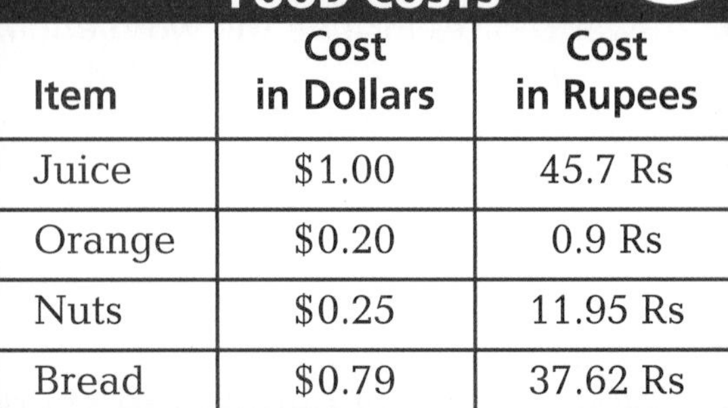

FOOD COSTS

Item	Cost in Dollars	Cost in Rupees
Juice	$1.00	45.7 Rs
Orange	$0.20	0.9 Rs
Nuts	$0.25	11.95 Rs
Bread	$0.79	37.62 Rs

CHECK WHAT YOU KNOW

Use this page to help you review and remember important skills needed for Chapter 26.

USE MONEY NOTATION (For Intervention, see p. H30.)

Count and write the amount.

1.    

2. 

3. 

4. 

NAME THE DECIMAL (For Intervention, see p. H31.)

Write each fraction or mixed number as a decimal.

5. $\frac{1}{10}$ **6.** $\frac{7}{10}$ **7.** $2\frac{6}{10}$ **8.** $\frac{2}{10}$ **9.** $\frac{5}{10}$

10. $1\frac{65}{100}$ **11.** $\frac{4}{100}$ **12.** $\frac{30}{100}$ **13.** $\frac{3}{100}$ **14.** $\frac{10}{100}$

COMPARE AND ORDER DECIMALS (For Intervention, see p. H31.)

Compare. Write < or > for each ●.

15. 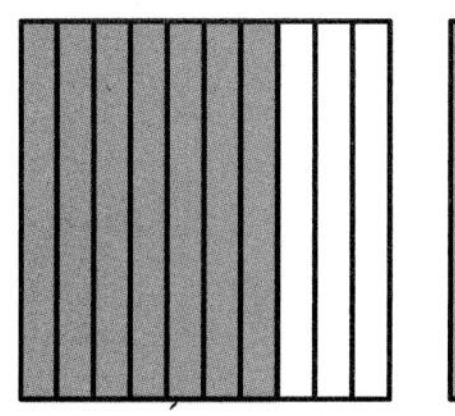 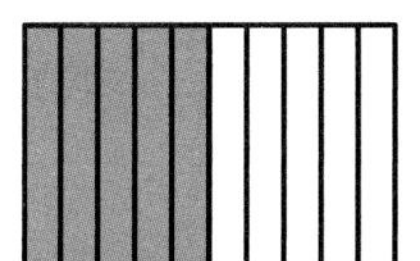

0.7 ● 0.5

16. 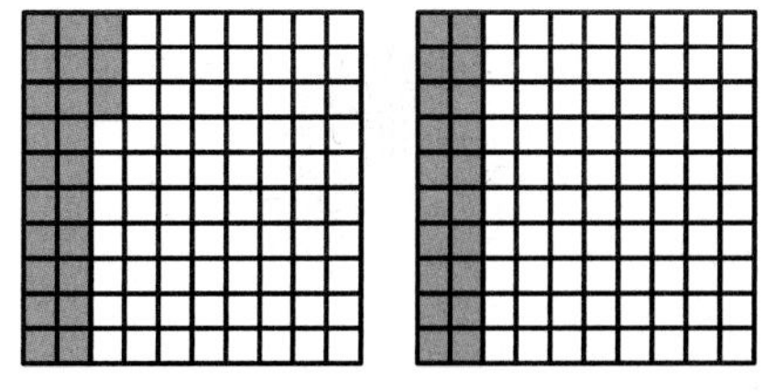

0.23 ● 0.20

Use the number line to order the decimals from least to greatest.

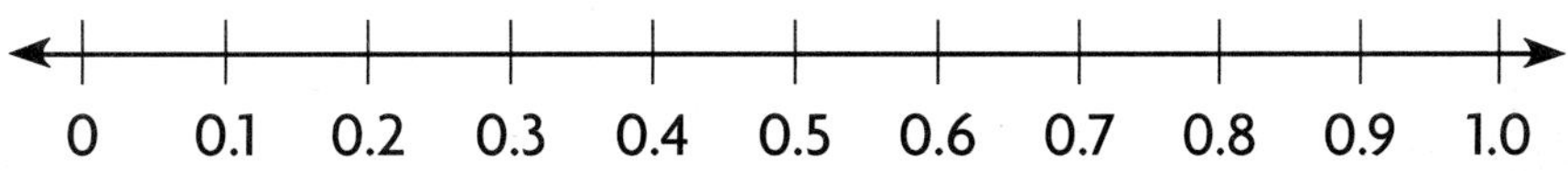

17. 0.7, 0.5, 0.9 **18.** 0.3, 0.2, 0.5 **19.** 0.1, 0, 0.6 **20.** 0.8, 0.1, 1.0

LESSON 1

Relate Fractions and Money

Quick Review

Write each decimal as a fraction.

1. 0.9 **2.** 0.1

3. 0.09 **4.** 0.75

5. 0.50

Learn

MILK MONEY Jesse needs half a dollar for milk at school. How many quarters does Jesse need?

4 quarters = 1 dollar = $\$1.00$

$0.25 $0.25 $0.25 $0.25

2 quarters are $\frac{2}{4}$ of a dollar. $\frac{2}{4} = \frac{1}{2}$

$\frac{1}{2}$ of a dollar = $\$0.50$

So, Jesse needs 2 quarters.

$0.25 $0.25 $0.25 $0.25

Three quarters are $\frac{3}{4}$ of a dollar, or $\$0.75$.

$\frac{3}{4}$ of a dollar = $\$0.75$

$0.25 $0.25 $0.25 $0.25

One quarter is $\frac{1}{4}$ of a dollar, or $\$0.25$.

$\frac{1}{4}$ of a dollar = $\$0.25$

Example

Write the amount of money shown. Then write the amount as a fraction of a dollar.

$\$0.25$, or $\frac{1}{4}$ of a dollar

Check

1. **Tell** how much money Anna will have left if she spends $\frac{3}{4}$ of a dollar on a frozen fruit bar.

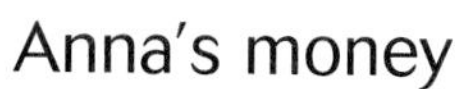
Anna's money

Write the amount of money shown. Then write the amount as a fraction of a dollar.

2. 

3. 

Practice and Problem Solving

Write the amount of money shown. Then write the amount as a fraction of a dollar.

4. 

5. 

6. 

7. 

8. **REASONING** What fraction of a dollar equals one dime?

9. **Write About It** Give two examples of how you could show $\frac{3}{4}$ of a dollar using quarters, dimes, nickels, or pennies.

Mixed Review and Test Prep

Add or subtract.

10. $\frac{3}{8} + \frac{2}{8}$ (p. 436)

11. $\frac{9}{10} - \frac{2}{10}$ (p. 442)

12. $\frac{5}{6} - \frac{5}{6}$ (p. 442)

13. $\frac{7}{8} - \frac{2}{8}$ (p. 442)

14. **TEST PREP** Leroy has a scrapbook with 8 pages. Each page has 7 stickers. How many stickers does he have in all? (p. 152)

A 52 **B** 54 **C** 56 **D** 64

EXTRA PRACTICE page H57, Set A

LESSON 2

Relate Decimals and Money

Quick Review

Write in expanded form.

1. 0.21
2. 0.45
3. 0.72
4. 0.53
5. 2.87

MATERIALS Play money: 100 pennies, 10 dimes

Explore

 = 

100 pennies = 1 dollar

 = $\frac{1}{100}$ or 0.01 of a dollar

 = 

10 dimes = 1 dollar

 = $\frac{1}{10}$ or 0.1 of a dollar

Use play money to connect money and decimals.

Activity

Copy Tables A and B.

STEP 1

Use pennies to show $\frac{31}{100}$ or 0.31 of a dollar. Record the number of pennies you used.

STEP 2

Repeat Step 1 for 0.02 of a dollar.

TABLE A

Decimal	Number of Pennies
0.31	■
0.02	■

STEP 3

Use dimes and pennies to show $\frac{31}{100}$ or 0.31 of a dollar. Use as few coins as possible. Record the coins you used.

STEP 4

Repeat Step 3 for 0.02 of a dollar.

TABLE B

Decimal	Number of Dimes	Number of Pennies
0.31	■	■
0.02	■	■

Try It

a. Write 53¢ or $\frac{53}{100}$ as a decimal.

b. What coins would you use to show 0.12 of a dollar?

Connect

You can think of dimes as tenths and pennies as hundredths.

0.49			
Ones	.	Tenths	Hundredths
0	.	4	9

$0.49 = 49$ hundredths

$0.49 = 4$ tenths 9 hundredths

$0.49			
Dollars	.	Dimes	Pennies
0	.	4	9

$0.49 = 49 pennies = 49 hundredths of a dollar

$0.49 = 4 dimes 9 pennies = 4 tenths 9 hundredths of a dollar

Practice and Problem Solving

Write the money amount for each fraction of a dollar.

1. $\frac{59}{100}$ **2.** $\frac{3}{100}$ **3.** $\frac{13}{100}$ **4.** $\frac{20}{100}$ **5.** $\frac{10}{100}$

Write the money amount.

6. 3 hundredths of a dollar

7. 24 hundredths of a dollar

8. 19 hundredths of a dollar

Write the missing numbers. Use the fewest coins possible.

9. $0.52 = ■ dimes ■ pennies $0.52 = ■ tenths ■ hundredths of a dollar

10. $0.80 = ■ dimes ■ pennies $0.80 = ■ tenths ■ hundredths of a dollar

11. $0.06 = ■ dimes ■ pennies $0.06 = ■ tenths ■ hundredths of a dollar

12. **Write About It** You know that 1 dime is 0.1 or 1 tenth of a dollar. Is 1 dime also 0.10 or 10 hundredths of a dollar? Explain.

13. Juan has $1.50 to spend at the school fair. He spends $\frac{3}{4}$ of a dollar on a hot dog and $\frac{1}{2}$ of a dollar on a drink. How much money does he have left over?

Mixed Review and Test Prep

Subtract 352 from each number. (p. 58)

14. 556 **15.** 398

16. 409 **17.** 500

18. **TEST PREP** How many sides does a rectangle have? (p. 314)

A 1 **C** 2

B 3 **D** 4

LESSON 3

Add and Subtract Decimals and Money

Quick Review

1. $\frac{1}{10} + \frac{3}{10}$
2. $\frac{2}{10} + \frac{5}{10}$
3. $\frac{5}{10} - \frac{1}{10}$
4. $\frac{7}{10} - \frac{4}{10}$
5. $\frac{9}{10} - \frac{3}{10}$

Learn

MOVING RIGHT ALONG Gary walked to Dan's house and then went to the playground. How far did Gary walk in all?

Example

One Way Use decimal models to add.

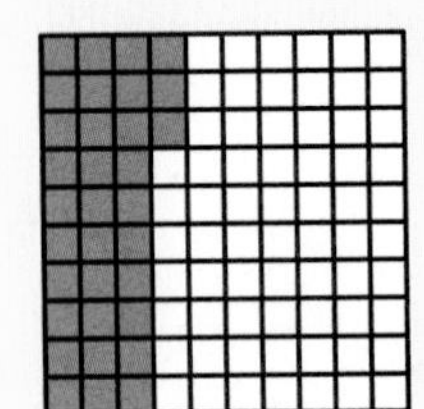

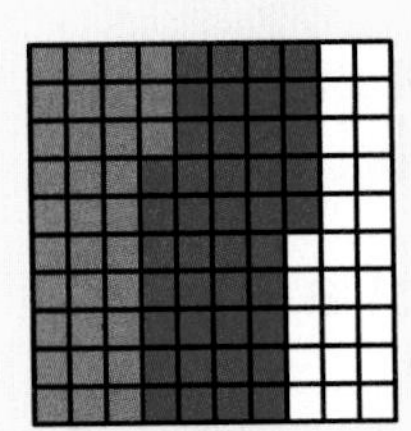

Shade 0.33 and 0.42. Add the shaded squares.

$$0.33 + 0.42 = 0.75$$

So, Gary walked 0.75 kilometer.

Another Way Add 0.33 and 0.42.

STEP 1

Line up the decimal points.

$$\begin{array}{r} 0.33 \\ +0.42 \\ \hline \end{array}$$

↑ decimal point

STEP 2

Add as you would add whole numbers. Regroup if necessary.

$$\begin{array}{r} 0.33 \\ +0.42 \\ \hline 0\ 75 \end{array}$$

STEP 3

Write the decimal point in the sum.

$$\begin{array}{r} 0.33 \\ +0.42 \\ \hline 0.75 \end{array}$$

More Examples

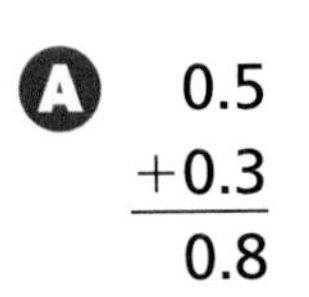

A
$$\begin{array}{r} 0.5 \\ +0.3 \\ \hline 0.8 \end{array}$$

B
$$\begin{array}{r} 1.33 \\ +0.06 \\ \hline 1.39 \end{array}$$

C
$$\begin{array}{r} ^{1} \\ \$0.38 \\ +\$0.45 \\ \hline \$0.83 \end{array}$$

D
$$\begin{array}{r} ^{1} \\ \$2.17 \\ +\$3.58 \\ \hline \$5.75 \end{array}$$

Example

One Way Use decimal models to subtract.

Find 0.73 − 0.50.

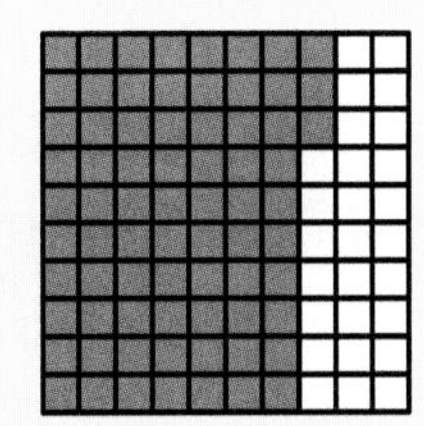

Shade 0.73 of a decimal model.

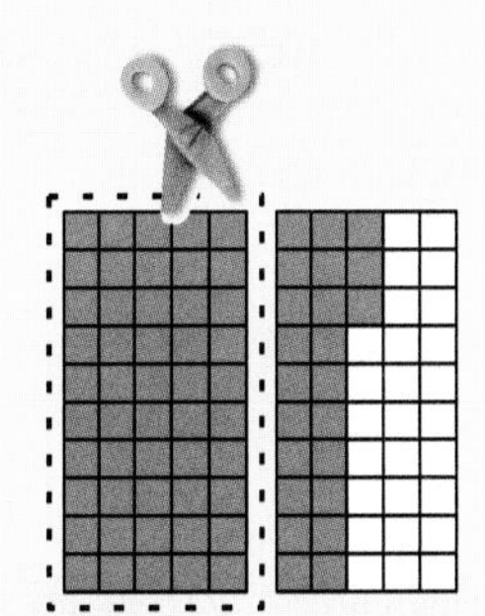

Take away 50 shaded squares. 23 shaded squares are left. So, 0.73 − 0.50 = 0.23.

Another Way Subtract 0.38 from 0.71.

STEP 1

Line up the decimal points.

$$\begin{array}{r} 0.71 \\ -0.38 \\ \hline \end{array}$$

↑ decimal point

STEP 2

Subtract as you would with whole numbers.

$$\begin{array}{r} \scriptsize{6\ 11} \\ 0.\not{7}\,\not{1} \\ -0.\,3\,8 \\ \hline 0\ \ 3\,3 \end{array}$$

STEP 3

Write the decimal point in the difference.

$$\begin{array}{r} \scriptsize{6\ 11} \\ 0.\not{7}\,\not{1} \\ -0.\,3\,8 \\ \hline 0.\,3\,3 \end{array}$$

More Examples

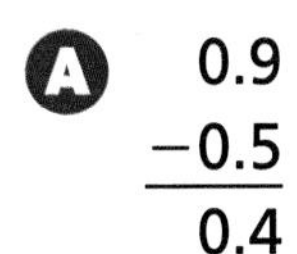

A
$$\begin{array}{r} 0.9 \\ -0.5 \\ \hline 0.4 \end{array}$$

B
$$\begin{array}{r} 2.75 \\ -1.33 \\ \hline 1.42 \end{array}$$

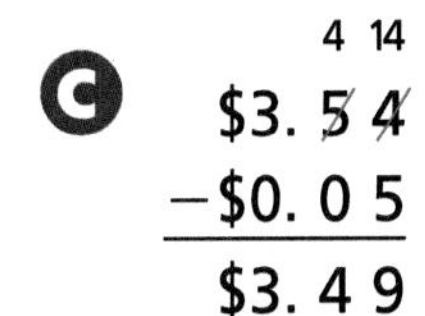

C
$$\begin{array}{r} \scriptsize{4\ 14} \\ \$3.\not{5}\,\not{4} \\ -\$0.\,0\,5 \\ \hline \$3.\,4\,9 \end{array}$$

D
$$\begin{array}{r} \scriptsize{5\ 13} \\ 0.\not{6}\,\not{3} \\ -0.\,3\,7 \\ \hline 0.\,2\,6 \end{array}$$

You can also use a calculator to find a sum or difference.

0.83 + 0.39 = ■

0.71 − 0.28 = ■

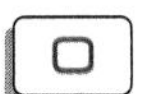 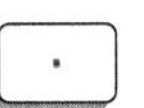      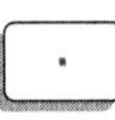   

LESSON CONTINUES ▶

Check

1. **Tell** why you write a 5 above the 6 tenths in Example D.

Add or subtract.

2. $\begin{array}{r} 0.46 \\ +0.19 \\ \hline \end{array}$

3. $\begin{array}{r} 0.75 \\ -0.56 \\ \hline \end{array}$

4. $\begin{array}{r} 0.29 \\ +0.33 \\ \hline \end{array}$

To learn more about adding and subtracting decimals and money, watch the Harcourt Math Newsroom Video, *Rare Coins.*

Practice and Problem Solving

Add or subtract.

5. $\begin{array}{r} \$7.73 \\ -\$4.55 \\ \hline \end{array}$

6. $\begin{array}{r} \$0.39 \\ +\$0.46 \\ \hline \end{array}$

7. $\begin{array}{r} 5.5 \\ -1.3 \\ \hline \end{array}$

8. $\begin{array}{r} 0.80 \\ -0.53 \\ \hline \end{array}$

9. $\begin{array}{r} 2.3 \\ -1.1 \\ \hline \end{array}$

10. $\begin{array}{r} 2.35 \\ +6.19 \\ \hline \end{array}$

11. $\begin{array}{r} \$0.13 \\ +\$2.41 \\ \hline \end{array}$

12. $\begin{array}{r} 0.73 \\ -0.32 \\ \hline \end{array}$

13. $\begin{array}{r} \$3.33 \\ -\$0.25 \\ \hline \end{array}$

14. $\begin{array}{r} 0.5 \\ +0.4 \\ \hline \end{array}$

15. $\begin{array}{r} 1.6 \\ -1.2 \\ \hline \end{array}$

16. $\begin{array}{r} 0.79 \\ +0.12 \\ \hline \end{array}$

17. **What's the Error?** Gina wrote 0.20 + 0.07 = 0.90. Describe her error. Give the correct sum.

18. **REASONING** Do you have to regroup to find 0.64 – 0.27? Explain.

19. Kendra bikes 0.55 kilometer to school and 0.34 kilometer to a friend's house. How many kilometers does she bike in all?

20. Rhea lives 0.92 kilometer from school. She has walked 0.38 kilometer so far this morning. How much farther must she walk to get to school?

21. **ALGEBRA** Use the price list. Conner bought popcorn and one other item with a $5 bill. His change was $4.05. What was the other item?

22. **Write a problem** about adding or subtracting money amounts. Use the price list. Trade with a partner and solve.

PRICE LIST

Item	Cost
Popcorn	$0.50
Juice	$0.45
Trail mix	$0.39

23. David went to the store with \$10.00. He bought milk, apples, and eggs. He had \$4.78 when he left the store. The milk was \$2.89, and the eggs were \$0.89. How much were the apples?

Mixed Review and Test Prep

Find each missing factor. (p. 142)

24. ■ × 8 = 32
25. 8 × ■ = 48
26. ■ × 4 = 28
27. 7 × ■ = 21

Find each quotient. (p. 216)

28. 12 ÷ 6 = ■
29. 54 ÷ 6 = ■
30. 24 ÷ 8 = ■
31. 49 ÷ 7 = ■

32. **TEST PREP** Find 309 − 87. (p. 58)

A 396
B 322
C 241
D 222

33. **TEST PREP** Lori had 5 carrots. She cut each carrot into 3 pieces. Then she ate 2 pieces. How many pieces of carrot did she have left? (p. 172)

F 10 G 12 H 13 J 15

PROBLEM SOLVING LINKUP ... to Reading

STRATEGY • MAKE PREDICTIONS You can use a graph to help you make a prediction. A prediction is a guess that is based on information.

USE DATA Emmett's sunflower should grow to be about 1 meter tall. Using the graph, predict whether the flower will be taller than 0.65 m at the end of the fifth week.

The sunflower has grown at least 0.15 m each week. It was 0.60 m tall at the end of the fourth week.

$$0.60 \text{ m} + 0.15 \text{ m} = 0.75 \text{ m}$$

So, the sunflower will probably be taller than 0.65 m at the end of the fifth week.

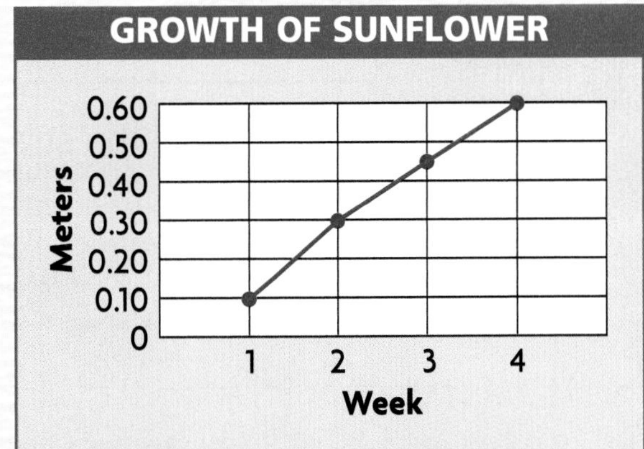

- Between week 1 and week 2 the flower grew 0.20 meter.
- Between week 2 and week 3 it grew 0.15 meter.
- Between week 3 and week 4 it grew 0.15 meter.

1. Predict whether the sunflower will be taller than 0.75 m at the end of the sixth week. Explain.

2. Between which two weeks was the sunflower 0.35 m tall?

LESSON 4

Problem Solving Strategy
Solve a Simpler Problem

Quick Review

1. 0.24 + 0.24
2. 0.3 + 0.3
3. 0.44 + 0.44
4. 0.2 + 0.2
5. 0.39 + 0.39

PROBLEM Alma has $5 to buy fruit. If she buys 1 pound of grapes and 2 pounds of bananas, how much money will she have left?

UNDERSTAND

- What are you asked to find?
- What information will you use?
- Is there information you will not use? If so, what?

PLAN

- What strategy can you use to solve the problem?

You can solve a simpler problem.

SOLVE

- How can you use the strategy to solve the problem?

Find the price for each fruit.
1 lb of grapes: $0.98
2 lb of bananas: $0.45 + $0.45

Add to find the total cost.

grapes:	$0.98
bananas:	+$0.90
	$1.88

Subtract the total cost of the fruit from $5.

$$\begin{array}{r} \$5.00 \\ -\$1.88 \\ \hline \$3.12 \end{array}$$

So, Alma will have $3.12 left.

CHECK

- Look at the Problem. Does your answer make sense for the problem? Explain.

Problem Solving Practice

Use the prices on page 476. ***Solve a simpler problem*** **to solve.**

1. **What if** Alma buys 2 pounds of grapes and 1 pound of bananas? How much change will she receive from $5?
2. Brandon has $3. Does he have enough money to buy 2 apples and 2 pounds of grapes? Explain.

PROBLEM SOLVING STRATEGIES

- Draw a Diagram or Picture
- Make a Model or Act It Out
- Make an Organized List
- Find a Pattern
- Make a Table or Graph
- Predict and Test
- Work Backward
- **Solve a Simpler Problem**
- Write a Number Sentence
- Use Logical Reasoning

Maddie bought 3 apples and 1 pound of grapes. She gave the clerk $2. How much change did she get? Use the prices on page 476.

3. Which number sentence could help you solve a simpler problem?
   - **A** $0.24 + $4 = $4.24
   - **B** $0.24 + $0.61 = $0.85
   - **C** $0.24 + $0.24 + $0.24 = $0.72
   - **D** $0.98 + $0.98 = $1.96
4. Which answers the question?
   - **F** $0.30
   - **G** $0.75
   - **H** $0.99
   - **J** $1.55

Mixed Strategy Practice

5. On Wednesday, Emily had $24.75. She earned $5.50 on Thursday. Then she spent $6.25. How much money does she have now?
6. Norm biked $\frac{7}{10}$ of a mile one Saturday. Joel biked $\frac{9}{10}$ of a mile. How much farther did Joel bike?
7. Mark is 1 year younger than Rick. Jaime is twice as old as Rick. Jaime is 12 years old. How old are Mark and Rick?
8. **What's the Question?** The perimeter of Kiyo's rectangular rabbit pen is 20 ft. The length is 6 ft. The answer is 4 ft.
9. Shelly worked on her science project for 45 minutes on Monday, and for 1 hour and 20 minutes on Tuesday. How much time did she work on her project in the two days?
10. Toni baby-sat from 4:30 P.M. until 8:30 P.M. She earned $3 per hour. How long did she baby-sit? How much did she earn?

CHAPTER 26

Review/Test

CHECK CONCEPTS

Write the money amount for each fraction of a dollar. (pp. 470–471)

1. $\frac{20}{100}$ **2.** $\frac{7}{100}$ **3.** $\frac{44}{100}$ **4.** $\frac{75}{100}$

Write the money amount.

5. 14 hundredths of a dollar

6. 10 hundredths of a dollar

7. 54 hundredths of a dollar

CHECK SKILLS

Write the amount of money shown. Then write the amount as a fraction of a dollar. (pp. 468–469)

8. 

9.

10.

Add or subtract. (pp. 472–475)

11. $\begin{array}{r} 0.6 \\ +0.2 \\ \hline \end{array}$ **12.** $\begin{array}{r} 2.5 \\ +7.4 \\ \hline \end{array}$ **13.** $\begin{array}{r} 0.06 \\ +0.19 \\ \hline \end{array}$ **14.** $\begin{array}{r} \$3.52 \\ +\$4.16 \\ \hline \end{array}$

15. $\begin{array}{r} 0.9 \\ -0.6 \\ \hline \end{array}$ **16.** $\begin{array}{r} 2.8 \\ -0.1 \\ \hline \end{array}$ **17.** $\begin{array}{r} \$1.70 \\ -\$0.05 \\ \hline \end{array}$ **18.** $\begin{array}{r} 0.61 \\ -0.59 \\ \hline \end{array}$

PROBLEM SOLVING

Solve. (pp. 476–477)

19. Felix bought a notebook for \$1.29 and two pens for \$0.41 each. He gave the clerk \$5.00. How much change did he receive?

20. Ginger wants to buy 2 pears for \$0.35 each and 1 cookie for \$1.05. She has \$2.00. Will she have enough money? Explain.

Standardized Test Prep

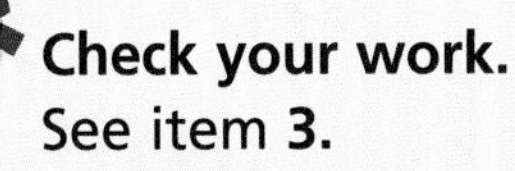

Check your work.
See item **3.**
Be sure you used the correct information from the table. You need to subtract to find how much change.
Also see problem **6,** p. H65.

For 1–7, choose the best answer.

1. $\$0.46 + \$0.38$

A $0.08 C $0.84
B $0.46 D NOT HERE

2. How many dimes equal one dollar?

F 5 H 20
G 10 J 50

3. Tim buys 1 pencil and 2 erasers. He gives the clerk $4.00. How much change should he receive?

ITEM	COST
pencil	$0.39
eraser	$0.69

A $2.23 C $1.77
B $1.93 D $0.77

4. Which fraction names the part of a dollar that is shown?

F $\frac{1}{9}$ G $\frac{1}{5}$ H $\frac{1}{4}$ J $\frac{1}{2}$

5. Which fraction names the shaded part?

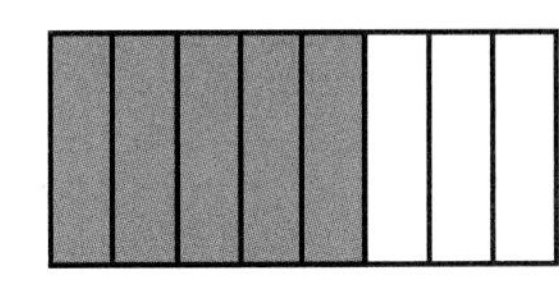

A $\frac{3}{8}$ B $\frac{1}{2}$ C $\frac{3}{5}$ D $\frac{5}{8}$

6. Nancy gave $\frac{1}{8}$ of her stickers to Jill and $\frac{3}{8}$ of her stickers to Laurie. What fraction of her stickers did she give away?

F $\frac{1}{8}$ G $\frac{2}{8}$ H $\frac{1}{2}$ J $\frac{5}{8}$

7. Gerald ran 0.4 mile to the library and then 0.3 mile to school. How far did he run in all?

A 1.2 mi C 0.1 mi
B 0.7 mi D 0.07 mi

Write What You Know

8. Find the sum. Explain how to find it.

$0.28 + 0.36$

9. Jerry had $0.61. He spent $0.25. What amount does he have left? Show your work. Write the decimal as a fraction.

PROBLEM SOLVING
MATH DETECTIVE

I Found the Whole Thing

If you know what a part of a whole looks like, can you figure out what the whole looks like? Use the clues below to solve each case!

Draw a picture of each clue to help you find the answer to the question.

Case 1

If $\frac{1}{8}$ = □, what does one whole equal?

Case 2

If $\frac{1}{4}$ = 🌼🌼🌼, what does one whole equal?

Case 3

If $\frac{2}{3}$ = ▭ ▭, what does one whole equal?

Case 4

If 2 = ⬡, what does one whole equal?

Case 5

If $\frac{3}{2}$ = ▽▽▽ ▽▽▽ ▽▽▽, what does one whole equal?

Case 6

If $\frac{3}{4}$ = 🍎🍎🍎 🍎🍎🍎, what does one whole equal?

STRETCH YOUR THINKING Draw a fraction as a picture. Have a classmate use your clues and draw a picture showing what the whole looks like.

Challenge

Fractions and Decimals on a Number Line

You can use decimal models and a number line to help you understand fractions and decimals.

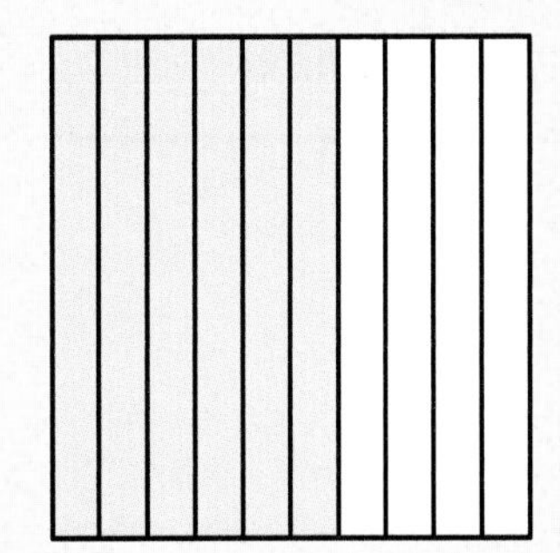

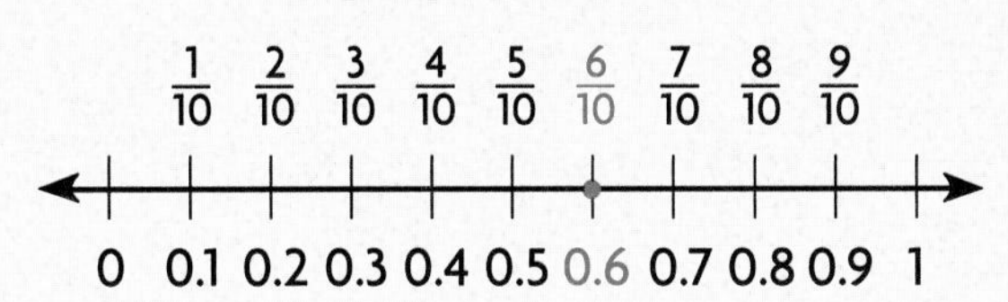

Since 0.6 and $\frac{6}{10}$ name 6 parts out of 10, they name the same amount, or are equivalent.

One whole can be divided into 100 equal parts on a number line.

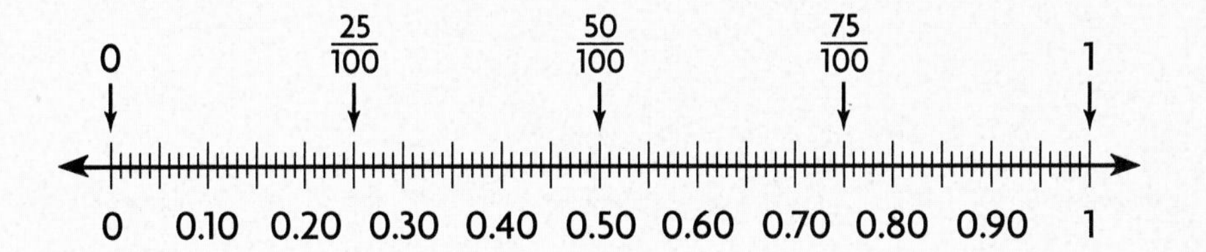

Practice

1. Find the point on the number line that is halfway between 0 and 1. Write a fraction name and a decimal name for that point.
2. Find the point on the number line that is one fourth of the way from 0 to 1. Write a fraction name and a decimal name for that point.

Use the number line. Write each as a fraction.

3. 0.30 **4.** 0.45 **5.** 0.75 **6.** 0.99

Remember

Equivalent fractions and decimals are different names for the same amount.

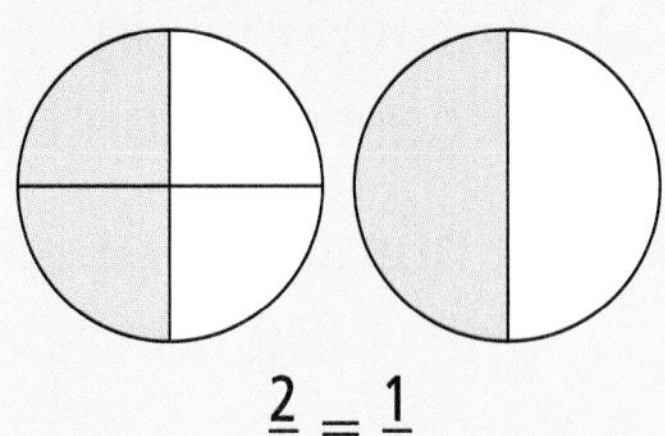

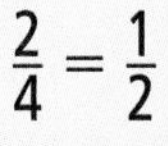

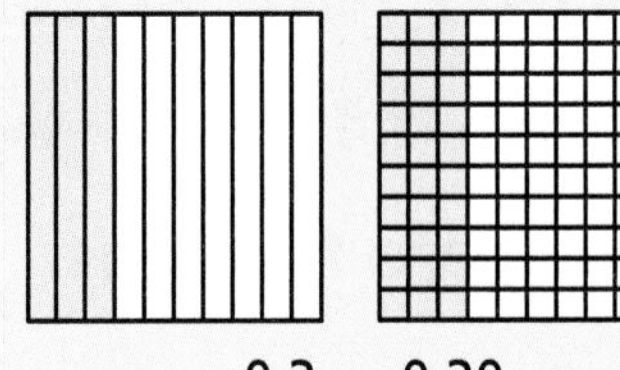

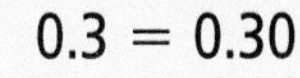

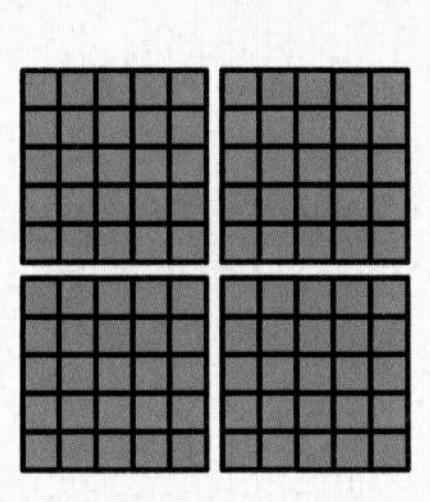

$\frac{25}{100} = \frac{1}{4}$

Study Guide and Review

VOCABULARY

Choose the best term from the box.

1. A number with one or more digits to the right of the decimal point is a _?_. (p. 452)
2. A fraction is in _?_ when a model of it uses the largest fraction bars possible. (p. 436)
3. A number that is represented by a whole number and a fraction is a _?_. (p. 428)

decimal
fraction
equivalent fractions
simplest form
mixed number

STUDY AND SOLVE

Chapter 23

Compare fractions.

Compare $\frac{1}{2}$ and $\frac{3}{5}$.

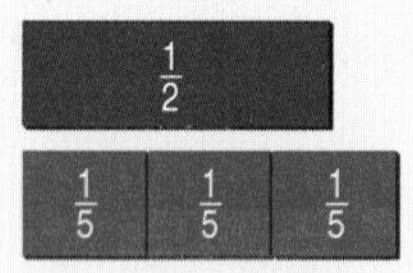

The bar for $\frac{1}{2}$ is shorter than the bars for $\frac{3}{5}$.

So, $\frac{1}{2} < \frac{3}{5}$ or $\frac{3}{5} > \frac{1}{2}$.

Compare. Write <, >, or = for each ●. (pp. 422–425)

4. 

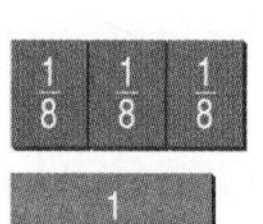

$\frac{3}{8}$ ● $\frac{1}{3}$

5. 

$\frac{2}{6}$ ● $\frac{5}{12}$

Chapter 24

Add and subtract like fractions.

Write the sum or difference in simplest form.

$\frac{3}{8} + \frac{1}{8} = \frac{4}{8}$

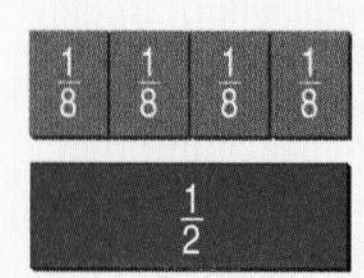

So, the sum, $\frac{4}{8}$, is $\frac{1}{2}$ in simplest form.

$\frac{4}{6} - \frac{2}{6} = \frac{2}{6}$

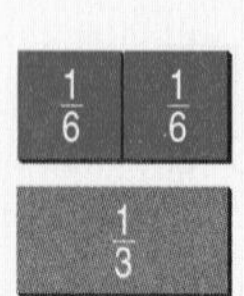

So, the difference, $\frac{2}{6}$, is $\frac{1}{3}$ in simplest form.

Use fraction bars to find the sum or difference in simplest form. (pp. 434–445)

6. $\frac{1}{8} + \frac{1}{8} = $ ■
7. $\frac{2}{4} + \frac{1}{4} = $ ■
8. $\frac{1}{6} + \frac{3}{6} = $ ■
9. $\frac{9}{12} - \frac{5}{12} = $ ■
10. $\frac{2}{5} - \frac{1}{5} = $ ■
11. $\frac{4}{10} - \frac{2}{10} = $ ■
12. $\frac{1}{12} + \frac{4}{12} = $ ■
13. $\frac{4}{4} - \frac{2}{4} = $ ■

Chapter 25

Relate fractions and decimals.

Write the fraction and the decimal for the shaded part.

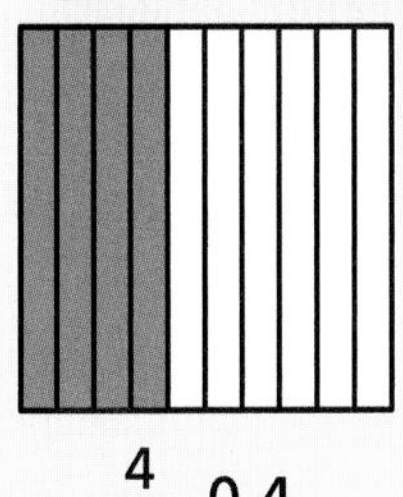

$\frac{4}{10}$, 0.4

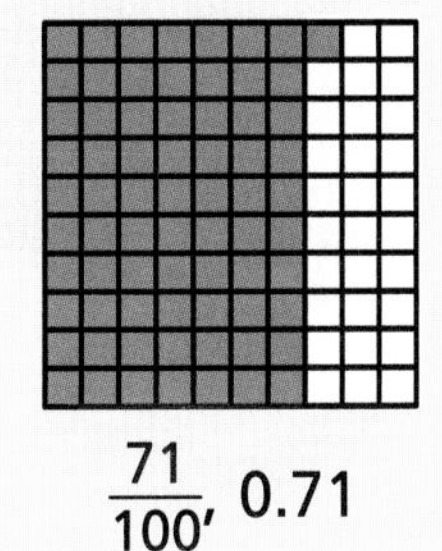

$\frac{71}{100}$, 0.71

Write the fraction and the decimal for the shaded part. (pp. 456–457)

14. 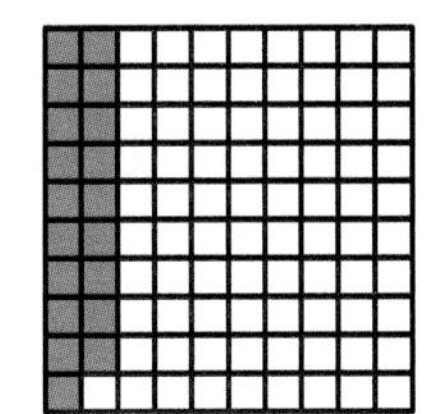

Write the decimal for each fraction. (pp. 452–457)

15. $\frac{6}{10}$ **16.** $\frac{12}{100}$ **17.** $\frac{55}{100}$

Chapter 26

Add and subtract decimals.

Subtract. \$0.83 − \$0.56

$$\begin{array}{r} \overset{7\,13}{\$0.\cancel{8}\cancel{3}} \\ -\$0.56 \\ \hline \$0.27 \end{array}$$

- Line up the decimal points.
- Subtract. Regroup if necessary.
- Write the decimal point in the difference.

Remember to write the dollar sign in your answer when you are adding or subtracting money amounts.

Find the sum or difference. (pp. 472–475)

18. $\begin{array}{r} \$0.23 \\ +\$0.44 \\ \hline \end{array}$ **19.** $\begin{array}{r} 0.78 \\ +0.04 \\ \hline \end{array}$

20. $\begin{array}{r} \$0.64 \\ +\$0.29 \\ \hline \end{array}$ **21.** $\begin{array}{r} 0.5 \\ -0.4 \\ \hline \end{array}$

22. $\begin{array}{r} \$0.72 \\ -\$0.19 \\ \hline \end{array}$ **23.** $\begin{array}{r} \$0.19 \\ -\$0.12 \\ \hline \end{array}$

PROBLEM SOLVING PRACTICE

Solve. (pp. 446–447, 476–477)

24. Cole and Jon are sharing a pizza. If they each eat $\frac{1}{3}$ of the pizza, is it reasonable to say that $\frac{1}{2}$ of the pizza will be left? Explain.

25. Mr. Jackson is training for a race. Last week, he ran 1.25 miles on each of two days. He ran 1.1 miles on each of two other days. How far did Mr. Jackson run last week?

PERFORMANCE ASSESSMENT

TASK A • Mmmm . . . BROWNIES!

Jeff made a pan of brownies to share with Carla and Keith. The brownies were cut into equal pieces. The three friends each ate a different number of brownies.

a. Make a drawing of the pan of brownies. Mark each piece with J, C, or K to show which child ate it.

b. Write each child's name and tell what fraction of the brownies he or she ate.

c. Who ate the most brownies?

d. Write a fraction in simplest form that tells what part of the brownies Jeff and Keith ate altogether.

TASK B • SCHOOL FAIR

Colton, Erica, and Whitney each have 100 tickets to sell for the school fair. Colton sold 45 tickets, and Whitney sold 70 tickets. Erica sold more tickets than Colton, but fewer tickets than Whitney.

a. Shade the decimal model to show the part of her 100 tickets that Erica may have sold.

b. Write a decimal and a fraction for the part of the model that you shaded.

c. Each ticket for the fair costs $\frac{1}{4}$ of a dollar. Erica's brother has \$2.00. Can he buy 5 tickets and still have $\frac{1}{2}$ of a dollar left to buy a pretzel?

Mighty Math Number Heroes • Fractions

You can use Mighty Math Number Heroes to show fractions.

- Click on *Fraction Fireworks*.
- Click . Choose Level A, B, C, or D.
- Answer at least 5 questions.
- Draw pictures to record your work.

Practice and Problem Solving

Click . Make each fraction.

For 1–6, draw a picture of each fraction you make.

1. Make a fraction that is equivalent to $\frac{1}{2}$.

2. Make a fraction that is equivalent to $\frac{2}{3}$.

3. Make a fraction that is equivalent to $\frac{3}{4}$.

4. Make a fraction that is equivalent to $\frac{2}{8}$.

5. Add $\frac{1}{6} + \frac{4}{6}$. Make the fraction that shows the sum.

6. Subtract $\frac{11}{12} - \frac{6}{12}$. Make the fraction that shows the difference.

Solve.

7. Click on . Then click 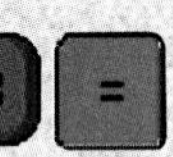.

Write the number sentence you see.

8. **Stretch Your Thinking** Click on .
Use it to subtract $\frac{2}{3} - \frac{1}{4}$.
What is the difference?

> **Multimedia Math Glossary** www.harcourtschool.com/mathglossary
>
> **9.** **Vocabulary** Look up *decimals* and *equivalent fractions* in the Multimedia Math Glossary. Give an example of two decimals that are equivalent.

Baby great horned owls in nest

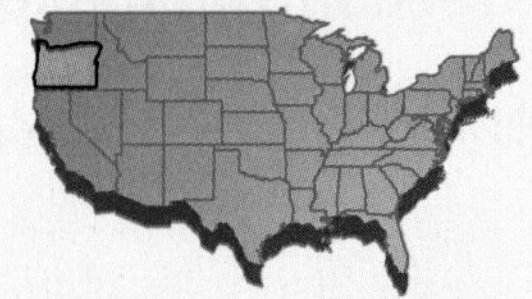

CORVALLIS, OREGON

Each year workers at the Chintimini Wildlife Rehabilitation Center in Corvallis, Oregon take care of about 700 animals. These baby great horned owls will leave the nest when they are very young. If you see one, its parents are probably close by.

OREGON BIRDS	
Bird	**Weight Range**
Owl	1.4–141.0 ounces
Swift	0.3–5.0 ounces
Grebe	4.6–49.0 ounces

USE DATA For 1–5, use the table.

1. Find the swift's least weight. Write this number as a fraction.
2. Use decimal models to show the smallest owl's weight.
3. Three grebes weigh 8.9 ounces, 9.8 ounces, and 8.19 ounces. Write these numbers in order from least to greatest.
4. A swift's egg weighs 0.04 ounces. What is the difference in weight between the egg and a swift that weighs 0.40 ounces?
5. **REASONING** A bird weighs 1.1 ounces. Is it most likely an owl, a swift, or a grebe? Explain your choice.
6. An owl's egg weighs 0.2 ounces. Do three eggs weigh more than or less than 0.5 ounce? Explain how you know.

THE WILLAMETTE VALLEY, OREGON

Many birds at the Wildlife Center, such as falcons, hawks, and eagles, live in this area. In spring and summer, peregrine falcons nest in the Willamette Valley. They return to Central America for the fall and winter.

You may use fraction bars for 1–5.

1. Peregrine falcons live in Oregon for 2 seasons each year. Write a fraction to show the part of the year they live in Oregon.
2. Write your answer to Exercise 1 in simplest form.
3. You have $10 to buy books from the wildlife center. A book about birds costs $7.15. You want to buy two copies of the book, one for yourself and one for your friend. How much more money do you need?
4. Only 2 out of every 10 peregrine falcon eggs will become adult birds. What fraction is that?
5. There are 7 continents in the world: Africa, Asia, Australia, Europe, North America, South America, and Antarctica. Peregrine falcons live on 6 of them. What fraction of the continents is that?

▲ **Adult peregrine falcons can fly at about 200 miles per hour!**

Multiply by 1-Digit Numbers

4-H is a club for young people. 4-H members living in rural areas often raise livestock, such as pigs, cows, and sheep. The animals are then shown at fairs. At the fairs, the pigs are kept in barns.

PROBLEM SOLVING Look at the diagram. If there is 1 pig in each stall and there are 4 barns like this one, how many pigs can be kept at the fair?

DATA LINK

PIG BARN

one stall

CHECK WHAT YOU KNOW

Use this page to help you review and remember important skills needed for Chapter 27.

COLUMN ADDITION (For Intervention, see p. H10.)

Find the sum.

1. $\begin{array}{r} 47 \\ 47 \\ +47 \\ \hline \end{array}$ **2.** $\begin{array}{r} 36 \\ 36 \\ +36 \\ \hline \end{array}$ **3.** $\begin{array}{r} 54 \\ 54 \\ +54 \\ \hline \end{array}$ **4.** $\begin{array}{r} 18 \\ 18 \\ +18 \\ \hline \end{array}$

5. $\begin{array}{r} 29 \\ 29 \\ +29 \\ \hline \end{array}$ **6.** $\begin{array}{r} 65 \\ 65 \\ +65 \\ \hline \end{array}$ **7.** $\begin{array}{r} 72 \\ 72 \\ +72 \\ \hline \end{array}$ **8.** $\begin{array}{r} 83 \\ 83 \\ +83 \\ \hline \end{array}$

REGROUP ONES AND TENS (For Intervention, see p. H20.)

Regroup. Write the missing number.

9. 6 tens 13 ones = ■ tens 3 ones

10. 2 tens 18 ones = 3 tens ■ ones

11. 4 tens ■ ones = 5 tens 8 ones

12. ■ tens 15 ones = 6 tens 5 ones

13. 6 tens 20 ones = 8 tens ■ ones

14. 7 tens 24 ones = ■ tens 4 ones

15. ■ tens 25 ones = 6 tens 5 ones

16. 5 tens ■ ones = 7 tens 3 ones

MULTIPLICATION FACTS (For Intervention, see p. H20.)

Find the product.

17. $6 \times 7 =$ ■ **18.** $5 \times 5 =$ ■ **19.** ■ $= 9 \times 4$ **20.** $9 \times 1 =$ ■

21. $7 \times 10 =$ ■ **22.** ■ $= 8 \times 6$ **23.** $5 \times 9 =$ ■ **24.** ■ $= 2 \times 3$

25. ■ $= 4 \times 5$ **26.** $4 \times 8 =$ ■ **27.** ■ $= 0 \times 6$ **28.** $8 \times 7 =$ ■

29. $\begin{array}{r} 5 \\ \times 7 \\ \hline \end{array}$ **30.** $\begin{array}{r} 8 \\ \times 4 \\ \hline \end{array}$ **31.** $\begin{array}{r} 4 \\ \times 9 \\ \hline \end{array}$ **32.** $\begin{array}{r} 10 \\ \times\ 3 \\ \hline \end{array}$ **33.** $\begin{array}{r} 8 \\ \times 8 \\ \hline \end{array}$

LESSON 1

Multiply 2-Digit Numbers

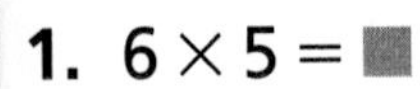

Quick Review

1. $6 \times 5 = ■$
2. $■ = 4 \times 3$
3. $10 \times 6 = ■$
4. $■ = 6 \times 8$
5. $4 \times 10 = ■$

MATERIALS
base-ten blocks

Explore

Students in the school chorus stand in 4 rows of 16. How many students are in the chorus?

$$4 \times 16 = ■$$

Activity

Model the problem using an array of base-ten blocks. Find the product.

STEP 1

Use 1 ten 6 ones to show 16. Make 4 rows of 16 to show 4×16.

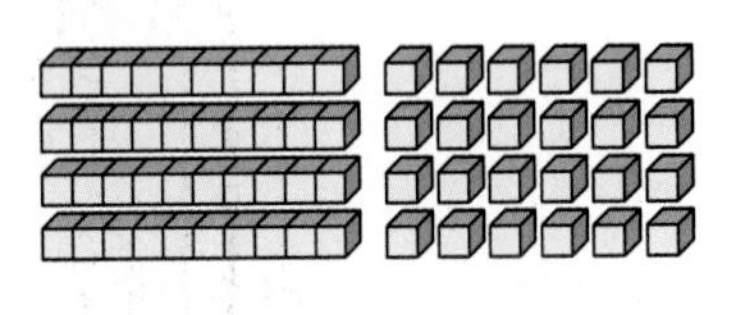

STEP 2

Combine the ones and the tens to find the product.

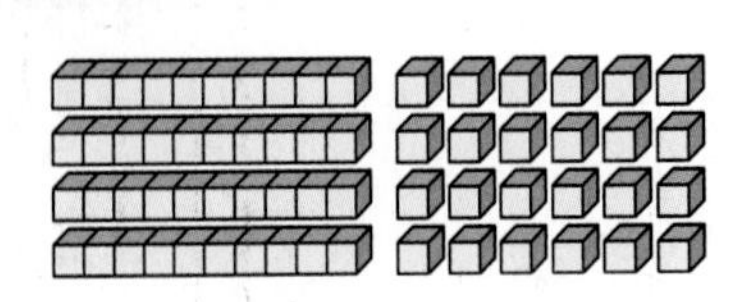

$4 \times 10 = 40 \quad 4 \times 6 = 24$

$40 + 24 = 64$

So, there are 64 students in the school chorus.

Remember

An array shows objects in rows and columns. An array with 3 rows of 5 shows 3×5.

- How did you combine the ones and tens to find the product?
- **REASONING** Why is it helpful to show 16 by using 1 ten 6 ones instead of 16 ones?

I have an array of 4 rows of 1 ten 3 ones. How do I find the product?

Try It

Use base-ten blocks to find each product.

a. $4 \times 13 = ■$ **b.** $3 \times 18 = ■$

Connect

You can use arrays on grid paper to multiply 2-digit numbers.

Technology Link
More Practice: Use E-Lab, *Modeling Multiplication.*
www.harcourtschool.com/elab2002

Find 3×17.

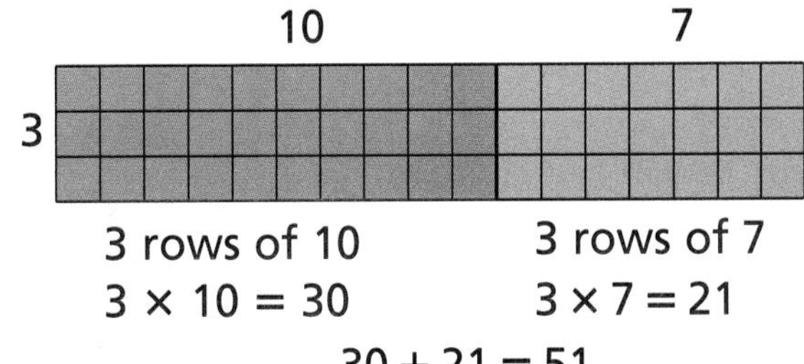

3 rows of 10 — $3 \times 10 = 30$

3 rows of 7 — $3 \times 7 = 21$

$30 + 21 = 51$

So, $3 \times 17 = 51$.

Practice and Problem Solving

Use the array to help find the product.

1. $2 \times 10 = 20$ $\quad$ $2 \times 9 = 18$

$2 \times 19 = \blacksquare$

2. 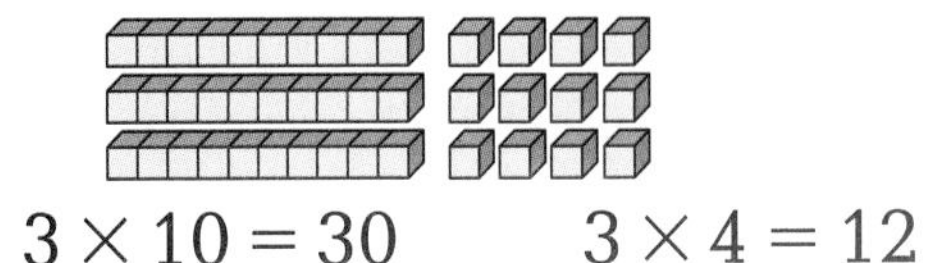

$3 \times 10 = 30$ $\quad$ $3 \times 4 = 12$

$3 \times 14 = \blacksquare$

3. 3 rows of 10 $\quad$ 3 rows of 5

$3 \times 10 = 30$ $\quad$ $3 \times 5 = 15$

$3 \times 15 = \blacksquare$

4. 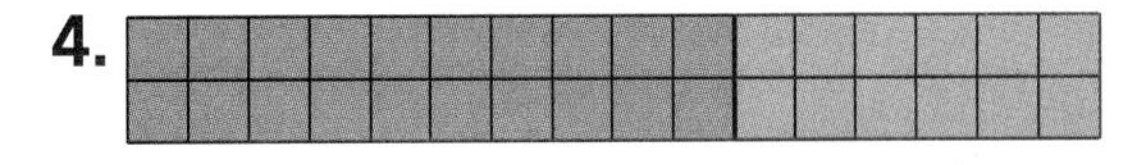

2 rows of 10 $\quad$ 2 rows of 6

$2 \times 10 = 20$ $\quad$ $2 \times 6 = 12$

$2 \times 16 = \blacksquare$

Use base-ten blocks or grid paper to find the product.

5. $6 \times 14 = \blacksquare$ $\quad$ **6.** $5 \times 13 = \blacksquare$ $\quad$ **7.** $6 \times 16 = \blacksquare$

8. **REASONING** Draw an array on grid paper to find the missing factor in $\blacksquare \times 18 = 72$.

9. **Write About It** Describe how to model 3×21. Explain how to find the product.

Mixed Review and Test Prep

10. $30 \div 10$ (p. 220) $\quad$ **11.** $90 \div 10$ (p. 220)

12. $\$1.14 + \$3.76 = \blacksquare$ (p. 88)

13. Round 24,515 to the nearest thousand. (p. 30)

14. **TEST PREP** Which polygon has 8 sides and 8 angles? (p. 314)

A rectangle $\quad$ **C** hexagon

B octagon $\quad$ **D** triangle

LESSON 2

Record Multiplication

Quick Review

1. $2 \times 4 = ■$
2. $5 \times 7 = ■$
3. $■ = 2 \times 10$
4. $10 \times 7 = ■$
5. $■ = 8 \times 3$

Learn

IN A HEARTBEAT The human heart exhibit at the Science Museum has 4 sections. Each section seats 30 students. How many students can sit in all 4 sections?

Example 1

$4 \times 30 = ■$ or $\begin{array}{r} 30 \\ \times\ 4 \\ \hline \end{array}$

Base-ten blocks can help you find the product.

STEP 1

Model 4 groups of 30.

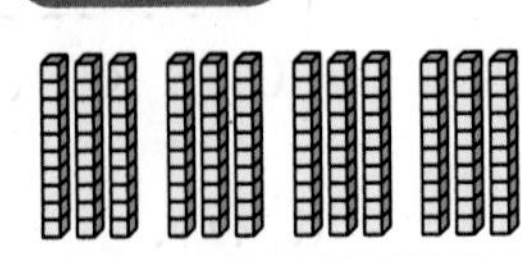

STEP 2

Combine the tens to find the product. Regroup 12 tens as 1 hundred 2 tens.

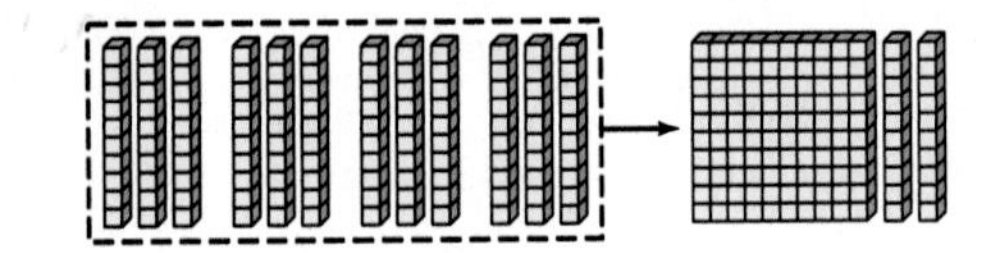

So, 120 students can sit in all 4 sections.

- What addition problem is the same as finding 4×30?

More Examples

A $3 \times 20 = 60$

B $2 \times 40 = 80$

C $\begin{array}{r} 50 \\ \times\ 2 \\ \hline 100 \end{array}$

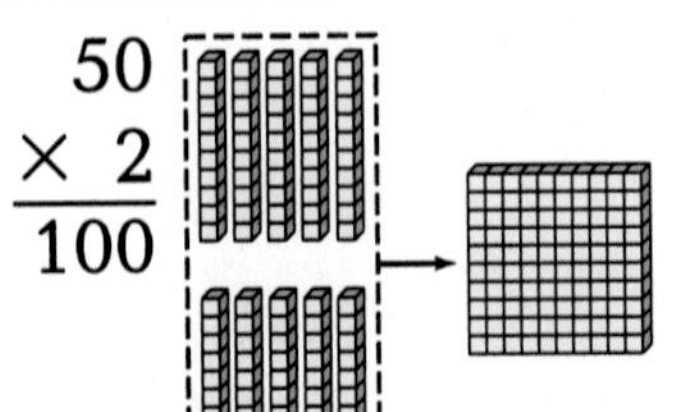

- What addition problem is the same as finding 2×50?

Example 2

For the field trip, there are 5 buses with 23 people on each bus. How many people in all are going on the field trip?

$5 \times 23 = \blacksquare$ or $\begin{array}{r} 23 \\ \times\ 5 \\ \hline \end{array}$

STEP 1

Model 5 groups of 23. Multiply the ones.

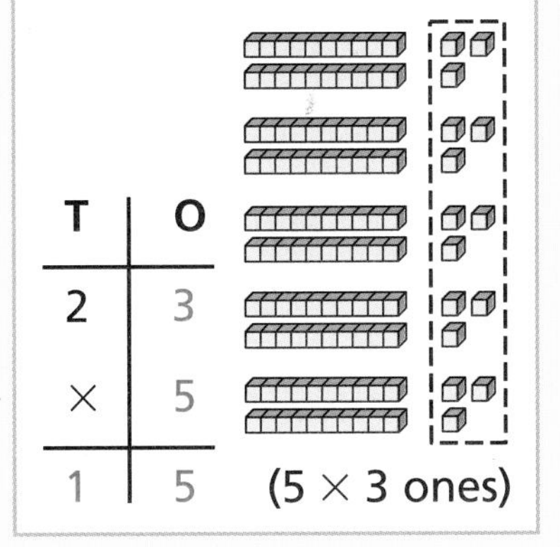

STEP 2

Multiply the tens.

H	T	O
	2	3
×		5
	1	5
1	0	0

(5 × 2 tens)

STEP 3

Add to find the product.

H	T	O
	2	3
×		5
	1	5
+ 1	0	0
1	1	5

So, 115 people are going on the field trip.

- How does knowing 5×2 help you multiply 5×20?

More Examples

A
$$\begin{array}{rl} 32 & \\ \times\ 4 & \\ \hline 8 & (4 \times 2 \text{ ones}) \\ +120 & (4 \times 3 \text{ tens}) \\ \hline 128 & \end{array}$$

B
$$\begin{array}{rl} 51 & \\ \times\ 3 & \\ \hline 3 & (3 \times 1 \text{ one}) \\ +150 & (3 \times 5 \text{ tens}) \\ \hline 153 & \end{array}$$

C
$$\begin{array}{rl} 73 & \\ \times\ 5 & \\ \hline 15 & (5 \times 3 \text{ ones}) \\ +350 & (5 \times 7 \text{ tens}) \\ \hline 365 & \end{array}$$

Check

1. **Model** 5×28 with base-ten blocks. Use paper and pencil to record what you did.

Technology Link

To learn more about multiplication by 1-digit numbers, watch the Harcourt Math Newsroom Video, *Milk in a Bag.*

Find the product.

2. $\begin{array}{r} 30 \\ \times\ 2 \\ \hline \end{array}$

3. $\begin{array}{r} 24 \\ \times\ 3 \\ \hline \end{array}$

4. $\begin{array}{r} 38 \\ \times\ 2 \\ \hline \end{array}$

Practice and Problem Solving

Find the product.

5. $\begin{array}{r} 20 \\ \times\ 4 \\ \hline \end{array}$

6. $\begin{array}{r} 22 \\ \times\ 4 \\ \hline \end{array}$

7. $\begin{array}{r} 67 \\ \times\ 2 \\ \hline \end{array}$

Find the product. You may wish to use base-ten blocks.

8. $\begin{array}{r} 10 \\ \times\ 3 \\ \hline \end{array}$
9. $\begin{array}{r} 31 \\ \times\ 5 \\ \hline \end{array}$
10. $\begin{array}{r} 40 \\ \times\ 6 \\ \hline \end{array}$
11. $\begin{array}{r} 19 \\ \times\ 8 \\ \hline \end{array}$

12. $\begin{array}{r} 18 \\ \times\ 2 \\ \hline \end{array}$
13. $\begin{array}{r} 41 \\ \times\ 3 \\ \hline \end{array}$
14. $\begin{array}{r} 17 \\ \times\ 3 \\ \hline \end{array}$
15. $\begin{array}{r} 50 \\ \times\ 5 \\ \hline \end{array}$

16. $\begin{array}{r} 14 \\ \times\ 9 \\ \hline \end{array}$
17. $\begin{array}{r} 21 \\ \times\ 4 \\ \hline \end{array}$
18. $\begin{array}{r} 28 \\ \times\ 4 \\ \hline \end{array}$
19. $\begin{array}{r} 15 \\ \times\ 5 \\ \hline \end{array}$

20. $5 \times 14 = \blacksquare$
21. $6 \times 18 = \blacksquare$
22. $3 \times 27 = \blacksquare$
23. $5 \times 36 = \blacksquare$

ALGEBRA For 24–27, use base-ten blocks to find the missing factor.

24. $6 \times \blacksquare = 132$
25. $4 \times \blacksquare = 56$
26. $\blacksquare \times 21 = 63$
27. $\blacksquare \times 25 = 100$

28. **Write a problem** There are 24 hours in a day. There are 7 days in a week. Write a problem using this information. Trade problems with a classmate and solve.

29. **What's the Error?** Lori says that the product 3×16 is the same as the sum $16 + 16$. Describe Lori's error and then find the product.

30. **REASONING** Brandon says that 5×18 is the same as $40 + 50$. Do you agree or disagree? Explain.

31. The sum of two numbers is 20. The product of the two numbers is 75. What are the numbers?

32. Julio bought 8 dozen bagels. How many bagels did he buy? Write a number sentence and solve. (HINT: 1 dozen = 12)

33. Tanya bought 2 pairs of jeans for \$18 each, a shirt for \$12, and a sweatshirt for \$8. How much did Tanya spend?

Mixed Review and Test Prep

Find the number that the variable stands for. (p. 188)

34. $2 \times a = 8$
35. $b \times 3 = 18$
36. $c \times 4 = 16$
37. $8 \times d = 56$
38. $5 \times a = 35$
39. $b \times 7 = 63$
40. $c \times 6 = 42$
41. $d \times 9 = 81$

42. **TEST PREP** Choose the unit you would use to measure the length of your finger. (p. 356)

A foot
B yard
C inch
D mile

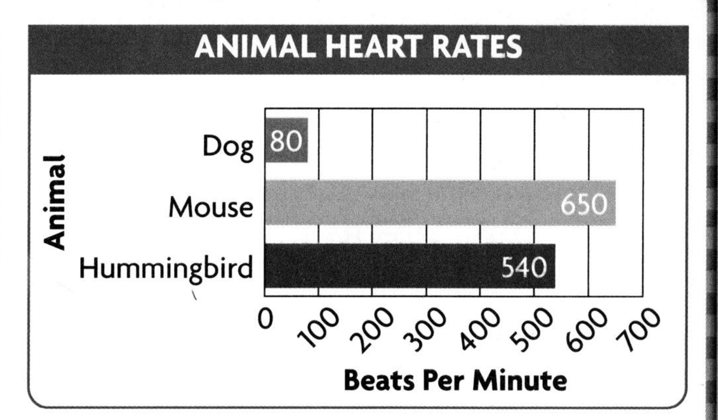

43. **TEST PREP** How many more times does a hummingbird's heart beat per minute than a dog's heart? (p. 254)

F 460 G 505 H 585 J 595

PROBLEM SOLVING LINKUP . . . to Art

Dr. John Biggers is known for his murals about African American and African culture. He visited Africa in 1957. While there he studied the traditions, culture, and people. In 1997 an exhibit showed almost 100 of John Biggers' drawings, prints, sculptures, paintings, and murals. The exhibit traveled to different art museums.

John Thomas Biggers *Starry Crown* 1987. Acrylic mixed media on masonite Dallas Museum of Art. Museum League Purchase Fund.

Starry Crown **is one of John Biggers' most famous works of art.**

1. **What if** 35 people see the John Biggers exhibit every hour? How many people see it in 8 hours?
2. A school field trip brings 3 buses with 57 students in each bus to the museum to see the exhibit. How many students are there in all?
3. Suppose the museum sells prints of John Biggers' paintings for $13 each. How much would 7 prints cost?
4. Suppose the museum sells John Biggers T-shirts for $19 each. How much would it cost to buy 5 T-shirts?

EXTRA PRACTICE page H58, Set A

LESSON 3

Practice Multiplication

Quick Review

1. $3 \times 9 = \blacksquare$
2. $\blacksquare = 4 \times 6$
3. $\blacksquare = 8 \times 9$
4. $10 \times 5 = \blacksquare$
5. $9 \times 10 = \blacksquare$

Learn

ROAD TRIP Mrs. Barrett drove from Portland to Seattle in 3 hours. She drove 56 miles each hour. How many miles did she drive?

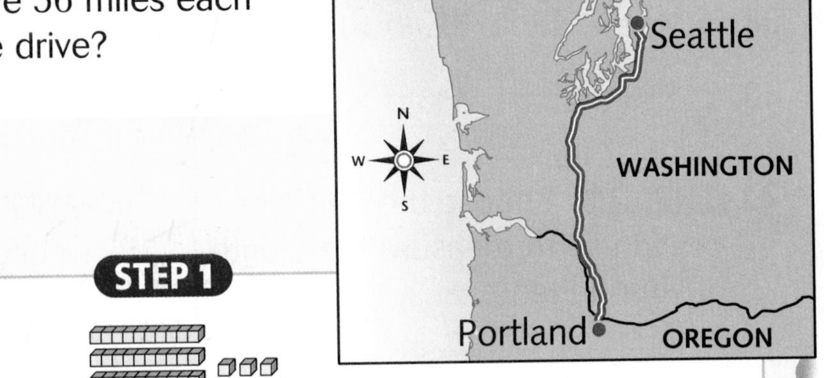

Example

$3 \times 56 = \blacksquare$ or $\begin{array}{r} 56 \\ \times\ 3 \\ \hline \end{array}$

STEP 1

Multiply the ones.
3×6 ones $= 18$ ones
Regroup 18 ones as 1 ten 8 ones.

	Hundreds	Tens	Ones
		1	
		5	6
×			3
			8

STEP 2

Multiply the tens. 3×5 tens $= 15$ tens
Add the 1 ten you regrouped. $(15 + 1)$ tens $= 16$ tens
Regroup 16 tens as 1 hundred 6 tens.

	Hundreds	Tens	Ones
		1	
		5	6
×			3
	1	6	8

You can also use a calculator to find the product.

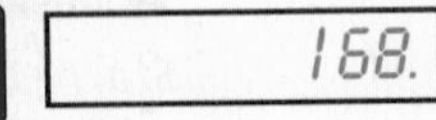

So, Mrs. Barrett drove 168 miles.

- The 8 ones were recorded in the ones column. How was the regrouped ten recorded? How was the regrouped hundred recorded?

Check

1. **Explain** how to use regrouping to find 5×34. Then find the product.

Find the product. Tell whether you need to regroup. Write *yes* or *no*.

2.

H	T	O
	1	3
×		3

3.

H	T	O
	1	4
×		7

4.

H	T	O
	3	4
×		2

5.

H	T	O
	3	1
×		5

Practice and Problem Solving

Find the product. Tell whether you need to regroup. Write *yes* or *no*.

6. $\begin{array}{r} 19 \\ \times\ 5 \\ \hline \end{array}$

7. $\begin{array}{r} 14 \\ \times\ 2 \\ \hline \end{array}$

8. $\begin{array}{r} 74 \\ \times\ 3 \\ \hline \end{array}$

9. $\begin{array}{r} 15 \\ \times\ 8 \\ \hline \end{array}$

10. $\begin{array}{r} 43 \\ \times\ 7 \\ \hline \end{array}$

Find the product.

11. $\begin{array}{r} 42 \\ \times\ 4 \\ \hline \end{array}$

12. $\begin{array}{r} 27 \\ \times\ 4 \\ \hline \end{array}$

13. $\begin{array}{r} 95 \\ \times\ 2 \\ \hline \end{array}$

14. $\begin{array}{r} 42 \\ \times\ 6 \\ \hline \end{array}$

15. $\begin{array}{r} 39 \\ \times\ 7 \\ \hline \end{array}$

16. $2 \times 49 = ■$

17. $5 \times 23 = ■$

18. $8 \times 67 = ■$

Write <, >, or = for each ●.

19. 2×25 ● 7×15

20. 6×27 ● 9×13

21. 8×64 ● 64×8

22. **REASONING** Explain how you know without finding the product that 4×82 will have more than 2 digits.

23. Write About It Hiro said that 8×12 and 8×32 have the same ones digit. Explain how he knew without finding each product.

Mixed Review and Test Prep

24. $\$5.00 - \$1.65 = ■$ (p. 88)

25. $48 + 484 = ■$ (p. 42)

26. $\$1.25 + \$2.55 = ■$ (p. 42)

27. $5{,}324 - 2{,}908 = ■$ (p. 62)

28. **TEST PREP** Which fraction is equivalent to $\frac{1}{2}$ (p. 418)

A $\frac{2}{3}$ **B** $\frac{3}{5}$ **C** $\frac{4}{6}$ **D** $\frac{4}{8}$

EXTRA PRACTICE page H58, Set B

LESSON 4

Problem Solving Skill
Choose the Operation

UNDERSTAND › PLAN › SOLVE › CHECK

Quick Review

1. 15×6
2. 72×8
3. $232 + 125$
4. $495 - 208$
5. $18 \div 9 = \blacksquare$

GRAND CHAMP Steve will work 25 hours each week to get his cattle ready for the State Fair. The fair is in 4 weeks. How many hours will Steve work to get his cattle ready?

This chart can help you decide which operation to use.

Add	• Join groups of different amounts.
Subtract	• Take away. • Compare amounts.
Multiply	• Join equal groups.
Divide	• Separate into equal groups. • Find the number in each group.

Since you are joining equal groups, you multiply.

$4 \times 25 = 100$ or $\begin{array}{r} 25 \\ \times\ 4 \\ \hline 100 \end{array}$

So, Steve will work 100 hours to get his cattle ready.

MATH IDEA Before you solve a problem, decide what operation to use.

Talk About It

- What other operation could you use to solve the problem? Explain.
- **What if** Steve will work 8 more hours during the week before the fair? How many hours will he work to get ready?

Problem Solving Practice

Write whether you would *add, subtract, multiply,* or *divide*. Then solve.

1. Tamara spends 30 minutes each day grooming her horse. How many minutes does she spend grooming her horse in 5 days?

2. Drew paid $\$12.99$ for a State Fair T-shirt and $\$10.95$ for a cap. How much more did the T-shirt cost than the cap?

Mieko bought 3 flowers for 75¢ each. How much did she spend for the flowers?

3. Which is the best number sentence to use to solve the problem?
   A $75¢ + 3¢ = ■$
   B $75¢ - 3¢ = ■$
   C $3 \times 75¢ = ■$
   D $75¢ \div 3 = ■$

4. How much did Mieko spend for the flowers?
   F 25¢
   G 75¢
   H $\$1.50$
   J $\$2.25$

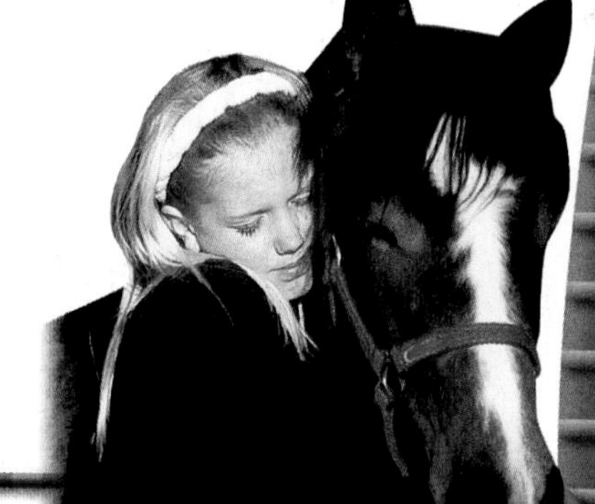

Mixed Applications

USE DATA For 5-9, use the table.

ANIMAL CONTEST ENTRIES	
Animals	**Number of Entries**
Horses	60
Hogs	55
Sheep	50
Rabbits	30

5. Six rabbits were entered in a special contest. The others were put into 3 equal groups for the Largest Rabbit Contest. How many rabbits were in each of the 3 groups?

6. The horses are shown in groups of 6. How many groups of horses will there be?

7. Make a bar graph using the data in the table. Be sure to write a title and labels.

8. Llamas were added to the contest. They were put into 2 equal groups. Do you have enough information to find the number of llamas in each group? Explain.

9. **? What's the Question?** The answer is 10 more animals. What is the question? What operation would you use?

Review/Test

CHECK CONCEPTS

Use the array to help find the product. (pp. 488–489)

1. $2 \times 10 = 20$ $2 \times 7 = 14$
   $2 \times 17 = ■$

2. 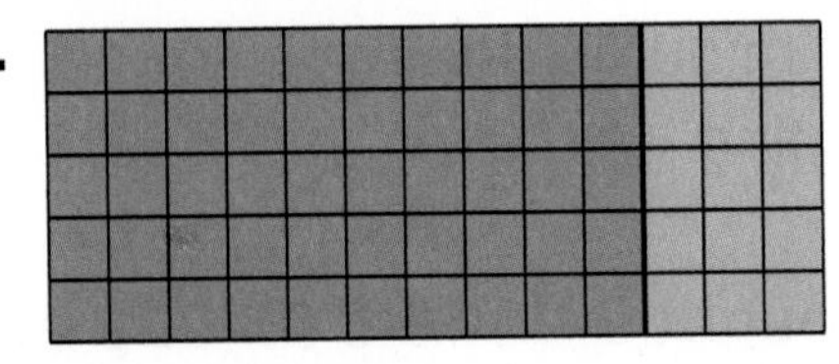

   5 rows of 10 5 rows of 3
   $5 \times 10 = 50$ $5 \times 3 = 15$
   $5 \times 13 = ■$

CHECK SKILLS

Find the product. (pp. 490–495)

3. 18×4
4. 34×3
5. 40×7
6. 12×6
7. 27×5
8. 14×2
9. 27×3
10. 36×6
11. 20×4
12. 56×8
13. $9 \times 40 = ■$
14. $4 \times 35 = ■$
15. $7 \times 83 = ■$
16. $6 \times 18 = ■$

CHECK PROBLEM SOLVING

Write whether you would *add, subtract, multiply,* or *divide*. Then solve. (pp. 496–497)

17. Caitlin practices the flute for 45 minutes each day. How many minutes does she practice in 4 days?

18. Manny paid $2.99 for a magazine and $1.09 for a bottle of apple juice. How much money did he spend?

19. Eva sold 7 plates of cookies at a bake sale. There were 18 cookies on each plate. How many cookies did she sell?

20. Lee has 64 photos in an album. There are 8 pages with the same number of photos on each page. How many photos are on each page?

Standardized Test Prep

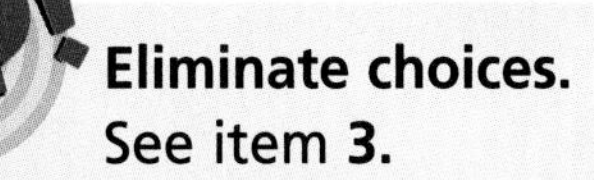

Eliminate choices.
See item **3**.

Estimate your answer and eliminate any unreasonable choices. If your answer does not match any of the choices, check the numbers you used and your computation.

Also see problem **5**, p. H64.

For 1–6, choose the best answer.

1. $\begin{array}{r} 57 \\ \times\ 7 \\ \hline \end{array}$

A 64 **C** 389
B 349 **D** NOT HERE

---

2. Jean collects seashells. She has 19 large shells, 37 medium shells, and 54 small shells. How many shells does she have in all?

F 90 **H** 110
G 100 **J** 120

---

3. Ken earns \$12 each week caring for his neighbor's dog. How much money will he earn in 8 weeks?

A \$20 **C** \$96
B \$86 **D** \$106

4. Shea has 4 boxes of markers. Each box holds 16 markers. How many markers does Shea have in all?

F 4 **H** 54
G 20 **J** 64

For 5–6, use the line plot.

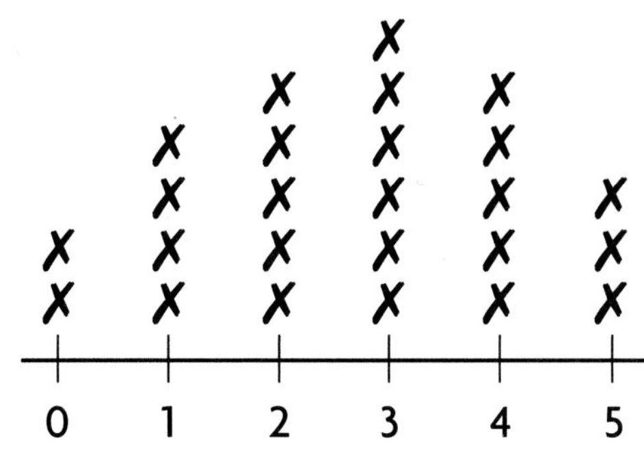

Number of lunches students bought last week

5. How many students bought a school lunch 3 times last week?

A 3 **C** 8
B 6 **D** 12

6. How many students bought a school lunch 4 or 5 days last week?

F 8 **G** 5 **H** 4 **J** 3

Write What You Know

7. Alan used tiles to make an array with 3 rows of 18. Draw a picture of the array. Write an expression that finds how many tiles he used. Solve.

8. Show and explain the steps for finding the product.

$42 \times 9 = \blacksquare$

Divide by 1-Digit Numbers

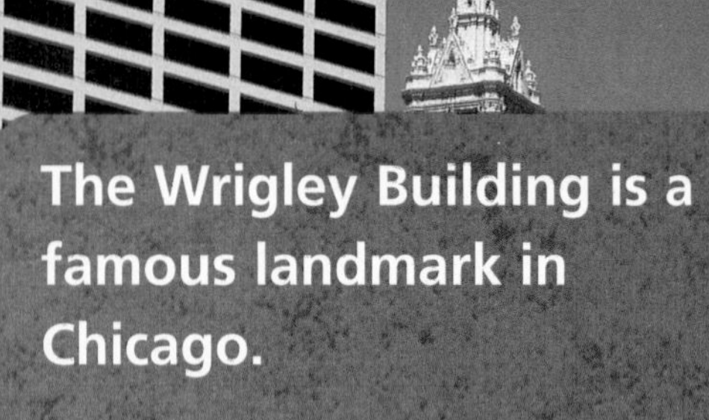

The Wrigley Building is a famous landmark in Chicago.

PROBLEM SOLVING Suppose the classes shown on the bar graph are going to see the Wrigley Building. If the same number of students ride on each of 4 buses, how many students will be on each bus?

DATA LINK

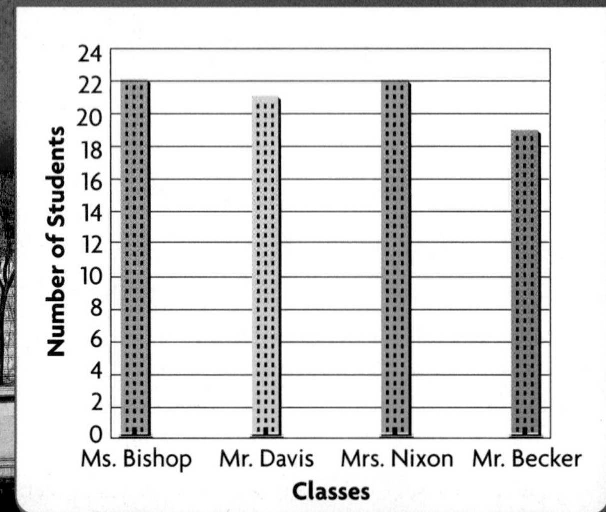

The Wrigley Building

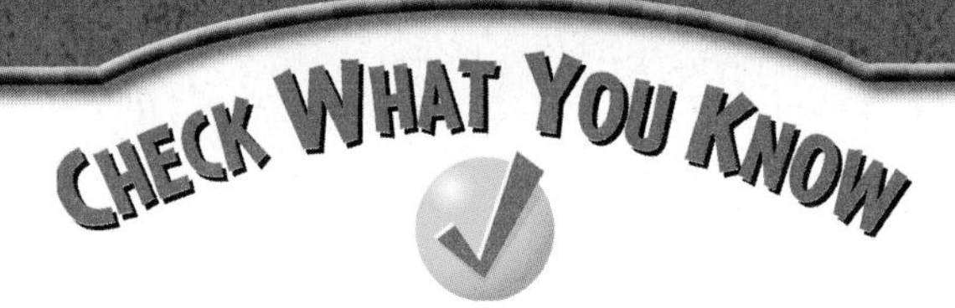

Use this page to help you review and remember important skills needed for Chapter 28.

VOCABULARY

Choose the best term from the box.

1. In the example, the number 8 is the ___?___.
2. In the example, the number 32 is the ___?___.
3. In the example, the number 4 is the ___?___.

Example:

$$\begin{array}{r} 4 \\ 8\overline{)32} \end{array}$$

dividend
divisor
quotient
sum

SUBTRACTION FACTS (For Intervention, see p. H6.)

Subtract.

4. $7 - 5 = ■$ **5.** $12 - 5 = ■$ **6.** $14 - 6 = ■$ **7.** $■ = 13 - 5$

8. $8 - 2 = ■$ **9.** $■ = 15 - 9$ **10.** $18 - 9 = ■$ **11.** $8 - 7 = ■$

DIVISION FACTS (For Intervention, see p. H21.)

Divide.

12. $81 \div 9 = ■$ **13.** $21 \div 3 = ■$ **14.** $■ = 48 \div 8$ **15.** $18 \div 6 = ■$

16. $24 \div 6 = ■$ **17.** $■ = 25 \div 5$ **18.** $16 \div 8 = ■$ **19.** $12 \div 3 = ■$

20. $2\overline{)14}$ **21.** $5\overline{)40}$ **22.** $6\overline{)42}$ **23.** $7\overline{)56}$

24. $9\overline{)36}$ **25.** $4\overline{)16}$ **26.** $3\overline{)24}$ **27.** $7\overline{)49}$

REGROUP HUNDREDS AND TENS (For Intervention, see p. H22.)

Regroup.

28. 50 = 4 tens ■ ones **29.** 60 = ■ tens 10 ones

30. 73 = 5 tens ■ ones **31.** 3 tens ■ ones = 44

32. 5 hundreds ■ tens = 610 **33.** 2 hundreds ■ tens = 340

LESSON 1

Divide with Remainders

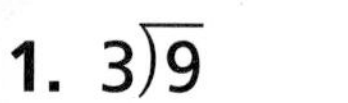

Quick Review

1. $3\overline{)9}$
2. $6\overline{)12}$
3. $4\overline{)24}$
4. $5\overline{)35}$
5. $7\overline{)28}$

VOCABULARY
remainder

MATERIALS
counters

Explore

Noah collected all 19 dinosaur toys from Crispy Crunch cereal. He wants to keep an equal number of toys in each of 3 shoe boxes. Can Noah divide them equally among the 3 boxes? Why or why not?

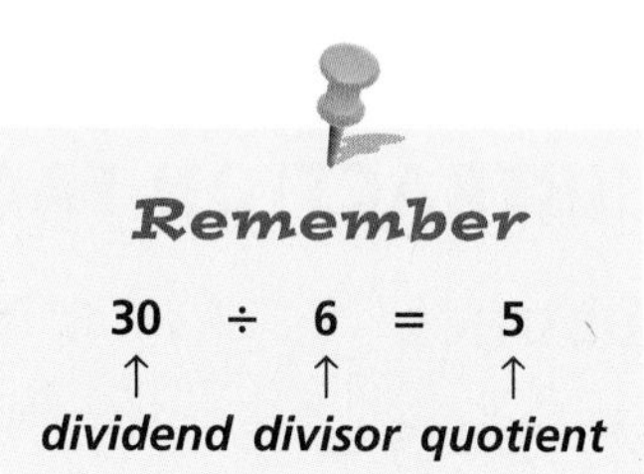

Activity

Use counters to find 19 ÷ 3.

STEP 1

Use 19 counters. Draw 3 circles.

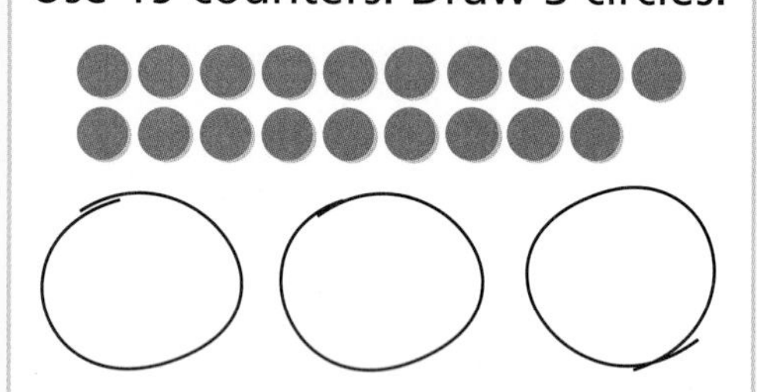

STEP 2

Divide the 19 counters into 3 equal groups by putting them in the circles.

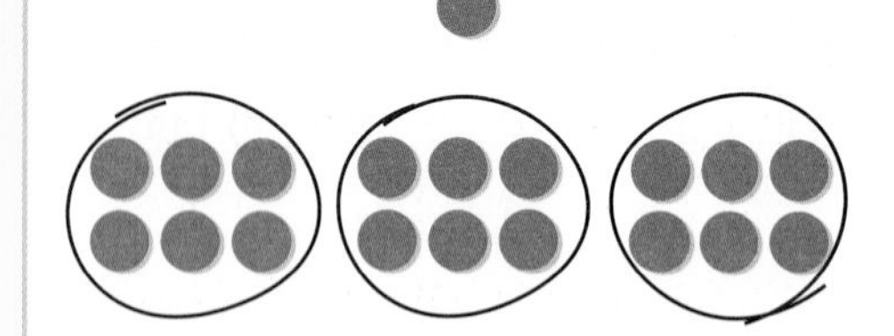

So, there will be 6 toys in each shoe box, with 1 toy left over.

I'm putting 11 counters into 2 equal groups. Will there be any left over ?

Try It

Use counters to make equal groups. Draw a picture of the model you made.

a. 11 ÷ 2 **b.** 15 ÷ 4 **c.** 26 ÷ 3

Connect

In division, the amount left over when a number cannot be divided evenly is called the **remainder**.

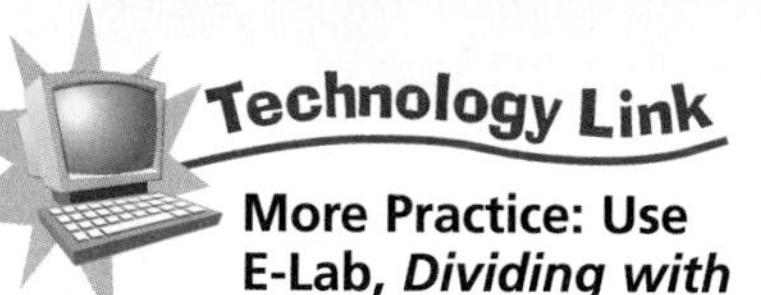

Technology Link

More Practice: Use E-Lab, *Dividing with Remainders.*

www.harcourtschool.com/elab2002

Activity

Find 20 ÷ 6.

STEP 1

Use 20 counters. Draw 6 circles.

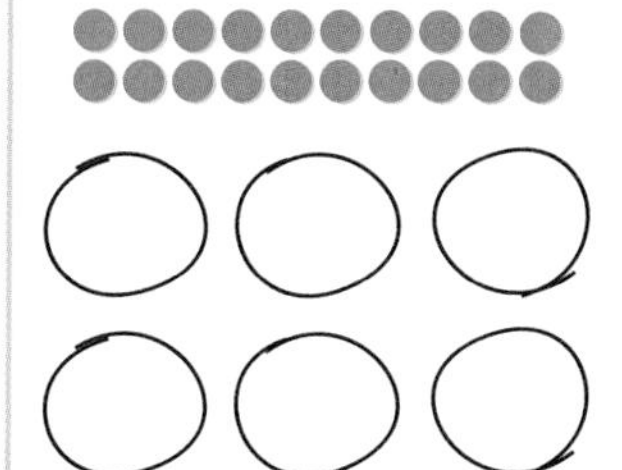

STEP 2

Divide the 20 counters into 6 equal groups.

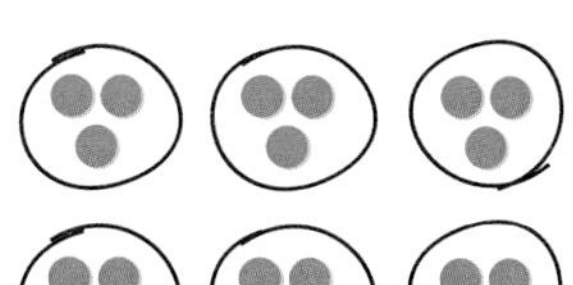

The quotient is 3.
The remainder is 2.

Practice and Problem Solving

Use counters to find the quotient and remainder.

1. $9 \div 2 = $ ■ **2.** $17 \div 4 = $ ■ **3.** $10 \div 3 = $ ■ **4.** $20 \div 3 = $ ■

5. $4\overline{)18}$ **6.** $5\overline{)19}$ **7.** $5\overline{)27}$ **8.** $6\overline{)21}$

Find the quotient and remainder. You may use counters or draw a picture to help.

9. $16 \div 3 = $ ■ **10.** $21 \div 5 = $ ■ **11.** $2\overline{)15}$ **12.** $4\overline{)22}$

13. **Write About It** What is the greatest remainder you could have when you divide by 5? Explain.

14. **What's the Question?** Darnell has 12 Beastie Buddies in his collection. He puts an equal number in each of 3 drawers. The answer is 4 Beastie Buddies.

Mixed Review and Test Prep

Multiply. (p. 490)

15. $2 \times 13 = $ ■ **16.** $3 \times 16 = $ ■

17. $5 \times 14 = $ ■ **18.** $4 \times 19 = $ ■

19. **TEST PREP** Judy has 2 rolls of pennies with 50 pennies in each roll. She also has 4 dimes and 1 quarter. How much money does she have in all? (p. 80)

A $1.65 **B** $1.40 **C** $1.00 **D** $0.65

LESSON 2

Model Division of 2-Digit Numbers

Quick Review

1. $16 \div 4 = \blacksquare$
2. $14 \div 7 = \blacksquare$
3. $15 \div 3 = \blacksquare$
4. $32 \div 4 = \blacksquare$
5. $18 \div 6 = \blacksquare$

Learn

COOL COLLECTIONS Suppose you have a collection of 63 hockey cards. You display an equal number of cards on each of 5 shelves. How many cards are on each shelf? How many cards are left over?

Activity

$63 \div 5 = \blacksquare$ $\quad 5\overline{)63}$

Materials: base-ten blocks

STEP 1

Divide the tens first. Put an equal number of tens in each group.

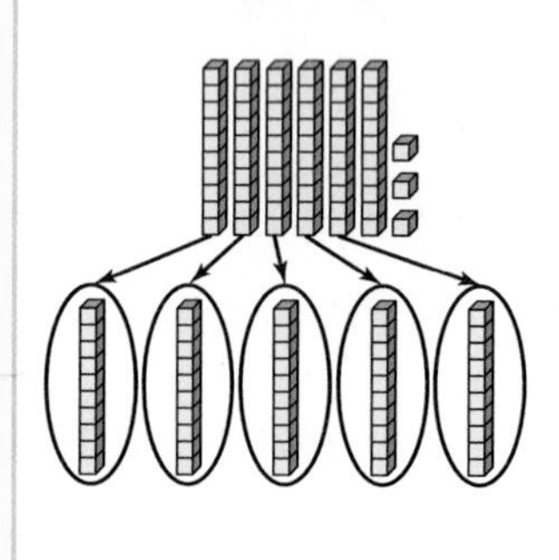

STEP 2

One ten is left. Regroup it as 10 ones. Put an equal number of ones in each group.

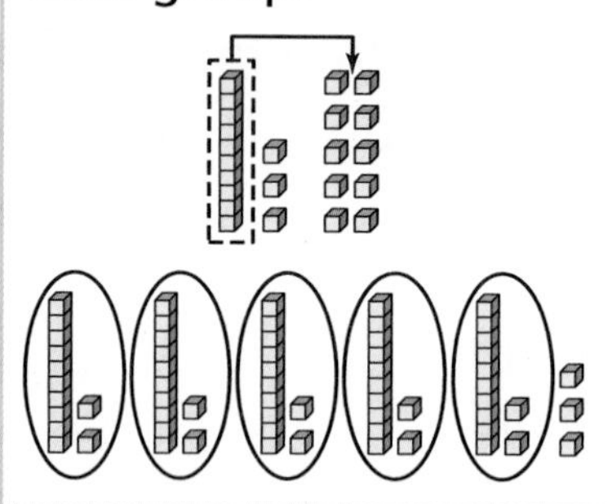

STEP 3

Use the letter *r* to show the remainder.

$63 \div 5 = 12\ r3$

↑ quotient ↑ remainder

So, there are 12 cards on each shelf. There are 3 cards left over.

- In Step 2, why was the 1 ten regrouped into 10 ones?

Check

1. **Explain** why $63 \div 5$ has a remainder.

Use the model. Write the quotient and remainder.

2. $21 \div 2$

3. $44 \div 3$

Practice and Problem Solving

Use the model. Write the quotient and remainder.

4. $62 \div 4$

5. $71 \div 5$

Divide. You may use base-ten blocks to help.

6. $23 \div 2 = \blacksquare$
7. $25 \div 2 = \blacksquare$
8. $39 \div 3 = \blacksquare$
9. $43 \div 4 = \blacksquare$
10. $2\overline{)37}$
11. $3\overline{)42}$
12. $2\overline{)51}$
13. $4\overline{)55}$

14. Matt needs 42 cans of juice for a party. How many 6-packs of juice should he buy?

15. ALGEBRA Find the missing digit.

$$\begin{array}{r} 15 \text{ r2} \\ 3\overline{)4\blacksquare} \end{array}$$

16. **What's the Error?** This is how Cheryl modeled $32 \div 3$.

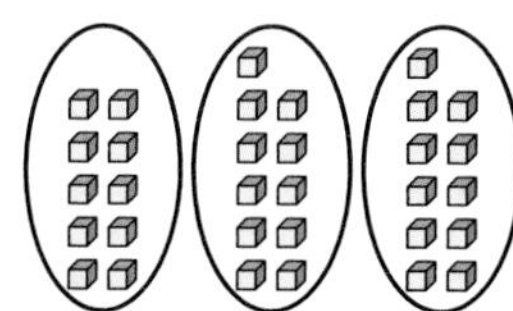

Describe Cheryl's mistake. Draw the correct model.

17. Mitsugi has 4 shelves of hockey cards. Each shelf has 13 cards. How many hockey cards does he have?

Mixed Review and Test Prep

For 18–20, find the product. (p. 490)

18. 12×9
19. 14×3
20. 27×2

21. $9{,}427 - 5{,}613 = \blacksquare$ (p. 62)

22. **TEST PREP** Amy has 4 boxes with 8 toys in each box. She gives 3 toys to her sister. How many toys does Amy have left? (p. 172)

A 35
B 33
C 29
D 25

EXTRA PRACTICE page H59, Set A

LESSON 3

Record Division of 2-Digit Numbers

Quick Review

1. $2\overline{)16}$
2. $6\overline{)48}$
3. $5\overline{)25}$
4. $9\overline{)63}$
5. $6\overline{)24}$

Learn

TAKE YOUR PICK Brad picked 58 apples at an orchard. He put an equal number of apples in 3 bags. How many were in each bag? How many were left over?

Example 1

$58 \div 3 = ■$

Read: 58 divided by 3 **Write:** $3\overline{)58}$

STEP 1

Divide the 5 tens first. $3\overline{)5}$

Think: $3 \times 1 = 3$

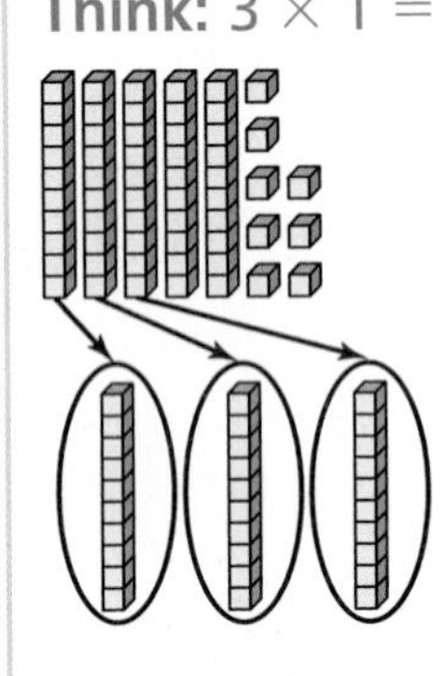

$$\begin{array}{r} 1 \\ 3\overline{)58} \\ -3 \\ \hline 2 \end{array}$$

1 ten in each group.

Divide. $3\overline{)5}$
Multiply. 3×1
Subtract. $5 - 3$
Compare. $2 < 3$

The difference, 2, must be less than the divisor, 3.

STEP 2

Bring down the 8 ones.

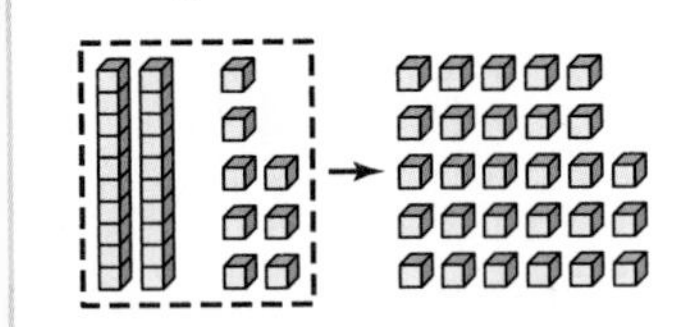

$$\begin{array}{r} 1 \\ 3\overline{)58} \\ -3\downarrow \\ \hline 28 \end{array}$$

Regroup 2 tens 8 ones as 28 ones.

STEP 3

Divide the 28 ones. $3\overline{)28}$ **Think:** $3 \times 9 = 27$

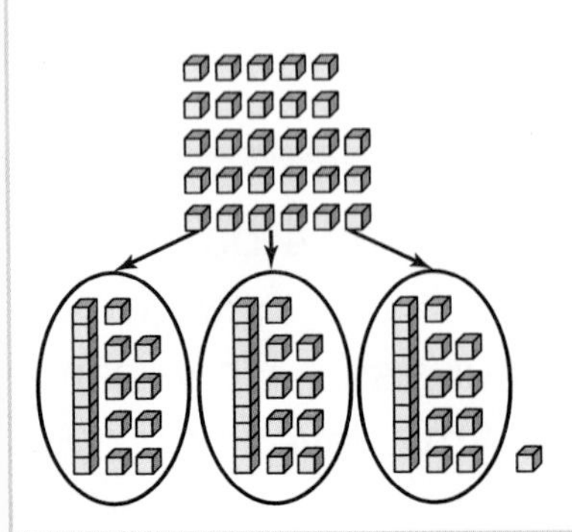

$$\begin{array}{r} 19 \text{ r1} \\ 3\overline{)58} \\ -3 \\ \hline 28 \\ -27 \\ \hline 1 \end{array}$$

9 ones in each group.

Divide. $3\overline{)28}$
Multiply. 3×9
Subtract. $28 - 27$
Compare. $1 < 3$

STEP 4

Multiply to check your answer.

$$\begin{array}{r} 19 \\ \times\ 3 \\ \hline 57 \end{array}$$

19 ← quotient
3 ← divisor

$$\begin{array}{r} 57 \\ +\ 1 \\ \hline 58 \end{array}$$

1 ← Add the remainder.
58 ← This should equal the dividend.

So, there were 19 apples in each bag and 1 apple was left over.

Example 2

Tyler bought 35 pumpkins for the school's Fall Festival. He puts 4 pumpkins in each box. How many boxes does he use? How many pumpkins are left over?

$35 \div 4 = \blacksquare$

STEP 1

Since there are not enough tens to divide, start with ones.

$4\overline{)35}$ $4 > 3$, so place the first digit in the ones place.

STEP 2

Divide the 35 ones.

$$\begin{array}{r} 8\text{ r}3 \\ 4\overline{)35} \\ -32 \\ \hline 3 \end{array}$$

Divide. $4\overline{)35}$
Multiply. 4×8
Subtract. $35 - 32$
Compare. $3 < 4$

STEP 3

Multiply to check your answer.

$$\begin{array}{r} 8 \\ \times 4 \\ \hline 32 \end{array}$$

8 ← quotient
×4 ← divisor

$$\begin{array}{r} 32 \\ +\ 3 \\ \hline 35 \end{array}$$

Add the remainder. 35 ← dividend

So, Tyler uses 8 boxes. There are 3 pumpkins left over.

More Examples

A

$$\begin{array}{r} 14 \\ 7\overline{)98} \\ -7 \\ \hline 28 \\ -28 \\ \hline 0 \end{array}$$

Check:

$$\begin{array}{r} 14 \\ \times\ 7 \\ \hline 98 \end{array}$$

B

$$\begin{array}{r} 22\text{ r}1 \\ 3\overline{)67} \\ -6 \\ \hline 07 \\ -6 \\ \hline 1 \end{array}$$

Check:

$$\begin{array}{r} 22 \\ \times\ 3 \\ \hline 66 \end{array} \quad \begin{array}{r} 66 \\ +\ 1 \\ \hline 67 \end{array}$$

MATH IDEA The steps to record divison are divide, multiply, subtract, compare.

- Why couldn't you write a 6 in the ones place in Step 2 above?

Check

1. **Describe** why you do not need to add a remainder when you check Example A.

Divide and check.

2. $29 \div 2 = \blacksquare$ 3. $26 \div 3 = \blacksquare$ 4. $52 \div 4 = \blacksquare$ 5. $38 \div 3 = \blacksquare$

LESSON CONTINUES

Practice and Problem Solving

Divide and check.

6. $33 \div 2 = ■$ **7.** $45 \div 4 = ■$ **8.** $49 \div 5 = ■$ **9.** $41 \div 3 = ■$

10. $42 \div 2 = ■$ **11.** $50 \div 3 = ■$ **12.** $67 \div 6 = ■$ **13.** $61 \div 5 = ■$

14. $38 \div 4 = ■$ **15.** $84 \div 7 = ■$ **16.** $23 \div 4 = ■$ **17.** $76 \div 3 = ■$

18. $4\overline{)77}$ **19.** $6\overline{)52}$ **20.** $5\overline{)73}$ **21.** $3\overline{)66}$

22. $5\overline{)48}$ **23.** $2\overline{)71}$ **24.** $7\overline{)92}$ **25.** $9\overline{)89}$

26. $2\overline{)87}$ **27.** $7\overline{)34}$ **28.** $8\overline{)99}$ **29.** $6\overline{)39}$

Write the *check* step for each division problem.

30. $\begin{array}{r} 5\text{ r2} \\ 4\overline{)22} \end{array}$ **31.** $\begin{array}{r} 7\text{ r1} \\ 6\overline{)43} \end{array}$ **32.** $\begin{array}{r} 28\text{ r2} \\ 3\overline{)86} \end{array}$ **33.** $\begin{array}{r} 27\text{ r1} \\ 2\overline{)55} \end{array}$

Decide whether you start by dividing the tens or the ones. Write *tens* or *ones*.

34. $5\overline{)22}$ **35.** $6\overline{)81}$ **36.** $2\overline{)15}$ **37.** $3\overline{)32}$

38. $9\overline{)72}$ **39.** $4\overline{)34}$ **40.** $7\overline{)61}$ **41.** $2\overline{)52}$

42. **What's the Error?** Lou began to solve a problem like this:

$$\begin{array}{r} 1 \\ 2\overline{)53} \\ -2 \\ \hline 3 \end{array}$$

Describe how he must use the *compare* step to fix his error. Solve the problem.

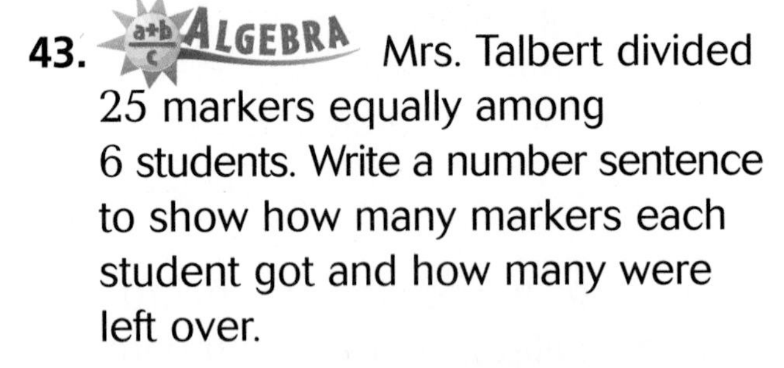

43. **ALGEBRA** Mrs. Talbert divided 25 markers equally among 6 students. Write a number sentence to show how many markers each student got and how many were left over.

44. **REASONING** Alyssa divided 79 jelly beans equally into 6 jars. She kept 1 jar and the leftover jelly beans. How many jelly beans did she keep?

45. Marshal bought 16 packs of gum. Each pack holds 3 baseball cards. How many cards did he get?

46. Hitoshi has 97 strawberries. He puts an equal number of strawberries into 3 baskets. How many strawberries are in each basket? How many are left over?

47. It takes Grant 12 minutes to get to his friend's house. It takes him 4 times as long to get to his grandfather's house. How long does it take to get to his grandfather's?

Mixed Review and Test Prep

Find the product. (p. 494)

48. $\begin{array}{r} 21 \\ \times\ 2 \\ \hline \end{array}$ **49.** $\begin{array}{r} 34 \\ \times\ 4 \\ \hline \end{array}$ **50.** $\begin{array}{r} 55 \\ \times\ 3 \\ \hline \end{array}$

51. TEST PREP Mrs. Martin kept a lunch box she bought in 1971. She was 9 years old when she bought the lunch box. How old was she in 2000? (p. 62)

A 38 **B** 28 **C** 19 **D** 9

Compare. Write <, >, or = for each ●.

52. 3×4 ● 6×2 (p. 138)

53. $54 \div 9$ ● $18 \div 2$ (p. 222)

54. TEST PREP Elise cut a pie into 8 equal pieces. She gave 2 pieces to her teacher. Elise ate 1 piece. What fraction of the pie was left? (p. 442)

A $\frac{5}{8}$ **B** $\frac{3}{8}$ **C** $\frac{2}{8}$ **D** $\frac{1}{8}$

PROBLEM SOLVING LINKUP ... to Reading

STRATEGY • USE GRAPHIC AIDS

The steps for making crayons have not changed much since 1903. Wax in different colors is heated to 240°F. Workers pour the hot wax across crayon molds. In $7\frac{1}{2}$ minutes, the wax cools into 72 rows of 8 crayons.

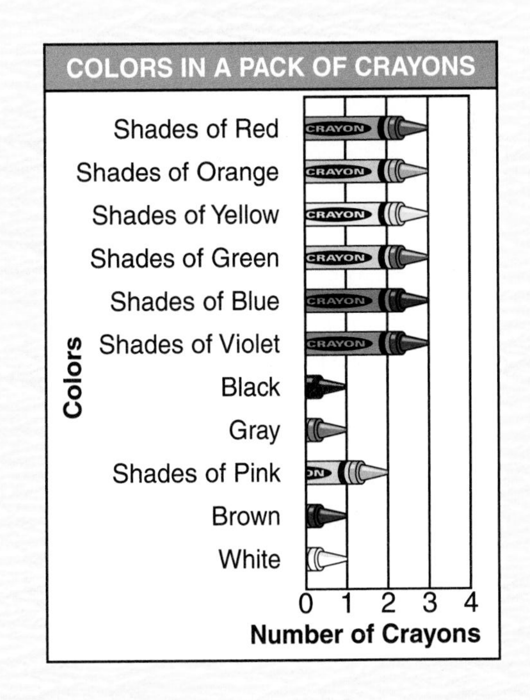

USE DATA For 1–2, use the bar graph.

1. How many crayons are in the pack shown in the bar graph?
2. **Write a problem** about the bar graph. Use division. Trade with a partner and solve.
3. How many crayons are made in the 72 rows of 8 crayon molds?
4. The 64-pack of crayons has 4 equal rows. How many crayons are in each row?

LESSON 4

Practice Division

Quick Review

1. $12 \div 2 = \blacksquare$
2. $21 \div 3 = \blacksquare$
3. $40 \div 5 = \blacksquare$
4. $12 \div 4 = \blacksquare$
5. $54 \div 6 = \blacksquare$

Learn

TOO MANY TOYS Ellen has collected 87 plastic toys. She puts an equal number of toys on 6 shelves in her room. How many toys are on each shelf? How many toys are not on shelves?

Example

$87 \div 6 = \blacksquare$

Read: 87 divided by 6 **Write:** $6\overline{)87}$

STEP 1

Divide the 8 tens.

$$\begin{array}{r} 1 \\ 6\overline{)87} \\ -6 \\ \hline 2 \end{array}$$

Divide. $6\overline{)8}$
Multiply. 6×1
Subtract. $8 - 6$
Compare. $2 < 6$

STEP 2

Bring down the 7 ones. Divide the 27 ones.

$$\begin{array}{r} 14 \text{ r}3 \\ 6\overline{)87}\phantom{\text{ r}3} \\ -6\downarrow\phantom{\text{ r}3} \\ \hline 27\phantom{\text{ r}3} \\ -24\phantom{\text{ r}3} \\ \hline 3\phantom{\text{ r}3} \end{array}$$

Divide. $6\overline{)27}$
Multiply. 6×4
Subtract. $27 - 24$
Compare. $3 < 6$

STEP 3

Multiply to check your answer.

$$\begin{array}{rl} 14 & \leftarrow \text{quotient} \\ \times\ 6 & \leftarrow \text{divisor} \\ \hline 84 & \end{array}$$

$$\begin{array}{rl} 84 & \\ +\ 3 & \leftarrow \text{remainder} \\ \hline 87 & \leftarrow \text{dividend} \end{array}$$

So, there are 14 toys on each shelf and 3 toys not on shelves.

- Could you begin by writing a 1 over the ones place in Step 1? Why or why not?

Check

1. **Tell** why you must multiply 6×4 in Step 2.

Divide.

2. $38 \div 3 = \blacksquare$
3. $29 \div 4 = \blacksquare$
4. $35 \div 2 = \blacksquare$
5. $65 \div 5 = \blacksquare$

Practice and Problem Solving

Divide.

6. $26 \div 5 = ■$ **7.** $52 \div 2 = ■$ **8.** $55 \div 4 = ■$ **9.** $49 \div 3 = ■$

10. $6\overline{)69}$ **11.** $4\overline{)77}$ **12.** $5\overline{)72}$ **13.** $3\overline{)64}$

14. $3\overline{)80}$ **15.** $7\overline{)91}$ **16.** $4\overline{)65}$ **17.** $9\overline{)79}$

Divide and check.

18. $70 \div 8 = ■$ **19.** $77 \div 2 = ■$ **20.** $81 \div 3 = ■$ **21.** $88 \div 5 = ■$

22. $5\overline{)78}$ **23.** $4\overline{)49}$ **24.** $8\overline{)36}$ **25.** $6\overline{)45}$

26. $4\overline{)50}$ **27.** $3\overline{)65}$ **28.** $2\overline{)34}$ **29.** $4\overline{)13}$

30. $34 \div 5$ **31.** $45 \div 3$ **32.** $29 \div 7$ **33.** $55 \div 7$

Write the *check* step for each division problem.

34. $\begin{array}{r} 24 \\ 4\overline{)96} \end{array}$ **35.** $\begin{array}{r} 45 \text{ r1} \\ 2\overline{)91} \end{array}$ **36.** $\begin{array}{r} 12 \text{ r4} \\ 6\overline{)76} \end{array}$ **37.** $\begin{array}{r} 6 \text{ r2} \\ 3\overline{)20} \end{array}$

38. What if Ellen put 87 toys on 7 shelves? How many toys would be on each shelf? How many toys would be left over?

39. MENTAL MATH How can you use the basic fact $12 \div 4 = 3$ to find $13 \div 4$ in your head?

40. ALGEBRA Ida had 64 marbles. She gave an equal number to 4 friends. Write a number sentence to show how many each friend got.

41. REASONING Suppose you find that $42 \div 3$ has no remainder. Can you tell without dividing whether $45 \div 3$ has a remainder? Explain.

Mixed Review and Test Prep

Find the product. (p. 170)

42. $3 \times 3 \times 5 = ■$

43. $4 \times 2 \times 6 = ■$

Describe the lines. Write *parallel* or *intersecting* (p. 304)

44.

45.

46. At the zoo, Cathy counted 74 animals. Sam counted half that number. How many animals did the two friends count? (p. 504)

A 74 **B** 111 **C** 148 **D** 174

47. TEST PREP Find the number that the variable stands for.

$6 \times r = 48 \quad r = \underline{\ ?\ }$ (p. 188)

F 5 **G** 6 **H** 8 **J** 10

EXTRA PRACTICE page H59, Set C

LESSON 5

Problem Solving Skill
Interpret the Remainder

UNDERSTAND > PLAN > SOLVE > CHECK

Quick Review

1. $3\overline{)55}$
2. $4\overline{)18}$
3. $2\overline{)61}$
4. $5\overline{)60}$
5. $7\overline{)32}$

PICNIC PLANS Clare needs to take 50 cans of juice to the picnic. The juice is sold in packages of 6 cans. How many packages must she buy?

Since you need to know how many groups of 6 are in 50, you divide.

Find $50 \div 6$.

$$\begin{array}{r} 8 \text{ r}2 \\ 6\overline{)50} \\ -48 \\ \hline 2 \end{array}$$

The 8 packages of juice will have only 48 cans.

If Clare buys 8 packages of juice, she will still need 2 more cans of juice. So, Clare must buy 9 packages of juice.

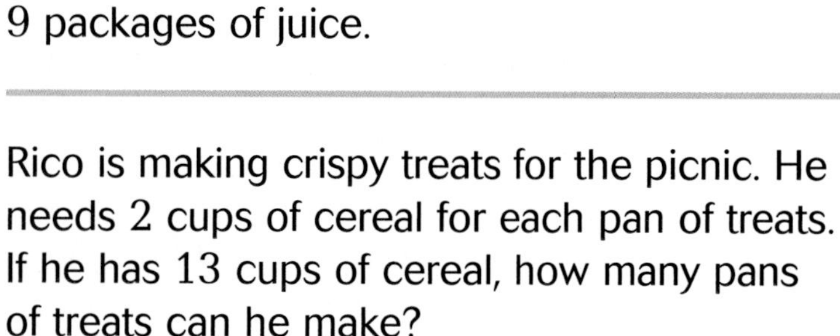

Rico is making crispy treats for the picnic. He needs 2 cups of cereal for each pan of treats. If he has 13 cups of cereal, how many pans of treats can he make?

Find $13 \div 2$.

$$\begin{array}{r} 6 \text{ r}1 \\ 2\overline{)13} \\ -12 \\ \hline 1 \end{array}$$

Making 6 pans of treats uses 12 cups of cereal.

If Rico makes 6 pans of treats, he will have only 1 cup of cereal left. This is not enough for another pan of treats. So, he can make 6 pans of treats.

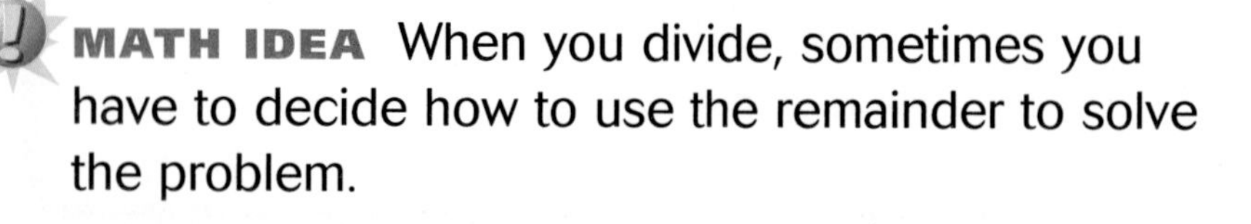

MATH IDEA When you divide, sometimes you have to decide how to use the remainder to solve the problem.

Problem Solving Practice

1. **What if** Clare had to take 56 cans of juice? If there are 6 cans in each package, how many packages would she need to buy?

2. Simon has 63 bird stamps in a collection. He can fit 8 stamps on a page. How many pages does he need?

3. Beth is making bows to decorate a float in a parade. It takes 5 feet of ribbon to make a bow. She has 89 feet of ribbon. How many bows can Beth make?

4. Dimitri worked on a math puzzle for 32 minutes. If each part of the puzzle took 5 minutes to solve, how many parts did he solve?

A class of 28 students is going on a field trip. Each van can hold 8 students. How many vans does the class need?

5. Which number sentence can help solve the problem?

   A $28 - 28 = 0$
   B $28 \times 8 = 224$
   C $28 \div 8 = 3$ r4
   D $8 \div 8 = 1$

6. How many vans does the class need?

   F 3 vans
   G 4 vans
   H 8 vans
   J 28 vans

Problem Solving Skill

Mixed Applications

USE DATA For 7–9, use the price list.

7. Tomas bought 1 lb of peanuts and 1 lb of cashews. He paid with \$5. How much change did he get?

8. Sheila has \$5.36. Does she have enough to buy 1 pound of raisins, 1 pound of peanuts, and 1 pound of cashews? Explain.

9. Cole bought 2 pounds of raisins. The checker charged him \$1.19. Is this price reasonable? Explain.

PRICE LIST	
Raisins	\$0.85 per lb
Peanuts	\$1.99 per lb
Cashews	\$2.50 per lb

Review/Test

CHECK VOCABULARY AND CONCEPTS

Choose the best term from the box.

divisor
remainder

1. In division, the amount left over is called the __?__. (p. 503)

Find the quotient and remainder. You may use counters or draw a picture to help. (pp. 502–503)

2. $11 \div 2 = ■$
3. $26 \div 4 = ■$
4. $38 \div 5 = ■$
5. $14 \div 5 = ■$

CHECK SKILLS

Use the model. Write the quotient and remainder. (pp. 504–505)

6. $27 \div 2 = ■$

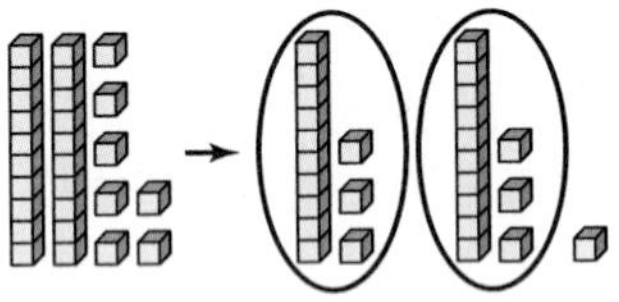

7. $41 \div 3 = ■$

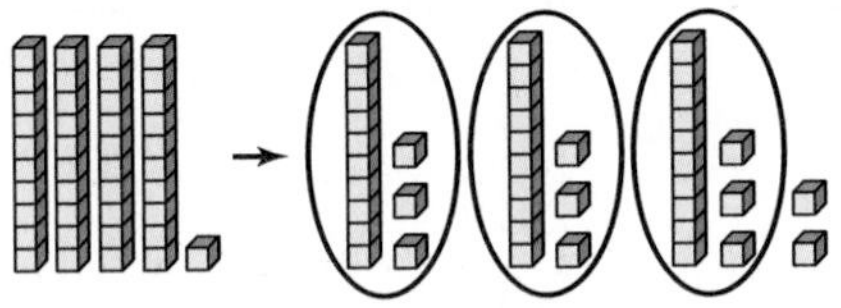

Divide. (pp. 506–511)

8. $22 \div 3 = ■$
9. $35 \div 2 = ■$
10. $50 \div 4 = ■$
11. $59 \div 5 = ■$
12. $2\overline{)53}$
13. $5\overline{)72}$
14. $6\overline{)82}$
15. $3\overline{)78}$
16. $8\overline{)69}$
17. $3\overline{)94}$
18. $9\overline{)95}$
19. $4\overline{)71}$

Write the *check* step for each division problem. (pp. 506–511)

20. $\begin{array}{r} 8 \text{ r1} \\ 3\overline{)25} \end{array}$
21. $\begin{array}{r} 7 \text{ r1} \\ 6\overline{)43} \end{array}$
22. $\begin{array}{r} 12 \text{ r3} \\ 4\overline{)51} \end{array}$
23. $\begin{array}{r} 22 \\ 2\overline{)44} \end{array}$

CHECK PROBLEM SOLVING

Solve. (pp. 512–513)

24. A class of 33 students is going to the science museum. If each van holds 9 students, how many vans does the class need?

25. Caroline needs 4-inch pieces of yarn for an art project. She has 58 inches of yarn. How many 4-inch pieces can she make?

Standardized Test Prep

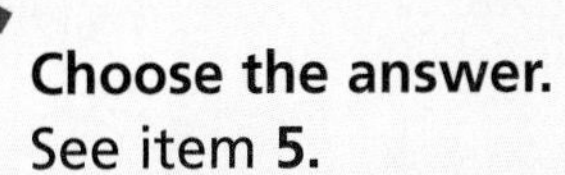

Choose the answer.
See item **5.**

All the answer choices show expressions with multiplication. Think how the Order Property of Multiplication relates to equivalent expressions.

Also see problem **6,** p. H65.

For 1–9, choose the best answer.

1. $8\overline{)92}$

A 10 r2 **B** 11 r2 **C** 11 r4 **D** 12

2. There are 47 students going by car to visit the Space Museum. Each car can carry 4 students. How many cars are needed?

F 7 **G** 10 **H** 11 **J** 12

3. $\begin{array}{r} \$0.47 \\ +\$0.35 \\ \hline \end{array}$

A \$0.82 **B** \$0.72 **C** \$0.12 **D** NOT HERE

4. $30 \div 4 = \blacksquare$

F 7 r2 **G** 8 r2 **H** 9 **J** 10

5. Which expression is equivalent to 38×9?

A 8×39 **B** 9×38 **C** 9×39 **D** 18×9

6. What is the name of this solid figure?

F cube
G cone
H square pyramid
J sphere

7. Mr. King's class painted 26 pictures to decorate 3 bulletin boards. Each bulletin board has space for 9 pictures. How many spaces will there be left to fill?

A 0 **B** 1 **C** 3 **D** 9

8. John's brother Jake is 3 feet tall. How many inches tall is he?

F 48 inches **G** 42 inches **H** 36 inches **J** NOT HERE

9. Find the number that the variable stands for.

$7 \times a = 56 \qquad a = \underline{?}$

A 6 **B** 7 **C** 8 **D** 9

Write What You Know

10. Write a division sentence that would have a remainder. Show how to find the quotient.

11. After dividing some pennies among 4 children, Mrs. Cruz had 2 pennies left. Each child got 8 pennies. Write a division sentence that shows what Mrs. Cruz did. Explain how you found the division sentence.

CHAPTER 29

Multiply Greater Numbers

Airlines must decide how many flights to schedule, and which types of planes to use.

PROBLEM SOLVING For example, could they carry more passengers on four 757s or on five 727s? Write a problem using the information in the table.

PLANES AND PASSENGERS

Plane	Number of Passengers
727	149
737	119
757	190
767	254

CHECK WHAT YOU KNOW

Use this page to help you review and remember important skills needed for Chapter 29.

REGROUP ONES AND TENS (For Intervention, see p. H20.)

Regroup. Write the missing number.

1. 5 tens 15 ones = ■ tens 5 ones
2. 4 tens 17 ones = 5 tens ■ ones
3. 3 tens ■ ones = 4 tens 4 ones
4. ■ tens 12 ones = 6 tens 2 ones
5. 6 tens 20 ones = 8 tens ■ ones
6. 7 tens 24 ones = ■ tens 4 ones
7. ■ tens 20 ones = 5 tens 0 ones
8. 5 tens ■ ones = 7 tens 1 one

MULTIPLY 2-DIGIT NUMBERS (For Intervention, see p. H21.)

Find the product.

9. 14×3
10. 27×4
11. 56×5
12. 23×6
13. 71×9
14. 15×7
15. 43×3
16. 92×6
17. 85×2
18. 44×4
19. 58×8
20. 25×3
21. 62×5
22. 81×9
23. 64×6
24. 18×9
25. 53×6
26. 36×5
27. 75×2
28. 60×9
29. 45×3
30. 97×3
31. 33×8
32. 47×6

LESSON 1

Mental Math: Patterns in Multiplication

Quick Review

1. $6 \times 4 = \blacksquare$
2. $\blacksquare = 7 \times 2$
3. $\blacksquare = 8 \times 7$
4. $3 \times 8 = \blacksquare$
5. $10 \times 6 = \blacksquare$

Learn

PRODUCT PATTERNS Jena found different patterns in these multiplication exercises.

Jena

The product starts with the same digit as the first factor. The number of zeros in the second factor and the product increase by 1.

3 x 1 = 3	3 x 1 = 3
3 x 10 = 30	3 x 10 = 30
3 x 100 = 300	3 x 100 = 300
3 x 1,000 = 3,000	3 x 1,000 = 3,000

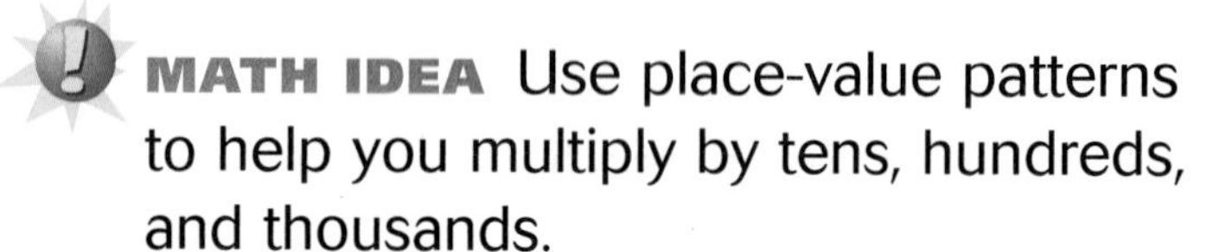

MATH IDEA Use place-value patterns to help you multiply by tens, hundreds, and thousands.

				THOUSANDS	HUNDREDS	TENS	ONES
2×4	=	2×4 ones	=				8
2×40	=	2×4 tens	=			8	0
2×400	=	2×4 hundreds	=		8	0	0
$2 \times 4{,}000$	=	2×4 thousands	=	8,	0	0	0

Examples

A
$4 \times 1 = 4$
$4 \times 10 = 40$
$4 \times 100 = 400$
$4 \times 1{,}000 = 4{,}000$

B
$3 \times 2 = 6$
$3 \times 20 = 60$
$3 \times 200 = 600$
$3 \times 2{,}000 = 6{,}000$

C
$2 \times 5 = 10$
$2 \times 50 = 100$
$2 \times 500 = 1{,}000$
$2 \times 5{,}000 = 10{,}000$

- Why are there more zeros in the products in Example C than in Examples A and B?

Check

1. **Explain** how to use the basic fact $6 \times 4 = 24$ to help you find 6×40.

Copy and complete. Use patterns and mental math to help.

2. $6 \times 1 = ■$	**3.** $3 \times 7 = ■$	**4.** $4 \times 5 = ■$
$6 \times 10 = ■$	$3 \times 70 = ■$	$4 \times 50 = ■$
$6 \times 100 = ■$	$3 \times 700 = ■$	$4 \times 500 = ■$
$6 \times 1{,}000 = ■$	$3 \times 7{,}000 = ■$	$4 \times 5{,}000 = ■$

Practice and Problem Solving

Copy and complete. Use patterns and mental math to help.

5. $5 \times 1 = ■$	**6.** $7 \times 5 = ■$	**7.** $10 \times 1 = ■$
$5 \times 10 = ■$	$■ \times 50 = 350$	$■ \times 10 = 100$
$5 \times 100 = ■$	$7 \times ■ = 3{,}500$	$10 \times 100 = ■$
$5 \times 1{,}000 = ■$	$7 \times 5{,}000 = ■$	$10 \times ■ = 10{,}000$

Use mental math and basic facts to find the product.

8. $2 \times 70 = ■$	**9.** $■ = 6 \times 400$	**10.** $4 \times 8{,}000 = ■$
11. $■ = 5 \times 300$	**12.** $5 \times 4{,}000 = ■$	**13.** $■ = 3 \times 90$

14. **REASONING** What basic multiplication fact could you use to help you find 5×600? Explain.

15. Emilio had 8 rolls of 50 pennies each. How many dollars is this?

16. **What's the Question?** A case of staples contains 80 boxes. The answer is 480 boxes.

Mixed Review and Test Prep

Divide. (p. 222)

17. $4\overline{)32}$ **18.** $7\overline{)28}$ **19.** $5\overline{)40}$

Add or subtract. (p. 472)

20. $5.2 + 3.1$ **21.** $3.02 - 1.01$ **22.** $7.9 - 5.8$

23. **TEST PREP** A parking lot can fit 132 cars. There are 84 cars in the lot. How many more cars can park in the lot? (p. 58)

A 36 **B** 46 **C** 48 **D** 58

EXTRA PRACTICE page H60, Set A

LESSON 2

Problem Solving Strategy
Find a Pattern

Quick Review

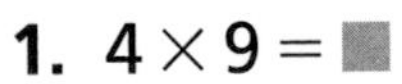

1. $4 \times 9 = ■$
2. $■ = 6 \times 7$
3. $3 \times 6 = ■$
4. $■ = 10 \times 8$
5. $2 \times 9 = ■$

PROBLEM A recycling container holds 200 pounds of newspaper. How many pounds of newspaper will 4 containers hold?

UNDERSTAND

- What are you asked to find?
- What information will you use?

PLAN

- What strategy can you use to solve the problem?

You can *find a pattern* to solve the problem.

SOLVE

- How can you use the strategy to solve the problem?

4	×	200	=	■
↑		↑		↑
number of containers		pounds in each container		pounds in all containers

Use a basic fact and place-value patterns to find the product.

$4 \times 2 = 8$
$4 \times 20 = 80$
$4 \times 200 = 800$

So, 4 containers will hold 800 pounds of newspaper.

CHECK

- What other strategy could you use to solve the problem?

Problem Solving Practice

Find a pattern to solve.

1. **What if** there are 6 containers to fill? How many pounds of newspaper will 6 containers hold?

2. A ferry boat carries 300 people on each trip. How many people can it carry on 6 trips?

A movie theater complex has 8 theaters. Each theater can seat 200 people. How many people can the theater complex seat?

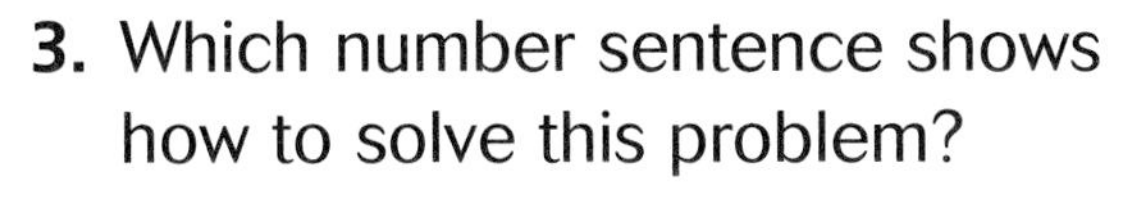

3. Which number sentence shows how to solve this problem?

   A $8 \times 2 = 16$
   B $2 \times 80 = 160$
   C $2 + 800 = 802$
   D $8 \times 200 = 1,600$

4. **What if** the complex builds 2 more theaters that seat 500 people each? How many people could the theater complex seat in all?

   F 520
   G 1,000
   H 2,600
   J 4,000

PROBLEM SOLVING STRATEGIES

- Draw a Diagram or Picture
- Make a Model or Act It Out
- Make an Organized List
- **Find a Pattern**
- Make a Table or Graph
- Predict and Test
- Work Backward
- Solve a Simpler Problem
- Write a Number Sentence
- Use Logical Reasoning

Mixed Strategy Practice

USE DATA For 5–6, use the table.

5. Make a bar graph of the data in the table. Be sure to write a title and labels.

6. How much longer is a blue whale than a whale shark? Write a number sentence to solve.

7. The difference between 2 numbers is 24. The product of the numbers is 112. What are the numbers?

8. Yuji wants to buy 5 boxes of crackers. Each box costs $3. How much will all of the boxes cost? Tell what operation you would use. Then solve the problem.

SEA ANIMAL LENGTHS

Animal	Length
Blue whale	70 feet
Bottlenose dolphin	10 feet
Killer whale	25 feet
Whale shark	45 feet

9. Timothy wrote this number pattern.
   7, 14, 9, 18, 13, 26, 21, 42, 37
   What is the rule and the next 4 numbers in his pattern?

LESSON

3 Estimate Products

Quick Review

1. $30 \times 9 = \blacksquare$
2. $\blacksquare = 6 \times 60$
3. $2 \times 70 = \blacksquare$
4. $\blacksquare = 100 \times 3$
5. $50 \times 4 = \blacksquare$

Learn

BUNCHES OF BISCUITS There are 28 dogs at the pet fair. If each dog gets 4 biscuits, about how many biscuits will be needed?

$4 \times 28 = \blacksquare$ or $\begin{array}{r} 28 \\ \times\ 4 \\ \hline \end{array}$

MATH IDEA When you don't need an exact answer, you can estimate.

Remember

To round a number:

- Decide on the place to be rounded.
- Look at the digit to its right.
- If the digit is less than 5, the digit being rounded stays the same.
- If the digit is 5 or more, the digit being rounded is increased by 1.

Example

STEP 1

Round the first factor to the nearest ten.

$\begin{array}{r} 28 \\ \times\ 4 \\ \hline \end{array} \rightarrow \begin{array}{r} 30 \\ \times\ 4 \\ \hline \end{array}$

STEP 2

Find the estimated product.

$\begin{array}{r} 30 \\ \times\ 4 \\ \hline 120 \end{array}$

So, about 120 biscuits are needed.

- Is the actual answer less than or greater than 120? How do you know?

More Examples

A Round to the nearest ten.

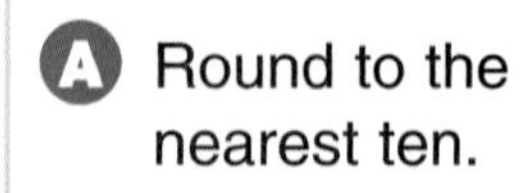

$\begin{array}{r} 52 \\ \times\ 7 \\ \hline \end{array} \rightarrow \begin{array}{r} 50 \\ \times\ 7 \\ \hline 350 \end{array}$

B Round to the nearest hundred.

$\begin{array}{r} 213 \\ \times\ \ 4 \\ \hline \end{array} \rightarrow \begin{array}{r} 200 \\ \times\ \ 4 \\ \hline 800 \end{array}$

Check

1. **Explain** how to estimate 7×65.

Estimate the product.

2. $\begin{array}{r} 47 \\ \times\ 3 \\ \hline \end{array}$

3. $\begin{array}{r} 58 \\ \times\ 4 \\ \hline \end{array}$

4. $\begin{array}{r} 396 \\ \times\ \ 5 \\ \hline \end{array}$

5. $\begin{array}{r} 64 \\ \times\ 2 \\ \hline \end{array}$

Practice and Problem Solving

Estimate the product.

6. $\begin{array}{r} 38 \\ \times\ 5 \\ \hline \end{array}$

7. $\begin{array}{r} 73 \\ \times\ 8 \\ \hline \end{array}$

8. $\begin{array}{r} 44 \\ \times\ 3 \\ \hline \end{array}$

9. $\begin{array}{r} 89 \\ \times\ 4 \\ \hline \end{array}$

10. $\begin{array}{r} 169 \\ \times\ \ 6 \\ \hline \end{array}$

11. $\begin{array}{r} 228 \\ \times\ \ 7 \\ \hline \end{array}$

12. $\begin{array}{r} 514 \\ \times\ \ 4 \\ \hline \end{array}$

13. $\begin{array}{r} 682 \\ \times\ \ 8 \\ \hline \end{array}$

14. $\begin{array}{r} 37 \\ \times\ 2 \\ \hline \end{array}$

15. $\begin{array}{r} 83 \\ \times\ 4 \\ \hline \end{array}$

16. $\begin{array}{r} 286 \\ \times\ \ 7 \\ \hline \end{array}$

17. $\begin{array}{r} 267 \\ \times\ \ 5 \\ \hline \end{array}$

18. $3 \times 29 = ■$

19. $■ = 3 \times 78$

20. $5 \times 173 = ■$

21. $■ = 7 \times 731$

22. $■ = 9 \times 395$

23. $6 \times 419 = ■$

24. $4 \times 59 = ■$

25. $■ = 8 \times 612$

26. **REASONING** Gretchen says that $7 \times 43 = 3{,}001$. Estimate to decide if you agree or disagree. Explain.

27. **REASONING** Rodrigo said that 2×413 is the same as $800 + 20 + 6$. Do you agree or disagree? Explain.

28. Mr. Cory has 29 students and Miss Jan has 27 students. Each student needs 4 jars for an experiment. About how many jars are needed for all students?

29. **? What's the Error?** Tui said that the product 7×488 is less than 2,800. Describe her error and give a more reasonable estimate.

Mixed Review and Test Prep

Subtract. (p. 58)

30. $\begin{array}{r} 942 \\ -258 \\ \hline \end{array}$

31. $\begin{array}{r} 604 \\ -465 \\ \hline \end{array}$

32. $\begin{array}{r} 731 \\ -283 \\ \hline \end{array}$

33. How many feet are in 3 yards? (p. 364)

34. **TEST PREP** A rabbit has a mass of 3 kg. How many grams is that? (p. 380)

A 3 g
B 30 g
C 300 g
D 3,000 g

EXTRA PRACTICE page H60, Set B

LESSON 4

Choose a Method

Quick Review

Find each product.

1. $\begin{array}{r} 32 \\ \times\ 8 \\ \hline \end{array}$ 2. $\begin{array}{r} 53 \\ \times\ 6 \\ \hline \end{array}$ 3. $\begin{array}{r} 36 \\ \times\ 4 \\ \hline \end{array}$

4. $\begin{array}{r} 24 \\ \times\ 7 \\ \hline \end{array}$ 5. $\begin{array}{r} 16 \\ \times\ 3 \\ \hline \end{array}$

Learn

You can find a product by using paper and pencil, a calculator, or mental math.

LUNCH TIME The airline catering service is preparing lunches for 2 flights. There are 188 passengers on each flight. How many lunches will the service prepare?

$2 \times 188 = \blacksquare$

Estimate. $2 \times 200 = 400$

Use Paper and Pencil The problem involves regrouping. So, paper and pencil is a good choice.

STEP 1

Multiply the ones.
2×8 ones = 16 ones
Regroup 16 ones as 1 ten 6 ones.

Hundreds	Tens	Ones
	1	
1	8	8
×		2
		6

STEP 2

Multiply the tens.
2×8 tens = 16 tens
16 tens + 1 ten = 17 tens
Regroup 17 tens as 1 hundred 7 tens.

Hundreds	Tens	Ones
1	1	
1	8	8
×		2
	7	6

STEP 3

Multiply the hundreds.
2×1 hundred = 2 hundreds
2 hundreds + 1 hundred = 3 hundreds

Hundreds	Tens	Ones
1	1	
1	8	8
×		2
3	7	6

So, the catering service will prepare 376 lunches.
Since 376 is close to 400, the product is reasonable.

Use a Calculator

$7 \times 975 = \blacksquare$

The product will be large. The problem involves regrouping. So, a calculator is a good choice.

7 × 9 7 5 = 6825.

Use Mental Math

$7 \times 31 = \blacksquare$

Since there is no regrouping, mental math is a good choice.

Think: 7 times 3 tens, or $7 \times 30 = 210$.
7 times 1 one, or $7 \times 1 = 7$.
So, $31 \times 7 = 210 + 7$, or 217.

More Examples

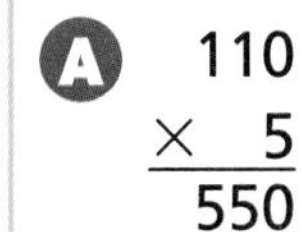 

A	B	C
$\begin{array}{r} 110 \\ \times\ 5 \\ \hline 550 \end{array}$	$\begin{array}{r} 382 \\ \times\ 3 \\ \hline 1{,}146 \end{array}$	$\begin{array}{r} 406 \\ \times\ 2 \\ \hline 812 \end{array}$

- Which problems can you solve by using mental math? Explain.
- Which method would you use to solve Example B?

MATH IDEA You can find a product by using paper and pencil, a calculator, or mental math. Choose the method that works best with the numbers in the problem.

Check

1. **Explain** what method you would use to multiply 742 by 2.

Tell what method you used. Then find the product.

2. $\begin{array}{r} 124 \\ \times\ 3 \\ \hline \end{array}$
3. $\begin{array}{r} 275 \\ \times\ 4 \\ \hline \end{array}$
4. $\begin{array}{r} 461 \\ \times\ 5 \\ \hline \end{array}$
5. $\begin{array}{r} 514 \\ \times\ 2 \\ \hline \end{array}$
6. $\begin{array}{r} 643 \\ \times\ 2 \\ \hline \end{array}$

LESSON CONTINUES

Practice and Problem Solving

Multiply. Tell what method you used.

7. 203×7 **8.** 411×9 **9.** 381×3 **10.** 428×5

11. 326×4 **12.** 187×3 **13.** 202×7 **14.** 423×3

Use mental math. Find the product.

15. 202×3 **16.** 413×3 **17.** 120×6 **18.** 112×4

19. 121×8 **20.** 131×3 **21.** 403×4 **22.** 241×5

Use a calculator. Find the product.

23. $4 \times 345 = \blacksquare$ **24.** $2 \times 499 = \blacksquare$ **25.** $\blacksquare = 6 \times 322$ **26.** $3 \times 279 = \blacksquare$

27. $5 \times 118 = \blacksquare$ **28.** $\blacksquare = 4 \times 421$ **29.** $8 \times 123 = \blacksquare$ **30.** $\blacksquare = 2 \times 391$

USE DATA For 31–35, use the table.

31. Diesta and her family took a train to Washington, D.C. How many seats were on the train in all?

32. How many coach seats are there on 4 trains?

33. **What's the Question?** The answer is 192 seats.

34. Claudia orders 1 pillow for each seat in the sleeper cars for 3 trains and in the deluxe car for 1 train. How many does she order?

35. **What's the Error?** George says there are 620 coach seats on two trains. Describe George's error. What is the correct answer?

TRAIN

Type of Car	Total Seats
Coach	360
Sleeper	168
Deluxe	90

36. Write a problem that can be solved using mental math. Explain your method.

37. **REASONING** Is 3×672 greater than or less than 4×415? Explain.

Mixed Review and Test Prep

Find the sum or difference.

38. $\begin{array}{r} 275 \\ +376 \\ \hline \end{array}$ (p. 42)

39. $\begin{array}{r} 382 \\ -158 \\ \hline \end{array}$ (p. 58)

40. $\begin{array}{r} 822 \\ -568 \\ \hline \end{array}$ (p. 58)

41. $\begin{array}{r} 936 \\ +826 \\ \hline \end{array}$ (p. 42)

42. $\frac{1}{5} + \frac{2}{5} = \blacksquare$ (p. 436)

43. $\frac{9}{10} - \frac{7}{10} = \blacksquare$ (p. 442)

44. **TEST PREP** What is the perimeter of a square when the length of one side is 3 cm? (p. 388)

A 3 cm
B 6 cm
C 9 cm
D 12 cm

45. **TEST PREP** What is 5,976 rounded to the nearest thousand? (p. 30)

F 5,000
G 5,900
H 5,980
J 6,000

PROBLEM SOLVING LINKUP . . . Math History

Lattice multiplication was used in Europe during the fourteenth and fifteenth centuries. You can use basic multiplication facts and a grid to find 7×465.

A Draw 1 row of 3 squares with a diagonal line in each square. Then write the factors as shown.

4 6 5
7

B Multiply each pair of digits.

$7 \times 5 = 35$
$7 \times 6 = 42$
$7 \times 4 = 28$

Write each product as shown.

4 6 5
2/8 4/2 3/5 7

C Start at the right. Add the digits in each diagonal. Regroup if needed.

So, $7 \times 465 = 3,255$.

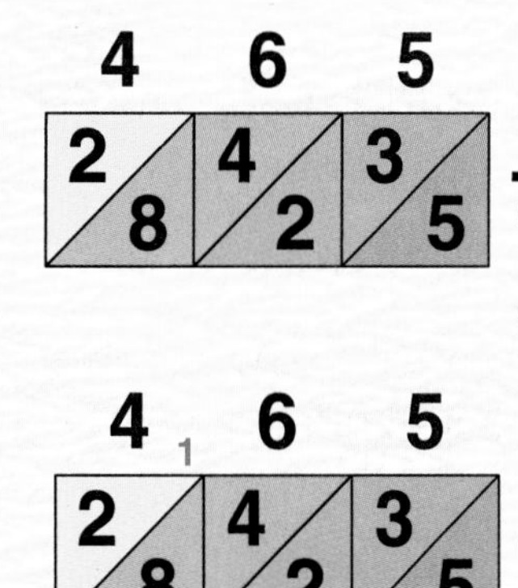

Use a grid to find the product.

1. $6 \times 722 = \blacksquare$
2. $8 \times 346 = \blacksquare$
3. $4 \times 296 = \blacksquare$
4. $3 \times 435 = \blacksquare$

EXTRA PRACTICE page H60, Set C

LESSON 5

Find Products Using Money

Quick Review

Find each product.

1. 57×6
2. 35×4
3. 27×5
4. 59×2
5. 27×7

Learn

PLANE FUN Mr. Haines and his son bought 3 puzzle books for their flight. Each book cost \$3.85. How much did the books cost in all?

Example

Estimate. \$3.85 rounded to the nearest dollar is \$4.00. 3 × \$4.00 = \$12.00

STEP 1

Write the problem in cents.

Think:

\$3.85 = 385 cents

$$\begin{array}{r} \$3.85 \\ \times \quad 3 \\ \hline \end{array} \rightarrow \begin{array}{r} 385 \\ \times \quad 3 \\ \hline \end{array}$$

STEP 2

Multiply to find the product in cents.

$$\begin{array}{r} {}^{2\,1} \\ 385 \\ \times \quad 3 \\ \hline 1{,}155 \end{array}$$

STEP 3

Write the product in dollars and cents.

1,155 cents = \$11.55

Since \$11.55 is close to \$12.00, the answer is reasonable.

So, the puzzle books cost \$11.55.

Remember

When you write money amounts, separate the dollars and cents with a decimal point and write a dollar sign on the left.

decimal point ↓
dollar sign → \$4.78

- How can estimation help you decide if your answer is reasonable?

MATH IDEA Multiply money amounts the same way you multiply whole numbers. Then write the product in dollars and cents.

More Examples

A
$$\begin{array}{r} {}^{2} \\ \$2.60 \\ \times \quad 4 \\ \hline \$10.40 \end{array}$$

B
$$\begin{array}{r} {}^{4\,4} \\ \$2.78 \\ \times \quad 6 \\ \hline \$16.68 \end{array}$$

C
$$\begin{array}{r} {}^{3} \\ \$4.16 \\ \times \quad 5 \\ \hline \$20.80 \end{array}$$

Check

1. **Explain** why you do not need to regroup when you multiply 5×1 in Example C.

Technology Link

More Practice: Use Calculating Crew, *Superhero Superstore,* Levels I and R.

Find the product in dollars and cents. Estimate to check.

2. $\$2.95 \times 6$ **3.** $\$1.38 \times 8$ **4.** $\$4.76 \times 2$ **5.** $\$3.75 \times 3$

Practice and Problem Solving

Find the product in dollars and cents. Estimate to check.

6. $\$5.09 \times 9$ **7.** $\$4.68 \times 5$ **8.** $\$3.29 \times 7$ **9.** $\$3.82 \times 2$

Find the product in dollars and cents.

10. $\$6.75 \times 4$ **11.** $\$2.87 \times 8$ **12.** $\$1.67 \times 9$ **13.** $\$8.43 \times 6$

14. $\$4.71 \times 3$ **15.** $\$7.65 \times 4$ **16.** $\$2.48 \times 9$ **17.** $\$1.82 \times 6$

18. $4 \times \$5.08 = ■$ **19.** $■ = 7 \times \$3.94$ **20.** $3 \times \$6.48 = ■$ **21.** $■ = 5 \times \$7.31$

22. $8 \times \$2.65 = ■$ **23.** $■ = 9 \times \$3.84$ **24.** $■ = 2 \times \$4.73$ **25.** $6 \times \$9.24 = ■$

26. Mario buys 3 dozen pens. Each dozen costs \$7.85. How much change will he get if he pays with \$25?

27. **Write a problem** that uses $4 \times \$5.72$. Solve the problem.

Mixed Review and Test Prep

Divide. (p. 222)

28. $7\overline{)49}$ **29.** $6\overline{)48}$ **30.** $5\overline{)45}$

31. Joey ate $\frac{2}{5}$ of a pie. Elise ate $\frac{1}{5}$. What fraction of the pie is left? (p. 442)

32. **TEST PREP** Trevor bought a book for \$12.95 and a video for \$15.95. How much did he spend in all? (p. 46)

A \$27.80 **C** \$28.80
B \$27.90 **D** \$28.90

EXTRA PRACTICE page H60, Set D

LESSON 6

Practice Multiplication

Quick Review

Find each product.

1. 46×3
2. 73×4
3. 28×5
4. 37×2
5. 89×6

Learn

ON THE ROAD AGAIN Mr. Wilson drove 1,853 miles from Orlando to Denver. Three days later, he drove back to Orlando along the same route. How many miles did Mr. Wilson drive in all?

Example

$2 \times 1{,}853 = \blacksquare$

Estimate. Round 1,853 to the nearest thousand. $2 \times 2{,}000 = 4{,}000$

STEP 1

Multiply the ones.

$$\begin{array}{r} 1{,}853 \\ \times \quad 2 \\ \hline 6 \end{array}$$

STEP 2

Multiply the tens.

$$\begin{array}{r} \scriptstyle 1 \quad \\ 1{,}853 \\ \times \quad 2 \\ \hline 06 \end{array}$$

STEP 3

Multiply the hundreds.

$$\begin{array}{r} \scriptstyle 1\ 1 \quad \\ 1{,}853 \\ \times \quad 2 \\ \hline 706 \end{array}$$

STEP 4

Multiply the thousands.

$$\begin{array}{r} \scriptstyle 1\ 1 \quad \\ 1{,}853 \\ \times \quad 2 \\ \hline 3{,}706 \end{array}$$

So, Mr. Wilson drove 3,706 miles. The estimate was 4,000, so the product is reasonable.

More Examples

A

$$\begin{array}{r} \scriptstyle 1\ \ 1 \quad \\ 1{,}625 \\ \times \quad 3 \\ \hline 4{,}875 \end{array}$$

B

$$\begin{array}{r} \scriptstyle 2\ 4\ 1 \quad \\ 1{,}482 \\ \times \quad 5 \\ \hline 7{,}410 \end{array}$$

C

$$\begin{array}{r} \scriptstyle 3\ 3 \quad \\ \$20.89 \\ \times \quad 4 \\ \hline \$83.56 \end{array}$$

Check

1. **Tell** why there are no tens to regroup in Example A.

Find the product. Estimate to check.

2. $1,186 \times 8$
3. $3,245 \times 3$
4. $1,514 \times 6$
5. $4,692 \times 2$

Practice and Problem Solving

Find the product. Estimate to check.

6. $1,231 \times 3$
7. $2,843 \times 6$
8. $2,418 \times 4$
9. $\$30.22 \times 3$
10. $4,395 \times 2$
11. $1,587 \times 6$
12. $\$31.25 \times 3$
13. $3,974 \times 2$
14. $1,279 \times 7$
15. $\$43.25 \times 2$
16. 746×4
17. 642×9
18. $\blacksquare = 5 \times 1,653$
19. $\blacksquare = 2 \times 3,742$

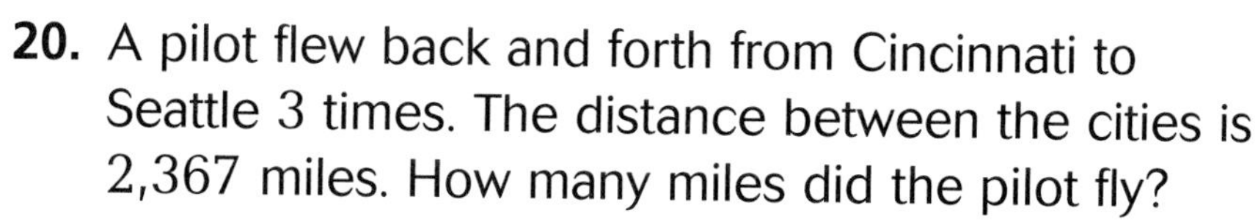

20. A pilot flew back and forth from Cincinnati to Seattle 3 times. The distance between the cities is 2,367 miles. How many miles did the pilot fly?

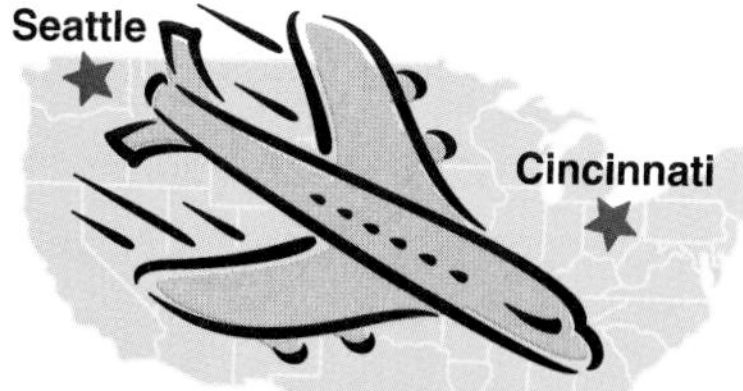

21. **Write About It** Find the missing digit. Explain how you found it.

$$\begin{array}{r} 2,1\blacksquare5 \\ \times \quad 4 \\ \hline 8,700 \end{array}$$

22. **ALGEBRA** Use estimation to find the missing factor. Then explain how you found the missing number. $\$9.30 \times \blacksquare = \$37.20$

Mixed Review and Test Prep

23. Write a rule for the table. Then copy and complete the table. (p. 168)

Tricycles	1	2	3	4
Wheels	3	6	■	■

Write the value of each blue digit. (p. 6)

24. 1,073 25. 4,861 26. 9,524

27. **TEST PREP** What is the area of this figure? (p. 394)

A 4 square units
B 8 square units
C 10 square units
D 12 square units

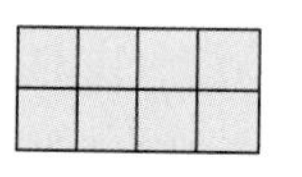

EXTRA PRACTICE page H60, Set E

CHAPTER 29

Review/Test

CHECK CONCEPTS

Copy and complete. Use patterns and mental math to help. (pp. 518–519)

1. $6 \times 1 = ■$
$6 \times 10 = ■$
$6 \times 100 = ■$
$6 \times 1{,}000 = ■$

2. $5 \times 3 = ■$
$5 \times 30 = ■$
$5 \times 300 = ■$
$5 \times 3{,}000 = ■$

3. $8 \times 4 = ■$
$8 \times 40 = ■$
$8 \times ■ = 3{,}200$
$■ \times 4{,}000 = 32{,}000$

CHECK SKILLS

Estimate the product. (pp. 522–523)

4. 45×6

5. 323×3

6. 386×5

7. 24×8

8. 592×7

Find the product. (pp. 524–531)

9. 201×4

10. 63×5

11. $\$2.07 \times 9$

12. $\$9.31 \times 7$

13. $1{,}285 \times 8$

14. 64×2

15. 531×4

16. $\$7.49 \times 3$

17. $\$21.96 \times 4$

18. $4{,}238 \times 5$

19. $4 \times 2{,}052 = ■$ **20.** $8 \times 527 = ■$ **21.** $3 \times \$6.15 = ■$ **22.** $6 \times \$42.17 = ■$

CHECK PROBLEM SOLVING

USE DATA For 23–25, use the bar graph. (pp. 520–521)

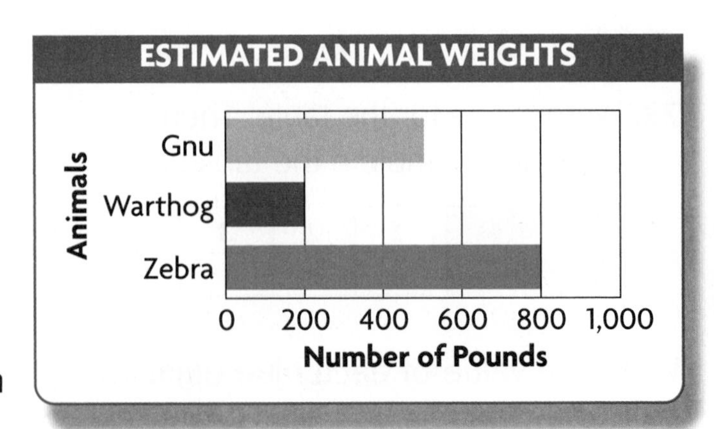

23. About how much would 4 warthogs weigh?

24. About how much would 3 zebras weigh?

25. Would 2 zebras weigh more than or less than 4 gnu? Explain.

Standardized Test Prep

TIP! **Look for important words.**
See item **3.**

An important word is *rounded.* When you are asked to round a number, it is important to find the place to which you need to round.

Also see problem **2,** p. H63.

For 1–8, choose the best answer.

1. $$\begin{array}{r} \$2.52 \\ \times \quad 9 \\ \hline \end{array}$$

A $23.68 B $22.68 C $22.58 D $22.50

2. What is the best name for this quadrilateral?

F square G rhombus H rectangle J parallelogram

3. What is 1,456 rounded to the nearest hundred?

A 2,000 B 1,500 C 1,400 D 1,000

4. How many inches are in 3 feet?

F 12 G 24 H 36 J NOT HERE

5. Alex biked 2.5 km on Monday and 3.2 km on Wednesday. How many kilometers did he bike on both days?

A 5.02 km B 5.52 km C 5.6 km D 5.7 km

6. Which is the best estimate for this product?

$$\begin{array}{r} 178 \\ \times \ 6 \\ \hline \end{array}$$

F 1,200 G 1,000 H 800 J 600

7. A cruise ship can carry 2,000 passengers on each cruise. There are 4 cruises in a month. How many passengers can the ship carry in a month?

A 4,000 B 6,000 C 8,000 D NOT HERE

8. Which angle is less than a right angle?

F

G

H 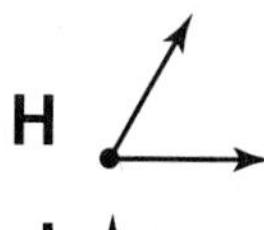

J 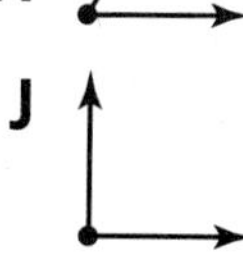

Write What You Know

9. Show and explain the steps for finding the product.

$$\begin{array}{r} 128 \\ \times \ 5 \\ \hline \end{array}$$

10. Mario had $3.25. He spent $1.97. Then he earned $2.50 for feeding the neighbor's cat. Explain how you would find out how much money Mario has now.

Divide Greater Numbers

Soccer is the world's most popular sport. A game like soccer was played in China about 2,500 years ago.

PROBLEM SOLVING

Suppose a soccer coach orders 6 pairs of shinguards. How much does each shinguard cost? Explain how to find the answer.

DATA LINK

SOCCER SUPPLIES

Item	Quantity	Cost
Corner flags	8	$64
Goal nets	4	$120
Soccer balls	9	$144
Air pumps	3	$60
Shinguards	6 pairs	$72

CHECK WHAT YOU KNOW

Use this page to help you review and remember important skills needed for Chapter 30.

MULTIPLICATION AND DIVISION FACTS (For Intervention, see p. H22.)

Find the product.

1. $6 \times 6 = ■$ **2.** $■ = 3 \times 9$ **3.** $6 \times 8 = ■$ **4.** $7 \times 3 = ■$

5. $6 \times 7 = ■$ **6.** $5 \times 5 = ■$ **7.** $■ = 2 \times 9$ **8.** $8 \times 8 = ■$

9. $■ = 9 \times 7$ **10.** $6 \times 4 = ■$ **11.** $5 \times 6 = ■$ **12.** $■ = 9 \times 9$

13. $8 \times 6 = ■$ **14.** $7 \times 7 = ■$ **15.** $4 \times 9 = ■$ **16.** $8 \times 7 = ■$

Find the quotient.

17. $12 \div 6 = ■$ **18.** $63 \div 7 = ■$ **19.** $■ = 24 \div 3$ **20.** $30 \div 6 = ■$

21. $■ = 35 \div 5$ **22.** $48 \div 6 = ■$ **23.** $16 \div 4 = ■$ **24.** $■ = 81 \div 9$

25. $40 \div 8 = ■$ **26.** $■ = 20 \div 5$ **27.** $42 \div 7 = ■$ **28.** $64 \div 8 = ■$

29. $8\overline{)56}$ **30.** $9\overline{)45}$ **31.** $4\overline{)32}$ **32.** $7\overline{)49}$

33. $6\overline{)54}$ **34.** $5\overline{)40}$ **35.** $6\overline{)42}$ **36.** $8\overline{)48}$

REGROUP HUNDREDS AND TENS (For Intervention, see p. H22.)

Write the missing number to regroup.

37. 7 hundreds 18 tens = ■ hundreds 8 tens

38. 5 hundreds 23 tens = 7 hundreds ■ tens

39. 1 hundred ■ tens = 2 hundreds 3 tens

40. 4 hundreds 34 tens = ■ hundreds 4 tens

LESSON 1

Mental Math: Patterns in Division

Quick Review

1. 2×3
2. 20×3
3. 200×3
4. 300×2
5. What number times 3 equals 900?

Learn

PATTERN SEARCH Ms. Ward showed these problems to her students. She asked them to look for a pattern.

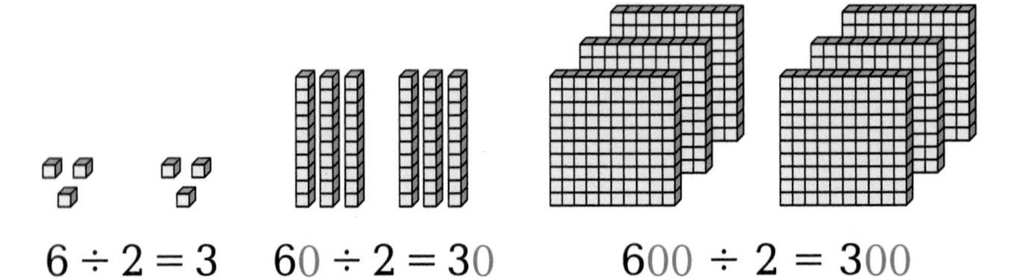

$6 \div 2 = 3$ $60 \div 2 = 30$ $600 \div 2 = 300$

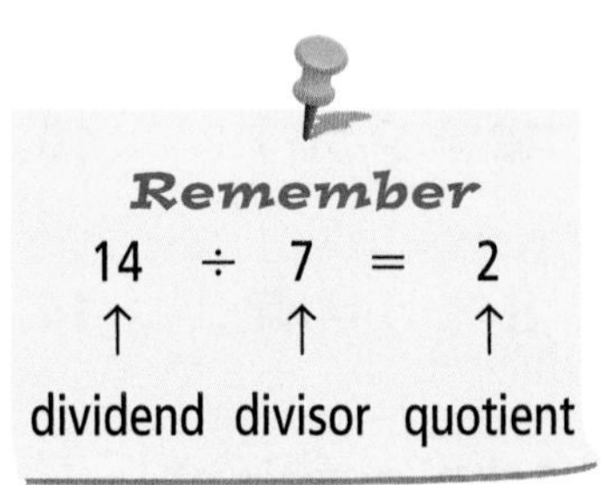

Mallory said the number of zeros in the dividend is the same as in the quotient. Do you agree?

Examples

A	B	C
$25 \div 5 = 5$	$12 \div 6 = 2$	$40 \div 8 = 5$
$250 \div 5 = 50$	$120 \div 6 = 20$	$400 \div 8 = 50$
$2{,}500 \div 5 = 500$	$1{,}200 \div 6 = 200$	$4{,}000 \div 8 = 500$

- In Example C, why is there one more zero in the dividend than in the quotient?

MATH IDEA You can use mental math and patterns to help you divide multiples of 10 and 100.

Check

1. **Explain** what happens to the number of zeros in the quotient as the number of zeros in the dividend increases.

Copy and complete. Use patterns and mental math.

2. $8 \div 8 = ■$
$80 \div 8 = ■$
$800 \div 8 = ■$

3. $12 \div 3 = ■$
$120 \div 3 = ■$
$1{,}200 \div 3 = ■$

4. $20 \div 4 = ■$
$200 \div 4 = ■$
$2{,}000 \div 4 = ■$

Practice and Problem Solving

Copy and complete. Use patterns and mental math.

5. $35 \div 5 = ■$
$350 \div 5 = ■$
$3{,}500 \div 5 = ■$

6. $28 \div ■ = 4$
$280 \div 7 = ■$
$■ \div 7 = 400$

7. $30 \div 6 = ■$
$■ \div 6 = 50$
$3{,}000 \div ■ = 500$

Use mental math and a basic fact to find the quotient.

8. $700 \div 7$
9. $360 \div 6$
10. $1{,}600 \div 4$
11. $800 \div 2$
12. $4{,}800 \div 8$
13. $400 \div 5$
14. $1{,}200 \div 4$
15. $30 \div 3$
16. $420 \div 7$

USE DATA For 17–18, use the graph.

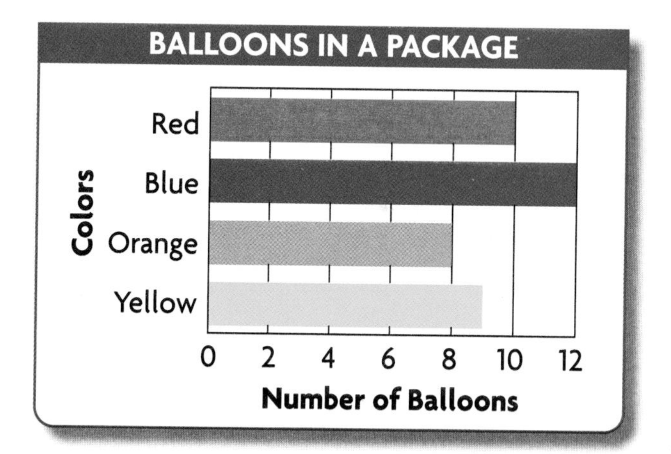

17. Grace bought a package of balloons. If each balloon costs 5 cents, how much does the package cost?

18. Jorge said he likes the red and blue balloons best. How many more red and blue balloons are in the package than orange and yellow?

19. ALGEBRA A grapefruit is 3 times as heavy as an apple. Together they weigh 24 ounces. What does each one weigh?

Mixed Review and Test Prep

20. $16 \div 2 = ■$
21. $64 \div 8 = ■$
22. $63 \div 9 = ■$
23. $45 \div 5 = ■$ (p. 222)

24. TEST PREP Leah started eating lunch at 11:55 A.M. It took her 30 minutes. When did Leah finish eating? (p. 98)

A 12:25 P.M.
B 12:30 P.M.
C 12:35 P.M.
D 12:55 P.M.

EXTRA PRACTICE page H61, Set A

LESSON 2

Estimate Quotients

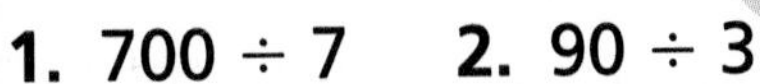

Quick Review

1. $700 \div 7$
2. $90 \div 3$
3. $200 \div 4$
4. $1{,}800 \div 2$
5. $120 \div 3$

Learn

PUNCH LINE Mrs. Allison is making 135 cups of punch. If she divides the punch evenly among 3 punch bowls, about how many cups will be in each bowl?

When you do not need an exact answer, you can *estimate* to find the quotient.

Example

Estimate. $135 \div 3$ or $3\overline{)135}$

STEP 1

Look at the first two digits.

$135 \div 3 = ■$

STEP 2

Think of a basic fact that is close to $13 \div 3$.

$12 \div 3 = 4$

STEP 3

Then use the pattern to estimate.

$12 \div 3 = 4$

$120 \div 3 = 40$

So, about 40 cups of punch will be in each bowl.

- Will the exact answer be more than or less than 40 cups? Explain.

Check

1. **Tell** what estimate you would get if you used $15 \div 3 = 5$ as the basic fact to find the number of cups of punch.

Estimate each quotient. Write the basic fact you used to find the estimate.

2. $181 \div 2 = ■$
3. $501 \div 5 = ■$
4. $374 \div 6 = ■$
5. $8\overline{)490}$
6. $3\overline{)268}$
7. $7\overline{)223}$

▶ Practice and Problem Solving

Estimate each quotient. Write the basic fact you used to find the estimate.

8. $143 \div 5 = ■$

9. $174 \div 9 = ■$

10. $161 \div 2 = ■$

11. $253 \div 6 = ■$

Estimate the quotient.

12. $365 \div 5 = ■$

13. $116 \div 4 = ■$

14. $493 \div 8 = ■$

15. $618 \div 6 = ■$

16. $784 \div 2 = ■$

17. $407 \div 4 = ■$

18. $3\overline{)876}$

19. $7\overline{)644}$

20. $9\overline{)716}$

21. $5\overline{)442}$

22. $8\overline{)331}$

23. $8\overline{)552}$

For 24–26, find an estimate to solve each problem.

24. The students raised $226 to buy 3 new trees for their school. About how much can they spend for each tree?

25. Kathy put 102 photos into her album. She put 3 photos on each page. About how many pages have photos?

26. There are 4 large tables at the sports banquet. There are 163 team members. About how many team members will be at each table?

27. **REASONING** Masao saved $1.50 more this week than he saved last week. He saved $2.25 last week. How much money in all has he saved in the two weeks?

28. **? What's the Error?** A clerk has 196 shirts to display on 6 racks. She wants to display about the same number of shirts on each rack. She says she can put about 20 shirts on each rack. What's her error?

Mixed Review and Test Prep

Complete. (p. 188)

29. $■ \times 3 = 30 \div 5$

30. $24 \div ■ = 4 \times 2$

31. $2 \times 2 = ■ \div 4$

32. $14 \div 2 = 1 \times ■$

33. **TEST PREP** Tara's quilt is 3 feet long and 4 feet wide. What is the perimeter of her quilt? (p. 390)

A 16 feet
B 14 feet
C 12 feet
D 10 feet

EXTRA PRACTICE page H61, Set B

LESSON 3

Place the First Digit in the Quotient

Quick Review

1. $18 \div 3$
2. $24 \div 8$
3. $30 \div 5$
4. $42 \div 6$
5. How many sixes are in 36?

Learn

ROLL CALL There are 6 classes of third graders. Each class has the same number of students. There are 192 third graders in all. How many students are in each class?

Since each class has an equal number of students, you can divide to find the answer.

Example

Divide. $192 \div 6 = \blacksquare$ or $6\overline{)192}$

Estimate. $192 \div 6 \rightarrow 180 \div 6 = 30$

STEP 1

Decide where to place the first digit in the quotient.

$6\overline{)192}$ $1 < 6$, so look at the tens.

$\blacksquare$
$6\overline{)192}$ $19 > 6$, so use 19 tens.

STEP 2

Divide the 19 tens by 6.

$$\begin{array}{r} 3 \\ 6\overline{)192} \\ -18 \\ \hline 1 \end{array}$$

Multiply.
Subtract.
Compare. $1 < 6$
The difference must be less than the divisor.

STEP 3

Bring down the 2 ones. Divide the 12 ones.

$$\begin{array}{r} 32 \\ 6\overline{)192} \\ -18\downarrow \\ \hline 12 \\ -12 \\ \hline 0 \end{array}$$

Multiply.
Subtract.
Compare. $0 < 6$

So, each class has 32 students.

Since 32 is close to the estimate of 30, the quotient is reasonable.

- How can you use multiplication to check your answer?

MATH IDEA When you divide, you must decide where to place the first digit in the quotient.

Check

1. **Explain** how an estimate can help you decide how many digits will be in the quotient.

Copy. Place an X where the first digit in the quotient should be.

2. $5\overline{)65}$ 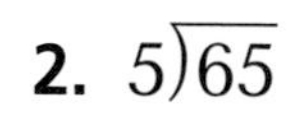 **3.** $3\overline{)324}$ **4.** $4\overline{)56}$ **5.** $6\overline{)138}$

Practice and Problem Solving

Copy. Place an X where the first digit in the quotient should be.

6. $4\overline{)92}$ **7.** $8\overline{)104}$ **8.** $6\overline{)650}$ **9.** $7\overline{)98}$

Find the quotient.

10. $5\overline{)155}$ **11.** $2\overline{)72}$ **12.** $3\overline{)84}$ **13.** $4\overline{)92}$

14. $8\overline{)680}$ **15.** $4\overline{)136}$ **16.** $7\overline{)294}$ **17.** $5\overline{)755}$

18. $3\overline{)72}$ **19.** $4\overline{)472}$ **20.** $9\overline{)819}$ **21.** $6\overline{)294}$

22. $7\overline{)154}$ **23.** $2\overline{)78}$ **24.** $5\overline{)90}$ **25.** $9\overline{)558}$

26. Write a problem in which the dividend has 3 digits and the first digit of the quotient is in the tens place.

27. **What's the Question?** Ruth's dogs eat 5 pounds of dog food in a week. The answer is 12 weeks.

28. Bonnie has $2.35 in pennies. John has $1.20 less than Bonnie. How much money do they have altogether?

Mixed Review and Test Prep

What time does each clock show? (p. 94)

29.

30.

31.

32.

33. **TEST PREP** Max's team had practice 3 days a week for 9 weeks. Max missed 2 practices. How many practices did he go to? (p. 122)

A 3 **B** 24 **C** 25 **D** 27

EXTRA PRACTICE page H61, Set C

LESSON

4 Practice Division of 3-Digit Numbers

Quick Review

1. $56 \div 7$
2. $45 \div 5$
3. $24 \div 6$
4. $72 \div 9$
5. How many twos are in 15? What is left over?

Learn

FAIR TRADE Roger and his sister have 358 trading cards. They want to divide them evenly. How many cards does each one get?

Example

Divide. $358 \div 2 = ■$ or $2\overline{)358}$

Estimate. $358 \div 2 \rightarrow 400 \div 2 = 200$

STEP 1

Decide where to place the first digit in the quotient.

■
2)358 — $3 > 2$, so divide the hundreds.

STEP 2

Divide 3 hundreds by 2.

```
   1
2)358
 -2      Multiply.
  1      Subtract.
         Compare.
         1 < 2
```

STEP 3

Bring down the 5 tens. Divide 15 tens.

```
   17
2)358
 -2↓     Multiply.
  15     Subtract.
 -14     Compare.
   1     1 < 2
```

STEP 4

Bring down the 8 ones. Divide 18 ones.

```
   179
2)358
 -2
  15
 -14↓    Multiply.
   18    Subtract.
  -18    Compare.
    0    0 < 2
```

So, each one gets 179 trading cards.

• Is the quotient reasonable? Explain.

More Examples

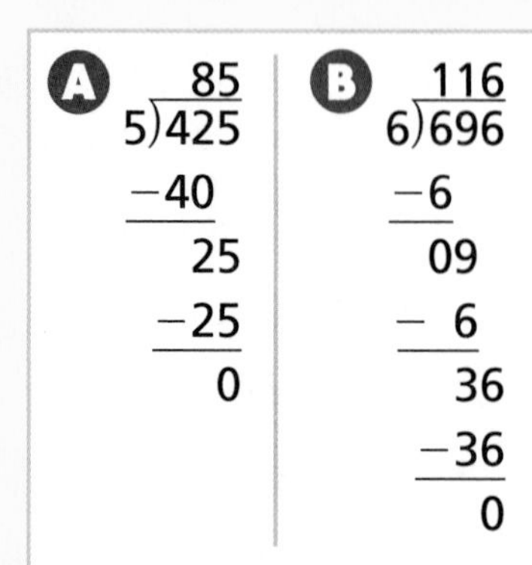

C You can also use a calculator to divide greater numbers.

$161 \div 7 = ■$

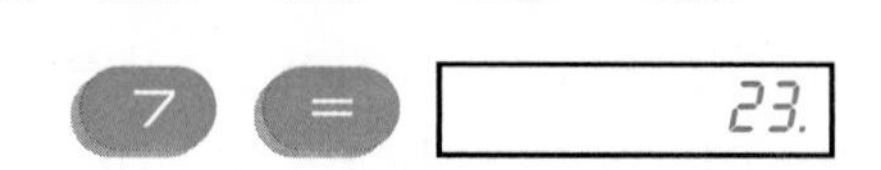

Check

1. **Discuss** how you can decide where to place the first digit in a quotient.

Technology Link

More Practice: Use Mighty Math Calculating Crew, *Nick Knack*, Level Q.

Find the quotient.

2. $5\overline{)365}$ 3. $3\overline{)231}$ 4. $4\overline{)336}$ 5. $2\overline{)568}$

Practice and Problem Solving

Find the quotient.

6. $7\overline{)448}$ 7. $6\overline{)276}$ 8. $5\overline{)335}$ 9. $2\overline{)384}$

10. $5\overline{)145}$ 11. $7\overline{)847}$ 12. $6\overline{)486}$ 13. $4\overline{)644}$

14. $6\overline{)324}$ 15. $9\overline{)819}$ 16. $7\overline{)294}$ 17. $4\overline{)904}$

18. Katie delivers 497 newspapers in 7 days. She delivers the same number each day. How many papers does she deliver each day?

19. **Write About It** Explain why it is important to know the basic division facts when you are dividing 3-digit dividends.

20. **What's the Error?** Raul wrote this problem to show his sister. What did he do wrong? Correct his error.

$$\begin{array}{r} 122 \text{ r4} \\ 4\overline{)492} \\ -4 \\ \hline 09 \\ -\ 8 \\ \hline 12 \\ -\ 8 \\ \hline 4 \end{array}$$

21. **REASONING** Julia bought some trading cards for $3.00. She paid for them with the same number of dimes as nickels. How many of each coin did she use?

Mixed Review and Test Prep

Add. (p. 436)

22. $\frac{1}{5} + \frac{2}{5} = \blacksquare$ 23. $\frac{2}{4} + \frac{2}{4} = \blacksquare$

24. $\frac{2}{9} + \frac{3}{9} = \blacksquare$ 25. $\frac{1}{3} + \frac{1}{3} = \blacksquare$

26. **TEST PREP** Coach Sloan bought 8 baseballs for $32. What was the cost for one baseball? (p. 226)

A $8 C $4
B $6 D $2

LESSON 5

Divide Amounts of Money

Quick Review

1. $7 ÷ 7
2. $20 ÷ 4
3. $9 ÷ 3
4. $8 ÷ 2
5. $18 ÷ 6

MATERIALS
play money

Explore

Rafi spent $3.74 for 2 toy cars. Each car was the same price. How much did each car cost?

Activity

Use division to find the cost of one car.

Divide. $\$3.74 \div 2 = \blacksquare$ or $2\overline{)\$3.74}$

Estimate. $\$3.74 \div 2 \rightarrow \$4.00 \div 2 = \$2.00$

STEP 1

Model the amount Rafi spent.

Dollars

Dimes

Pennies

STEP 2

Divide the dollars into 2 groups. Regroup 1 dollar as 10 dimes.

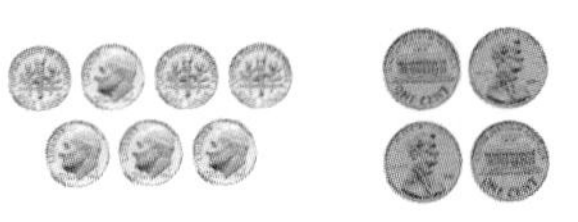

STEP 3

Divide the dimes. Regroup 1 dime as 10 pennies.

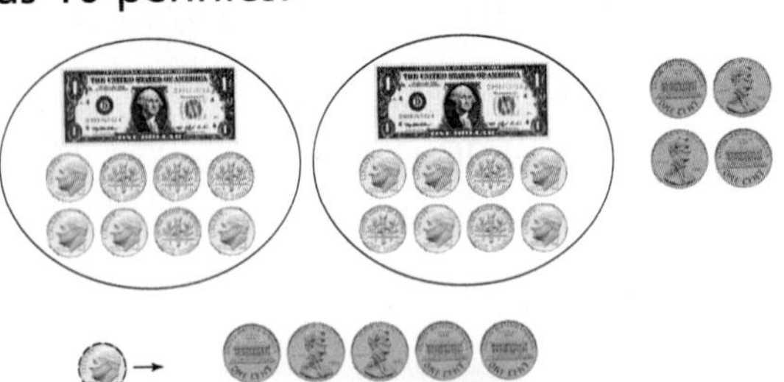

STEP 4

Divide the pennies.

$1.87 $1.87

So, Rafi spent $1.87 on each car.

Try It

Use play money to model each problem. Record your steps.

a. $\$5.26 \div 2 = \blacksquare$

b. $\$3.64 \div 4 = \blacksquare$

Connect

You can divide money amounts like whole numbers. Place the dollar sign and decimal point in the quotient.

Example

Divide. $4.92 ÷ 3 = ■ or $3\overline{)\$4.92}$

STEP 1

Divide the 4 dollars.

$$\begin{array}{r} 1 \\ 3\overline{)\$4.92} \\ -3 \\ \hline 1 \end{array}$$

Multiply.
Subtract.
Compare.
$1 < 3$

STEP 2

Bring down the 9 dimes. Divide 19 dimes.

$$\begin{array}{r} 16 \\ 3\overline{)\$4.92} \\ -3\downarrow \\ \hline 19 \\ -18 \\ \hline 1 \end{array}$$

Multiply.
Subtract.
Compare.
$1 < 3$

STEP 3

Bring down the 2 pennies. Divide the 12 pennies.

$$\begin{array}{r} 164 \\ 3\overline{)\$4.92} \\ -3 \\ \hline 19 \\ -18\downarrow \\ \hline 12 \\ -12 \\ \hline 0 \end{array}$$

Multiply.
Subtract.
Compare.
$0 < 3$

STEP 4

Write the quotient with a dollar sign and a decimal point.

$$\begin{array}{r} \$1.64 \\ 3\overline{)\$4.92} \\ -3 \\ \hline 19 \\ -18 \\ \hline 12 \\ -12 \\ \hline 0 \end{array}$$

- Why is it important to keep each digit in the quotient in the correct place-value position?

Practice and Problem Solving

Find the quotient.

1. $2\overline{)\$1.76}$
2. $3\overline{)\$8.04}$
3. $6\overline{)\$8.28}$
4. $2\overline{)\$9.26}$
5. $9\overline{)\$5.13}$
6. $8\overline{)\$9.36}$
7. $3\overline{)\$1.95}$
8. $4\overline{)\$8.48}$
9. **REASONING** Five ounces of cherries cost $1.95. Ali will buy the cherries if they are less than $0.50 per ounce. Should Ali buy them? Explain.
10. **? What's the Error?** Bob paid $3.00 for a bag of popcorn that he shared with 4 friends. Bob said each friend owes him $0.06. What's his error?

Mixed Review and Test Prep

Find the missing addend. (p. 68)

11. 25 + ■ = 53
12. ■ + 32 = 47
13. ■ + 19 = 38
14. 84 + ■ = 101

15. **TEST PREP** What is the missing factor in 3 × ■ × 4 = 24? (p. 142)

A 5 **B** 4 **C** 3 **D** 2

LESSON 6

Problem Solving Strategy
Solve a Simpler Problem

Quick Review

1. $54 \div 6$
2. $40 \div 5$
3. $72 \div 9$
4. $63 \div 7$
5. $48 \div 8$

PROBLEM Sunrise Elementary School is having its annual field day. There are 640 students in the school. The students are placed in 8 groups to participate in the events. How many students are in each group?

UNDERSTAND

- What do you know?
- What are you asked to find?

PLAN

- What strategy can you use to solve the problem?

  You can use *solve a simpler problem.*

SOLVE

- How can you use the strategy *solve a simpler problem?*

  Think: $640 \div 8 = \blacksquare$.
  You can use the basic fact $64 \div 8$ to find $640 \div 8$.
  $64 \div 8 = 8$, so $640 \div 8 = 80$.

So, the students are in 8 groups of 80.

CHECK

- How can you decide if your answer is correct?

Problem Solving Practice

For 1–4, *solve a simpler problem*.

1. **What if** there were 720 students in the school and they needed to be placed into 8 groups? How many students would be in each group?
2. Betty has $4.00 in nickels. How many nickels does she have?

José has $12.00. He wants to buy model plane kits that cost $3.89 each. How many kits can he buy?

3. What simpler problem can José use to help him decide how many kits he can buy?

   A $12 - 3 = 9$ **C** $12 \times 3 = 36$
   B $3 + 3 = 6$ **D** $12 \div 4 = 3$

4. How many kits can José buy?

   F 2 **G** 3 **H** 4 **J** 5

PROBLEM SOLVING STRATEGIES

- Draw a Diagram or Picture
- Make a Model or Act It Out
- Make an Organized List
- Find a Pattern
- Make a Table or Graph
- Predict and Test
- Work Backward
- **Solve a Simpler Problem**
- Write a Number Sentence
- Use Logical Reasoning

Mixed Strategy Practice

USE DATA For 5–7, use the table.

HAPPY HOLLOW CAMP
Daily Rentals
Canoe $25.00
Raft $6.25
Fishing pole $3.50

5. The Thomas family rented 1 canoe and 2 fishing poles for 3 days each. How much did they spend in all?
6. The Davis family rented a canoe and a raft for 4 days. How much did they spend?
7. **REASONING** The Murray family rented a canoe and 3 fishing poles. They paid $71.00 in all. For how many days did they rent the equipment?
8. The Clarks arrived home at 5:00 P.M. They had spent 2 hours fishing, 45 minutes driving to the camp, and 45 minutes driving home. At what time did they leave home?

CHAPTER 30

Review/Test

CHECK CONCEPTS

Copy and complete. Use patterns and mental math to help. (pp. 536–537)

1. $6 \div 6 = ■$
$60 \div 6 = ■$
$600 \div 6 = ■$

2. $9 \div 3 = ■$
$■ \div 3 = 30$
$900 \div ■ = 300$

3. $■ \div 4 = 3$
$120 \div 4 = ■$
$1{,}200 \div ■ = 300$

CHECK SKILLS

Estimate the quotient. (pp. 538–539)

4. $325 \div 4 = ■$
5. $573 \div 7 = ■$
6. $268 \div 3 = ■$

7. $8\overline{)497}$
8. $5\overline{)509}$
9. $9\overline{)362}$

Find the quotient. (pp. 540–545)

10. $2\overline{)24}$
11. $6\overline{)360}$
12. $8\overline{)312}$

13. $2\overline{)180}$
14. $3\overline{)213}$
15. $5\overline{)255}$

16. $4\overline{)252}$
17. $9\overline{)747}$
18. $3\overline{)744}$
19. $6\overline{)678}$

20. $5\overline{)\$2.40}$
21. $8\overline{)\$9.52}$
22. $2\overline{)\$9.78}$
23. $3\overline{)\$8.82}$

CHECK PROBLEM SOLVING

Solve. (pp. 546–547)

24. Ms. Jones stopped after work to buy apples for school lunches. She had $6.00 and wanted to buy 8 apples that cost $0.69 each. Did she have enough money? Explain.

25. The hobby shop ordered 120 model cars. The models come packaged with 4 cars in a box. How many boxes of cars will the shop receive?

Standardized Test Prep

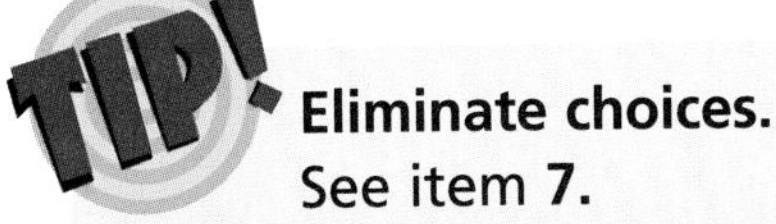

Eliminate choices.
See item **7**.
After you estimate your answer, you can eliminate unreasonable choices. Only one answer choice is greater than $1.00. Use your estimate to decide if this is the most reasonable amount.

Also see problem **5,** p. H64.

For 1–8, choose the best answer.

1. $7\overline{)945}$

A 135
B 145
C 149
D NOT HERE

2. $28 \div 3 = \blacksquare$

F 8 r3
G 9
H 9 r1
J 9 r2

3. $8\overline{)184}$

A 20
B 24
C 25
D NOT HERE

4. $5\overline{)\$9.35}$

F $1.05
G $1.80
H $1.85
J $1.87

5. Jerry had 33 feet of fencing. How many yards is that?

A 11 yards
B 15 yards
C 99 yards
D 103 yards

6. Akira has 270 pennies. Suzanne has the same amount of money in nickels. How many nickels does Suzanne have?

F 27
G 54
H 135
J 270

7. Brandon needs to buy 18 erasers for some friends. Each eraser costs 9 cents. How much money does Brandon need?

A $0.02
B $0.09
C $0.27
D $1.62

8. An airplane has 140 seats. Each row of the plane has 4 seats. How many rows of seats does the airplane have?

F 35
G 40
H 60
J 560

9. Greg says $1{,}800 \div 6 = 300$. Is he correct? Explain.

10. Estimate the quotient. Explain how you found your estimate. $8\overline{)736}$

PROBLEM SOLVING MATH DETECTIVE

Mystery Signs

REASONING Use the clues to help you find the mystery signs. Some cases may have more than one solution.

Hints: Use each number or operation only once.

Complete the operation in ( ) first.

Example

A Write +, −, ×, or ÷ for each $\bullet$ so the number sentence is true.

$(7 \bullet 3) \bullet 8 = 29$

Answer:

$(7 \times 3) + 8 = 29$

B Write 3, 42, or 2 for each $\blacksquare$ to make the number sentence true.

$(\blacksquare \div \blacksquare) + \blacksquare = 16$

Answer:

$(42 \div 3) + 2 = 16$

Cases 1–6

Write +, −, ×, or ÷ for each $\bullet$ so the number sentence is true.

1. $(32 \bullet 4) \bullet 1 = 7$
2. $(40 \bullet 5) \bullet 11 = 211$
3. $(9 \bullet 39) \bullet 8 = 6$
4. $(90 \bullet 6) \bullet 1 = 15$
5. $49 \bullet (4 \bullet 3) = 7$
6. $5 \bullet (50 \bullet 10) = 25$

Cases 7–9

Write a number for each $\blacksquare$ to make the number sentence true.

7. $(\blacksquare \times \blacksquare) + \blacksquare = 26$ Use 2, 4, and 6.
8. $(\blacksquare - \blacksquare) \div \blacksquare = 25$ Use 2, 8, and 58.
9. $\blacksquare + (\blacksquare \times \blacksquare) = 65$ Use 2, 3, and 21.

STRETCH YOUR THINKING **Make the number sentence true. Write + or × for each $\bullet$. Write 6, 7, or 5 for each $\blacksquare$.**

$(\blacksquare \bullet \blacksquare) \bullet \blacksquare = 41$

CASE CLOSED

Challenge

Prime Numbers

You can show factors of a number by using square tiles.

Use 16 square tiles. Make arrays to show all the factors of 16.

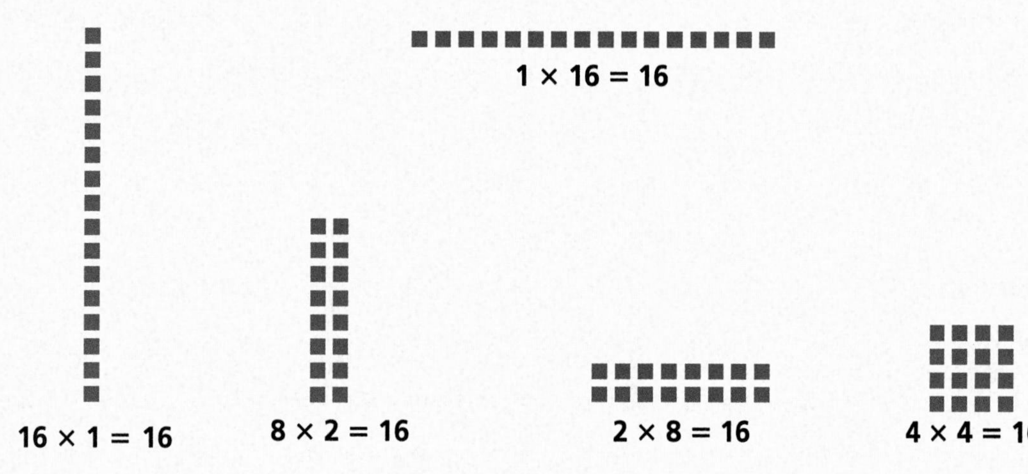

So, the number 16 has five factors: 1, 2, 4, 8, and 16.

The number 16 is a **composite number** because it has more than two factors.

Use 7 square tiles. Make arrays to show all the factors of 7.

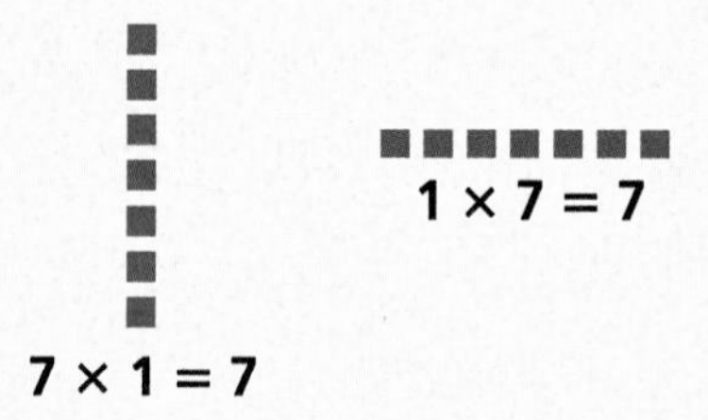

The number 7 has only two factors: 1 and 7. So, 7 is a prime number. **Prime numbers** have only 2 factors, 1 and the number itself.

Try It

Use square tiles. Make arrays to show all the factors for each. Write *prime* or *composite*.

1. 6 **2.** 3 **3.** 15 **4.** 20 **5.** 13

Study Guide and Review

VOCABULARY

Choose the best term from the box.

array
divisor
remainder

1. In division, the amount left over is called the __?__. (p. 502)

STUDY AND SOLVE

Chapter 27

Multiply 2-digit numbers by 1-digit numbers.

Find the product. $9 \times 36 = \blacksquare$

$$\begin{array}{r} {}^{5} \\ 36 \\ \times\ 9 \\ \hline 324 \end{array}$$

- Multiply the ones. Regroup into tens and ones.
- Multiply the tens. Add the regrouped tens.
- Regroup the tens as hundreds and tens.

So, $9 \times 36 = 324$.

Find each product. (pp. 490–495)

2. $\begin{array}{r} 15 \\ \times\ 2 \\ \hline \end{array}$
3. $\begin{array}{r} 26 \\ \times\ 5 \\ \hline \end{array}$
4. $\begin{array}{r} 38 \\ \times\ 4 \\ \hline \end{array}$
5. $\begin{array}{r} 42 \\ \times\ 8 \\ \hline \end{array}$
6. $\begin{array}{r} 67 \\ \times\ 6 \\ \hline \end{array}$
7. $\begin{array}{r} 88 \\ \times\ 9 \\ \hline \end{array}$
8. There are 24 hours in a day. How many hours are there in 9 days?

Chapter 28

Divide 2-digit dividends.

Divide and check. $97 \div 8 = \blacksquare$

$$\begin{array}{r} 12 \text{ r1} \\ 8\overline{)97} \\ -8\downarrow \\ \hline 17 \\ -16 \\ \hline 1 \end{array}$$

- Divide the 9 tens.
- Bring down the 7 ones.
- Divide the 17 ones.
- Write the remainder beside the quotient.

So, $97 \div 8 = 12$ r1.

Check: $8 \times 12 = 96$; $96 + 1 = 97$

Divide and check. (pp. 506–511)

9. $54 \div 6 = \blacksquare$
10. $17 \div 5 = \blacksquare$
11. $25 \div 4 = \blacksquare$
12. $42 \div 3 = \blacksquare$
13. $2\overline{)37}$
14. $6\overline{)90}$
15. $4\overline{)45}$
16. $5\overline{)74}$
17. $3\overline{)99}$
18. $8\overline{)67}$

Chapter 29

Multiply 3-digit numbers by 1-digit numbers.

Find the product. $7 \times 482 = \blacksquare$

$$\begin{array}{r} {}^{51} \\ 482 \\ \times \quad 7 \\ \hline 3{,}374 \end{array}$$

- Multiply the ones. Regroup if needed.
- Multiply the tens. Regroup if needed.
- Multiply the hundreds. Regroup if needed.

So, $7 \times 482 = 3{,}374$.

Find each product. (pp. 524–531)

19. $\begin{array}{r} 450 \\ \times \quad 2 \\ \hline \end{array}$

20. $\begin{array}{r} \$2.16 \\ \times \quad 5 \\ \hline \end{array}$

21. $\begin{array}{r} 412 \\ \times \quad 3 \\ \hline \end{array}$

22. $\begin{array}{r} \$6.55 \\ \times \quad 4 \\ \hline \end{array}$

23. $\begin{array}{r} 428 \\ \times \quad 7 \\ \hline \end{array}$

24. $\begin{array}{r} 391 \\ \times \quad 1 \\ \hline \end{array}$

Chapter 30

Divide 3-digit dividends.

$147 \div 3 = \blacksquare$

$$\begin{array}{r} 49 \\ 3\overline{)147} \\ -12\downarrow \\ \hline 27 \\ -27 \\ \hline 0 \end{array}$$

So, $147 \div 3 = 49$.

$318 \div 2 = \blacksquare$

$$\begin{array}{r} 159 \\ 2\overline{)318} \\ -2\downarrow \\ \hline 11 \\ -10\downarrow \\ \hline 18 \\ -18 \\ \hline 0 \end{array}$$

So, $318 \div 2 = 159$.

Divide. (pp. 540–545)

25. $2\overline{)166}$

26. $3\overline{)291}$

27. $5\overline{)\$2.45}$

28. $7\overline{)434}$

29. $6\overline{)306}$

30. $8\overline{)736}$

31. $4\overline{)\$4.68}$

32. $8\overline{)544}$

33. $9\overline{)801}$

34. $3\overline{)\$7.23}$

PRACTICE AND PROBLEM SOLVING

Solve. (pp. 496–497, 512–513)

35. Paul bought 3 slices of pizza for $1.49 each. He had a $10 bill. What operations would you use to find out how much change Paul received? What was Paul's change?

36. Mrs. Ramsey bakes pies at a bakery. She has 52 cups of apples. If 6 cups of apples are needed for one pie, how many pies can she make? Explain.

PERFORMANCE ASSESSMENT

TASK A • TABLETOP TILES

Evan has 90 square tiles to make a rectangular design for a tabletop. So far, his design has 4 rows with 18 tiles in each row.

a. How many tiles has Evan used so far? Draw a model and write a multiplication sentence to show your answer is correct.

b. How many rows of 18 tiles can Evan make with 90 tiles?

c. Draw a model and write a multiiplication sentence to show a different way to use 90 tiles to make a rectangular design.

TASK B • SHELL ART

The table shows the number of shells Jarrod, Gina, and Tim have collected. They will use the shells to make flower pictures. It takes 4 shells to make a small flower and 8 shells to make a large flower.

SHELLS WE COLLECTED	
Jarrod	62
Gina	45
Tim	60

a. Jarrod wants to make all large flowers. Draw a model to show how he can use his shells. How many large flowers can he make? How many shells will he have left over?

b. Gina wants to make all small flowers. Draw a model to show how she can use her shells. How many small flowers can she make? How many shells will she have left over?

c. Tim wants to use all his shells to make some large flowers and some small flowers. Draw a model to show how he can use his shells. How many large flowers and how many small flowers can he make?

E-Lab • Modeling Multiplication

Craig gives helicopter tours. On each trip, he can take 5 passengers in his helicopter. He makes 36 trips each week. How many passengers can Craig take in his helicopter each week?

You can use E-Lab to model multiplication.

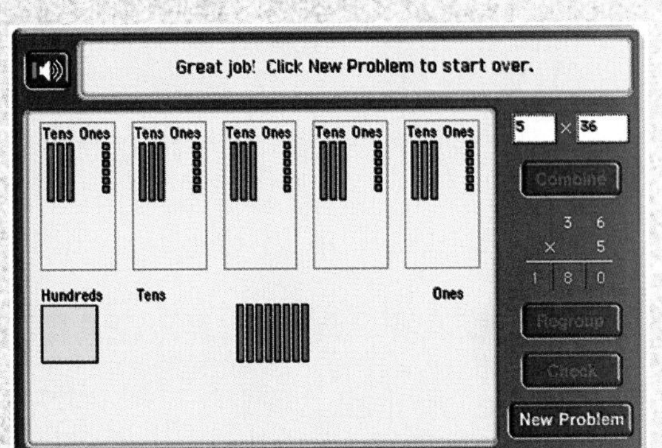

- Click *Modeling Multiplication.*
- Click *New Problem* to begin.
- Type 5. Press *Enter.*
- Type 36. Press *Enter.*
- Click *Combine.*
- Click *Regroup* 3 times to regroup ones as tens.
- Click *Regroup* to regroup tens as hundreds.
- Click *Check.*
- Record the product.

So, Craig can take 180 passengers in his helicopter each week.

Why did you need to click *Regroup* 3 times to regroup ones as tens?

Practice and Problem Solving

Use E-Lab to multiply.

1. 3×76
2. 4×29
3. 5×93

Solve.

4. 2×87
5. 6×38
6. 5×56

7. **REASONING** What would the E-Lab model for multiplying 4×124 look like? Draw a picture. Tell how many times you would need to click *Regroup*.

8. **Write a Problem** that involves multiplying 4 by a factor between 55 and 99. Use E-Lab to solve the problem.

Multimedia Math Glossary www.harcourtschool.com/mathglossary

9. **Vocabulary** Look up *remainder* in the Multimedia Math Glossary. Write three examples of division problems that have remainders.

PROBLEM SOLVING ON LOCATION

in Kansas

If you take a tour of the Konza Prairie, you can spot many wildflowers.

THE KONZA PRAIRIE RESEARCH NATURAL AREA

Most of the original prairie land in the United States is gone. The Konza Prairie is the last original piece of tallgrass prairie in the country.

Kelly looked for wildflowers at the Konza Prairie. The graph shows the number of different kinds of flowers she saw.

SPRING WILDFLOWERS	
Purple coneflower	❀ ❀
Butterfly milkweed	❀ ❀ ❀
Missouri evening primrose	❀ ❀ ❀ ❀
White prairie clover	❀ ❀ ❀ ❀ ❀ ❀
Spider milkweed	❀ ❀ ❀ ❀ ❀
KEY: Each ❀ = 25 flowers.	

For 1–5, use the graph.

1. Native Americans used the purple coneflower for toothaches. How many coneflowers did Kelly find?
2. Which flower did Kelly see the most of? How many did she see?
3. The Missouri evening primrose is a flower that opens at night. How many of these flowers did Kelly see?
4. The graph shows that Kelly found two kinds of milkweed. How many milkweed flowers did she find in all? Explain.
5. Luis studied prairie flowers. He drew 7 flowers on his graph to show the spider milkweeds he found. In his graph, each flower symbol stands for 18 flowers. Compare the number of spider milkweeds each student found.

ON THE PRAIRIE

About 200 years ago, there were millions of buffalo on the prairie. About 100 years ago, there were only about 550 buffalo. There are about 130,000 buffalo in North America now. Many of these buffalo live in special fenced areas where they are protected.

1. A buffalo eats about 900 pounds of grass each month. About how much grass would a buffalo eat during June, July, and August?
2. Buffalo can run up to 45 miles in 1 hour. If a buffalo could run at this speed for 3 hours, could it cover 120 miles? How can you tell?
3. **What if** the 210 buffalo are divided into 7 groups for checkups? How many buffalo would be in each group?
4. About 60 buffalo calves are born each year. About every third calf is male. About how many calves are male?
5. A hectare is a unit used to measure a large area of land, and is equal to about $2\frac{1}{2}$ acres. There is 1 buffalo for every 6 hectares on the Konza Prairie. How many buffalo live on 90 hectares?

About 210 buffalo are in the Konza Prairie herd.

STUDENT HANDBOOK

PREREQUISITE SKILLS REVIEW Do you have the math skills needed to start a new chapter? Use this list of skills to review and remember your skills from last year.

TROUBLESHOOTING

ORDINAL NUMBERS

Ordinal numbers tell order or position.

★ first ★ second ★ third ★ fourth ★ fifth ★ sixth ★ seventh ★ eighth ★ ninth ★ tenth

Example

Write the position of the yellow marble.

The yellow marble is second.

▶ **Practice**

Write the position of each marble.

1. purple **2.** white **3.** red **4.** green

READ AND WRITE ONES, TENS, HUNDREDS

You can use base-ten blocks to show numbers.

Example

Write the number.

3 tens 4 ones = 34

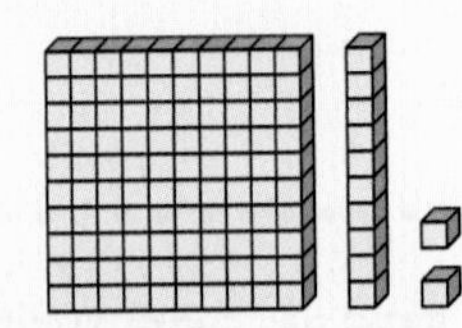

1 hundred 1 ten 2 ones = 112

▶ **Practice**

Write the number.

1. 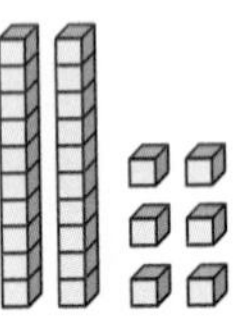

2. 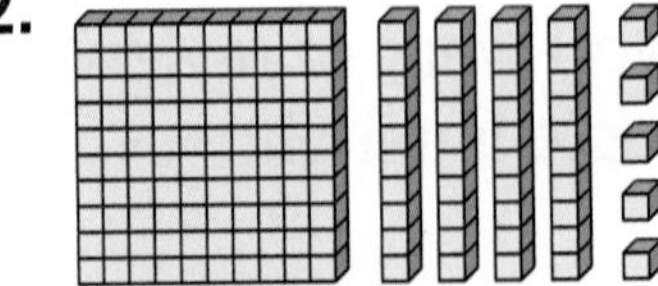

3. 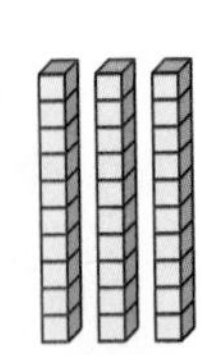

4. 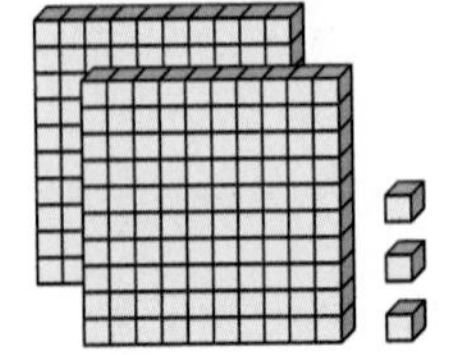

PLACE VALUE WITH ONES, TENS, HUNDREDS

You can use a place-value chart to understand numbers.

Hundreds	Tens	Ones
5	8	2

582 = 5 hundreds 8 tens 2 ones

▶ Practice

Write the number.

1. 3 hundreds 7 tens 9 ones

2. 8 tens 1 one

3. 6 tens 5 ones

4. 1 hundred 8 ones

5. 7 hundreds 5 tens 1 one

6. 4 hundreds 2 tens

COMPARE 2-DIGIT NUMBERS

You can use base-ten blocks to compare numbers.

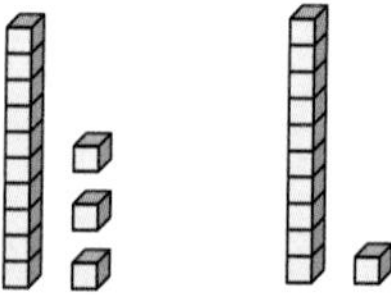

13 is **greater than** 11.

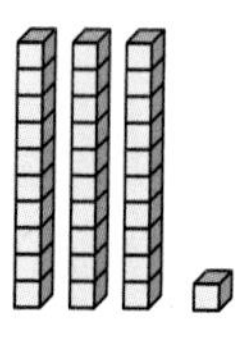

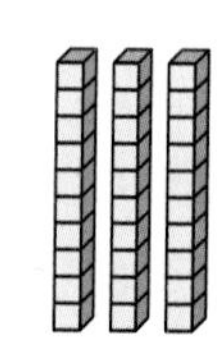

31 is **less than** 32.

Examples

Write the words *greater than* or *less than*.

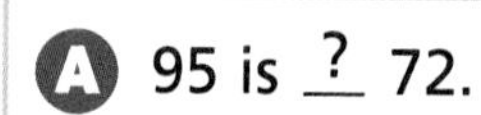

A 95 is ? 72.

Think: 95 has more tens than 72.

So, 95 is *greater than* 72.

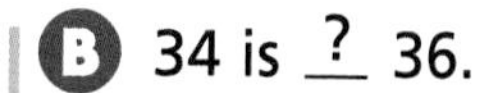

B 34 is ? 36.

Think: 34 and 36 have the same number of tens. 34 has fewer ones than 36.

So, 34 is *less than* 36.

▶ Practice

Write the words *greater than* or *less than*.

1. 15 is ? 51.

2. 77 is ? 68.

3. 84 is ? 83.

4. 99 is ? 97.

5. 55 is ? 65.

6. 12 is ? 21.

7. 23 is ? 33.

8. 14 is ? 40.

9. 90 is ? 80.

TROUBLESHOOTING

ORDER NUMBERS

You can write numbers in order.

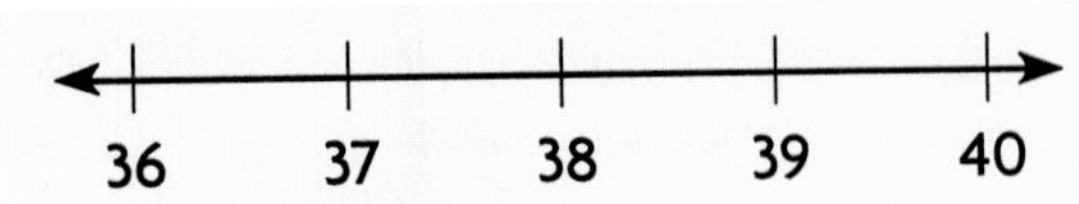

37 is just **after** 36.
38 is just **before** 39.
39 is **between** 38 and 40.

▶ Practice

Write the number that is just after, just before, or between.

1. 29, ■ **2.** ■, 179 **3.** 60, ■

4. 506, ■, 508 **5.** 552, ■ **6.** ■, 13

7. 27, ■, 29 **8.** ■, 60 **9.** 439, ■, 441

ADDITION FACTS

The answer to an addition problem is the **sum**. If you don't remember a fact, you can think about how to **make a ten** to help find the sum.

Example Find 7 + 5.

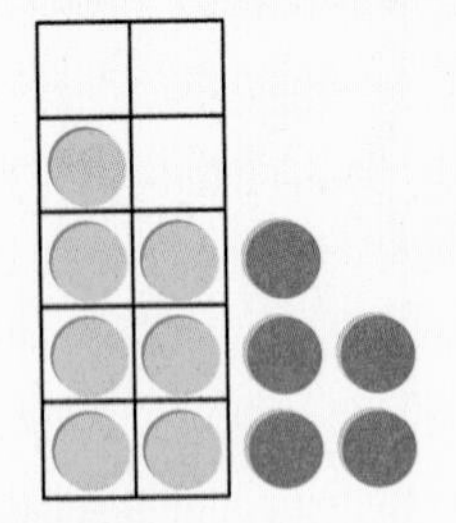

Show 7 + 5.

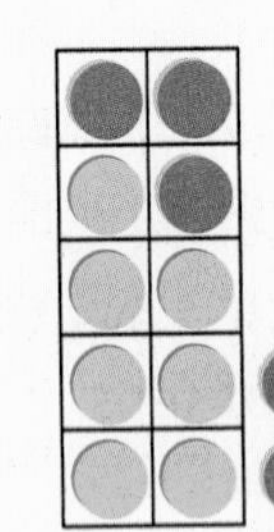

Move 3 counters to fill the ten frame. 7 + 3 = 10

There are 2 more counters to add. 10 + 2 = 12

So, 7 + 5 = 12.

▶ Practice

Add. You may wish to make a ten.

1. 5 + 6

2. 8 + 5

3. 6 + 4

4. 7 + 7

5. 9 + 6

2-DIGIT ADDITION

Sometimes you must regroup to add 2-digit numbers.

Find 14 + 17.

STEP 1

Add the ones.
4 + 7 = 11

tens	ones
1	4
+ 1	7

STEP 2

Regroup.
11 ones = 1 ten 1 one

tens	ones
1	
1	4
+ 1	7
	1

STEP 3

Add the tens.
1 + 1 + 1 = 3

tens	ones
1	
1	4
+ 1	7
3	1

▶ Practice

Add.

1. 62 + 23

2. 31 + 49

3. 58 + 18

4. 73 + 17

MENTAL MATH: ADD 2-DIGIT NUMBERS

You can use mental math to add.

Examples

A Find 32 + 46 = ■.

- Add the tens. 30 + 40 = 70
- Add the ones. 2 + 6 = 8
- Add the sums. 70 + 8 = 78

So, 32 + 46 = 78.

B Find 19 + 21 = ■.

- Add the tens. 10 + 20 = 30
- Add the ones. 9 + 1 = 10
- Add the sums. 30 + 10 = 40

So, 19 + 21 = 40.

▶ Practice

Use mental math to add.

1. 12 + 11 = ■ **2.** 45 + 19 = ■ **3.** 59 + 31 = ■ **4.** 46 + 47 = ■

5. 27 + 13 = ■ **6.** 32 + 29 = ■ **7.** 64 + 28 = ■ **8.** 39 + 43 = ■

9. 76 + 25 = ■ **10.** 89 + 32 = ■ **11.** 94 + 46 = ■ **12.** 49 + 83 = ■

TROUBLESHOOTING

SUBTRACTION FACTS

The answer to a subtraction problem is the **difference**.
If you can't remember a fact, you can **count back** to help find the difference.

Example

Find 7 − 2.

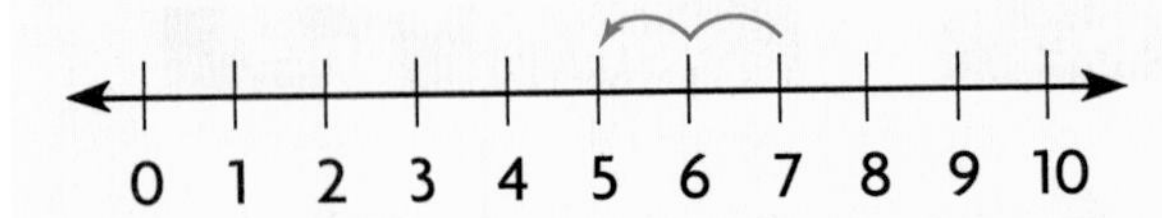

Say 7. Count back 2.
6, 5

The difference is 5.
So, 7 − 2 = 5.

▶ **Practice**

Subtract.

1. $11 - 2$
2. $11 - 5$
3. $13 - 7$
4. $15 - 8$
5. $18 - 9$

2-DIGIT SUBTRACTION

Check whether you must regroup to subtract 2-digit numbers.

Find 32 − 15.

STEP 1

Since 5 > 2, you must regroup.

tens	ones
3	2
− 1	5

STEP 2

Regroup.
3 tens 2 ones = 2 tens 12 ones.
Subtract the ones.

tens	ones
2	12
~~3~~	~~2~~
− 1	5
	7

STEP 3

Subtract the tens.

tens	ones
2	12
~~3~~	~~2~~
− 1	5
1	7

▶ **Practice**

Subtract.

1. $57 - 22$
2. $73 - 34$
3. $66 - 18$
4. $31 - 17$
5. $80 - 27$

MENTAL MATH: SUBTRACT 2-DIGIT NUMBERS

You can use mental math to subtract.

Use mental math to find 95 − 19.

STEP 1

Add more to make the smaller number a ten.

$$\begin{array}{r} 95 \\ -\ 19 \\ \hline \blacksquare \end{array}$$

Think: 19 + 1 = 20

STEP 2

Add the same number to the larger number.

95 + 1 = 96
19 + 1 = 20

STEP 3

Subtract your answers.

$$\begin{array}{r} 96 \\ -\ 20 \\ \hline 76 \end{array}$$

So, 95 − 19 = 76.

▶ Practice

Use mental math to subtract.

1. 41 − 19 = ■ **2.** 78 − 29 = ■ **3.** 37 − 18 = ■ **4.** 56 − 28 = ■

MONEY: COUNT BILLS AND COINS

1 dollar, or $1.00

1 quarter = 25¢, or $0.25

1 dime = 10¢, or $0.10

1 nickel = 5¢, or $0.05

1 penny = 1¢, or $0.01

Count on to find the value of a group of coins.

Example

Count and write the amount.

$1.00

$1.25

$1.50

$1.60

$1.65

$1.66

So, the value of the bill and coins is $1.66.

▶ Practice

Count and write the amount.

1.    

2.        

TROUBLESHOOTING

TELL TIME

The short hand on a clock is the **hour hand**. The long hand on a clock is the **minute hand**.

You can **count by fives** to find the minutes after the hour.

This clock shows 9:35.

Examples

Write the time.

A 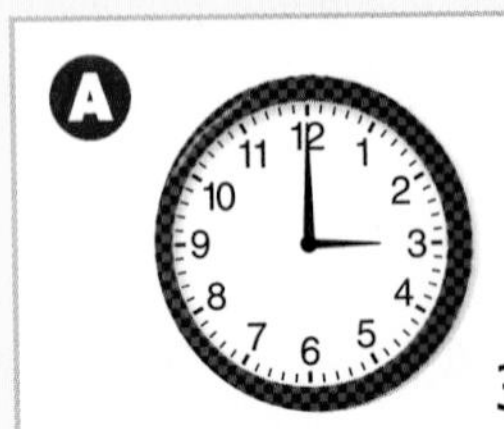 3:00

B 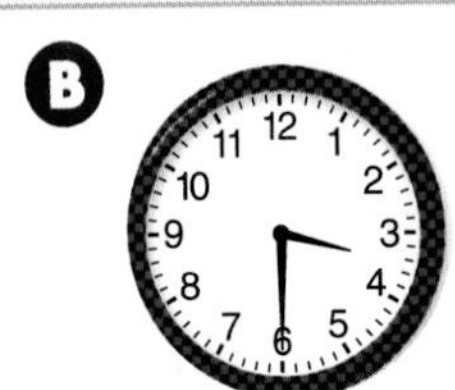 3:30

C  3:45

▶ Practice

Write the time.

1. 

2. 

3. 

CALENDAR

You can use a calendar to keep track of days, weeks, and months.

There are 5 Wednesdays in October. The third Friday in this month is October 18.

OCTOBER						
Sun	Mon	Tue	Wed	Thu	Fri	Sat
		1	2	3	4	5
6	7	8	9	10	11	12
13	14	15	16	17	18	19
20	21	22	23	24	25	26
27	28	29	30	31		

▶ Practice

Use the calendar.

1. The second Tuesday in October is _?_.
2. The last day in October is a _?_.
3. There are _?_ Mondays in October.
4. The fifth Wednesday in October is _?_.

SKIP-COUNT

To skip-count, start with any number and add or subtract the same number.

Example

Skip-count to find the missing numbers. 6, 9, 12, ■, ■

Think: $6 + 3 = 9$ $9 + 3 = 12$

So, skip-count by threes to find the missing numbers.

$12 + 3 = 15$ $15 + 3 = 18$

So, the numbers are 6, 9, 12, 15, 18.

▶ Practice

Skip-count to find the missing numbers.

1. 8, 10, 12, ■ **2.** 25, 30, 35, ■ **3.** 100, 90, 80, ■, ■

4. 36, 38, 40, ■ **5.** 12, 15, 18, ■ **6.** 35, 40, 45, ■, ■

EQUAL GROUPS

When you have **equal groups**, you can skip-count to find how many in all.

Example

Write how many there are in all.

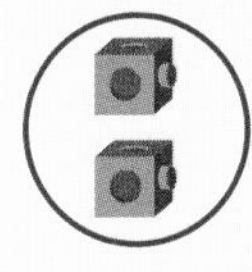  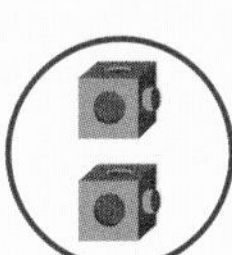 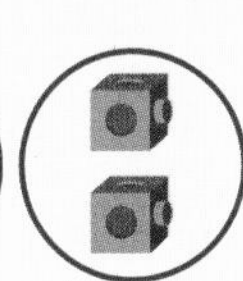

Think: Since there are 2 in each group, you can skip-count by twos: 2, 4, 6, 8.

So, 4 groups of $2 = 8$.

▶ Practice

Write how many there are in all.

1.   

3 groups of $2 =$ ■

2. 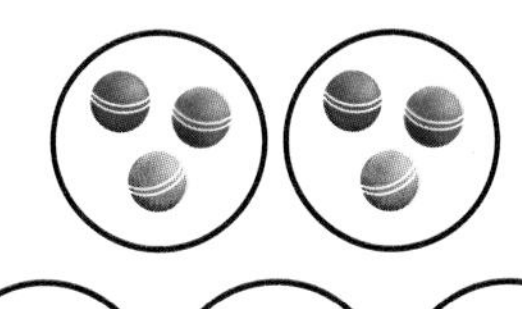 

5 groups of $3 =$ ■

3. 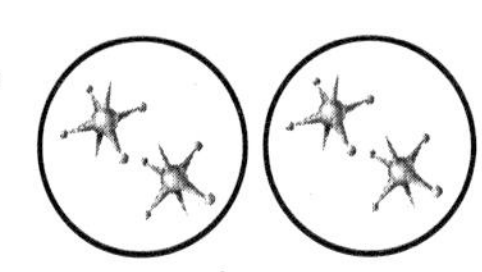

2 groups of $2 =$ ■

TROUBLESHOOTING

COLUMN ADDITION

To add more than two numbers, choose two numbers to add first. Look for facts you know.

Example

$$\begin{array}{r} 2 \\ 5 \\ +\ 8 \\ \hline 15 \end{array} \quad 2 + 8 = 10,\ 10 + 5 = 15$$

$$\begin{array}{r} 2 \\ 5 \\ +\ 8 \\ \hline 15 \end{array} \quad 2 + 5 = 7,\ 7 + 8 = 15$$

$$\begin{array}{r} 2 \\ 5 \\ +\ 8 \\ \hline 15 \end{array} \quad 5 + 8 = 13,\ 13 + 2 = 15$$

▶ Practice

Find the sum.

1. $\begin{array}{r} 4 \\ 4 \\ +\ 2 \\ \hline \end{array}$ **2.** $\begin{array}{r} 6 \\ 2 \\ +\ 3 \\ \hline \end{array}$ **3.** $\begin{array}{r} 9 \\ 1 \\ +\ 2 \\ \hline \end{array}$ **4.** $\begin{array}{r} 7 \\ 3 \\ +\ 5 \\ \hline \end{array}$ **5.** $\begin{array}{r} 2 \\ 7 \\ +\ 4 \\ \hline \end{array}$ **6.** $\begin{array}{r} 1 \\ 8 \\ +\ 1 \\ \hline \end{array}$

ADDITION

When addends are the same, you can skip-count to find the sum. Any number plus 0 equals that same number.

Examples

A Find the sum.

$3 + 3 + 3 + 3 + 3 = ■$

Think: Since each addend is 3, skip-count by threes.
3, 6, 9, 12, 15

So, $3 + 3 + 3 + 3 + 3 = 15$.

B Find the missing number.

$7 + ■ = 7$

Think: What number plus 7 equals 7?

So, $7 + 0 = 7$.

▶ Practice

Find the sum or missing number.

1. $5 + 5 + 5 = ■$ **2.** $■ + 9 = 9$ **3.** $4 + 4 + 4 + 4 = ■$

4. $2 + 2 + 2 = ■$ **5.** $0 + ■ = 8$ **6.** $■ + 3 = 3$

7. $3 + 3 + 3 + 3 = ■$ **8.** $■ + 1 = 1$ **9.** $7 + 7 + 7 = ■$

ORDER PROPERTY OF ADDITION

Changing the order of addends does not change the sum.

Example

Find the missing number.

$2 + \blacksquare = 3 + 2$ **Think:** $3 + 2 = 5$
So, I must solve $2 + \blacksquare = 5$.
The missing number is 3.

So, $2 + 3 = 3 + 2$.

▶ Practice

Find the missing number.

1. $4 + 1 = 1 + \blacksquare$
2. $7 + 6 = 6 + \blacksquare$
3. $\blacksquare + 5 = 5 + 8$
4. $9 + \blacksquare = 5 + 9$
5. $\blacksquare + 2 = 2 + 4$
6. $3 + \blacksquare = 0 + 3$

MULTIPLICATION FACTS

The answer to a multiplication problem is the **product**.

Examples

A Find 2×6.

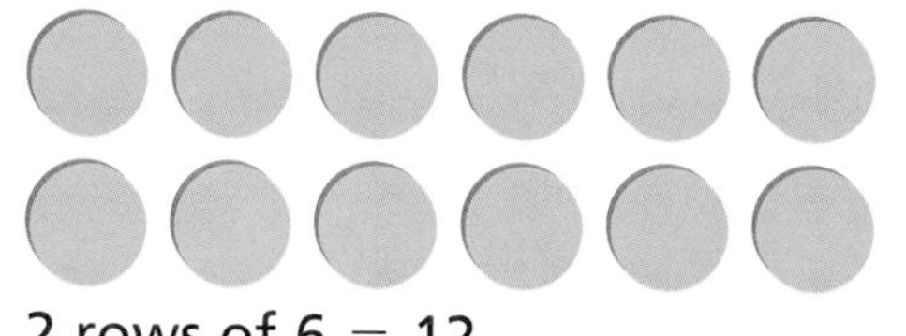

2 rows of 6 = 12
So, $2 \times 6 = 12$.

B Find 3×5.

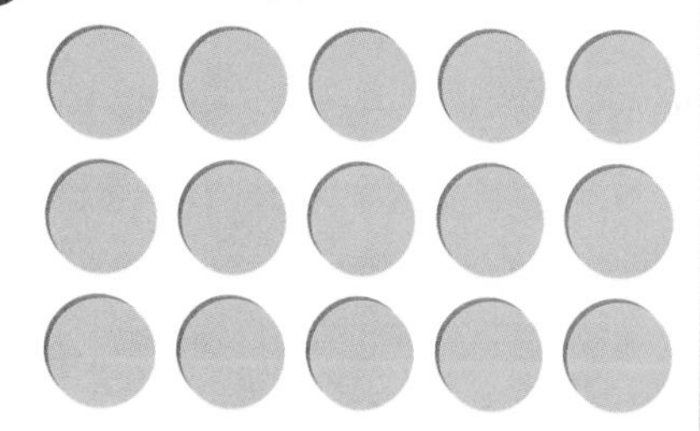

3 rows of 5 = 15
So, $3 \times 5 = 15$.

▶ Practice

Find the product.

1. $2 \times 2 = \blacksquare$
2. $4 \times 3 = \blacksquare$
3. $5 \times 2 = \blacksquare$
4. $\blacksquare = 5 \times 5$
5. $3 \times 2 = \blacksquare$
6. $\blacksquare = 5 \times 9$
7. $8 \times 3 = \blacksquare$
8. $\blacksquare = 7 \times 5$
9. $3 \times 7 = \blacksquare$
10. $3 \times 9 = \blacksquare$
11. $2 \times 8 = \blacksquare$
12. $5 \times 1 = \blacksquare$

TROUBLESHOOTING

ADD EQUAL GROUPS

When you multiply, you add equal groups.

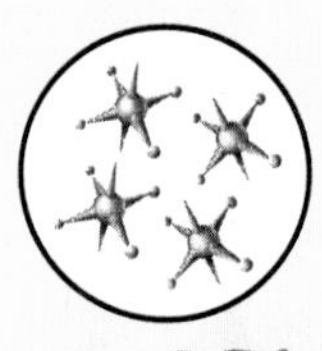 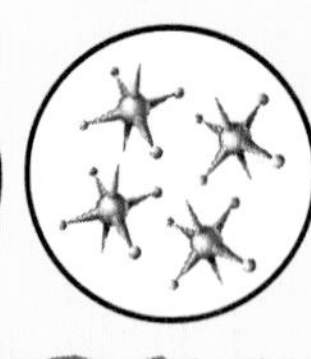 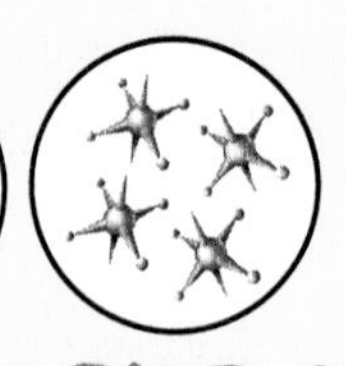

$4 + 4 + 4 = 12$
So, 3 groups of $4 = 12$.

▶ Practice

Write how many there are in all.

1. 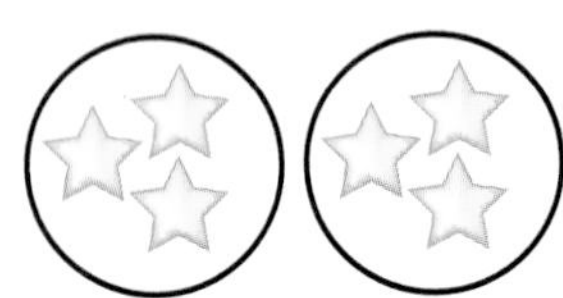

2 groups of 3 = ■

2. 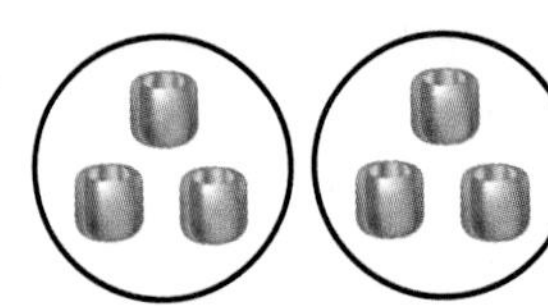 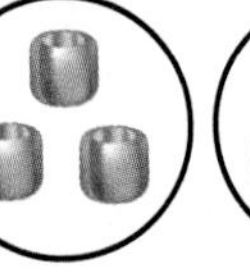 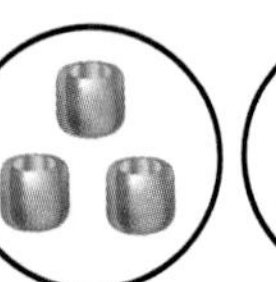 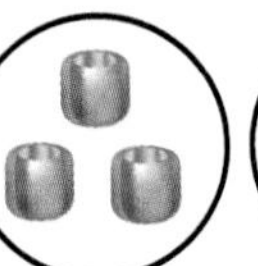 

5 groups of 3 = ■

3. 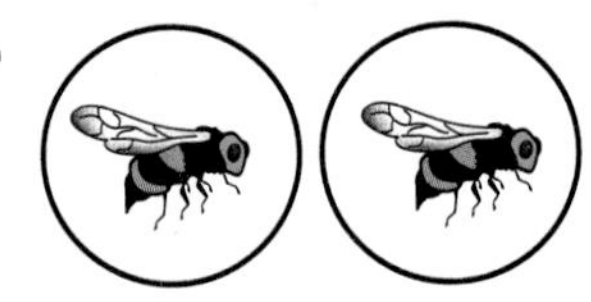

2 groups of 1 = ■

4.    

4 groups of 4 = ■

MULTIPLICATION FACTS THROUGH 5

When one of the factors of a problem is even, you can double a fact you already know.

Example

Find 4×7.

Think: To find a 4's fact, you can double a 2's fact.

- First find the 2's fact. $\mathbf{2 \times 7 = 14}$
- Double the product. $\mathbf{14 + 14 = 28}$

So, $4 \times 7 = 28$.

▶ Practice

Find the product.

1. $8 \times 2 =$ ■ **2.** $4 \times 4 =$ ■ **3.** $4 \times 6 =$ ■ **4.** $6 \times 5 =$ ■

5. $3 \times 6 =$ ■ **6.** ■ $= 4 \times 7$ **7.** $9 \times 4 =$ ■ **8.** $7 \times 6 =$ ■

9. $8 \times 3 =$ ■ **10.** $4 \times 5 =$ ■ **11.** ■ $= 6 \times 2$ **12.** $8 \times 4 =$ ■

SKIP-COUNT BY TENS

You can use a pattern to skip-count by tens.

Example

Skip-count by tens to find the missing numbers.

18, 28, 38, ■, ■, 68

1	2	3	4	5	6	7	8	9	10
11	12	13	14	15	16	17	18	19	20
21	22	23	24	25	26	27	28	29	30
31	32	33	34	35	36	37	38	39	40
41	42	43	44	45	46	47	48	49	50
51	52	53	54	55	56	57	58	59	60
61	62	63	64	65	66	67	68	69	70

38 + **10** = 48
48 + **10** = 58

So, the numbers are 18, 28, 38, 48, 58, 68.

▶ Practice

Skip-count to find the missing numbers.

1. 15, 25, 35, 45, ■, 65

2. 11, 21, 31, ■, 51, 61

3. 54, 44, 34, ■, ■

4. 16, 26, 36, ■, 56, ■

MULTIPLICATION FACTS THROUGH 8

Break apart an array to find a multiplication fact.

Find 4 × 8.

STEP 1

Show 4 rows of 8.

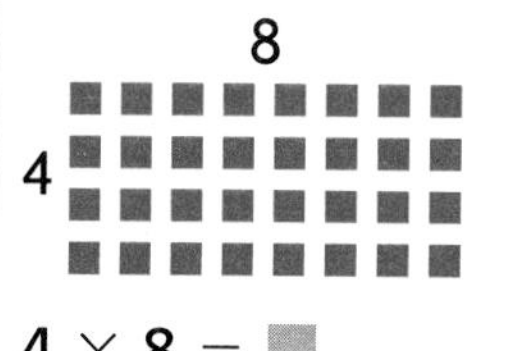

4 × 8 = ■

STEP 2

Break apart the array.

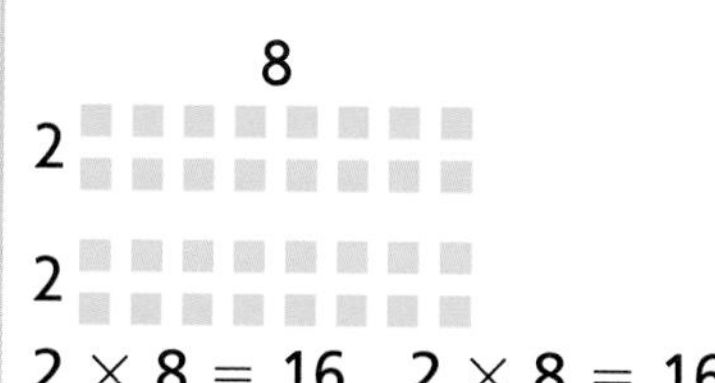

2 × 8 = 16   2 × 8 = 16

STEP 3

Add the products.

$$\begin{array}{r} 16 \\ +\ 16 \\ \hline 32 \end{array}$$

So, 4 × 8 = 32.

▶ Practice

Find the product.

1. 7 × 6 = ■ **2.** 7 × 4 = ■ **3.** 8 × 6 = ■ **4.** 8 × 9 = ■

5. 9 × 6 = ■ **6.** 9 × 4 = ■ **7.** 7 × 7 = ■ **8.** 8 × 8 = ■

9. 4 × 6 = ■ **10.** 3 × 8 = ■ **11.** 8 × 4 = ■ **12.** 6 × 3 = ■

TROUBLESHOOTING

MEANING OF MULTIPLICATION

When you multiply, you add equal groups.

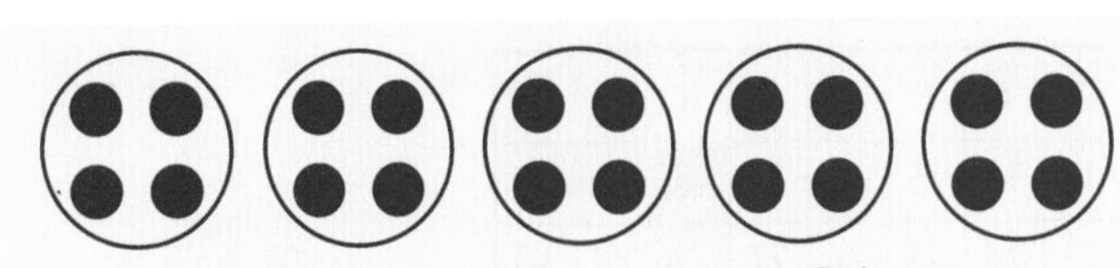

5 groups of 4

$4 + 4 + 4 + 4 + 4 = 20$

So, $5 \times 4 = 20$.

▶ Practice

Copy and complete.

1.

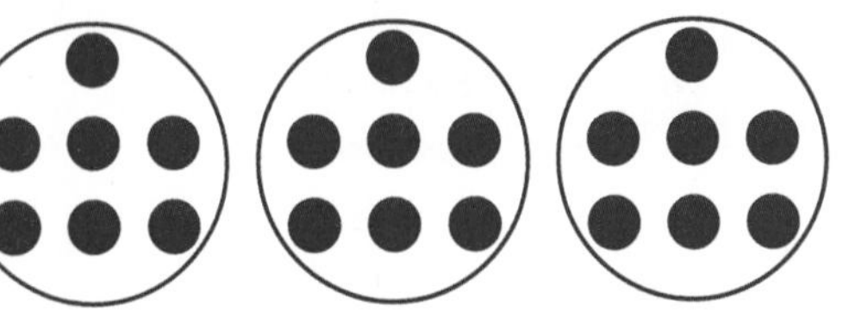

a. ■ groups of ■

b. ■ + ■ + ■ = ■

c. ■ × ■ = ■

2.

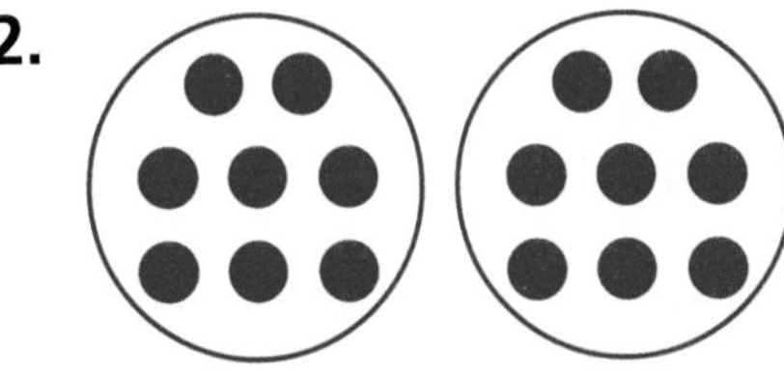

a. ■ groups of ■

b. ■ + ■ = ■

c. ■ × ■ = ■

MULTIPLICATION FACTS THROUGH 10

You can use an array to find a multiplication fact.

Examples

A Find $5 \times 9 =$ ■.

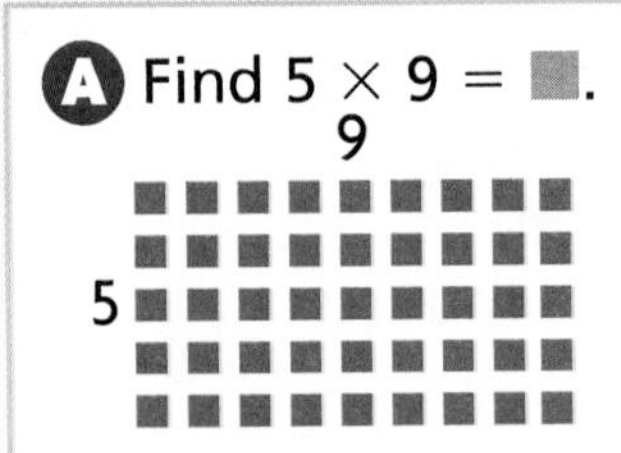

So, $5 \times 9 = 45$.

B Find $4 \times 6 =$ ■.

6

4

So, $4 \times 6 = 24$.

▶ Practice

Find the product.

1. $9 \times 7 =$ ■ **2.** $8 \times 10 =$ ■ **3.** $6 \times 9 =$ ■ **4.** $5 \times 9 =$ ■

5. $10 \times 7 =$ ■ **6.** $9 \times 9 =$ ■ **7.** $7 \times 7 =$ ■ **8.** $10 \times 6 =$ ■

9. $9 \times 6 =$ ■ **10.** $3 \times 8 =$ ■ **11.** $3 \times 10 =$ ■ **12.** $6 \times 8 =$ ■

13. $9 \times 8 =$ ■ **14.** $9 \times 4 =$ ■ **15.** $9 \times 10 =$ ■ **16.** $8 \times 8 =$ ■

SUBTRACTION

To find a missing number in a subtraction sentence, think about fact families. Remember how numbers are related.

$5 + 7 = 12$ $7 + 5 = 12$ $12 - 5 = 7$ $12 - 7 = 5$

Examples

Find the missing number.

A $12 - ■ = 7$

To solve $12 - ■ = 7$,
find $12 - 7 = ■$.
$12 - 7 = 5$
So, $12 - 5 = 7$.

B $32 - ■ = 28$

To solve $32 - ■ = 28$,
find $32 - 28 = ■$.
$32 - 28 = 4$
So, $32 - 4 = 28$.

▶ Practice

Find the missing number.

1. $21 - ■ = 15$ **2.** $32 - ■ = 29$ **3.** $43 = 51 - ■$ **4.** $98 - ■ = 92$

5. $31 - ■ = 17$ **6.** $40 - ■ = 25$ **7.** $50 = 70 - ■$ **8.** $67 - ■ = 11$

DIVISION FACTS THROUGH 5

The answer in a division problem is a **quotient**. You can think about a related multiplication fact to find a quotient.

Example

Find $3\overline{)15}$.

Read: 15 divided by 3
Write: $15 \div 3 = ■$

Think: 15 counters in 3 groups
How many in each group?

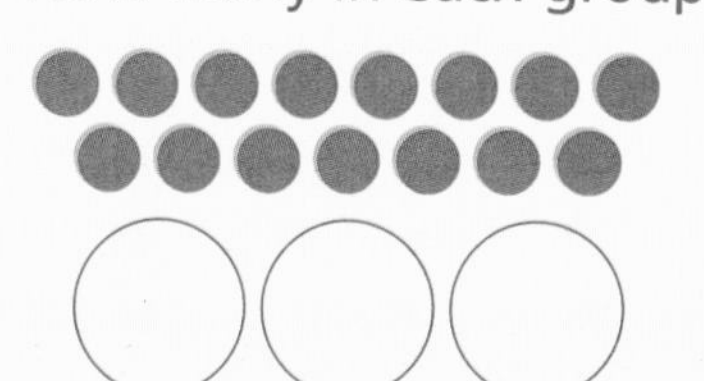

5 in each group

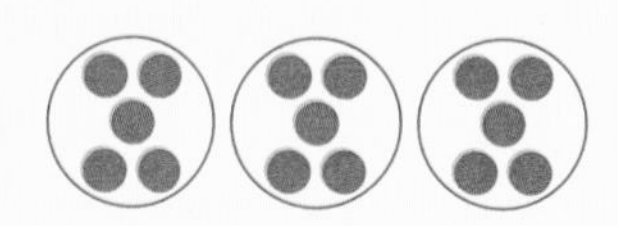

So, $15 \div 3 = 5$, or $3\overline{)15}$ with quotient 5.

▶ Practice

Find the quotient.

1. $25 \div 5 = ■$ **2.** $30 \div 5 = ■$ **3.** $24 \div 3 = ■$ **4.** $14 \div 2 = ■$

5. $40 \div 5 = ■$ **6.** $36 \div 4 = ■$ **7.** $28 \div 4 = ■$ **8.** $45 \div 5 = ■$

TROUBLESHOOTING

✓ ORDER PROPERTY OF MULTIPLICATION

Two numbers can be multiplied in any order. The product will be the same.

Example

3 groups of 5 = 15
So, $3 \times 5 = 15$.

5 groups of 3 = 15
So, $5 \times 3 = 15$.

If you can't remember 3×5, try thinking of 5×3.

So, $3 \times 5 = 5 \times 3$.

▶ Practice

Use the Order Property of Multiplication to help you find each product.

1. $4 \times 5 = \blacksquare$ **2.** $8 \times 4 = \blacksquare$ **3.** $7 \times 8 = \blacksquare$

4. $9 \times 6 = \blacksquare$ **5.** $7 \times 3 = \blacksquare$ **6.** $6 \times 3 = \blacksquare$

✓ MISSING FACTORS

You can use division to find a missing factor.

Examples

Find the missing factor.

A $4 \times \blacksquare = 36$

Think: $36 \div 4 = 9$

So, the missing factor is 9.
$4 \times 9 = 36$

B $5 \times \blacksquare = 40$

Think: $40 \div 5 = 8$

So, the missing factor is 8.
$5 \times 8 = 40$

▶ Practice

Find the missing factor.

1. $\blacksquare \times 3 = 18$ **2.** $4 \times \blacksquare = 20$ **3.** $3 \times \blacksquare = 24$

4. $\blacksquare \times 4 = 32$ **5.** $\blacksquare \times 5 = 15$ **6.** $4 \times \blacksquare = 16$

7. $4 \times \blacksquare = 40$ **8.** $\blacksquare \times 5 = 35$ **9.** $\blacksquare \times 3 = 27$

READ A CHART

A chart is used to organize information.

FAVORITE BREAKFAST	
Type	Students
Cereal	12
Pancakes	6
Eggs	7
Fruit	3

← Title

The data shows the number of students who chose each type of breakfast food.

This chart shows that

- Cereal had the most votes.
- 7 students chose eggs.
- 28 students in all voted.

▶ Practice

For 1–3, use the information in the chart above.

1. Which breakfast food had the fewest votes?
2. How many students chose pancakes?
3. How many students in all chose cereal or eggs?

TALLY DATA

A **tally table** has tally marks to record data.
Tally marks are grouped by fives. (5 = 𝍸)

FAVORITE SPORT	
Sport	Tally
Baseball	𝍸 I
Basketball	𝍸 𝍸
Hockey	IIII
Soccer	𝍸 IIII

← 6 students chose baseball.

← 9 students chose soccer.

▶ Practice

For 1–2, use the tally table.

1. How many students chose basketball as their favorite sport?
2. How many students did NOT choose soccer as their favorite sport?

TROUBLESHOOTING

✓ READ PICTOGRAPHS

A **pictograph** shows data by using pictures.

The **key** in this pictograph shows that each ◆ equals 2 votes.

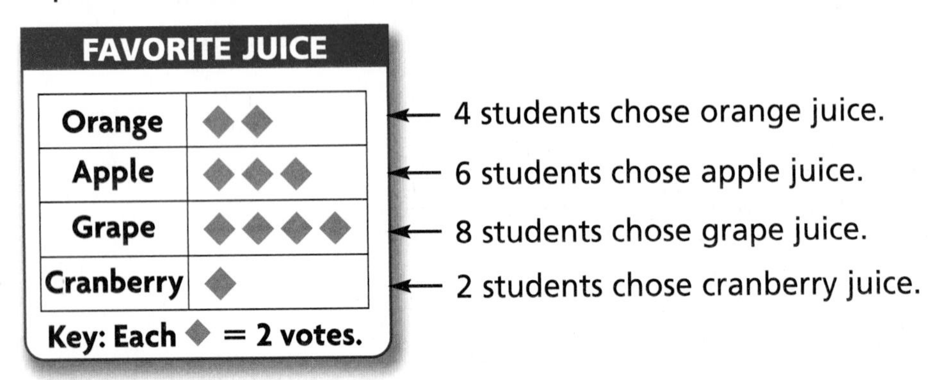

FAVORITE JUICE	
Orange	◆◆
Apple	◆◆◆
Grape	◆◆◆◆
Cranberry	◆
Key: Each ◆ = 2 votes.	

← 4 students chose orange juice.
← 6 students chose apple juice.
← 8 students chose grape juice.
← 2 students chose cranberry juice.

▶ Practice

Use the value of the symbol to find the missing number.

1. If ✱ = 2, then ✱ + ✱ + ✱ = ■.
2. If ✳ = 4, then ✳ + ✳ = ■.
3. If ★ = 10, then ★ + ★ + ★ + ★ = ■.
4. If ✿ = 5, then ✿ + ✿ + ✿ = ■.

✓ IDENTIFY PARTS OF A WHOLE

A **fraction** names a part of a whole.

The top number tells how many parts are being used.

The bottom number tells how many equal parts are in the whole.

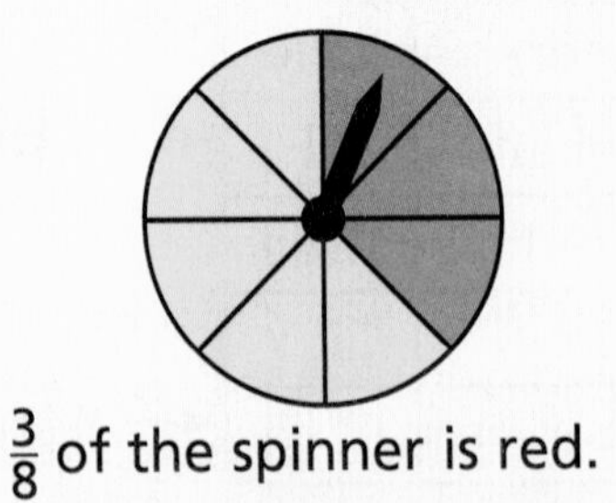

$\frac{3}{8}$ of the spinner is red.

▶ Practice

Write a fraction that names the red part of the spinner.

1. 
2. 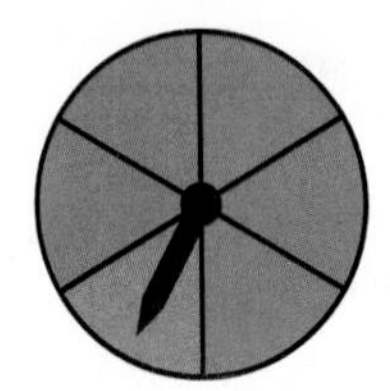
3. 
4. 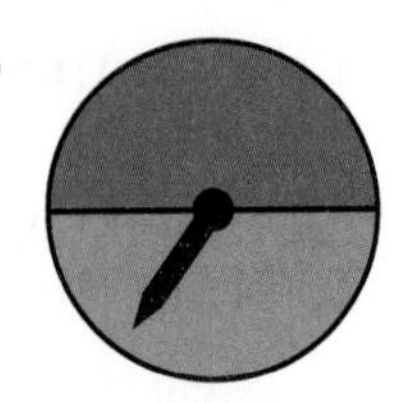

COMPARE PARTS OF A WHOLE

You compare fractions by comparing parts of a whole.

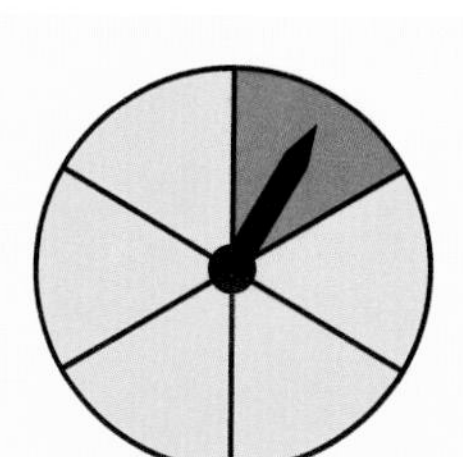

$\frac{1}{6}$ of the spinner is blue.

$\frac{5}{6}$ of the spinner is yellow.

The largest part of the spinner is yellow.

▶ Practice

Write the color shown by the largest part of each spinner.

1. 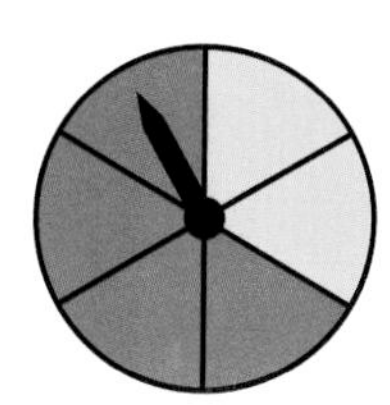
2. 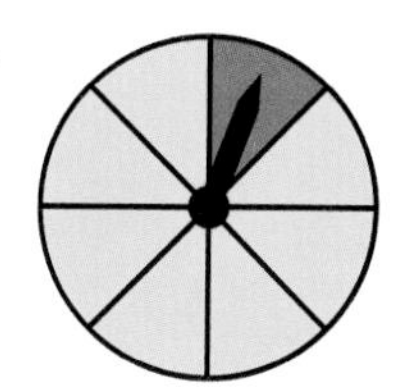
3. 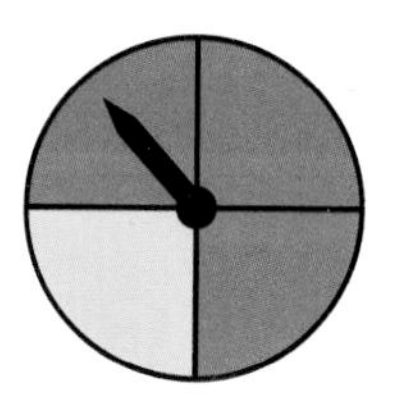
4. 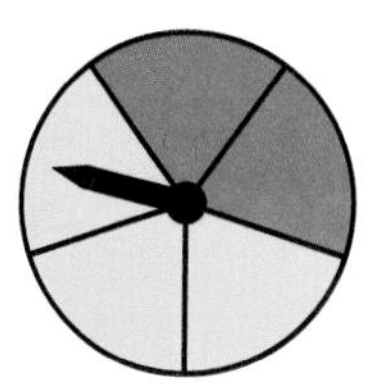

USE A TALLY TABLE

You can record the number of times a pointer stops on each color by using a tally table.

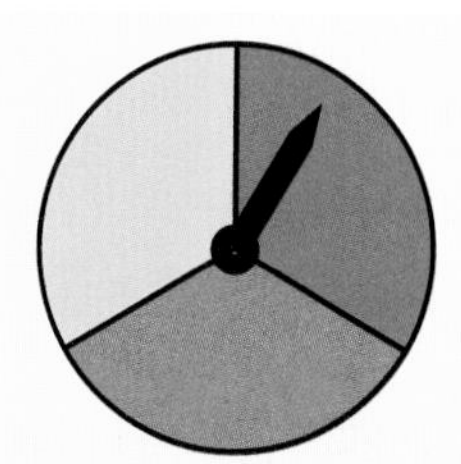

SPINNER RESULTS	
Color	**Tally**
Yellow	|||
Red	卌
Blue	||||

This spinner was used 12 times.
The pointer stopped on yellow 3 times.

▶ Practice

Use the tally table.

1. How many times did the pointer stop on red?
2. Which color was landed on the fewest times?
3. How many more times was red landed on than blue?

TROUBLESHOOTING

REGROUP ONES AND TENS

You can regroup 10 ones as 1 ten.

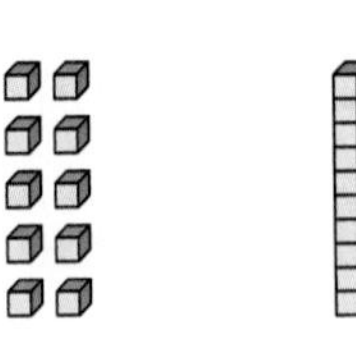

10 ones = 1 ten

Example

Regroup 2 tens 16 ones.

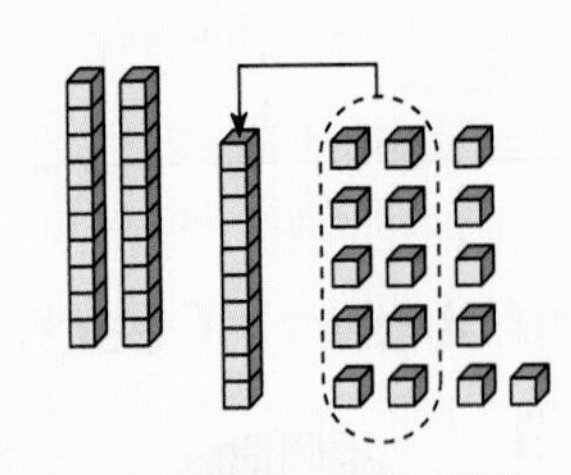

So, 2 tens 16 ones = 3 tens 6 ones, or 36.

▶ Practice

Regroup.

1. 6 tens 12 ones = ■ tens 2 ones

2. 3 tens 15 ones = ■ tens 5 ones

3. 4 tens 25 ones = ■ tens 5 ones

4. 4 tens ■ ones = 5 tens 8 ones

MULTIPLICATION FACTS

There are many ways you have learned to remember multiplication facts.

Example

Find 4 × 9 = ■.

- Skip-count by equal groups.
- Use an array.
- Break apart an array.
- If one of the factors is even, use doubling.
- If you can't remember a fact, try changing the order of the factors.

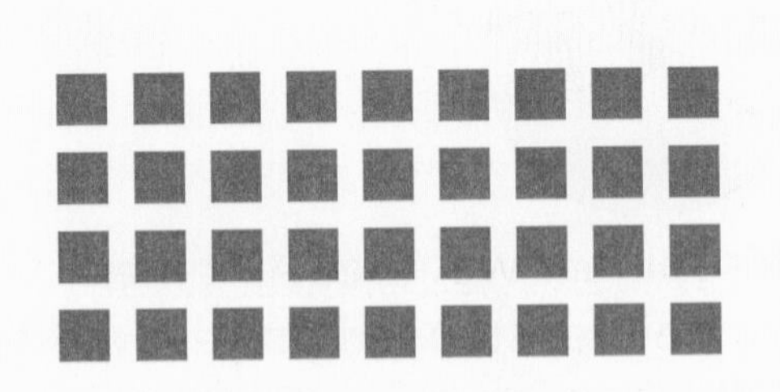

So, 4 × 9 = 36.

▶ Practice

Find the product.

1. 6 × 8 = ■ **2.** 3 × 9 = ■ **3.** 5 × 7 = ■ **4.** 6 × 7 = ■

5. 3 × 5 = ■ **6.** 7 × 7 = ■ **7.** 5 × 8 = ■ **8.** 2 × 9 = ■

MULTIPLY 2-DIGIT NUMBERS

You can use base-ten blocks to multiply 2-digit numbers.

Find $3 \times 16 = \blacksquare$.

STEP 1

Model 3 groups of 16.
Multiply the ones.

$$\begin{array}{r} 16 \\ \times\ 3 \\ \hline 18 \end{array} \quad (3 \times 6 \text{ ones})$$

STEP 2

Multiply the tens.

$$\begin{array}{r} 16 \\ \times\ 3 \\ \hline 18 \\ 30 \\ \hline \end{array} \quad (3 \times 1 \text{ ten})$$

STEP 3

Add to find the product.

$$\begin{array}{r} 16 \\ \times\ 3 \\ \hline 18 \\ +\ 30 \\ \hline 48 \end{array}$$

So, $3 \times 16 = 48$.

▶ Practice

Find the product.

1. $18 \times 8 = \blacksquare$ **2.** $27 \times 4 = \blacksquare$ **3.** $23 \times 6 = \blacksquare$ **4.** $81 \times 9 = \blacksquare$

5. $64 \times 4 = \blacksquare$ **6.** $25 \times 3 = \blacksquare$ **7.** $62 \times 5 = \blacksquare$ **8.** $85 \times 2 = \blacksquare$

DIVISION FACTS

You have learned many ways to remember division facts.

Example Find $18 \div 3 = \blacksquare$.

A Use counters.

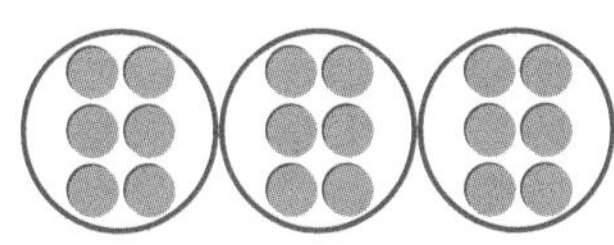

B Think about an array.

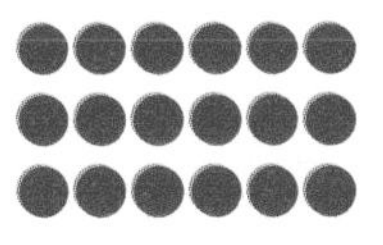

C Use repeated subtraction.

$$\begin{array}{rrrrrr} 18 & 15 & 12 & 9 & 6 & 3 \\ -\ 3 & -\ 3 & -\ 3 & -\ 3 & -\ 3 & -\ 3 \\ \hline 15 & 12 & 9 & 6 & 3 & 0 \end{array}$$

So, $18 \div 3 = 6$.

D Think about fact families.

$3 \times 6 = 18$ $18 \div 6 = 3$

$6 \times 3 = 18$ $18 \div 3 = 6$

▶ Practice

Divide.

1. $20 \div 5 = \blacksquare$ **2.** $36 \div 6 = \blacksquare$ **3.** $27 \div 3 = \blacksquare$ **4.** $42 \div 6 = \blacksquare$

5. $72 \div 9 = \blacksquare$ **6.** $24 \div 3 = \blacksquare$ **7.** $56 \div 8 = \blacksquare$ **8.** $63 \div 7 = \blacksquare$

TROUBLESHOOTING

REGROUP HUNDREDS AND TENS

You can regroup 10 tens as 1 hundred and 10 ones as 1 ten.

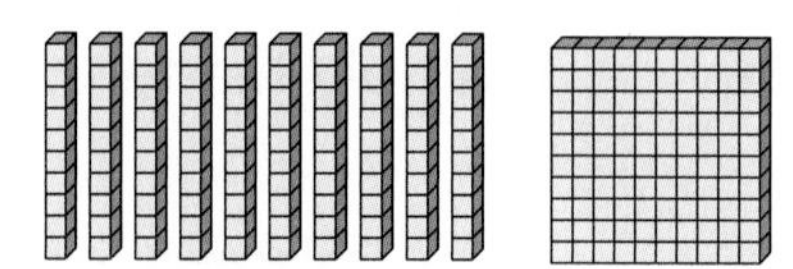

10 ones = 1 ten 10 tens = 1 hundred

Example

Regroup 3 hundreds 12 tens.

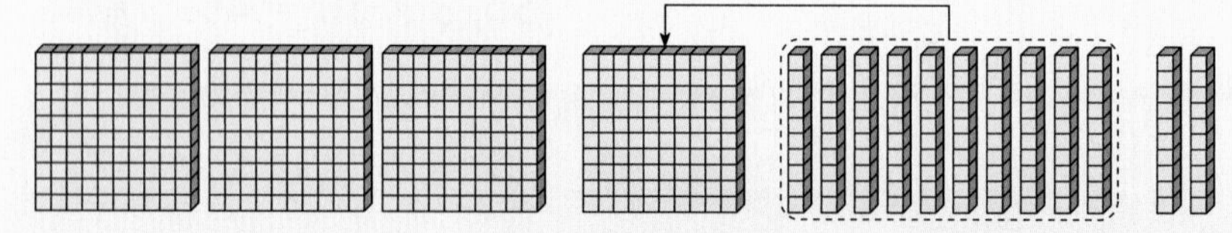

So, 3 hundreds 12 tens = 4 hundreds 2 tens, or 420.

▶ Practice

Regroup.

1. 2 hundreds ■ tens = 350

2. 910 = ■ hundreds 11 tens

3. 1 hundred 17 tens = 2 hundreds ■ tens

MULTIPLICATION AND DIVISION FACTS

Fact families can help you remember multiplication and division facts.

Examples

A Find $21 \div 3 =$ ■.

Think of a related multiplication fact.

If you know $3 \times 7 = 21$,
then you know $21 \div 3 = 7$.

B Find $4 \times 6 =$ ■.

Think of a related division fact.

If you know $24 \div 4 = 6$,
then you know $4 \times 6 = 24$.

▶ Practice

Find the product or quotient.

1. $6 \times 7 =$ ■ **2.** $9 \times 3 =$ ■ **3.** $56 \div 7 =$ ■ **4.** $6 \times 6 =$ ■

5. $45 \div 5 =$ ■ **6.** $3 \times 8 =$ ■ **7.** $42 \div 6 =$ ■ **8.** $72 \div 9 =$ ■

IDENTIFY SOLID FIGURES

These are solid figures.

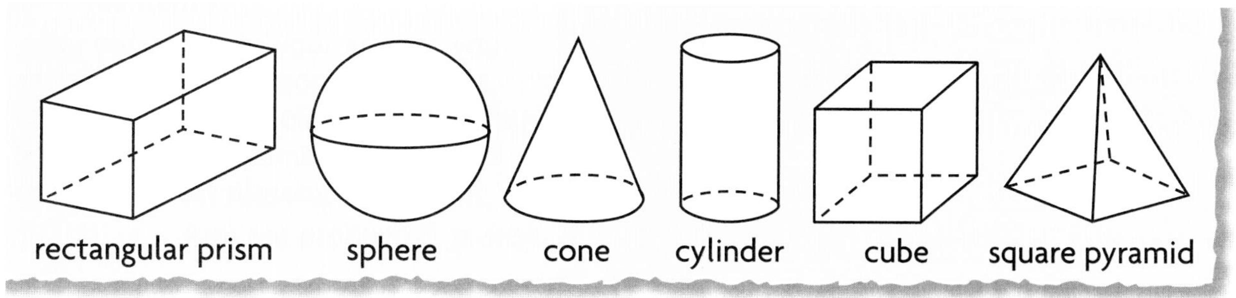

▶ Practice

Write the name of each solid figure.

1.

2. 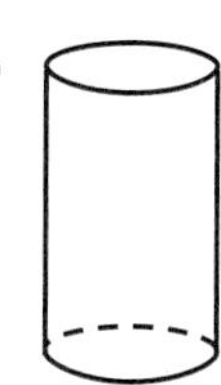

3. 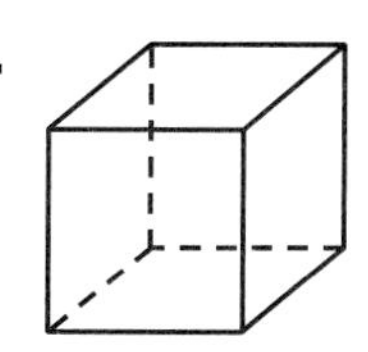

4. 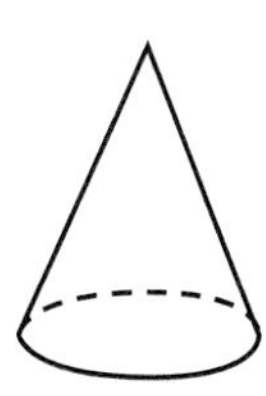

IDENTIFY PLANE FIGURES

These are plane figures.

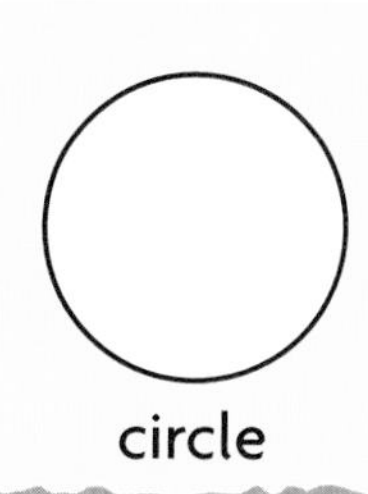
circle

rectangle

square

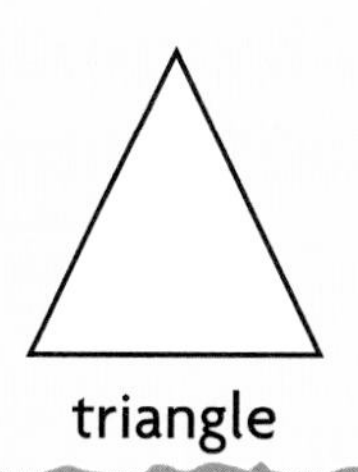
triangle

▶ Practice

Write the name of each plane figure.

1.

2.

3.

4.

TROUBLESHOOTING

SIDES AND CORNERS

A rectangle has 4 sides and 4 corners.

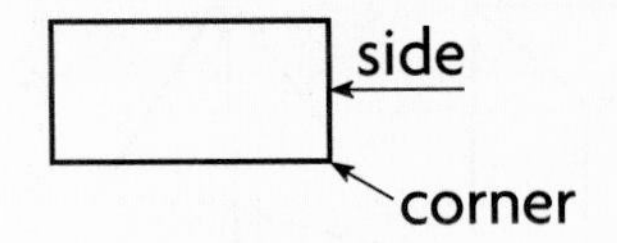

A triangle has 3 sides and 3 corners.

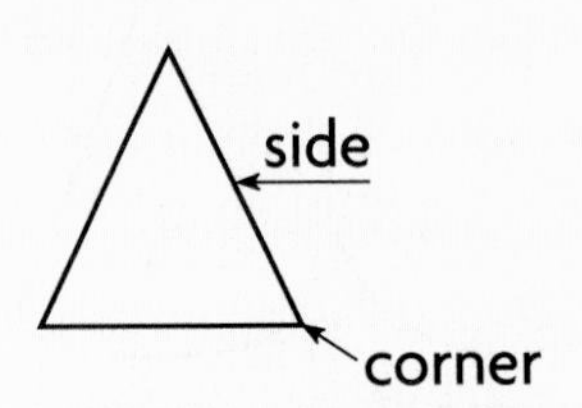

▶ Practice

Write the number of sides and corners in each figure.

1. 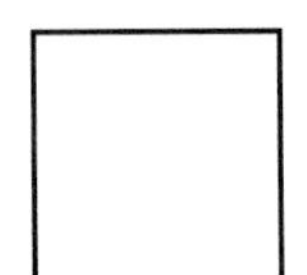

2. 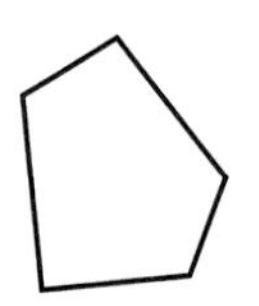

3. 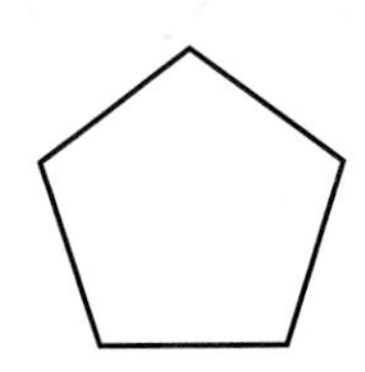

4. 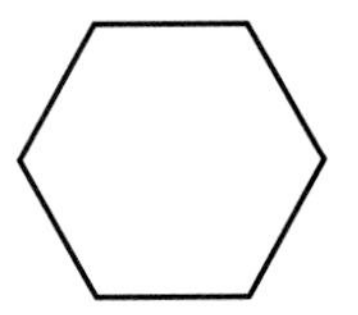

GEOMETRIC PATTERNS

You can find a pattern by looking for the place where a figure, or group of figures, repeats.

Example

Name the missing figure in the pattern. Describe the pattern.

○△□○△□○ _?_ □○△□

Think: The pattern has 3 figures: ○△□.

These figures repeat.

So, the missing figure is a triangle.

▶ Practice

Describe the pattern. Name the missing figure in the pattern.

1. ○△○△○△ _?_ △○△

2. ▯△△▯△△▯ _?_ △▯△△

SAME SIZE, SAME SHAPE

Figures that are the same size and shape are **congruent**.

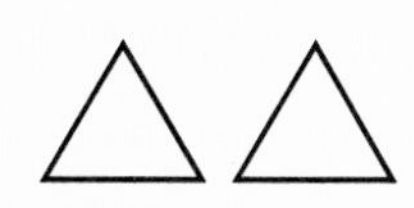

These figures have the same size and shape. They are **congruent**.

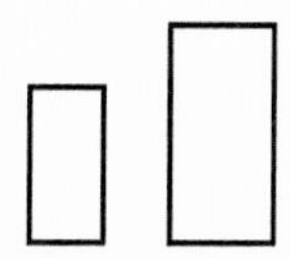

These figures have the same shape, but they are not the same size. They are **not congruent**.

▶ Practice

Tell whether the two figures are congruent. Write *yes* or *no*.

1. 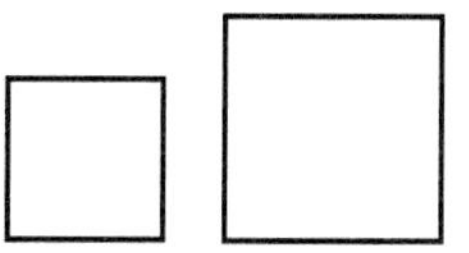
2. 
3. 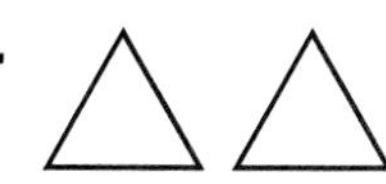
4. 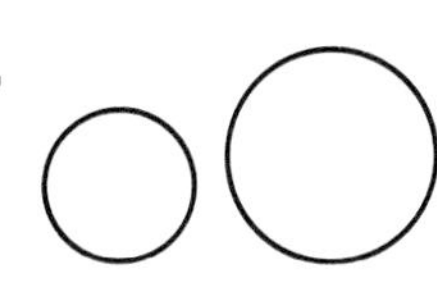
5. 
6. 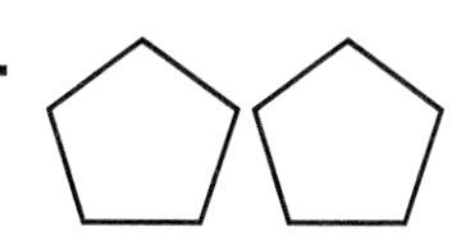

TYPES OF ANGLES

A **right angle** is an angle that forms a square corner.

This is a **right angle**.

This angle is **less than** a right angle.

This angle is **greater than** a right angle.

▶ Practice

Write if each angle is a *right angle, greater than* a right angle, or *less than* a right angle.

1. 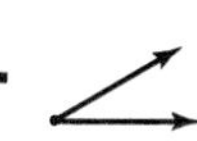
2. 
3. 
4. 
5. 
6. 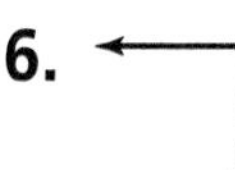

TROUBLESHOOTING

USE A CUSTOMARY RULER

An **inch (in.)** is used to measure length.
Use an inch ruler to measure.

- The red crayon is 2 inches long.

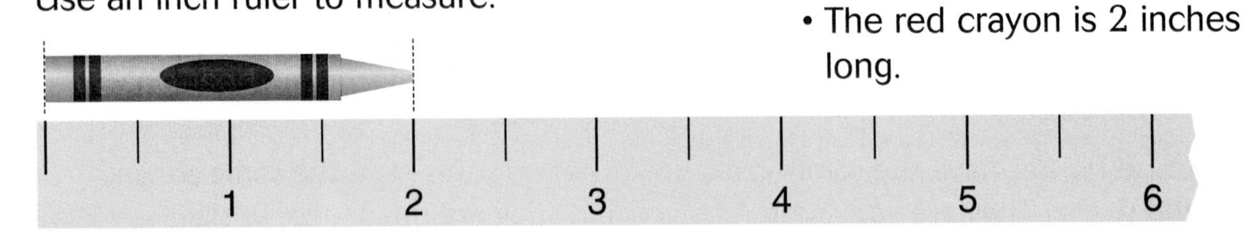

▶ Practice

For 1–2, use the drawing below.

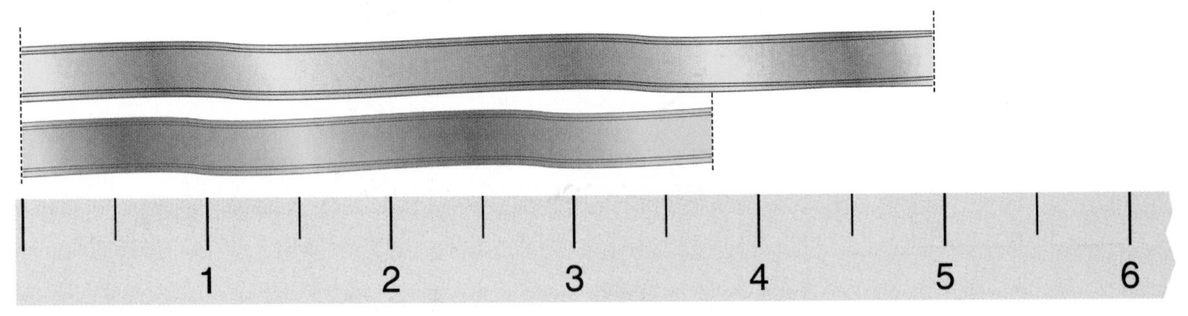

1. Which ribbon is about 5 inches long?
2. Which ribbon is between 3 and 4 inches long?

MEASURE TO THE NEAREST INCH

Example

Write the length to the nearest inch.

1 2 3

To measure to the nearest inch:

- Line up one end of the object with the left end of the ruler.
- Find the inch mark nearest the other end of the object.

The pencil is 3 inches long to the nearest inch.

▶ Practice

Write the length to the nearest inch.

1.

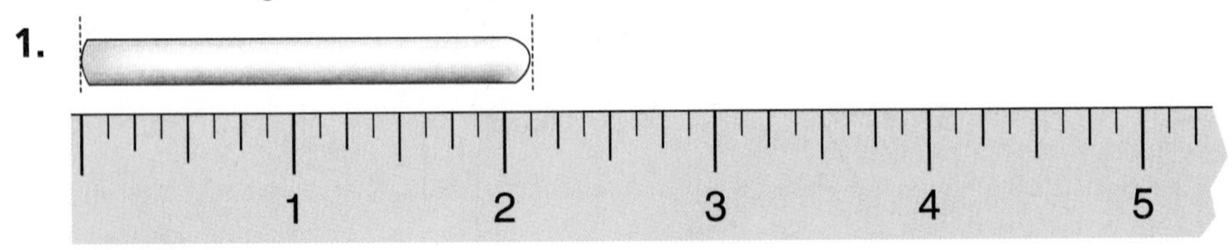

FIND A RULE

You can write a rule to describe a number pattern.

Example

Write a rule for the table. Then complete the table.

Tricycles	1	2	3	4	5	6
Wheels	3	6	9	12	■	■

Pattern: The number of wheels equals the number of tricycles times 3.

Rule: Multiply the number of tricycles by 3.

So, the missing numbers are 15 and 18.

▶ Practice

Write a rule for the table. Then complete the table.

1.

Packs	1	2	3	4	5	6
Cookies	6	12	18	24	■	■

2.

Chairs	3	6	2	5	1	4
Legs	12	24	8	20	■	■

USE A METRIC RULER

A **centimeter (cm)** is a metric unit used to measure length.

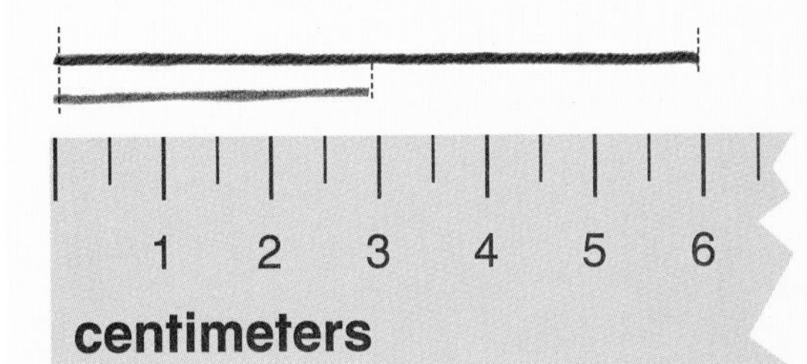

- The blue string is about 6 cm long.
- The red string is about 3 cm long.

▶ Practice

For 1–3, use the drawing below.

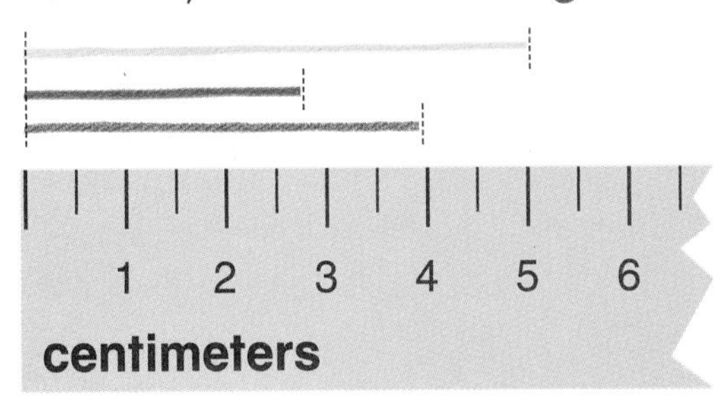

1. Which string is about 5 cm long?
2. Which string is about 1 cm longer than the pink string?
3. Which string is less than 3 cm long?

TROUBLESHOOTING

MEASURE TO THE NEAREST CENTIMETER

To measure to the nearest centimeter:

- Line up one end of the object with the left end of the ruler.
- Find the cm mark nearest the other end of the object.

Example

Write the length to the nearest centimeter.

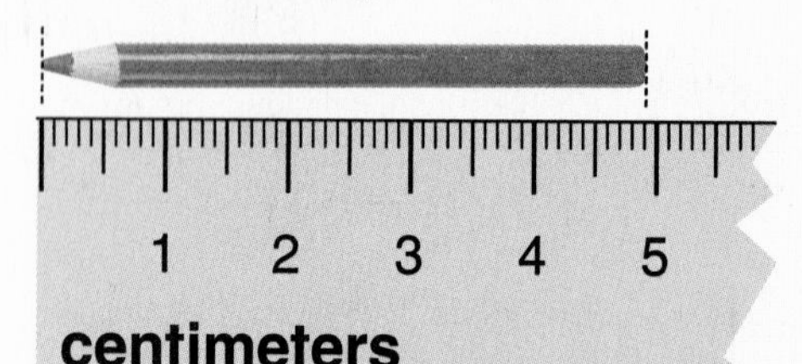

The pencil is 5 centimeters long to the nearest centimeter.

▶ **Practice**

Write the length to the nearest centimeter.

1.

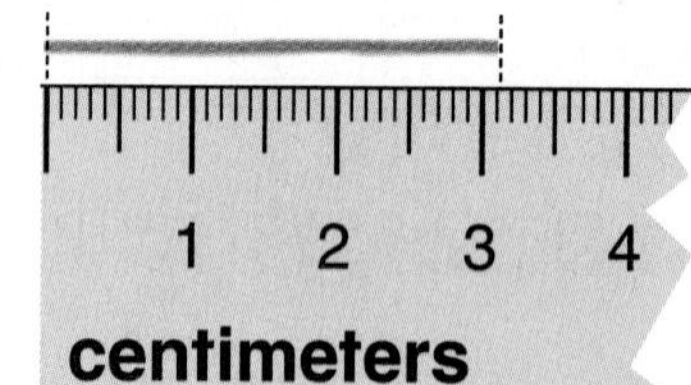

2.

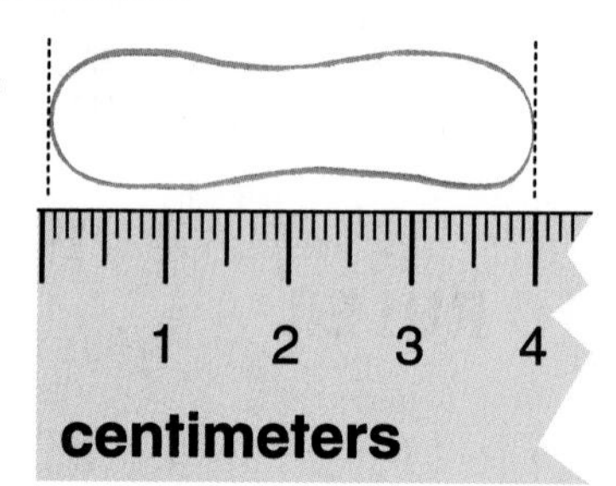

MODEL PARTS OF A WHOLE

This figure has 5 equal parts.

- 3 equal parts are shaded.
- 2 equal parts are *not* shaded.

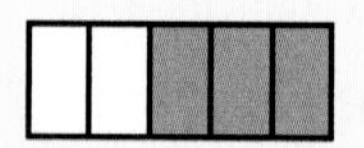

▶ **Practice**

Write how many equal parts are in the whole figure. Then write how many parts are shaded.

1.

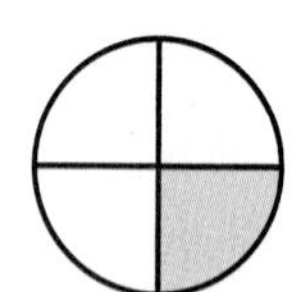

2.

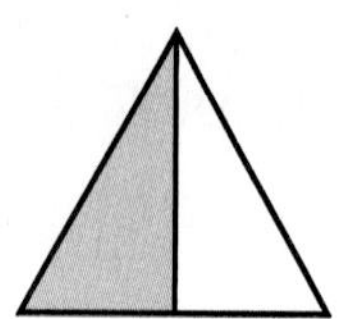

3.

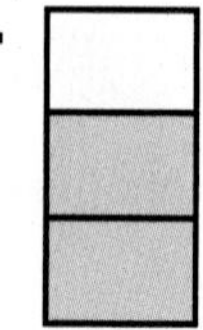

MODEL PARTS OF A GROUP

This group has 6 squares.

- 1 square is yellow.
- 5 squares are *not* yellow.

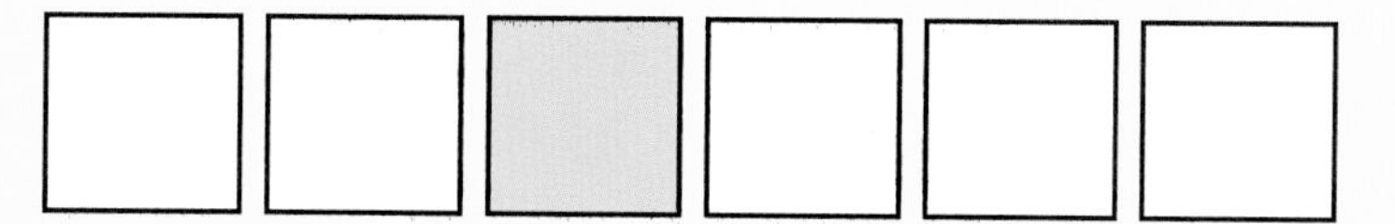

▶ Practice

Write the number in each group. Then write the number that are *not* yellow in each group.

1. ○○○ **2.** □□□□□ **3.** △△△△

NAME THE FRACTION

A number that names part of a whole or part of a group is called a **fraction**.

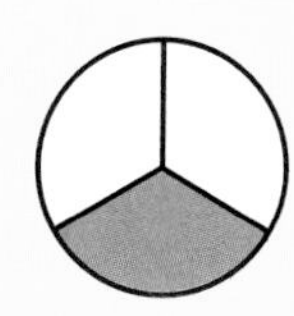

- 1 part is shaded.
- There are 3 parts in all.

So, $\frac{1}{3}$ of the circle is shaded.

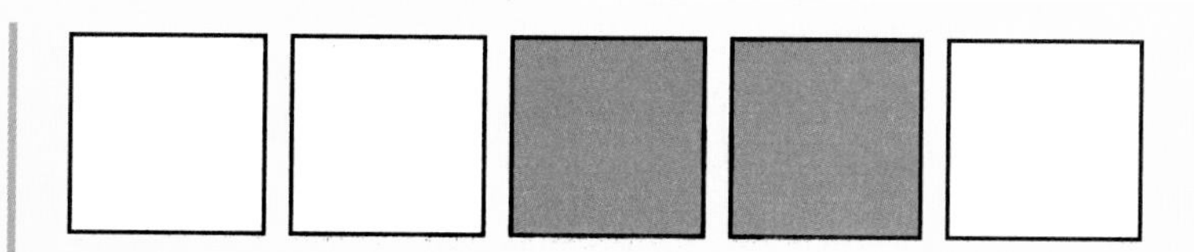

- 2 squares are blue.
- There are 5 squares in all.

So, $\frac{2}{5}$ of the squares are blue.

▶ Practice

Write the fraction for the part that is shaded.

1. 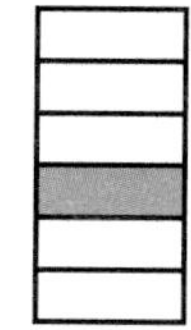

2. 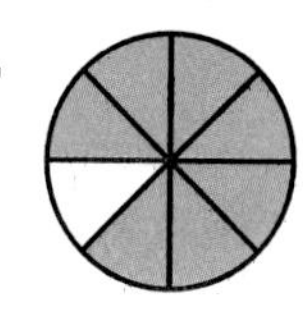

3. 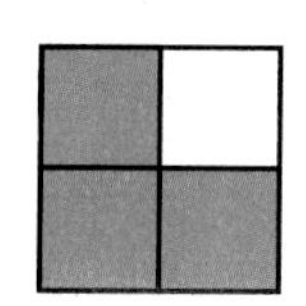

Write the fraction for the part that is blue.

4. 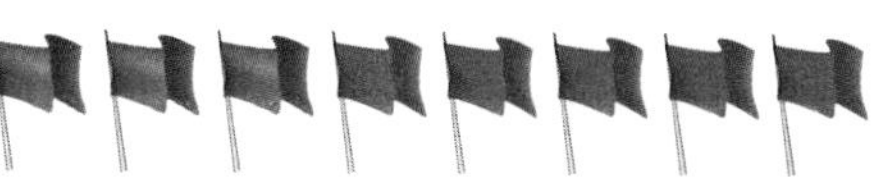

5. 

6. 

TROUBLESHOOTING

COMPARE FRACTIONS

As you compare fractions, remember the symbols you have used.

$<$ means *is less than.*
$1 < 3$

$>$ means *is greater than.*
$3 > 1$

Examples

A Compare $\frac{2}{4}$ and $\frac{3}{4}$.

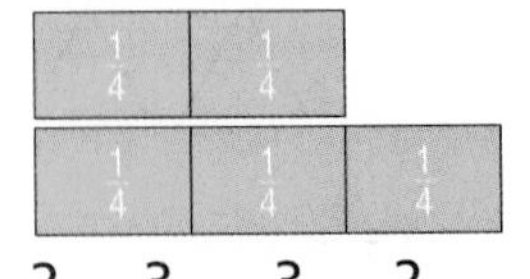

$\frac{2}{4} < \frac{3}{4}$ or $\frac{3}{4} > \frac{2}{4}$

B Compare $\frac{3}{5}$ and $\frac{4}{10}$.

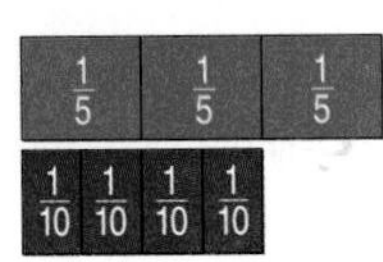

$\frac{3}{5} > \frac{4}{10}$ or $\frac{4}{10} < \frac{3}{5}$

▶ Practice

Compare. Write $<$ or $>$ for each ●.

1. 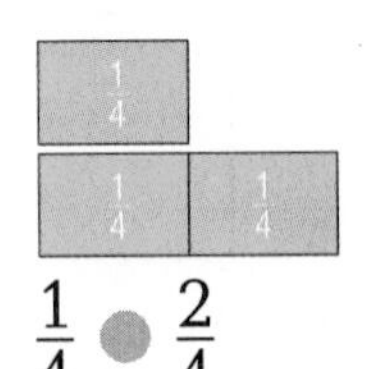

$\frac{1}{4}$ ● $\frac{2}{4}$

2. 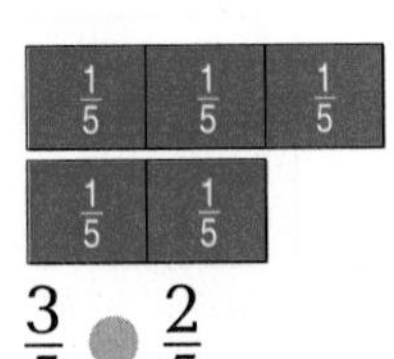

$\frac{3}{5}$ ● $\frac{2}{5}$

3. 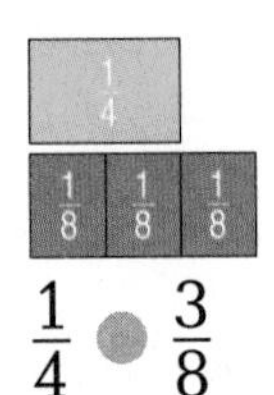

$\frac{1}{4}$ ● $\frac{3}{8}$

USE MONEY NOTATION

Use a **dollar sign** and a **decimal point** to write money amounts.

Example

Count and write the amount.

$0.25

$0.35

$0.40

$0.45

$0.46

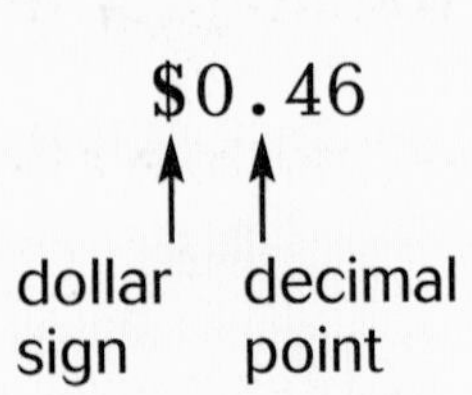

So, the value of the coins is $0.46.

▶ Practice

Count and write the amount.

1.          

2.     

NAME THE DECIMAL

You can use decimals to show tenths and hundredths.

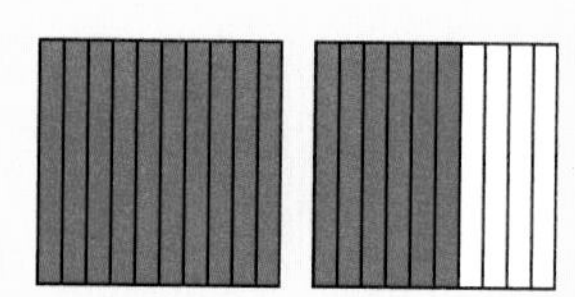

Mixed number: $1\frac{6}{10}$

Decimal: 1.6

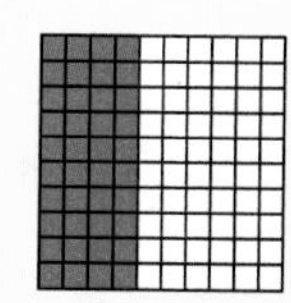

Fraction: $\frac{40}{100}$

Decimal: 0.40

▶ Practice

Write each fraction or mixed number as a decimal.

1. $\frac{3}{10}$ **2.** $\frac{9}{10}$ **3.** $\frac{45}{100}$ **4.** $1\frac{5}{10}$

5. $\frac{2}{100}$ **6.** $\frac{50}{100}$ **7.** $1\frac{7}{10}$ **8.** $1\frac{75}{100}$

COMPARE AND ORDER DECIMALS

Use a number line to compare and order decimals.

Example

Write 0.9, 0.5, and 0.8 in order from least to greatest.

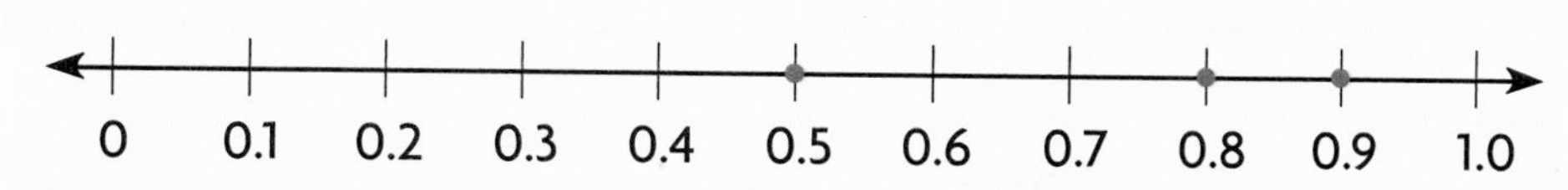

0.5 is to the left of 0.8 0.8 is to the left of 0.9

So, the numbers from least to greatest are 0.5, 0.8, 0.9.

▶ Practice

Use the number line above. Write < or > for each ●.

1. 0.6 ● 0.5 **2.** 0.3 ● 0.6 **3.** 0.7 ● 0.4 **4.** 0.9 ● 1.0

Use the number line to order the decimals from least to greatest.

5. 0.1, 0.7, 0.4 **6.** 1.0, 0.2, 0.1 **7.** 0.9, 0.8, 0.7 **8.** 0.8, 0.4, 1.0

Use the number line to order the decimals from greatest to least.

9. 0.4, 0.9, 0.2 **10.** 0.1, 0.8, 1.0 **11.** 0.7, 0.3, 0.5 **12.** 0.6, 0.2, 0.8

CHAPTER 1

EXTRA PRACTICE

Set A (pp. 4–5)

Write each number in standard form.

1. 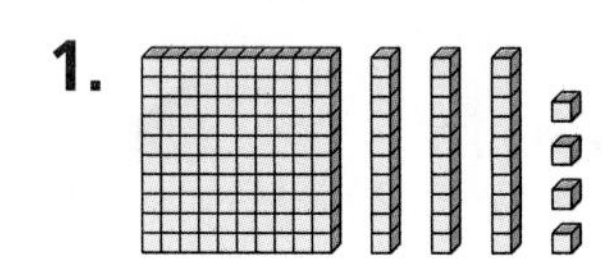

2. 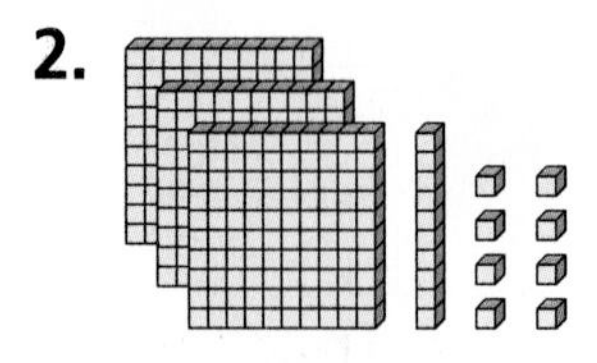

3. 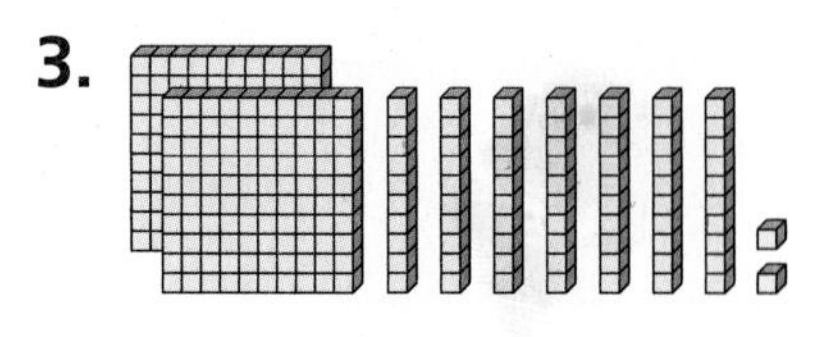

4. 500 + 60 + 6
5. 700 + 4
6. 800 + 10 + 4
7. four hundred seventy-six
8. nine hundred ninety-one

Write the value of the blue digit.

9. 346
10. 872
11. 13
12. 554

Set B (pp. 6–9)

Write in standard form.

1. 1,000 + 900 + 40 + 2
2. 5,000 + 700 + 80 + 3
3. 3,000 + 5
4. 9,000 + 900 + 10 + 4
5. two thousand, four hundred sixty-seven
6. eight thousand, eighteen

Write in expanded form.

7. 5,487
8. 6,055
9. 6,170
10. 7,796

Set C (pp. 10–11)

Write in standard form.

1. 10,000 + 6,000 + 900 + 60 + 5
2. 50,000 + 3,000 + 400 + 80 + 6
3. 30,000 + 6,000 + 1
4. 90,000 + 5,000 + 900 + 10 + 3
5. fifty-one thousand, four hundred
6. twenty-two thousand, eighteen
7. seventy thousand, forty-nine
8. ninety-nine thousand, four

Write in expanded form.

9. 65,487
10. 76,055
11. 36,173
12. 47,796

Set A (pp. 18–19)

For 1–2, choose a benchmark of 10, 100, or 500 to estimate each.

1. the number of pieces in a small bag of dog food
2. the number of teeth in your mouth
3. There are 25 students in Ken's class. There are 4 third-grade classes. About how many students are in the third grade?

Set B (pp. 20–23)

Compare the numbers. Write <, >, or = for each ●.

1. 400 ● 12
2. 646 ● 600
3. 741 ● 741
4. 57 ● 75
5. 165 ● 164
6. 313 ● 515

Set C (pp. 24–25)

Write the numbers in order from least to greatest.

1. 124; 562; 347
2. 102; 89; 157
3. 702; 546; 212

Write the numbers in order from greatest to least.

4. 1,466; 1,365; 1,988
5. 2,218; 3,010; 2,115
6. 5,010; 5,100; 5,310

Set D (pp. 28–29)

Round to the nearest hundred and the nearest ten.

1. 414
2. 888
3. 502
4. 635
5. 157
6. 733
7. 374
8. 498

Set E (pp. 30–31)

Round to the nearest thousand.

1. 3,345
2. 8,866
3. 5,533
4. 6,500
5. 9,457
6. 1,168
7. 7,662
8. 2,220

CHAPTER 3

EXTRA PRACTICE

Set A (pp. 36–37)

Use the Grouping Property to find the sum.

1. 7 + 8 + 3
2. 7 + 13 + 9
3. 16 + 14 + 27
4. 12 + 37 + 18
5. 31 + 19 + 76

6. 6 + 6 + 4 = ■
7. 27 + 10 + 13 = ■
8. 3 + 7 + 18 = ■

9. Ron bought 8 goldfish, 15 guppies, and 12 minnows. How many fish did he buy?

Set B (pp. 38–39)

Estimate the sum.

1. 11 + 23
2. 24 + 17
3. 63 + 27
4. 165 + 220
5. 298 + 456
6. 645 + 594
7. 584 + 248
8. 2,375 + 4,082
9. 560 + 439
10. 6,757 + 4,446

Set C (pp. 42–43)

Find the sum. Estimate to check.

1. 465 + 521
2. 978 + 200
3. 789 + 123
4. $3.12 + $5.29
5. 919 + 453

6. 327 + 246 = ■
7. 299 + 403 = ■
8. 544 + 671 = ■

Set D (pp. 46–49)

Find the sum. Estimate to check.

1. 3,120 + 5,771
2. 2,899 + 901
3. 6,247 + 2,319
4. 7,231 + 1,191
5. 3,485 + 9,856

6. Estimate to decide if the sum of 1,874 + 3,205 is greater than 4,000. Explain your answer.

7. ALGEBRA Write the missing addend. 4,000 + ■ = 5,100

Set A (pp. 54–55)

Estimate the difference.

1. $63 - 49$
2. $54 - 14$
3. $98 - 55$
4. $595 - 227$
5. $873 - 475$
6. $795 - 309$
7. $7,850 - 2,187$
8. $8,026 - 4,826$
9. $5,575 - 4,746$
10. $2,521 - 1,779$

Set B (pp. 58–61)

Find the difference. Estimate to check.

1. $205 - 67$
2. $608 - 409$
3. $500 - 165$
4. $900 - 198$
5. $402 - 317$
6. ALGEBRA The sum of 885 and another number is 901. What is the other number?
7. **MENTAL MATH** How might you use mental math to find $500 - 199$?

Set C (pp. 62–65)

Find the difference. Tell what method you used.

1. $4,561 - 4,327$
2. $8,135 - 3,645$
3. $7,608 - 5,810$
4. $4,005 - 3,318$
5. $8,922 - 902$
6. Lindsey paddled 4,033 feet in her canoe. Ronald paddled 2,077 feet in his. How much farther did Lindsey paddle than Ronald?
7. **REASONING** The library was built 6 years before the post office was built. The post office was built in 1971. How old was the library in 1988?

Set D (pp. 68–69)

Write + or – to make the number sentence true.

1. 8 ● $15 = 23$
2. 95 ● $16 = 79$
3. 517 ● $483 = 1,000$
4. $35 = 29$ ● 6
5. 42 ● $9 = 51$
6. 27 ● $4 = 23$
7. Shelby had $18.50. She spent $7.25 at the car wash. Write an expression that shows how much she has left.

CHAPTER 5

EXTRA PRACTICE

Set A (pp. 84–85)

Use $<$ or $>$ to compare the amounts of money.

1. a. 

b. 

2. a. 

b. 

3. a. 

b. 

4. Maria has 3 quarters, 3 dimes, and 1 nickel. She wants to buy a slice of pie for $1.10. Does she have enough money? Explain.

5. Ronnie has 1 quarter, 4 dimes, and 3 nickels. Lydia has 3 quarters. Who has more money? Explain.

Set B (pp. 88–89)

Find the sum or difference. Estimate to check.

1.	2.	3.	4.	5.
$3.35 +$2.14	$8.56 −$4.45	$7.08 −$5.35	$6.45 +$5.77	$4.07 +$1.96

6.	7.	8.	9.	10.
$8.00 −$4.44	$9.05 +$8.96	$5.40 +$2.95	$9.42 −$3.58	$10.00 −$ 5.65

USE DATA For 11–12, use the table.

11. How much more does a pound of boysenberries cost than an apricot?

12. Ezra buys 2 plums and a pound of grapes with a $5 bill. The clerk gives him two $1 bills, 2 quarters, 3 dimes, and 7 pennies. Is the change correct? Explain.

FREIDA'S FRUIT STAND

Fruit	Price
Apricots	$0.65 each
Plums	$0.45 each
Grapes	$1.33 each pound
Apples	$0.55 each
Boysenberries	$1.79 each pound

Set A (pp. 96–97)

Write two ways you can read each time. Then write the time, using A.M. or P.M.

1. 
recess

2. play at park

3. 
plant flowers

4. 1:49
go to library

Set B (pp. 100–101)

Copy and complete the schedule.

	MONDAY NIGHT ON CHANNEL 8		
	Program	**Time**	**Elapsed Time**
1.	Game Show	4:00 P.M. – 5:00 P.M.	■
2.	Evening News	5:00 P.M. – 5:45 P.M.	■
3.	Basketball Game	5:45 P.M. – ■	2 hours 30 minutes
4.	Mystery Theater	■ – 10:00 P.M.	1 hour 45 minutes
5.	Nighttime News	10:00 P.M. – 10:30 P.M.	■

Set C (pp. 102–103)

Solve. Use the calendars.

November						
Sun	Mon	Tue	Wed	Thu	Fri	Sat
					1	2
3	4	5	6	7	8	9
10	11	12	13	14	15	16
17	18	19	20	21	22	23
24	25	26	27	28	29	30

December						
Sun	Mon	Tue	Wed	Thu	Fri	Sat
1	2	3	4	5	6	7
8	9	10	11	12	13	14
15	16	17	18	19	20	21
22	23	24	25	26	27	28
29	30	31				

1. Mr. Burns went on a trip from November 24 to December 6. How many days was he gone?

2. Enya went on a trip for 3 weeks and 3 days. She returned December 4. When did Enya leave?

3. What date is 2 weeks before November 19?

4. What day of the week is 3 days after December 14?

CHAPTER 7

EXTRA PRACTICE

Set A (pp. 116–117)

Copy and complete.

1.

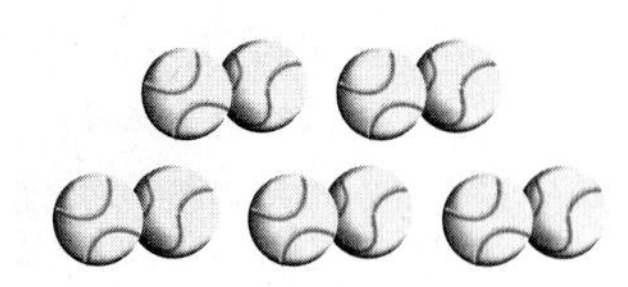

a. ■ groups of ■ = ■

b. ■ + ■ + ■ + ■ + ■ = ■

c. ■ × ■ = ■

2.

a. ■ groups of ■ = ■

b. ■ + ■ + ■ + ■ = ■

c. ■ × ■ = ■

For 3–8, find how many in all.

3. 5 groups of 3

4. $4 + 4 + 4 + 4$

5. $7 + 7 + 7$

6. 5×4

7. 4 groups of 5

8. 5×5

9. Ana bought 6 packages of 5 cards each. How many cards did she buy?

Set B (pp. 118–119)

Find the product.

1. $7 \times 2 =$ ■

2. $5 \times 2 =$ ■

3. $7 \times 5 =$ ■

4. $8 \times 2 =$ ■

5. $4 \times 5 =$ ■

6. $9 \times 5 =$ ■

7. ■ $= 6 \times 5$

8. ■ $= 2 \times 2$

9. ■ $= 5 \times 5$

10. Keith bought 5 packages of toy cars. Each package has 3 cars in it. David has 16 toy cars. Who has more cars? How do you know?

Set C (pp. 122–125)

Find the product.

1. $7 \times 3 =$ ■

2. $5 \times 3 =$ ■

3. ■ $= 3 \times 9$

4. $4 \times 5 =$ ■

5. ■ $= 3 \times 3$

6. $8 \times 5 =$ ■

7. $\begin{array}{r} 6 \\ \times 3 \\ \hline \end{array}$

8. $\begin{array}{r} 5 \\ \times 6 \\ \hline \end{array}$

9. $\begin{array}{r} 3 \\ \times 8 \\ \hline \end{array}$

10. $\begin{array}{r} 3 \\ \times 9 \\ \hline \end{array}$

11. $\begin{array}{r} 5 \\ \times 9 \\ \hline \end{array}$

Set A (pp. 132–133)

Find the product.

1. $0 \times 5 = \blacksquare$
2. $3 \times 7 = \blacksquare$
3. $1 \times 7 = \blacksquare$
4. $4 \times 3 = \blacksquare$
5. $\blacksquare = 6 \times 3$
6. $\blacksquare = 8 \times 5$
7. $\blacksquare = 0 \times 9$
8. $\blacksquare = 1 \times 1$
9. Is the product of 3 and 0 *greater than, less than,* or *equal to* the product of 0 and 6? Explain.

Set B (pp. 134–135)

Find the product.

1. 8×4
2. 4×0
3. 4×6
4. 4×4
5. 4×9
6. $3 \times 4 = \blacksquare$
7. $\blacksquare = 2 \times 4$
8. $\blacksquare = 4 \times 7$
9. $5 \times 4 = \blacksquare$
10. Mario has 4 packs of 8 stickers. He also has 19 loose stickers. How many stickers does he have in all?

Set C (pp. 138–141)

Find the product.

1. $4 \times 6 = \blacksquare$
2. $5 \times 3 = \blacksquare$
3. $8 \times 0 = \blacksquare$
4. $9 \times 5 = \blacksquare$
5. $5 \times 8 = \blacksquare$
6. $3 \times 6 = \blacksquare$
7. $7 \times 4 = \blacksquare$
8. $3 \times 9 = \blacksquare$
9. $8 \times 2 = \blacksquare$
10. $3 \times 7 = \blacksquare$
11. $6 \times 5 = \blacksquare$
12. $4 \times 8 = \blacksquare$
13. A movie is shown 5 times each day. How many times is that movie shown in one week?

Set D (pp. 142–143)

Find the missing factor.

1. $\blacksquare \times 9 = 18$
2. $5 \times \blacksquare = 20$
3. $\blacksquare \times 1 = 8$
4. $\blacksquare \times 9 = 9$
5. $2 \times \blacksquare = 14$
6. $4 \times \blacksquare = 16$
7. $3 \times \blacksquare = 21$
8. $\blacksquare \times 8 = 32$
9. Jill has 9 baskets with an equal number of eggs in each. If she has 36 eggs in all, how many are in each basket?

EXTRA PRACTICE

Set A (pp. 148–149)

Find each product.

1. $5 \times 6 = ■$ **2.** $6 \times 2 = ■$ **3.** $0 \times 6 = ■$ **4.** $6 \times 7 = ■$

5. $6 \times 4 = ■$ **6.** $6 \times 3 = ■$ **7.** $■ = 8 \times 6$ **8.** $4 \times 7 = ■$

9. $8 \times 2 = ■$ **10.** $■ = 6 \times 6$ **11.** $6 \times 1 = ■$ **12.** $6 \times 9 = ■$

Complete.

13. $■ \times 4 = 16$ **14.** $■ \times 5 = 35$ **15.** $2 \times ■ = 8$ **16.** $8 \times ■ = 24$

Set B (pp. 150–151)

Find each product.

1. $1 \times 7 = ■$ **2.** $■ = 7 \times 5$ **3.** $6 \times 7 = ■$ **4.** $3 \times 7 = ■$

5. $2 \times 7 = ■$ **6.** $3 \times 3 = ■$ **7.** $7 \times 7 = ■$ **8.** $4 \times 7 = ■$

9. $8 \times 1 = ■$ **10.** $0 \times 7 = ■$ **11.** $■ = 7 \times 8$ **12.** $9 \times 7 = ■$

13. Alex has 8 bags of rocks. Each bag has 7 rocks. Vera takes 3 bags. How many rocks does Alex have left?

Set C (pp. 152–153)

Find each product.

1. $7 \times 8 = ■$ **2.** $8 \times 0 = ■$ **3.** $8 \times 3 = ■$ **4.** $6 \times 8 = ■$

5. $■ = 8 \times 4$ **6.** $■ = 2 \times 5$ **7.** $8 \times 1 = ■$ **8.** $8 \times 8 = ■$

9. $8 \times 2 = ■$ **10.** $5 \times 8 = ■$ **11.** $2 \times 2 = ■$ **12.** $8 \times 9 = ■$

13. Dolores has 8 bags of 4 apples. Ruth has 3 bags of 8 apples. How many more apples does Dolores have?

Set D (pp. 156–159)

Find each product.

1. $8 \times 7 = ■$ **2.** $6 \times 7 = ■$ **3.** $■ = 6 \times 4$ **4.** $7 \times 3 = ■$

5. $9 \times 8 = ■$ **6.** $8 \times 6 = ■$ **7.** $7 \times 7 = ■$ **8.** $3 \times 5 = ■$

9. $9 \times 7 = ■$ **10.** $■ = 6 \times 5$ **11.** $8 \times 8 = ■$ **12.** $9 \times 6 = ■$

Set A (pp. 164–167)

Find the product.

1. $9 \times 7 = ■$
2. $6 \times 9 = ■$
3. $6 \times 10 = ■$
4. $10 \times 3 = ■$
5. $3 \times 9 = ■$
6. $8 \times 9 = ■$
7. $10 \times 7 = ■$
8. $9 \times 5 = ■$
9. $9 \times 4 = ■$
10. $10 \times 5 = ■$
11. $8 \times 10 = ■$
12. $9 \times 2 = ■$

Find the missing factor.

13. $■ \times 9 = 27$
14. $90 = 9 \times ■$
15. $7 \times ■ = 56$
16. $21 = ■ \times 7$

Set B (pp. 168–169)

Write a rule for the table. Then copy and complete the table.

1.

PACKS	1	2	3	4	5	6
CARDS	4	8	12	■	■	■

2.

BAGS	1	2	3	4	5	6
ORANGES	6	12	18	■	■	■

3.

CANS	1	2	3	4	5	6
TENNIS BALLS	3	6	9	■	■	■

4.

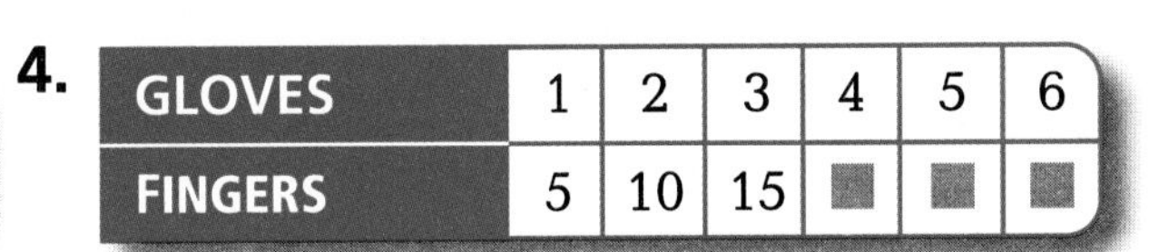

GLOVES	1	2	3	4	5	6
FINGERS	5	10	15	■	■	■

For 5–6, use the table in Exercise 3.

5. If you had 18 tennis balls, how many cans would you have?
6. Suppose you had 7 cans. How many tennis balls would you have?

Set C (pp. 170–171)

Find each product.

1. $(4 \times 1) \times 3 = ■$
2. $(3 \times 2) \times 3 = ■$
3. $(5 \times 1) \times 5 = ■$
4. $4 \times (2 \times 2) = ■$
5. $10 \times (3 \times 3) = ■$
6. $(6 \times 1) \times 7 = ■$

Find the missing factor.

7. $(4 \times ■) \times 1 = 16$
8. $5 \times (2 \times ■) = 20$
9. $■ \times (7 \times 1) = 49$
10. Mason made 2 sandwiches for each of 3 friends. He used 2 slices of bread for each sandwich. How many slices of bread did he use?

CHAPTER 11

EXTRA PRACTICE

Set A (pp. 186–187)

Write a division sentence for each.

1. $18 - 6 = 12$; $12 - 6 = 6$; $6 - 6 = 0$

2. $36 - 9 = 27$; $27 - 9 = 18$; $18 - 9 = 9$; $9 - 9 = 0$

Use subtraction to solve.

3. $15 \div 5 = ■$ **4.** $18 \div 3 = ■$ **5.** $12 \div 4 = ■$ **6.** $16 \div 4 = ■$

7. $7\overline{)14}$ **8.** $5\overline{)20}$ **9.** $3\overline{)24}$ **10.** $8\overline{)40}$

Set B (pp. 188–191)

Complete each number sentence. Draw an array to help.

1. $4 \times ■ = 8$ $8 \div 4 = ■$ **2.** $6 \times ■ = 30$ $30 \div 6 = ■$

3. $8 \times ■ = 32$ $32 \div 8 = ■$ **4.** $4 \times ■ = 12$ $12 \div 4 = ■$

5. What division sentence could you write for an array that shows $5 \times 8 = 40$?

6. How can you use $5 + 5 + 5 + 5 = 20$ to help you find $20 \div 5$?

Find the number that the variable stands for.

7. $2 \times a = 12$
$a =$ ___?___

8. $b \times 4 = 36$
$b =$ ___?___

9. $20 \div 4 = c$
$c =$ ___?___

Set C (pp. 192–195)

Write the fact family.

1. 2, 3, 6 **2.** 3, 7, 21 **3.** 3, 9, 27 **4.** 3, 6, 18

5. 4, 6, 24 **6.** 4, 8, 32 **7.** 5, 5, 25 **8.** 3, 8, 24

Find the quotient or the missing divisor.

9. $6 \div ■ = 3$ **10.** $18 \div 6 = ■$ **11.** $■ = 12 \div 3$ **12.** $20 \div ■ = 4$

13. $30 \div 6 = ■$ **14.** $3 = 15 \div ■$ **15.** $9 \div ■ = 3$ **16.** $16 \div 4 = ■$

17. Jerome made 30 cookies. He ate 3 and divided the rest equally among 3 friends. How many cookies did each friend get?

Set A (pp. 202–203)

Find each missing factor and quotient.

1. $2 \times \blacksquare = 10 \qquad 10 \div 2 = \blacksquare$
2. $5 \times \blacksquare = 30 \qquad 30 \div 5 = \blacksquare$

Find each quotient.

3. $15 \div 5 = \blacksquare$
4. $\blacksquare = 16 \div 2$
5. $\blacksquare = 45 \div 5$
6. $10 \div 5 = \blacksquare$
7. $2\overline{)2}$
8. $5\overline{)20}$
9. $2\overline{)18}$
10. $5\overline{)25}$
11. $2\overline{)12}$
12. Divide 20 by 2.
13. Divide 35 by 5.
14. Divide 6 by 2.

Set B (pp. 204–205)

Write the multiplication fact you can use to find the quotient. Then write the quotient.

1. $18 \div 3 = \blacksquare$
2. $32 \div 4 = \blacksquare$
3. $9 \div 3 = \blacksquare$

Find each quotient.

4. $28 \div 4 = \blacksquare$
5. $12 \div 3 = \blacksquare$
6. $\blacksquare = 27 \div 3$
7. $\blacksquare = 8 \div 4$
8. $4\overline{)16}$
9. $3\overline{)15}$
10. $4\overline{)24}$
11. $3\overline{)21}$
12. $4\overline{)12}$
13. Divide 30 by 3.
14. Divide 20 by 4.
15. Divide 36 by 4.

Set C (pp. 206–207)

Find each quotient.

1. $0 \div 4 = \blacksquare$
2. $\blacksquare = 3 \div 3$
3. $\blacksquare = 8 \div 1$
4. $10 \div 10 = \blacksquare$
5. $7\overline{)7}$
6. $8\overline{)0}$
7. $1\overline{)4}$
8. $3\overline{)0}$
9. $9\overline{)9}$
10. Divide 8 by 8.
11. Divide 0 by 6.
12. Divide 9 by 1.

Set D (pp. 208–209)

Write an expression to describe each problem.

1. Four friends share 28 stickers equally. How many stickers does each friend get?
2. Melinda had \$15. She buys slippers for \$8. How much money does she have now?

CHAPTER 13

EXTRA PRACTICE

Set A (pp. 216–219)

Find the missing factor and quotient.

1. $8 \times ■ = 32$ $32 \div 8 = ■$
2. $7 \times ■ = 35$ $35 \div 7 = ■$

Find the quotient.

3. $42 \div 6 = ■$
4. $■ = 24 \div 4$
5. $64 \div 8 = ■$
6. $■ = 21 \div 7$
7. $7\overline{)49}$
8. $2\overline{)2}$
9. $6\overline{)36}$
10. $5\overline{)40}$
11. $8\overline{)48}$
12. Divide 63 by 7.
13. Divide 80 by 8.
14. Divide 15 by 3.

Set B (pp. 220–221)

Find the quotient.

1. $■ = 36 \div 9$
2. $20 \div 10 = ■$
3. $■ = 20 \div 5$
4. $54 \div 6 = ■$
5. $3\overline{)12}$
6. $10\overline{)70}$
7. $9\overline{)72}$
8. $4\overline{)16}$
9. $9\overline{)27}$
10. Divide 56 by 8.
11. Divide 60 by 6.
12. Divide 100 by 10.

Set C (pp. 222–225)

Find the quotient.

1. $6 \div 6 = ■$
2. $■ = 7 \div 1$
3. $■ = 0 \div 5$
4. $28 \div 7 = ■$
5. $9\overline{)81}$
6. $3\overline{)18}$
7. $6\overline{)24}$
8. $2\overline{)20}$
9. $7\overline{)0}$
10. Divide 32 by 4.
11. Divide 40 by 10.
12. Divide 16 by 8.

Set D (pp. 226–227)

PRICE LIST	
Mugs	$4
Aprons	$8

USE DATA For 1–4, use the price list at the right to find the cost of each number of items.

1. 3 aprons
2. 5 mugs
3. 6 aprons
4. 8 mugs

Find the cost of one of each item.

5. 2 pizzas cost $14.
6. 4 tapes cost $32.
7. 5 books cost $25.
8. 6 pens cost $12.
9. 7 balls cost $21.
10. 3 shirts cost $27.

Set A (pp. 242–243)

For 1–3, use the tally table.

FAVORITE HOBBY	
Hobby	**Tally**
Collecting stamps	||||
Collecting sports cards	𝍸 𝍸
Collecting coins	||
Reading	𝍸 |||
Drawing	𝍸

1. How many people answered the survey?
2. What is the most popular hobby?
3. How many fewer people chose collecting coins than chose reading?

For 4–6, use the frequency table.

4. How many people answered the survey?
5. Did more people choose peas or carrots?
6. How many more people chose corn than chose broccoli?

FAVORITE VEGETABLE	
Type	**Number**
Carrots	6
Peas	6
Beans	3
Corn	12
Broccoli	1

Set B (pp. 244–245)

For 1–3, use the table.

FAVORITE BREAKFAST FOOD				
	Bacon and Eggs	**French Toast**	**Cereal**	**Muffins**
Boys	9	6	4	1
Girls	4	8	5	3

1. How many girls were surveyed?
2. How many boys liked French toast the best?
3. How many students chose bacon and eggs?

For 4–6, use the table.

4. How many students are wearing blue shoes?
5. How many students are wearing white or black shoes?
6. How many more students are wearing blue shoes than red shoes?

SHOE COLOR				
	White	**Black**	**Red**	**Blue**
Boys	9	4	3	5
Girls	8	3	2	4

EXTRA PRACTICE

Set A (pp. 254–255)

For 1–4, use the bar graph.

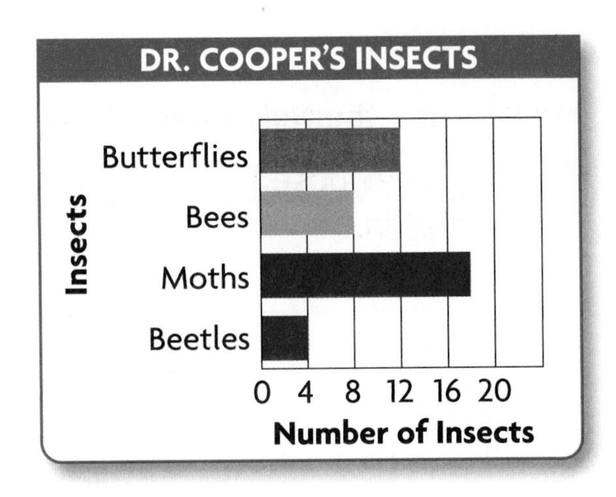

1. How many moths are there?
2. What is the total number of insects?
3. How many more butterflies are there than beetles?
4. If Dr. Cooper sold five of his bees, how many bees would be left?

Set B (pp. 258–261)

For 1–3, use the line plot.

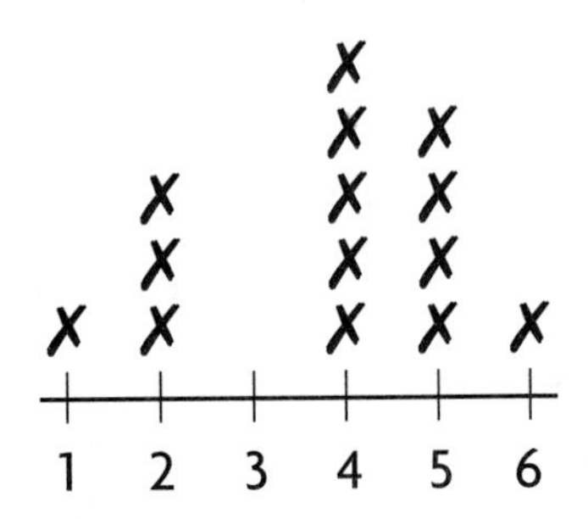

1. What is the range of the data?
2. What is the mode for this data? Explain.
3. How many students were surveyed?

Set C (pp. 262–263)

For 1–6, use the grid. Write the letter of the point named by the ordered pair.

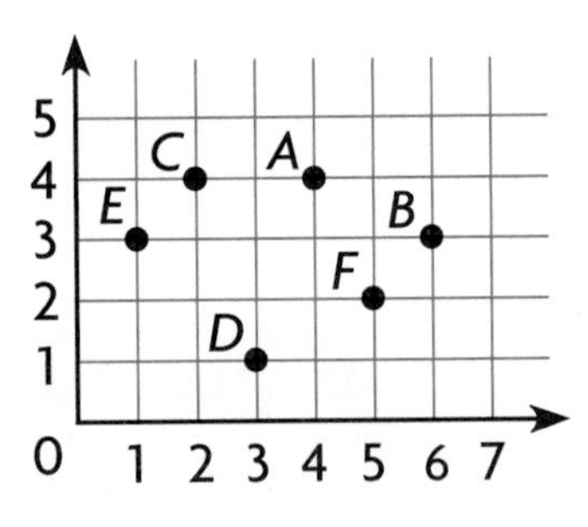

1. (3,1)
2. (6,3)
3. (5,2)
4. (1,3)
5. (4,4)
6. (2,4)

Set D (pp. 264–265)

For 1–2, use the line graph.

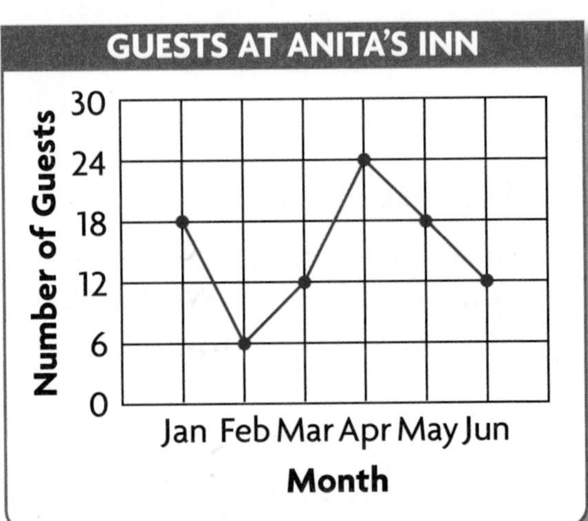

1. In what month did Anita's Inn have the most guests? In what month were there the fewest guests?
2. How many guests stayed at Anita's Inn from January through June?

Set A (pp. 270–271)

Tell whether each event is *certain* or *impossible*.

1. You can choose a nickel from these coins.

2. You can spin red or blue on this spinner.

Set B (pp. 272–273)

1. Name the color that you are most likely to spin.

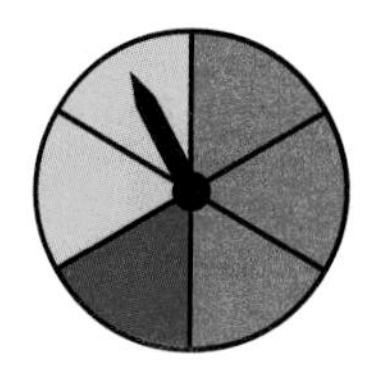

2. Name the color marble that is least likely to be pulled.

Set C (pp. 276–279)

1. Tenesha pulled animal crackers from the box. She made the graph below of the outcomes. Which animals are equally likely to be pulled from the box?

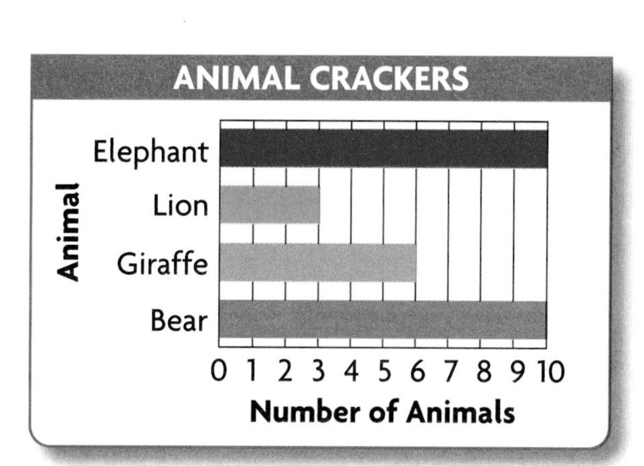

2. Ned pulled marbles from the bag. He made the graph below of the outcomes. Which color marble is most likely to be pulled from the bag? Which is least likely to be pulled?

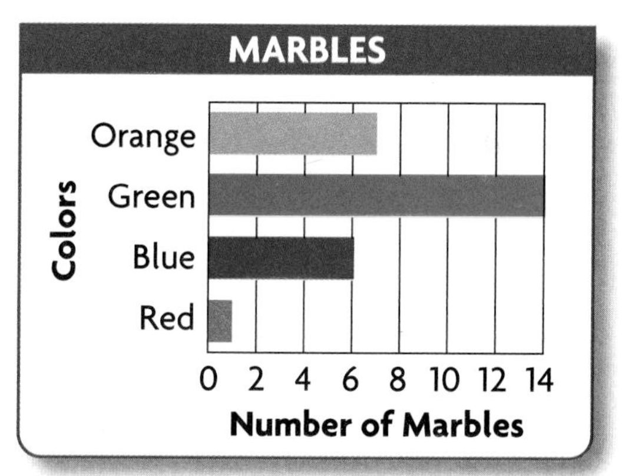

Set D (pp. 280–281)

1. This line plot shows how often each number on the number cube was tossed. Which number is most likely to be tossed next?

```
X X X X X X
X X X X X X
X X X X X X
+-+-+-+-+-+
1 2 3 4 5 6
```

2. This tally table shows the color tiles that were pulled from a bag. Which color tile is most likely to be pulled?

COLOR	TALLIES
green	卌 |||
blue	|
red	|||
yellow	||

EXTRA PRACTICE

Set A (pp. 294–297)

Name the solid figure that each object looks like.

1. 

2. 

3. 

4. 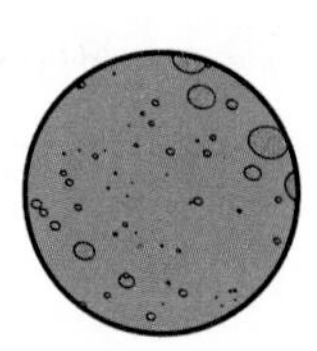

Set B (pp. 298–299)

Name the solid figures used to make each object.

1. 

2. 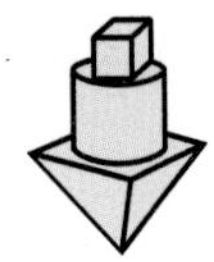

3. 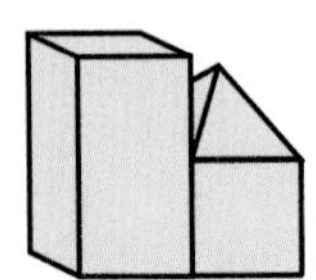

Set C (pp. 300–303)

Name each figure.

1.

2. 

3.

4. 

Write whether each angle is a *right angle, greater than* a right angle, or *less than* a right angle.

5. 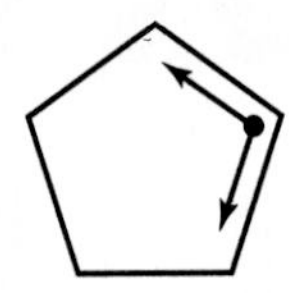

6. 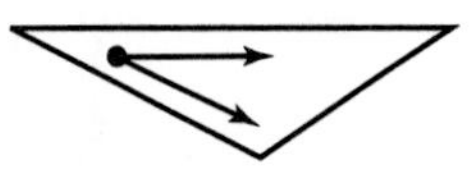

7. 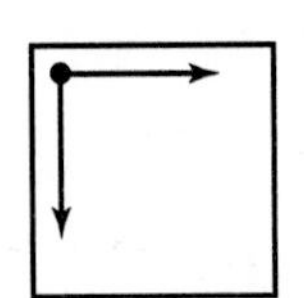

Set D (pp. 304–305)

Describe the lines. Write *parallel* or *intersecting*.

1. 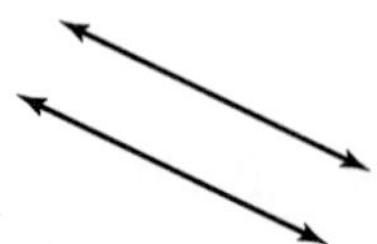

2. 

3. 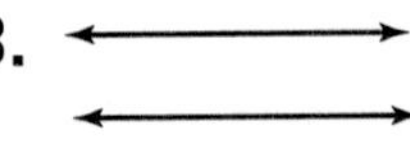

Describe the intersecting lines. Write *form right angles* or *do not form right angles*.

4. 

5. 

6. 

Set A (pp. 314–315)

Write the number of sides and angles each polygon has. Then name the polygon.

1. 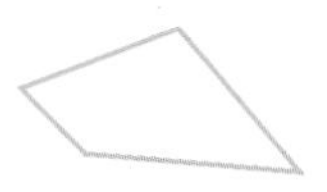
2. 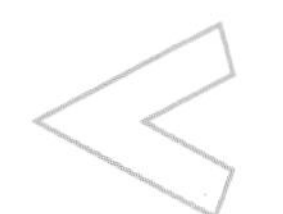
3. 
4. 
5. 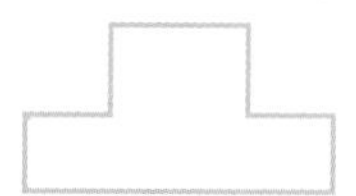

6. Melanie was drawing polygons for homework. She decided that squares and rectangles are quadrilaterals. Do you agree? Explain.

Set B (pp. 316–319)

Name each triangle. Write *equilateral, isosceles,* or *scalene*.

1. 1 cm, 4 cm, 3 cm
2. 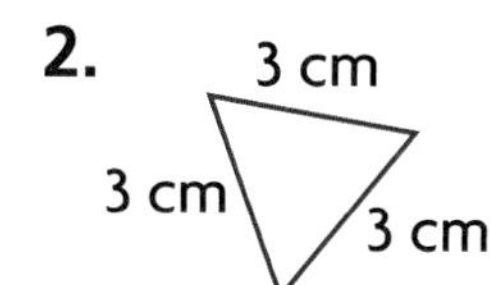
3 cm, 3 cm, 3 cm
3. 1 cm, 2 cm, 2 cm
4. 3 cm, 3 cm, 2 cm
5. 4 cm, 2 cm, 3 cm

Set C (pp. 320–323)

Write as many names for each quadrilateral as you can.

1. 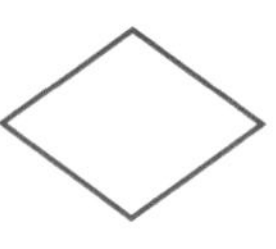
2. 
3. 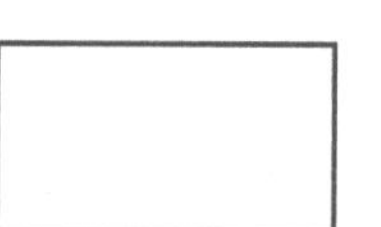
4. 
5. 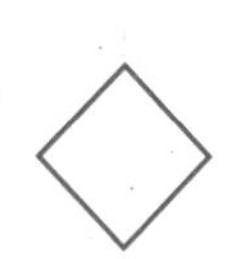

Set D (pp. 324–325)

Tell if each figure will tessellate. Write *yes* or *no*.

1. 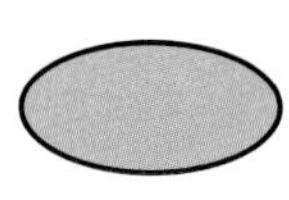
2. 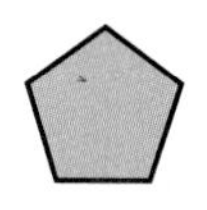
3. 
4. 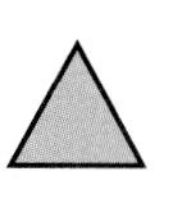

5. Use grid paper and pattern blocks to make a design. Repeat your design to make a tessellation.

EXTRA PRACTICE

Set A (pp. 334–335)

Write how many lines of symmetry each object has.

1. 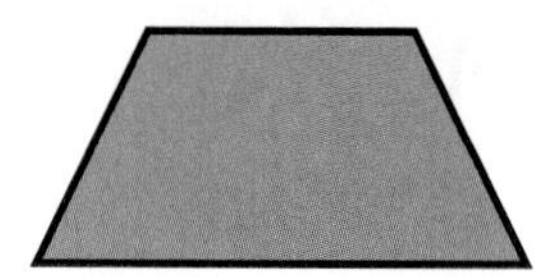

2. 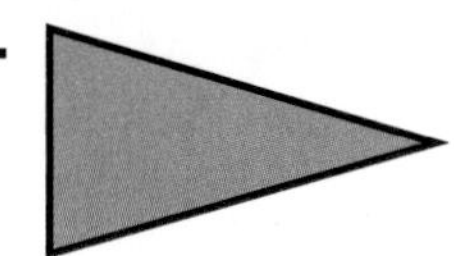

3. 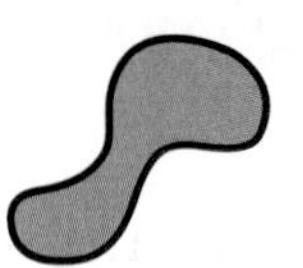

4. 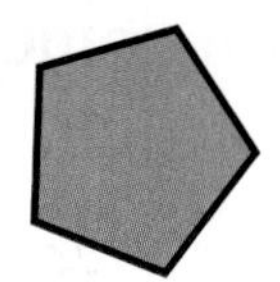

5. Look at the letters in the word MATH. Which letters have one line of symmetry? Do any of the letters have more than one line of symmetry? Explain.

Tell if the blue line is a line of symmetry. Write *yes* or *no*.

6. 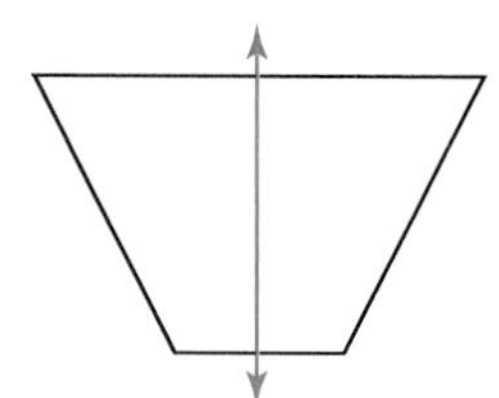

7. 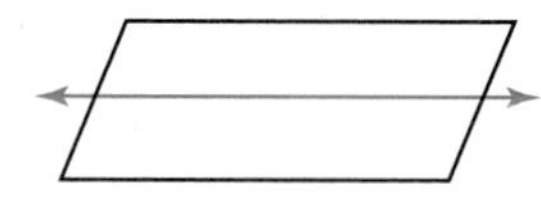

8. 

Set B (pp. 336–337)

Tell if the figures are similar. Write *yes* or *no*.

1. 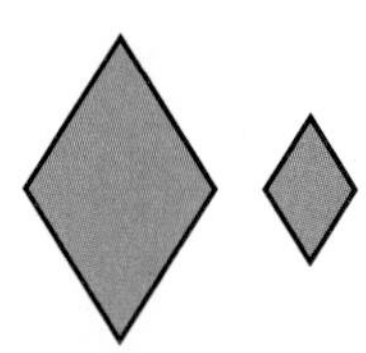

2. 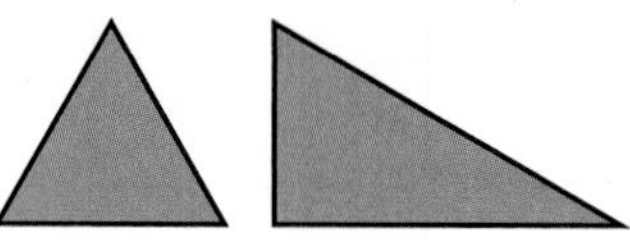

3. 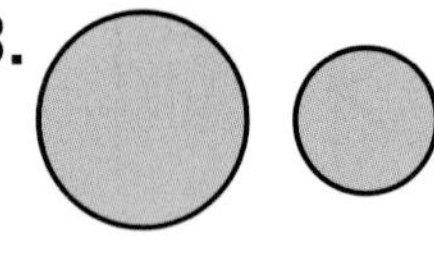

4. Draw a design on 1-cm grid paper. Then draw the design on 1-inch grid paper. Are the designs similar? Explain.

Set A (pp. 352–355)

Measure the length to the nearest inch.

1.

2.

Measure the length to the nearest half inch.

3. 

4. 

5. 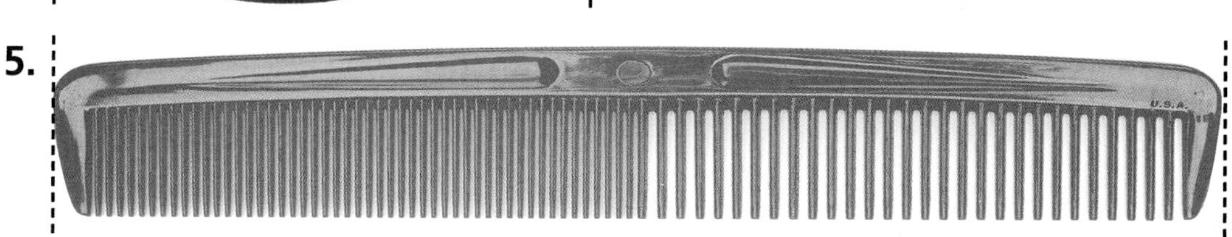

Set B (pp. 356–357)

Choose the unit you would use to measure each. Write *inch, foot, yard,* or *mile.*

1. the height of a 2-story house
2. the distance between 2 towns
3. the length of a hockey stick
4. the length of your thumb

Set C (pp. 362–363)

Change the units. Use the Table of Measures on page 362 to help.

1. ■ cups = 1 quart

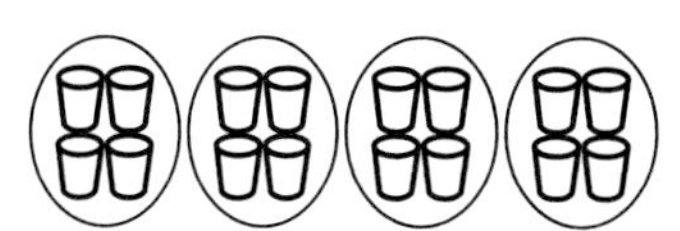

16 cups = ■ quarts

2. ■ feet = 1 yard

feet	3	6
yards	1	2

■ feet = 4 yards

Set D (pp. 364–365)

Write a rule and change the units. You may make a table to help. (2 pints = 1 quart)

1. How many pints are in 9 quarts?

   ■ pints = 9 quarts

2. How many quarts are in 6 pints?

   ■ quarts = 6 pints

EXTRA PRACTICE

Set A (pp. 372–375)

Estimate the length in centimeters. Then use a ruler to measure to the nearest centimeter.

1.
2.
3.
4.

5.

Choose the unit you would use to measure each. Write *cm, m,* or *km*.

6. the length of a marker
7. the height of a two-story building
8. the distance you can kick a ball
9. the distance you can walk in half an hour

Choose the better estimate.

10. Jana's math book is almost 3 _?_ long.
    - **A** meters
    - **B** decimeters
11. Ed walked 1 _?_ to get to Jeff's house.
    - **A** kilometer
    - **B** centimeter
12. The tree is about 6 _?_ high.
    - **A** kilometers
    - **B** meters
13. The ant is 1 _?_ long.
    - **A** decimeter
    - **B** centimeter

Solve.

14. Mathias has a piece of yarn that measures 13 centimeters. Is that more or less than 2 decimeters? Explain.
15. Alice had a poster board 1 m long. She cut 12 cm off of the poster board. How many centimeters long is the poster board now?

Set A (pp. 390–393)

Find the perimeter.

1.

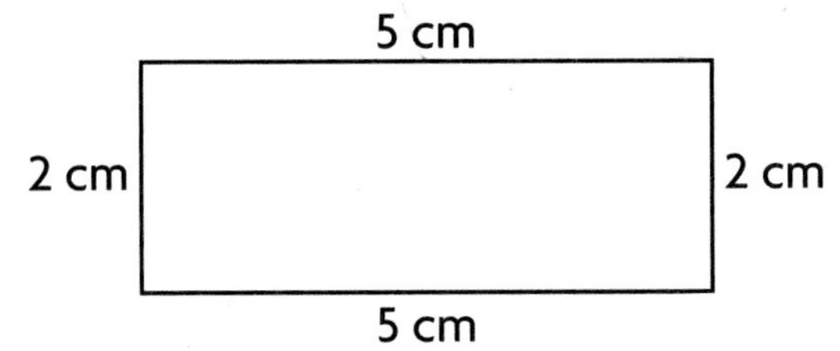

2.

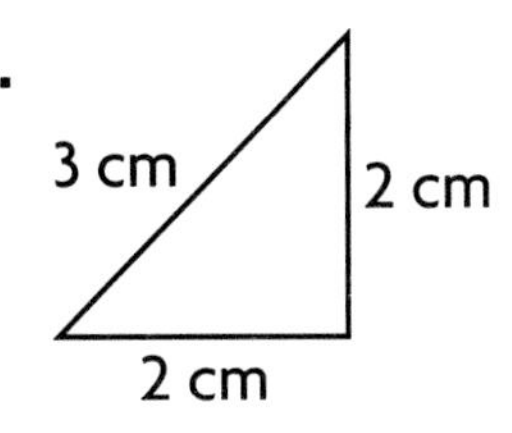

Use your centimeter ruler to find the perimeter.

3.

4.

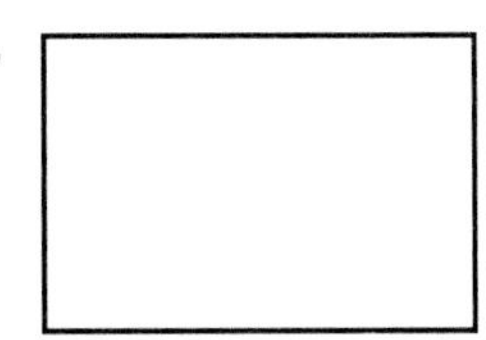

Set B (pp. 398–401)

Find the volume of each solid. Write the volume in cubic units.

1.

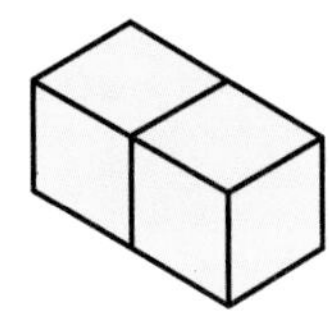

2.

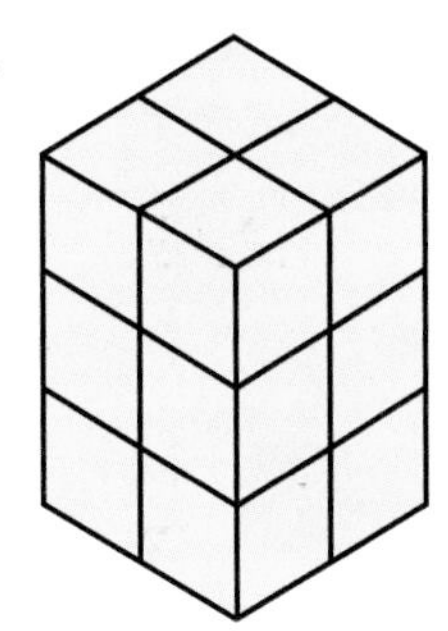

3.

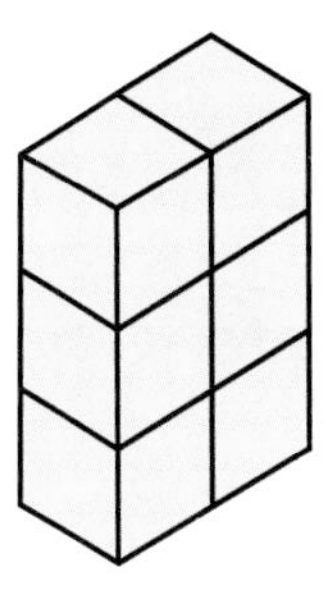

4.

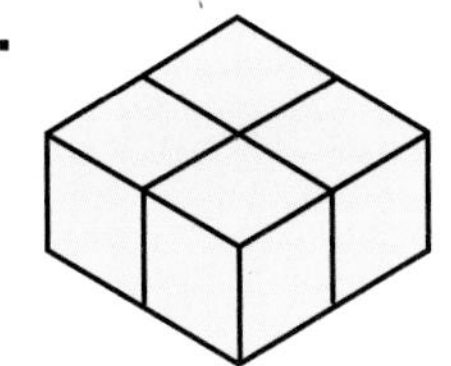

5.

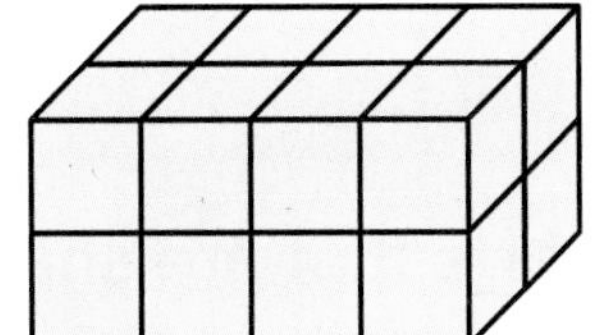

6.

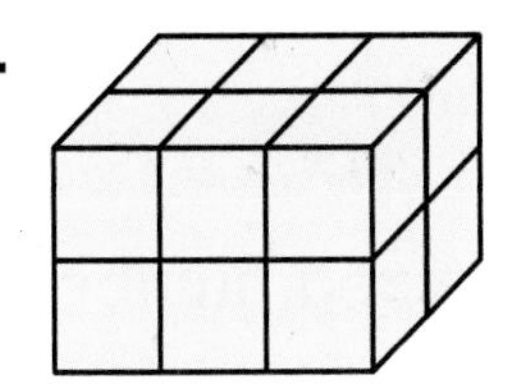

CHAPTER 23 EXTRA PRACTICE

Set A (pp. 412–415)

Write a fraction in numbers and words that names the shaded part.

1. 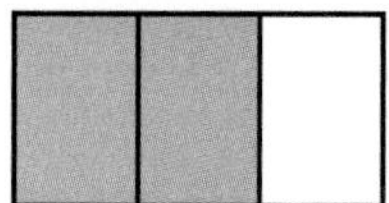

2. 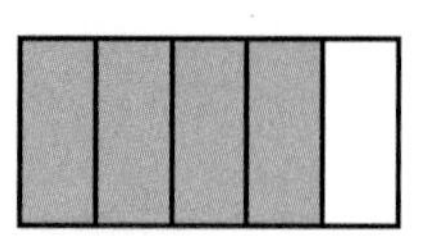

3. 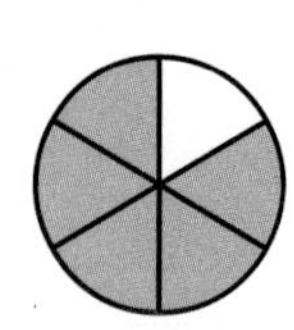

Write the fraction, using numbers.

4. one eighth
5. four out of seven
6. two divided by five

Set B (pp. 416–417)

1. Draw 4 nickels. Circle $\frac{3}{4}$ of them.
2. Draw 5 rectangles. Circle $\frac{1}{5}$ of them.
3. Draw 8 triangles. Circle $\frac{1}{2}$ of them.

Set C (pp. 418–421)

Find an equivalent fraction. Use fraction bars.

1. 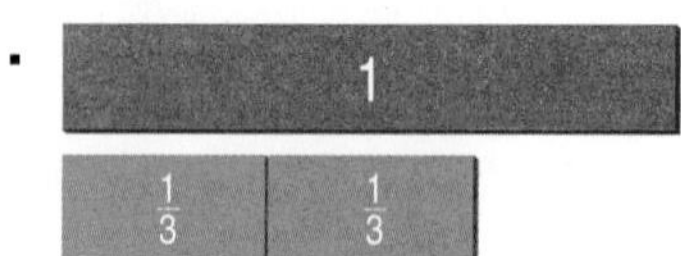

2. 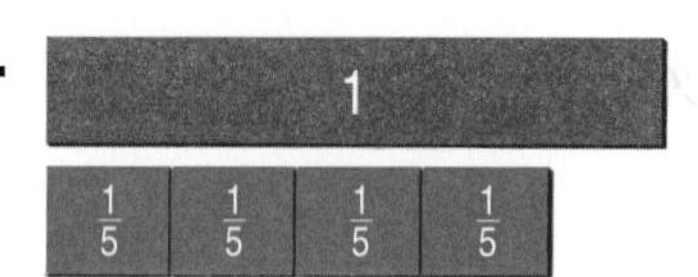

3. 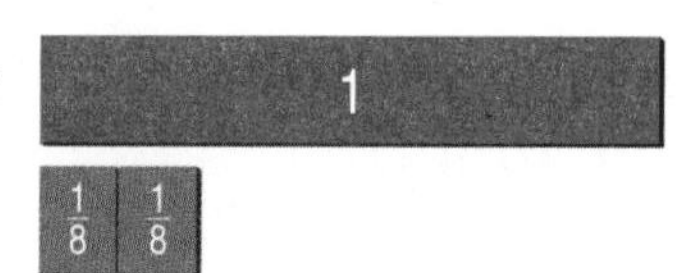

Set D (pp. 422–425)

Compare. Write <, >, or = for each ●.

1. 

$\frac{1}{8}$ ● $\frac{3}{10}$

2. 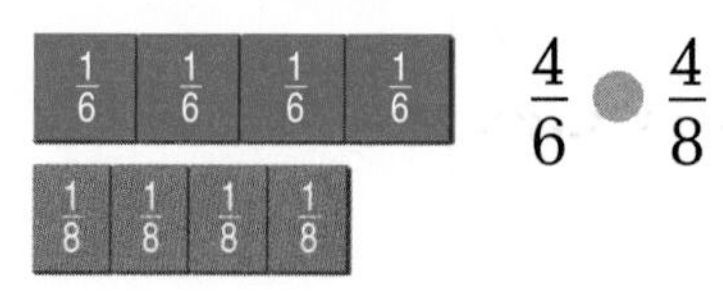

$\frac{4}{6}$ ● $\frac{4}{8}$

3. 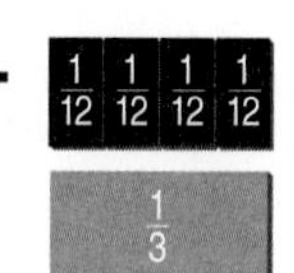

$\frac{4}{12}$ ● $\frac{1}{3}$

4. Use fraction bars to order $\frac{1}{2}$, $\frac{3}{10}$, and $\frac{2}{3}$ from greatest to least.

Set E (pp. 428–429)

Write a mixed number for the parts that are shaded.

1. 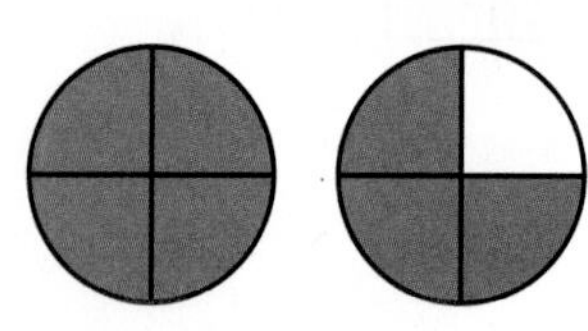

2. 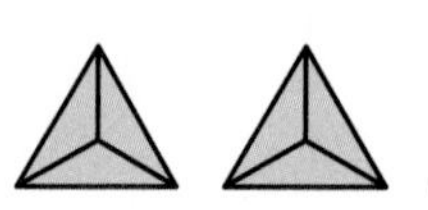

3. 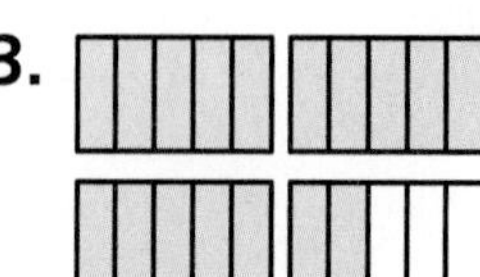

Set A (pp. 436–439)

Find the sum. Write the answer in simplest form. Use fraction bars.

1. $\frac{2}{4} + \frac{1}{4} = \blacksquare$
2. $\frac{2}{5} + \frac{1}{5} = \blacksquare$
3. $\frac{4}{8} + \frac{1}{8} = \blacksquare$
4. $\frac{4}{10} + \frac{2}{10} = \blacksquare$
5. $\frac{1}{3} + \frac{1}{3} = \blacksquare$
6. $\frac{6}{12} + \frac{1}{12} = \blacksquare$
7. $\frac{3}{6} + \frac{2}{6} = \blacksquare$
8. $\frac{1}{4} + \frac{2}{4} = \blacksquare$
9. $\frac{1}{8} + \frac{2}{8} = \blacksquare$
10. $\frac{6}{10} + \frac{3}{10} = \blacksquare$
11. $\frac{2}{6} + \frac{2}{6} = \blacksquare$
12. $\frac{1}{12} + \frac{1}{12} = \blacksquare$
13. Miles gave $\frac{1}{5}$ of the cake to Molly and $\frac{2}{5}$ to Sarah. What fraction of the cake did Miles give away?
14. Gwen says that the simplest form of $\frac{6}{12}$ is $\frac{3}{6}$. Is she correct? Explain.

Set B (pp. 442–445)

Compare. Find the difference. Write the answer in simplest form.

1. 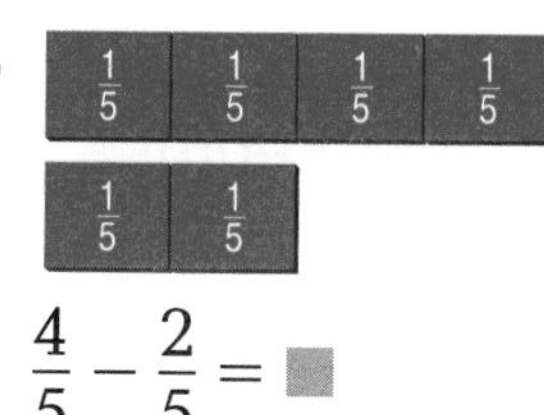

$\frac{4}{5} - \frac{2}{5} = \blacksquare$

2. 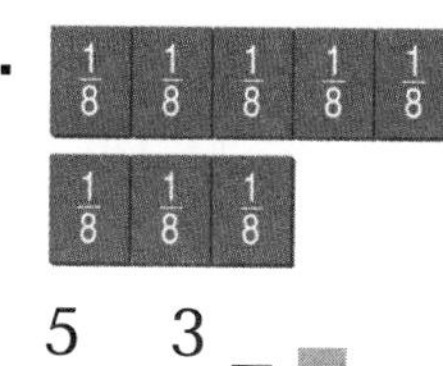

$\frac{5}{8} - \frac{3}{8} = \blacksquare$

3. 

$\frac{8}{12} - \frac{2}{12} = \blacksquare$

Find the difference. Write the answer in simplest form. Use fraction bars.

4. $\frac{2}{3} - \frac{1}{3} = \blacksquare$
5. $\frac{5}{6} - \frac{2}{6} = \blacksquare$
6. $\frac{3}{4} - \frac{2}{4} = \blacksquare$
7. $\frac{5}{10} - \frac{3}{10} = \blacksquare$
8. $\frac{3}{4} - \frac{1}{4} = \blacksquare$
9. $\frac{4}{5} - \frac{1}{5} = \blacksquare$
10. $\frac{7}{8} - \frac{5}{8} = \blacksquare$
11. $\frac{10}{12} - \frac{1}{12} = \blacksquare$

Solve.

12. Nate and Sam ate a candy bar. Nate ate $\frac{5}{10}$ of the candy, and Sam ate $\frac{5}{10}$ of the candy. How much of the candy bar did the boys eat in all?
13. Valencia and Dora ate some jelly beans. Valencia ate $\frac{3}{8}$ of them, and Dora ate $\frac{5}{8}$ of them. What fraction tells how much more of the jelly beans Dora ate than Valencia?

CHAPTER 25

EXTRA PRACTICE

Set A (pp. 452–453)

Write the fraction and decimal for the shaded part.

1. 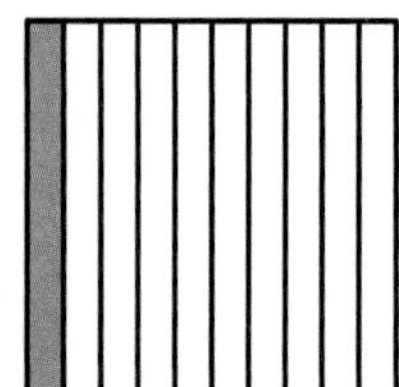

2. 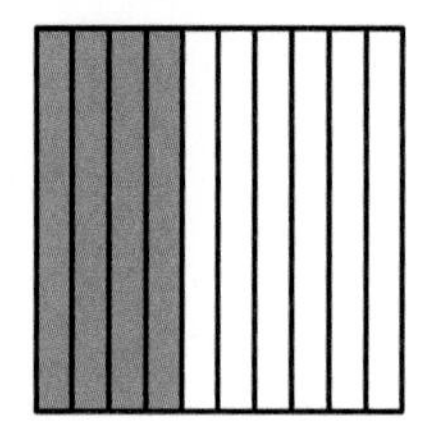

3. 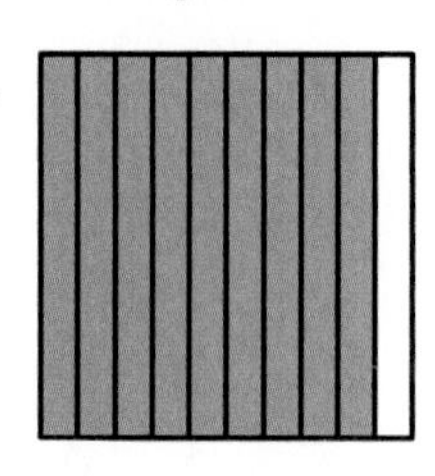

4. 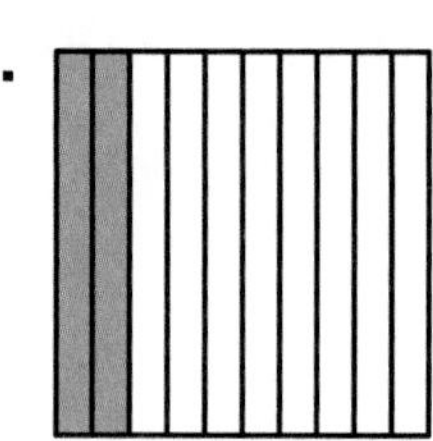

5. What decimal names the amount in Exercise 1 that is **NOT** shaded?

6. How many more parts would you shade in Exercise 2 to show 0.6?

Set B (pp. 458–459)

Write the word form and expanded form for each decimal.

1.

ONES	•	TENTHS	HUNDREDTHS
0	•	1	8

2.

ONES	•	TENTHS	HUNDREDTHS
1	•	3	3

Write the missing number.

3. 0.45 = ■ tenths 5 hundredths

4. 0.16 = 1 tenth ■ hundredths

Set C (pp. 460–461)

Compare. Write < or > for each ●.

1.

ONES	•	TENTHS
2	•	2
2	•	4

2.2 ● 2.4

2.

ONES	•	TENTHS	HUNDREDTHS
1	•	0	9
1	•	0	3

1.09 ● 1.03

3. 2.31 ● 1.32 **4.** 5.1 ● 1.5 **5.** 0.09 ● 0.90 **6.** 1.10 ● 0.10

For 7–10, use the number line to order the decimals from least to greatest.

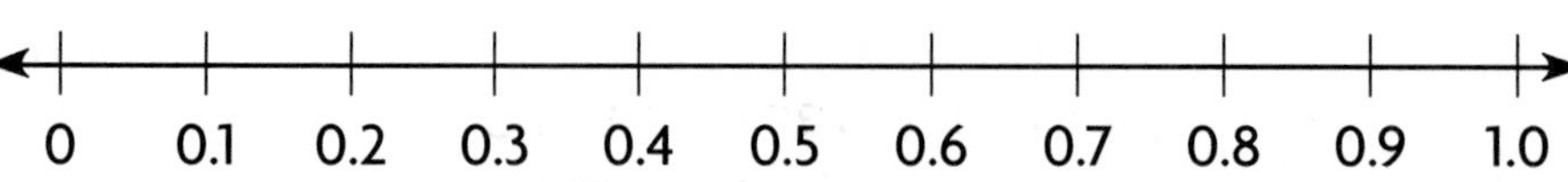

7. 0.2, 0.1, 0.3 **8.** 0.7, 0.1, 0.4 **9.** 0.9, 0.6, 0.3 **10.** 0.4, 0.2, 0.6

Set A (pp. 468–469)

Write the amount of money shown. Then write the amount as a fraction of a dollar.

1. 

2. 

3. 

4. 

5. Ian had 2 quarters, 2 dimes, and 5 pennies. He gave $0.50 to Lucia. What fraction of a dollar does he have left?

6. Give two examples of how you could model $\frac{1}{4}$ of a dollar using dimes, nickels, and pennies.

Set B (pp. 472–475)

Add or subtract.

1. 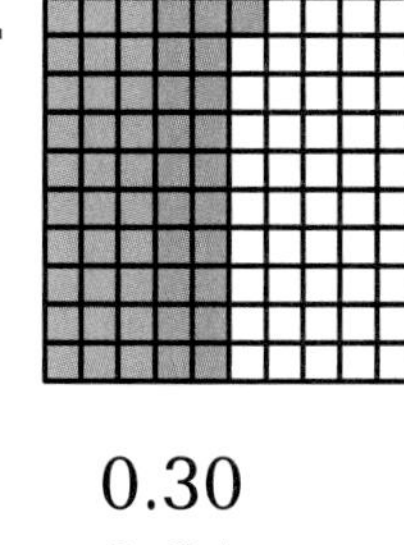

$$\begin{array}{r} 0.30 \\ +0.21 \\ \hline \end{array}$$

2. 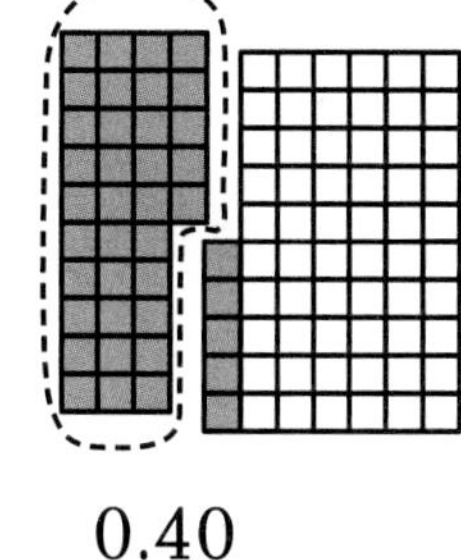

$$\begin{array}{r} 0.40 \\ -0.35 \\ \hline \end{array}$$

3. $\begin{array}{r} 0.1 \\ +0.6 \\ \hline \end{array}$

4. $\begin{array}{r} 3.09 \\ +2.51 \\ \hline \end{array}$

5. $\begin{array}{r} 1.7 \\ -0.2 \\ \hline \end{array}$

6. $\begin{array}{r} \$0.70 \\ -\$0.54 \\ \hline \end{array}$

7. $\begin{array}{r} 0.88 \\ +0.01 \\ \hline \end{array}$

8. $\begin{array}{r} \$4.56 \\ -\$2.38 \\ \hline \end{array}$

9. $\begin{array}{r} 0.5 \\ +0.3 \\ \hline \end{array}$

10. $\begin{array}{r} 8.43 \\ -7.22 \\ \hline \end{array}$

11. Mrs. Saguchi is training for a bike race. She rode 2.5 kilometers on Tuesday, and 2.7 kilometers on Thursday. How far did she ride on those two days?

CHAPTER 27

EXTRA PRACTICE

Set A (pp. 490–493)

Find the product.

1. 15 × 3

2. 27 × 2

Find the product. You may wish to use base-ten blocks.

3. 16 × 2 **4.** 32 × 3 **5.** 13 × 6 **6.** 25 × 5 **7.** 47 × 4

8. 34 × 8 **9.** 73 × 2 **10.** 23 × 7 **11.** 45 × 5 **12.** 99 × 3

13. $6 \times 15 = ■$ **14.** $9 \times 21 = ■$ **15.** $7 \times 33 = ■$ **16.** $4 \times 51 = ■$

17. Jill has 3 boxes of nature magazines. There are 28 magazines in each box. How many nature magazines does Jill have in all?

Set B (pp. 494–495)

Find the product. Tell whether you need to regroup. Write *yes* or *no*.

1. 21 × 4 **2.** 23 × 9 **3.** 38 × 5 **4.** 44 × 3 **5.** 33 × 3

Find the product.

6. 14 × 6 **7.** 29 × 5 **8.** 37 × 7 **9.** 18 × 2 **10.** 22 × 3

11. 38 × 4 **12.** 25 × 8 **13.** 72 × 7 **14.** 49 × 8 **15.** 53 × 9

16. $9 \times 19 = ■$ **17.** $4 \times 91 = ■$ **18.** $3 \times 12 = ■$ **19.** $8 \times 88 = ■$

20. Angelo sells hot dogs for $3 each. If he sells 56 hot dogs, how much money will he make?

Set A (pp. 504–505)

Use the model. Write the quotient and remainder.

1. 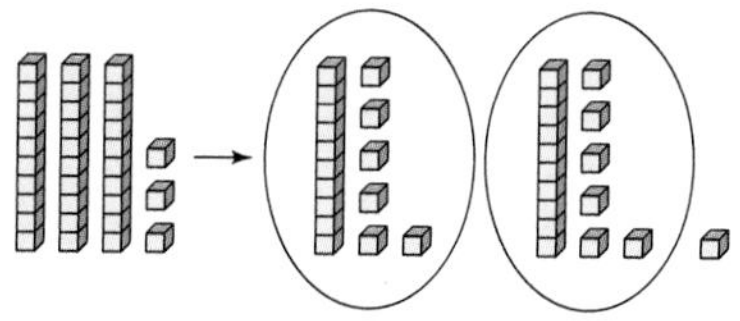 $33 \div 2 = \blacksquare$

2. 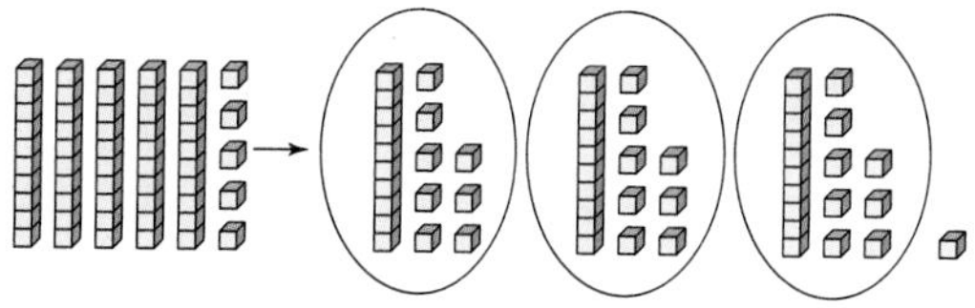 $55 \div 3 = \blacksquare$

Divide. You may use base-ten blocks to help.

3. $29 \div 2 = \blacksquare$
4. $45 \div 4 = \blacksquare$
5. $54 \div 3 = \blacksquare$
6. $68 \div 6 = \blacksquare$
7. $3\overline{)58}$
8. $2\overline{)23}$
9. $3\overline{)57}$
10. $4\overline{)59}$
11. $7\overline{)66}$
12. $8\overline{)51}$
13. $4\overline{)63}$
14. $5\overline{)54}$

Set B (pp. 506–509)

Divide and check.

1. $15 \div 7 = \blacksquare$
2. $47 \div 5 = \blacksquare$
3. $68 \div 5 = \blacksquare$
4. $38 \div 2 = \blacksquare$
5. $9\overline{)16}$
6. $8\overline{)74}$
7. $5\overline{)70}$
8. $4\overline{)89}$
9. $3\overline{)56}$
10. $9\overline{)32}$
11. $3\overline{)47}$
12. $2\overline{)27}$
13. Ms. Payne has 65 crayons. She divides them equally among 9 students. How many crayons does each student get? How many are left over?
14. Jackie has 35 crackers. She divides them equally among 5 plates. How many crackers are on 2 plates?

Set C (pp. 510–511)

Divide.

1. $28 \div 5 = \blacksquare$
2. $45 \div 7 = \blacksquare$
3. $15 \div 8 = \blacksquare$
4. $39 \div 4 = \blacksquare$
5. $52 \div 2 = \blacksquare$
6. $78 \div 9 = \blacksquare$
7. $21 \div 6 = \blacksquare$
8. $54 \div 3 = \blacksquare$
9. $5\overline{)59}$
10. $2\overline{)25}$
11. $9\overline{)28}$
12. $6\overline{)32}$
13. $7\overline{)79}$
14. $9\overline{)59}$
15. $8\overline{)67}$
16. $4\overline{)98}$

EXTRA PRACTICE

Set A (pp. 518–519)

Copy and complete. Use patterns and mental math.

1. $8 \times 3 = ■$
   $8 \times 30 = ■$
   $8 \times 300 = ■$
   $8 \times 3{,}000 = ■$

2. $4 \times 7 = ■$
   $4 \times 70 = ■$
   $4 \times ■ = 2{,}800$
   $4 \times 7{,}000 = ■$

3. $■ \times 6 = 54$
   $9 \times ■ = 540$
   $9 \times 600 = ■$
   $9 \times ■ = 54{,}000$

Use mental math and basic multiplication facts to find the product.

4. $2 \times 900 = ■$
5. $5 \times 4{,}000 = ■$
6. $3 \times 80 = ■$
7. $7 \times 300 = ■$

Set B (pp. 522–523)

Estimate the product.

1. 15×7
2. 46×6
3. 92×9
4. 130×8
5. 683×5

Set C (pp. 524–527)

Find the product. Tell what method you used.

1. 141×4
2. 718×2
3. 455×4
4. 257×7
5. 609×6

6. $3 \times 252 = ■$
7. $5 \times 279 = ■$
8. $9 \times 191 = ■$
9. $8 \times 348 = ■$

Set D (pp. 528–529)

Find the product in dollars and cents.

1. $\$3.32 \times 5$
2. $\$6.24 \times 4$
3. $\$1.99 \times 8$
4. $\$7.13 \times 6$
5. $\$5.58 \times 2$

6. $7 \times \$1.37 = ■$
7. $3 \times \$8.30 = ■$
8. $9 \times \$4.59 = ■$
9. $4 \times \$2.46 = ■$

Set E (pp. 530–531)

Find the product. Estimate to check.

1. $2{,}481 \times 3$
2. $5{,}082 \times 2$
3. $\$12.39 \times 7$
4. $8{,}469 \times 5$
5. $\$35.94 \times 4$

Set A (pp. 536–537)

Copy and complete. Use patterns and mental math.

1. $9 \div 9 = ■$
   $90 \div 9 = ■$
   $900 \div 9 = ■$
2. $7 \div 7 = ■$
   $70 \div 7 = ■$
   $700 \div 7 = ■$
3. $42 \div ■ = 7$
   $■ \div 6 = 70$
   $4{,}200 \div ■ = 700$

Use mental math and a basic fact to find the quotient.

4. $400 \div 5 = ■$
5. $360 \div 9 = ■$
6. $1{,}800 \div 2 = ■$
7. $5\overline{)450}$
8. $4\overline{)3{,}600}$
9. $8\overline{)4{,}800}$

Set B (pp. 538–539)

Estimate each quotient. Write the basic fact you used to find the estimate.

1. $155 \div 3 = ■$
2. $639 \div 7 = ■$
3. $374 \div 6 = ■$

Estimate the quotient.

4. $4\overline{)318}$
5. $5\overline{)212}$
6. $2\overline{)801}$
7. $3\overline{)291}$
8. $8\overline{)653}$
9. $9\overline{)551}$

Set C (pp. 540–541)

Find the quotient.

1. $2\overline{)68}$
2. $3\overline{)108}$
3. $5\overline{)250}$
4. $9\overline{)279}$
5. $6\overline{)552}$
6. $4\overline{)296}$
7. $7\overline{)119}$
8. $8\overline{)624}$
9. Jan has 384 jelly beans. How many can she put in each of 6 bags?
10. Tyrone has 423 cards. How many can he give to each of his 9 friends?

Set D (pp. 542–543)

Find the quotient.

1. $2\overline{)200}$
2. $4\overline{)128}$
3. $6\overline{)672}$
4. $3\overline{)549}$
5. $5\overline{)495}$
6. $9\overline{)369}$
7. $8\overline{)808}$
8. $7\overline{)623}$
9. The product of 7 and a certain number is 238. What is the other number?
10. Marian has 516 sunflower seeds. How many seeds can she put in each of 3 bags?

Tips for Taking Math Tests

Being a good test-taker is like being a good problem solver. When you answer test questions, you are solving problems. Remember to **UNDERSTAND, PLAN, SOLVE,** and **CHECK.**

UNDERSTAND

Read the problem.

- Look for math terms and recall their meanings.
- Reread the problem and think about the question.
- Use the details in the problem and the question.

1. The sum of the digits of a number is 14. Both the digits are odd. The ones digit is 4 less than the tens digit. What is the number?

A 59 **C** 86
B 77 **D** 95

TIP! Understand the problem.
Remember the meanings of *sum, digits,* and *odd*. Reread the problem to compare the details to the answer choices. Since all choices have a sum of 14, look for the odd digits. Then look for a ones digit that is 4 less than the tens digit. The answer is D.

- Each word is important. Missing a word or reading it incorrectly could cause you to get the wrong answer.
- Pay attention to words that are in **bold** type, all CAPITAL letters, or *italics*.
- Some other words to look for are round, about, only, best, or least to greatest.

2. Mr. Karza drew a diagram of 8 squares. He made 3 squares red, 2 squares blue, 2 squares yellow, and 1 square green. What fraction of the squares was **not** green?

F $\frac{7}{8}$ **H** $\frac{1}{2}$
G $\frac{5}{8}$ **J** $\frac{1}{8}$

TIP! Look for important words.
The word not is important. Without the word not, the answer would be $\frac{1}{8}$. Find the number of squares that were not green. The answer is F.

PLAN

Think about how you can solve the problem.

- See if you can solve the problem with the information given.
- Pictures, charts, tables, and graphs may have the information you need.
- You may need to think about information you already know.
- The answer choices may have information you need.

3. Soccer practice started at 12:00. The clock shows the time practice ended. How long did soccer practice last?

A 10 minutes **C** 35 minutes
B 20 minutes **D** 50 minutes

TIP! Get the information you need.
Use the clock to find how long soccer practice lasted. You can find out how much time passed by counting by fives. The answer is D.

- You may need to write a number sentence and solve it.
- Some problems have two steps or more.
- In some problems you need to look at relationships instead of computing an answer.
- If the path to the solution isn't clear, choose a problem solving strategy and use it to solve the problem.

4. June always has 30 days. Mary takes swimming lessons every three days in June, starting on June 3. How many times will she have lessons?

F 5 **H** 12
G 10 **J** 30

TIP! Decide on a plan.
Lessons every three days sounds like a pattern. Use the strategy *find a pattern*. Count by 3 beginning with June 3 until you reach 30. You need to count 10 numbers, so the answer is G.

SOLVE

Follow your plan, working logically and carefully.

- Estimate your answer. Look for unreasonable answer choices.
- Use reasoning to find the most likely choices.
- Solve all steps needed to answer the problem.
- If your answer does not match any answer choice, check your numbers and your computation.

5. The cafeteria served 76 lunches each day for a week. How many lunches were served in 5 days?

A 76 **C** 380

B 353 **D** 1,380

TIP! Eliminate choices.

Estimate the product (5 × 80). The only reasonable answers are B and C. Since 5 times the ones digit 6 is 30, the answer must end in zero. If you are still not certain, multiply and check your answer against B and C. The answer is C.

- If your answer still does not match, look for another form of the number, such as a decimal instead of a fraction.
- If answer choices are given as pictures, look at each one by itself while you cover the other three.
- If the answer choices include NOT HERE and your answer is not given, make sure your work is correct and then mark NOT HERE.
- Read answer choices that are statements and relate them to the problem one by one.
- If your strategy isn't working, try a different one.

6. Mr. Rodriguez is putting a wallpaper border around a room. The room is 9 feet wide and 12 feet long. How many feet of border does he need?

F 21 ft **H** 108 ft

G 84 ft **J** NOT HERE

TIP! Choose the answer.

The border goes around all four walls, two that are 9 feet and two that are 12 feet. Answer choices are given using the abbreviation for feet (ft). Add the lengths of the four walls (9 + 9 + 12 + 12). That total is not given. So, mark J for NOT HERE.

CHECK

Take time to catch your mistakes.

- Be sure you answered the question asked.
- Check for important words you might have missed.
- Did you use all the information you needed?
- Check your computation by using a different method.
- Draw a picture when you are unsure of your answer.

7. Katy is buying 3 books. Their prices are $4.95, $3.25, and $7.49. What is the total cost of the books?

A $14.59 **C** $15.59
B $14.69 **D** $15.69

TIP! **Check your work.**
To check column addition, write the numbers in a different order. Then you will be adding different basic facts. For example, add $7.49 + $3.25 + $4.95. The answer is D.

Don't Forget!

Before the test

- Listen to the teacher's directions and read the instructions.
- Write down the ending time if the test is timed.
- Know where and how to mark your answers.
- Know whether you should write on the test page or use scratch paper.
- Ask any questions you may have before the test begins.

During the test

- Work quickly but carefully. If you are unsure how to answer a question, leave it blank and return to it later.
- If you cannot finish on time, read the questions that are left. Answer the easiest ones first. Then answer the others.
- Fill in each answer space carefully. Erase completely if you change an answer. Erase any stray marks.
- Check that the answer number matches the question number, especially if you skip a question.

ADDITION FACTS TEST

	K	L	M	N	O	P	Q	R
A	$3 + 2$	$0 + 6$	$2 + 4$	$5 + 9$	$6 + 1$	$2 + 5$	$3 + 10$	$4 + 4$
B	$8 + 9$	$0 + 7$	$3 + 5$	$9 + 6$	$6 + 7$	$2 + 8$	$3 + 3$	$7 + 10$
C	$4 + 6$	$9 + 0$	$7 + 8$	$4 + 10$	$3 + 7$	$7 + 7$	$4 + 2$	$7 + 5$
D	$5 + 7$	$3 + 9$	$8 + 1$	$9 + 5$	$10 + 5$	$9 + 8$	$2 + 6$	$8 + 7$
E	$7 + 4$	$0 + 8$	$3 + 6$	$6 + 10$	$5 + 3$	$2 + 7$	$8 + 2$	$9 + 9$
F	$2 + 3$	$1 + 7$	$6 + 8$	$5 + 2$	$7 + 3$	$4 + 8$	$10 + 10$	$6 + 6$
G	$8 + 3$	$7 + 2$	$7 + 0$	$8 + 5$	$9 + 1$	$4 + 7$	$8 + 4$	$10 + 8$
H	$7 + 9$	$5 + 6$	$8 + 10$	$6 + 5$	$8 + 6$	$9 + 4$	$0 + 9$	$7 + 1$
I	$4 + 3$	$5 + 5$	$6 + 4$	$10 + 2$	$7 + 6$	$8 + 0$	$6 + 9$	$9 + 2$
J	$5 + 8$	$1 + 9$	$5 + 4$	$8 + 8$	$6 + 2$	$6 + 3$	$9 + 7$	$9 + 10$

SUBTRACTION FACTS TEST

	K	L	M	N	O	P	Q	R
A	9 − 1	10 − 4	7 − 2	6 − 4	20 − 10	7 − 0	8 − 3	13 − 9
B	9 − 9	13 − 4	7 − 1	11 − 5	9 − 7	6 − 3	15 − 10	6 − 2
C	10 − 2	8 − 8	16 − 8	6 − 5	18 − 10	8 − 7	13 − 3	15 − 6
D	11 − 7	9 − 5	12 − 8	8 − 1	15 − 8	18 − 9	14 − 10	9 − 4
E	9 − 2	7 − 7	10 − 3	8 − 5	16 − 9	11 − 9	14 − 8	12 − 6
F	7 − 3	12 − 10	17 − 9	6 − 0	9 − 6	11 − 8	10 − 9	12 − 2
G	15 − 7	8 − 4	13 − 6	7 − 5	11 − 2	12 − 3	14 − 6	11 − 4
H	7 − 6	13 − 5	12 − 9	10 − 5	13 − 8	11 − 3	16 − 10	14 − 7
I	5 − 0	10 − 8	11 − 6	9 − 3	14 − 5	5 − 4	7 − 7	14 − 9
J	15 − 9	9 − 8	13 − 7	8 − 2	7 − 4	13 − 10	10 − 6	16 − 7

MULTIPLICATION FACTS TEST

	K	L	M	N	O	P	Q	R
A	2×7	0×6	6×6	9×2	8×3	3×4	2×8	6×1
B	7×7	5×9	2×2	7×5	2×3	10×8	4×10	8×4
C	4×5	5×1	7×0	6×3	3×5	6×8	7×3	9×9
D	0×9	6×4	6×10	1×6	9×8	4×4	3×2	9×3
E	0×7	9×4	1×7	9×7	2×5	7×9	5×6	5×8
F	4×3	6×9	1×9	7×6	7×10	6×0	2×9	10×3
G	5×3	1×5	7×1	3×8	3×6	8×10	3×9	6×7
H	7×4	7×2	3×7	2×4	7×8	4×7	5×10	8×6
I	4×6	5×5	5×7	3×3	9×6	8×0	4×9	8×8
J	8×9	6×2	4×8	9×5	5×4	0×5	10×6	9×10

DIVISION FACTS TEST

	K	L	M	N	O	P	Q	R
A	$1\overline{)1}$	$3\overline{)9}$	$2\overline{)6}$	$2\overline{)4}$	$1\overline{)6}$	$3\overline{)12}$	$5\overline{)15}$	$7\overline{)21}$
B	$6\overline{)24}$	$8\overline{)56}$	$5\overline{)40}$	$6\overline{)18}$	$6\overline{)30}$	$7\overline{)42}$	$9\overline{)81}$	$5\overline{)45}$
C	$5\overline{)30}$	$2\overline{)16}$	$3\overline{)21}$	$7\overline{)35}$	$3\overline{)15}$	$9\overline{)9}$	$8\overline{)16}$	$9\overline{)63}$
D	$4\overline{)32}$	$9\overline{)90}$	$4\overline{)8}$	$8\overline{)48}$	$9\overline{)54}$	$3\overline{)18}$	$10\overline{)50}$	$6\overline{)48}$
E	$7\overline{)28}$	$3\overline{)0}$	$5\overline{)20}$	$4\overline{)24}$	$7\overline{)14}$	$3\overline{)6}$	$5\overline{)50}$	$10\overline{)60}$
F	$9\overline{)18}$	$4\overline{)36}$	$5\overline{)25}$	$7\overline{)63}$	$1\overline{)5}$	$8\overline{)32}$	$9\overline{)45}$	$6\overline{)54}$
G	$2\overline{)14}$	$8\overline{)24}$	$4\overline{)4}$	$5\overline{)40}$	$3\overline{)9}$	$4\overline{)12}$	$7\overline{)56}$	$8\overline{)72}$
H	$5\overline{)35}$	$1\overline{)4}$	$8\overline{)64}$	$5\overline{)10}$	$8\overline{)40}$	$2\overline{)12}$	$6\overline{)42}$	$10\overline{)70}$
I	$7\overline{)49}$	$9\overline{)27}$	$10\overline{)90}$	$3\overline{)27}$	$9\overline{)36}$	$4\overline{)20}$	$9\overline{)72}$	$8\overline{)80}$
J	$8\overline{)0}$	$4\overline{)28}$	$2\overline{)10}$	$7\overline{)70}$	$1\overline{)3}$	$10\overline{)80}$	$6\overline{)60}$	$10\overline{)100}$

MULTIPLICATION AND DIVISION FACTS TEST

	K	L	M	N	O	P	Q	R
A	$2\overline{)18}$	8×4	$5\overline{)15}$	10×6	8×1	$3\overline{)24}$	$6\overline{)12}$	5×8
B	8×2	$7\overline{)56}$	$9\overline{)81}$	4×10	7×9	$1\overline{)6}$	$8\overline{)80}$	4×9
C	$6\overline{)36}$	8×5	7×7	$10\overline{)90}$	$5\overline{)45}$	6×7	$8\overline{)16}$	9×9
D	10×2	$4\overline{)32}$	$9\overline{)54}$	7×8	9×3	$9\overline{)90}$	$6\overline{)54}$	9×4
E	8×10	7×6	$8\overline{)64}$	$2\overline{)20}$	9×0	10×10	$3\overline{)36}$	$10\overline{)100}$
F	$4\overline{)40}$	8×3	8×6	9×6	$7\overline{)49}$	$9\overline{)45}$	10×3	9×7
G	$8\overline{)48}$	$6\overline{)60}$	9×2	5×9	$7\overline{)42}$	$4\overline{)36}$	5×10	9×8
H	6×5	8×8	$9\overline{)72}$	$5\overline{)50}$	6×9	8×5	$9\overline{)36}$	$7\overline{)63}$
I	$8\overline{)56}$	$10\overline{)80}$	7×8	10×9	$5\overline{)50}$	9×5	10×8	$10\overline{)80}$
J	6×8	10×9	$4\overline{)40}$	$7\overline{)35}$	3×6	$8\overline{)56}$	9×8	7×5

TABLE OF MEASURES

METRIC	CUSTOMARY
Length	
1 decimeter (dm) = 10 centimeters	1 foot (ft) = 12 inches (in.)
1 meter (m) = 100 centimeters	1 yard (yd) = 3 feet, or 36 inches
1 meter (m) = 10 decimeters	1 mile (mi) = 1,760 yards, or 5,280 feet
1 kilometer (km) = 1,000 meters	
Mass/Weight	
1 kilogram (kg) = 1,000 grams (g)	1 pound (lb) = 16 ounces (oz)
Capacity	
1 liter (L) = 1,000 milliliters (mL)	1 pint (pt) = 2 cups (c)
	1 quart (qt) = 2 pints
	1 gallon (gal) = 4 quarts

TIME

1 minute (min) = 60 seconds (sec)
1 hour (hr) = 60 minutes
1 day = 24 hours
1 week (wk) = 7 days
1 year (yr) = 12 months (mo), or about 52 weeks
1 year = 365 days
1 leap year = 366 days

MONEY

1 penny = 1 cent (¢)
1 nickel = 5 cents
1 dime = 10 cents
1 quarter = 25 cents
1 half dollar = 50 cents
1 dollar ($) = 100 cents

SYMBOLS

$<$ is less than
$>$ is greater than
= is equal to
°F degrees Fahrenheit
°C degrees Celsius
(2,3) ordered pair

GLOSSARY

Pronunciation Key

a	**add**, **map**	f	**fit**, **half**	n	**nice**, **tin**	p	**pit**, **stop**	yōō	**fuse**, **few**
ā	**ace**, **rate**	g	**go**, **log**	ng	**ring**, **song**	r	**run**, **poor**	v	**vain**, **eve**
â(r)	**care**, **air**	h	**hope**, **hate**	o	**odd**, **hot**	s	**see**, **pass**	w	**win**, **away**
ä	**palm**, **father**	i	**it**, **give**	ō	**open**, **so**	sh	**sure**, **rush**	y	**yet**, **yearn**
b	**bat**, **rub**	ī	**ice**, **write**	ô	**order**, **jaw**	t	**talk**, **sit**	z	**zest**, **muse**
ch	**check**, **catch**	j	**joy**, **ledge**	oi	**oil**, **boy**	th	**thin**, **both**	zh	**vision**, **pleasure**
d	**dog**, **rod**	k	**cool**, **take**	ou	**pout**, **now**	t͟h	**this**, **bathe**		
e	**end**, **pet**	l	**look**, **rule**	o͝o	**took**, **full**	u	**up**, **done**		
ē	**equal**, **tree**	m	**move**, **seem**	o͞o	**pool**, **food**	û(r)	**burn**, **term**		

ə the schwa, an unstressed vowel representing the sound spelled *a* in **above**, *e* in **sicken**, *i* in **possible**, *o* in **melon**, *u* in **circus**

Other symbols:
- • separates words into syllables
- ′ indicates stress on a syllable

acute angle [ə•kyo͞ot′ ang′gəl] An angle that has a measure less than a right angle (p. 317)

acute triangle [ə•kyo͞ot′ trī′ang•gəl] A triangle that has three angles less than a right angle (p. 317)

addend [a′dend] Any of the numbers that are added (p. 36)
Example: 2 + 3 = 5
↑ ↑
addend addend

addition [ə•dish′ən] The process of finding the total number of items when two or more groups of items are joined; the opposite operation of subtraction (p. 36)

A.M. [ā em] Between midnight and noon (p. 96)

angle [ang′gəl] The figure formed when two rays share the same endpoint (p. 301)
Example:

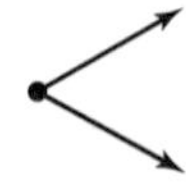

area [âr′ē•ə] The number of square units needed to cover a flat surface (p. 394)
Example:

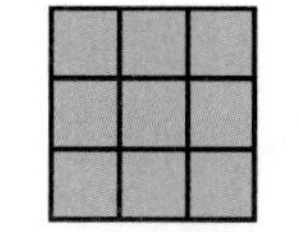

area = 9 square units

array [ə•rā′] An arrangement of objects in rows and columns (p. 120)
Example:

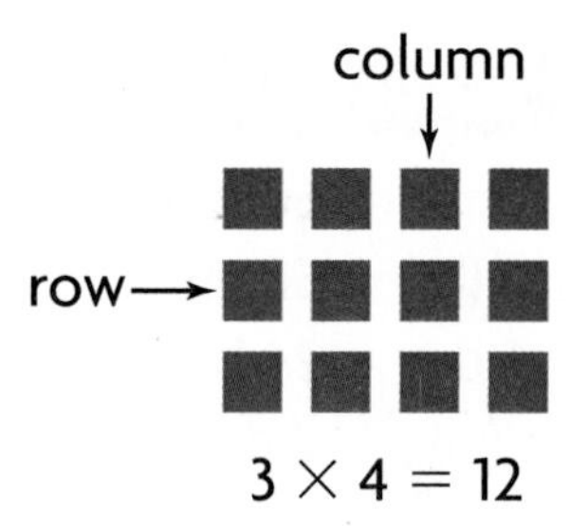

bar graph [bär graf] A graph that uses bars to show data (p. 254)
Example:

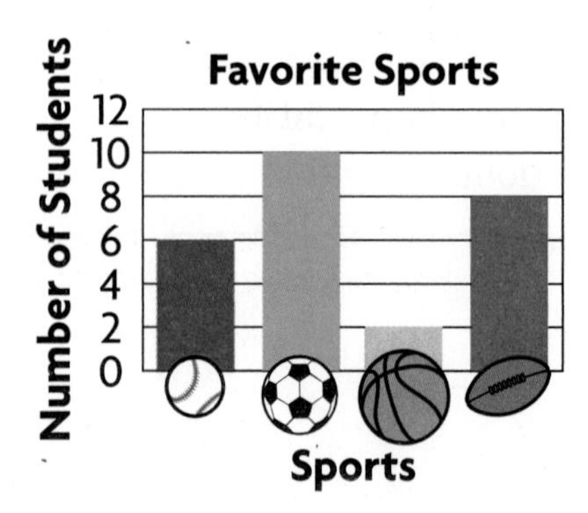

benchmark numbers [bench′märk num′bərz] Numbers that help you estimate the number of objects without counting them, such as 25, 50, 100, 1,000 (p. 18)

calendar [ka′lən•dər] A table that shows the days, weeks, and months of a year (p. 102)
Example:

July						
Sun	Mon	Tue	Wed	Thu	Fri	Sat
	1	2	3	4	5	6
7	8	9	10	11	12	13
14	15	16	17	18	19	20
21	22	23	24	25	26	27
28	29	30	31			

capacity [kə•pa′sə•tē] The amount a container can hold (p. 358)

center [sen′tər] A point in the middle of a circle that is the same distance from anywhere on the circle (p. 307)
Example:

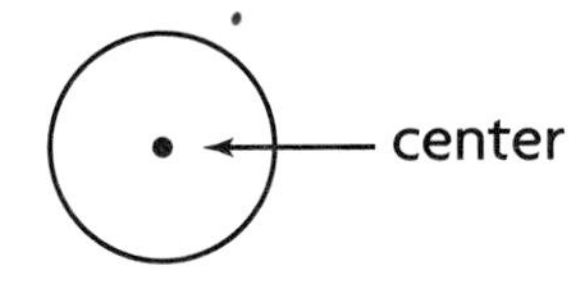

centimeter (cm) [sen′tə•mē•tər] A metric unit that is used to measure length (p. 372)
Example:

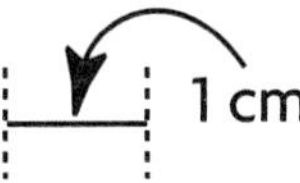

certain [sûr′tən] An event is certain if it will always happen. (p. 270)

classify [kla′sə•fī] To group pieces of data according to how they are the same; for example, you can classify data by size, color, or shape. (p. 244)

closed figure [klōzd fi′•gyər] A shape that begins and ends at the same point (p. 314)
Examples:

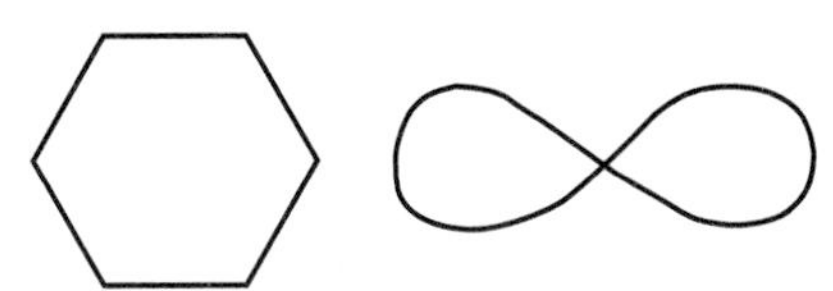

compare [kəm•pâr′] To describe whether numbers are equal to, less than, or greater than each other (p. 20)

cone [kōn] A solid, pointed figure that has a flat, round base (p. 294)
Example:

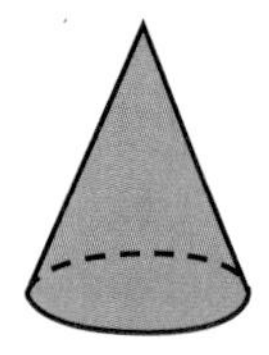

congruent [kən•grōō′ənt] Figures that have the same size and shape (p. 332)
Example:

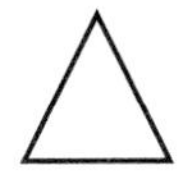

counting back [koun′ting bak] A way to find the difference when you subtract 1, 2, or 3
Example: 8 − 3 = ■ Count: 8 . . . 7, 6, 5

counting on [koun′ting on] A way to find the sum when one of the addends is 1, 2, or 3
Example: 5 + 2 = ■ Count: 5 . . . 6, 7

counting up [koun′ting up] A way to find the difference by beginning with the smaller number
Example: 7 − 4 = ■

Count: 4 . . . 5, 6, 7 ← 3 is the difference.

cube [kyōōb] A solid figure with six congruent square faces (p. 294)
Example:

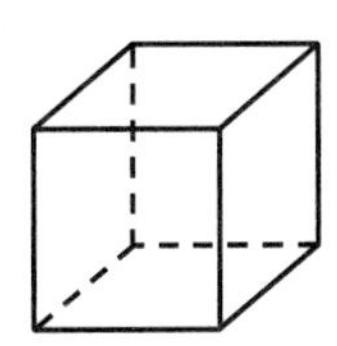

cubic unit [kyōō′bik yōō′nət] A cube with a side length of one unit; used to measure volume (p. 398)

cup (c) [kup] A customary unit used to measure capacity (p. 358)

cylinder [sil′in•dər] A solid or hollow object that is shaped like a can (p. 294)
Example:

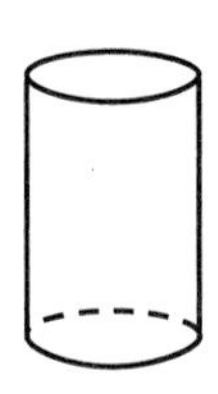

D

data [dā′tə] Information collected about people or things (p. 240)

decimal [de′sə•məl] A number with one or more digits to the right of the decimal point (p. 452)

decimal point [de′sə•məl point] A symbol used to separate dollars from cents in money and to separate the ones place from the tenths place in decimals (p. 452)
Example: 4.5
↑ decimal point

decimeter (dm) [de′sə•mē•tər] A metric unit that is used to measure length; 1 decimeter = 10 centimeters (p. 372)

degrees Celsius (°C) [di•grēz′ sel′sē•əs] A unit for measuring temperature in the metric system (p. 382)

degrees Fahrenheit (°F) [di•grēz′ far′ən•hīt] A unit for measuring temperature in the customary system (p. 382)

denominator [di•nä′mə•nā•tər] The part of a fraction that tells how many equal parts are in the whole (p. 412)
Example: $\frac{3}{4}$ ← denominator

diameter [dī•a′mə•tər] A line segment that passes through the center of a circle and whose endpoints are on the circle (p. 307)
Example:

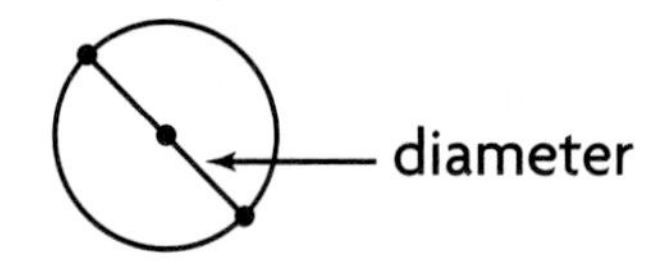

difference [dif′rən(t)s] The answer in a subtraction problem
Example: 6 − 4 = 2
↑ difference

digits [di′jəts] The symbols 0, 1, 2, 3, 4, 5, 6, 7, 8, and 9 (p. 4)

divide [di•vīd′] To separate into equal groups; the opposite operation of multiplication (p. 184)

dividend [di′və•dend] The number that is to be divided in a division problem (p. 188)
Example: 35 ÷ 5 = 7
↑ dividend

divisor [də•vī′zər] The number that divides the dividend (p. 188)
Example: 35 ÷ 5 = 7
↑ divisor

E

edge [ej] A line segment formed where two faces meet (p. 294)
Example:

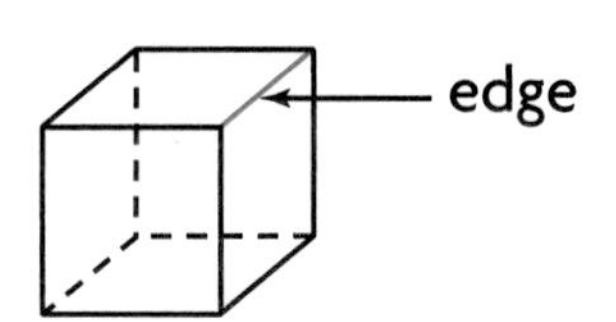

elapsed time [i•lapst′ tīm] The amount of time that passes from the start of an activity to the end of that activity (p. 98)

equal sign (=) [ē′kwəl sīn] A symbol used to show that two numbers have the same value (p. 20)
Example: 384 = 384

equally likely [ē′kwəl•lē lī′klē] Having the same chance of happening (p. 274)

equilateral triangle [ē•kwə•lat′ər•əl trī′ang•gəl] A triangle that has all sides equal (p. 317)

equivalent [ē•kwiv′ə•lənt] Two or more sets that name the same amount are equivalent. (p. 80)

equivalent fractions [ē•kwiv′ə•lənt frak′shənz] Two or more fractions that name the same amount (p. 418)
Example:

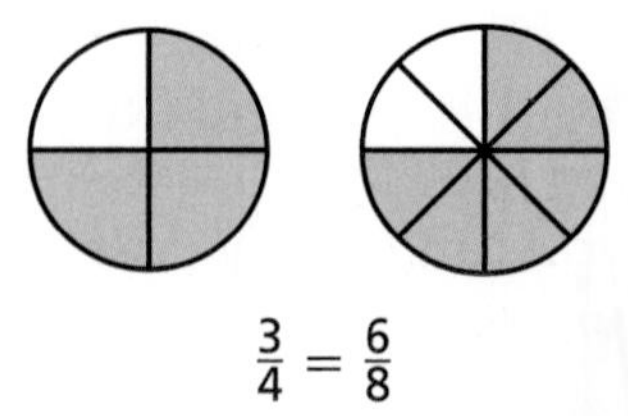

$\frac{3}{4} = \frac{6}{8}$

estimate [es′tə•māt] To find about how many or how much (p. 38)

even [ē′vən] A whole number that has a 0, 2, 4, 6, or 8 in the ones place is even (p. 2)

event [i•vent′] Something that happens (p. 270)

expanded form [ik•spand′id fôrm] A way to write numbers by showing the value of each digit (p. 4)
Example: 7,201 = 7,000 + 200 + 1

experiment [ik•sper′ə•mənt] A test that is done in order to find out something (p. 276)

expression [ik•spre′shən] The part of a number sentence that combines numbers and operation signs, but doesn't have an equal sign (p. 68)
Example: 5×6

face [fās] A flat surface of a solid figure (p. 294)
Example:

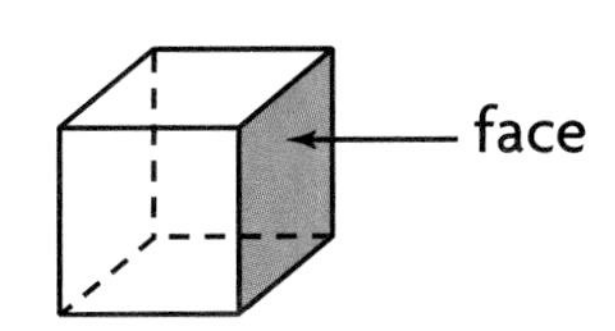

fact family [fakt fam′ə•lē] A set of related multiplication and division, or addition and subtraction, number sentences (p. 192)
Example:

$4 \times 7 = 28$ $28 \div 7 = 4$
$7 \times 4 = 28$ $28 \div 4 = 7$

factor [fak′tər] A number that is multiplied by another number to find a product (p. 118)
Example: $3 \times 8 = 24$
↑ ↑
factor factor

fair [fâr] A game is fair if every player has an equal chance to win. (p. 282)

flip [flip] A movement of a figure to a new position by flipping the figure over a line. (p. 338)

foot (ft) [fo͝ot] A customary unit used to measure length or distance; 1 foot = 12 inches (p. 356)

fraction [frak′shən] A number that names part of a whole or part of a group (p. 412)
Example:

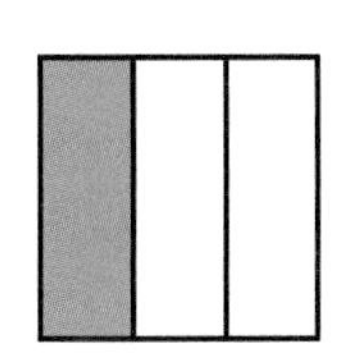

$\frac{1}{3}$

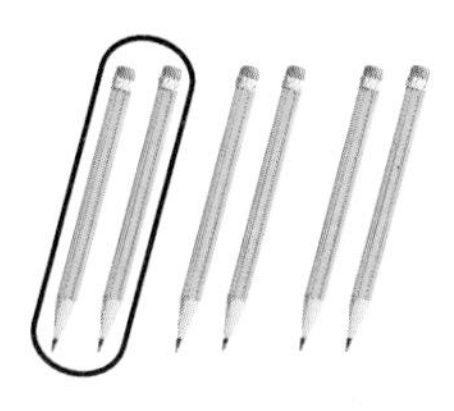

frequency table [frē′kwen•sē tā′bəl] A table that uses numbers to record data (p. 240)

gallon (gal) [ga′lən] A customary unit for measuring capacity; 4 quarts = 1 gallon (p. 359)

gram (g) [gram] A metric unit that is used to measure mass (p. 380)

greater than (>) [grā′tər t͟han] A symbol used to compare two numbers, with the larger number given first (p. 20)
Example: $6 > 4$

grid [grid] Horizontal and vertical lines on a map (p. 262)

Grouping Property of Addition [gro͞o′ping prä′pər•tē əv ə•di′shən] A rule stating that you can group addends in different ways and still get the same sum (p. 36)
Example:

$4 + (2 + 5) = 11$ and
$(4 + 2) + 5 = 11$

Grouping Property of Multiplication [gro͞o′ping prä′pər•tē əv məl•tə•plə•kā′shən] A rule stating that when the grouping of factors is changed, the product remains the same (p. 170)
Example:

$3 \times (4 \times 1) = 12$ and
$(3 \times 4) \times 1 = 12$

hexagon [hek′sə•gän] A polygon with six sides and six angles (p. 314)
Example:

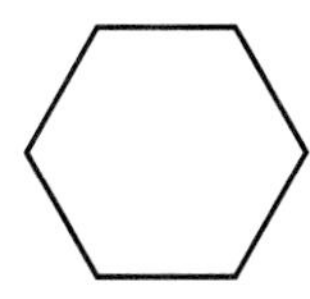

horizontal bar graph [hôr•ə•zän′təl bär graf] A bar graph in which the bars go from left to right (p. 254)

hour (hr) [our] A unit used to measure time; in one hour, the hour hand on a clock moves from one number to the next. 1 hour = 60 minutes (p. 94)

hour hand [our hand] The short hand on an analog clock (p. 94)

hundredth [hən′drədth] One of one hundred equal parts (p. 456)
Example:

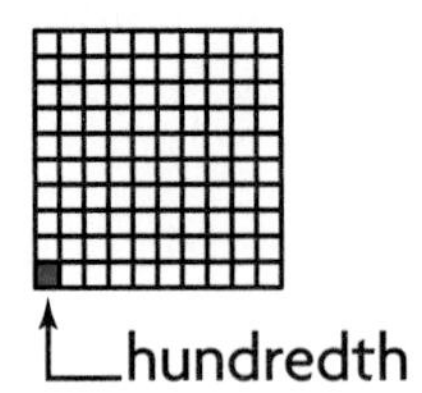

impossible [im•pä′sə•bəl] An event is impossible if it will never happen. (p. 270)

inch (in.) [inch] A customary unit used to measure length (p. 352)
Example:

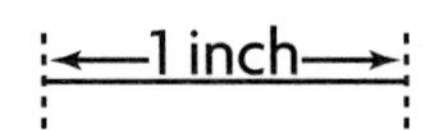

intersecting lines [in•tər•sek′ting līnz] Lines that cross (p. 304)
Example:

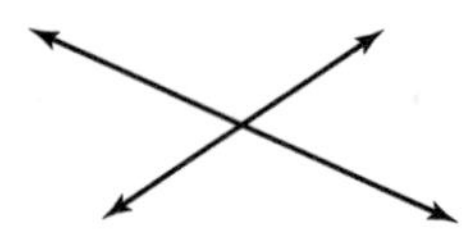

inverse operations [in′vərs ä•pə•rā′shənz] Opposite operations, or operations that undo each other, such as addition and subtraction or multiplication and division (p. 188)

isosceles triangle [ī•sos′ə•lēz trī′ang•gəl] A triangle that has two equal sides (p. 317)
Example:

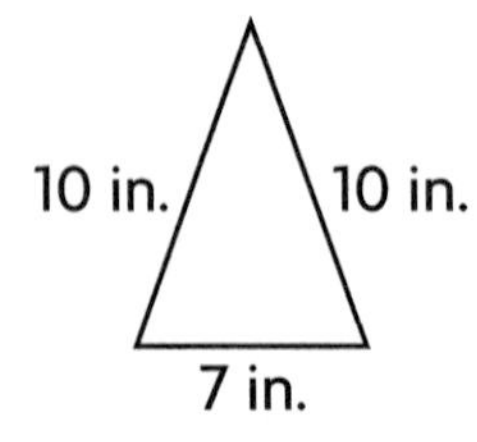

kilogram (kg) [kil′ə•gram] A metric unit that is used to measure mass; 1 kilogram = 1,000 grams (p. 380)

kilometer (km) [kə•lä′mə•tər] A metric unit that is used to measure length and distance; 1 kilometer = 1,000 meters (p. 372)

less than (<) [les than] A symbol used to compare two numbers, with the lesser number given first (p. 20)
Example: $3 < 7$

like fractions [līk frak′shənz] Fractions that have the same denominator (p. 434)

likely [līk′lē] Having a good chance of happening (p. 272)

line [līn] A straight path extending in both directions with no endpoints (p. 300)
Example:

line graph [līn graf] A graph that uses a line to show how something changes over time (p. 264)
Example:

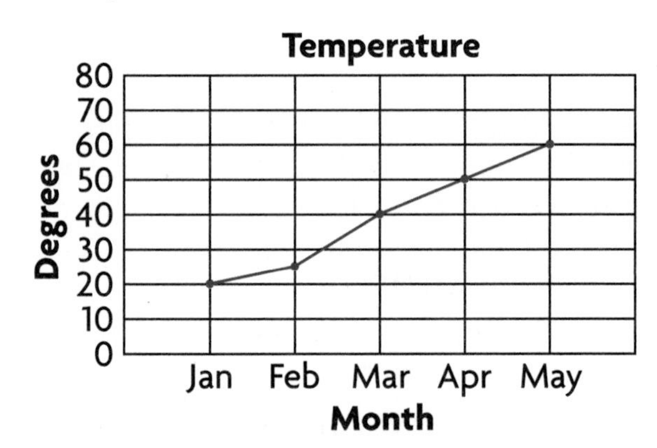

line of symmetry [līn əv sim′ə•trē] An imaginary line that divides a figure into two congruent parts (p. 334)
Example:

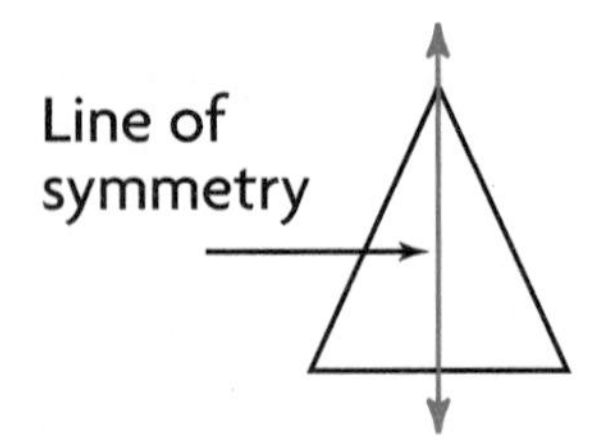

line plot [līn plot] A diagram that records each piece of data on a number line (p. 258)
Example:

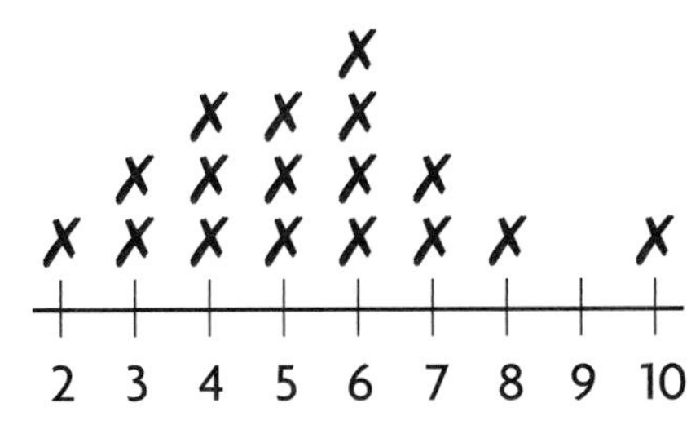

line segment [līn seg′mənt] A part of a line that extends between two points, called endpoints (p. 300)
Example:

liter (L) [lē′tər] A metric unit that is used to measure capacity; 1 liter = 1,000 milliliters (p. 378)

mass [mas] The amount of matter in an object (p. 380)

meter (m) [mē′tər] A metric unit that is used to measure length and distance; 1 meter = 100 centimeters (p. 372)

midnight [mid′nīt] 12:00 at night (p. 96)

mile (mi) [mīl] A customary unit used to measure length and distance; 1 mile = 5,280 feet (p. 356)

milliliter (mL) [mi′lə•lē•tər] A metric unit that is used to measure capacity (p. 378)

minute (min) [mi′nət] A unit used to measure short amounts of time; in one minute, the minute hand moves from one mark to the next. (p. 94)

minute hand [mi′nət hand] The long hand on an analog clock (p. 94)

mixed number [mikst nəm′bər] A number represented by a whole number and a fraction (p. 428)
Example: $4\frac{1}{2}$

mode [mōd] The number found most often in a set of data (p. 258)

multiply [mul′tə•plī] When you combine equal groups, you can multiply to find how many in all; the opposite operation of division. (p. 116)

multistep problem [məl′tē•step prä′bləm] A problem with more than one step (p. 172)

noon [no͞on] 12:00 in the day (p. 96)

number sentence [num′bər sen′təns] A sentence that includes numbers, operation symbols, and a greater than or less than symbol or an equal sign (p. 68)
Example:

5 + 3 = 8 is a number sentence.

numerator [no͞o′mə•rā•tər] The part of a fraction above the line, which tells how many parts are being counted (p. 412)
Example: $\frac{3}{4}$ ←numerator

obtuse angle [əb•t(y)o͞os′ ang′gəl] An angle that has a measure greater than a right angle (p. 317)

obtuse triangle [əb•t(y)o͞os′ trī′ang•gəl] A triangle that has 1 angle greater than a right angle (p. 317)

octagon [äk′tə•gän] A polygon with eight sides and eight angles (p. 314)
Example:

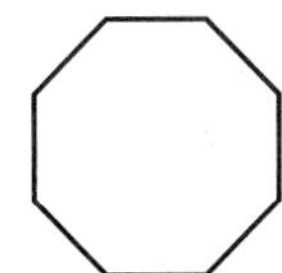

odd [od] A whole number that has a 1, 3, 5, 7, or 9 in the ones place is odd (p. 2)

Order Property of Addition [ôr′dər prä′pər•tē əv ə•dish′ən] A rule stating that you can add two numbers in any order and get the same sum (p. xxvi)

Order Property of Multiplication [ôr′dər prä′pər•tē əv məl•tə•plə•kā′shən] A rule stating that you can multiply two factors in any order and get the same product (p. 121)
Example: $4 \times 2 = 8$
$2 \times 4 = 8$

ordered pair [ôr′dərd pâr] A pair of numbers that names a point on a grid (p. 262)
Example: **(3,4)**

ounce (oz) [ouns] A customary unit used to measure weight (p. 360)

outcome [out′kum′] A possible result of an experiment (p. 272)

parallel lines [par′ə•lel līnz] Lines that never cross (p. 304)
Example:

parallelogram [par•ə•lel′ə•gram] A quadrilateral with 2 pairs of parallel sides and 2 pairs of equal sides (p. 321)
Example:

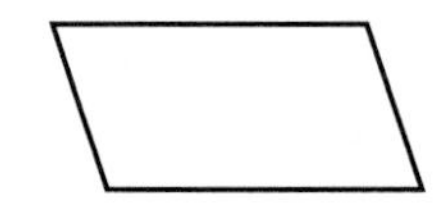

pentagon [pen′tə•gän] A polygon with five sides and five angles (p. 314)
Example:

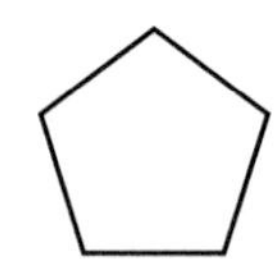

perimeter [pə•ri′mə•tər] The distance around a figure (p. 388)
Example:

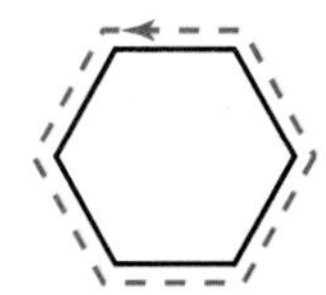

pictograph [pik′tə•graf] A graph that uses pictures to show and compare information (p. 252)
Example:

HOW WE GET TO SCHOOL

Walk	❋ ❋ ❋
Ride a Bike	❋ ❋ ❋ ❋
Ride a Bus	❋ ❋ ❋ ❋ ❋ ❋
Ride in a Car	❋ ❋

Key: Each ❋ = 10 students.

pint (pt) [pīnt] A customary unit for measuring capacity; 1 pint = 2 cups (p. 358)

place value [plās val′yōō] The value of each digit in a number, based on the location of the digit (p. 4)

P.M. [pē em] Between noon and midnight (p. 96)

point [point] An exact position or location (p. 300)

polygon [pol′ē•gän] A closed plane figure with straight sides; each side is a line segment. (p. 314)
Examples:

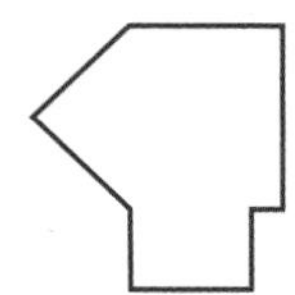

possible outcome [pos′ə•bəl out′kəm] Something that has a chance of happening (p. 274)

pound (lb) [pound] A customary unit used to measure weight; 1 pound = 16 ounces (p. 360)

predict [pri•dikt′] To make a reasonable guess about what will happen (p. 274)

product [prä′dəkt] The answer in a multiplication problem (p. 118)
Example: 3 × 8 = 24
↑ product

pyramid [pir′ə•mid] A solid, pointed figure with a flat base that is a polygon (p. 294)
Example:

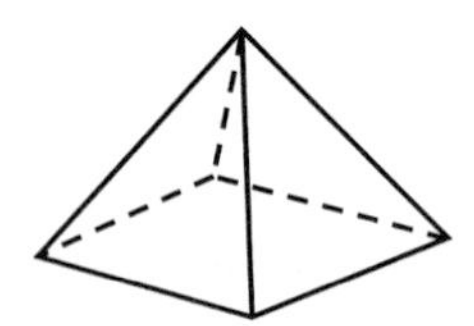

quadrilateral [kwa•drə•lat′ər•əl] A polygon with four sides and four angles (p. 314)
Example:

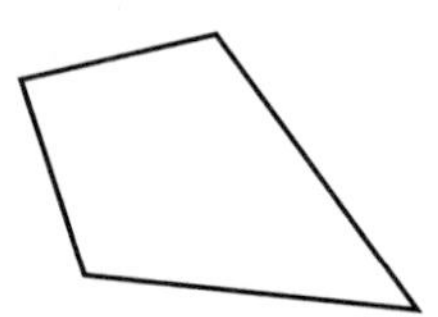

quart (qt) [kwôrt] A customary unit for measuring capacity; 1 quart = 2 pints (p. 358)

quotient [kwō′shənt] The number, not including the remainder, that results from dividing (p. 188)
Example: 8 ÷ 4 = 2
↑ quotient

radius [rā′dē•əs] A line segment whose endpoints are the center of a circle and any point on the circle (p. 307)
Example:

range [rānj] The difference between the greatest number and the least number in a set of data (p. 258)

ray [rā] A part of a line, with one endpoint, that is straight and continues in one direction (p. 301)
Example:

rectangle [rek′tang•gəl] A quadrilateral with 2 pairs of parallel sides, 2 pairs of equal sides, and 4 right angles (p. 321)
Example:

rectangular prism [rek•tan′gyə•lər pri′zəm] A solid figure with six faces that are all rectangles (p. 294)
Example:

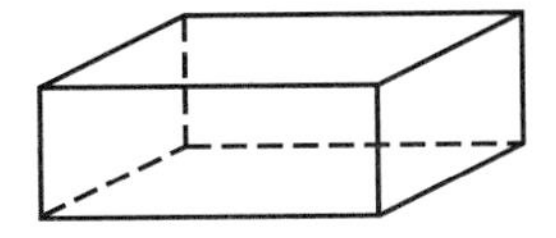

regroup [rē•gro͞op′] To exchange amounts of equal value to rename a number (p. 40)
Example: 5 + 8 = 13 ones or 1 ten 3 ones

remainder [ri•mān′dər] The amount left over when a number cannot be divided evenly (p. 503)

results [ri•zults′] The answers from a survey (p. 242)

rhombus [räm′bəs] A quadrilateral with 2 pairs of parallel sides and 4 equal sides (p. 321)
Example:

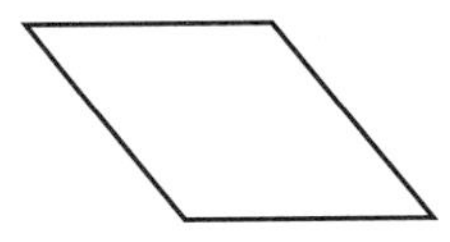

right angle [rīt ang′gəl] A special angle that forms a square corner (p. 301)
Example:

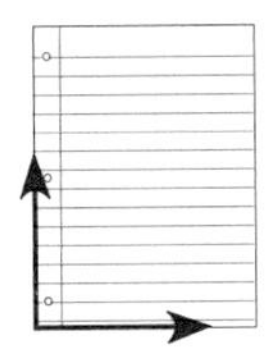

right triangle [rīt trī′ang•gəl] A triangle with one right angle (p. 317)
Example:

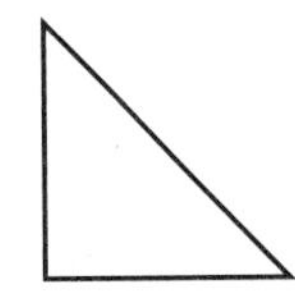

rounding [roun′ding] One way to estimate (p. 28)

scale [skāl] The numbers on a bar graph that help you read the number each bar shows (p. 254)

scalene triangle [skā′lēn trī′ang•gəl] A triangle in which no sides are equal (p. 317)
Example:

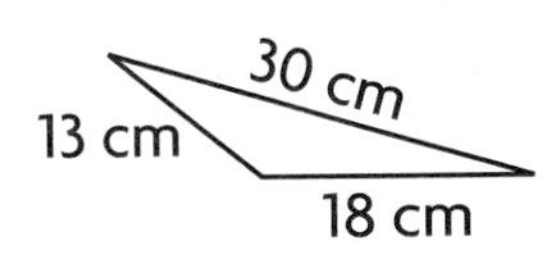

schedule [ske′•jōōl] A table that lists activities or events and the times they happen (p. 100)

sequence [sē′kwəns] To write events in order (p. 104)

similar [si′mə•lər] Having the same shape and the same or different size (p. 336)
Example:

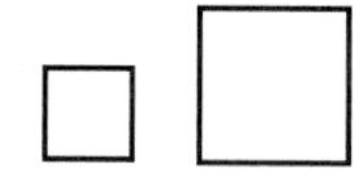

simplest form [sim′pləst fôrm] When a fraction is modeled with the largest fraction bar or bars possible (p. 436)

slide [slīd] A movement of a figure to a new position without turning or flipping it (p. 338)

sphere [sfir] A solid figure that has the shape of a round ball (p. 294)
Example:

square [skwâr] A quadrilateral with 2 pairs of parallel sides, 4 equal sides, and 4 right angles (p. 321)
Example:

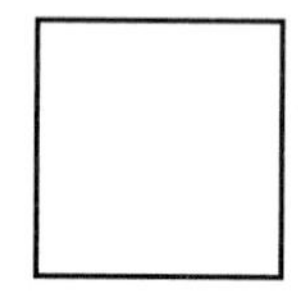

square number [skwâr nəm′bər] A product of two factors that are the same (p. 159)
Example: $4 \times 4 = 16$;
16 is a square number.

square unit [skwâr yōō′nət] A square with a side length of one unit; used to measure area (p. 394)

standard form [stan′dərd fôrm] A way to write numbers by using the digits 0–9, with each digit having a place value (p. 4)
Example: 345 ← standard form

subtraction [səb•trak′shən] The process of finding how many are left when a number of items are taken away from a group of items; the process of finding the difference when two groups are compared; the opposite operation of addition (p. 54)

sum [səm] The answer to an addition problem (p. 36)

survey [sər′vā] A question or set of questions that a group of people are asked (p. 242)

symmetry [sim′mə•trē] When one half of a figure looks like the mirror image of the other half (p. 334)

tally table [ta′lē tā′bəl] A table that uses tally marks to record data (p. 240)

tenth [tenth] One of ten equal parts (p. 452)
Example:

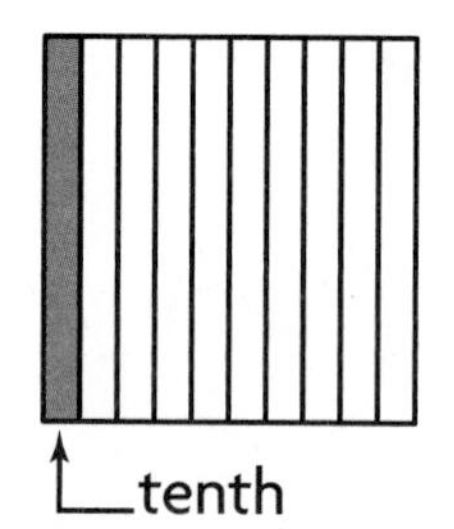

tessellate [tes′ə•lāt] To combine plane figures so they cover a surface without overlapping or leaving any space between them (p. 324)

tessellation [te•sə•lā′shən] A repeating pattern of closed figures that covers a surface with no gaps and no overlaps (p. 324)
Example:

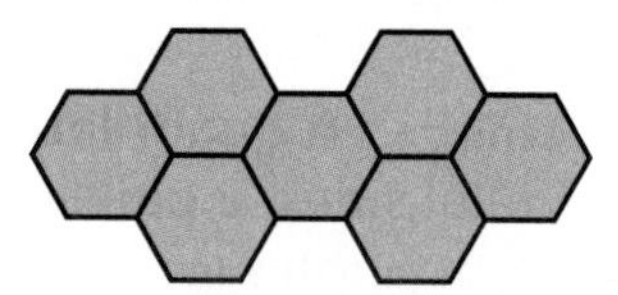

time line [tīm līn] A schedule of events, or an ordered list of historic moments (p. 109)

trapezoid [trap′ə•zoid] A quadrilateral with one pair of parallel sides (p. 326)
Example:

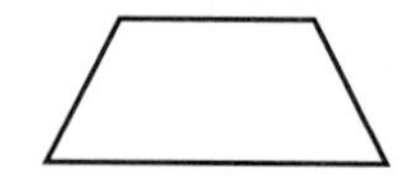

triangle [trī′ang•gəl] A polygon with three sides and three angles (p. 314)
Example:

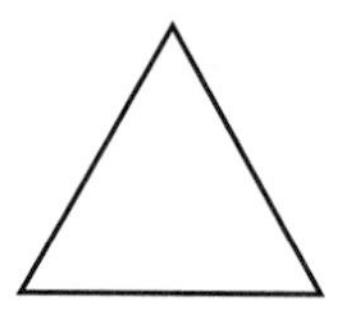

turn [tûrn] A movement of a figure to a new position by rotating the figure around a point (p. 338)

unit cost [yōō′nit kôst] The cost of one item when several items are sold for a single price (p. 237)

unlikely [ən•lī′klē] An event is unlikely if it does not have a good chance of happening. (p. 272)

variable [vâr′ē•ə•bəl] A symbol or a letter that stands for an unknown number (p. 189)

vertex [vûr′teks] In a solid figure, a corner where three or more edges meet (p. 294)
Example:

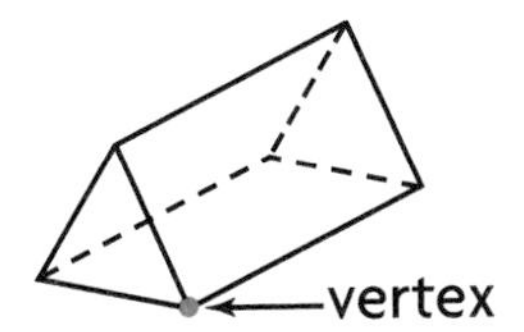

vertical bar graph [vûr′ti•kəl bär graf] A bar graph in which the bars go up from bottom to top (p. 254)

volume [väl′yəm] The amount of space a solid figure takes up (p. 398)

whole number [hōl nəm′bər] One of the numbers 0, 1, 2, 3, 4, ... The set of whole numbers goes on without end.

word form [wûrd form] A way to write numbers by using words (p. 4)
Example: The word form of 212 is two hundred twelve.

yard (yd) [yärd] A customary unit used to measure length or distance; 1 yard = 3 feet (p. 356)

Zero Property of Addition [zē′rō prä′pər•tē əv ə•di′shən] A rule stating that when you add zero to, or subtract zero from, a number, the result is that number

Zero Property of Multiplication [zē′rō prä′pər•tē əv məl•tə•plə•kā′shən] A rule stating that the product of zero and any number is zero (p. 132)

INDEX

C

D

E

F

G

M

N

O

P

S

T

PHOTO CREDITS

Photography Credits
Page placement key: (t) top, (c) center, (b) bottom, (l) left, (r) right, (bg) background, (i) inset.
v Laurie Campbell/Stone; xii Stephen J. Krasemann/DRK; xiv Mike Severn/Stone; xix Bill Bachmann/PhotoEdit; xx Tony Freeman/PhotoEdit; 1 Buzz Binzen/International Stock Photography; 4 (b) Larry Lefever/Grant Heilman Photography; 16 Darrell Gulin/Dembinsky Photo Associates; 18 (b) Mugshots/The Stock Market; 23 John Elk III/Bruce Coleman, Inc.; 24 Jeff Greenberg/Photri; 26 Ric Ergenbright/Corbis; 28 AFP/Corbis; 29 Kevin Schafer; 30 Joe McDonald/PictureQuest; 31 Ronn Maratea/International Stock; 34 Lawrence Migdale/Photo Researchers; 36 JC Carton/Bruce Coleman, Inc.; 37 (t) Orion/International Stock; 37 (b) Patricia Doyle/Stone; 38 Index Stock; 45 John Daniels/Bruce Coleman, Inc.; 52 D. Muench/H. Armstrong Roberts; 54 Luiz C. Marigo/Peter Arnold, Inc.; 55 Fred Bavendam/Peter Arnold, Inc.; 58 Doug Perrine/Innerspace Visions; 59 Mike Severns/Stone; 61 Ingrid Visser/Innerspace Visions; 62 Superstock; 63 Bob Burch/Bruce Coleman, Inc.; 64 Mark E. Gibson; 66 David Young Wolff/Stone; 68 Jackie Pirret; 69 Steven Needham/Envision; 77a (b) Scott Barrow, Inc.; 77b (b) June Evelyn Atwood/Contact Photos/PictureQuest; 78 J. Scott Applewhite/AP/Wide World Photos; 89 (t) Patricia Doyle/Stone; 89 (b) Hans Reinhard/Stone; 92 R. Kord/H. Armstrong Roberts; 101 Mark Newman/International Stock Photography; 105 Mark E. Gibson; 109 (tr) Myrleen Ferguson/PhotoEdit; 109 (bl) Robert Rubic, Photographer/New York Public Library; 109 (br) NASA; 109 (tl) The Granger Collection, New York; 113a (t) Jean Higgins/Envision; 113b (t) Jim Pickerell/Stock Connection/PictureQuest; 113b (b) National Gallery of Art Credit Mark C. Burnett/Stock, Boston/PictureQuest; 114 Thayer Syme/FPG International; 118 (t) Don Mason/The Stock Market; 118 (b) Charles D. Winters/Photo Researchers; 118 (bl) Carolina Biological Supp/Phototake; 118 (bc) Stephen J. Krasemann/Photo Researchers; 118 (br) Kelvin Aitken/Peter Arnold, Inc.; 130 Dan Feicht/Cedar Point; 132 Comstock; 133 Vicki Silbert/PhotoEdit; 138 Chuck Mason/International Stock; 139 Chris Sorensen; 146 John Troha/Black Star/Harcourt; 148 Myrleen Ferguson/PhotoEdit; 150 Chromo Sohm/Sohm/Visions of America; 156 Tony Freeman/PhotoEdit; 157 Bill Bachmann/PhotoEdit; 162 Kim Heacox/Peter Arnold, Inc.; 164 Pat & Tom Leeson/Photo Researchers; 167 Zefa/The Stock Market; 181A Bob Krist/Corbis; 181b (t) S. J. Krasemann/Peter Arnold, Inc.; 181B (b) Steve Gettle/ENP Images; 182 Mug Shots/The Stock Market; 191 H. Mark Weidman; 200 Charles Gupton/The Stock Market; 210 Len Rue Jr./Photo Researchers; 211 Laurie Campbell/Stone; 214 Paul Chauncey/The Stock Market; 225 Ted C. Hilliard; 237A Kelly-Mooney Photography/Corbis; 237B Churchill & Klehr; 238 AP/Wide World Photos; 238 (inset) AFP/Corbis; 242 Bob Firth/International Stock; 243 Bill Tocker/International Stock; 250 Mark J. Thomas/Dembinsky Photo Associates; 253 David Stoecklein/The Stock Market; 254 Erwin & Peggy Bauer/Bruce Coleman, Inc.; 255 Robert Winslow/Animals Animals; 256 (t) Anup Shah/DRK; 257 Stephen J. Krasemann/DRK; 260 (b) Orion Press/Stone; 261 Barbara Kreye/International Stock; 262 Christoph Burki/Stone; 263 David Northcott/DRK; 264 Curt Maas/AGStock USA; 268 Terry Donnelly/Dembinsky Photo Associates; 279 NASA/Peter Arnold, Inc.; 291a Effigy Mounds State Park, Iowa; 291b (t) Scott Leonhart/Positive Images; 291b (b) Effigy Mounds State Park, Iowa; 292 Paula Lerner/Woodfin Camp & Associates; 294 (bl) Dick Durrance/Woodfin Camp & Associates; 296 #6 Dick Durrance/Woodfin Camp & Associates; 296 #9 Robert Stottlemeyer/International Stock; 298 Tom McCrathy/Unicorn Stock Photography; 312 (bg) Jose L. Pelaez/The Stock Market; 312 (i) Aneal Vohra/Unicorn Stock Photos; 312 (i) Jim Shippe/Unicorn Stock Photos; 312 (i) Aneal Vohra/Unicorn Stock Photos; 319 Eric A. Wessman/Stock, Boston; 324 Cordon Art-Baarn-Holland; 330 Stephen Frink/Waterhouse Stock Photo; 349A Sandy Felsentha/Corbis; 349B Richard A. Cooke/Corbis; 350 NASA; 357 (c) J H. Robinson/Photo Researchers; 370 Larry Ulrich/DRK; 381 (t) Tetsu Yamazaki/International Stock Photography; 381 (#3) Oliver Strewe/Tony Stone Images; 384 ({no}9) Rod Planck/Tony Stone Images;385 Peter Vadnai/The Stock Market; 386 Lefever/Grushow/Grant Heilman Photography; 393 (b) R. Lautman/Poplar Forest; 409A American Maze Company's Amazing Maize Maze at Lebanon Valley College, Annville, PA 1993 *www.AmericanMaze.com* Maze design Team; Don Frantz, Ian Marshall, Adrian Fisher, and Rich Whorl; 409B (l) American Maze Company's Amazing Maize Maze at Mountain Creek, Vernon, NJ 2000 *www.AmericanMaze.com* Maze design Team; Don Frantz, Ian Marshall, Adrian Fisher, and Rich Whorl; 409B (r) American Maze Company's Amazing Maize Maze at Cherry-Crest Farm, Paradise, PA 1996 *www.AmericanMaze.com* Maze design Team; Don Frantz, Ian Marshall, Adrian Fisher, and Rich Whorl; 410 Bruce Davidson/Animals Animals/Earth Scenes; 416 M. Gibbs/Animals Animals; 417 (l) B. Von Hoffmann/H. Armstrong Roberts, Inc.; 417 (r) Carolyn A. McKeone/Photo Researchers; 421 (t) Robert & Linda Mitchell; 421 (c) Norman Owen Tomalin/Bruce Coleman, Inc.; 421 (b) Dwight R. Kuhn; 424 Wes Thompson/The Stock Market; 425 Steven Needham/Envision; 432 Index Stock Photography; 447 Superstock; 450 VCG/FPG International; 460 Ryan Williams/International Stock; 462 Nancy Sheehan/PhotoEdit; 466 Nicholas DeVore/Stone; 475 Scott Nielsen/Bruce Coleman, Inc.; 485a Daniel Waggoner/Envision; 485b (b) Steve Bly/Dave G. Houser; 486 Scott Smith/ Animals Animals Earth Scenes; 490 Gail Mooney/Kelly/Mooney Photography; 493 John Thomas Biggers, Starry Crown 1987. Acrylic, mixed media on masonite. Dallas Museum of Art. Museum League Purchase Fund; 497 Ed Harp/ Unicorn Stock Photos; 500 TravelPix/FPG International; 506 R. Hutchings/PhotoEdit; 507 Richard Hutchings/Photo Researchers; 516 Kevin Horan/Stock, Boston; 522 Ron Kimball; 523 Barbara Reed/Animals Animals; 524 Chris Sorensen; 526 Chris Kapolka/Stone; 534 Rudi Von Briel/PhotoEdit; 540 Myrleen Ferguson Cate/PhotoEdit; 541 American Images, Inc./FPG International; 542 D. Young-Wolfe/ PhotoEdit; 546 Rudi Von Briel/PhotoEdit; 554 George Lepp/Corbis; 555a (l) Lynn M. Stone; 555a (r) Ted Levin; 555b Francis & Donna Caldwell/Affordable Photo Stock.
All other photographs by Harcourt photographers listed below,
© Harcourt: Weronica Ankarorn, John Bateman, Victoria Bowen, Ken Kinzie, Ron Kunzman, Allan Maxwell, Sheri O'Neal, Quebecor Digital Imaging, Sonny Senser, Terry Sinclair.